ANNALS OF
THE NEW YORK ACADEMY
OF SCIENCES

Volume 1042

EDITORIAL STAFF

Director, Publishing and New Media
SARAH GREENE

Managing Editor
JUSTINE CULLINAN

Associate Editor
MARION L. GARRY

The New York Academy of Sciences
2 East 63rd Street
New York, New York 10021

THE NEW YORK ACADEMY OF SCIENCES
(Founded in 1817)

BOARD OF GOVERNORS, September 2004 – September 2005

TORSTEN N. WIESEL, *Chairman of the Board*
GERALD D. FISCHBACH, *Vice Chairman*
MICHAEL SCHMERTZLER, *Treasurer*
ELLIS RUBINSTEIN, *Chief Executive Officer* [ex officio]

Honorary Life Governors
WILLIAM T. GOLDEN   JOSHUA LEDERBERG

Governors

KAREN E. BURKE	VIRGINIA W. CORNISH	PETER B. CORR
R. BRIAN FERGUSON	RONALD L. GRAHAM	MARNIE IMHOFF
WENDY EVANS JOSEPH	JACQUELINE LEO	ROBERT W. LUCKY
PAUL MARKS	BRUCE McEWEN	RONAY MENSCHEL
JOHN T. MORGAN	JOHN F. NIBLACK	SANDRA PANEM
PETER RINGROSE		DAVID D. SABATINI

VICTORIA BJORKLUND, *Counsel* [ex officio]   LARRY R. SMITH, *Secretary* [ex officio]

THE ROLE OF THE MITOCHONDRIA IN HUMAN AGING AND DISEASE

FROM GENES TO CELL SIGNALING

ANNALS OF THE NEW YORK ACADEMY OF SCIENCES

Volume 1042

THE ROLE OF THE MITOCHONDRIA IN HUMAN AGING AND DISEASE

FROM GENES TO CELL SIGNALING

Edited by Yau-Huei Wei, Horng-Mo Lee, and Chung Y. Hsu

The New York Academy of Sciences
New York, New York
2005

Copyright © 2005 by the New York Academy of Sciences. All rights reserved. Under the provisions of the United States Copyright Act of 1976, individual readers of the Annals are permitted to make fair use of the material in them for teaching or research. Permission is granted to quote from the Annals provided that the customary acknowledgment is made of the source. Material in the Annals may be republished only by permission of the Academy. Address inquiries to the Permissions Department (editorial@nyas.org) at the New York Academy of Sciences.

Copying fees: For each copy of an article made beyond the free copying permitted under Section 107 or 108 of the 1976 Copyright Act, a fee should be paid through the Copyright Clearance Center, Inc., 222 Rosewood Drive, Danvers, MA 01923 (www.copyright.com).

∞ The paper used in this publication meets the minimum requirements of the American National Standard for Information Sciences—Permanence of Paper for Printed Library Materials, ANSI Z39.48-1984.

Library of Congress Cataloging-in-Publication Data

Asian Society for Mitochondrial Research and Medicine. Scientific
   Meeting (2nd : 2004 : Taipei, Taiwan)
  The role of the mitochondria in human aging and disease : from
  genes to cell signaling / edited by Yau-Huei Wei, Horng-Mo Lee,
  and Chung Y. Hsu.
     p. ; cm. – (Annals of the New York Academy of Sciences, ISSN
  0077-8923 ; v. 1042)
   This volume is the result of the Second Scientific Meeting of the
  Asian Society for Mitochondrial Research and Medicine, held April
  1–2, 2004 in Taipei, Taiwan.
   Includes bibliographical references and index.
     ISBN 1-57331-541-9 (cloth : alk. paper) – ISBN 1-57331-542-7
  (pbk. : alk. paper)
     1. Mitochondrial pathology–Congresses. 2. Mitochondria–Congresses. 3. Mitochondrial DNA–Congresses.   I. Wei, Y.-H. (Yau
-Huei) II. Lee, Horng-Mo. III. Hsu, Chung Y. IV. New York
Academy of Sciences. V. Title. VI. Series.
   [DNLM: 1. Mitochondrial Diseases–genetics–Congresses.
2. Mitochondrial Diseases–physiopathology–Congresses. 3. Cell
Aging–physiology–Congresses. 4. Cell Death–genetics–Congresses. 5. Disease Models, Animal–Congresses. 6. Oxidative
Stress–genetics–Congresses.   W1 AN626YL v.1042 2005 /
WD 200.5.M6 A825r 2005]
Q11.N5 vol. 1042
[RB147.5]
500 s–dc22
[616.07]
                                                              2005010196

GYAT/PCP
Printed in the United States of America
ISBN 1-57331-541-9 (cloth)
ISBN 1-57331-542-7 (paper)
ISSN 0077-8923

ANNALS OF THE NEW YORK ACADEMY OF SCIENCES
Volume 1042
May 2005

THE ROLE OF THE MITOCHONDRIA IN HUMAN AGING AND DISEASE

FROM GENES TO CELL SIGNALING

Editors and Conference Organizers
YAU-HUEI WEI, HORNG-MO LEE, AND CHUNG Y. HSU

This volume is the result of the Second Scientific Meeting of the Asian Society for Mitochondrial Research and Medicine, held April 1–2, 2004, in Taipei, Taiwan.

CONTENTS

Preface. *By* YAU-HUEI WEI, HORNG-MO LEE, AND CHUNG Y. HSU xiii

Part I. Mitochondrial Disease Models

Mitochondria-Based Model for Fetal Origin of Adult Disease and Insulin Resistance. *By* HONG KYU LEE, KYONG SOO PARK, YOUNG MIN CHO, YUN YONG LEE, AND YOUNGMI KIM PAK 1

Evolutional Analysis in Determining Pathogenic versus Nonpathogenic Mutations of ATPase 6 in Human Mitochondriopathy. *By* CHIN-YUAN TZEN AND TSU-YEN WU 19

Restoration of Mitochondrial Function in Cells with Complex I Deficiency. *By* YIDONG BAI, JEONG SOON PARK, JIAN-HONG DENG, YOUFEN LI, AND PEIQING HU 25

Hearing Loss in Mitochondrial Disorders. *By* CHANG-HUNG HSU, HAEYOUNG KWON, CHERNG-LIH PERNG, REN-KUI BAI, PU DAI, AND LEE-JUN C. WONG 36

Brain Single Photon Emission Computed Tomography in Patients with A3243G Mutation in Mitochondrial DNA tRNA. *By* PETERUS THAJEB, MING-CHE WU, BING-FU SHIH, CHIN-YUAN TZEN, MING-FU CHIANG, AND REY-YUE YUAN 48

Upregulation of Matrix Metalloproteinase 1 and Disruption of Mitochondrial Network in Skin Fibroblasts of Patients with MERRF Syndrome. *By* YI-SHING MA, YIN-CHIU CHEN, CHING-YOU LU, CHUN-YI LIU, AND YAU-HUEI WEI 55

Increased Oxidative Damage with Altered Antioxidative Status in Type 2 Diabetic Patients Harboring the 16189 T to C Variant of Mitochondrial DNA. *By* TSU-KUNG LIN, SHANG-DER CHEN, PEI-WEN WANG, YAU-HUEI WEI, CHENG-FENG LEE, TZU-LING CHEN, YAO-CHUNG CHUANG, TENG-YEOW TAN, KU-CHOU CHANG, AND CHIA-WEI LIOU   64

Alteration of the Copy Number of Mitochondrial DNA in Leukocytes of Patients with Hyperlipidemia. *By* CHIN-SAN LIU, CHING-LING KUO, WEN-LING CHENG, CHING-SHAN HUANG, CHENG-FENG LEE, AND YAU-HUEI WEI   70

A New Noninvasive Test to Detect Mitochondrial Dysfunction of Skeletal Muscles in Progressive Supranuclear Palsy. *By* YUNG-YEE CHANG, CHIANG-HSUAN LEE, MIN-YU LAN, HSIU-SHAN WU, CHIUNG-CHIH CHANG, AND JIA-SHOU LIU   76

High Prevalence of the COII/tRNA^{Lys} Intergenic 9-bp Deletion in Mitochondrial DNA of Taiwanese Patients with MELAS or MERRF Syndrome. *By* CHIN-SAN LIU, WEN-LING CHENG, YI-YUN CHEN, YI-SHING MA, CHENG-YOONG PANG, AND YAU-HUEI WEI   82

Part II. Mitochondrial Genome

Analysis of Proteome Bound to D-Loop Region of Mitochondrial DNA by DNA-Linked Affinity Chromatography and Reverse-Phase Liquid Chromatography/Tandem Mass Spectrometry. *By* YON-SIK CHOI, BO-KYUNG RYU, HYE-KI MIN, SANG-WON LEE, AND YOUNGMI KIM PAK   88

Mitochondrial Transcription Factor A in the Maintenance of Mitochondrial DNA: Overview of Its Multiple Roles. *By* DONGCHON KANG AND NAOTAKA HAMASAKI   101

Mitochondrial Genome Instability and mtDNA Depletion in Human Cancers. *By* HSIN-CHEN LEE, PEN-HUI YIN, JIN-CHING LIN, CHENG-CHUNG WU, CHIH-YI CHEN, CHEW-WUN WU, CHIN-WEN CHI, TSENG-NIP TAM, AND YAU-HUEI WEI   109

Frequent Occurrence of Mitochondrial Microsatellite Instability in the D-Loop Region of Human Cancers. *By* YUE WANG, VINCENT W.S. LIU, HEXTAN Y.S. NGAN, AND PHILLIP NAGLEY   123

Analysis of Heteroplasmy in Hypervariable Region II of Mitochondrial DNA in Maternally Related Individuals. *By* MEI-CHEN LO, HORNG-MO LEE, MING-WEI LIN, AND CHIN-YUAN TZEN   130

Deleted Mitochondrial DNA in Human Luteinized Granulosa Cells. *By* HENG-KIEN AU, SHYH-HSIANG LIN, SHIH-YI HUANG, TIEN-SHUN YEH, CHII-RUEY TZENG, AND RONG-HONG HSIEH   136

Identification of Human-Specific Adaptation Sites of *ATP6*. *By* BEY-LIING MAU, HORNG-MO LEE, AND CHIN-YUAN TZEN   142

Part III. Mitochondrial Function

Repeated Ovarian Stimulations Induce Oxidative Damage and Mitochondrial DNA Mutations in Mouse Ovaries. *By* HSIANG-TAI CHAO, SHU-YU LEE, HORNG-MO LEE, TIEN-LING LIAO, YAU-HUEI WEI, AND SHU-HUEI KAO   148

Calcium Stimulates Mitochondrial Biogenesis in Human Granulosa Cells. *By* TIEN-SHUN YEH, JAU-DER HO, VIVIAN WEI-CHUNG YANG, CHII-RUEY TZENG, AND RONG-HONG HSIEH 157

Dynamics of Mitochondria and Mitochondrial Ca^{2+} near the Plasma Membrane of PC12 Cells: A Study by Multimode Microscopy. *By* DE-MING YANG, CHUNG-CHIH LIN, HSIA YU LIN, CHIEN-CHANG HUANG, DIN PING TSAI, CHIN-WEN CHI, AND LUNG-SEN KAO 163

Propofol Specifically Inhibits Mitochondrial Membrane Potential but Not Complex I NADH Dehydrogenase Activity, Thus Reducing Cellular ATP Biosynthesis and Migration of Macrophages. *By* GONG-JHE WU, YU-TING TAI, TA-LIANG CHEN, LI-LING LIN, YUNE-FANG UENG, AND RUEI-MING CHEN 168

Abnormal Mitochondrial Structure in Human Unfertilized Oocytes and Arrested Embryos. *By* HENG-KIEN AU, TIEN-SHUN YEH, SHU-HUEI KAO, CHII-RUEY TZENG, AND RONG-HONG HSIEH 177

Oxidative Damage and Mitochondrial DNA Mutations with Endometriosis. *By* SHU-HUEI KAO, HSIENG-CHIANG HUANG, RONG-HONG HSIEH, SU-CHEE CHEN, MING-CHUAN TSAI, AND CHII-REUY TZENG 186

Neuroprotective Role of Coenzyme Q10 against Dysfunction of Mitochondrial Respiratory Chain at Rostral Ventrolateral Medulla during Fatal Mevinphos Intoxication in the Rat. *By* F.C.H. LI, H.P. TSENG, AND A.Y.W. CHANG 195

Part IV. Oxidative Stress

Mitochondrial Dysfunction and Oxidative Stress as Determinants of Cell Death/Survival in Stroke. *By* PAK H. CHAN 203

Oxidative Damage in Mitochondrial DNA Is Not Extensive. *By* KOK SEONG LIM, KANDIAH JEYASEELAN, MATTHEW WHITEMAN, ANDREW JENNER, AND BARRY HALLIWELL 210

Enhanced Generation of Mitochondrial Reactive Oxygen Species in Cybrids Containing 4977-bp Mitochondrial DNA Deletion. *By* MEI-JIE JOU, TSUNG-I PENG, HONG-YUEH WU, AND YAU-HUEI WEI 221

S-Nitrosoglutathione and Hypoxia-Inducible Factor–1 Confer Chemoresistance against Carbamoylating Cytotoxicity of BCNU in Rat C6 Glioma Cells. *By* DING-I YANG, SHANG-DER CHEN, JIU-HAW YIN, AND CHUNG Y. HSU 229

Celecoxib Induces Heme-Oxygenase Expression in Glomerular Mesangial Cells. *By* CHUN-CHENG HOU, SU-LI HUNG, SHU-HUEI KAO, TSO HSIAO CHEN, AND HORNG-MO LEE 235

Oxidative Stress–Induced Depolymerization of Microtubules and Alteration of Mitochondrial Mass in Human Cells. *By* CHENG-FENG LEE, CHUN-YI LIU, RONG-HONG HSIEH, AND YAU-HUEI WEI 246

Blood Lipid Peroxides and Muscle Damage Increased following Intensive Resistance Training of Female Weightlifters. *By* JEN-FANG LIU, WEI-YIN CHANG, KUEI-HUI CHAN, WEN-YEE TSAI, CHEN-LI LIN, AND MEI-CHIEH HSU 255

Anti-Inflammatory and Antioxidative Effects of Propofol on Lipopolysaccharide-Activated Macrophages. *By* RUEI-MING CHEN, TYNG-GUEY CHEN, TA-LIANG CHEN, LI-LING LIN, CHIA-CHEN CHANG, HWA-CHIA CHANG, AND CHIH-HSIUNG WU 262

Protective Effect of α-Keto-β-Methyl-n-Valeric Acid on BV-2 Microglia under Hypoxia or Oxidative Stress. *By* HSUEH-MEEI HUANG, HSIO-CHUNG OU, HUAN-LIAN CHEN, ROLIS CHIEN-WEI HOU, AND KEE CHING G. JENG 272

Oxidative Toxicity in BV-2 Microglia Cells: Sesamolin Neuroprotection of H_2O_2 Injury Involving Activation of p38 Mitogen-Activated Protein Kinase. *By* ROLIS CHIEN-WEI HOU, CHIA-CHUAN WU, JING-RONG HUANG, YUH-SHUEN CHEN, AND KEE-CHING G. JENG 279

Angiotensin II Stimulates Hypoxia-Inducible Factor 1α Accumulation in Glomerular Mesangial Cells. *By* TSO-HSIAO CHEN, JIN-FONG WANG, PAUL CHAN, AND HORNG-MO LEE 286

Effects of Glucose and α-Tocopherol on Low-Density Lipoprotein Oxidation and Glycation. *By* CHUN-JEN CHANG, RONG-HONG HSIEH, HUI-FANG WANG, MEI-YUN CHIN, AND SHIH-YI HUANG 294

Identification of Three Mutations in the Cu,Zn-Superoxide Dismutase (Cu,Zn-SOD) Gene with Familial Amyotrophic Lateral Sclerosis: Transduction of Human Cu,Zn-SOD into PC12 Cells by HIV-1 TAT Protein Basic Domain. *By* CHIH-MING CHOU, CHANG-JEN HUANG, CHWEN-MING SHIH, YI-PING CHEN, TSANG-PAI LIU, AND CHIEN-TSU CHEN ... 303

Kainic Acid–Induced Oxidative Injury Is Attenuated by Hypoxic Preconditioning. *By* CHENG-HAO WANG, ANYU CHANG, MAY-JYWAN TSAI, HENRICH CHENG, LI-PING LIAO, AND ANYA MAAN-YUH LIN 314

Antioxidant *N*-Acetylcysteine Blocks Nerve Growth Factor–Induced H_2O_2/ERK Signaling in PC12 Cells. *By* LIANG-YO YANG, WUN-CHANG KO, CHUN-MAO LIN, JIA-WEI LIN, JEN-CHINE WU, CHIEN-JU LIN, HUEY-HWA CHENG, AND CHWEN-MING SHIH 325

Part V. Antioxidants

Effect of Enhanced Prostacyclin Synthesis by Adenovirus-Mediated Transfer on Lipopolysaccharide Stimulation in Neuron-Glia Cultures. *By* MAY-JYWAN TSAI, SONG-KUN SHYUE, CHING-FENG WENG, YING CHUNG, DANN-YING LIOU, CHI-TING HUANG, HUAI-SHENG KUO, MENG-JEN LEE, PEI-TEH CHANG, MING-CHAO HUANG, WEN-CHENG HUANG, K.D. LIOU, AND HENRICH CHENG 338

Potential Mechanism of Blood Vessel Protection by Resveratrol, a Component of Red Wine. *By* HUEI-MEI HUANG, YU-CHIH LIANG, TZU-HURNG CHENG, CHENG-HEIEN CHEN, AND SHU-HUI JUAN 349

Pravastatin Attenuates Ceramide-Induced Cytotoxicity in Mouse Cerebral Endothelial Cells with HIF-1 Activation and VEGF Upregulation. *By* SHANG-DER CHEN, CHAUR-JONG HU, DING-I YANG, ABDULLAH NASSIEF, HONG CHEN, KEJIE YIN, JAN XU, AND CHUNG Y. HSU 357

Alleviation of Oxidative Damage in Multiple Tissues in Rats with Streptozotocin-Induced Diabetes by Rice Bran Oil Supplementation. *By* RONG-HONG HSIEH, LI-MING LIEN, SHYH-HSIANG LIN, CHIA-WEN CHEN, HUEI-JU CHENG, AND HSING-HSIEN CHENG 365

Curcumin Inhibits ROS Formation and Apoptosis in Methylglyoxal-Treated Human Hepatoma G2 Cells. *By* WEN-HSIUNG CHAN, HSIN-JUNG WU, AND YAN-DER HSUUW 372

Prevention of Cellular Oxidative Damage by an Aqueous Extract of *Anoectochilus formosanus*. *By* LENG-FANG WANG, CHUN-MAO LIN, CHWEN-MING SHIH, HUI-JU CHEN, BORCHERNG SU, CHENG-CHUANG TSENG, BAO-BIH GAU, AND KUR-TA CHENG 379

Inhibitory Effects of a Rice Hull Constituent on Tumor Necrosis Factor α, Prostaglandin E2, and Cyclooxygenase-2 Production in Lipopolysaccharide-Activated Mouse Macrophages. *By* SHENG-TUNG HUANG, CHIEN-TSU CHEN, KUR-TA CHIENG, SHIH-HAO HUANG, BEEN-HUANG CHIANG, LENG-FANG WANG, HSIEN-SAW KUO, AND CHUN-MAO LIN 387

Raffinee in the Treatment of Spinal Cord Injury: An Open-Labeled Clinical Trial. *By* HSIN-YING CHEN, JER-MIN LIN, HUNG-YI CHUANG, AND WEN-TA CHIU 396

Part VI. Apoptosis

Induction of Thioredoxin and Mitochondrial Survival Proteins Mediates Preconditioning-Induced Cardioprotection and Neuroprotection. *By* CHUANG C. CHIUEH, TSUGUNOBU ANDOH, AND P. BOON CHOCK 403

Mitochondrion-Targeted Photosensitizer Enhances the Photodynamic Effect–Induced Mitochondrial Dysfunction and Apoptosis. *By* TSUNG-I PENG, CHENG-JEN CHANG, MEI-JIN GUO, YU-HUAI WANG, JAU-SONG YU, HONG-YUEH WU, AND MEI-JIE JOU 419

Attenuation of UV-Induced Apoptosis by Coenzyme Q_{10} in Human Cells Harboring Large-Scale Deletion of Mitochondrial DNA. *By* CHENG-FENG LEE, CHUN-YI LIU, SHU-MEI CHEN, MARIANNA SIKORSKA, CHEN-YU LIN, TZU-LING CHEN, AND YAU-HUEI WEI 429

Antisense RNA to Inducible Nitric Oxide Synthase Reduces Cytokine-Mediated Brain Endothelial Cell Death. *By* DING-I YANG, SHAWEI CHEN, UTHAYASHANKER R. EZEKIEL, JAN XU, YINGJI WU, AND CHUNG Y. HSU 439

2,6-Diisopropylphenol Protects Osteoblasts from Oxidative Stress-Induced Apoptosis through Suppression of Caspase-3 Activation. *By* RUEI-MING CHEN, GONG-JHE WU, HWA-CHIA CHANG, JUE-TAI CHEN, TZENG-FU CHEN, YI-LING LIN, AND TA-LIANG CHEN 448

Nitric Oxide Induces Osteoblast Apoptosis through a Mitochondria-Dependent Pathway. *By* WEI-PIN HO, TA-LIANG CHEN, WEN-TA CHIU, YU-TING TAI, AND RUEI-MING CHEN 460

Bcl-2 Gene Family Expression in the Brain of Rat Offspring after Gestational and Lactational Dioxin Exposure. *By* SHWU-FEN CHANG, YU-YO SUN, LIANG-YO YANG, SSU-YAO HU, SHIH-YING TSAI, WEN-SEN LEE, AND YI-HSUAN LEE 471

IL-5 Inhibits Apoptosis by Upregulation of c-myc Expression in Human Hematopoietic Cells. *By* SHU-HUI JUAN, JEFFREY JONG-YOUNG YEN, HORNG-MO LEE, AND HUEI-MEI HUANG 481

Detection of Apoptosis and Necrosis in Normal Human Lung Cells Using ^{1}H NMR Spectroscopy. *By* CHWEN-MING SHIH, WUN-CHANG KO, LIANG-YO YANG, CHIEN-JU LIN, JUI-SHENG WU, TSUI-YUN LO, SHWU-HUEY WANG, AND CHIEN-TSU CHEN 488

Cadmium Toxicity toward Caspase-Independent Apoptosis through the Mitochondria–Calcium Pathway in mtDNA-Depleted Cells. *By* YUNG-LUEN SHIH, CHIEN-JU LIN, SHENG-WEI HSU, SHENG-HAO WANG, WEI-LI CHEN, MEI-TSU LEE, YAU-HUEI WEI, AND CHWEN-MING SHIH 497

Pectinesterase Inhibitor from Jelly Fig (*Ficus awkeotsang* Makino) Achene Induces Apoptosis of Human Leukemic U937 Cells. *By* JIA-HUEI CHANG, YUH-TAI WANG, AND HUNG-MIN CHANG 506

Enhancement of Cisplatin-Induced Apoptosis and Caspase 3 Activation by Depletion of Mitochondrial DNA in a Human Osteosarcoma Cell Line. *By* HSIU-CHUAN YEN, YI-CHIA TANG, FAN-YI CHEN, SHIH-WEI CHEN, AND HIDEYUKI J. MAJIMA 516

Thallium Acetate Induces C6 Glioma Cell Apoptosis. *By* CHEE-FAH CHIA, SOUL-CHIN CHEN, CHIN-SHYANG CHEN, CHUEN-MING SHIH, HORNG-MO LEE, AND CHIH-HSIUNG WU 523

Effects of Gonadotrophin-Releasing Hormone Agonists on Apoptosis of Granulosa Cells. *By* NU-MAN TSAI, RONG-HONG HSIEH, HENG-KIEN AU, MING-JER SHIEH, SHIH-YI HUANG, AND CHII-RUEY TZENG 531

Index of Contributors 539

Financial assistance was received from:

Co-sponsors
- NATIONAL SCIENCE COUNCIL, TAIWAN
- MINISTRY OF EDUCATION, TAIWAN
- ASIAN SOCIETY FOR MITOCHONDRIAL RESEARCH AND MEDICINE (ASMRM)

Major Funder
- MITOCON LTD.
- PFIZER INC.
- SERONO SINGAPORE PTE LTD.

Contributors
- BECKMAN COULTER INC.
- EAST WIND LIFE SCIENCE CO., LTD.
- FMC CO.
- LIN TRADING CO., LTD.
- ROCHE PRODUCTS LTD.
- YUANYU INDUSTRY CO., LTD.
- SANOFI-SYNTHELABO INC.
- UNITED CORPS CO., LTD.
- BECTON, DICKINSON AND COMPANY
- AMERSHAM PHARMACIA BIOTECH CO., LTD.
- CHIH CHIN H&W ENTERPRISE CO., LTD.
- DOUBLE ENGLE ENTERPRISE CO., LTD.
- YOUNG DEAR CO., LTD.
- ANATECH CO., LTD.
- BIOMAN SCIENTIFIC CO., LTD.
- CHINHSIN ENTERPRISE CO., LTD.
- PROTECH TECHNOLOGY ENTERPRISE CO., LTD.
- RAINBOW BIOTECHNOLOGY CO., LTD.
- SMARTEC SCIENTIFIC CO., LTD.
- NATURE OPERA BIOTECHNOLOGY INC.

The New York Academy of Sciences believes it has a responsibility to provide an open forum for discussion of scientific questions. The positions taken by the participants in the reported conferences are their own and not necessarily those of the Academy. The Academy has no intent to influence legislation by providing such forums.

Preface

In early February 2003, a group of scholars, researchers, and physicians in Asia organized the 1st Scientific Meeting of the Asian Society for Mitochondrial Research and Medicine (ASMRM) in Seoul, Korea, with the belief that understanding mitochondria is essential in solving the mysteries of life and illness in humans. Under the leadership of Dr. Hong Kyu Lee, President of ASMRM, the 1st Scientific Meeting of the ASMRM was a great success. This international conference brought together basic and clinical investigators interested in mitochondrial research. There were more than 200 participants from more than 10 countries. The proceedings were published as Volume 1011 of the *Annals of the New York Academy of Sciences*.

The 2nd ASMRM Conference was held on April 1–2, 2004 in Taipei, Taiwan. The scientific program featured one keynote address, three plenary sessions, four special lectures, and more than 110 free papers contributed by investigators from Hong Kong, Japan, Korea, Singapore, Taiwan, and the United States, as well as other countries. This volume contains review articles by invited speakers and selected original papers presented at the 2nd ASMRM Conference. These papers cover advances on various topics of mitochondrial research and medicine in a timely and comprehensive manner.

—YAU-HUEI WEI
—HORNG-MO LEE
—CHUNG Y. HSU

Mitochondria-Based Model for Fetal Origin of Adult Disease and Insulin Resistance

HONG KYU LEE,[a] KYONG SOO PARK,[a] YOUNG MIN CHO,[a] YUN YONG LEE,[b] AND YOUNGMI KIM PAK[c]

[a]*Department of Internal Medicine, Seoul National University College of Medicine, Seoul, Korea*

[b]*Department of Internal Medicine, Korea Cancer Center Hospital, Seoul, Korea*

[c]*Department of Biochemistry, Ulsan University College of Medicine, Seoul, Korea*

> ABSTRACT: Insulin resistance has been recognized as the fundamental underlying metabolic defect in the pathogenesis of metabolic syndrome, a clustering of cardiovascular risk factors such as diabetes, hypertension, dyslipidemia, and obesity. Recent studies established that mitochondrial dysfunction is involved in insulin resistance in general and fetal origin of this state in particular. Because genes are the fundamental molecular basis of inheritance—and thus the cornerstones of evolution—a model explaining insulin resistance is based at the gene level at best. Since a certain mtDNA polymorphism, 16189T>C, is associated with insulin resistance, mtDNA has to be a basic component of the gene-based model. We developed a mitochondria-based model that explains insulin resistance in terms of quantitative and qualitative change of the mitochondrion and its DNA. This model can accommodate several important hypotheses, such as thrifty genotype hypothesis, thrifty phenotype hypothesis, fetal insulin hypothesis, contribution of metabolic imprinting by epigenetic changes, and many other features associated with insulin resistance. We will discuss mechanisms that indicate why the perturbed initial condition of mitochondrial function should lead to the reduced insulin sensitivity.
>
> KEYWORDS: mitochondrial DNA; mitochondrial dysfunction; insulin resistance; fetal malnutrition; thrifty phenotype

THRIFTY GENOTYPE, THRIFTY PHENOTYPE, AND INSULIN RESISTANCE

Insulin resistance has been recognized as the fundamental metabolic defect of metabolic syndrome, a clustering of cardiovascular risk factors such as diabetes, hypertension, dyslipidemia, and obesity.[1,2] Many epidemiological studies have revealed links between various indices of reduced intrauterine and early postnatal growth, and susceptibility to insulin resistance syndrome in adult life (for review,

Address for correspondence: Hong Kyu Lee, M.D., Ph.D., Department of Internal Medicine, Seoul National University College of Medicine, 28 Yongon-Dong, Chongno-Gu, Seoul, 110-744, Korea. Voice: +82-2-760-2266; fax: +82-2-762-7966.
  hkleemd@snu.ac.kr

see Ref. 3). Barker and Hales suggested that fetal origin of this "thrifty phenotype," a physical condition programmed by poor nutritional condition in early life, leads to insulin resistance in later life.[4] Altered programming of the body, such as hypothalamo-pituitary-adrenal axis, is suggested to cause decreased insulin sensitivity.[5,6] Compared to the nutritionally induced changes in the genetic program, genes were suggested responsible for insulin resistance. A contrasting concept, the "thrifty genotype" hypothesis of Neel[7] was originally proposed to explain the very high prevalence of obesity and diabetes in some American Indians such as the Pima. He suggested that native Indians might have accumulated genes that are beneficial for survival under famine conditions, but are detrimental when society becomes affluent. Both hypotheses are well accepted; the predisposition to insulin resistance is likely to be the result of both genetic and fetal environmental factors. Although Hattersley and his co-workers suggested that altered fetal growth may be a phenotype of a genotype—in other words, the thrifty phenotype is the result of a thrifty genotype[8,9]—one cannot explain all the thrifty phenotypes with genotypes, because experimental study using the same strain of animals induced insulin resistance in offspring.

The "fetal insulin" hypothesis of Hattersley emphasizes the insulin secreted by the fetal pancreas in response to maternal glucose concentrations as a key factor. There is strong evidence supporting the fetal insulin hypothesis. For example, monogenic diseases that impair sensing of glucose, such as glucokinase gene mutations, lower insulin secretion, or increase insulin resistance, as in IGF-1 gene polymorphism, are associated with impaired fetal growth.[10,11] Polygenic influences resulting in insulin resistance in the normal population are therefore likely to result in lower birth weight.

Further study is needed to determine whether common gene variants can explain the association between reduced birth weight and increased risk of type 2 diabetes or insulin resistance. Considering recent advances in epigenetics and genomic imprinting, these hypotheses were criticized in that they omitted the important contribution of these mechanisms in the pathogenesis of insulin resistance.[12,13]

Recent studies established without doubt that mitochondrial dysfunction in general, or dysfunction of mitochondrial genome, might be a major abnormality of this state. Because genes are the fundamental molecular basis of inheritance—and thus cornerstones of evolution—explaining insulin resistance is based at the gene level at best.

As a certain mtDNA polymorphism, 16189T>C, is associated with insulin resistance,[14–19] mtDNA has to be a fundamental component of the model. Here we present a mitochondria (and its DNA)–based model that is versatile enough to accommodate all the components of previously proposed hypotheses and other unexplained features such as imprinting.

MITOCHONDRIA-BASED MODEL OF INSULIN RESISTANCE—INTRODUCTION

By expanding three major hypotheses presented above and taking into account several observations that show (1) a certain mitochondrial DNA polymorphism, 16189T>C variant, is associated with insulin resistance, (2) mtDNA density in pe-

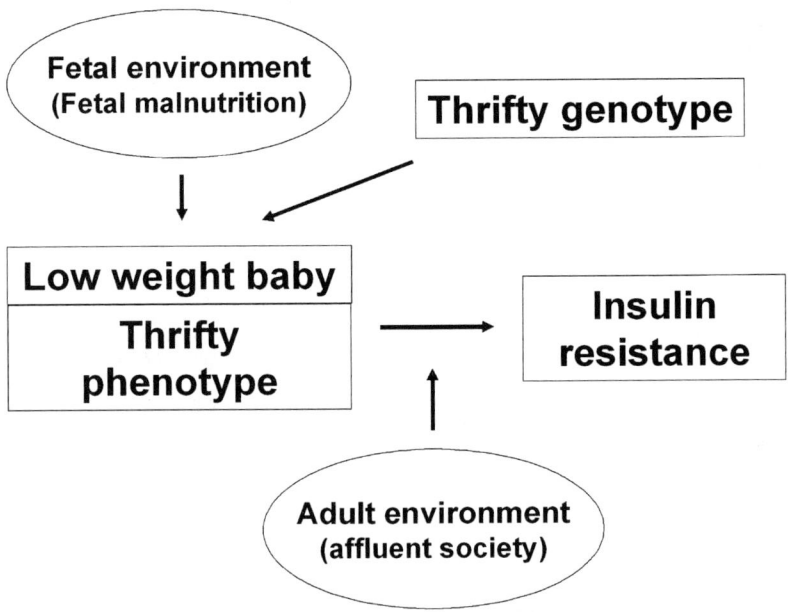

FIGURE 1. Diagram of the thrifty genotype/phenotype hypotheses.

ripheral blood is associated with insulin sensitivity and predictive of development of diabetes at population level, (3) mtDNA density is lower in offspring of a diabetic parent, (4) mitochondrial structure and mtDNA density are altered in offspring of malnutrition induced dams, and (5) much other supporting evidence, we developed a model based on mitochondria or the mitochondrial genome, which is shown in FIG. 1. This model is based on our early idea,[20] which was revised recently.[21]

This model accommodates three critical elements newly identified: first, taurine is an important structural component of mitochondrial tRNA, and its deficiency during early life causes poor fetal development and impaired insulin secretion in offspring, and, most importantly, a lowered fetal plasma taurine level from a low-protein diet is the main predictor of the fetal plasma insulin level.[22] Second, epigenetic mechanisms play a critical role at the nexus between nutrition and genome; and finally that highly active antiretroviral therapy (HAART) for AIDS patients causes mitochondrial toxicity and insulin resistance.[23–25] Details on metabolic imprinting will not be discussed here (for two recent reviews, see Refs. 26 and 27).

Since the mitochondrion has its own genome from its symbiotic origin and is under the control of the nuclear genome, it plays the dutiful servant role of generating ATP for the cell. On the other hand, the mitochondrion also plays a role as a commander, to protect and to supply substrates for cell survival. From the mitochondrial perspective, one can incorporate nuclear gene effects without making a model highly complex, which is otherwise inevitable. This synthesis bring us to a view that one of the defects underlying insulin resistance is a quantitative or qualitative change of mitochondria or their DNA, which might result from the thrifty genes the subjects have, as well as the programming effect of poor nutritional conditions during early life and the environmental effect during development. This model considers the mitochondrial genome as a thrifty gene that is rapidly evolving to adapt to the human environment,[28,29] and (thrifty) nuclear genes are considered as controlling elements controlling mitochondrial function.

BASES OF MITOCHONDRIA-BASED MODEL—BIOCHEMICAL ASPECT

Mitochondrial Abnormalities Are Linked to Diabetes and Insulin Resistance

mtDNA mutations are well known to cause diabetes by affecting insulin secretion from pancreatic beta cells.[30] Glucose enters the cell through a specific transporter, followed by phosphorylation of glucose, activation of glycolysis, and stimulation of mitochondrial oxidative phosphorylation, resulting in an increase of intracellular ATP. This leads to the closure of the ATP-sensitive K^+ channel, the opening of the Ca^{2+} channel, and increased intracellular Ca^{2+}, which eventually triggers insulin secretion. It is thus reasonable that pancreatic beta cells with abnormal mitochondria would show a poor insulin-secretory response to glucose stimulation.

In addition to reducing insulin secretion, mitochondrial abnormalities can also cause insulin resistance. Poulton *et al.* reported that a polymorphism in the first hypervariable region of the mtDNA control region (16189 T>C) is associated with insulin resistance in the English population.[14] The finding was confirmed in other populations, including Koreans.[15–19]

Our laboratory found that the decreased mtDNA density in peripheral blood preceded the development of diabetes,[31] suggesting that quantitative mitochondrial abnormalities could be primary events. Inverse correlations were noted between mtDNA content and components of the metabolic syndrome such as blood pressure, fasting glucose level, and waist-to-hip circumference ratio,[31] suggesting for the first time that mitochondrial abnormality might be associated with insulin resistance. Song *et al.* extended this observation to diabetic offspring and found that mtDNA density was indeed associated with insulin sensitivity in the offspring of type 2 diabetic patients.[32]

Petersen *et al.* established that there is deranged mitochondrial function in insulin resistance.[33,34] They reported that elderly people were insulin resistant compared with young controls matched for lean body mass and fat mass, and this resistance was attributable to reduced insulin-stimulated muscle glucose metabolism. These changes were associated with increased fat accumulation in muscle and with an approximately 40% reduction in mitochondrial oxidative and phosphorylation activity. This group extended the study to the insulin-resistant diabetic offspring, and found that they have a 80% increase in intramyocellular lipid content and a 30% reduction in mitochondrial phosphorylation activity compared with insulin-sensitive, age-, height-, and weight-matched control subjects.[34]

As peripheral blood mtDNA density was correlated positively with fat oxidation during hyperinsulinemic clamp,[35] and as insulin resistance in the elderly is related to the increase in intramyocellular fatty acid metabolites, these two observations might be corollaries to a common phenotype. This possibility is consistent with the fact that the whole-body oxygen consumption rate correlates with insulin sensitivity.[36,37]

Park et al.[38] investigated the effects of mtDNA depletion on glucose metabolism at the cellular level. When the human hepatoma SK-Hep1 cells were treated with repeated sublethal doses of ethidium bromide, they lost mtDNA (ρ^0 cells) and the cells failed to hyperpolarize their mitochondrial membrane potential in response to glucose stimulation. Intracellular ATP content, glucose-stimulated ATP production, glucose uptake, steady-state mRNA, and protein levels of the glucose transporters, and cellular activities of the glucose metabolizing enzymes, including hexokinase, decreased. These results suggest that quantitative reduction of mtDNA suppresses the expression of nuclear DNA-encoded glucose transporters and enzymes of glucose metabolism.

Insulin Resistance and Mitochondrial Dysfunction Are Induced by Malnutrition in Animal Models

Several metabolic abnormalities that lead to insulin resistance in protein-malnourishment models also have been reported. The muscle and liver organ weight were reduced in these models,[39] and the activities and gene expression of insulin-sensitive hepatic enzymes changed. In addition, the glucokinase activity was reduced and phosphoenolpyruvate carboxykinase (PEPCK) increased, both resulting in increased hepatic glucose production.[40,41] The ability of insulin to inhibit glucagon-stimulated glucose output from the perfused liver was lost and indeed reversed.[42] Proteome analysis of the fetal pancreas to examine the intrauterine programming of β cell gene expression showed that the expressions of 70 proteins were changed by fetal protein malnutrition.[43] They include the proteins related to mitochondrial energy transfer, glucose metabolism, RNA and DNA metabolism, protein synthesis and metabolism, the cell cycle and differentiation, cellular structure, and cellular defense. The glucose tolerance of the offspring of low-protein–fed dams was markedly age dependent. The younger of such offspring had improved glucose tolerance at 12 weeks of age when compared with controls,[44] which is explained by adaptation of the peripheral tissue through increased whole-body insulin sensitivity.[45–47] These age-dependent changes are compatible with ever-decreasing mitochondrial function along the aging process.[48]

Ogata et al.[49,50] have developed a rat model using utero-placental insufficiency as a cause of intrauterine growth retardation by inducing ischemia to the fetus by partially ligating the placental blood supply. This model exhibits marked insulin resistance early in life, characterized by blunted whole-body glucose disposal in response to insulin,[51] and impaired insulin suppression of hepatic glucose output.[52] Mitochondria of intrauterine growth retardation (IUGR) rats have not been well studied but, in skeletal muscle from IUGR, rats exhibited significantly decreased rates of stage 3 oxygen consumption with pyruvate, glutamate, α-ketoglutarate, and succinate. Such a defect in mitochondria leads to a chronic reduction in the supply of ATP available from oxidative phosphorylation, and the authors of the study concluded that impaired ATP synthesis in muscle compromises energy-dependent glucose transport and glycogen synthesis, which in turn contribute to the insulin resistance and hyperglycemia of type 2 diabetes.[53]

In the offspring of dams fed a low-protein diet during gestation and lactation, the mtDNA content of the liver and skeletal muscle was reduced and did not recover until 20 weeks of age, despite restoration of nutrition after weaning.[54] The reduced mtDNA content was accompanied by a decrease in mitochondrial DNA-encoded gene expression.[54] Park et al.[55] found that these rats also have decreased mtDNA levels in the pancreas at 25 weeks of age, accompanied by decreased pancreatic β cell mass, and reduced insulin secretory responses to glucose load. These findings indicate that poor nutrition in early life causes long-lasting changes in the mitochondria and their mtDNA densities.

Mitochondria-Based Model Accommodates Nuclear Gene Effects

The replication of mtDNA is under the relaxed control of the cell cycle. All of the proteins required for replication and transcription of mtDNA are nuclear encoded, suggesting the importance of mitochondrial and nuclear interaction in mtDNA replication. The detailed process of mtDNA replication is well reviewed by Clayton et al.[56] and Lecrenier et al.[57] The control of mitochondrial biogenesis is extremely complex, involving the coordinated expression of hundreds of genes. The nuclear respiratory factors NRF-1 and NRF-2 are transcriptional regulators for the genes of subunits of the oxidative phosphorylation system, as well as for many genes involved in mtDNA replication such as the Tfam gene.[58] The fact that numerous genes are regulated by NRFs suggests that the NRF-dependent genes are involved in the control of mitochondrial biogenesis in general, which is a possible link between external stimuli and mtDNA content.[59] The mechanisms of how malnutrition in early life induces mtDNA reduction that lasts until the adult period without recovery despite restoration of nutrition are not clear yet. Further understanding of nuclear gene effects on insulin resistance could be incorporated to this model as molecular biology advances.

Taurine Is a Key Component of Mitochondrial tRNA, and Taurine Availability Is a Key Determinant for the Development of Insulin Resistance

In an isocaloric, low-protein fetal-malnutrition model, although basal blood sugar and plasma insulin were not modified, the amino acid profile was disturbed in the maternal and fetal plasma as well as in the amniotic fluid.[60] The levels of essential, branched, and sulfur-containing amino acids were reduced. The most affected amino acid in maternal and fetal plasma, amniotic fluid, and fetal islets is taurine. Surprisingly, supplementation of the maternal low-protein diet by 2.5% taurine in the drinking water completely restored islet cell proliferation as well as the insulin-like growth factor 2 (IGF-2) and vascular endothelial growth factor content in the islets of fetuses and suckling pups. Taurine supplementation also restored fetal islet vascularization by increasing the number of blood vessels.[61] Moreover, taurine supplementation of the mother's diet normalized the insulin secretion of the fetal islets.[62] Taurine is an amino acid that does not participate in protein synthesis, but has a function in cholesterol excretion, as a neurotransmitter and as a potent antioxidant (for review, see Ref. 63). Taurine is not considered an essential amino acid for humans because it can be synthesized from cysteine in the liver.[64] However, the plasma concentration of taurine in the fetus is 1.5-fold that of maternal blood, and this level is mostly dependent on transport from the maternal blood through the placenta because the bioregulatory systems for taurine in the fetus are not fully developed.[65] Reduced activity of placental taurine transporters results in low fetal-taurine levels and IUGR fetuses.[66] Most importantly, Bertin et al. reported that a lowered fetal-plasma taurine level from a low-protein diet was the main predictor of the fetal-plasma insulin level.[22]

A recent report provides unequivocal evidence that taurine critically affects mitochondrial function. Suzuki et al.[67] found two novel taurine-containing modified uridines, 5-taurinomethyluridine and 5-taurinomethyl-2-thiouridine, in mtDNA, and they showed that taurine was a constituent of mitochondrial tRNAs. Modification of the taurine-containing uridines has been found to be lacking in mutant mitochondrial tRNAs for Leu (UUR) and Lys in pathogenic cells of the mitochondrial encephalomyopathies, mitochondrial myopathy, encephalopathy, lactic acidosis and stroke-like episode (MELAS), and myoclonus epilepsy associated with ragged red fibers (MERRF), respectively. In addition, these modification deficiencies of mutant tRNAs cause defective translation due to weak codon–anticodon interactions, which might significantly contribute to the defective mitochondrial function in mitochondrial diseases. It is likely that low taurine in the fetus may induce a deficiency of modification of nucleosides that leads to defective translation activity and mitochondrial function, and may result in impaired insulin secretion and insulin resistance.

Furthermore, there is an interesting study by Nakaya et al.[68] that reports that abdominal fat accumulation, hyperglycemia, and insulin resistance were significantly ameliorated by taurine supplementation in Otsuka Long-Evans Tokushima Fatty (OLETF) rats, a genetic model of spontaneous development of type 2 diabetes, which exhibits hyperglycemia, obesity, and insulin resistance, similar to that seen in humans.

HAART in HIV-Infected Subjects Causes Insulin Resistance Syndrome by Mitochondrial Toxicity

Another example is the nucleoside-analogue reverse transcriptase inhibitors that are used as therapy for human immunodeficiency virus (HIV) infection, which act by inhibiting HIV replication. After phosphorylation in the cell, these agents are internalized into nascent mtDNA by their substitution for the natural base, which leads to mtDNA damage. The human DNA polymerase gamma is also inhibited,[23] thereby inhibiting replication of mitochondrial DNA, thus leading to depletion of mitochondrial DNA.[24,25] A syndrome almost identical to clinical features of metabolic syndrome, type 2 diabetes, central obesity, increased intra-abdominal fat, hyperlipidemia, and insulin resistance is caused by mtDNA depletion resulting from the toxicity of these drugs.

Genomic Imprinting Is Changed in Malnutrition

The long-term programming effect of early malnutrition can be partly explained by an epigenetic mechanism, because inheritance of the thrifty phenotype shows characteristics of genetic imprinting. Once an offspring is affected by malnutrition, it takes several generations to fully recover from it.[69] This phenomenon can only be explained on the basis of imprinting of the genes. Genomic imprinting is an epigenetic phenomenon in which only a single allele of a gene is expressed in a parent-of-origin–dependent manner (for review, see Ref. 26). In mammals, the imprinted genes are particularly implicated in the regulation of fetal growth and development, cancer, and aging. The DNA modification for imprinting includes DNA methylation, histone modifications, and differences in chromatin structure. The mice that lack methylation because of a deficiency in DNA-methyltransferase-1 die at early post-implantation. Some genetic imprints are already lost at this stage, suggesting a crucial role for DNA methylation in determining the imprinted status of genes.[70]

Transient nutritional stimuli at critical ontogenic stages can yield lasting influences on the expression of genes by interacting epigenetic mechanisms.[27] If certain genomic regions such as imprinted domains are especially labile to such influences, early nutritional influence on these genomic components could have a long-lasting impact on phenotype. Indeed, because nutrition can affect imprinted growth factors such as IGF-2, insulin has been established in animal experiments.[71] Several studies revealed that early nutritional changes, by changing the culture media used during *in vitro* manipulations of early mouse embryos, alter allelic methylation and expression of imprinted genes,[72,73] and lower birth weight.[73] Recent observations have identified that *in vitro* manipulation of human embryos also induces imprinting alterations that lead to congenital disorders such as Angelman's syndrome[74] and Beckwith-Wiedermann syndrome.[75] Data from animal models have indicated that the epigenetic lability of imprinted genes is not limited to the early embryonic period. Hu *et al.*[76] treated mice with 5-azacytidine—an inhibitor of DNA methylation—at postnatal days 11 and 14, and found dramatic alteration in allelic expression of the IGF-2 gene. Waterland and Garza reported that two of 10 differentially expressed genes in the islets of undernourished rats during the suckling period were imprinted genes.[77]

Mammalian one-carbon metabolism, which ultimately synthesizes S-adenosylmethionine, providing the methyl group for all biological methylation reactions, is highly dependent on dietary methyl donors and cofactors.[78] Dietary methionine and choline are the major source of one-carbon units, and folic acid, vitamin B12, and pyridoxal phosphate are critical cofactors in methyl metabolism. The availability of dietary methyl donors and cofactors during critical ontogenic periods therefore might influence DNA methylation patterns.[79] For example, the coat-color distribution of A^{vy}/a offspring (the A^{vy} mutation resulted from the insertion of retrotransposon into an exon of the agouti gene; a is the loss of function mutation) was shifted when their mothers' diets were supplemented with methyl donors and cofactors.[80] The coat-color shift in the offspring caused by methyl donor supplementation of the dam was revealed to be caused by altered methylation status of the agouti promoter in the offspring,[81] suggesting the epigenetic effects of dietary nutrition.

Last, it will be important to note the unpublished data of Ng *et al.*, who found that mtDNA density is more affected by the paternal history of diabetes,[82] suggesting that imprinting in paternal genes might be directly involved in mtDNA density determination and inheritance of diabetes mellitus.

Vicious Cycle of Mitochondrial Dysfunction and Oxidative Stress

Protein malnutrition is associated with depressed antioxidant defense systems and increased oxidative stress.[83] Proteome analysis of fetal protein-malnourished pancreas revealed that antioxidant protein 2, which protects the pancreas against oxidative injury[84] by reducing hydrogen peroxide (H_2O_2), was downregulated.[84] Because of the role of oxidative stress in mtDNA damage,[85] we can speculate that oxidative stress might be involved in malnutrition-associated mitochondrial changes.

The mitochondrion is a major organelle for free radical production and also most vulnerable to free radical damage. Once mitochondrial function is deranged, there is more free radical production and more rapid mitochondrial function decline. Thus, it is likely once the mitochondrial capacity of the whole body is set low, a vicious cycle will operate, leading to early exhaustion. This mechanism is one of the most plausible hypotheses explaining aging.[86] Because the antioxidant system is also deficient in offspring of malnourished animals, one can explain the development of insulin resistance or the metabolic syndrome in later life by persistent decline of mitochondrial system through a vicious cycle, a feature of the mitochondria-based gene model.

BASES OF MITOCHONDRIA-BASED MODEL—BIOPHYSICAL ASPECT

Mitochondrial Function of Unit Cell and Tissue Is Quantitatively Related to Whole-Body VO_2max and Insulin Resistance

In 1956, Smith reported that the mitochondrial density of liver correlated with whole-body energy expenditure, and the relative amount of mitochondria in any given tissue was suggested to be the controlling factor in determining the regression of oxygen utilization on total body size of the species.[87] Recently, Rasmussen et al.[88] examined whether parameters of isolated mitochondria could account for the *in vivo* maximum oxygen uptake (VO_2max) of human skeletal muscle. VO_2max and work performance of the quadriceps muscle of six volunteers were measured in the knee extensor model (range 10–18 mmol $O_2 \times min^{-1} \times kg^{-1}$ at work rates of 22–32 W/kg). Mitochondria were isolated from the same muscle at rest. Strong correlations were obtained between VO_2max and a number of mitochondrial parameters (mitochondrial protein, cytochrome aa3, citrate synthase, and respiratory activities). The activities of citrate synthase, succinate dehydrogenase, and pyruvate dehydrogenase measured in isolated mitochondria corresponded, respectively, to 15, 3, and 1.1 times the rates calculated from VO_2max. Fully coupled *in vitro* respiration, which is limited by the rate of ATP synthesis, could account for, at most, 60% of the VO_2max. It is thus not surprising that mtDNA density correlates with oxidative capacity in animal and man, as discussed previously.[89] Furthermore, because this capacity is related to insulin sensitivity,[90] one can appreciate that mitochondrial function at the cell or tissue level is quantitatively related to the whole-body insulin resistance.[20,91]

Biophysical Law between Metabolic Power and Body Mass— Allometric Scaling Law

Recent reformulation of allometric scaling laws by West and his group[92] provides reasons why mitochondrial density at cell or tissue level should correlate with whole-body insulin sensitivity. This group generated a general model for the origin of allometric scaling law from three broad principles required in the living organ-

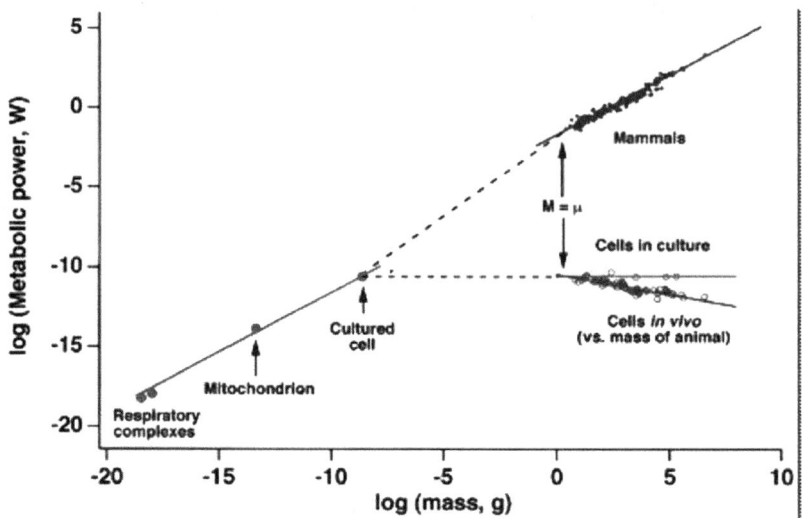

FIGURE 2. A logarithmic plot of metabolic power as a function of mass (from West et al.[94]). The entire range is shown, covering 27 orders of magnitude from a cytochrome oxidase molecule and respiratory complex through a mitochondrion and a single cell *in vitro*, up to whole mammals. The *solid lines* through the corresponding dots are the $M^{3/4}$ predictions. The *dashed line* is the linear extrapolation from $M = \mu$, the approximate mass predicted and observed for the smallest mammal to an isolated mammalian cell. (From West et al.[94] Reprinted, with permission, from the *Proceedings of the National Academy of Sciences USA*.)

isms: (1) a space-filling, fractal-like branching pattern is necessary to supply the organism with what it needs; (2) the final unit of this branching pattern is a size-invariant unit; and (3) the energy required to distribute resources is minimized. The metabolic rate of organisms is known to be proportional to body mass raised to the power of three-quarters developed from Kleiber's original analysis.[93] West et al.[94] extended the relation between body mass and metabolic power, which covers 27 orders of magnitude of body size, from elephant, mice, and mitochondrion to electron transfer chain enzymes (FIG. 2).

It is illuminating that each cultured cell has identical metabolic power (universal metabolic rate), and *in vivo* the metabolic power decreases as weight of the animal increases. Body temperature is a key constraint of metabolic scale law.[95] They argued that this relationship is inevitable to keep core body temperature optimal, to which all the enzyme systems were adaptively evolved. If heat production of each cell or tissue is not changed according to body size, core temperature will go up too high or too low and damage cellular function. As West et al. discuss in their findings, thus if one knows the scale of power generation at the molecular level, it will be sufficient to predict the metabolic rate of individual mitochondria and cells (whether *in vitro* or *in vivo*) as well as intact mammals. Since this biophysical law of metabolic scaling is true, it predicts that the reverse statement will be true. In other words, if the unit cellular metabolic rate is decreased, increased body mass will occur adaptively; in people who were to develop obesity, metabolic power of the unit cell should be lower. This prediction is compatible with data collected on Pima Indians (to be discussed below).

Physiologic Mechanisms for Allometric Scale Law—Adaptive Thermogenesis

Adaptive thermogenesis is the physiological process whereby energy is dissipated in the form of heat in response to external stimuli such as exposure to cold and ingestion of high-calorie diets. Adaptive thermogenesis has been regarded as a physiological defense against obesity,[96] where brown fat plays a major role.[97] In humans, there are great differences in how individuals metabolize an intentional caloric overload; some people store most of this energy as fat, while others dissipate much of it through altered energy expenditure, including adaptive thermogenesis.[98,99] Adaptive thermogenesis occurs primarily in the mitochondria of brown fat and skeletal muscle. Since brown fat is hardly found in large animals including humans, skeletal muscle is thought to be the site of primary importance for this process.

The uncoupling proteins (UCPs) are small intramembranous mitochondrial proteins that are expressed in a tissue-selective manner and play a key role in thermogenesis.[100,101] Cold and high-calorie diets stimulate the sympathetic nervous system leading to the release of norepinephrine. Norepinephrine triggers the activation of the β-adrenergic receptors (AR) resulting in the elevation of intracellular cAMP and inducing the expression of PGC-1. PGC-1 activates the expression of the subunits of the respiratory chain and mtTFA through the induction of the expression of NRFs and the coactivation of NRF-1–mediated transcription. mtTFA subsequently translocates into the mitochondrion and directly activates the transcription and replication of mtDNA. PGC-1–induced mitochondrial biogenesis is accompanied by an enhanced capacity for energy production, which is made possible by the increased expression of enzymes acting in fatty acid oxidation, the tricarboxylic acid (TCA) cycle, and oxidative phosphorylation.[102–105] Interestingly, PGC-1 can modulate mitochondrial activity in a cell-type–specific manner. In adipocytes, it induces the inner mitochondrial membrane UCP-1, which can uncouple fuel oxidation from ATP production and generate heat.[102–105] In muscle, it induces the UCP-2 and stimulates both uncoupled (heat-producing) and coupled (ATP-producing) respiration.[103] In cardiac myocytes, PGC-1 does not activate the UCPs and stimulates only coupled, ATP-producing respiration.[104] These findings suggest that PGC-1 gears mitochondria to meet tissue-specific metabolic needs. It is possible that, within a given tissue, the PGC-1–induced response (e.g., coupled versus uncoupled respiration) also depends on the physiological state of the organism. The suggested reason involves the need to increase the overall rate of fuel oxidation, allowing for the preservation of normal cellular ATP/ADP ratios, while "wasting" a significant fraction of metabolic energy in the form of heat.

The compensatory mechanism to the decreased metabolic power was studied in detail in human. Leibel *et al.*[106] found maintenance of a body weight at a level 10% or more below the initial weight was associated with a mean ($\pm$ SD) reduction in total energy expenditure of 6 ± 3 kcal per kilogram of fat-free mass per day in the subjects who had never been obese ($P < 0.001$) and 8 ± 5 kcal per kilogram per day in the obese subjects ($P < 0.001$). Maintenance of body weight at a level 10% above the usual weight was associated with an increase in total energy expenditure of 9 ± 7 kcal

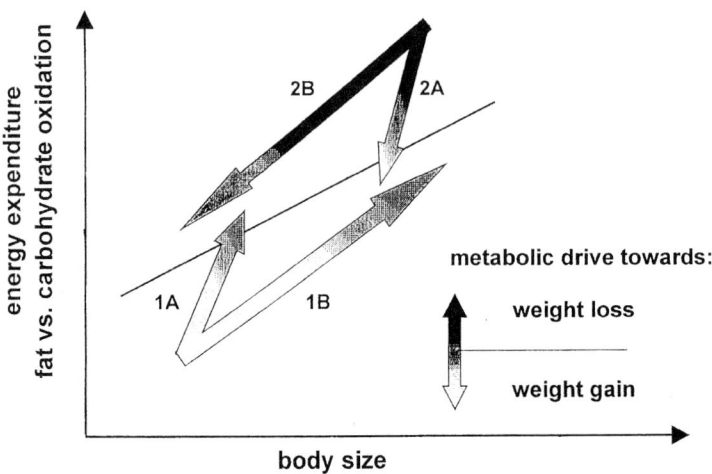

FIGURE 3. Energy expenditure increase after the weight gain in the Pima Indians. (From Weyer et al.[107] Reprinted, with permission, from the *Journal of Clinical Endocrinology & Metabolism*.)

per kilogram of fat-free mass per day in the subjects who had never been obese ($P < 0.001$) and 8 ± 4 kcal per kilogram per day in the obese subjects ($P < 0.001$). The thermic effect of feeding and non-resting energy expenditure increased by approximately 1–2 and 8–9 kcal per kilogram of fat-free mass per day, respectively, after weight gain. Maintenance of a reduced or elevated body weight is associated with compensatory changes in energy expenditure, which oppose the maintenance of a body weight that is different from the usual weight. These compensatory changes are consistent with the prediction of allometric scaling law.

Consistent with this law, the metabolic rate of individuals increased after a certain period in Pima Indian subjects whose rate was initially low (arrow 1A and 1B in FIG. 3).[107] Resultant changes in body scale and metabolic rate fall to the line of scale law relation. The weight gain was highest among those subjects with lowest metabolic rate.[108] These findings support the concept that lowered metabolic rate is compensated by increased body mass.

CONCLUSION

Many epidemiologic findings suggest that intrauterine growth retardation is linked to the risk of developing both type 2 diabetes and insulin resistance syndrome. In the light of advances made in the last few years, including development of the thrifty animal model, the thrifty phenotype hypothesis as a possible explanation for these links has been clarified. However, its explanation remains conflicting. Then, the vast amount of work from mitochondrial research and in the area of the pathophysiology of type 2 diabetes provided evidence that mitochondrial dysfunction may

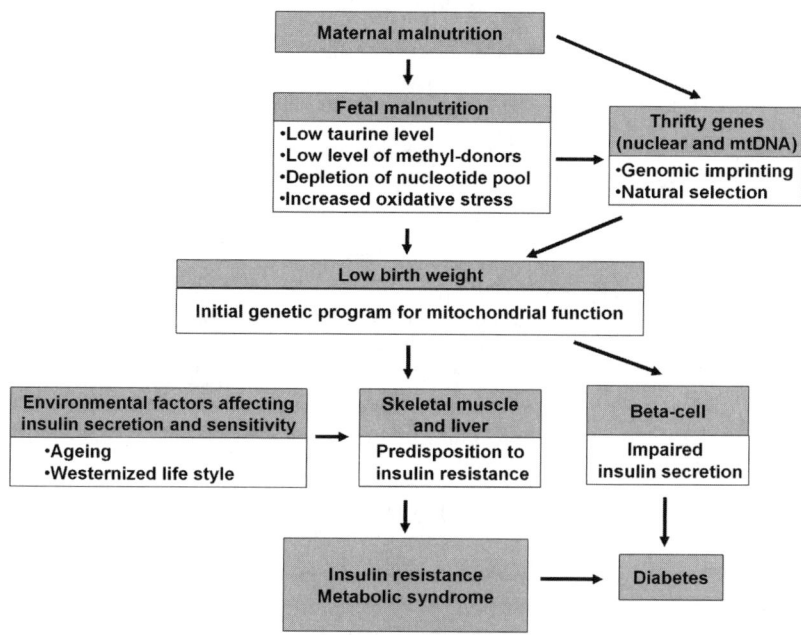

FIGURE 4. General outline of the mitochondrial hypothesis.

be a link between malnutrition in early life and insulin resistance syndrome in adult life. However, its explanation remains conflicting. Recent works on the pathophysiology of type 2 diabetes provided evidence that mitochondrial dysfunction may be a link between fetal malnutrition and insulin resistance syndrome. Here we propose a mitochondria-based model for the fetal origin of adult diseases or insulin resistance syndrome, in which a set point of mtDNA content might change according to the nutritional status of early life and might be inheritable by an imprinting mechanism (FIG. 4).

The mitochondria-based model is compatible with a biophysical law of allometric scaling. It might be simply thought as a special case of this law existing between the metabolic power of mitochondrial unit and the body mass. Among the many predictions this model makes, it is suggested it will be fruitful to identify genes whose expression is persistently altered by early malnutrition, both in genomic and epigenetic approaches, especially the genes that control mitochondrial replication.

ACKNOWLEDGMENTS

This study was supported by a grant (02-PJ1-PG1-CH04-0001) from the Korea Health 21 R&D Project, Ministry of Health & Welfare, Republic of Korea.

REFERENCES

1. REAVEN, G.M. 1988. Banting lecture 1988. Role of insulin resistance in human disease. Diabetes **37**: 1590–1607.
2. GARG, A. & S.M. HAFFNER. 1996. Insulin resistance and atherosclerosis. Diabetes Care **19**: 274.
3. MCCANCE, D.R. et al. 1994. Birth weight and non-insulin dependent diabetes: thrifty genotype, thrifty phenotype, or surviving small baby genotype? Br. Med. J. **308**: 942–945.
4. HALES, C.N. & D.J.P. BARKER. 1992. Type 2 (non-insulin-dependent) diabetes mellitus: the thrifty phenotype hypothesis. Diabetologia **35**: 595–601.
5. LUCAS, A. 1994. Role of nutritional programming in determining adult morbidity. Arch.Dis. Child. **71**: 288–290.
6. CLARK, P.M. 1998. Programming of the hypothalamo-pituitary-adrenal axis and the fetal origins of adult disease hypothesis. Eur. J. Pediatr. **157**: S7–10.
7. NEEL, J.V. 1962. Diabetes mellitus: a "thrifty" genotype rendered detrimental by "progress"? Am. J. Hum. Genet. **14**: 353–362.
8. HATTERSLEY, A.T. & J.E. TOOKE. 1999. The fetal insulin hypothesis: an alternative explanation of the association of low birthweight with diabetes and vascular disease. Lancet **353**: 1789–1792.
9. FRAYLING, T.M. & A.T. HATTERSLEY. 2001. The role of genetic susceptibility in the association of low birth weight with type 2 diabetes. Br. Med. Bull. **60**: 89–101.
10. VELHO, G., A.T. HATTERSLEY & P. FROGUEL. 2000. Maternal diabetes alters birth weight in glucokinase-deficient (MODY2) kindred but has no influence on adult weight, height, insulin secretion or insulin sensitivity. Diabetologia **43**: 1060–1063.
11. ARENDS, N. et al. 2002. Polymorphism in the IGF-I gene: clinical relevance for short children born small for gestational age (SGA). J. Clin. Endocrinol. Metab. **87**: 2720.
12. YOUNG, L.E. 2001. Imprinting of genes and Barker hypothesis. Twin Res. **4**: 307–317.
13. WATERLAND, R.A. & R.L. JIRTLE. 2004. Early nutrition, epigenetic changes at transposons and imprinted genes, and enhanced susceptibility to adult chronic diseases. Nutrition **20**: 63–68.
14. POULTON, J. et al. 1998. A common mitochondrial DNA variant is associated with insulin resistance in adult life. Diabetologia **41**: 54–58.
15. POULTON, J. et al. 2002. Type 2 diabetes is associated with a common mitochondrial variant: evidence from a population-based case-control study. Hum. Mol. Genet. **11**: 1581–1583.
16. CASTEELS, K. et al. 1999. Mitochondrial 16189 variant, thinness at birth, and type-2 diabetes. ALSPAC study team. Avon Longitudinal Study of Pregnancy and Childhood. Lancet **353**: 1499–1500.
17. POULTON, J. et al. 2002. The presence of a common mitochondrial DNA variant is associated with fasting insulin levels in Europeans in Auckland. Diabet. Med. **19**: 969–971.
18. KIM, J.H. et al. 2002. The prevalence of the mitochondrial DNA 16189 variant in non-diabetic Korean adults and its association with higher fasting glucose and body mass index. Diabet. Med. **19**: 681–684.
19. JI, L., L. GAO & X. HAN. 2001. Association of 16189 variant (T→C transition) of mitochondrial DNA with genetic predisposition to type 2 diabetes in Chinese populations. Zhonghua Yi. Xue. Za Zhi. **81**: 711–714.
20. LEE, H.K. 1999. Evidence that the mitochondrial genome is the thrifty genome. Diabetes Res. Clin. Pract. **45**: 127–135.
21. LEE, Y.Y. et al. 2005. The role of mitochondrial DNA in the development of type 2 diabetes caused by fetal malnutrition. J. Nutr. Biochem. In press.
22. BERTIN, E. et al. 2002. Development of beta-cell mass in fetuses of rats deprived of protein and/or energy in last trimester of pregnancy. Am. J. Physiol. Regul. Integr. Comp. Physiol. **283**: R623–630.
23. MARTIN, J.L. et al. 1994. Effects of antiviral nucleoside analogs on human DNA polymerases and mitochondrial DNA synthesis. Antimicrob. Agents Chemother. **38**: 2743–2749.

24. LEWIS, W. & M.C. DALAKAS. 1995. Mitochondrial toxicity of antiviral drugs. Nat. Med. **1:** 417–422.
25. BRINKMAN, K. et al. 1998. Adverse effects of reverse transcriptase inhibitors: mitochondrial toxicity as common pathway. AIDS **12:** 1735–1744.
26. REIK, W. & J. WALTER. 2001. Genomic imprinting: parental influence on the genome. Nat. Rev. Genet. **2:** 21–32.
27. JAENISCH, R. & A. BIRD. 2003. Epigenetic regulation of gene expression: how the genome integrates intrinsic and environmental signals. Nat. Genet. **33:** 245–254.
28. WALLACE, D.C. 1982. Structure and evolution of organelle genomes. Microbiol. Rev. **46:** 208–240.
29. WALLACE, D.C. et al. 1997. Ancient mtDNA sequences in the human nuclear genome: a potential source of errors in identifying pathogenic mutations. Proc. Natl. Acad. Sci. USA **94:** 14900–14905.
30. MAECHLER, P. & C.B. WOLLHEIM. 2001. Mitochondrial function in normal and diabetic beta-cells. Nature **414:** 807–812.
31. LEE, H.K. et al. 1998. Decreased mitochondrial DNA content in peripheral blood precedes the development of non-insulin-dependent diabetes mellitus. Diabetes Res. Clin. Pract. **42:** 161–167.
32. SONG, J. et al. 2001. Peripheral blood mitochondrial DNA content is related to insulin sensitivity in offspring of type 2 diabetic patients. Diabetes Care **24:** 865–869.
33. PETERSEN, K.F. et al. 2003. Mitochondrial dysfunction in the elderly: possible role in insulin resistance. Science **300:** 1140–1142.
34. PETERSEN, K.F. et al. 2004. Impaired mitochondrial activity in the insulin-resistant offspring of patients with type 2 diabetes. N. Engl. J. Med. **350:** 664–671.
35. PARK, K.S. et al. 1999. Peripheral blood mitochondrial DNA content correlates with lipid oxidation rate during euglycemic clamps in healthy young men. Diabetes Res. Clin. Pract. **46:** 149–154.
36. DENGEL, D.R. et al. 1996. Distinct effects of aerobic exercise training and weight loss on glucose homeostasis in obese sedentary men. J. Appl. Physiol. **81:** 318–325.
37. THAMER, C. et al. 2003. Reduced skeletal muscle oxygen uptake and reduced beta-cell function: two early abnormalities in normal glucose-tolerant offspring of patients with type 2 diabetes. Diabetes Care **26:** 2126–2132.
38. PARK, K.S. et al. 2001. Depletion of mitochondrial DNA alters glucose metabolism in SK-Hep1 cells. Am. J. Physiol. **280:** E1007–1014.
39. HALES, C.N. 1997. Metabolic consequences of intrauterine growth retardation. Acta Paediatr. Suppl. **423:** 184–187.
40. DESAI, M. et al. 1997. Programming of hepatic insulin-sensitive enzymes in offspring of rat dams fed a protein-restricted diet. Am. J. Physiol. **272):** G1083–1090.
41. BURNS, S.P. et al. 1997. Gluconeogenesis, glucose handling, and structural changes in livers of the adult offspring of rats partially deprived of protein during pregnancy and lactation. J. Clin. Invest. **100:** 1768–1774.
42. OZANNE, S.E. et al. 1996. Altered regulation of hepatic glucose output in the male offspring of protein-malnourished rat dams. Am. J. Physiol. **270:** E559–564.
43. SPARRE, T. et al. 2003. Intrauterine programming of fetal islet gene expression in rats—effects of maternal protein restriction during gestation revealed by proteome analysis. Diabetologia **46:** 1497–1511.
44. HALES, C.N. et al. 1996. Fishing in the stream of diabetes: from measuring insulin to the control of fetal organogenesis. Biochem. Soc. Trans. **24:** 341–350.
45. HOLNESS, M.J., L.G. FRYER & M.C. SUGDEN. 1999. Protein restriction during early development enhances insulin responsiveness but selectively impairs sensitivity to insulin at low concentrations in white adipose tissue during a later pregnancy. Br. J. Nutr. **81:** 481–489.
46. OZANNE, S.E. et al. 1996. Altered muscle insulin sensitivity in the male offspring of protein-malnourished rats. Am. J. Physiol. **271:** E1128–1134.
47. SHEPHERD, P.R. et al. 1997. Altered adipocyte properties in the offspring of protein malnourished rats. Br. J. Nutr. **78:** 121–129.
48. MIQUEL, J. et al. 1980. Mitochondrial role in cell aging. Exp. Gerontol. **15:** 575–591.

49. OGATA, E.S. *et al.* 1985. Altered growth, hypoglycemia, hypoalaninemia, and ketonemia in the young rat: postnatal consequences of intrauterine growth retardation. Pediatr. Res. **19:** 32–37.
50. OGATA, E.S., M.E. BUSSEY & S. FINLEY. 1986. Altered gas exchange, limited glucose and branched chain amino acids, and hypoinsulinism retard fetal growth in the rat. Metabolism **35:** 970–977.
51. SIMMONS, R.A., L.J. TEMPLETON & S.J. GERTZ. 2001. Intrauterine growth retardation leads to the development of type 2 diabetes in the rat. Diabetes **50:** 2279–2286.
52. VUGUIN, P. *et al.* 2002. Abberent insulin-mediated hepatic glucose production in growth-retarded rats. Pediatr. Res. **51:** 129A
53. SELAK, M.A. *et al.* 2003. Impaired oxidative phosphorylation in skeletal muscle of intrauterine growth-retarded rats. Am. J. Physiol. Endocrinol. Metab. **285:** E130–137.
54. PARK, K.S. *et al.* 2003. Fetal and early postnatal protein malnutrition cause long-term changes in rat liver and muscle mitochondria. J. Nutr. **133:** 3085–3090.
55. PARK, H.K. *et al.* 2004. Changes of mitochondrial DNA content in the male offspring of protein-malnourished rats. Ann. N.Y. Acad. Sci. **1011:** 205–216.
56. CLAYTON, D.A. 2000. Transcription and replication of mitochondrial DNA. Hum. Reprod. **15:** 11–17.
57. LECRENIER, N. & F. FOURY. 2000. New features of mitochondrial DNA replication system in yeast and man. Gene **246:** 37–48.
58. SCARPULLA, R.C. 1997. Nuclear control of respiratory chain expression in mammalian cells. J. Bioenerg. Biomembr. **29:** 109–119.
59. WU, Z. *et al.* 1999. Mechanisms controlling mitochondrial biogenesis and respiration through the thermogenic coactivator PGC-1. Cell **98:** 115–124.
60. REUSENS, B. *et al.* 1995. Long-term consequences of diabetes and its complications may have fetal origin: experimental and epidemiological evidences. *In* Nestle' Nutrition Workshop Series. R.M. Cowett, Ed. Vol 25: 187–198. Raven Press. New York.
61. BOUJENDAR, S. *et al.* 2003. Taurine supplementation of a low protein diet fed to rat dams normalizes the vascularization of the fetal endocrine pancreas. J. Nutr. **133:** 2820–2825.
62. CHERIF, H. *et al.* 1998. Effects of taurine on the insulin secretion of rat fetal islets from dams fed a low-protein diet. J. Endocrinol. **159:** 341–348.
63. HANSEN, S.H. 2001. The role of taurine in diabetes and the development of diabetic complications. Diabetes Metab. Res. Rev. **17:** 330–346.
64. SPAETH, D.G., D.L. SCHNEIDER & H.P. SARETT. 1974. Taurine synthesis, concentration, and bile salt conjugation in rat, guinea pig, and rabbit. Proc. Soc. Exp. Biol. Med. **147:** 855–858.
65. STURMAN, J.A. *et al.* 1986. Feline maternal taurine deficiency: effect on mother and offspring. J. Nutr. **116:** 655–667.
66. NORBERG, S., T.L. POWELL & T. JANSSON. 1998. Intrauterine growth restriction is associated with a reduced activity of placental taurine transporters. Pediatr. Res. **44:** 233–238.
67. SUZUKI, T. *et al.* 2002. Taurine as a constituent of mitochondrial tRNAs: new insights into the functions of taurine and human mitochondrial diseases. EMBO J. **21:** 6581–6589.
68. NAKAYA, Y. *et al.* 2000. Taurine improves insulin sensitivity in the Otsuka Long-Evans Tokushima fatty rat, a model of spontaneous type 2 diabetes. Am. J. Clin. Nutr. **71:** 54–58.
69. MARTIN, J.F. *et al.* 2000. Nutritional origins of insulin resistance: a rat model for diabetes-prone human populations. J. Nutr. **130:** 741–744.
70. LI, E., C. BEARD & R. JAENISCH. 1993. Role for DNA methylation in genomic imprinting. Nature **366:** 362–365.
71. BARKER, D.J. & P.M. CLARK. 1997. Fetal undernutrition and disease in later life. Rev. Reprod. **2:** 105–112.
72. DOHERTY, A.S. *et al.* 2000. Differential effects of culture on imprinted H19 expression in the preimplantation mouse embryo. Biol. Reprod. **62:** 1526–1535.

73. KHOSLA, S. et al. 2001. Culture of preimplantation mouse embryos affects fetal development and the expression of imprinted genes. Biol. Reprod. **64:** 918–926.
74. COX, G.F. et al. 2002. Intracytoplasmic sperm injection may increase the risk of imprinting defects. Am. J. Hum. Genet. **71:** 162–164.
75. DEBAUN, M.R., E.L. NIEMITZ & A.P. FEINBERG. 2003. Association of in vitro fertilization with Beckwith-Wiedemann syndrome and epigenetic alterations of LIT1 and H19. Am. J. Hum. Genet. **72:** 156–160.
76. HU, J.F. et al. 1997. Modulation of Igf2 genomic imprinting in mice induced by 5-azacytidine, an inhibitor of DNA methylation. Mol. Endocrinol. **11:** 1891–1898.
77. WATERLAND, R.A. & C. GARZA. 2002. Early postnatal nutrition determines adult pancreatic glucose-responsive insulin secretion and islet gene expression in rats. J. Nutr. **132:** 357–364.
78. VAN DEN VEYVER, I.B. 2002. Genetic effects of methylation diets. Annu. Rev. Nutr. **22:** 255–282.
79. WATERLAND, R.A. & C. GARZA. 1999. Potential mechanisms of metabolic imprinting that lead to chronic disease. Am. J. Clin. Nutr. **69:** 179–197.
80. WOLFF, G.L. et al. 1998. Maternal epigenetics and methyl supplements affect agouti gene expression in Avy/a mice. FASEB J. **12:** 949–957.
81. WATERLAND, R.A. & R.L. JIRTLE. 2003. Transposable elements: targets for early nutritional effects on epigenetic gene regulation. Mol. Cell. Biol. **23:** 5293–5300.
82. NG, M.C.Y. et al. 2001. Qualitative and quantitative analysis of mitochondrial DNA in Chinese patients with type 2 diabetes. Diabetes **50:** A493.
83. HUANG, C.J. & M.L. FWU. 1993. Degree of protein deficiency affects the extent of the depression of the antioxidative enzyme activities and the enhancement of tissue lipid peroxidation in rats. J. Nutr. **123:** 803–810.
84. BAST, A. et al. 2002. Oxidative and nitrosative stress induces peroxiredoxins in pancreatic beta cells. Diabetologia **45:** 867–876.
85. SASTRE, J., F.V. PALLARDO & J. VINA. 2000. Mitochondrial oxidative stress plays a key role in aging and apoptosis. IUBMB Life **49:** 427–435.
86. WEI, Y.H. et al. 1998. Oxidative damage and mutation to mitochondrial DNA and age-dependent decline of mitochondrial respiratory function. Ann. N. Y. Acad. Sci. **854:** 155–170.
87. SMITH, R.E. 1956. Quantitative relations between liver mitochondria metabolism and total body weight in mammals. Ann. N.Y. Acad. Sci. **62:** 403–422
88. RASMUSSEN, U.F. et al. 2001. Aerobic metabolism of human quadriceps muscle: in vivo data parallel measurements on isolated mitochondria. Am. J. Physiol. Endocrinol. Metab. **280:** E301–307.
89. WANG, H. et al. 1999. Relationship between muscle mitochondrial DNA content, mitochondrial enzyme activity and oxidative capacity in man: alterations with diseases. Eur. J. Appl. Physiol. **80:** 22–27.
90. LIM, S. et al. 2000. Effect of exercise on the mitochondrial DNA content of peripheral blood in healthy women. Eur. J. Appl. Physiol. **82:** 407–412.
91. LEE, H.K. 2001. Method of proof and evidences for the concept that mitochondrial genome is a thrifty genome. Diabetes Res. Clin. Pract. **54:** S57–S63.
92. WEST, G.B., J.H. BROWN & B.J. ENQUIST. 1997. A general model for the origin of allometric scaling laws in biology. Science **276:** 122–126.
93. KLEIBER, M. 1932. Body size and metabolism. Hilgardia **6:** 315–353.
94. WEST, G.B., W.H. WOODRUFF & J.H. BROWN. 2002. Allometric scaling of metabolic rate from molecules and mitochondria to cells and mammals. Proc. Natl. Acad. Sci. USA **99:** 2473–2478.
95. GILLOOLY, J.F. et al. 2001. Effects of size and temperature on metabolic rate. Science **293:** 2248–2251.
96. HAMANN, A., J.S. FLIER & B.B. LOWELL. 1996. Decreased brown fat markedly enhances susceptibility to diet-induced obesity, diabetes, and hyperlipidemia. Endocrinology **137:** 21–29.
97. HIMMS-HAGEN, J. 1989. Role of thermogenesis in the regulation of energy balance in relation to obesity. Can. J. Physiol. Pharmacol. **67:** 394–401.

98. BOUCHARD, C. *et al.* 1990. The response to long-term overfeeding in identical twins. N. Engl. J. Med. **322:** 1477–1482.
99. LEVINE, J.A., N.L. EBERHARDT & M.D. JENSEN. 1999. Role of nonexercise activity thermogenesis in resistance to fat gain in humans. Science **283:** 212–214.
100. FLIER, J.S. & B.B. LOWELL. 1997. Obesity research springs a proton leak. Nat. Genet. **15:** 223–224.
101. RICQUIER, D. & F. BOUILLAUD. 1997. The mitochondrial uncoupling protein: structural and genetic studies. Prog. Nucleic Acid Res. Mol. Biol. **56:** 83–108.
102. PUIGSERVER, P. *et al.* 1998. A cold-inducible coactivator of nuclear receptors linked to adaptive thermogenesis. Cell **92:** 829–839.
103. WU, Z. *et al.* 1999. Mechanisms controlling mitochondrial biogenesis and respiration through the thermogenic coactivator PGC-1. Cell **98:** 115–124.
104. LEHMAN, J.J. *et al.* 2000. Peroxisome proliferator-activated receptor gamma coactivator-1 promotes cardiac mitochondrial biogenesis. J. Clin. Invest. **106:** 847–856.
105. NEDERGAARD, J. *et al.* 2001. UCP1: the only protein able to mediate adaptive nonshivering thermogenesis and metabolic inefficiency. Biochim. Biophys. Acta **1504:** 82–106.
106. LEIBEL, R.L., M. ROSENBAUM & J. HIRSCH. 1995. Changes in energy expenditure resulting from altered body weight. N. Engl. J. Med. **332:** 621–628.
107. WEYER, C. *et al.* 2000. Energy expenditure, fat oxidation, and body weight regulation: a study of metabolic adaptation to long-term weight change. J. Clin. Endocrinol. Metab. **85:** 1087–1094.
108. RAVUSSIN, E. *et al.* 1988. Reduced rate of energy expenditure as a risk factor for body-weight gain. N. Engl. J. Med. **318:** 467–472.

Evolutional Analysis in Determining Pathogenic versus Nonpathogenic Mutations of ATPase 6 in Human Mitochondriopathy

CHIN-YUAN TZEN[a,b,c,d] AND TSU-YEN WU[e]

Departments of Pathology[a] and Medical Research,[e]
Mackay Memorial Hospital, Taipei, Taiwan
[b]Taipei Medical University, Taipei, Taiwan
[c]National Taipei College of Nursing, Taipei, Taiwan
[d]Mackay Medicine, Nursing and Management College, Taipei, Taiwan

ABSTRACT: Because mitochondrial ATPase 6 plays an important role in ATP synthesis, mutations affecting ATPase 6 can undoubtedly cause human diseases. In contrast, the ATPase 6 gene is known to be a fast-evolving gene and has generated enough polymorphisms to allow identity investigation for forensic casework. To investigate these seemingly opposite views, we analyzed amino acid sequences of ATPase 6 in at least 1,266 humans, 102 mammals, and 213 vertebrates. The result showed that the amino acids of human ATPase 6 could be divided into the following four groups. Amino acid residue 192 (affected by alteration at nt 9101) and 79 other residues were variable, and therefore substitutions of these residues would not be pathogenic. Amino acid residue 156 (affected by alteration at nt 8993) and 93 other residues were conserved in *Homo sapiens*, but not in Mammalia. Therefore, they were potentially pathogenic if altered. Function studies would be necessary to confirm their role in pathogenesis. Amino acid residue 217 (affected by alteration at nt 9176) and 9 other residues were conserved across all species, including *S. cerevisiae* and *E. coli*. Mutations involving these residues would be pathogenic, some of which might even be life threatening. The remainder (42 residues) were conserved in Mammalia, but not in yeast and *E. coli*. They were probably pathogenic if mutated. The classification proposed in this study may, therefore, provide an algorithm for a diagnostic approach when a newly identified change of ATPase 6 is suspected for human mitochondriopathy.

KEYWORDS: mitochondriopathy; pathogenic mutation; ATPase 6; evolutional analysis

INTRODUCTION

In the respiratory chain, the pumping of hydrogen ion into the mitochondrial intermembrane space generates a proton-motive force, which is used by the ATP syn-

Address for correspondence: Chin-Yuan Tzen, M.D., Ph.D., Department of Pathology, Mackay Memorial Hospital, 45, Minsheng Road, Tamshui, Taipei 251, Taiwan. Voice: +8862-2809-4661 ext. 2491; fax: +8862-2809-3385.
jeffrey@ms2.mmh.org.tw

thase, also called F_1F_0ATPase, to form ATP. Nearly identical protein subunits are found in eukaryotic mitochondria and bacteria, and they all operate by converting the electromotive force into a rotary torque. In *Escherichia coli*, the F_0 portion consists of three transmembrane subunits: a_6, b_2, and c_{12}. Human ATPase 6 is equivalent to subunit *a* of *E. coli* F_0-ATPase.

Because ATPase 6 plays an important role in forming ATP, it would be understandable that mutation of this gene can cause human diseases. For example, mutation at nucleotide (nt) 9176 was reported in familial bilateral striated necrosis[1] and Leigh syndrome,[2–4] nt 9101 in Leber hereditary optic neuroretinopathy,[5] and nt 8993 in subacute necrotizing encephalopathy,[6] Leigh syndrome,[7–10] neurogenic muscle weakness, ataxia, retinitis pigmentosa syndrome,[11,12] and some types of mitochondrial myopathy with variable combination of neurological manifestation.[13]

However, from the forensic point of view, we reported that the mitochondrial DNA (mtDNA) coding such an important enzyme contained many polymorphic sites.[14] In 2003, Lutz-Bonengel *et al.* confirmed our observation and further reported that some DNA polymorphisms resulted in changes of amino acids of ATPase 6.[15] In addition, Poetsch *et al.* recently applied these findings in their forensic casework.[16] Taken together, these observations suggest that human ATPase 6 contains many variable amino acids, implying that mutations involving these amino acid residues may not be involved in significant functions. Therefore, it may cast doubt on correlating syndromes with nucleotide alternations of the ATPase 6 gene.

To understand the above seemingly opposite viewpoints, we examined ATPase 6 protein sequences in humans, mammals, and vertebrates. From the alignment of the amino acid sequences, we determined the conserved sites of human ATPase 6 and classified these sites based on how conserved they were across the species. These conserved sites might be important in the ATP formation, and mutations involving these sites were likely to be pathogenic. From this perspective, we re-examined the reported ATPase 6 mutations associated with mitochondriopathy and confirmed the usefulness of the proposed classification of ATPase 6 amino acid residues.

MATERIALS AND METHODS

The amino acid sequences of human ATPase 6 were deduced from unpublished sequences of Taiwanese Chinese, published data of Lutz-Bonengel *et al.*[15] and the public databases—NCBI[17] and Pfam.[18] The amino acid sequences of mammalian and vertebral ATPase 6 were retrieved from Pfam[18] and NCBI,[17] respectively. Vector NTI Suite 8.0 (InforMax, Inc., MD) was used for sequence alignment and analysis.

RESULTS AND DISCUSSION

Analysis of Mammalian and Vertebral ATPase 6

The length of ATPase 6 in 916 sequences of vertebrate from the available database[17] ranged from 84 to 347 amino acid residues in length. Among them, 534 sequences had 227 amino acids and 335 sequences had 226 amino acids. The species with ATPase 6 of 227 amino acids were primarily fishes and birds, whereas those of 226 amino acids were seen in many species, including 221 warm-blooded and 5 cold-blooded animals.

Analysis of 102 species of Mammalia (2 species in Monotremata, 95 species in Eutheria, and 5 species in Metatheria) showed that mammalian ATPase 6 was 226 amino acids in length except in shrew (*Sortex unguiculatus*), which had 227 amino acids.[18] The alignment among 102 species of mammal showed that 66 amino acids were conserved in mammalian ATPase 6. Among them, 43 residues were conserved in 213 vertebral animals with ATPase 6 of 226 amino acids.

Conserved Amino Acids of ATPase 6 in Homo Sapiens

Theoretical topography (FIG. 1) of human ATPase 6 was constructed according to SwissPfam.[19] After comparing 136 amino acid sequences of Taiwanese Chinese (unpublished data) and 384 amino acid sequences from three database,[15,17,18] plus the collected polymorphic residues reported in two websites,[20,21] which had collected data from more than 746 individuals, we identified the conserved and variable sites of ATPase 6. As depicted in FIGURE 1, there were 80 varied amino acid residues (labeled □), 94 amino acids conserved in humans but not other mammals (labeled ▨), and 52 amino acids conserved in both humans and mammals (labeled ■).

Nonpathogenic Substitutions in ATPase 6

Based on the above analysis, the amino acid residues of human ATPase 6 can be divided into four groups from the perspective of causing human diseases. The first group included 80 polymorphic sites as follows: N^2, F^6, A^7, S^8, F^9, I^{10}, A^{11}, T^{13}, I^{14}, G^{16}, A^{20}, A^{21}, V^{22}, I^{24}, F^{26}, I^{31}, T^{33}, Y^{36}, L^{37}, N^{39}, N^{40}, R^{41}, Q^{47}, L^{52}, T^{53}, T^{59}, M^{60}, H^{61}, W^{68}, S^{74}, L^{75}, I^{76}, I^{77}, A^{80}, H^{90}, T^{93}, M^{104}, A^{105}, I^{106}, L^{108}, A^{110}, G^{111}, T^{112}, M^{115}, F^{117}, A^{126}, H^{127}, G^{132}, I^{138}, V^{142}, T^{146}, I^{147}, M^{154}, A^{155}, T^{161}, I^{164}, T^{165}, A^{166}, H^{172}, S^{176}, A^{177}, T^{178}, S^{182}, I^{184}, N^{185}, T^{189}, I^{191}, I^{192}, F^{193}, T^{194}, I^{195}, I^{201}, I^{204}, A^{205}, V^{206}, A^{207}, I^{209}, V^{213}, S^{219}, and D^{224}. According to the so-called Anderson Sequence, T^{112} was the most variable site that occurred in 99.6% of individuals. The second most common variable site (T^{59}) accounted for 23.6% of polymorphism (FIG. 1).

Alterations of amino acids in these variable sites, ranging from 0.1 to 99.6% in frequency, should not be pathogenic. From this point of view, the previously reported "mutation" at nt 9101, affecting residue I^{192} in a family with Leber hereditary optic neuoretinopathy,[5] was probably not pathogenic.

Potentially Pathogenic Mutations in ATPase 6

The second group included the 94 residues that conserved in *Homo sapiens*, but not in Mammalia. Because these amino acids were conserved among more than 1,266 individuals, substitutions involving these residues are potentially pathogenic. However, function assay may be required. This can be exemplified by mutation at nt 8993, which affects amino acids L^{156} and has been repeatedly reported in Leigh syndrome and other diseases.[6-13] Baracca et al.[22] explained that alteration at L^{156} (equivalent to L^{207} of the subunit *a* in *E. coli*) could impede the rotation of the F_0 portion (through the subunit *c* in *E. coli*). Recently, the pathogenesis of mutation at nt 8993 have been proved by function study.[23]

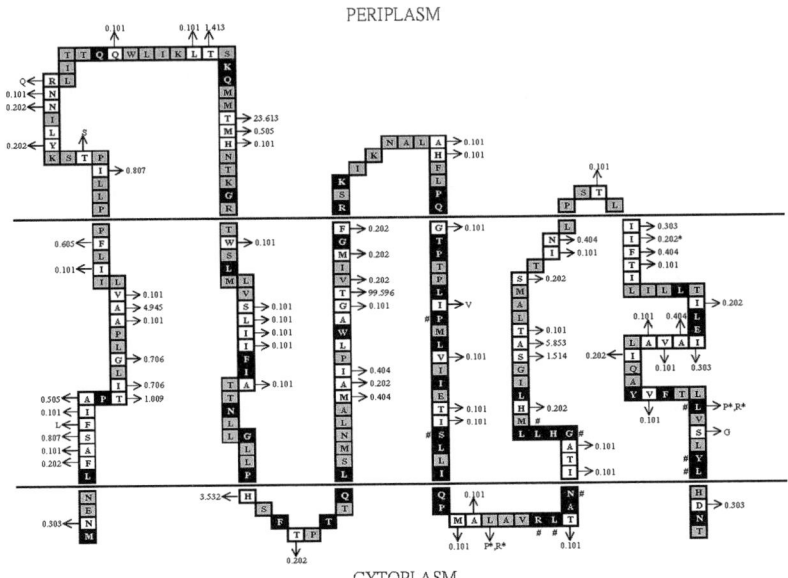

FIGURE 1. Theoretical secondary structure of the mammalian ATPase 6 with indicated conserved and variable amino acid residues. Predicted two-dimensional structure of the mammalian ATPase 6 molecule based on SwissPfam is illustrated with boxes containing one-letter codes for amino acids. The sequences between two horizontal lines are transmembranous domains. Codes in the filled boxes (■) and empty boxes (□) represent the amino acid residues in the conserved and variable positions among mammalian ATPase 6 proteins, respectively. The codes in the shaded boxes (▨) are amino acids conserved only in *Homo sapiens*. There are 10 amino acids conserved among Mammalia, *S. cerevisiae*, and *E. coli* (■#). The arrowed (→) numbers are the calculated frequencies of amino acid polymorphism while the arrowed codes are altered amino acids without knowing their frequencies. The asterisks (*) indicate reported cases associated with mitochondriopathy.

Probable Pathogenic Mutations in ATPase 6

Among 52 conserved residues in Mammalia, 10 amino acids were conserved in *S. cerevisiae* and *E. coli*. The rest of the 42 amino acids conserved in mammals were as follows: M^1, L^5, P^{12}, Q^{46}, K^{55}, Q^{56}, G^{65}, L^{70}, F^{78}, I^{79}, N^{83}, G^{86}, P^{89}, F^{92}, T^{95}, Q^{97}, L^{98}, W^{109}, G^{116}, R^{118}, K^{110}, P^{131}, Q^{131}, T^{133}, P^{134}, L^{137}, L^{141}, I^{144}, L^{149}, I^{151}, Q^{152}, P^{153}, A^{162}, H^{168}, L^{170}, L^{173}, L^{199}, L^{202}, E^{203}, Y^{212}, F^{214}, and N^{225}. These 42 residues constituted the third group, and their mutations were probably pathogenic because they are conserved among Mammalia. However, cases in this group remained to be reported.

Pathogenic Mutations in ATPase 6

The 10 amino acids conserved across the species from human to yeast and *E. coli* were as follows: P^{139}, S^{148}, R^{159}, L^{160}, N^{163}, G^{167}, L^{168}, L^{217}, Y^{221}, and L^{222}. Because these 10 amino acids were conserved across all species, they must play a critical role in ATP production. For example, the residue R^{159} of human ATPase is equivalent to R^{210} in the subunit c of the *E. coli*, which is believed to play an essential role in proton transport.[24–27] According to Hatch,[26] protons from the intermembrane space would move to subunit c through the involvement of E^{194} and H^{168}.

From this perspective, mutations affecting any of these highly conserved residues would cause severe diseases in humans. For example, substitution at nt 9176 affecting amino acid L^{217} was reported in Leigh syndrome and other diseases.[1–4] Recently, the pathogenesis of mutation at nt 9176, affecting amino acid L^{217} (equivalent to *E. coli* L^{259}), was proved by function study.[28]

Because L^{217} was conserved across all species, whereas L^{156} was conserved only in Mammalia, the former should be more important than the latter in the ATP synthesis. This may explain why mutation at nt 9176 can be life threatening[4] and is rare, whereas mutation at nt 8993 might be associated with a milder and more chronic course and is more commonly reported.[29]

ACKNOWLEDGMENTS

This work was supported by grants MMH-E-94002 from Mackay Memorial Hospital and NSC 91-2314-B-195-026 from the National Science Council, Taiwan, to C.-Y.T. The authors thank Ms. Ching-Hui Chen for secretarial expertise.

REFERENCES

1. THYAGARAJAN, D. *et al.* 1995. A novel mitochondrial ATPase 6 point mutation in familial bilateral striatal necrosis. Ann. Neurol. **38:** 468–472.
2. CAMPOS, Y. *et al.* 1997. Leigh syndrome associated with the T9176C mutation in the ATPase 6 gene of mitochondrial DNA. Neurology **49:** 595–597.
3. MAKINO, M. *et al.* 1998. Confirmation that a T-to-C mutation at 9176 in mitochondrial DNA is an additional candidate mutation for Leigh's syndrome. Neuromuscul. Disord. **8:** 149–151.
4. DIONISI-VICI, C. *et al.* 1998. Fulminant Leigh syndrome and sudden unexpected death in a family with the T9176C mutation of the mitochondrial ATPase 6 gene. J. Inherit. Metab. Dis. **21:** 2–8.
5. LAMMINEN, T. *et al.* 1995. A mitochondrial mutation at nt 9101 in the ATP synthase 6 gene associated with deficient oxidative phosphorylation in a family with Leber hereditary optic neuroretinopathy. Am. J. Hum. Genet. **56:** 1238–1240.
6. SHOFFNER, J.M. *et al.* 1992. Subacute necrotizing encephalopathy: oxidative phosphorylation defects and the ATPase 6 point mutation. Neurology **42:** 2168–2174.
7. VRIES, D.D. *et al.* 1993. A second missense mutation in the mitochondrial ATPase 6 gene in Leigh's syndrome. Ann. Neurol. **34:** 410–412.
8. YOSHINAGA, H. *et al.* 1993. A T-to-G mutation at nucleotide pair 8993 in mitochondrial DNA in a patient with Leigh's syndrome. J. Child. Neurol. **8:** 129–133.
9. SANTORELLI, F.M. *et al.* 1993. The mutation at nt 8993 of mitochondrial DNA is a common cause of Leigh's syndrome. Ann. Neurol. **34:** 827–834.
10. PASTORES, G.M. *et al.* 1994. Leigh syndrome and hypertrophic cardiomyopathy in an infant with a mitochondrial DNA point mutation (T8993G). Am. J. Med. Genet. **50:** 265–271.
11. MAKELA-BENGS, P. *et al.* 1995. Correlation between the clinical symptoms and the proportion of mitochondrial DNA carrying the 8993 point mutation in the NARP syndrome. Pediatr. Res. **37:** 634–639.
12. FRYER, A. *et al.* 1994. Mitochondrial DNA 8993 (NARP) mutation presenting with a heterogeneous phenotype including 'cerebral palsy'. Arch. Dis. Child. **71:** 419–422.

13. HOLT, I.J. *et al.* 1990. A new mitochondrial disease associated with mitochondrial DNA heteroplasmy. Am. J. Hum. Genet. **46:** 428–433.
14. TZEN, C.Y. *et al.* 2001. Sequence polymorphism in the coding region of mitochondrial genome encompassing position 8389–8865. Forensic Sci. Int. **120:** 204–209.
15. LUTZ-BONENGEL, S. *et al.* 2003. Sequence polymorphisms within the human mitochondrial genes MTATP6, MTATP8 and MTND4. Int. J. Legal Med. **117:** 133–142.
16. POETSCH, M. *et al.* 2003. The impact of mtDNA analysis between positions nt8306 and nt9021 for forensic casework. Mitochondrion **3:** 133–137.
17. http://www.ncbi.nlm.nih.gov/entrez/guery.fcgi
18. http://pfam.wustl.edu/cgi-bin/getdesc?name=ATP-suy_A
19. http://pfam.wustl.edu/cgi-bin/getswisspfam?key=ATP6_HUMAN
20. http://www.mitomap.org/
21. http://www.genpat.uu.se/mtDB/polysites.html
22. BARACCA, A. *et al.* 2000. Catalytic activities of mitochondrial ATP synthase in patients with mitochondrial DNA T8993G mutation in the ATPase 6 gene encoding subunit α. J. Biol. Chem. **275:** 4177–4182.
23. MANFREDI, G. *et al.* 2002. Rescue of a deficiency in ATP synthesis by transfer of MTATP6, a mitochondrial DNA-encoded gene, to the nucleus. Nat. Genet. **30:** 394–399.
24. FILLINGAME, R.H. 1997. Coupling H+ transport and ATP synthesis in F1F0-ATP synthases: glimpses of interacting parts in a dynamic molecular machine. J. Exp. Biol. **200:** 217–224.
25. CAIN, B.D. & R.D. SIMONI. 1989. Proton translocation by the F1F0ATPase of *Escherichia coli*. Mutagenic analysis of the α subunit. J. Biol. Chem. **264:** 3292–3300.
26. HATCH, L.P. *et al.* 1995. The essential arginine residue at position 210 in the alpha subunit of the *Escherichia coli* ATP synthase can be transferred to position 252 with partial retention of activity. J. Biol. Chem. **270:** 29407–29412.
27. VIK, S.B. & B.J. ANTONIO. 1994. A mechanism of proton translocation by F1F0 ATP synthases suggested by double mutants of the α subunit. J. Biol. Chem. **269:** 30364–30369.
28. CARROZZO, R. *et al.* 2000. The T9176G mutation of human mtDNA gives a fully assembled but inactive ATP synthase when modeled in *Escherichia coli*. FEBS Lett. **486:** 297–299.
29. MAKINO, M. *et al.* 2000. Mitochondrial DNA mutations in Leigh syndrome and their phylogenetic implications. J. Hum. Genet. **45:** 69–75.

Restoration of Mitochondrial Function in Cells with Complex I Deficiency

YIDONG BAI, JEONG SOON PARK, JIAN-HONG DENG, YOUFEN LI, AND PEIQING HU

Department of Cellular and Structural Biology, University of Texas Health Science Center at San Antonio, San Antonio, Texas 78229, USA

ABSTRACT: The mammalian mitochondrial NADH dehydrogenase (complex I) is the major entry point for the electron transport chain. It is the largest and most complicated respiratory complex consisting of at least 46 subunits, 7 of which are encoded by mitochondrial DNA (mtDNA). Deficiency in complex I function has been associated with various human diseases including neurodegenerative diseases and the aging process. To explore ways to restore mitochondrial function in complex I–deficient cells, various cell models with mutations in genes encoding subunits for complex I have been established. In this paper, we discuss various approaches to recover mitochondrial activity, the complex I activity in particular, in cultured cells.

KEYWORDS: complex I deficiency; mtDNA mutation; restoration; revertant; NDI1

Although mitochondria were found to contain their own DNA about 40 years ago,[1] and abnormal mitochondrial function was related to human diseases even earlier,[2] mutations in mtDNA were not firmly linked to human diseases until 1988.[3,4] Over the last 16 years, large numbers of mtDNA mutations have been associated with human diseases, including seizure, ataxia, cortical blandness, dystonia, exercise intolerance, opthalmoplegia, optic atrophy, cataracts, diabetes mellitus, short stature, cardiomyopathy, sensorineural hearing loss, and kidney failure.[5,6] More than 50 point mutations in mtDNA, including changes in protein-encoding genes for subunits of complex I, III, IV, and V, rRNA and tRNA genes, and large rearrangements or deletions of the mitochondrial genome, have been linked to a variety of clinical disorders.[6] Accumulation of mtDNA mutations has also been suggested to play a major role in the aging process and in the development of various aging-related degenerative diseases.[7] Interestingly, high levels of mitochondrial DNA mutations have also been found in many tumors.[8–10]

Complex I deficiency is the most frequently observed enzyme defect among mitochondrial diseases,[11] and has been associated with Leigh syndrome, fatal infantile lactic acidosis, macrocephaly with progressive leukodystrophy, and encephalomy-

---

Address for correspondence: Dr. Yidong Bai, Department of Cellular and Structural Biology, University of Texas Health Science Center at San Antonio, San Antonio, TX 78229. Voice: 210-567-0561; fax: 210-567-3800.
 baiy@uthscsa.edu

opathy.[11] Complex I defects were also associated with neurodegenerative diseases such as Parkinson's disease, dystonia, and Leber's hereditary optical neuropathy (LHON).[12] Moreover, declined complex I activity was also found in several tissues of aged animals.[13]

Mitochondrial DNA mutations are estimated responsible for 5–10% of all complex I deficiencies. LHON, which is defined by subacute, bilateral visual loss,[14] is one of first human diseases linked to the mtDNA mutations. First described in the last century,[15] LHON is found to be inherited onlythrough the female.[16] Since mtDNA is transmitted maternally, it was long hypothesized as responsible for the diseases.[16] Wallace and colleagues identified a point mutation at complex I subunit to be ND4 gene from 11 families with LHON from Finland and the United States.[4] Subsequent studies has identified two more primary mutations in ND1 and ND6 genes.[17–19] Moreover, mutations on ND4 and ND6 were found to cause LHON and dystonia, a rapidly progressive, neurodegenerative disease associated with symmetric basal ganglia lesions.[20,21] Besides LHON disease, mutations on the ND1[22] and ND5 gene were associated with mitochondrial encephalomyopathy, lactic acidosis, and stroke-like episodes (MELAS).[23–25] A nonsense mutation in the ND4 gene was found to cause defective complex I activity and lifelong exercise intolerance.[26] Among the common diseases, about 1–2% of type 2, or non-insulin-dependent, diabetes were related to mutations in mtDNA.[27] While point mutations in the mitochondrial tRNA gene are the primary cause, mutation on the ND1 gene were also reported as associated with diabetes.[27,28] High levels of mitochondrial DNA mutations, including those on the genes encoding complex I subunits, have also been found in several tumor cells.[8–10]

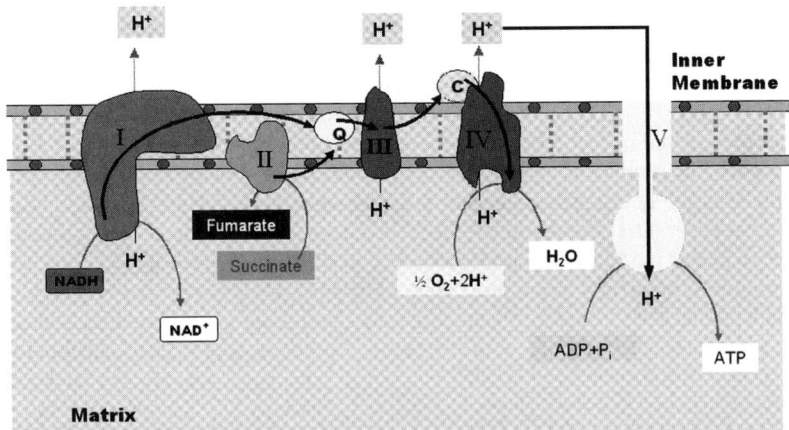

FIGURE 1. The mitochondrial oxidative phosphorylation system. The electrons transfer along the protein complexes on the inner membrane of mitochondria. Protons are pumped out through complexes I, III, and IV, and they flow back through complex V with concomitant production of ATP.

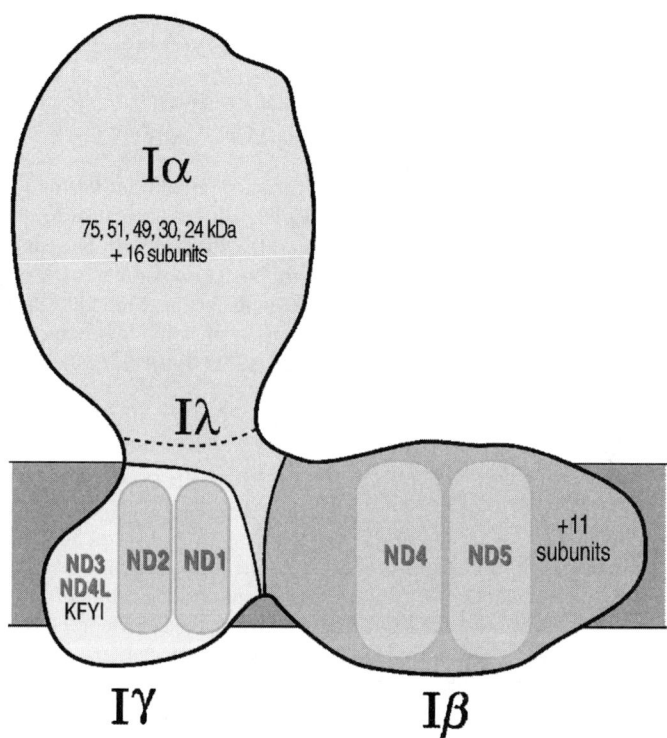

FIGURE 2. L-Structure of complex I. The locations of subcomplexes Ia, Ib, Il, and Ig are indicated. The catalytic center is at Ia, intruding to the matrix. All mtDNA-encoded subunits are at the membrane arm. (Modified from Sazanov et al. 2000. Biochemistry **39:** 7229.)

Complex I is the major entry point for electrons to get into the electron transport chain (FIG. 1). It is embedded in the inner membrane of mitochondria, and is capable of dehydrogenating NADH and shuttling electrons to ubiquinone in a rotenone-sensitive pathway.[29] The mammalian complex I consists about 46 subunits,[30] and 7 of them are encoded by the mitochondrial genes.[31,32] Complex I is the only remaining complex in the mammalian oxidative phosphorylative chain whose crystallographic data is still to be worked out.[29] A series of biochemical, electro-microscopical, and functional investigations have led to a model of the structure of the complex I in *Neurospora crassa,* which contains 32 subunits.[33–36] It is suggested to have an L-shaped structure consisting two complementary moieties: one membrane domain and a peripheral arm protruding to the matrix space. Since the mammalian complex I has a subunit composition similar to that of the Neurospora enzyme, the L-shaped structure is considered to approximate the mammalian complex I structure as well.[37,38] All the mtDNA-encoded hydrophobic subunits are located on the membrane arm, including the ubiquinone-binding subunit, ND1[39] (FIG. 2), while the catalytic center is on the matrix arm.

Various cell lines carrying mtDNA mutations in genes encoding complex I subunits and/or deficient in complex I activity have been established. mtDNA from LHON patients carrying mutations in ND1, ND4, and ND6 genes has been transferred cells depleted of their own mitochondrial genomes.[40,41] An efficient method to isolate mtDNA mutations in complex I genes in culture cells has also been developed.[42,43]

With cell models deficient in complex I function available, various approaches have been explored to restore the mitochondrial function. First, administration of substrates, cofactors and antioxidants are generally utilized with patients with mitochondrial diseases. Revertants have been sought based on some markers selected for cells with functional mitochondria. Wild-type genes are introduced into the mutant cell lines to replace the defective products. Further, proteins that can enhance mitochondrial function or bypass the defective components of the mitochondrial machinery have also been brought into the host cells.

MONITOR THE COMPLEX I GENE EXPRESSION AND FUNCTION

Seven of 46 of complex I subunits are encoded by the mitochondrial genome. Analysis of the protein synthesis of the mtDNA-encoded genes has been achieved by the development of a gradient electrophoresis system and the usage of a cytoplasmic-specific translation inhibitor.[44] Cells are grown in a medium-depleted methionine and supplemented with [^{35}S]methionine for 30 min in the presence of emetine, to inhibit cytoplasmic protein synthesis. The labeled cells then are trypsinized, washed, and lysed in 1% SDS. Samples containing 30–50 μg of protein are electrophoresed through an SDS-polyacrylamide gel (15–20% exponential gradient), and the mitochondrial translational products are analyzed by phosphorImaging.[44] The reference patterns for human and mouse mitochondrial translation products are available in the literature.[45,46]

The mammalian mitochondrial complex I is one of the most complicated membrane complexes, and its proper assembly is essential for the functional complex I.[47,48] To investigate the assembly of other mtDNA-encoded subunits of complex I, immunoprecipitation experiments can be carried out using antibodies against the nuclear-encoded 49k subunit for complex I.[45] Immunocomplexes are bound to formaldehyde-fixed *Staphyococcus aureus* (Zysorbin; Zymed Laboratories, San Francisco, CA),[31] pelleted, and washed repeatedly by centrifugation and resuspension. The proteins are eluted from the immunoabsorbent in the final pellets, and electrophoresed through an SDS-polyacrylamide gel (15–20% exponential gradient). The integrity of the whole complex I can also be studied by Blue-Native PAGE (BN-PAGE) analysis.[49] Mitochondrial proteins, prepared from cells by treatment with of digitonin,[50] are separated on 6–15% polyacrylamide gradient gels. In-gel qualitative assays for different complexes are carried out by histochemical staining.[51]

The primary function of mitochondria is to generate ATP through the oxidative phosphorylation process. ATP synthesis is coupled with electron transfers along the respiratory complexes to oxygen. Complex I function analysis can be achieved by measurement of the respiratory capacity. In particular, a combination of malate and glutamate are used to as respiratory substrates for measuring the complex I activity. As controls, the succinate/glycerol-3-phosphate (G-3-P)–driven respiration, which usually reflects the activity of complex III, and the N,N,N',N'-tetramethyl-*p*-phenylenediamine (TMPD)/ascorbate-driven respiration, which reflects the activity of complex IV, are analyzed.

Respiration measurements *in situ* are usually complemented by enzyme assays. These are carried out on isolated mitochondrial membranes, with sonication and detergent solubilization.[41,45,52,53] For complex I, the oxidoreductase activities are monitored by absorbance measurements at 275 nm for the reduction of Q_1, a ubiquinone analogue ($\varepsilon = 12,250$ $M^{-1}cm^{-1}$).[45] As controls, complex III activities are determined for cytochrome *c* reduction in the presence of reduced decylubiquinone (DBH$_2$) by monitoring the increase in absorbance at 550 minus 540 nm ($\varepsilon = 19,000$ $M^{-1}cm^{-1}$);[54] complex IV activities are measured by following the oxidation of reduced cytochrome *c* at 550 minus 540 nm ($\varepsilon = 19,000$ $M^{-1}cm^{-1}$) in a buffer containing potassium phosphate and cytochrome *c*.[54]

The efficiency of oxidative phosphorylation of cells can be measured as the coupling of mitochondrial respiration to ATP production. This is represented by the P/O ratio, defined as nanomoles of ATP produced per nanoatoms of oxygen consumed during ATP-stimulated respiration.[46,54] Complex I, as well as complexes III and IV are involved in proton pumping, and the proton gradient generated across the inner mitochondrial membrane is utilized to synthesize ATP. The measurements are carried out by an *in situ* method in digitonin-permeabilized cells[46] using a Hansatech oxygraph with an adjustable chamber volume. Malate and glutamate (for complex I coupled P/O) are used as respiratory substrates. ADP is added for each P/O measurement. The decrease in the P/O number indicates a compromised proton-pumping ability.

CYBRIDS CARRYING LHON SPECIFIC MUTATIONS

Since mitochondria are under dual genetic control—nuclear and mitochondrial—alterations in either genome could cause enzyme defects observed in the patients with mitochondrial diseases. To focus on the contribution from the mtDNA, the mtDNA-less (ρ^0) cell repopulation approach has been utilized.[55] Three primary mtDNA mutations for LHON—3460A in ND1, 11778A in ND4, and 14484C in ND6—were transferred in cybrid experiments.[40,41]

Complex I–dependent oxygen consumption analysis showed that 3460A and 11778A mutations significantly reduced the ADP-stimulated (state III) respiration.[40,41] Complex I enzyme assay also revealed that 3460A and 11778A mutations resulted in a severe reduction in specific activities.[40,41] These results indicated that cybrids containing 3460A and 11778A mutations could serve as good cell models to investigate mtDNA mutations associated with human diseases.

CELL LINES ISOLATED BASED ON RESISTANCE TO RESPIRATORY INHIBITOR

An approach for isolation of mtDNA mutants defective in complex I based on resistance to high concentration of rotenone, a complex I–specific respiration inhibitor, has been successfully developed with both human[42,56] and mouse cells.[43] It was shown that the growth capacity of the wild-type cells was significantly affected when the rotenone concentration reached 0.6 mM, with almost complete inhibition by 1.2 mM. The procedure was as follows: about 5 million wild-type cells in each dish were treated with 0.8 mM rotenone for about 3 weeks, individual rotenone resistant colonies were picked, and then those clones were subjected to stepwise increases in the concentration of rotenone, from 1.0–1.2 mM. Those cell lines adapted to grow in the presence of 1.2 mM were candidates for having mutations in mtDNA.

An analysis of the respiratory capacity of these rotenone-resistant cell lines was carried out. In the screen with mouse cells, it was revealed that all rotenone-resistant clones exhibited a 43–83% decrease in activities of complex I, and, furthermore, 7 of the 10 cell lines exhibited a 14–64% decrease in overall O_2 consumption.[45] In order to investigate the genetic origin of the respiratory defects in the isolated rotenone-resistant cells, two mouse mtDNA-less (ρ^0) cell lines derived from two different mouse cell lines have been established. Mouse LL/2 cells and A9 cells were exposed to 5 mg of ethidium bromide per milliliter for 11 to 12 weeks in medium supplemented with 50 mg of uridine per milliliter. Two clones, ρ^0 LL/2-m21 and ρ^0 A9-m34, were isolated and tested for the presence of mtDNA by Southern blot hybridization of a mouse mtDNA- specific probe and mtDNA-specific PCR amplification. Both clones showed complete absence of mtDNA. Advantage was then taken of the mtDNA-less (ρ^0) cell repopulation approach.[55] For example, mitochondria from respiratory defective A9 derivative cell were transferred into the ρ^0 LL/2-m21 cells by fusion of the latter with a population of enucleated cytoplasts of the former. Transformants were obtained; these were expected to contain the respiration defective A9-derivatives mtDNA, but in the LL/2 nuclear background. We found that both the reductions in overall respiratory capacity and the specific complex I defects of the two clone cells were transferred into ρ^0 cells, together with their mtDNA.[43]

Examination of mitochondrial translation products followed with DNA sequencing analysis that identified one frameshift mutation in ND6 gene[45] and a nonsense mutation in ND5 gene.[53]

PHARMACOLOGICAL ADMINISTRATION ON MITOCHONDRIAL DEFICIENCY

Tremendous progress has been made in the diagnosis of mitochondrial diseases. However, therapeutic advancements for these are not impressive.[57,58] Up until now, most of the attempts to improve mitochondrial function have used vitamins, coenzymes, and metabolic intermediates. Among all the agents used to treat mitochondrial diseases, coenzyme Q10 (CoQ10) has been used most extensively.[59] CoQ10 is a fat-soluble quinone involving the transfer of electron from complex I to complex III. It was also reported that CoQ10 stabilizes the oxidative phosphorylation complexes in the inner membrane of mitochondria and detoxifies the free radicals.[60] An analogue of CoQ10, idebenone, was used with a LHON patient carrying the 11778 mutation, and a mild improvement of visual acuity was reported.[61] Vitamin B2, riboflavin, functions as a cofactor for electron transfer in complex I after conversion to flavin monophosphate and flavin adenine dinucleotide. Improved excise capacity was recorded in a mitochondrial myopathy patient with complex I defects following the taking of riboflavin.[62] As one kind of vitamin B, nicotinamide has been used mainly to treat niacin deficiency.[63] The biochemistry activity of nicotinamide is based on the conversion into nicotinamide adenine dinucleotide (NAD), which in turn is linked with complex I respiration. In a global ischemia-reperfusion mouse model, the administration of nicotinamide resulted in higher levels of NADH and restored the abolished complex I function.[64]

INTRODUCTION OF A SYNTHETIC COMPLEX I SUBUNIT

The G11778A mutation in LHON was the first mtDNA mutation linked to human diseases.[4] The mutation results in an arginine-to-histidine substitution at amino acid 340 in ND4 gene, and is one of three primary mutations for LHON, accounting for 50% of LHON cases. One of the conventional approaches to restore mitochondrial function is to introduce the wild-type gene into the cells carrying the mutation. Currently, it is still impossible to bring in exogenous genes to the mammalian mitochondrial genome; thus, an alternative approach—allotropic expression—was utilized in which a nuclear-encoded version of ND4 specifies the subunit in the cytoplasm that is then imported into the mitochondria. A synthetic ND4 construct compatible with the universal genetic code was generated and tagged with a mitochondrial targeting sequence, and it was then transfected into the cybrids carrying the 11778 mutation.[65] The growth capacity of the transfectants in the galactose medium was greatly improved, and the ATP synthesis based on the complex I-specific substrates showed a threefold increase, compared with the original 11778A cybrids.[65]

BYPASSING THE DEFECTIVE COMPLEX I

A rather radical method to correct the mitochondrial deficiency caused by a defective complex I is to bypass it altogether (FIG. 3). In contrast to the multisubunit

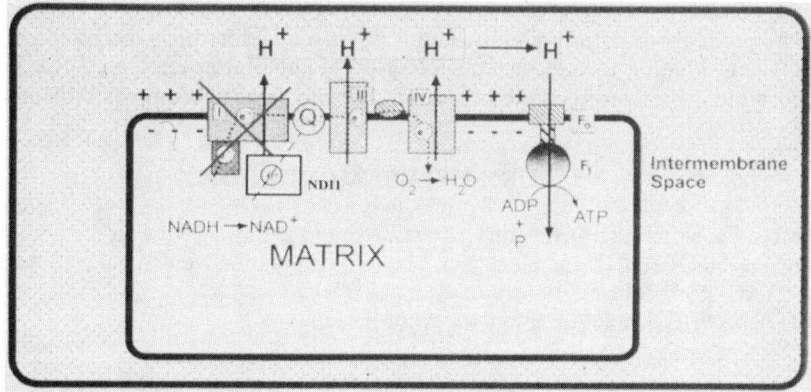

FIGURE 3. Bypassing the defective complex I. The yeast NDI1 protein is integrated into human oxidative phosphorylation chain and transfers electrons from NADH to ubiquinone. (Modified from Scheffler. 1999. Mitochondria. Wiley-Liss, Inc.)

enzyme of mammalian cells, the mitochondrial NADH-Q oxidoreductase of *Saccharomyces cerevisiae,* NDI1, is a simple peptide of 513 amino acid residues, including the NH_2-terminal 26-residue signal sequence which guides its import into mitochondria.[66,67] Furthermore, NDI1, in contrast to mammalian complex I, contains no proton-translocating site and is rotenone insensitive.

C4T is a human cell line with a homoplasmic frameshift mutation in the mitochondrial ND4 gene.[56] The abolishment of the protein synthesis of ND4 subunit caused by the mutation resulted in disruption of assembly of mtDNA-encoded complex I subunits and consequently a complete loss of its respiratory function and enzyme activity.[56] The yeast NDI1 gene was successfully introduced to C4T cells, and two transformants, C4T-Ca and C4T-AAV, were isolated. The NDI1 protein was further shown predominantly localized in the host mitochondria in both transformants.

The endogenous respiration rates and the complex I–dependent respiration in the permeabilized cells in intact C4T cells and their NDI1 transformants were measured; NDI1 restored respiration in the transformants, the extent of which restoration corresponded to the different expression levels of the NDI1 gene. The restored activities in the NDI1 transformants were insensitive to the complex I inhibitor, rotenone, but sensitive to flavone, a known inhibitor to NDI1 enzyme. NDI1-dependent respiration was also shown to be sensitive to mammalian complex III inhibitor, antimycin, indicating that the yeast NDI1 has integrated into the endogenous respiratory chain.[46]

Recently, the yeast NDI1 gene has also been introduced into a cybrid carrying the LHON 11778A mutation. It will be interesting to know whether NDI1 can be functional with the existence of a partially active complex I, and to explore further its potential as a tool for gene therapy.

Mitochondrial respiratory complex I appears to be the rate-limiting step of the overall respiratory activity and oxidative phosphorylation in mammalian systems,[53,68] and it is particularly vulnerable to oxidative damages.[68,69] Defective complex I is regarded as one of the major contributors to mitochondrial diseases.[68,70] Although impressive progress has been achieved during the past several years in the effort to restore mitochondrial function in complex I–deficient cells, we are still far from developing an effective therapeutic method. Nevertheless, the authors expect that the genetic screening approach by overexpression or downregulation of genes involving mitochondrial function (which is not discussed in this paper) will make a major contribution toward understanding the regulation of complex I function and development of new methods to recover the mitochondrial complex I function.

ACKNOWLEDGMENTS

Part of the work described in this paper was supported by a UMDF grant, an AHA grant (AHA 0430303N), and an Ellison Medical Foundation New Scholar in Aging Award (AG-NS-0183-02).

We thank Ning Wang for help with FIGURE 1.

REFERENCES

1. NASS, S. & M. NASS. 1963. Intramitochondrial fibers with DNA characteristics. J. Cell. Biol. **19:** 593–629.

2. ERNSTER, L., D. IKKOS & R. LUFT. 1959. Enzymic activities of human skeletal muscle mitochondria: a tool in clinical metabolic research. Nature **184:** 1851–854.
3. HOLT, I.J., A.E. HARDING & J.A. MORGAN-HUGHES. 1988. Deletions of muscle mitochondrial DNA in patients with mitochondrial myopathies. Nature **331:** 717–719.
4. WALLACE, D.C. et al. 1988. Mitochondrial DNA mutation associated with Leber's hereditary optic neuropathy. Science **242:** 1427–1430.
5. WALLACE, D.C. 1997. Mitochondrial DNA in aging and disease. Sci. Am. **277:** 40–47.
6. DIMAURO, S. & E.A. SCHON. 2001. Mitochondrial DNA mutations in human disease. Am. J. Med. Genet. **106:** 18–26.
7. WALLACE, D.C. 2001. A mitochondrial paradigm for degenerative diseases and ageing. Novartis Found. Symp. **235:** 247–263; discussion 263–266.
8. POLYAK, K. et al. 1998. Somatic mutations of the mitochondrial genome in human colorectal tumours. Nat. Genet. **20:** 291–293.
9. PENTA, J. S. et al. 2001. Mitochondrial DNA in human malignancy. Mutat. Res. **488:** 119–133.
10. HOCHHAUSER, D. 2000. Relevance of mitochondrial DNA in cancer. Lancet **356:** 181–182.
11. LOEFFEN, J.L. et al. 2000. Isolated complex I deficiency in children: clinical, biochemical and genetic aspects. Hum. Mutat. **15:** 123–134.
12. SCHAPIRA, A.H. 1998. Human complex I defects in neurodegenerative diseases. Biochim. Biophys. Acta **1364:** 261–270.
13. LENAZ, G. et al. 1997. Mitochondrial complex I defects in aging. Mol. Cell. Biochem. **174:** 329–333.
14. HUOPONEN, K. 2001. Leber hereditary optic neuropathy: clinical and molecular genetic findings. Neurogenetics **3:** 119–125.
15. LEBER, T. 1871. Über hereditare und congenital-angelegte Sehnervenleiden. Graefe's Arch. Ophthalmol. **17:** 249–291.
16. CHALMERS, R.M. & A.H. SCHAPIRA. 1999. Clinical, biochemical and molecular genetic features of Leber's hereditary optic neuropathy. Biochim. Biophys. Acta **1410:** 147–158.
17. HOWELL, N. et al. 1992. Mitochondrial gene segregation in mammals: is the bottleneck always narrow? Hum. Genet. **90:** 117–120.
18. HUOPONEN, K. et al. 1991. A new mtDNA mutation associated with Leber hereditary optic neuroretinopathy. Am. J. Hum. Genet. **48:** 1147–1153.
19. MACKEY, D. & N. HOWELL. 1992. A variant of Leber hereditary optic neuropathy characterized by recovery of vision and by an unusual mitochondrial genetic etiology. Am. J. Hum. Genet. **51:** 1218–1228.
20. DE VRIES, D.D. et al. 1996. Genetic and biochemical impairment of mitochondrial complex I activity in a family with Leber hereditary optic neuropathy and hereditary spastic dystonia. Am. J. Hum. Genet. **58:** 703–711.
21. JUN, A.S., M.D. BROWN & D.C. WALLACE. 1994. A mitochondrial DNA mutation at nucleotide pair 14459 of the NADH dehydrogenase subunit 6 gene associated with maternally inherited Leber hereditary optic neuropathy and dystonia. Proc. Natl. Acad. Sci. USA **91:** 6206–6210.
22. CAMPOS, Y. et al. 1997. Bilateral striatal necrosis and MELAS associated with a new T3308C mutation in the mitochondrial ND1 gene. Biochem. Biophys. Res. Commun. **238:** 323–325.
23. SANTORELLI, F.M. et al. 1997. Identification of a novel mutation in the mtDNA ND5 gene associated with MELAS. Biochem. Biophys. Res. Commun. **238:** 326–328.
24. PENISSON-BESNIER, I. et al. 2000. Recurrent brain hematomas in MELAS associated with an ND5 gene mitochondrial mutation. Neurology **55:** 317–318.
25. CORONA, P. et al. 2001. A novel mtDNA mutation in the ND5 subunit of complex I in two MELAS patients. Ann. Neurol. **49:** 106–110.
26. ANDREU, A.L. et al. 1999. Exercise intolerance due to a nonsense mutation in the mtDNA ND4 gene. Ann. Neurol. **45:** 820–823.
27. MAECHLER, P. & C.B. WOLLHEIM. 2001. Mitochondrial function in normal and diabetic beta-cells. Nature **414:** 807–812.
28. KALININ, V.N. et al. 1995. A new point mutation in the mitochondrial gene ND1, detected in a patient with type II diabetes. Genetika **31:** 1180–1182.

29. YAGI, T. *et al.* 2001. NADH dehydrogenases: from basic science to biomedicine. J. Bioenerg. Biomembr. **33:** 233–242.
30. FEARNLEY, I.M., *et al.* 2001. Grim-19, a cell death regulatory gene product, is a subunit of bovine mitochondrial nadh:ubiquinone oxidoreductase (complexI). J. Biol. Chem. **276:** 38345–38348.
31. CHOMYN, A. *et al.* 1985. Six unidentified reading frames of human mitochondrial DNA encode components of the respiratory-chain NADH dehydrogenase. Nature **314:** 592–597.
32. CHOMYN, A. *et al.* 1986. URF6, last unidentified reading frame of human mtDNA, codes for an NADH dehydrogenase subunit. Science **234:** 614–618.
33. GUENEBAUT, V. *et al.* 1997. Three-dimensional structure of NADH-dehydrogenase from *Neurospora crassa* by electron microscopy and conical tilt reconstruction. J. Mol. Biol. **265:** 409–418.
34. FRIEDRICH, T. *et al.* 1993. Attempts to define distinct parts of NADH:ubiquinone oxidoreductase (complex I). J. Bioenerg. Biomembr. **25:** 331–337.
35. HOFHAUS, G., H. WEISS & K. LEONARD. 1991. Electron microscopic analysis of the peripheral and membrane parts of mitochondrial NADH dehydrogenase (complex I). J. Mol. Biol. **221:** 1027–1043.
36. WEISS, H. *et al.* 1991. The respiratory-chain NADH dehydrogenase (complex I) of mitochondria. Eur. J. Biochem. **197:** 563–576.
37. GRIGORIEFF, N. 1999. Structure of the respiratory NADH:ubiquinone oxidoreductase (complex I). Curr. Opin. Struct. Biol. **9:** 476–483.
38. GRIGORIEFF, N. 1998. Three-dimensional structure of bovine NADH:ubiquinone oxidoreductase (complex I) at 22 A in ice. J. Mol. Biol. **277:** 1033–1046.
39. EARLEY, F.G. *et al.* 1987. Photolabelling of a mitochondrially encoded subunit of NADH dehydrogenase with [3H]dihydrorotenone. FEBS Lett. **219:** 108–112.
40. BROWN, M.D. *et al.* 2000. Functional analysis of lymphoblast and cybrid mitochondria containing the 3460, 11778, or 14484 Leber's hereditary optic neuropathy mitochondrial DNA mutation. J. Biol. Chem. **275:** 39831–39836.
41. HOFHAUS, G. *et al.* 1996. Respiration and growth defects in transmitochondrial cell lines carrying the 11778 mutation associated with Leber's hereditary optic neuropathy. J. Biol. Chem. **271:** 13155–13161.
42. HOFHAUS, G. & G. ATTARDI. 1995. Efficient selection and characterization of mutants of a human cell line which are defective in mitochondrial DNA-encoded subunits of respiratory NADH dehydrogenase. Mol. Cell. Biol. **15:** 964–974.
43. BAI, Y. *et al.* 2004. Genetic and functional analysis of mitochondrial DNA-encoded complex I genes. Ann. N.Y. Acad. Sci. **1011:** 272–283.
44. CHOMYN, A. 1996. In vivo labeling and analysis of human mitochondrial translation products. Methods Enzymol. **264:** 197–211.
45. BAI, Y. & G. ATTARDI. 1998. The mtDNA-encoded ND6 subunit of mitochondrial NADH dehydrogenase is essential for the assembly of the membrane arm and the respiratory function of the enzyme. EMBO J. **17:** 4848–4858.
46. BAI, Y. *et al.* 2001. Lack of complex I activity in human cells carrying a mutation in MtDNA-encoded ND4 subunit is corrected by the *Saccharomyces cerevisiae* NADH-quinone oxidoreductase (NDI1) gene. J. Biol. Chem. **276:** 38808–38813.
47. VOGEL, R. *et al.* 2004. Complex I assembly: a puzzling problem. Curr. Opin. Neurol. **17:** 179–186.
48. CHOMYN, A. 2001. Mitochondrial genetic control of assembly and function of complex I in mammalian cells. J. Bioenerg. Biomembr. **33:** 251–257.
49. SCHAGGER, H. & G. VON JAGOW. 1991. Blue native electrophoresis for isolation of membrane protein complexes in enzymatically active form. Anal. Biochem. **199:** 223–231.
50. KLEMENT, P. *et al.* 1995. Analysis of oxidative phosphorylation complexes in cultured human fibroblasts and amniocytes by blue-native-electrophoresis using mitoplasts isolated with the help of digitonin. Anal. Biochem. **231:** 218–224.
51. ZERBETTO, E., L. VERGANI & F. DABBENI-SALA. 1997. Quantification of muscle mitochondrial oxidative phosphorylation enzymes via histochemical staining of blue native polyacrylamide gels. Electrophoresis **18:** 2059–2064.

52. GUAN, M.X. *et al.* 1998. The deafness-associated mitochondrial DNA mutation at position 7445, which affects tRNASer(UCN) precursor processing, has long-range effects on NADH dehydrogenase subunit ND6 gene expression. Mol. Cell. Biol. **18:** 5868–5879.
53. BAI, Y., R.M. SHAKELEY & G. ATTARDI. 2000. Tight control of respiration by NADH dehydrogenase ND5 subunit gene expression in mouse mitochondria. Mol. Cell. Biol. **20:** 805–815.
54. TROUNCE, I.A. *et al.* 1996. Assessment of mitochondrial oxidative phosphorylation in patient muscle biopsies, lymphoblasts, and transmitochondrial cell lines. Methods Enzymol. **264:** 484–509.
55. KING, M.P. & G. ATTARDI. 1989. Human cells lacking mtDNA: repopulation with exogenous mitochondria by complementation. Science **246:** 500–503.
56. HOFHAUS, G. & G. ATTARDI. 1993. Lack of assembly of mitochondrial DNA-encoded subunits of respiratory NADH dehydrogenase and loss of enzyme activity in a human cell mutant lacking the mitochondrial ND4 gene product. EMBO J. **12:** 3043–3048.
57. DIMAURO, S., M. MANCUSO & A. NAINI. 2004. Mitochondrial encephalomyopathies: therapeutic approach. Ann. N.Y. Acad. Sci. **1011:** 232–245.
58. DIMAURO, S., S. TAY & M. MANCUSO. 2004. Mitochondrial encephalomyopathies: diagnostic approach. Ann. N.Y. Acad. Sci. **1011:** 217–231.
59. CHOW, C.K. 2004. Dietary coenzyme Q10 and mitochondrial status. Methods Enzymol. **382:** 105–112.
60. LANDI, L. *et al.* 1984. Antioxidative effect of ubiquinones on mitochondrial membranes. Biochem. J. **222:** 463–466.
61. MASHIMA, Y., Y. HIIDA & Y. OGUCHI. 1992. Remission of Leber's hereditary optic neuropathy with idebenone. Lancet **340:** 368–369.
62. ARTS, W.F. *et al.* 1983. NADH-CoQ reductase deficient myopathy: successful treatment with riboflavin. Lancet **2:** 581–582.
63. YANG, J., L.K. KLAIDMAN & J.D. ADAMS. 2002. Medicinal chemistry of nicotinamide in the treatment of ischemia and reperfusion. Mini Rev. Med. Chem. **2:** 125–134.
64. KLAIDMAN, L. *et al.* 2003. Nicotinamide offers multiple protective mechanisms in stroke as a precursor for NAD+, as a PARP inhibitor and by partial restoration of mitochondrial function. Pharmacology **69:** 150–157.
65. GUY, J. *et al.* 2002. Rescue of a mitochondrial deficiency causing Leber hereditary optic neuropathy. Ann. Neurol. **52:** 534–542.
66. DE VRIES, S. & L.A. GRIVELL. 1988. Purification and characterization of a rotenone-insensitive NADH:Q6 oxidoreductase from mitochondria of *Saccharomyces cerevisiae*. Eur. J. Biochem. **176:** 377–384.
67. DE VRIES, S. *et al.* 1992. Primary structure and import pathway of the rotenone-insensitive NADH-ubiquinone oxidoreductase of mitochondria from *Saccharomyces cerevisiae*. Eur. J. Biochem. **203:** 587–592.
68. PAPA, S. *et al.* 2004. Respiratory complex I in brain development and genetic disease. Neurochem. Res. **29:** 547–560.
69. LENAZ, G. *et al.* 1998. Oxidative stress, antioxidant defences and aging. Biofactors **8:** 195–204.
70. SMEITINK, J., L. VAN DEN HEUVEL & S. DIMAURO. 2001. The genetics and pathology of oxidative phosphorylation. Nat. Rev. Genet. **2:** 342–352.

Hearing Loss in Mitochondrial Disorders

CHANG-HUNG HSU,[a,b] HAEYOUNG KWON,[a] CHERNG-LIH PERNG,[b] REN-KUI BAI, PU DAI,[c] AND LEE-JUN C. WONG

Institute for Molecular and Human Genetics, Georgetown University Medical Center, Washington, DC 2007, USA 20007

ABSTRACT: Hearing loss is a common clinical feature in mitochondria-syndrome disorders. The underlining molecular etiology of hearing loss has not been fully investigated. In this study, 83 patients with mitochondrial syndromic hearing loss were evaluated clinically and their blood and tissue samples were examined molecularly. Using modified Walker's criteria, 31, 31, 14, and 7 patients had been classified as having definite, probable, possible, and unlikely diagnosis of mitochondrial disease, respectively. Deleterious mtDNA point mutations and/or abnormal mtDNA content or multiple deletions were identified in 20 patients with definite diagnosis and 2 patients with probable diagnosis. In addition to known, undisputed pathogenic mutations, several novel mutations believed to be clinically significant were found. Furthermore, abnormal mtDNA content and mtDNA deletions were found in some of the cases. Evaluation of clinical and diagnostic features associated with hearing loss revealed that cardiomyopathy, lactic acidosis, deficient respiratory chain enzyme complex activities, histochemical and ultrastructural abnormalities in mitochondria, and abnormal brain imaging results occurred significantly more frequently in patients with mtDNA alterations than in those without. This study revealed that the majority of the mtDNA defects in patients with mitochondrial syndromic hearing loss affect the overall mitochondrial gene expression.

KEYWORDS: multisystemic mitochondrial disorder; deafness; maternally inherited deafness; syndromic hearing loss

INTRODUCTION

Hearing loss is the most common sensory disorder in human[1,2] with an incidence of approximately 1 in 1,000 children.[1] Hereditary deafness is extremely genetically heterogeneous—there are more than 40 autosomal dominant (DFNA), more than 30 autosomal recessive (DNFB), and more than six X-linked (DFN) genes for non-

---

[a]C-H.H. and H.K. contributed equally to this study.
[b]Current address: Department of Neurology (C-H.H.) and Division of Clinical Pathology (C-L.P.), Tri-Service General Hospital, National Defense Medical Center, Taipei, Taiwan.
[c]Current address: Department of Otolaryngology, 301 General Hospital, Beijing, People's Republic of China.
Address for correspondence: Lee-Jun C. Wong, Ph.D., Institute for Molecular and Human Genetics, Georgetown University Medical Center, M4000, 3800 Reservoir Rd., NW, Washington, DC 20007. Voice: 202-444-0760; fax: 202-444-1770.
wonglj@georgetown.edu

syndromic deafness that account for 60–70% of inherited hearing impairment.[3] The most common cause for non-syndromic hearing loss is mutation in Connexin 26, a gap junction protein encoded by the autosomal recessive *GJB2* gene.[4–8] About 30% of hereditary deafness presents syndromic hearing loss with other clinical features in addition to hearing impairment. More than 30 nuclear genes, including those encoded for transcription factors and gap junction proteins in Waardenburg syndrome and Usher syndrome, have been identified as responsible for syndromic hearing loss.[3] Although most hereditary hearing loss is due to nuclear gene defects, in recent years, it has become clear that mitochondrial genes also play an important role. Mutations in mitochondrial DNA (mtDNA) can cause both non-syndromic hearing loss[2,9,10] and syndromic hearing loss. The most well studied is the A1555G mutation in 12S rRNA that causes the non-syndromic hypersensitivity to ototoxic effects of aminoglycosides by increased binding of aminoglycosides to mitochondrial ribosomes, thus disrupting mitochondrial protein synthesis.[11,12] Several mutations in tRNA ser(AGY) also cause maternally inherited hearing loss (MIHL) by disturbing the tRNA structure and function.[13] Syndromic sensorineural hearing loss is also one of the most prevalent clinical features of mitochondrial disorders.[2,14] Mitochondrial respiratory chain disorders caused by mtDNA mutation are clinically heterogeneous, with nerve, brain, and muscle being the most susceptible tissues, and often manifest as a multisystemic dysfunction that can include hearing loss. The most well known are the A3243G mutation in MELAS (mitochondrial encephalopathy, lactic acidosis, and stroke-like episodes) syndrome and large mtDNA deletion in Kearns-Sayre syndrome (KSS). Mutations in other regions of mtDNA may also cause syndromic deafness. Comprehensive molecular analysis of mtDNA mutations in patients with mitochondrial syndromic hearing loss is important in that it will provide information on the prevalence of pathogenic mtDNA mutations causing syndromic hearing loss and facilitate genetic counseling to affected families. It will also assist in monitoring the affected individual and family members for disease complications associated with the syndrome, so that appropriate patient care can be administered. In this paper, we report the results of clinical and molecular analysis of patients presenting mitochondrial disorders with deafness.

MATERIAL AND METHODS

Patients and Specimens

Approximately 1,500 patient samples sent to our laboratory (Molecular Genetics Laboratory, Institute for Molecular and Human Genetics, at Georgetown University Medical Center, Washington, DC) were evalutated for molecular diagnosis of mitochondrial DNA disorders from January 1, 1998 to December 31, 2002. DNA was extracted from blood or muscle specimens according to published procedures.[15,16] Clinical evaluation of the patients was performed by referring physicians who were asked to fill out a checklist of clinical indicators (TABLE 1). The clinical history of the patients was reviewed, and patients were classified using modified Walker's criteria for the diagnosis of mitochondrial disorders.[17,18] A definite diagnosis is defined as the identification of either two major criteria or one major plus two minor criteria (TABLE 2).[18] A probable diagnosis is defined as either one major plus one mi-

TABLE 1. Clinical and laboratory evaluators for mitochondrial disorders

Central Nervous System
- Developmental delay/mental retardation
- Hypotonia/floppy baby
- Autistic features
- Dementia/encephalopathy
- Headaches/migraine
- Stroke, ischemic episodes
- Ataxia
- Episodic coma
- Seizures
- Myoclonus or myoclonic siezures
- Perinatal insult
- Pyramidal signs
- Hemiparesis
- Intractable seizure, refractory
- Spasticity
- Dystonia
- Chorea

Neuromuscular
- Peripheral neuropathy
- Exercise intolerance
- Muscle weakness
- Cramps after exercise
- Easy fatigability
- Cardiomyopathy
- Heart block
- Arrhythmia
- Ophthalmoparesis, CPEO
- Abnormal EMG/NCV
- Ptosis

Visceral
- Gastrointestinal reflux
- Delayed gastric emptying
- Pancreatitis
- Diarrhea
- Constipation
- Cyclic vomiting
- Pseudoobstruction
- Hepatic failure
- Elevated transaminases
- Renal tubular disease
- Apnea/hypoventilation
- Respiratory deficiency/failure

TABLE 1. (*continued*) **Clinical and laboratory evaluators for mitochondrial disorders**

Metabolites/Metabolic
    Ketosis
    Dicarboxylic aciduria
    Lactic acidosis
    High cerebrospinal fluid lactate
    Organic aciduria
    Low plasma carnitine
    Abnormal CPK
    Elevated pyruvate
    Elevated cerebrospinal fluid protein
    Elevated alanine

Sensory
    Retinitis pigmentosa
    Optic atrophy
    Cataract
    Sensorineural hearing loss
    Tortuous retinal vessels

Endocrine
    Diabetes
    Exocrine/pancreas deficiencies
    Gonadal failure
    Hypothyroidism
    Hypoparathyroidism
    Hypo/hyperadrenal function
    Short stature

Other Clinical Features
    Failure to thrive
    Microcephaly
    SIDS/unexplained death
    Congenital anomalies
    Dysmorphic features
    Immunodeficiency
    Macrocytic anemia
    Pancytopenia/Bone marrow failure
    Neutropenia

Electrophysiology (Brain Stem)
    Abnormal BAERS
    Abnormal VERS
    Abnormal EEG

Imaging Studies/Other Studies
    Increased signal basal ganglia
    Delay myelination
    Cerebellar atrophy

TABLE 1. (*continued*) **Clinical and laboratory evaluators for mitochondrial disorders**

	Posterior stroke
	Leukodystrophy
	MRS/lactate peak
	Abnormal MRI
Muscle Biopsy	
	Abnormal histology
	Abnormal ultrastructure (electron miscroscopic)
	Abnormal respiratory enzymes
	Large mitochondrial/proliferation
	COX deficiency
	Ragged red fiber
	Hair/skin finings
	Rashes with hypopigmentation
	Hypertrichosis
	Alopecia
	Acrocyanosis
Family history	
	Identified mutation
	Evidence of maternal inheritance

CPEO: chronic progressive external ophthalmoplegia; EMG: electromyogram; NCV: nerve conduction velocity; BAERS: brainstem auditory evoked potential; VERS: visual evoked potential; EEG: electroencephalogram; MRI: magnetic resonance imaging; MRS: magnetic resonance spectroscopy; COX: cytochrome *c* oxidase.

nor criterion or at least three minor criteria. A possible diagnosis is defined as either a single major criterion or two minor criteria, one of which must be clinical.

Mutational Analysis of Common mtDNA Mutations

The DNA samples were analyzed for the presence of large deletions and 11 common point mutations; A3243G, T3271C, G3460A, A8344G, T8356C, G8363A, T8993G, T8993C, G11778A, G14459A, and T14484C, by Southern blot and allele-specific oligonucleotide (ASO) dot blot analysis, respectively, according to published procedures.[19,20]

Mutational Analysis of mtDNA Mutations Responsible for Non-Syndromic Maternally Inherited Hearing Loss

A1555G, T7445C, 7472insC, T7510C, T7511C, and T7512C mutations responsible for non-syndromic MIHL were analyzed. Two primer pairs—mtF1351/mtR1762 and mtF7234/mtR7921 (the number stands for the 5′ nucleotide position of the 20-mer primer)—were used in multiplex PCR/ASO analysis for point mutations. The PCR products were dotted onto nylon membrane followed by hybridization with wild type or mutant ASO probes as described previously.[19,20]

Deletion Analysis by PCR

Multiple deletions not detectable by Southern analysis were investigated by PCR using mtF8295/mtR13738 and mtF8295/mtR14499 primer pairs followed by agarose gel analysis. Deletions were also confirmed by real-time quantitative PCR (RT qPCR) analysis (see below).

Whole Mitochondrial Genome Mutational Analysis

Mutations in the entire mitochondrial genome were studied by temporal temperature gradient gel electrophoresis (TTGE) analysis, using 32 pairs of overlapping primers according to published procedures.[21] To identify the nucleotide sequence alterations, TTGE-positive DNA fragments were sequenced by direct DNA sequencing of the PCR product using a BigDye terminator cycle sequencing kit (Applied Biosystem, Foster City, CA) and an ABI 377 (Applied Biosystem) automated sequencer.

Real-Time Quantitative Analysis of mtDNA Deletions and Content

Multiple deletions and mtDNA depletion were analyzed simultaneously using real-time quantitative PCR (RT qPCR) method. For mtDNA, two regions, np3212–3319 and np12093–12170, were amplified for the measurement of mtDNA content. Region np3212–3319 is almost always present in all mtDNA molecules including the deletion mutants. Therefore these regions can be probed for the total mtDNA quantification. Region np12093–12170 is deleted in >95% of the deletion molecules that have been reported. Thus, probe in this region is used to measure the amount of non-deletion molecules. The difference in the amount of total and non-deleted mtDNA molecules is the amount of deletion mtDNA molecules. The $\beta2$ microglobulin (B2M) gene is used as the nuclear gene (nDNA) normalizer for the calculation of mtDNA to nDNA ratio. The target sequences were detected by using TaqMan probes 6FAM-5′TTACCGGGTCCTGCCATCT3′-TAMRA, 6FAM-5′CATCATTACCGGGTTTT-CCTCTTGTA3′-TAMRA, and VIC-5′TTGCTCCACAGGTAGCTCTAGGAGG3′-TAMRA for mtDNA regions np3212–3319 and np12093–12170, and B2M, respectively.[22] The 10 μL PCR reaction contains 1× TaqMan Universal PCR Master Mix (ABI P/N 4304437), 500 nM of each primer, 200 nM of TaqMan probe, 0.2–2 ng of total genomic DNA extract. PCR conditions are 2 min at 50°C, 10 min at 95°C, followed by 40 cycles of 15 s of denaturation at 95°C and 60 s of annealing/extension at 60°C. RT qPCR analysis was performed on Sequence Detector System ABI-Prism 7700.[22,23]

RESULTS

Clinical Evaluation and Fitting Diagnostic Criteria of Mitochondrial Disorders in Patients with Syndromic Hearing Loss

Among approximately 1,500 patients referred for molecular diagnosis of mitochondrial respiratory-chain disorders, 102 (6.8%) indicated hearing impairment as one of the clinical complaints. Only 83 with syndromic hearing loss had DNA available for further analysis. The female-to-male ratio was 1.03, and the average age of the overall patients was 17.4 years (range from 0.4 to 83). According to the information given in the evaluation list (TABLE 1), the majority of the patients (73/83 = 88%) had at least central nervous system (CNS) and motor system involvement in addition to hearing loss. Ten had CNS involvement only. Diagnosis using published modified criteria (TABLE 2)[17,18] revealed that 31 patients had definite and another 31 patients had probable diagnosis of mitochondrial disorder (TABLE 3). Only 21 patients had muscle biopsies. Since most of the patients did not have muscle specimens for analysis, the histological, biochemical, and ultrastructural results were lacking. Thus, the majority of them (52/83 = 62.7%) did not meet criteria for definite diagnosis of mitochondrial disease.

Detection of Point Mutations in mtDNA

Analysis of common point mutations detected A3243G mutation in four patients and A8344G and G14459A mutation in one patient each (TABLE 1). One patient had A15257G, A15812G, and T4216C mutations responsible for Leber hereditary optic neuropathy, as well as significantly reduced mtDNA content (25% of normal mean) and deletion. Analysis of six point mutations known to cause non-syndromic hearing loss identified a homoplasmic A1555G mutation in one patient who had multisystemic CNS and neuromuscular motor dysfunction in addition to diabetes and hearing loss. Screening of the entire mitochondrial genome identified six mutations: delT961insC, 5783G>A, 12258C>A, 5553T>C, 3505A>G, and 9022G>A. The delT961insC mutation was known to cause aminoglycoside-induced ototoxicity.[24,25] The remaining five mutations were novel. The G5783A mutation in tRNA cys occurs in the stem region of the T arm of tRNA cys. The mutation disrupts a base pair in an evolutionarily highly conserved region. The C12258A mutation in tRNA ser(AGY) disrupts a base pair at the acceptor stem region. The 5553T>C mutation occurred at the anti-codon stem region of tRNA trp. The 9022G>A missense mutation changes a highly conserved alanine to threonine in ATPase 6 subunit. The 3505A>G mutation in ND1 changes a mildly conserved threonine residue to alanine. The clinical significance of this mutation is unclear. However, the patient also had significant mtDNA proliferation and respiratory chain enzyme deficiency.

mtDNA Deletion, Depletion, and Amplification

Southern analysis did not detect any large deletions or rearrangement. Regular PCR/agarose gel and real-time qPCR analyses were performed on muscle specimens to detect low percentage of multiple mtDNA deletions and to determine the mtDNA content. Only when both methods showed consistent results was the mtDNA deletion scored. The RT qPCR analysis using three probes was designed to allow the detection of deletions and measurement of mtDNA content. Five patients had more than 20% multiple deletion mtDNA molecules. Two of them also had elevated mtDNA content (overreplication) in addition to multiple deletions. Three patients had mtDNA content below 30% of age-matched mean. All these patients manifested multisystemic disorders that were strongly supported by histological, ultrastructural, and biochemical evidences.

TABLE 2. Modified Walker criteria for the diagnosis of mitochondrial disorders[18]

	Major criteria	Minor criteria
Clinical	Complete RC encephalopathy (Leigh, Alpers, Person, Kearns-Sayre, MELAS, MERRF, NARP, MNGIE, LHON), or mitochondrial cytopathy defined as all three of the following: • Multisystemic symptoms for a RC disorder • A progressive clinical course with episodes of exacerbation or a strong family history • Other possible metabolic or nonmetabolic disorders have been excluded by appropriate testing	Symptoms compatible with a RC defect
Histology	>2% ragged red fibers in skeletal muscle	Smaller numbers of ragged red fibers depending on age, mitochondrial accumulation, or widespread EM abnormalities
Enzymology	>2% COX-negative fibers if <50 years old >5% COX-negative fibers if >50 years old <20% activity of any RC complex in tissue <30% activity of any RC complex in cell line <30% activity RC complex in ≥2 tissues	Antibody demonstration of RC defect 20–30% activity of any RC complex in tissue 30–40% activity RC complex in cell line 30–40% activity RC complex in ≥2 tissues
Functional	Skin fibroblast ATP synthesis rate >3 SD below mean	Skin fibroblast ATP synthesis rate 2–3 SD below mean
Molecular	Identified undisputed pathogenic mutation	MtDNA mutation of probable pathogenicity
Metabolic		>1 metabolic indicators of impaired RC function

MELAS: mitochondrial encephalopathy, lactic acidosis, and stroke-like episodes; MERRF: myoclonus epilepsy, ragged red fibers; KSS: Kearns-Sayre syndrome; LHON: Leber hereditary optic neuropathy; NARP: neuropathy, ataxia, and retinitis pigmentosa; MNGIE: mitochondrial neuro-gastro-intestinal encephalomyopathy; COX: cytochrome *c* oxidase; RC: respiratory chain; SD: standard deviation; EM: electron microscopy.

TABLE 3. Specimen type and meeting diagnosis criteria for mitochondrial disorders

Specimen type	Undisputed mtDNA Defect	Diagnosis of mitochondrial disorders				
		Definite	Probable	Possible	Unlikely	Total
Muscle	yes	14	0	0	0	14
	no	6	1	0	0	7
Blood	yes	5	2	0	0	7
	no	5	28	14	6	53
Kidney	yes	1	0	0	0	1
Skin fibroblast	no	0	0	0	1	1
Total		31	31	14	7	83

TABLE 4. Summary of mtDNA alterations in patients with mitochondrial syndromic hearing loss

	mtDNA Alteration	Patients (N)
1	Undisputed mtDNA point mutations	15
2	Reduced or elevated level of mtDNA content, mtDNA deletion, or combination of defects (muscle specimens only)	7
3	Mutations or variations with unknown significance	7
4	Respiratory chain enzyme complex deficiency without identifiable mtDNA mutations or quantitative alteration	4
5	Alterations not found	50
	Total	83

Significance of Molecular Defects in Syndromic Hearing Loss with Mitochondrial Disorder

The molecular defects were classified into four categories: (1) undisputed deleterious mutations, (2) severely reduced mtDNA content, high percentage of deletion mutant mtDNA molecules, highly elevated mtDNA content, or combination of any of these and/or point mutations of unclear clinical consequences, (3) mutations or variations with unknown significance, and (4) respiratory-chain enzyme complex deficiency without identifiable mtDNA mutations or quantitative alteration (TABLE 4). Category 3 includes samples with mtDNA content 50-80% of age-matched mean or mtDNA elevated 120–150%, or deletion mutant load less than 20%. Overall, 22/83 (27%) unrelated patients (categories 1 and 2) with mitochondrial syndromic hearing loss had identifiable molecular defects in mtDNA. Twenty of these 22 patients met definite diagnosis of mitochondrial disorders based on modified Walker's criteria[17,18] and two had probable diagnosis (TABLE 3). Seven patients (category 3) had mtDNA defects of unknown significance (data not shown). Four patients had respiratory-chain enzyme complex deficiency and abnormal mitochondrial structure without identifiable molecular defects (category 4). The remaining 50 patients did not have mutations of known pathogenic significance, and only 5 of them met the definite diagnosis of mitochondrial disorder. None of the patients with possible or unlikely diagnosis had identifiable molecular defects in mtDNA (TABLE 3).

Clinical Correlation

All patients were evaluated for the presence of clinical complications listed in TABLE 1. The clinical findings were grouped into the following categories: CNS, cardiac, muscular motor function, visceral systems including renal and hepatic functions, metabolites, ophthalmologic, endocrine, brain imaging examination, respiratory chain enzymes, histology and ultrastructure, and others including dysmorphic features. In addition, stroke, lactic acidosis, and diabetes mellius were evaluated and analyzed separately. It was found that CNS, endocrine dysfunction, abnormal brain MRI findings, lactic acidosis, deficient respiratory enzyme activities, and abnormal muscle histology and ultrastructure, but not stroke or diabetes, occur at a significantly higher frequency in the patients with identifiable mtDNA mutations (categories 1 and 2) when compared to patient without identifiable molecular defects in mtDNA (category 4).

DISCUSSION AND CONCLUSION

Sensorineural hearing loss is a common feature of the clinically heterogeneous mitochondrial disorders. This paper reports the results of comprehensive clinical and molecular studies of patients with mitochondrial syndromic hearing loss. Upon review of available clinical information, only 37% of the patients with hearing loss met definite diagnoses of mitochondrial disorders. This low percentage is probably due to the fact that the majority of the patients did not have muscle biopsies for analysis. Thus the lack of biochemical, histochemical, or ultrastructural evidence reduced the scores to fit the diagnostic criteria.

The most common mutation 1555A>G for aminoglycoside-induced non-syndromic deafness was found in only one patient who had multisystemic mitochondrial disorder. Mutations in tRNA ser(UCN) responsible for non-syndromic hearing loss were not detected in our patients, which would be expected since most of our patients have clinical features consistent with multisystemic mitochondrial encephalomyopathies.

Most of the molecular defects causing mitochondrial syndromic hearing loss are in tRNA and rRNA, or mtDNA deletions and depletions that affect the overall mitochondrial protein translation and gene expression. Recently, a nuclear encoded mitochodrial protein, DDP (deafness/dystonia peptide), has been shown to be responsible for the X-linked dystonia and deafness syndrome,[26] in which patients may also have visual disability and mental deficiency.[27] We screened the DDP gene for mutations in 43 males and sequenced 8 patients with both hearing loss and optic atrophy. Mutations in DDP gene were not detected.

In conclusion, comprehensive analysis of mitochondrial genome on patients presenting with mitochondrial disorders and hearing loss identified several novel mutations. Approximately 50% of the molecular defects in mtDNA responsible for mitochondrial disorder associated hearing loss are mutations in tRNAs and rRNAs. About 36% are due to mtDNA deletions and abnormal mtDNA content, either depletion or proliferation. For molecular diagnosis of patients with mitochondrial syndromic hearing loss, the common mutations for MELAS and MERRF syndromes, as well as A1555G, delT961insc, and other mutations in tRNA ser(UCN) known to cause maternally inherited deafness, should be analyzed. In addition, if muscle specimen is available, evaluation by real-time quantitative PCR for low percentage of multiple deletions and either depletion or proliferation of mtDNA content will be useful. Identification of mtDNA mutations can assist in genetic counseling and anticipation of possible disease complications.

ACKNOWLEDGMENTS

The authors would like to thank referring physicians and patients for participation in this study. This study was partly supported by a grant from Muscular Dystrophy Association to L-J.C.W.

REFERENCES

1. GORLIN, R.J., H.V. TORIELLO & M.M. COHEN. 1994. Hereditary Hearing Loss and Its Syndromes. Oxford University Press. Oxford, UK.
2. HUTCHIN, T.P. & G.A. CORTOPASSI. 2000. Mitochondrial defects and hearing loss. Cell Mol. Life Sci. **57:** 1927–1937.
3. BITNER-GLINDZICZ, M. 2002. Hereditary deafness and phenotyping in humans. Br. Med. Bull. **63:** 73–94.
4. DENOYELLE, F. et al. 1997. Prelingual deafness: high prevalence of a 30delG mutation in the connexin 26 gene. Hum. Mol. Genet. **6:** 2173–2177.
5. ESTIVILL, X. et al. 1998. Connexin-26 mutations in sporadic and inherited sensorineural deafness. Lancet **351:** 394–398.
6. ABE, S. et al. 2000. Prevalent connexin 26 gene (GJB2) mutations in Japanese. J. Med. Genet. **37:** 41–43.
7. LENCH, N. et al. 1998. Connexin-26 mutations in sporadic non-syndromal sensorineural deafness. Lancet **351:** 415.
8. MORELL, R.J. et al. 1998. Mutations in the connexin 26 gene (GJB2) among Ashkenazi Jews with nonsyndromic recessive deafness. N. Engl. J. Med. **339:** 1500–1505.
9. VAN CAMP, G. & R.J. SMITH. 2000. Maternally inherited hearing impairment. Clin. Genet. **57:** 409–414.
10. FISCHEL-GHODSIAN, N. 1998. Mitochondrial genetics and hearing loss: the missing link between genotype and phenotype. Proc. Soc. Exp. Biol. Med. **218:** 1–6.
11. NOLLER, H.F. 1991. Ribosomal RNA and translation. Annu. Rev. Biochem. **60:** 191–227.
12. HARPUR, E.S. 1982. The pharmacology of ototoxic drugs. Br. J. Audiol. **16:** 81–93.
13. GUAN, M.X. et al. 1998. The deafness-associated mitochondrial DNA mutation at position 7445, which affects tRNASer(UCN) precursor processing, has long-range effects on NADH dehydrogenase subunit ND6 gene expression. Mol. Cell. Biol. **18:** 5868–5879.
14. GOLD, M. & I. RAPIN. 1994. Non-Mendelian mitochondrial inheritance as a cause of progressive genetic sensorineural hearing loss. Int. J. Pediatr. Otorhinolaryngol. **30:** 91–104.
15. LAHIRI, D.K. & J.I. NURNBERGER, JR. 1991. A rapid non-enzymatic method for the preparation of HMW DNA from blood for RFLP studies. Nucleic Acids Res. **19:** 5444.

16. WONG, L-J.C. & C. LAM. 1997. Alternative, noninvasive tissues for quantitative screening of mutant mitochondrial DNA. Clin. Chem. **43:** 1241–1243.
17. WOLF, N.I. & J.A. SMEITINK. 2002. Mitochondrial disorders: a proposal for consensus diagnostic criteria in infants and children. Neurology **59:** 1402–1405.
18. BERNIER, F.P. et al. 2002. Diagnostic criteria for respiratory chain disorders in adults and children. Neurology **59:** 1406–1411.
19. WONG, L-J.C. & D. SENADHEERA. 1997. Direct detection of multiple point mutations in mitochondrial DNA. Clin. Chem. **43:** 1857–1861.
20. LIANG, M.H. & L. J. WONG. 1998. Yield of mtDNA mutation analysis in 2,000 patients. Am. J. Med. Genet. **77:** 395–400.
21. WONG, L-J.C. et al. 2002. Comprehensive scanning of the whole mitochondrial genome for mutations. Clin. Chem. **48:** 1901–1912.
22. WONG, L-J.C. & R. BAI. 2002. Real time quantitative PCR analysis of mitochodnrial DNA in patients with mitochondrial disease. Am. J. Hum. Genet. Suppl. **71:** 501.
23. WONG, L-J.C. et al. 2003. Compensatory amplification of mitochondrial DNA in a patient with a novel deletion/duplication and high mutant load. J. Med. Genet. **40:** 125.
24. BACINO, C. et al. 1995. Susceptibility mutations in the mitochondrial small ribosomal RNA gene in aminoglycoside induced deafness. Pharmacogenetics **5:** 165–172.
25. CASANO, R.A. et al. 1999. Inherited susceptibility to aminoglycoside ototoxicity: genetic heterogeneity and clinical implications. Am. J. Otolaryngol. **20:** 151–156.
26. KOEHLER, C.M. et al. 1999. Human deafness dystonia syndrome is a mitochondrial disease. Proc. Natl. Acad. Sci. USA **96:** 2141–2146.
27. TRANEBJAERG, L. et al. 1995. A new X linked recessive deafness syndrome with blindness, dystonia, fractures, and mental deficiency is linked to Xq22. J. Med. Genet. **32:** 257–263.

Brain Single Photon Emission Computed Tomography in Patients with A3243G Mutation in Mitochondrial DNA tRNA

PETERUS THAJEB,[a,e,f] MING-CHE WU,[b] BING-FU SHIH,[b] CHIN-YUAN TZEN,[c,e] MING-FU CHIANG,[d] AND REY-YUE YUAN[f]

Departments of [a]Neurology, [b]Nuclear Medicine, [c]Pathology, [d]Neurosurgery, and [e]Medical Research, Mackay Memorial Hospital, Taipei, Taiwan, Republic of China

[f]Section of Neurology, Taipei Medical University Hospital, Taipei, Taiwan, Republic of China

ABSTRACT: Brain single photon emission computed tomography (SPECT) studies were conducted in three patients with A3243G mutation of the mitochondrial (mt) DNA tRNA. All were born to mothers suffering from chronic progressive external ophthalmoplegia (CPEO) with the same A3243G point mutation of the mtDNA tRNA. The first case manifested clinically with MELAS, the second case manifested with CPEO, and third case was characterized by recurrent migraine-like headache, tremor, and epilepsy. Brain SPECT of all patients, regardless of whether they had or had not suffered from stroke-like episodes, showed multiple areas of asymmetrical decreased perfusion, particularly in the posterior and lateral head regions, especially the temporal lobes. Crossed-cerebellar diaschisis may occur. Conventional brain magnetic resonance images failed to show some of the lesions. Decreased regional cerebral blood flow, rather than previously proposed hyperemia, is likely to be the cause. We conclude that mitochondrial vasculopathy with regional cerebral hypoperfusion may be seen on brain SPECT in patients with mitochondrial disorders and A3243G mutations, regardless of whether they have or have not suffered from stroke-like episodes.

KEYWORDS: mitochondrial disease; A3243G mutation; MELAS; brain SPECT; CPEO; diaschisis; hypoperfusion; migraine

INTRODUCTION

Mitochondrial (mt) diseases are notorious for their heterogeneous genotypes and phenotypic expressions. Dozen of clinical syndromes have been recognized, such as chronic progressive external ophthalmoplegia (CPEO), mitochondrial myopathy with encephalopathy, lactic acidosis, and stroke-like episode (MELAS), Kearn-Sayre syndrome (KSS), mitochondrial myopathy with myoclonic epilepsy and

Address for correspondence: Dr. P. Thajeb, Department of Neurology and Medical Research, Mackay Memorial Hospital, P.O. Box Nei-hu 6-30, Taipei 11499, Taiwan, ROC. Voice: +8862-26477666; fax: +8862-25433642.

thajebp@hotmail.com

TABLE 1. Demographic characteristic of three patients with various phenotypes of mitochondrial disorders and A3243G mutation in the mtDNA tRNA Leu(UUR) gene

Case/ Age (yr)/ Sex	Phenotype	Maternal phenotype	Genotype	Lactate/ Pyruvate (mg/dL)	T2WI and PDI of brain MRI	Brain SPECT
1/32/M	MELAS	CPEO	A3243G	53.5/1.9	Infarctions/ ischemia	Hypoperfusion
2/18/M	CPEO	CPEO	A3243G	22.5/1.0	Negative	Hypoperfusion
3/15/M	Migraine, tremor, epilepsy	CPEO	A3243G	20.2/0.8	Negative	Hypoperfusion

ragged-red fiber (MERRF), Leber hereditary optic neuropathy (LHON), overlap syndrome, Leigh's disease, and so forth.[1–4] Even in a single family with the same genotype, the affected siblings and mother may have different clinical features (phenotypes).[5,6] The most common mt DNA mutation found in Taiwanese/Chinese patients with these disorders is an A-to-G substitution at nucleotide position 3243 (A3243G) of the mtDNA tRNA Leu(UUR).[6–12] Other single-point mutations or double mutations have also been reported in Taiwanese patients.[13] As a consequence of the alterations of the mitochondrial machinery, oxidative phosphorylation uncouples and cellular energy deficit ensues, and that deficit is reflected in the brain of patients with MELAS. However, brain perfusion studies of patients with mitochondrial diseases other than MELAS are rarely reported.[14–17] We herein report the brain single photon emission computed tomography (SPECT) of three different phenotypes of mitochondrial disease with the same A3243G mutation.

PATIENTS AND METHODS

The diagnosis of mitochondrial diseases is based on the clinical manifestations with positive family history (maternal relative with CPEO), the A3243G genotyping performed in the peripheral white blood cells and/or muscles, and electron microscopic examinations of the muscle biopsy. Demographic characteristics of the three patients are shown in TABLE 1. Conventional magnetic resonance imaging (MRI) of the brain, and technetium-99m ethyl cysteinate dimer (^{99m}Tc-ECD) SPECT of the brain were performed in all patients within 7 days of onset. Brief case history are described in case reports.

Case Reports

Patient 1

A 32-year-old man presented with MELAS since the age of 26 years. Six years after initial presentation he was demented. He was born to a mother with CPEO. Biopsy of left biceps muscle showed typical paracrystalline inclusions with "parking lot" appearance of the mitochondria on electron microscopic examinations. Se-

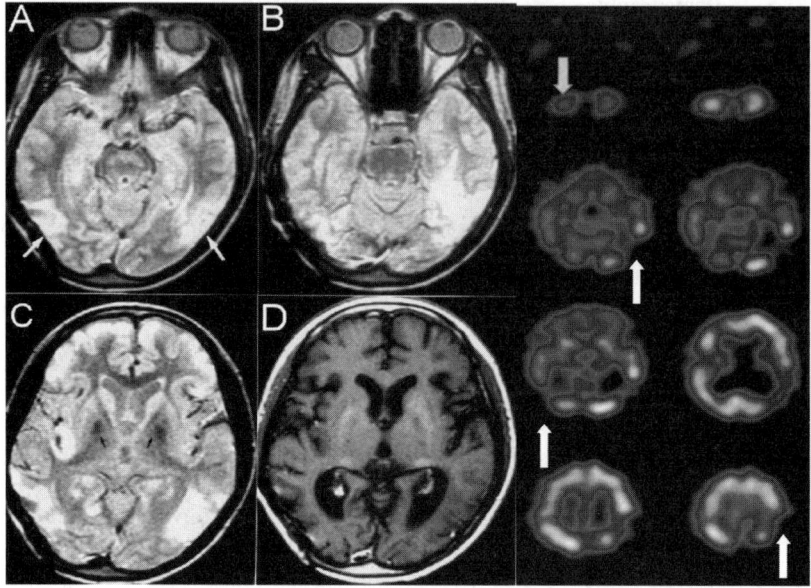

FIGURE 1. Brain MRI shows areas of increase signal intensity at bilateral posterior temporal lobes (*white arrows*) (**A–C**) proton density images (PDI). Symmetrical low signals on PDI (**C**) and high signals on T1WI (**D**) at bilateral basal ganglia suggest calcification. Brain SPECT (*right panel*) shows multiple regions of hypoperfusion at bilateral posterior temporal lobes (*white arrows*), left parietal lobe, and right frontal lobe. Crossed-cerebellar diaschisis is striking (*short arrow*).

quencing of the whole mitochondrial genomic DNA of the patient's white blood cells and muscles showed heteroplasmic A3243AG. Brain MRI (FIG. 1) showed multiple areas of increased signal intensities in the cortical areas crossing the vascular territories on proton density images (PDI) and T2-weighted images (T2WI). The cerebellum appeared normal on MRI. Brain SPECT showed asymmetric regions of multifocal hypoperfusion in bilateral temporal lobes, left parietal lobe, and, to a lesser degree, the frontal lobes. The most prominent region of decreased perfusion was at the left posterior temporal lobe. Crossed-cerebellar diaschisis (CCD) was noted (FIG. 1). Transcranial doppler ultrasonographic examinations showed high resistance flow profile of the middle (MCA) and posterior cerebral arteries (PCA) with elevated pulsatility index, especially in the left PCA (FIG. 2). MR angiography showed segmental narrowing of bilateral PCAs and MCAs (FIG. 2). Patient was put on co-enzyme Q therapy, and he remained stable in the past 3 years of follow-up.

Patient 2

An 18-year-old man who manifested with slowly progressive external ophthamoplegia without stroke-like episode, which began at 11 years of age. Complete external ophthalmoplegia with fixed eyes and ptosis were striking. Elevated plasma lactate/pyruvate (L/P) ratio, molecular diagnosis of A3243G, and electron micro-

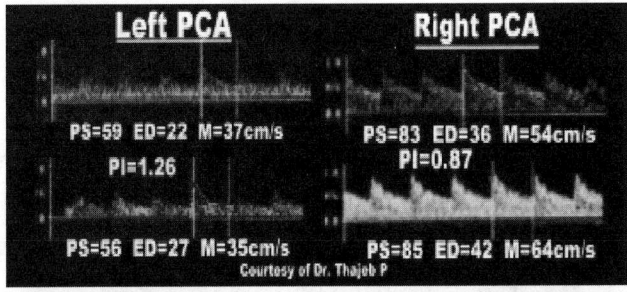

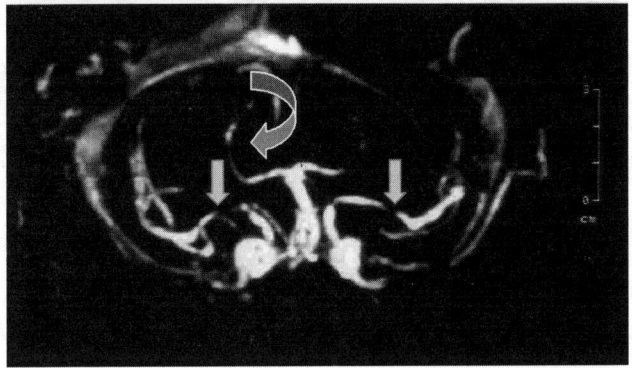

FIGURE 2. Transcranial doppler ultrasonography (*upper panel*) shows increased in pulsatility index of left posterior cerebral artery (PCA) suggesting increased downstream flow resistance, and elevated mean flow velocity of the contralateral PCA. Brain MR angiography [TR54/TI0/TE9/TOF 250] (*lower panel*) shows multiple segmental stenosis of bilateral PCAs (*curved arrow*) and MCAs (*short arrows*) suggestive of the mitochondrial vasculopathy.

scopic examinations of muscle biopsy confirmed the diagnosis of mitochondrial disease. The mother of this patient, who had undergone eye surgery twice for drooping eyelids, had CPEO. Conventional brain MRI was unrevealing. However, brain SPECT showed asymmetric areas of hypoperfusion in the posterior head regions, especially the temporal lobes, and to a greater extent than the frontal region (FIG. 3). He was placed on co-enzyme Q therapy and his CPEO did not improved.

Patient 3

The third patient was a 15-year-old boy with one-year history of recurrent migraine-like headache and limb tremors. He has had epilepsy since 13 years of age and has continued on carbamazepine therapy. Brain MRI was unrevealing. Transient homonymous hemianopsia was recorded within 24 h of onset of the migraine-like headache attacks. The interictal brain SPECT showed asymmetric areas of reduced perfusion, most remarkable in the left temporal lobe (FIG. 3). There was no CCD. He was born to a mother with CPEO. He himself did not have ophthalmoplegia. How-

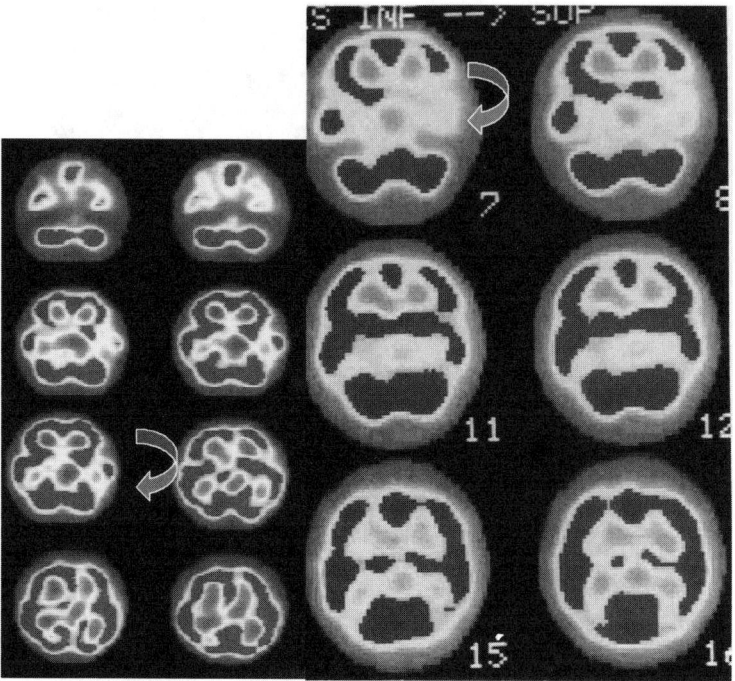

FIGURE 3. Brain SPECT of Patient 2 (*left*) and Patient 3 (*right*) show most remarkable decreased radiotracer uptake at left temporal lobe (*curved arrow*). Decreased perfusions are also seen in the right temporal lobe, bilateral parietal lobe, and the frontal lobes. The left temporal lobe hypoperfusion in Patient 3 corresponded to the clinical sign of transient homonymous right hemianopsia after the migraine-like attack.

ever, an elevated plasma level of creatine kinase (320 IU/dL), and lactate/pyruvate ratio (TABLE 1) suggested the presence of myopathy. Needle electromyographic examinations of left quadriceps muscles showed polyphasic potentials with mean amplitude of 291 ± 126 μV and mean duration of 6.3 ± 2.8 ms, consistent with myopathic change. Mitochondrial DNA sequencing of the white blood cells of the patient and his mother confirmed A3243G mutation. The manifestations of recurrent migraine-like headache, limb tremor, epilepsy, myopathy, and A3243G mutation suggested a mitochondria disease. Coenzyme Q10 and carbamazepine were continued until his death 2 years later because of a status epilepticus. Autopsy was not performed.

DISCUSSION

Stroke-like lesions of MELAS are characterized by multifocal areas of increased signals on T2- and diffusion-weighted MRI, predominantly involving the cortical regions crossing the vascular territories, the occipital lobes,[17] the parieto-occipital, and cerebellar regions.[15,16] ^{99m}Tc-HMPAO brain SPECT was reported to show remarkably increased tracer uptake (hyperperfusion, or hyperemia) in the abnormal regions revealed by MRI.[15–17] In contrast, our observations showed that the brain lesions of the A3243G mitochondrial diseases appear to be multiple regions of decreased tracer uptake (hypoperfusion) rather than increased tracer uptake (hyperperfusion) on brain SPECT. Similar patterns of regional perfusion defects were also seen in patients with either A3243G mutation and CPEO (without stroke-like episode) or A3243G mutation and migraine-like headache (with transient homonymous hemianopsia). Transcranial doppler ultrasonographic and MR angiographic studies of Patient 1 suggested that the regional brain hypoperfusion may be related to the increased downstream flow resistance due to a vasospasm or a segmental stenosis of the intracranial arteries (mitochondrial vasculopathy) (FIG. 2).

What factors determine the regional brain perfusion defects and vulnerability in patients with mitochondrial disorders remain to be elucidated. The most common site of focal brain perfusion defects is the posterior head region. We speculate that differences in the quotient of regional oxygen requirement and cellular metabolism, the regional activity of oxidative phosphorylation, or the so-called mitochondrial vasculopathy might contribute to the pathogenesis. Likewise, regional cerebellar hypoperfusion can either be caused by these mechanisms or by a well-known phenomenon called crossed-cerebellar diaschisis (CCD). The latter has been reported to occur in patients with cerebral infarctions and herpes simplex encephalitis.

Recently, a new MR technique, apparent diffusion coefficient mapping (ADC), has been advocated to help differentiating the stroke-like episodes of MELAS from acute ischemic stroke.[17] ADC is decreased in acute ischemic stroke, but is normal[17] or increased[14] in stroke-like episodes of MELAS. Ohshita and colleagues[14] has reported the time sequence of the reversible initial high signal lesions on DWI and ADC after stroke-like episode in MELAS. These lesions may not be seen on conventional MRI. High ADC may last 30 days after the stroke-like episodes, return to normal, and lesions may disappear completely with clinical improvement. They suggested that early increases of ADC in the acute and subacute phase of stroke-like episodes of MELAS reflect vasogenic rather than cytotoxic edema. The latter is characterized by ADC decrease and decreased diffusion.[18–20]

The non-neuromuscular features of A3243G mutations are common. Among 160 patients with deficiency of the respiratory chain enzyme of oxidative phosphorylation, 40% of cases manifested with neuromuscular symptoms and 60% with non-neuromuscular diseases.[21] The non-muscular features, migraine-like headache, and tremor, of our third patient are interesting. Different mutant load and distinct expression thresholds of different tissues may contribute to this phenomenon. But this speculation has to be proven. Taken together, we conclude that A3243G mutation of the mtDNA tRNA can have different phenotypes ranging from the classic MELAS, CPEO, to recurrent migraine-like headache and tremor. In the absence of stroke-like episodes, brain SPECT may be helpful to demonstrate the subclinical hypoperfusion brain lesions.

REFERENCES

1. PAVLAKIS, S.G. *et al.* 1984. Mitochondrial myopathy, encephalopathy, lactic acidosis, and stroke-like episodes: a distinctive clinical syndrome. Ann. Neurol. **16:** 481–488.
2. DIMAURO, S. & E. BONILLA. 1997. Mitochondrial encephalomyopathies. *In* The Molecular and Genetic Basis of Neurological Disease. R.N. Rosenberg *et al.*, Eds.: 201. Butterworth-Heinemann. Boston.

3. WALLACE, D.C., M.T. LOTT, M.D. BROWN, *et al.* 2001. Mitochondria and neuroophthalmologic diseases. *In* The Metabolic and Molecular Bases of Inherited Disease. 8th edit. C.R. Scriver *et al.*, Eds. **2:** 2425–2509.
4. HOLT, I.J., A.E. HARDING & J.A. MORGAN-HUGHES. 1988. Deletions of muscle mitochondrial DNA in patients with mitochondrial myopathies. Nature **331:** 717.
5. KOGA, Y., A. KOGA, R. IWANAGA, *et al.* 2000. Single-fiber analysis of mitochondrial A3243G mutation in four different phenotypes. Acta Neuropathol. **99:** 186–190.
6. THAJEB, P., H.C. LEE, C.Y. PANG, *et al.* 2000. Phenotypic heterogeneity in a Chinese family with mitochondrial disease and A3243G mutation of mitochondrial DNA. Chin. Med. J. (Taipei) **63:** 71–76.
7. FANG, W., C.C. HUANG, C.C. LEE, *et al.* 1993. Ophthalmologic manifestations in MELAS syndrome. Arch. Neurol. **50:** 977–980.
8. ZHANG, Y., J.F. LI, F.Y. WANG, *et al.* 2001. The study of A3243G and G13513A mitochondria DNA point mutation in patients with cerebral infarction. Chin. Med. J. (Beijing) **114:** 129-135.
9. HUANG, C.C., R.S. CHEN, C.M. CHEN, *et al.* 1994. MELAS syndrome with mitochondrial tRNA Leu (UUR) gene mutation in a Chinese family. J. Neurol. Neurosurg. Psychiatry **57:** 586–589.
10. LIOU, C.W., C.C. HUANG, E.C. CHEE, *et al.* 1994. MELAS syndrome: correlation between clinical features and molecular genetic analysis. Acta Neurol. Scand. **90:** 354–359.
11. HSU, C.C., Y.H. CHUANG, J.L. TSAI, *et al.* 1995. CPEO and carnitine deficiency overlapping in MELAS syndrome. Acta Neurol. Scand. **92:** 252–255.
12. THAJEB, P., K.M. HUANG, E.Y. CHI, *et al.* 1997. Koshevnikov syndrome in a patient with MELAS plus syndrome: electron microscopic and neuroimage studies. Chin. Med. J. (Beijing) **110:** 726–730.
13. TZEN, C.Y., P. THAJEB, T.Y. WU, *et al.* 2003. MELAS with point mutations involving tRNALeu (A3243G) and tRNAGlu (A14693G). Muscle Nerve **28:** 575–581.
14. OHSHITA, T., M. OKA, Y. IMON, *et al.* 2000. Serial diffusion-weighted imaging in MELAS. Neuroradiology **42:** 651–656.
15. PARRY, A. & P.M. MATTHEWS. 2003. Roles for imaging in understanding the pathophysiology, clinical evaluation, and management of patients with mitochondrial disease. J. Neuroimaging **13:** 293–302.
16. SUE, C.M., D.S. CRIMMINS, Y.S. SOO, *et al.* 1998. Neuroradiological features of six kindreds with MELAS tRNA(Leu) A3243G point mutation: implications for pathogenesis. J. Neurol. Neurosurg. Psychiatry **65:** 233–240.
17. YONEMURA, K., Y. HASEGAWA, K. KIMURA, *et al.* 2001. Diffusion-weighted MR imaging in a case of mitochondrial myopathy, encephalopathy, lactic acidosis, and stroke-like episodes. Am. J. Neuroradiol. **22:** 269–272.
18. SEVICK, R.J., F, KANDA, J. MINTOROVITCH, *et al.* 1992. Cytotoxic brain edema: assessment with diffusion-weighted MR imaging. Radiology **185:** 687–690.
19. CHIEN, D., K.K. KWONG, D.R. GRESS, *et al.* 1992. MR diffusion imaging of cerebral infarction in humans. Am. J. Neuroradiol. **13:** 1097–1102.
20. PIERPAOLI, C., A. RIGHINI, LINFANTE, *et al.* 1993. Histopathologic correlates of abnormal water diffusion in cerebral ischemia: diffusion-weighted MR imaging and light and electron microscopic study. Radiology **189:** 439–448.
21. MUNNICH, A., A. ROTIG, V. CORMIER, *et al.* 2001. Clinical presentation of respiratory chain deficiency. *In* The Metabolic and Molecular Bases of Inherited Disease. 8th edit. C.R. Scriver, *et al.*, Eds. **2:** 2261–2274.

Upregulation of Matrix Metalloproteinase 1 and Disruption of Mitochondrial Network in Skin Fibroblasts of Patients with MERRF Syndrome

YI-SHING MA,[a] YIN-CHIU CHEN,[a] CHING-YOU LU, CHUN-YI LIU, AND YAU-HUEI WEI

Department of Biochemistry and Center for Cellular and Molecular Biology, National Yang-Ming University, Taipei 112, Taiwan

ABSTRACT: By using cDNA microarray and RT-PCR techniques, we investigated the genome-wide alteration of gene expression in skin fibroblasts from patients with myoclonic epilepsy and ragged-red fibers (MERRF) syndrome. By screening for the genes with altered levels of expression, we first discovered that matrix metalloproteinase 1 (MMP1) was highly induced in the primary culture of skin fibroblasts of a female patient in a four-generation family with MERRF syndrome. This phenomenon was confirmed in skin fibroblasts from three other MERRF patients harboring about 85% of mtDNA with A8344G mutation. A further study revealed that the expression of MMP1 could be further induced by treatment of the skin fibroblasts with 200 μM hydrogen peroxide (H_2O_2) and inhibited by 1 mM N-acetylcysteine. Moreover, the intracellular level of H_2O_2 in skin fibroblasts of the female MERRF patient was higher than those of the asymptomatic family members and age-matched healthy controls. These findings imply that the increase in the expression of MMP1 may represent one of the responses to the increased oxidative stress in the skin fibroblasts of MERRF patients. We suggest that in affected tissues the oxidative stress–elicited overexpression of MMP1, and probably other matrix metalloproteinases involved in cytoskeleton remodeling, may play an important role in the pathogenesis and progression of mitochondrial encephalomyopathies such as MERRF syndrome.

KEYWORDS: MERRF syndrome; mitochondrial disease; matrix metalloproteinase; reactive oxygen species; oxidative stress

INTRODUCTION

In the past decade, molecular genetic studies have revealed that not only mutations of mitochondrial DNA (mtDNA) but also nuclear gene defects are involved in

---

[a]Y-S.M and Y-C.C. contributed equally to this work.

Address for correspondence: Professor Yau-Huei Wei, Department of Biochemistry and Molecular Biology, School of Life Science, National Yang-Ming University, Taipei 112, Taiwan. Voice: +886-2-2826-7118; fax: +886-2-2826-4843.

joeman@ym.edu.tw

a wide spectrum of human diseases associated with oxidative phosphorylation disorders.[1,2] The A8344G transition in the tRNA^{Lys} gene of mtDNA has been found in the affected tissues of about 80% of patients with myoclonic epilepsy and ragged-red fibers (MERRF) syndrome.[1,3] However, the severity and age of onset of the disease are poorly correlated with the proportion of mutant mtDNA in the affected tissue. These findings indicate that the etiology of mitochondrial diseases may involve other unidentified factors that affect the structure and function of mitochondria.[4] In recent years, we have demonstrated that pathogenic mtDNA mutations not only impair mitochondrial respiratory function but also enhance oxidative stress in human cells.[3] Based on this finding and other lines of evidence, we hypothesized that oxidative stress–elicited alteration in the expression of proteins and enzymes involved in oxidative metabolism may be another important consequence of pathogenic mtDNA mutation. This may also contribute to the pathogenesis and progression of mitochondrial diseases such as MERRF syndrome. To test this hypothesis, we launched a genome-wide study of the effect of A8344G mtDNA mutation on gene expression using Agilent Human 1 cDNA microarray. The results revealed that the expression of several clusters of genes involved in oxidative stress response, inflammatory response, and cytoskeleton remodeling were induced in skin fibroblasts of MERRF patients. One of the most notable findings is that the gene expression of matrix metalloproteinases (MMPs) and several marker enzymes of oxidative stress were upregulated in skin fibroblasts of a female patient with MERRF syndrome as compared with normal subjects and asymptomatic family members of the patient, who also harbored similar levels (approximately 80%) of mtDNA with A8344G mutation in their blood samples. It has been well documented that the expression of matrix metalloproteinases is regulated at the transcriptional level by growth factors, hormones, cytokines, oxidative stress, and agents that induce cell transformation.[5,6] An increase in the intracellular level of reactive oxygen species (ROS) may be due to inflammatory cytokines, defects in mitochondrial electron transport, and/or decrease in the capacity of antioxidant defense systems (e.g., superoxide dismutase [SOD] and glutathione peroxidase).[7] In light of these observations, we have been interested in the elucidation of the role of oxidative stress in the upregulation of these target genes. Here we report on the analysis of the expression levels of MMP1 in skin fibroblasts of the MERRF patient and her healthy siblings and of another three MERRF patients and three healthy controls.

MATERIALS AND METHODS

We first investigated the gene expression profiles of skin fibroblasts from a 15-year-old female patient and her maternal relatives of a four-generation MERRF family (FIG. 1), which was reported previously.[8] Skin fibroblasts cultured from skin biopsies of the patient (IV-1) and her asymptomatic siblings (IV-2 and IV-3) were used for genome-wide analysis of gene expression by cDNA microarray. The proportion of mtDNA with A8344G mutation in skin fibroblasts was determined by PCR-RFLP as described previously.[8] The proportion of mtDNA with A8344G mutation was 85% for skin fibroblasts of the female patient. Skin fibroblasts from three unrelated MERRF patients harboring mtDNA with A8344G mutation at an average proportion of 85.8 ± 6.9% were also recruited for this study. All the microarrays used in this

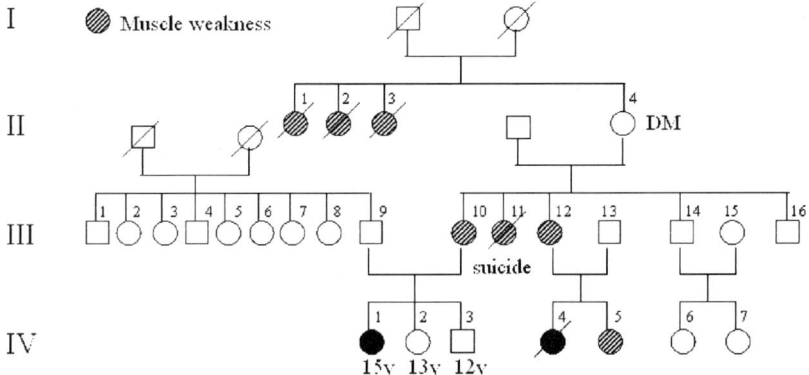

FIGURE 1. Illustration of the four-generation pedigree of a MERRF family investigated in this study. *Filled circle* indicates the proband, *open circles* and *squares* indicate asymptomatic family members.

study were Agilent's Human 1 cDNA microarrays spotted with 12,814 unique clones (Agilent Technologies, Palo Alto, CA). The genes with altered expression in the patient's fibroblasts were confirmed by RT-PCR with primers specific for the corresponding genes. The expression level of the MMP1 protein of skin fibroblasts from each MERRF patient was determined by Western blot. The activity of secreted MMP1 was measured by an ELISA method with a human active MMP-1 kit (R&D Systems, Minneapolis, MN). The intracellular concentration of H_2O_2 in skin fibroblasts was measured by using 2′,7′-dichlorofluorescein diacetate (DCFH-DA; Molecular Probes, Eugene, OR), which is converted to green fluorescent 2′,7′-dichlorofluorescein (DCF) after reacting with H_2O_2.[9] To observe the morphological changes of mitochondria, skin fibroblasts were stained with Mito Tracker Red (Molecular Probes), and imaged using a Leica TCS-SP2 laser-scanning confocal microscope system as previously described.[10]

RESULTS

We first screened by cDNA microarray the genes exhibiting more than twofold induction or suppression in the primary culture of skin fibroblasts of the female patient (IV-1) in the previously reported four-generation MERRF family.[8] When compared with the expression levels of the normal skin fibroblasts, we found that among the 12,814 unique genes in the Agilent Human 1 cDNA microarray 2.8% were upregulated ($P<0.05$) and 2.7% were downregulated ($P<0.05$), respectively. These genes were assigned and clustered into five specific groups as shown in TABLE 1. Some of these genes were further verified by RT-PCR (FIG. 2). The expression level of MMP1 in skin fibroblasts of the MERRF patient (IV-1) was significantly higher than that of normal control as revealed by Western blot (FIG. 3A). ELISA assay showed that the mean amplitude of increase in MMP1 activity in the MERRF fibroblasts was about 1.9-fold compared with that of control fibroblasts ($P<0.05$,

TABLE 1. Global view of transcriptional changes in the MEERF patient compared with normal subject

Induction	Suppression
Stress response	**Cytoskeleton protein**
Antioxidant enzymes	Intracellular cytoskeleton proteins: actin,
Heat shock proteins	myosin, elastin, tubulin
Inflammatory response	Extracellular matrix proteins
Cytokines	**Protein synthesis**
Complement components	Ribosomal proteins
Inflammatory marker proteins	Translation initiation factors
Interferon induced proteins	Translation elongation factors
Cytoskeleton remodeling	
MMP proteins	
Cathepsin S	

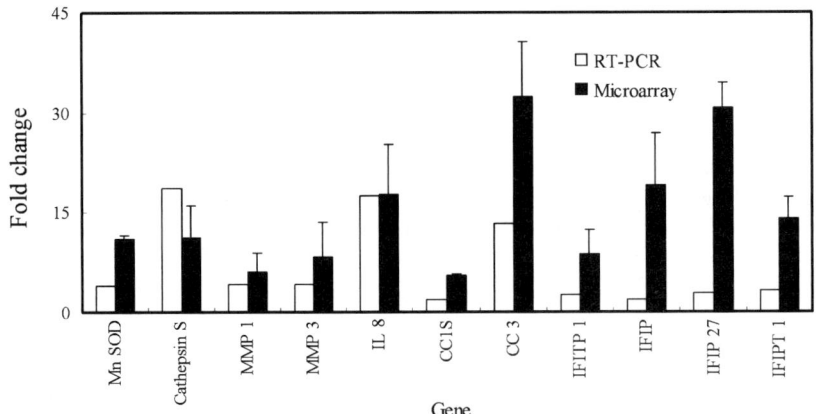

FIGURE 2. A comparison between microarray and RT-PCR analysis for several differentially expressed genes between the skin fibroblasts from a normal subject and the MERRF patient IV-1 in the four-generation MERRF family. The data of cDNA microarray were the mean ± S.D. of the results from two independent experiments. *Abbreviations*: IL 8, interleukin 8; CC1S, complement component 1s; CC3, complement component 3; IFIT1, interferon-induced transmembrane protein 1; IFIP, interferon α-inducible protein; IFIP27, interferon α-inducible protein 27; IFIPTI, interferon-induced protein with tetratricopeptide repeats 1.

FIG. 3B). Intracellular level of H_2O_2 in skin fibroblasts of the female MERRF patient IV-1 was 3.4-fold higher than those of controls ($P<0.05$, FIG. 3B). We also detected significant increase (165± 49% of control) of intracellular concentration of H_2O_2 in the skin fibroblasts of the other three MERRF patients. After 14-h treatment with 200 μM H_2O_2, the mean amplitude of increase in the activity of secreted MMP1 of all skin fibroblasts in the MERRF family was approximately 30% relative to the untreated skin fibroblasts (without H_2O_2) (FIG. 4A). It was noticed that the increase of

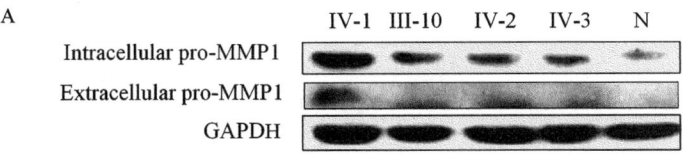

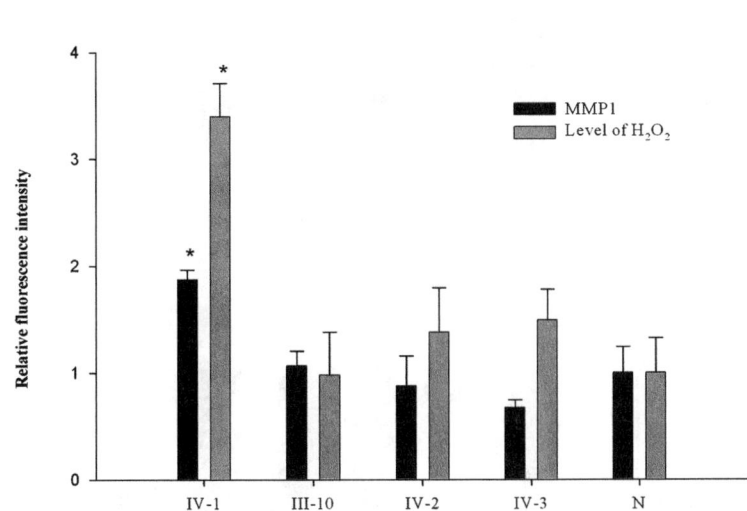

FIGURE 3. Comparison of the levels of matrix metalloproteinase 1 expression and intracellular concentration of hydrogen peroxide. (**A**) MMP1 expression analyzed by Western blot. (**B**) Relative levels of secreted MMP1 measured by ELISA and intracellular levels of hydrogen peroxide measured by flow cytometry. *Significantly different from healthy subjects, $P < 0.05$ ($N = 3$). N, normal control.

secreted MMP1 was sustained only in the skin fibroblasts of the female MEERF patient IV-1. The activity of secreted MMP1 was significantly reduced by pretreatment of skin fibroblasts for 2 h with 1 mM N-acetylcysteine (FIG. 4B). Confocal images showed that mitochondria assumed the fragmented and small tubular structures in skin fibroblasts of the patient IV-1 (FIG. 5A), whereas a continuous reticulum of mitochondria were seen in skin fibroblasts of her healthy sibling (IV-3) (data not shown) and control (FIG. 5H). Interestingly, 4 h after the pretreatment of skin fibroblasts of MERRF patient IV-1 with 1 mM N-acetylcysteine, the fragmented mitochondria were converted back to the continuous reticulum (FIG. 5).

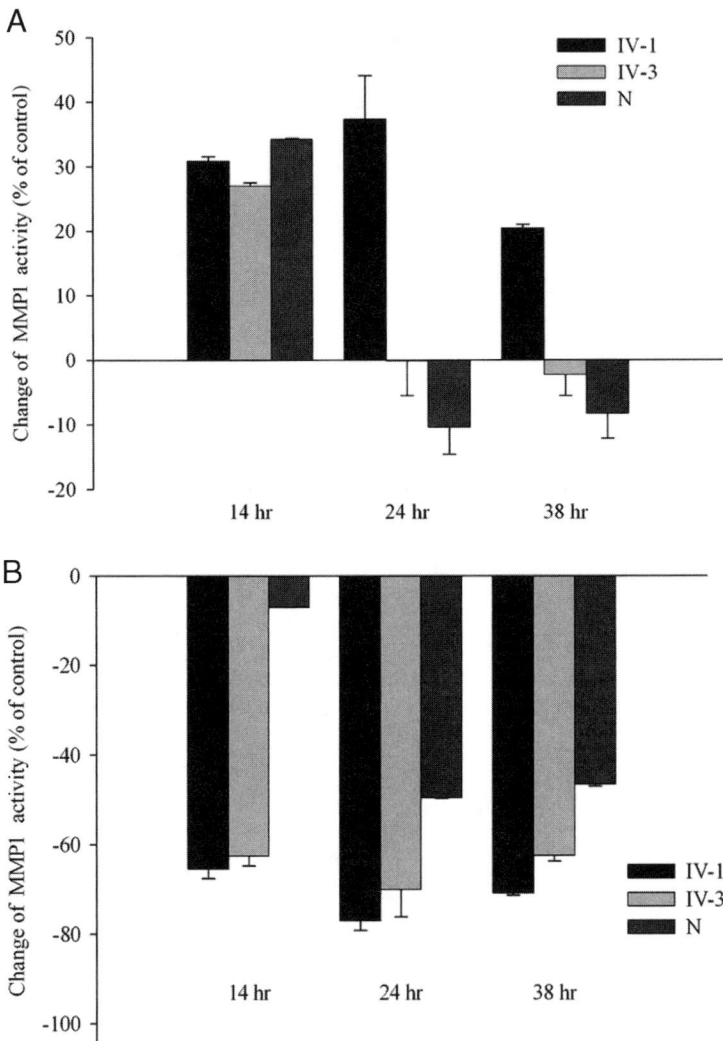

FIGURE 4. Suppression of the expression of MMP1 by treatment of skin fibroblasts with N-acetylcysteine. (**A**) Relative amplitude of secreted MMP1 at 14, 24, 38 h after a 1-h treatment of 200 μM H_2O_2. (**B**) Inhibition of secreted MMP1 by N-acetylcysteine (NAC). The MMP1 activities of three fibroblasts were assayed at 14, 24, and 38 h after 2-h treatment of 1 mM NAC. IV-1 and IV-3: the female MERRF patient and her young brother; N: an age-matched healthy subject.

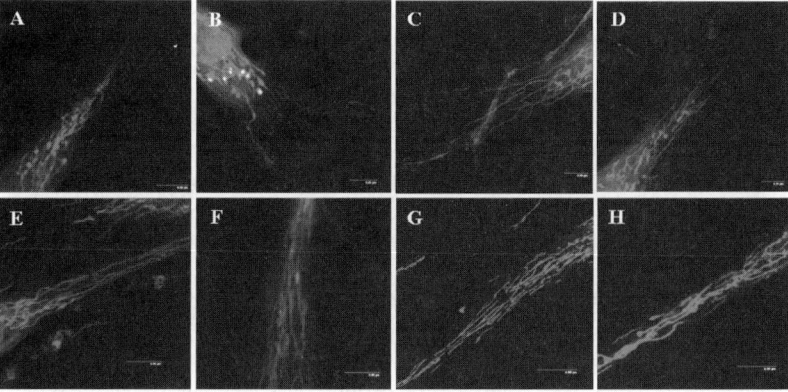

FIGURE 5. Effect of N-acetylcysteine on the subcellular organization and distribution of mitochondria in skin fibroblasts. (**A–F**) Confocal images of MitoTracker Red–stained skin fibroblasts of the female MERRF patient were obtained before (**A**) and after the treatment of 1 mM N-acetylcysteine (NAC) for 2 (**B**), 4 (**C**), 6 (**D**), 8 (**E**), and 10 (**F**) h, respectively. (**G** and **H**) Confocal images of MitoTracker Red–stained skin fibroblasts of a healthy subject were obtained before (**G**) and after (**H**) treatment of the cells with 1 mM NAC for 12 h. The *bar* in each image represents 8 μm in scale.

DISCUSSION

Although the suggestion that apoptosis is involved in the degeneration of muscle fibers in patients with mitochondrial myopathies has received great attention in the past few years, the contribution of apoptosis to the pathogenesis and progression of this group of disease remains controversial.[11–13] It has been proposed that increased expression of MnSOD and GSH may be considered as an early sign of mitochondrial dysfunction in myocytes of patients with mitochondrial diseases.[14] To investigate the involvement of free radical scavenging enzymes in the pathophysiology of mitochondrial disease, we analyzed the expression of these enzymes and found that the activity level of MnSOD, but not catalase and glutathione peroxidase, in skin fibroblasts of the three patients with MERRF syndrome was significantly higher (1.6-fold) than those of control. The mRNA and activity levels of MnSOD of the female patient (IV-1) in the MERRF family were increased to 3.5- and 1.5-fold, respectively. By contrast, the other antioxidant enzymes did not show significant changes. This has provided an explanation for the observation that the intracellular level of H_2O_2 was increased to 3.4-fold in skin fibroblasts of the female MERRF patient (IV-1) as compared with those of the asymptomatic members in the four-generation MERRF family and age-matched controls (FIG. 3). It has been recently reported that the levels of MMPs were increased in cell lines overexpressing MnSOD.[15,16] Moreover, Siwik and colleagues[17] demonstrated that ROS causes a decrease in fibrillar collagen synthesis and an increase in MMP activity in the myocardial cells.[17] In this study, we showed that the expression of MMP1 was inhibited by treatment of skin fibroblasts with 1 mM N-acetylcysteine. It has been shown in the TR9-7 cells that the frequency of apoptosis was increased after p53 induction and a concurrent increase in the expression of MnSOD and GPx, which were suppressed by N-acetylcysteine and a decrease in intracellular ROS level.[18] Thus, our findings are consistent with previous reports that excess generation of mitochondrial H_2O_2 by increased expression of MnSOD is the cause of the overexpression of MMP proteins.[15,16] Interestingly, the expression levels of MMP1 in skin fibroblasts of three different MERRF patients were consistently higher than those of controls (increased to 2.2–3.4 fold).

Surprisingly, the mitochondrial network of the skin fibroblasts of the MERRF patient was dramatically disrupted as revealed by confocal microscopy (FIG. 5). Furthermore, fragmented mitochondria of the skin fibroblasts of the MERRF patient IV-1 reverted to continuous reticulum with the treatment of antioxidant N-acetylcysteine. It has been reported that the decomposition of mitochondrial reticulum to numerous single mitochondria is due to cellular pathology such as depletion of mtDNA or heat shock.[19,20] In 1994, Hayashi and colleagues[21] reported that mitochondria in living cells were significantly swollen only when they contained a higher level of a predominant pathogenic mtDNA mutation (e.g., A3243G transition at tRNA$^{Leu(UUR)}$ gene or A4269G transition at tRNA^{Ile} gene). Abnormalities in mitochondrial morphology were also observed in the cybrids established from patients with sporadic Parkinson's and Alzheimer's disease.[22] Furthermore, abnormality in the structure of cytoskeleton has been proposed as one of the pathogenic factors for some mitochondrial diseases such as MELAS syndrome.[23] Based on the observed increase in the activity of extracellular matrix-degrading metalloproteinases and decrease in the expression levels of cytoskeleton and matrix proteins, we propose that the syndrome of muscle weakness and muscle wasting in MERRF patients may be associated with malfunction of the networks of cytoskeleton that results in the disruption of interconnected mitochondrial filaments and thus in defects in the energy transmission process.

ACKNOWLEDGMENTS

This work was supported by a grant (NSC92-2320-B010-037) from the National Science Council, Taiwan.

REFERENCES

1. ZEVIANI, M. & V. CARELLI. 2003. Mitochondrial disorders. Curr. Opin. Neurol. **16:** 585–594.
2. WALLACE, D.C. 1999. Mitochondrial diseases in man and mouse. Science **283:** 1482–1488.
3. WEI, Y.H. & H.C. LEE. 2003. Mitochondrial DNA mutations and oxidative stress in mitochondrial diseases. Adv. Clin. Chem. **37:** 83–128.
4. ESPOSITO, L.A., S. MELOV, A. PANVO, et al. 1999. Mitochondrial disease in mouse results in increased oxidative stress. Proc. Natl. Acad. Sci. USA **96:** 4820–4825.
5. NAGASE, H. & J.F. WOESSNER, JR. 1999. Matrix metalloproteinases. J. Biol. Chem. **274:** 21491–21494.
6. BRENNEISEN, P., K. BRIVIBA, M. WLASCHEK, et al. 1997. Hydrogen peroxide (H_2O_2) increases the steady-state mRNA levels of collagenase/MMP-1 in human dermal fibroblasts. Free Radic. Biol. Med. **22:** 515–524.
7. LU, C.Y., H.C. LEE, H.J. FAHN & Y.H. WEI. 1999. Oxidative damage elicited by imbalance of free radical scavenging enzymes is associated with large-scale mtDNA deletions in aging human skin. Mutat. Res. **423:** 11–21.

8. Lu, C.Y., D.J. Tso, T. Yang, et al. 2002. Detection of DNA mutations associated with mitochondrial diseases by Agilent 2100 bioanalyzer. Clin. Chim. Acta **318:** 97–105.
9. Lee, H.C., P.H. Yin, C.W. Chi & Y.H. Wei. 2002. Increase in mitochondrial mass in human fibroblasts under oxidative stress and during replicative cell senescence. J. Biomed. Sci. **9:** 517–526.
10. Liu, C.Y., C.F. Lee, C.H. Hong & Y.H. Wei. 2004. Mitochondrial DNA mutation and depletion increase the susceptibility of human cells to apoptosis. Ann. N.Y. Acad. Sci. **1011:** 133–145.
11. Sciacco, M., G. Fagiolari, C. Lamperti, et al. 2001. Lack of apoptosis in mitochondrial encephalomyopathies. Neurology **56:** 1070–1074.
12. Mirabella, M., S. Di Giovanni, G. Silvestri, et al. 2000. Apoptosis in mitochondrial encephalomyopathies with mitochondrial DNA mutations: a potential pathogenic mechanism. Brain **123:** 93–104.
13. Ikezoe, K., M. Nakagawa, C. Yan, et al. 2002. Apoptosis is suspended in muscle of mitochondrial encephalomyopathies. Acta Neuropathol. **103:** 531–540.
14. Filosto, M., P. Tonin, G. Vattemi, et al. 2002. Antioxidant agents have a different expression pattern in muscle fibers of patients with mitochondrial diseases. Acta Neuropathol. **103:** 215–220.
15. Wenk, J., P. Brenneisen, M. Wlaschek, et al. 1999. Stable overexpression of manganese superoxide dismutase in mitochondria identifies hydrogen peroxide as a major oxidant in the AP-1-mediated induction of matrix-degrading metalloprotease-1. J. Biol. Chem. **274:** 25869–25876.
16. Ranganathan, A.C., K.K. Nelson, A.M. Rodriguez, et al. 2001. Manganese superoxide dismutase signals matrix metalloproteinase expression via H_2O_2-dependent DRK1/2 activation. J. Biol. Chem. **276:** 14264–14270.
17. Siwik, D.A., P.J. Pagano & W.S. Colucci. 2001. Oxidative stress regulates collagen synthesis and matrix metalloproteinase activity in cardiac fibroblasts. Am. J. Physiol. Cell. Physiol. **280:** C53–C60.
18. Hussain, S.P., P. Amstad, P. He, et al. 2004. p53-induced up-regulation of MnSOD and GPx but not catalase increases oxidative stress and apoptosis. Cancer Res. **64:** 2350–2356.
19. Gilkerson, R.W., D.H. Margineantu, R.A. Capaldi & J.M. Selker. 2000. Mitochondrial DNA depletion causes morphological changes in the mitochondrial reticulum of cultured human cells. FEBS Lett. **474:** 1–4.
20. Collier, N.C., M.P. Sheetz & M.J. Schlesinger. 1993. Concomitant changes in mitochondria and intermediate filaments during heat shock and recovery of chicken embryo fibroblasts. J. Cell. Biochem. **52:** 297–307.
21. Hayashi, J., S. Ohta, Y. Kagawa, et al. 1994. Functional and morphological abnormalities of mitochondria in human cells containing mitochondrial DNA with pathogenic point mutations in tRNA genes. J. Biol. Chem. **269:** 19060–19066.
22. Trimmer, P.A., R.H. Swerdlow, J.K. Parks, et al. 2000. Abnormal mitochondrial morphology in sporadic Parkinson's and Alzheimer's disease cybrid cell lines. Exp. Neurol. **162:** 37–50.
23. Rusanen, H., J. Annunen, H. Yla-Outinen, et al. 2002. Cytoskeletal structure of myoblasts with the mitochondrial DNA 3243A→G mutation and of osteosarcoma cells with respiratory chain deficiency. Cell Motil. Cytoskeleton **53:** 231–238.

Increased Oxidative Damage with Altered Antioxidative Status in Type 2 Diabetic Patients Harboring the 16189 T to C Variant of Mitochondrial DNA

TSU-KUNG LIN,[a] SHANG-DER CHEN,[a] PEI-WEN WANG,[c] YAU-HUEI WEI,[b] CHENG-FENG LEE,[b] TZU-LING CHEN,[b] YAO-CHUNG CHUANG,[a] TENG-YEOW TAN,[a] KU-CHOU CHANG,[a] AND CHIA-WEI LIOU[a]

[a]*Department of Neurology and* [c]*Department of Metabolism, Chang Gung Memorial Hospital, Kaohsiung 833, Taiwan*

[b]*Department of Biochemistry and Center for Cellular and Molecular Biology, National Yang-Ming University, Taipei 112, Taiwan*

ABSTRACT: A transition of T to C at nucleotide position 16189 in mitochondrial DNA (mtDNA) has attracted biomedical researchers for its probable correlation with the development of diabetes mellitus in adult life. In diabetes, persistent hyperglycemia may cause high production of free radicals. Reactive oxygen species are thought to play a role in a variety of physiologic and pathophysiologic processes in which increased oxidative stress may play an important role in disease mechanisms. The aim of the present study was to clarify the degree of oxidative damage and plasma antioxidant status in diabetic patients and to see the potential influence of the 16189 variant of mtDNA on the oxidative status in these patients. An indicative parameter of lipid peroxidation, malondialdehyde (MDA), and total free thiols were measured from plasma samples of 165 type 2 diabetic patients with or without this variant and 168 normal subjects. Here we report an increase in the plasma levels of MDA and total thiols in type 2 diabetic patients compared with control subjects. The levels of plasma thiols in diabetic patients with the 16189 variant of mtDNA were not different from those in controls. These results suggest an increase in the oxidative damage and a compensatory higher antioxidative status in patients with type 2 diabetes. Harboring the 16189 mtDNA variant may impair the ability of a cell to respond properly to oxidative stress and oxidative damage.

KEYWORDS: mitochondrial DNA; T16189C polymorphism; antioxidative status; type 2 diabetes

Address for correspondence: Chia-Wei Liou, M.D., Department of Neurology, Chang Gung Memorial Hospital, 123, Ta-Pei Road, Niao-Sung Hsiang, Kaohsiung 833, Taiwan. Voice: +886-7-7317123 ext. 2283; fax: +886-7-7318762.
cwliou@ms22.hinet.net

INTRODUCTION

Mitochondria play a pivotal role in cell physiology not only in supplying energy to the cell but also in maintaining the redox potential,[1] in modulating intracellular calcium,[2] in generating free radicals,[3] and in initiating and executing apoptosis of the cell.[4,5] A mutation of mitochondrial DNA (mtDNA) can affect these functions and thus lead to clinical pathologies.[1] Some mtDNA mutations are strongly associated with diabetes,[6] with the most common mutation being the A3243G mutation in the mtDNA-encoded tRNA$^{Leu(UUR)}$ gene.[7,8] Variants in mtDNA could be associated with type 2 diabetes because ATP plays a critical role in the production and release of insulin.[9] Recently, a transition of T to C at nucleotide position (np) 16189 in the hypervariable D-loop region of mtDNA has attracted research interest for its probable correlation with the development of diabetes mellitus (DM) in adult life.[10,11] According to the Cambridge reference sequence, the first hypervariable segment of the human mtDNA control region contains a homopolymeric tract of cytosines between nt 16184 and 16193, interrupted at np 16189 by a thymine.[12] A variant commonly found in the general population is a T-to-C transition at np 16189, resulting in an uninterrupted homopolymeric cytosine tract. This common mtDNA variant has been shown to be positively correlated with blood fasting insulin and type 2 diabetes in a population-based case-control study in Cambridgeshire, UK.[13] In one of our previous studies a positive relationship between this mtDNA variant and cerebral infarction was established.[14]

However, few reports have explored oxidative stress in patients with this mtDNA variant. In this communication, we show that an increase in oxidative damage is associated with the occurrence of the 16189 C variant of mtDNA, and the implications of this sequence variation in the D-loop of mtDNA in the pathogenesis of type 2 diabetes are discussed.

MATERIAL AND METHODS

A total of 165 Taiwanese patients with type 2 diabetes and 168 normal subjects over age 40 were enrolled in this study. Venous blood samples were drawn in the fasting state (for at least 8 h) and processed immediately. Blood collected into tubes containing EDTA was centrifuged at 1,500 g for 10 min at 4°C, and mtDNA was extracted from peripheral leukocytes and the region of interest was amplified using polymerase chain reaction (PCR) techniques as described previously.[14] Two primer pairs were used; the forward primer consisted of np 15971-15990 of mtDNA and the reverse primer was an oligonucleotide spanning np 16471-16452 of mtDNA. The presence of the 16189 C variant was determined by using a combination of PCR and restriction fragment length polymorphism (RFLP) analysis with the restriction enzyme *Mnl* I. PCR products were digested with 1 U of the enzyme for at least 1 h at 37°C and subjected to electrophoresis with both positive and negative controls on a 2% agarose gel at 80 V for 45 min. DNA restriction fragments were visualized by UV transillumination of the gel stained with ethidium bromide.

The concentration of plasma thiobarbituric acid reactive substances (TBARS) was assessed based on the method of Ohkawa *et al.*[15] After centrifugation, the plasma samples were stored at −80°C for further analysis. Results are expressed as

TABLE 1. TBARS and thiol levels in DM and non-DM subjects

	N	Sex (male %)	Age (y ± SD)	TBARS (μmol/L)	Thiols (μmol/L)
DM	165	53.3	56.5 ± 8.8	1.69 ± 0.46	1.98 ± 0.58
Non-DM	168	54.8	56.3 ± 7.1	1.24 ± 0.42	1.74 ± 0.56
P	NS	NS	NS	<0.001*	0.003*

ABBREVIATIONS: DM, diabetic patients; non-DM, nondiabetic patients; TBARS, thiobarbituric acid reactive substances; NS, not significant.
*$P < 0.05$.

micromoles of TBARS per liter. A standard curve of TBARS was obtained by hydrolysis of 1,1,3,3-tetraethoxypropane (TEPP).

Plasma free thiols were determined by directly reacting thiols with 5,5-dithiobis 2-nitrobenzoic acid (DTNB) to form 5-thio-2-nitrobenzoic acid (TNB).[16] The amount of thiols in the sample was calculated from the absorbance determined using the extinction coefficient of TNB ($A_{412} = 13{,}600$ $M^{-1}cm^{-1}$).[16]

Statistical analysis was performed using Student's t test. Data are expressed as mean ± SD. A difference between groups with $P < 0.05$ is considered statistically significant.

RESULTS

RFLP analysis demonstrated the presence of the 16189 C variant of mtDNA in both study groups. The levels of plasma TBARS were 1.36 times higher in the diabetic group than in the controls (TABLE 1). The mean plasma TBARS levels of the diabetic groups with the 16189 variant of mtDNA (1.62 ± 0.46 μmol/L; $P < 0.05$) and without the variant (1.72 ± 0.47 μmol/L; $P < 0.05$) were higher than those of the control group (1.24 ± 0.42 μmol/L). No significant difference in plasma levels of TBARS was found between groups with the 16189 mtDNA variant and wild type (1.30 ± 0.45 μmol/L vs. 1.19 ± 0.36 μmol/L) in the control group. The analysis of antioxidant status revealed that the mean plasma level of free reduced thiols in diabetic patients without the 16189 mtDNA variant was significantly higher than that of the controls ($P < 0.05$) (TABLE 2). Plasma levels of free thiols were not different between diabetic patients with the 16189 variant of mtDNA and the controls.

DISCUSSION

Mitochondria are the major source of reactive oxygen species (ROS) in the cell and are susceptible to excessive oxidative damage.[17] Consequently, mitochondrial dysfunction has long been suggested to correlate with many human diseases and be linked with the process of aging.[18,19] The role of oxidative damage in the pathogenesis of the diabetic state has also been investigated extensively.[20,21] In the diabetic state, increased metabolic flux in the mitochondria due to high blood glucose, coupled with a reduction of NAD^+ to NADH, may result in increased formation of ROS

TABLE 2. TBARS and thiol levels in DM and non-DM subjects with and without mtDNA 16189 variant

	N	Sex (male %)	Age (y ± SD)	TBARS (μmol/L)	Thiols (μmol/L)
DM					
Variant	54	44.4	57.8 ± 10.2	1.62 ± 0.46	1.81 ± 0.48
Wild	111	57.7	55.8 ± 8.0	1.72 ± 0.47	2.07 ± 0.62
P		NS	NS	NS	0.022*
Non-DM					
Variant	71	53.5	55.7 ± 7.5	1.30 ± 0.45	1.82 ± 0.54
Wild	97	55.7	56.7 ± 6.8	1.19 ± 0.38	1.69 ± 0.58
P		NS	NS	NS	NS

ABBREVIATIONS: DM, diabetic patients; non-DM, nondiabetic patients; TBARS, thiobarbituric acid reactive substances; NS, not significant.
*$P < 0.05$.

such as superoxide anions, peroxinitrite, and highly reactive hydroxyl radicals.[6,22] These highly reactive ROS may cause membrane lipid peroxidation, nitration of proteins, and degradation of DNA, all of which could lead to tissue damage.[23,24] But the mechanism linking the 16189 mtDNA variant and type 2 diabetes is still not clear. It is therefore possible that the 16189 variant could result in a change in the mtDNA copy number, which might have a detrimental effect on respiratory chain function and thus lead to an increase in oxidative stress.[22] Therefore, we selected diabetic patients to examine whether oxidative status is altered among diabetic patients and whether this mtDNA variant is relevant to oxidative damage.

Malondialdehyde, measured as TBARS, is one of the well-known secondary products of lipid peroxidation and was used in this study as an indicator of oxidative damage.[25] A significant increase in the concentration of TBARS was noted in diabetic patients in comparison with healthy subjects (TABLE 1), and this finding is consistent with previous studies in which oxidative stress increased among diabetic patients.[20,24] In patients with or without the 16189 mtDNA variant, the degree of lipid peroxidation did not reveal a significant difference (TABLE 2). Although biochemical alterations in blood do not always reflect the clinical severity of the disease, these data showed a positive relationship between the lipid peroxidation index TBARS and diabetic patients. However, no significant difference in TBARS was noted between diabetic patients with or without the 16189 variant of mtDNA.

During the evolution of aerobic life, organisms have evolved several detoxifying mechanisms that protect themselves from the ill effects of oxygen by utilizing a series of antioxidant defense systems to directly react with ROS and disarm them.[1,3,17] Oxidative stress can arise when the production of harmful ROS overwhelms these antioxidant defenses. Consequences of this stress include modification of cellular macromolecules. Therefore, the ability of antioxidative defense in response to the increased oxidative damage was then evaluated by measuring the plasma level of total reduced thiols. Plasma thiols are physiological free radical scavengers.[26] Protein thiols may serve an antioxidant function as they may preemptively scavenge oxidants that initiate peroxidation, thus sparing biomolecules from oxidative damage.[21] Surprisingly, a higher plasma level of total reduced thiols was found in diabetic patients without the 16189 mtDNA variant (TABLE 2). As thiols play a central role in coordinating the antioxidant defense network in biological systems, these data might suggest that under a diabetic state, reduced thiols were increased to cope with higher oxidative damage as shown by TBARS data. This increased plasma level of total reduced thiols was not found in 16189 mtDNA variant carriers (TABLE 2). Thus, the compensatory antioxidatant defense ability might be impaired in patients with this mtDNA variant. Among the control group, both the plasma levels of lipid peroxidation and the total reduced thiols level did not reach a significant difference between people with and those without the 16189 mtDNA variant. Based on these data, the role of this common mtDNA variant probably involves the antioxidant defense system because of increased oxidative stress in diabetic patients.

In conclusion, the present study revealed a significant increase in oxidative damage and an increase in antioxidative activity in diabetic patients compared with controls. The levels of oxidative stress were not different among individuals with or without the 16189 mtDNA variant under hyperglycemic conditions. Diabetic patients carrying the 16189 variant of mtDNA did not respond to oxidative damage by increasing plasma thiols as seen in those with the wild type under a hyperglycemic state. Further studies are mandatory in the future to confirm this observation.

ACKNOWLEDGMENTS

This work was supported by research grants from the National Science Council, Executive Yuan (ROC), NSC-92-2314-B-182A-126, NSC-92-2314-B-182A-119, CMRPG8056. We gratefully acknowledge the assistance of Feng-Mei Huang and I-Ya Chen.

REFERENCES

1. WALLACE, D.C. 1999. Mitochondrial diseases in man and mouse. Science **283:** 1482–1488.
2. PACKER, M.A. & M.P. MURPHY. 1994. Peroxynitrite causes calcium efflux from mitochondria which is prevented by Cyclosporin A. FEBS Lett. **345:** 237–240.
3. WEI, Y.H. 1998. Oxidative stress and mitochondrial DNA mutations in human aging. Proc. Soc. Exp. Biol. Med. **217:** 53–63.
4. GREEN, D.R. & J.C. REED. 1998. Mitochondria and apoptosis. Science **281:** 1309–1312.
5. SCARLETT, J.L., P.W. SHEARD, G. HUGHES, et al. 2000. Changes in mitochondrial membrane potential during staurosporine-induced apoptosis in Jurkat cells. FEBS Lett. **475:** 267–272.
6. GREEN, K., M.D. BRAND & M.P. MURPHY. 2004. Prevention of mitochondrial oxidative damage as a therapeutic strategy in diabetes. Diabetes **53 (Suppl 1):** S110–118.
7. MORTEN, K.J., J. POULTON & B. SYKES. 1995. Multiple independent occurrence of the 3243 mutation in mitochondrial tRNA(leuUUR) in patients with the MELAS phenotype. Hum. Mol. Genet. **4:** 1689–1691.
8. LIOU, C.W., C.C. HUANG, C.F. LEE, et al. 2003. Low antioxidant content and mutation load in mitochondrial DNA A3243G mutation-related diabetes mellitus. J. Formos. Med. Assoc. **102:** 527–533.

9. MAASSEN, J.A., T.H. LM, E. VAN ESSEN, et al. 2004. Mitochondrial diabetes: molecular mechanisms and clinical presentation. Diabetes **53:** S103–109.
10. MARCHINGTON, D.R., J. POULTON, A. SELLAR, et al. 1996. Do sequence variants in the major non-coding region of the mitochondrial genome influence mitochondrial mutations associated with disease? Hum. Mol. Genet. **5:** 473–479.
11. POULTON, J., M.S. BROWN, A. COOPER, et al. 1998. A common mitochondrial DNA variant is associated with insulin resistance in adult life. Diabetologia **41:** 54–58.
12. GERBITZ, K.D., J.M. VAN DEN OUWELAND, J.A. MAASSEN, et al. 1995. Mitochondrial diabetes mellitus: a review. Biochim. Biophys. Acta **1271:** 253–260.
13. POULTON, J., J. LUAN, V. MACAULAY, et al. 2002. Type 2 diabetes is associated with a common mitochondrial variant: evidence from a population-based case-control study. Hum. Mol. Genet. **11:** 1581–1583.
14. LIOU, C.W., T.K. LIN, F.M. HUANG, et al. 2004. Association of the Mitochondrial DNA 16189 T to C Variant with Lacunar Cerebral Infarction: Evidence from a Hospital-Based Case-Control Study. Ann. N. Y. Acad. Sci. **1011:** 317–324.
15. OHKAWA, H., N. OHISHI & K. YAGI. 1979. Assay for lipid peroxides in animal tissues by thiobarbituric acid reaction. Anal. Biochem. **95:** 351–358.
16. ELLMAN, G. & H. LYSKO. 1979. A precise method for the determination of whole blood and plasma sulfhydryl groups. Anal. Biochem. **93:** 98–102.
17. FINKEL, T. & N.J. HOLBROOK. 2000. Oxidants, oxidative stress and the biology of ageing. Nature **408:** 239–247.
18. STUMP, C.S., K.R. SHORT, M.L. BIGELOW, et al. 2003. Effect of insulin on human skeletal muscle mitochondrial ATP production, protein synthesis, and mRNA transcripts. Proc. Natl. Acad. Sci. USA **100:** 7996–8001.
19. PETERSEN, K.F., S. DUFOUR, D. BEFROY, et al. 2004. Impaired mitochondrial activity in the insulin-resistant offspring of patients with type 2 diabetes. N. Engl. J. Med. **350:** 664–671.
20. MARTIN-GALLAN, P., A. CARRASCOSA, M. GUSSINYE, et al. 2003. Biomarkers of diabetes-associated oxidative stress and antioxidant status in young diabetic patients with or without subclinical complications. Free Radic. Biol. Med. **34:** 1563–1574.
21. TELCI, A., U. CAKATAY, S. SALMAN, et al. 2000. Oxidative protein damage in early stage Type 1 diabetic patients. Diabetes Res. Clin. Pract. **50:** 213–223.
22. LEE, H.K., J.H. SONG, C.S. SHIN, et al. 1998. Decreased mitochondrial DNA content in peripheral blood precedes the development of non-insulin-dependent diabetes mellitus. Diabetes Res. Clin. Pract. **42:** 161–167.
23. MELOV, S. 2000. Mitochondrial oxidative stress. Physiologic consequences and potential for a role in aging. Ann. N. Y. Acad. Sci. **908:** 219–225.
24. DINCER, Y., T. AKCAY, Z. ALADEMIR, et al. 2002. Assessment of DNA base oxidation and glutathione level in patients with type 2 diabetes. Mutat. Res. **505:** 75–81.
25. KARATAS, F., M. KARATEPE & A. BAYSAR. 2002. Determination of free malondialdehyde in human serum by high-performance liquid chromatography. Anal. Biochem. **311:** 76–79.
26. LIN, T.K., G. HUGHES, A. MURATOVSKA, et al. 2002. Specific modification of mitochondrial protein thiols in response to oxidative stress: a proteomics approach. J. Biol. Chem. **277:** 17048–17056.

Alteration of the Copy Number of Mitochondrial DNA in Leukocytes of Patients with Hyperlipidemia

CHIN-SAN LIU,[a,b] CHING-LING KUO,[a] WEN-LING CHENG,[a]
CHING-SHAN HUANG,[a] CHENG-FENG LEE,[c] AND YAU-HUEI WEI[c]

[a]*Vascular and Genomic Research Center and*
[b]*Department of Neurology, Changhua Christian Hospital,
Changhua 500, Taiwan*

[c]*Department of Biochemistry and Molecular Biology, and
Center for Cellular and Molecular Biology,
National Yang-Ming University, Taipei 112, Taiwan*

ABSTRACT: Lipid metabolism in leukocytes may be disturbed by mitochondrial dysfunction caused by depletion of mitochondrial DNA (mtDNA) in response to an increase of oxidative stress in blood circulation. It is possible that alteration in mtDNA copy number of the leukocyte is involved in the impairment of the scavenging of oxidatively modified plasma proteins such as oxidized low-density lipoprotein (oxLDL). To test this hypothesis, we recruited 91 healthy subjects and 63 patients with hyperlipidemia (LDL >130 mg/dL) for this study. The copy number of mtDNA in the leukocyte and the titer of oxLDL IgG autoantibody (oLAB) were determined as indices of the oxidative stress response of immune cells. The results revealed a significant higher level of plasma oxLDL, lower titer of oLAB, and decreased copy number of mtDNA in patients with hyperlipidemia ($P < 0.05$). In the analysis of partial correlations under age control, we found that an increase in the copy number of mtDNA was positively correlated with an increase in the level of oLAB ($P < 0.005, r = 0.3002$) and a decrease in the oxLDL level ($P < 0.05, r = -0.2654$) in healthy subjects but not in patients. Based on the results obtained from this case-control study, we conclude that the increase of mtDNA copy number might provide the leukocyte an increased capability of scavenging oxLDL, possibly by enhanced generation of oLAB in healthy subjects, but not in hyperlipidemic patients who had lower mtDNA copy numbers in their leukocytes. Taken together, these findings suggest that an alteration of mtDNA copy number in the leukocyte may be one of the risk factors for hyperlipidemia.

KEYWORDS: hyperlipidemia; oxidized LDL; autoantibody; mitochondrial DNA copy number

Address for correspondence: Professor Yau-Huei Wei, Department of Biochemistry and Molecular Biology, School of Life Science, National Yang-Ming University, Taipei 112, Taiwan. Voice: +886-2-28267118; fax: +886-2-28264843.
   joeman@ym.edu.tw

INTRODUCTION

The copy number of mitochondrial DNA (mtDNA) in the leukocyte may be changed in response to an increase of oxidative stress in blood plasma.[1] Dyslipidemia is one of the risk factors that contribute to the etiology of vascular diseases, and generation of oxidized low-density lipoprotein (oxLDL) is one of the major factors involved in the development of atherosclerosis.[2] The oxLDL molecules are immunogenic and may lead to the formation of oxLDL IgG autoantibody (oLAB) and immune complexes, which may play a role in the pathogenesis of atherosclerosis.[3] The pathologic role of oLAB was demonstrated in a study on the use of oLAB for the prediction of the severity of atherosclerosis.[4] However, other studies have failed to demonstrate any significant correlation between the level of oLAB and clinical or radiologic findings of vascular diseases.[5,6] By contrast, several studies demonstrated that immunization of animals with oxLDL induced higher levels of oLAB and decreased the incidence of atherosclerosis.[7,8] Therefore, the potential role that oLAB plays in the atherogenic process remains unsolved. It has been documented that mitochondrial dysfunction may disturb lipid metabolism in the human cell. In a previous study, we found that the mtDNA copy number in the leukocyte can be changed in response to the increase of oxidative stress in blood plasma.[1] This phenomenon may imply possible involvement of mitochondrial dysfunction of leukocytes in the impairment of the disposal of oxidatively modified plasma proteins such as oxLDL in patients with defective lipid metabolism. In this communication, we report on the alteration of mtDNA copy number in the leukocyte in relation to the plasma levels of lipids, lipoproteins, oxLDL, and oLAB in healthy subjects and hyperlipidemic patients.

MATERIALS AND METHODS

Subjects

Two hundred healthy subjects were recruited from the health clinic of Changhua Christian Hospital, Taiwan. Informed consent for study enrollment was obtained from each of the subjects. All of the human experimental procedures followed the medical ethics guidelines of Changhua Christian Hospital. General physical examination and blood biochemistry studies had been performed to exclude those with a systemic disease such as hypertension and diabetes. All enrolled subjects showed a normal distribution in leukocyte subpopulations, and a total of 91 healthy subjects and 63 patients with hyperlipidemia (LDL >130 mg/dL) were finally recruited in this study from year 2002 to 2003. The demographic characteristics, age, sex, body mass index (BMI), smoking history, clinical and biological characteristics of the study subjects are summarized in TABLE 1. Blood samples were drawn from each subject in the morning after overnight fasting. For each subject, 20 mL of whole blood was withdrawn from an antecubital vein and quickly delivered into an ethylenediamine tetraacetic acid (EDTA)-containing plastic tube. Plasma was collected by centrifugation of blood at $300 \times g$ for 10 min, divided into several aliquots, and stored in liquid nitrogen until analysis. Washed leukocytes were quickly separated from plasma as the buffy coat layer and immediately subjected to DNA isolation. The kit for

TABLE 1. Demographic and biochemical data of the subjects enrolled in this study

Characteristics of subjects	LDL <130[a] (n = 91)	LDL >130 (n = 63)	P
Age	49.65 ± 13.93	53.30 ± 12.70	0.003*
Sex	0.56 ± 0.498	0.54 ± 0.50	0.836
BMI	23.44 ± 3.37	24.01 ± 2.62	0.074
Smoking index	18.69 ± 14.59	17.66 ± 17.75	0.474
Cholesterol (mg/dL)	179.0 ± 29.51	235.7 ± 25.55	0.001*
Triglycerides (mg/dL)	133.5 ± 104.6	121.58 ± 55.22	0.453
HDL (mg/dL)	49.68 ± 18.38	51.65 ± 12.86	0.054
LDL (mg/dL)	99.67 ± 18.92	156.6 ± 20.69	0.001*
oLAB (mU/mL)	605.1 ± 651.9	479.6 ± 475.5	0.078
oxLDL (U/L)	47.63 ± 15.44	61.12 ± 19.75	0.001*
Log (mtDNA copy number)	2.91 ± 0.55	2.56 ± 0.47	0.074

ABBREVIATIONS: BMI, body mass index; HDL, high-density lipoprotein; LDL, low-density lipoprotein; oLAB, oxidized LDL IgG autoantibody; oxLDL, oxidized LDL; mtDNA, mitochondrial DNA.
[a]Arithmetic mean ± SD.
*$P < 0.05$.

determining the level of oxLDL was obtained from Mercodia (Sylveniusgatan, Sweden). The plasma oxLDL level was measured by a competitive ELISA utilizing a specific murine monoclonal antibody mAb-4E6. The kit for determination of the plasma level of oLAB was obtained from Biomedica (Wien, Austria). The plasma concentration of oLAB was determined by an enzyme immunoassay. Total cholesterol and triglycerides in serum were measured by enzymatic methods. Total cellular DNA of leukocytes was extracted by phenol/chloroform after lysis with proteinase K in an alkaline SDS solution.[1] The fluorescence-based quantitative PCR (QPCR) was used to determine the mtDNA copy number in the leukocyte.[1] Essentially, it was carried out by using the LightCycler-FastStart DNA Master SYBR Green I kit supplied by Roche Molecular Biochemicals (Pleasanton, CA). The Mann-Whitney test was applied in comparative analysis of data obtained from blood biochemistry and molecular biology studies between subjects with and those without hyperlipidemia. Partial correlation analysis was used to evaluate the correlation factors associated with the alteration of mtDNA copy number in the leukocyte under age control.

RESULTS

The biochemical and molecular analyses revealed significant higher levels of plasma oxLDL, lower titers of oLAB, and decreased copy numbers of mtDNA in patients with hyperlipidemia ($P <0.05$) (TABLE 1). In the analysis of partial correlations under age control, we found that the increase in mtDNA copy number was positively correlated with the increase in the plasma level of oLAB ($P <0.005, r = 0.300$) and a decrease in the oxLDL level ($P <0.05, r = -0.223$) in healthy subjects but not

TABLE 2. Correlation between log (mtDNA copy number) and various parameters

	LDL <130 ($n = 91$)		LDL >130 ($n = 63$)	
	r	P	r	P
Cholesterol (mg/dL)	0.142	0.176	−0.005	0.965
Triglycerides (mg/dL)	−0.027	0.796	−0.054	0.666
HDL (mg/dL)	0.238	0.022*	0.283	0.022*
LDL (mg/dL)	0.042	0.688	−0.164	0.190
oLAB (mU/mL)	0.300	0.003*	0.231	0.063
oxLDL (U/L)	−0.223	0.041*	−0.193	0.054

ABBREVIATIONS: HDL, high-density lipoprotein; LDL, low-density lipoprotein; oLAB, oxidized LDL IgG autoantibody; oxLDL, oxidized LDL.
*$P < 0.05$.

TABLE 3. Multiple regression models for variables associated with oxLDL in healthy subjects and patients with hyperlipidemia

	Healthy subjects[a]			Patients with hyperlipidemia[b]		
	Unstandardized coefficients			Unstandardized coefficients		
	Beta	SE	P	Beta	SE	P
LDL (mg/dL)	0.202	0.007	0.005[c]	0.253	0.002	0.016[c]
oLAB (mg/dL)	−0.005	0.002	0.007[c]	−0.0004	0.004	0.919

ABBREVIATIONS: LDL, low-density lipoprotein; oLAB, oxidized LDL IgG autoantibodies; oxLDL, oxidized low-density lipoprotein.
[a] $r^2 = 0.102$.
[b] $r^2 = 0.071$.
[c] $P < 0.05$.

in the patients (TABLE 2). However, a significantly positive correlation was found between mtDNA copy number in the leukocyte and the plasma level of lipoproteins in both subjects with hyperlipidemia ($P < 0.05$) and those without hyperlipidemia ($P < 0.05$) (TABLE 2). Both plasma LDL and oLAB levels could significantly contribute to the alteration of oxLDL (TABLE 3).

DISCUSSION

The immune system is one of the critical components in the elimination of oxLDL generated in the early stage of atherosclerosis. T cell and B cell lymphocytes, monocytes, and macrophages were all recruited in the process of scavenging of oxLDL.[10] Effective generation of oLAB may provide a possible way for efficient elimination of oxLDL in an individual. It has been speculated that these autoantibodies are involved in the protective process that may take place when plasma LDL is oxidized to oxLDL. Monoclonal autoantibodies of oxLDL have important biological properties that are critical in the blockade of the binding and degradation of oxLDL by macrophage.[11] The plasma oLAB titers of stroke patients were significantly lower than those of healthy subjects.[12] The plasma levels of autoantibodies and oxLDL had an opposite contribution to the development of carotid atherosclerosis.[13] Since the uptake of oxLDL by macrophages and formation of foam cells is one of the key events in the pathogenesis of atherosclerosis, oLAB may enhance the removal of oxLDL from plasma and prevent their entrance into the arterial wall.[14] The aforementioned observations can provide information to further our understanding of the anti-atherosclerosis effect of oLAB. One of the major findings (TABLE 3) of the present study is the significantly negative contribution of oLAB and positive contribution of LDL to the formation of oxLDL, which was only disclosed in healthy subjects but not in patients with hyperlipidemia. The inverse relationship between oxLDL and oLAB may indicate the protective role in the process of oxLDL formation only in healthy subjects. Mitochondria are important organelles in the leukocyte, and can provide the energy required for the execution of immune response. Recently, Lim and coworkers reported that mtDNA copy number in the leukocyte was negatively correlated with LDL, which may underscore the relationship between the mtDNA copy number in the leukocyte and the development of atherosclerosis.[15] In one of our previous studies, we showed that the copy number of mtDNA in human leukocytes may serve as a biomarker of oxidative stress in plasma.[1] As shown in TABLE 2, oLAB may be identified as an oxidative stress index, which is positively correlated with the mtDNA copy number and plasma level of lipid peroxides. Thus, we intend to believe that oxLDL-related oxidative stress can trigger the generation of oLAB via upregulation of mtDNA replication in healthy subjects but not in patients with hyperlipidemia, who had a lower copy number of mtDNA even under higher oxidative stress. Based on the results obtained from this case-control study, we conclude that the increase of mtDNA copy number in the leukocyte might be involved in the halting of atherosclerosis by an increase in the generation of oLAB and/or oxLDL scavenging in healthy subjects. However, in the patients with hyperlipidemia, the decrease in the mtDNA copy number of leukocytes might downregulate the humoral and cellular immunologic responses that are involved in the generation of oLAB. Further study is warranted to elucidate the relationship between inflammation-related atherosclerosis and alteration in the replication of mtDNA in the leukocytes of patients with hyperlipidemia.

ACKNOWLEDGMENTS

This work was supported by a grant (NSC92-2320-B010-037) from the National Science Council and partly by a grant (NHRI-EX93-9120BN) from the National Health Research Institutes, Taiwan.

REFERENCES

1. LIU, C.S., C.S. TSAI, C.L. KUO, et al. 2003. Oxidative stress-related alteration of the copy number of mitochondrial DNA in human leukocytes. Free Radic. Res. **37**: 1307–1317.
2. RAJMAN, I., M. KENDALL & R. CRAMB. 1994. The oxidation hypothesis of atherosclerosis. Lancet **12**: 1363–1364.

3. MIRONOVA, M., G. VIRELLA, I. VIRELLA-LOWELL, et al. 1997. Anti-modified LDL antibodies and LDL-containing immune complexes in IDDM patients and healthy controls. Clin. Immunol. Immunopathol. **85:** 73–82.
4. BERGMARK, C., R. WU, U. DE FAIRE, et al. 1995. Patients with early-onset peripheral vascular disease have increased levels of autoantibodies against oxidized LDL. Arterioscler. Thromb. Vasc. Biol. **15:** 441–445.
5. UUSITUPA, M.I., L. NISKANEN, J. LUOMA, et al. 1996. Autoantibodies against oxidized LDL do not predict atherosclerotic vascular disease in non-insulin-dependent diabetes mellitus. Arterioscler. Thromb. Vasc. Biol. **16:** 1236–1242.
6. VIRELLA, G., I. VIRELLA, R.B. LEMAN, et al. 1993. Anti-oxidized low-density lipoprotein antibodies in patients with coronary heart disease and normal healthy volunteers. Int. J. Clin. Lab. Res. **23:** 95–101.
7. ZHOU, X., G. CALIGIURI, A. HAMSTEN, et al. 2001. LDL immunization induces T-cell-dependent antibody formation and protection against atherosclerosis. Arterioscler. Thromb. Vasc. Biol. **21:** 108–114.
8. PALINSKI, W., E. MILLER & J.L. WITZTUM. 1995. Immunization of low density lipoprotein (LDL) receptor-deficient rabbits with homologous malondialdehyde- modified LDL reduces atherogenesis. Proc. Natl. Acad. Sci. USA **31:** 821–825.
9. LIU, C.S. & C.S. TSAI. 2002. Enhanced lipid peroxidation in epileptics with null genotype of glutathione S-transferase M1 and intractable seizure. Jpn. J. Pharmacol. **90:** 291–294.
10. HANSSON, G.K. 2001. Immune mechanisms in atherosclerosis. Arterioscler. Thromb. Vasc. Biol. **21:** 1876–1890.
11. HORKKO, S., D.A. BIRD, E. MILLER, et al. 1999. Monoclonal autoantibodies specific for oxidized phospholipids or oxidized phospholipid-protein adducts inhibit macrophage uptake of oxidized low-density lipoproteins. J. Clin. Invest. **103:** 117–128.
12. CHERUBINI, A., P. MECOCCI, U. SENIN, et al. 1997. Autoantibodies against oxidized low-density lipoproteins in older stroke patients. J. Am. Geriatr. Soc. **45:** 125.
13. KARVONEN, J., M. PAIVANSALO, Y.A. KESANIEMI, et al. 2003. Immunoglobulin M type of autoantibodies to oxidized low-density lipoprotein has an inverse relation to carotid artery atherosclerosis. Circulation **28:** 2107–2112.
14. WIKLUND, O., J.L. WITZTUM, T.E. CAREW, et al. 1987. Turnover and tissue sites of degradation of glycosylated low density lipoprotein in normal and immunized rabbits. J. Lipid Res. **28:** 1098–1109.
15. LIM, S., M.S. KIM, K.S. PARK, et al. 2001. Correlation of plasma homocysteine and mitochondrial DNA content in peripheral blood in healthy women. Atherosclerosis **158:** 399–405.

A New Noninvasive Test to Detect Mitochondrial Dysfunction of Skeletal Muscles in Progressive Supranuclear Palsy

YUNG-YEE CHANG,[a] CHIANG-HSUAN LEE,[b] MIN-YU LAN,[a] HSIU-SHAN WU,[a] CHIUNG-CHIH CHANG,[a] AND JIA-SHOU LIU[a]

[a]*Department of Neurology, Chang-Gung Memorial Hospital, Kaohsiung, Taiwan*

[b]*Department of Nuclear Medicine, Chang-Gung Memorial Hospital, Kaohsiung, Taiwan*

ABSTRACT: We present usage of technetium-99m methoxyisobutyl isonitrile (^{99m}Tc-sestamibi) single photon emission computed tomography (SPECT) as a novel noninvasive method to evaluate muscular mitochondrial function in patients with progressive supranuclear palsy (PSP). ^{99m}Tc-sestamibi SPECT revealed a statistically significant decrease in radionucleotide uptake in the quadriceps in PSP patients as compared with other neurodegenerative parkinsonism ($P < 0.05$) or control group ($P < 0.05$). This study demonstrates a remarkable deficit of skeletal muscle bioenergetics in patients with PSP. Our findings suggest a distinctive role of mitochondrial dysfunction in the pathogenesis of PSP. Furthermore, ^{99m}Tc-sestamibi SPECT provides a relatively simple, inexpensive, and noninvasive modality in further assessment of mitochondrial function and bioenergetic features in various muscular disorders.

KEYWORDS: progressive supranuclear palsy; parkinsonism; mitochondria; skeletal muscle; technetium-99m methoxyisobutyl isonitrile (^{99m}Tc-sestamibi); single photon emission computed tomography (SPECT)

INTRODUCTION

Clinically, progressive supranuclear palsy (PSP) presents as a progressive parkinsonian syndrome characterized by supranuclear gaze palsy, bulbar signs, dominant axial rigidity, retrocollis, and progressive gait disturbance with frequent falls.[1,2] Pathologically, it is classified as tauopathy leading to marked neuronal degeneration and gliosis. Tau-positive neurofibrillary tangles, neuropil treads, and tufted astrocytes are frequently observed in the subcortical and brainstem nuclei.[3] Although the molecular mechanism of PSP remains obscure, complex interplay between genetic predisposition, oxidative stress, mitochondrial dysfunction, and bioenergetic defects may contribute to accelerated neuronal degeneration.[4]

Address correspondence to: Jia-Shou Liu, Department of Neurology, Kaohsiung Chang-Gung Memorial Hospital, No. 123, Ta-Pei Road, Niao-Sung, Kaohsiung 833, Taiwan. Voice: +886-7-731-7123 ext. 3399; fax: +886-7-311-2516.

  josefliu@ms15.hinet.net

Several lines of evidence suggest that mitochondrial dysfunction with impaired energy production leading to increased free radical production with subsequent DNA and RNA damage may play a crucial role in the pathogenesis of PSP.[5-9] Previous studies demonstrated significantly decreased activities of mitochondrial enzyme complex and increased oxidative damage in the brains of PSP patients.[6,7] Defects in oxidative phosphorylation and ATP production have also been described in muscles from PSP patients.[9] Furthermore, using phosphorus magnetic resonance spectroscopy (^{31}P-MRS), Martinelli et al. have shown defective mitochondrial respiration in muscles of patients with PSP.[10] Given the characteristic manifestations of PSP such as ophthalmoparesis, resembling symptoms and signs of certain mitochondrial cytopathy, mitochondrial function should be evaluated in the clinical approach of patients with PSP.

Technetium-99m methoxyisobutyl isonitrile (^{99m}Tc-sestamibi) single photon emission computed tomography (SPECT) has been widely employed for the evaluation of myocardial perfusion.[11,12] Research has shown that the uptake, accumulation, and clearance of ^{99m}Tc-sestamibi appear to be related to mitochondrial integrity and cellular viability.[13,14] ^{99m}Tc-sestamibi SPECT has been used to evaluate muscle metabolic abnormalities in various muscular disorders, such as decreased uptake in patients with Duchenne muscular dystrophy, statin-induced myopathy, and myopathy related to uremia.[15-17] The aim of this study is to define whether mitochondrial function of skeletal muscles in PSP patients could be tracked in vivo and noninvasively by ^{99m}Tc-sestamibi SPECT.

MATERIALS AND METHODS

Six patients referred to the Neurological Outpatient Clinic at Kaohsiung Chang-Gung Memorial Hospital because of clinical symptoms and signs of PSP were examined. All patients fulfilled the diagnostic criteria of clinical probable PSP, according to the NINDS–SPSP (National Institute for Neurological Disorders and Stroke–Society for Progressive Supranuclear Palsy) criteria.[1] Six age-matched patients with other neurodegenerative parkinsonism, such as idiopathic Parkinson's disease, multiple system atrophy, or dementia with Lewy bodies, were concurrently enrolled in this study. In addition, seven patients suffering from other neurological disorders, malignant lymphoma, or thyroid cancer served as control subjects. The study was reviewed and approved by the Institutional Review Board of Chang-Gung Memorial Hospital. Written informed consent for this study has been obtained from each subject.

All patients and control subjects fasted overnight and rested for at least 60 min before radiotracer application. Early and delayed whole-body scans were performed 10 min and 4 h, respectively, after injection of 740 MBq of ^{99m}Tc-MIBI from the antecubital vein. A rectangular, large-field-of-view, double-head gamma camera (Millennium VG; General Electric Medical Systems, Milwaukee, WI) with low-energy, ultra-high resolution parallel collimator was used to obtain 128 × 128 pixel images. Radioactivity counts were semiquantitatively determined on the medial portion of the left thigh and knee (nonmuscle region), and the ratio between the thigh and knee (T/K ratio) regions of interest (ROI) counts was obtained. Mean ± standard deviation (SD) for each group was derived. Statistical comparisons were performed by the Wilcoxon test, and a P value less than .05 was considered significant.

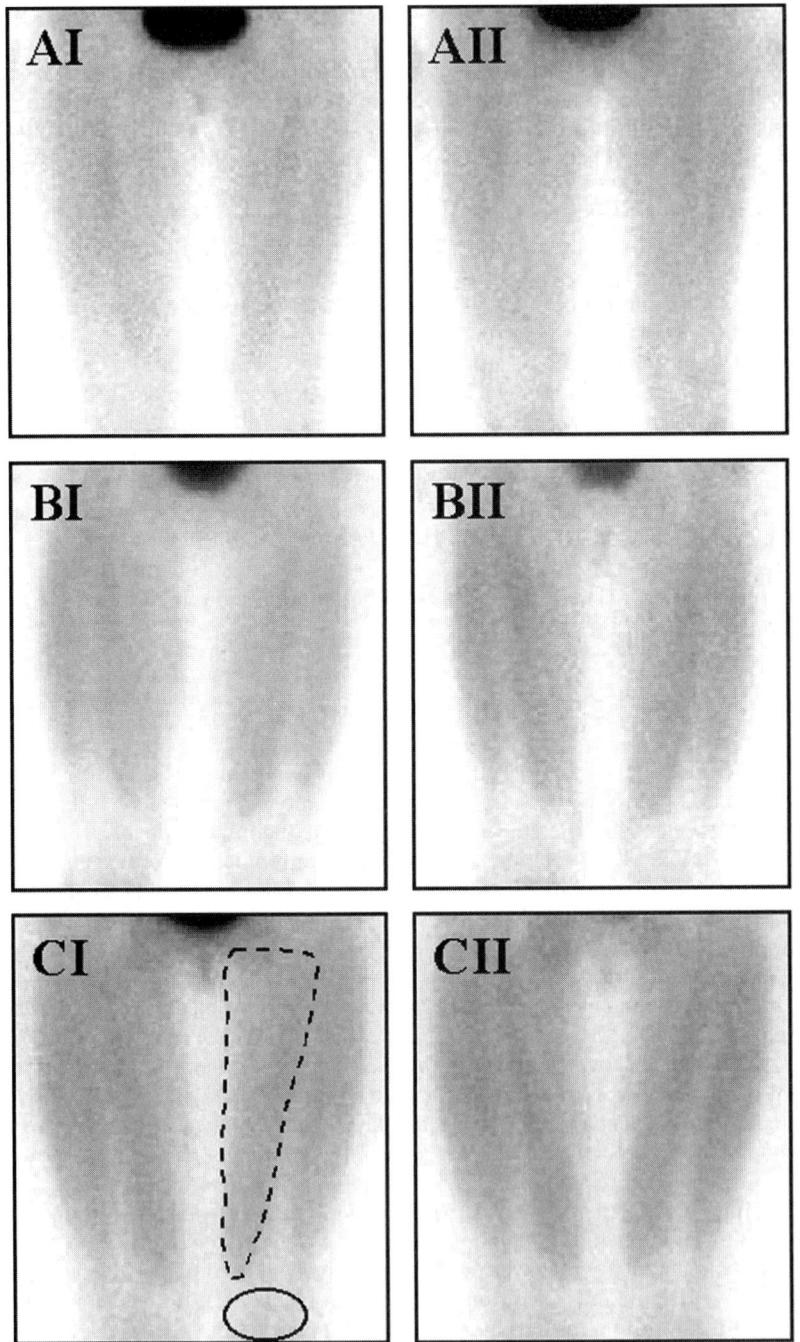

FIGURE 1. *See following page for legend.*

RESULTS

The mean age in the PSP group (three males, three females) was 68.5 ± 6.6 years, and in the non-PSP Parkinsonian group (three males, three females) was 65.8 ± 7.0 years ($P > 0.5$). The mean age of seven patients (four males, three females) in the control group was 43.6 ± 8.8 years.

Compared with other neurodegenerative parkinsonism or control, ^{99m}Tc-sestamibi SPECT revealed significant reduction of radionucleotide uptake in the quadriceps of PSP patients (FIG. 1).

The T/K ratio at 10 min for the PSP patients (1.76 ± 0.41) was significantly reduced when compared with the T/K ratios of non-PSP parkinsonism (4.07 ± 1.71) and control subjects (3.24 ± 1.11), respectively (FIG. 2A) ($P < 0.005$ in both cases). The T/K ratio at 4 h for the PSP patients (2.29 ± 0.63) was also lower in PSP patients when compared with the ratios of other neurodegenerative parkinsonism (4.96 ± 1.60; $P < 0.005$) and controls (4.21 ± 0.89; $P < 0.005$) (FIG. 2B). There was no association between the T/K uptake ratio and age.

DISCUSSION

In this study, we demonstrated that PSP patients sustained defective mitochondrial metabolism in skeletal muscles, a phenomenon not seen in other neurodegenera-

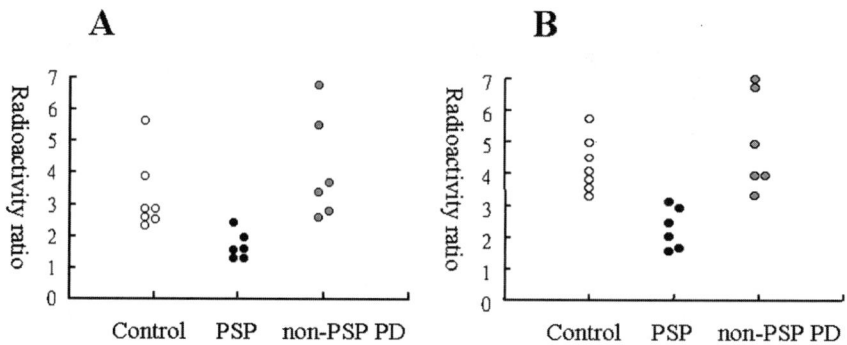

FIGURE 2. Distribution of ^{99m}Tc-sestamibi radioactivity ratio in patient groups and control subjects. **(A)** represents results 10 min after ^{99m}Tc-sestamibi injections, and **(B)** corresponds to 4 h after injections. PSP, progressive supranuclear palsy; non-PSP PD, other neurodegenerative parkinsonism.

FIGURE 1. Low body ^{99m}Tc-sestamibi scan. **Panels A, B,** and **C** represent progressive supranuclear palsy (PSP), idiopathic Parkinson's disease (IPD), and control, respectively. Subsection figure I shows imaging 10 min after ^{99m}Tc-sestamibi injections, and subsection II is 4 h after injections. Decrease in sestamibi uptake both at 10 min (AI) and 4 h (AII) session could be identified in patients with PSP. Radioactivity counts of regions of interest were semiquantitatively determined on the medial portion of the left thigh (*dotted area*) and knee (nonmuscle region, *oval circle*).

tive parkinsonism or control subjects. Although muscle metabolic function in PSP has been examined in biochemical[9] or ^{31}P-MRS studies,[10] to our knowledge, the current radionuclide scintigraphy has not been applied in this clinical setting. Our findings may suggest a distinctive role of mitochondrial dysfunction in the pathogenesis of PSP.

^{99m}Tc-sestamibi, a synthetic cationic imaging tracer, has been frequently used in the evaluation of myocardial perfusion and viability.[11,12] The cellular uptake of ^{99m}Tc-sestamibi is related mainly to its lipophilicity and charge distribution.[13,14] Studies have shown that in muscle cells, mitochondria are the subcellular organelles of sestamibi retention.[18,19] Piwnica-Worms *et al.* confirmed that at equilibrium state ^{99m}Tc-sestamibi is sequestered within mitochondria by large negative transmembrane potentials.[18] Pathological conditions causing cellular and mitochondrial calcium overload may lead to mitochondrial dysfunction with subsequent release of ^{99m}Tc-sestamibi.[19] Accordingly, ^{99m}Tc-sestamibi scintigraphy may act as a sensitive modality for myocardial cell viability. In addition, extended studies based on this modality have been done in muscular disorders such as Duchenne muscular dystrophy,[17] uremic myopathy,[15] and statin-induced myopathy.[16] Basically, ^{99m}Tc-sestamibi scintigraphy provides us an indirect *in vivo* measurement of mitochondrial function in muscle cells. With respect to energy levels, muscle and brain are both organs of high metabolic rate. Several reports have highlighted defective mitochondrial metabolism in the PSP brain and skeletal muscle.[5–7,9,10] Specifically, a transmitochondrial cytoplasmic hybrid (cybrid) study showed that complex I activity was reduced in PSP patients.[8] Decreased uptake of ^{99m}Tc-sestamibi represents mitochondrial dysfunction in the skeletal muscle in PSP patients, a finding not observed in non-PSP parkinsonism. We proposed that mitochondrial dysfunction with subsequent oxidation damage is the pathogenetic mechanism of the PSP brain. Unfortunately, imaging the central nervous system by ^{99m}Tc-sestamibi is not practical due to insufficient resolution, and the quantification of radiotracer is thus hampered by limited penetration of 99mTc-sestamibi into the brain. Further studies are needed in determining the role of mitochondrial dysfunction in uncovering the pathogenesis of PSP.

In summary, we demonstrate the presence of mitochondrial dysfunction in the skeletal muscle of PSP patients. To our knowledge, this is the first report on the application of ^{99m}Tc-sestamibi scintigraphy as a diagnostic modality for muscle abnormalities in patients with various parkinsonian presentations. In this study, significant mitochondrial abnormality has been observed in PSP patients. Further studies using scintigraphy for the assessment of disease course via pharmacological modification may offer more insights regarding the role of mitochondria in the pathogenesis of PSP. ^{99m}Tc-sestamibi scintigraphy is a simple, inexpensive, and noninvasive modality worthy of serial follow-up studies in this clinical entity.

REFERENCES

1. LITVAN, I., Y. AGID, D. CALNE, *et al.* 1996. Clinical research criteria for the diagnosis of progressive supranuclear palsy (Steele-Richardson-Olszewski syndrome): report of the NINDS-SPSP international workshop. Neurology **47**: 1–9.
2. BURN, D.J. & A.J. LEES. 2002. Progressive supranuclear palsy: where are we now? Lancet Neurol. **1**: 359–369.

3. SPILLANTINI, M.G. & M. GOEDERT. 1998. Tau protein pathology in neurodegenerative diseases. Trends Neurosci. **21:** 428–433.
4. ALBERS, D.S. & S.J. AUGOOD. 2001. New insights into progressive supranuclear palsy. Trends Neurosci. **24:** 347–353.
5. ALBERS, D.S. & M.F. BEAL. 2002. Mitochondrial dysfunction in progressive supranuclear palsy. Neurochem. Int. **40:** 559–564.
6. PARK, L.C., D.S. ALBERS, H. XU, et al. 2001. Mitochondrial impairment in the cerebellum of the patients with progressive supranuclear palsy. J. Neurosci. Res. **66:** 1028–1034.
7. ALBERS, D.S., S.G. AUGOOD, L.C. PARK, et al. 2000. Frontal lobe dysfunction in progressive supranuclear palsy: evidence for oxidative stress and mitochondrial impairment. J. Neurochem. **74:** 878–881.
8. SWERDLOW, R.H., L.I. GOLBE, J.K. PARKS, et al. 2000. Mitochondrial dysfunction in cybrid lines expressing mitochondrial genes from patients with progressive supranuclear palsy. J. Neurochem. **75:** 1681–1684.
9. DI MONTE, D.A., Y. HARATI, J. JANKOVIC, et al. 1994. Muscle mitochondrial ATP production in progressive supranuclear palsy. J. Neurochem. **62:** 1631–1634.
10. MARTINELLI, P., C. SCAGLIONE, R. LODI, et al. 2000. Deficit of brain and skeletal muscle bioenergetics in progressive supranuclear palsy shown *in vivo* by phosphorus magnetic resonance spectroscopy. Mov. Disord. **15:** 889–893.
11. WACKERS, F.J., D.S. BERMAN, J. MADDAHI, et al. 1989. Technetium-99m hexakis 2-methoxyisobutyl isonitrile: human biodistribution, dosimetry, safety, and preliminary comparison to thallium-201 for myocardial perfusion imaging. J. Nucl. Med. **30:** 301–311.
12. OKADA, R.D., D. GLOVER, T. GAFFNEY & S. WILLIAMS. 1988. Myocardial kinetics of technetium-99m-hexakis-2-methoxy-2-methylpropyl-isonitrile. Circulation **77:** 491–498.
13. BEANLANDS, R.S., F. DAWOOD, W.H. WEN, et al. 1990. Are the kinetics of technetium-99m methoxyisobutyl isonitrile affected by cell metabolism and viability? Circulation **82:** 1802–1814.
14. MAUBLANT, J.C., P. GACHON & N. MOINS. 1988. Hexakis (2-methoxy isobutylisonitrile) technetium-99m and thallium-201 chloride: uptake and release in cultured myocardial cells. J. Nucl. Med. **29:** 48–54.
15. SARIKAYA, A., S. SEN, T.F. CERMIK, et al. 2000. Evaluation of skeletal muscle metabolism and response to erythropoietin treatment in patients with chronic renal failure using 99Tcm-sestamibi leg scintigraphy. Nucl. Med. Commun. **21:** 83–87.
16. LUPATTELLI, G., B. PALUMBO & H. SINZINGER. 2001. Statin induced myopathy does not show up in MIBI scintigraphy. Nucl. Med. Commun. **22:** 575–578.
17. SCOPINARO, F., C. MANNI, A. MICCHELI, et al. 1996. Muscular uptake of Tc–99m MIBI and Tl-201 in Duchenne muscular dystrophy. Clin. Nucl. Med. **21:** 792–796.
18. PIWNICA-WORMS, D., J.F. KRONAUGE & M.L. CHIU. 1990. Uptake and retention of hexakis (2-methoxyisobutyl isonitrile) technetium (I) in cultured chick myocardial cells. Mitochondrial and plasma membrane potential dependence. Circulation **82:** 1826–1838.
19. CRANE, P., R. LALIBERTE, S. HEMINWAY, et al. 1993. Effect of mitochondrial viability and metabolism on technetium-99m-sestamibi myocardial retention. Eur. J. Nucl. Med. **20:** 20–25.

High Prevalence of the COII/tRNA^{Lys} Intergenic 9-bp Deletion in Mitochondrial DNA of Taiwanese Patients with MELAS or MERRF Syndrome

CHIN-SAN LIU,[a,b] WEN-LING CHENG,[a] YI-YUN CHEN,[a] YI-SHING MA,[c] CHENG-YOONG PANG,[c] AND YAU-HUEI WEI[c]

[a]*Vascular and Genomic Research Center, Changhua Christian Hospital, Changhua 500, Taiwan*

[b]*Department of Neurology, Changhua Christian Hospital, Changhua 500, Taiwan*

[c]*Department of Biochemistry and Molecular Biology, and Center for Cellular and Molecular Biology, National Yang-Ming University, Taipei 112, Taiwan*

ABSTRACT: The COII/tRNA^{Lys} intergenic 9-bp deletion (MIC9D) of mitochondrial DNA (mtDNA) has been established as a genetic polymorphism for Asian-Pacific populations. We investigated whether this small mtDNA deletion is co-transmitted with human diseases such as mitochondrial encephalomyopathy with lactic acidosis and stroke-like episodes (MELAS) and myoclonic epilepsy with ragged-red fibers (MERRF) syndromes. Forty unrelated Taiwanese families, including 12 families with MERRF and A8344G mtDNA mutation and 28 families with MELAS and A3243G mutation of mtDNA, respectively, were recruited in this study. In addition, 199 healthy subjects were recruited as control. We found that the frequency of occurrence of mtDNA with the MIC9D polymorphism in healthy subjects was 21% (41/199). However, the incidence of the MIC9D polymorphism was 67% (8/12) among the probands of all the families with MERRF syndrome ($P = 0.001$; OR = 8) and 39% (11/28) among the probands of the families with MELAS syndrome ($P = 0.038$; OR = 2). This finding indicates that the frequency of occurrence of mtDNA with the MIC9D polymorphism in patients with MERRF or MELAS syndrome is higher than that of healthy subjects. The prevalence of mitochondrial encephalomyopathies in relation to the MIC9D polymorphism of mtDNA in Taiwanese population is discussed.

KEYWORDS: mitochondrial disease; 9-bp deletion; MELAS; MERRF; Taiwanese

Address for correspondence: Yau-Huei Wei, Department of Biochemistry and Molecular Biology, School of Life Science, National Yang-Ming University, Taipei 112, Taiwan. Voice: +886-2-28267118; fax: +886-2-28264843.

joeman@ym.edu.tw

Ann. N.Y. Acad. Sci. 1042: 82–87 (2005). © 2005 New York Academy of Sciences.
doi: 10.1196/annals.1338.058

INTRODUCTION

Cann and Wilson[1] identified a length polymorphism in mitochondrial DNA (mtDNA) that involves insertion or deletion of a few base pairs. Wrischnik et al.[2] showed that there is a length mutation in region V of mtDNA caused by deletion of a 9-bp sequence (CCCCCTCTA) (MIC9D) in a small noncoding region between the genes coding for cytochrome c oxidase subunit II (COII) and tRNA^{Lys}, which has been established as a good anthropological marker for people of East Asia or Pacific origin.

Mitochondrial diseases are caused mostly by defects in the respiratory function of mitochondria in affected tissues and transmitted mainly through maternal lineages. The mtDNA mutations, which include point mutation, deletion, and duplication that affect transcription and translation of mtDNA, are the main etiology factor for most mitochondrial diseases.[3] It is plausible that point mutation of mtDNA associated with mitochondrial diseases may be co-transmitted with the MIC9D polymorphism of mtDNA. Here, we report the incidence of the MIC9D polymorphism of mtDNA in healthy Taiwanese subjects and compare it with that of patients with mitochondrial encephalomyopathy with lactic acidosis and stroke-like episodes (MELAS) or myoclonic epilepsy with ragged-red fibers (MERRF) syndrome, and we examine the correlation between the incidence of the MIC9D polymorphism of mtDNA and the prevalence rate of mitochondrial diseases.

MATERIALS AND METHODS

A total of 199 healthy Taiwanese subjects without any systemic or genetic diseases were recruited as control subjects from the health clinic of Changhua Christian Hospital, Taiwan. Forty unrelated Taiwanese families with mitochondrial diseases, including 12 families with MERRF syndrome and A8344→G mtDNA mutation and 28 families with MELAS syndrome and A3243→G mutation of mtDNA were recruited from five medical centers in Taiwan from 1991 to 2002. Written informed consent for participation in this study was obtained from each of the subjects. All of the human experimental procedures followed the medical ethics guidelines of the Department of Health of the Executive Yuan, Taiwan. Molecular analysis was performed to detect the A3243→G and A8344→G mutations and MIC9D polymorphism of mtDNA. Blood (10 mL) was drawn from an antecubital vein of each of the study subjects in the morning after overnight fasting and was quickly delivered into a plastic tube containing EDTA. Leukocytes were separated from plasma as the buffy coat and immediately subjected to DNA isolation. The A3243→G and A8344→G mutations of mtDNA in leukocytes of patients and their family members were detected by using polymerase chain reaction (PCR) and restriction fragment length polymorphism (RFLP) as described previously.[3,4] The MIC9D polymorphism of mtDNA was determined by direct sequencing of the mtDNA fragment encompassing the polymorphic region amplified by PCR using the primer pair of 5'-CCGGGGG-TATACTACGGTC-3' and 5'-GGGGCATTTCACTGTAAAGAGGT-3' according to the standard PCR procedure, which is illustrated in FIGURE 1.

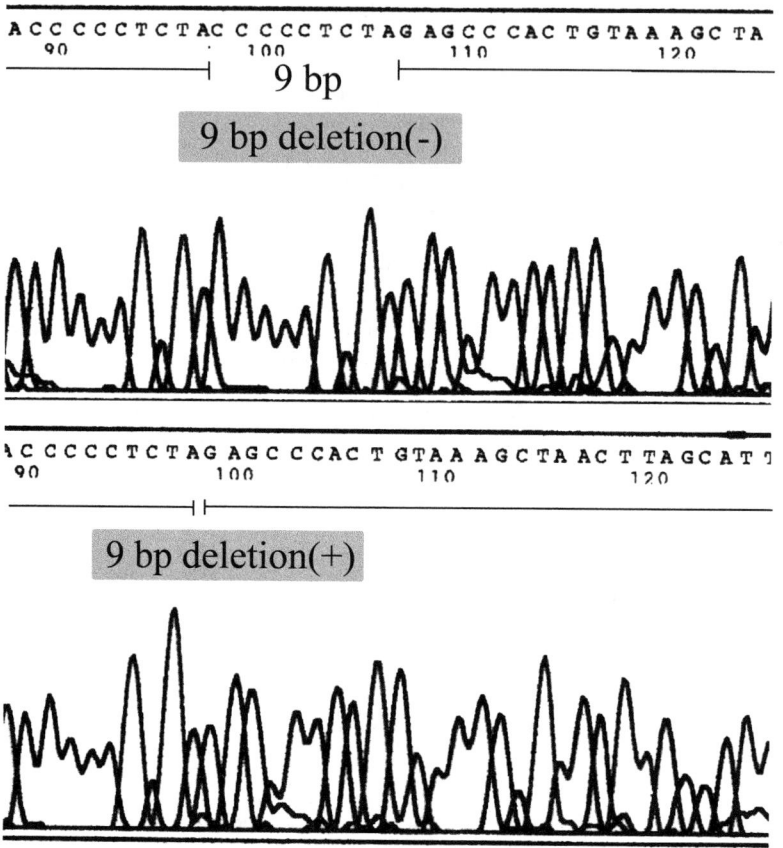

FIGURE 1. Direct sequencing of the PCR-amplified mtDNA fragment encompassing the COII/tRNA^{Lys} intergenic 9-bp deletion. *Upper panel* shows the sequence without the COII/tRNA^{Lys} intergenic 9-bp deletion; *lower panel* shows the sequence with the 9-bp deletion.

RESULTS

The frequency of occurrence of the MIC9D polymorphism of mtDNA was 21% in healthy subjects recruited in this study (FIG. 2). Higher prevalence of MIC9D polymorphism was found in the populations of the Asian-Pacific region including China (15%),[5] Vietnam (20%),[6] Thailand (29%),[7] Japan (20%),[2] and Polynesia (93%),[8] but lower prevalence was found in non–Asian-Pacific areas including New Britain (8%),[8] Russia (5%),[9] and Caucasus (0%).[10] This finding is consistent with previous reports that the MIC9D polymorphism of mtDNA in Asian-Pacific countries is more prevalent than that in non–Asian-Pacific countries or Caucasian populations. The frequency of occurrence of the MIC9D polymorphism of mtDNA was 67% (8/12) among the probands of the families with MERRF syndrome ($P = 0.001$; OR = 8), and it was 39% (11/28) among the probands of the families with MELAS

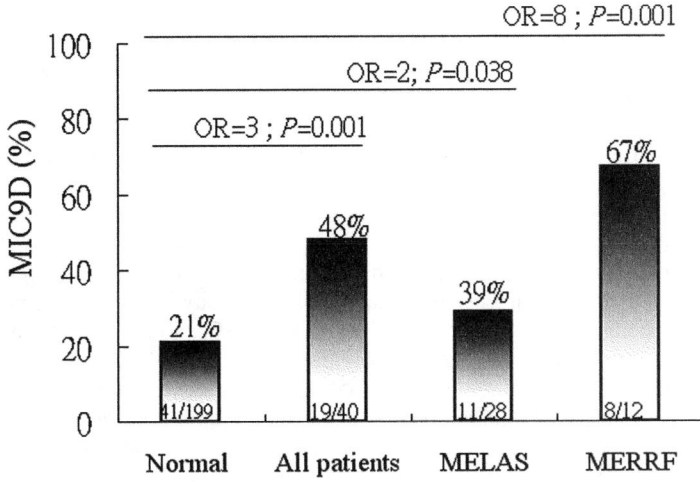

FIGURE 2. Frequency of occurrence within the Taiwanese population of the MIC9D polymorphism of mtDNA in healthy subjects and patients with MELAS or MERRF syndrome.

syndrome ($P = 0.038$; OR = 2), which are significantly higher than that of healthy subjects as revealed by chi-square test (FIG. 2). Similar patterns of MIC9D polymorphism were found in all asymptomatic matrilineal family members of the patients with either MERRF or MELAS syndrome.

DISCUSSION

Point mutation and large-scale deletion of mtDNA are comparatively common causes of neuromuscular diseases in both adults and children worldwide. However, the frequency of occurrence of MERRF syndrome with A8344→G mutation of mtDNA was very low in Western countries or Caucasian populations compared with that in Asian populations. In Northern Ostrobothnia of Finland, the prevalence rate of MERRF syndrome with A8344→G mutation of mtDNA was 0–1.5 per 100,000.[11] Only two families with MRRRF syndrome have been reported in Northern Finland. An epidemiological study conducted in the United Kingdom and western Sweden revealed that the prevalence rate of MERRF syndrome with A8344→G mtDNA mutation was only 0–0.25 per 100,000.[12,13] In 1999, we reported an 8-year molecular epidemiologic study of mtDNA mutations in patients with mitochondrial diseases in Taiwan. A total of 177 patients met the diagnostic criteria of mitochondrial disease and were recruited for the study. Among them, we found 32 patients carrying the A3243→G mtDNA mutation, 25 with MELAS syndrome, one with Kearns–Sayre syndrome, one with diabetes mellitus and deafness, and five with chronic progressive external ophthalmoplegia.[3] Nine patients with MERRF syndrome were found to harbor the A8344→G mutation of mtDNA.[3] In Taiwan, 4.3% of the patients with type II diabetes mellitus were found to harbor the A3243→G mutation of mtDNA.[14] In Japan, the incidence of mitochondrial disease was high, and it was estimated that 0.5–1.0% of the 6 million cases of diabetes were caused by the A3243→G mutation of mtDNA, and more than 100,000 subjects were affected by MELAS syndrome.[15] The A3243→G mutation of mtDNA has been established as one of the important genetic factors involved in the pathogenesis of type II diabetes mellitus in Taiwan and Japan.[3,15] However, there has been no satisfactory explanation for this relatively high prevalence of mitochondrial diabetes in the Taiwanese population. In this study, we found that the incidence of the MIC9D polymorphism was very high in patients with MERRF or MELAS syndrome compared with healthy subjects in the Taiwanese population. This finding implies that the epidemiology of MERRF and MELAS syndromes in Asian populations may be different from that of other ethnic groups such as Caucasians. The A8344→G and A3243→G mutations may be cotransmitted with the MIC9D polymorphism of mtDNA. Indeed, a linkage in the transmission of A8344→G, A3243→G, and MIC9D was found in this study of Taiwanese population. This notion is supported by the report of genomic instability in the COII/tRNA^{Lys} region of mtDNA in patients with mitochondrial disease.[16] However, Varlamov and coworkers[17] failed to find a correlation between the MIC9D polymorphism and mitochondrial diseases in the general population of Germany. This article mentioned that the MIC9D-related mitochondrial disease may fail to appear in the German population. In our study, a higher prevalence of MIC9D was found in forty families with MELAS or MERRF syndrome, which supported the notion that the incidence of mitochondrial disease varies in different ethnic groups. This finding suggests that mtDNA polymorphisms, such as MIC9D, may be cosegregated with some of the pathogenic mtDNA mutations in the long process of propagation and evolution of the human mitochondrial genome. However, more large-scale molecular epidemiologic studies of mitochondrial diseases are warranted to establish the relationship between the MIC9D polymorphism of mtDNA and the occurrence of mitochondrial diseases in Taiwanese and other Asian populations.

ACKNOWLEDGMENTS

This work was supported by a grant (NSC92-2320-B010-037) from the National Science Council and a grant (NHRI-EX93-9120BN) from National Health Research Institutes, Taiwan.

REFERENCES

1. CANN, R.L. & A.C. WILSON. 1983. Length mutations in human mitochondrial DNA. Genetics **104:** 699–711.
2. WRISCHNIK, L.A., R.G. HIGUCHI, M. STONEKING, et al. 1987. Length mutations in human mitochondrial DNA: direct sequencing of enzymatically amplified DNA. Nucleic Acids Res. **15:** 529–542.
3. PANG, C.Y., C.C. HUANG, M.Y. YEN, et al. 1999. Molecular epidemiologic study of mitochondrial DNA mutations in patients with mitochondrial diseases in Taiwan. J. Formos. Med. Assoc. **98:** 326–334.
4. HUANG, C.C., R.S. CHEN, C.M. CHEN, et al. 1994. MELAS syndrome with mitochondrial tRNA$^{Leu\ (UUR)}$ gene mutation in a Chinese family. J. Neurol. Neurosurg. Psychiatry **57:** 586–589.

5. YAO, Y.G., W.S. WATKINS & Y.P. ZHANG. 2000. Evolutionary history of the mtDNA 9-bp deletion in Chinese populations and its relevance to the peopling of east and southeast Asia. Hum. Genet. **107:** 504–512.
6. IVANOVA, R., A. ASTRINIDIS, V. LEPAGE, et al. 1999. Mitochondrial DNA polymorphism in the Vietnamese population. Eur. J. Immunogenet. **26:** 417–422.
7. FUCHAROEN, G., S. FUCHAROEN & S. HORAI. 2001. Mitochondrial DNA polymorphisms in Thailand. J. Hum. Genet. **46:** 115–125.
8. HERTZBERG, M., K.N. MICKLESON, S.W. SERJEANTSON, et al. 1989. An Asian-specific 9-bp deletion of mitochondrial DNA is frequently found in Polynesians. Am. J. Hum. Genet. **44:** 504–510.
9. DERENKO, M.V., I.K. DAMBUEVA, B.A. MALIARCHUK, et al. 1999. Structure and diversity of the mitochondrial gene pool of the aboriginal population of Tuva and Buriatia from restriction polymorphism data. Genetika **35:** 1706–1712.
10. WATKINS, W.S., M. BAMSHAD, M.E. DIXON, et al. 1999. Multiple origins of the mtDNA 9-bp deletion in populations of South India. Am. J. Phys. Anthropol. **109:** 147–158.
11. REMES, A.M., M. KARPPA, J.S. MOILANEN, et al. 2003. Epidemiology of the mitochondrial DNA 8344AG mutation for the myoclonus epilepsy and ragged-red fibres (MERRF) syndrome. J. Neurol. Neurosurg. Psychiatry **74:** 1158–1159.
12. CHINNERY, P.F., M.A. JOHNSON, T.M. WARDELL, et al. 2000. The epidemiology of pathogenic mitochondrial DNA mutations. Ann. Neurol. **48:** 188–193.
13. DARIN, N., A. OLDFORS, A.R. MOSLEMI, et al. 2001. The incidence of mitochondrial encephalomyopathies in childhood: clinical features and morphological, biochemical, and DNA abnormalities. Ann. Neurol. **49:** 377–383.
14. CHUANG, L.M., H.P. WU, K.C. CHIU, et al. 1995. Mitochondrial gene mutations in familial non-insulin-dependent diabetes mellitus in Taiwan. Clin. Genet. **48:** 251–254.
15. SUZUKI, S. 2004. Diabetes mellitus with mitochondrial gene mutations in Japan. Ann. N.Y. Acad. Sci. **1011:** 185–192.
16. PAUL, R., C. DESNUELLE, J. POUGET, et al. 2000. Importance of searching for associated mitochondrial DNA alterations in patients with multiple deletions. Eur. J. Hum. Genet. **8:** 331–338.
17. VARLAMOV, D.A., A.P. KUDIN, S. VIELHABER, et al. 2002. Metabolic consequences of a novel missense mutation of the mtDNA CO I gene. Hum. Mol. Genet. **11:** 1797–1805.

Analysis of Proteome Bound to D-Loop Region of Mitochondrial DNA by DNA-Linked Affinity Chromatography and Reverse-Phase Liquid Chromatography/Tandem Mass Spectrometry

YON-SIK CHOI,[a] BO-KYUNG RYU,[a] HYE-KI MIN,[b] SANG-WON LEE,[b] AND YOUNGMI KIM PAK[a]

[a]*Asan Institute for Life Sciences, College of Medicine, University of Ulsan, Seoul, 138-736, Korea*

[b]*Department of Chemistry and Center for Electro- & Photo-Responsive Molecules, Korea University, Seoul, 136-701, Korea*

ABSTRACT: Mitochondrial dysfunction has been suggested as a causal factor for insulin resistance and diabetes. Previously we have shown a decrease of mitochondrial DNA (mtDNA) content in tissues of diabetic patients. The mitochondrial proteins, which regulate the mitochondrial biogensis, including transcription and replication of mtDNA, are encoded by nuclear DNA. Despite the potential function of the proteins bound to the D-loop region of mtDNA in regulating mtDNA transcription/replication, only a few proteins are known to bind the D-loop region of mtDNA. The functional association of these known proteins with insulin resistance is weak. In this study, we applied proteomic analysis to identify a group of proteins (proteome) that physically bind to D-loop DNA of mtDNA. We amplified D-loop DNA (1.1 kb) by PCR and conjugated the PCR fragments to CNBr-activated sepharose. Mitochondria fractions were isolated by both differential centrifugation and Optiprep-gradient ultracentrifugation. The D-loop DNA binding proteome fractions were enriched via this affinity chromatography and analyzed by SDS-PAGE. The proteins on the gel were transferred onto PVDF membrane and the peptide sequences of each band were subsequently analyzed by capillary reverse-phase liquid chromatography/tandem mass spectrometry (RPLC/MS/MS). We identified many D-loop DNA binding proteins, including mitochondrial transcription factor A (mtTFA, Tfam) and mitochondrial single-stranded DNA binding protein (mtSSBP) which were known to bind to mtDNA. We also report the possibility of novel D-loop binding proteins such as histone family proteins and high-mobility group proteins.

KEYWORDS: mitochondria; D-loop; DNA binding proteome

Address for correspondence: Youngmi Kim Pak, Ph.D., Asan Institute for Life Sciences, College of Medicine, University of Ulsan, Songpa-Ku Pungnap-Dong 388-1, Seoul, 138-736, Korea. Voice: +82-2-3010-4191; fax: +82-2-3010-4182; or Sang-Won Lee, Ph.D., Department of Chemistry and Center for Electro- & Photo-Responsive Molecules, Korea University, Seoul, 136-701, Korea. Voice: +82-2-3290-3137; fax: +82-2-3290-3121.
ymkimpak@amc.seoul.kr or sw_lee@korea.ac.kr

Ann. N.Y. Acad. Sci. 1042: 88–100 (2005). © 2005 New York Academy of Sciences.
doi: 10.1196/annals.1338.009

INTRODUCTION

Mitochondrial DNA (mtDNA)[1] and mitochondrial dysfunction[2] might have important roles in developing insulin resistance and type 2 diabetes. Lee *et al.* reported that a decrease of mtDNA copy number occurred prior to developing type 2 diabetes,[3] and we reported that depletion of mtDNA impaired the glucose metabolism including glucose uptake and hexokinase activity in hepatoma cells.[4] These reports support that the decrease of mtDNA content may be one of the causes for the pathogenesis of type 2 diabetes.

The mammalian mtDNA, a closed-circular and double-strand DNA molecule of about 16 kb, is present in high cellular copy number (10^3–10^4 mtDNA per somatic cell), and their replication is unrestricted with regard to cell cycle.[5] The major part of the regulatory sequences in mtDNA replication and transcription is located in the D-loop region, a short, three-strand structure, which contains a heavy-strand promoter (HSP), a light-strand promoter (LSP), four mitochondrial transcription factor A (Tfam, mtTFA) binding sites, and three G+C rich conserved sequence blocks (CSB). The D-loop control region of mtDNA is the most variable portion of the human mtDNA genome, and it also contains heteroplasmic mutation(s) that are accumulated with the aging process.[6,7]

Since mitochondria are not self-supporting entities, the maintenance and expression of mtDNA depend on many nuclear-encoded gene products. Although some D-loop binding proteins such as Tfam, DNA polymerase gamma (polγ), and mitochondrial single-stranded DNA binding protein (mtSSB) have been identified and characterized,[5,8] the mammalian mtDNA replication machinery remains poorly defined.

To enhance the understanding of the mitochondrial replication/transcription machinery, we developed D-loop DNA-linked affinity chromatography to purify the D-loop DNA binding proteins. The resultant samples enriched with D-loop binding proteins were subsequently subjected to separation by 1D SDS-PAGE and proteomic analysis by capillary RPLC/MS/MS. Here, we report on various types of proteins bound to the control region of mtDNA and some DNA/RNA binding proteins, and we examine the possible candidates as important *trans*-acting factors for the maintenance of mtDNA.

MATERIALS AND METHODS

Cell Culture

HEK293 human kidney cells (ATCC #CRL-1573) were grown in Dulbecco's modified Eagle's medium supplemented with 10% fetal bovine serum.

Preparation of Anti-Human Tfam Antibody

His-tagged Tfam was expressed in *E. coli*, transformed with the recombinants of human Tfam coding sequence (GenBank # M62810) on pQE30 plasmids (Qiagen, Valencia, CA), a high-level bacterial expression vector of 6xHis-tagged protein. The overexpressed Tfam protein was purified by nickel column chromatography. For the immunization, a rabbit was used with the further purified Tfam protein from the SDS-PAGE gel band containing Tfam.[9] The rabbit was immunized three times at 3-

week intervals by subcutaneous injections of 300 μg of His-tagged Tfam dissolved in 0.5 mL of phosphate-buffered saline, 0.1 % SDS mixed with 1 vol of either complete or incomplete Freund's adjuvant (Sigma, St. Louis, MO). The sera obtained from the rabbit were tested for the presence of anti-human Tfam antibody by Western blot analysis.

Isolation of Mitochondria

Highly enriched mitochondria were prepared essentially as described by Meeusen et al.[10] The harvested cell pellets were resuspended in mitochondrial isolation buffer (MIB, 25 mM Tris pH 7.4, 0.25 M Sucrose, 1 mM EDTA) and homogenized using Dounce homogenizer (20 strokes in ice). After removing the nuclei pellets by centrifugation at $1,000 \times g$ for 10 min, the supernatant was centrifuged at $10,900 \times g$ for 10 min to obtain crude mitochondrial pellets. To isolate the highly enriched mitochondria, the pellets were resuspended in 50% w/v Optiprep (Axis-Shield, Oslo, Norway), 25 mM Tris pH 7.4, 0.25 M Sucrose, 1mM EDTA. The mixture was used as the bottom layer of Optiprep step gradients formed in SW28 centrifuge tubes (Beckman); the layers atop it had densities of 1.1 g/mL and 1.16 g/mL and were similarly osmotically balanced with sucrose. The gradients were centrifuged at $80,000 \times g$ for a minimum of 3 h in an SW28 rotor; mitochondria floated to the interface between the 1.1 and 1.6 g/mL layers. This interface was collected and diluted fivefold with MIB and centrifuged at $10,000 \times g$ for 10 min to pellet the mitochondria. The pellets were resuspended, centrifuged at $3,000 \times g$ for 5 min to pellet aggregated material; supernatants were centrifuged again at $10,000 \times g$ for 10 min, and the pellets were resuspended in minimal volume of MIB. The purity of mitochondria was confirmed by Western blot analysis using anti-poly (ADP-ribose) polymerase (PARP, Santa Cruz, CA, SC-7150), anti-hsp60 (Santa Cruz, CA, SC-13966), and anti-calregulin (Santa Cruz, CA, SC-11398) antibodies as nuclei, mitochondria, and endoplasmic reticulum (ER)/Golgi fraction markers, respectively.

Western Blot Analysis

The proteins (30 μg) were separated by 12.5% SDS-PAGE and transferred onto nitrocellulose membrane (Schleicher & Schuell, Dassel, Germany). The membrane was treated with 5% nonfat milk in Tris-buffered saline-Tween 0.05% for 1 h and incubated with the human Tfam antibody (1: 5000) or other organelle marker antibodies overnight at 4°C. Antigen-antibody complexes were detected by using horseradish peroxidase-conjugated secondary antibodies (Cell Signaling, Beverly, MA) diluted to 1:1000 in TBST (0.05% Tween-20 in Tris-buffered saline) supplemented with 5% nonfat milk and incubated for 1 h at room temperature. The chemiluminescence signals were detected using ECL Western blotting kit (Amersham Bioscience, Piscataway, NJ).

Preparation of DNA-Linked Affinity Column Containing the D-Loop Region of mtDNA

PCR synthesis of the D-loop region of mtDNA. The D-loop regions of mtDNA were produced by PCR amplification using the purified human genomic DNA (Promega, Madison, WI) as a template. The PCR was performed for 35 cycles of

94°C for 30 s, 56°C for 30 s, and 72 °C for 1 min using 5′-NH$_2$-CTG TTC TTT CAT GGG GAA GC-3′ and 5′-GAT GTG AGC CCG TCT AAA CA-3′, and the amine-modified PCR products were purified by phenol extraction and ethanol precipitation. The sequences of the purified 1.1 kb products were verified using the Autoread DNA sequencing kit and an autosequencer (Perkin-Elmer, Alameda, CA).

Coupling of the amine-modified PCR products to the CNBr-activated Sepharose. The CNBr-activated Sepharose Sigma, St. Louis, MO), previously swelled and washed with 1 mM HCl, was coupled to the amine group of the purified PCR products in 0.1M NaHCO3, 0.5M NaCl (pH 8.0) at room temperature for 2 h. To block remaining active groups, the gel matrices were transferred and incubated in 0.2 M glycine (pH 8.0) for 2 h. Then the products were washed with high (0.1 M NaHCO$_3$ pH 8.0, 0.5 M NaCl) and low (0.1 M sodium acetate, pH 4.0, 0.5 M NaCl) pH buffer solutions five times, to wash the excess uncoupled ligands. The coupling reaction of the D-loop region was confirmed by digesting a small aliquot of gel slurry with MspI or NcoI, separating the DNA fragment from gel matrix by brief centrifugation and analyzing the supernatant on gel electrophoresis.

Purification of the Mitochondrial Proteins through the D-Loop DNA-Affinity Chromatography

Mitochondrial lysates were prepared by diluting purified mitochondria to a protein concentration of 2 mg/mL with mitochondrial extraction buffer (MEB, 25 mM Tris (pH 7.9), 6.25 mM MgCl$_2$, 50 mM KCl, 0.5 mM EDTA, 0.5 mM DTT, 1 mM PMSF, 0.5% NP-40), followed by an incubation on ice for 30 min. The lysates were mixed with D-loop Sepharose and incubated at 4°C for 1 h. After the mixture was packed into a column, the gel matrix was washed with 10 column volumes of the same buffer in the absence of NP-40. The bound proteins were eluted with the buffer containing 1M KCl, pooled and concentrated to 1mg/mL. Each eluted protein fractions were separated by 15% 1D-SDS-PAGE, stained with 1% Coomassie blue. The protein bands were excised and the peptide profiles after tryptic digestion were analyzed by MALDI-TOF (Voyager-DETM STR Biospectrometry Workstation, Applied Biosystems) and MS-FIT (http://prospector.ucsf.edu/).

Electroblotting and Tryptic Digestion of the Protein Bands Transferred onto PVDF Membrane by Microcapillary LC/Ms/Ms

The protein aliquots eluted from D-loop-Sepharose were separated by 15% SDS-PAGE in Tricine buffer system using the method described by others.[11,12] Cathode buffer consists of 100 mM Tris, 100 mM Tricine at pH 8.25 containing 0.1% SDS and anode buffer consists of 200 mM Tris-HCl at pH 8.9. The gels were electroblotted onto PVDF membrane (Immobilon, pore size 0.45 μm, Millipore, Bedford, MA) using a semidry transfer method.[11,13] The PVDF membrane was stained with Ponceau S and 11 distinct protein bands were excised using clean razors. Each pieces of PVDF membrane was destained twice with 200 μL of 50% methanol for 10 min at room temperature. Then the pieces were washed twice in 200 μL of deionized water for 10 min. After removal of the supernatant, 100 of 100 mM ammonium bicarbonate containing 80% acetonitrile and 5 μL of 20 μg/mL trypsin (Promega) were added, and the mixture was incubated for 1 h at 37°C. Supernatant was collected and the membrane extracted with 50 μL of 80% ACN during 10 min with sonication to extract the peptides from the PVDF. The extract was pooled with the previous supernatant. After being dried in a speed-vac (Thermo Savant, Holbrook, NY), the digested samples were reconstituted in 5 μL of aqueous solution of 0.05% TFA, 0.2% acetic acid for LC/MS/MS experiments.

Capillary RPLC/MS/MS Analysis and Database Searching

A quadrupole ion-trap mass spectrometer (QIT-MS, LCQ, ThermoFinnigan, San Jose, CA) was used to analyze proteins bound to D-loop region. In brief, the mass spectrometer is equipped with a home-built nanoESI interface which allows minimizing the dead volume between the LC and the mass spectrometer by directly connecting the separation column to the electrospray emitter via a stainless steel union. In this experiment, we used a high-pressure capillary RPLC system similar to the one that was previously described elsewhere.[14]

The MS and MS/MS data from QIT-MS experiments were analyzed using the SEQUEST program against a database that was constructed by combining the IPI human database (ftp://ftp.ebi.ac.uk/pub/databases/IPI/current) and a common contaminant database (http://www.ncbi.nih.gov). A home-built application written using Microsoft Visual Basic was used to filter the identified peptide sequences using criteria of Xcorr ≥ 1.8 for +1 charge state, ≥ 2.5 for +2 charge state, and ≥ 3.5 for +3. The normalized difference between first and second match scores (ΔCn) higher than 0.08 was used in the filtering process as well. In order to minimize false positive identification, only proteins with two or more peptides were classified as positive protein identification. When a protein is identified by a single peptide, the corresponding MS/MS spectrum is manually inspected for positive identification.

RESULTS

Preparation of DNA-Linked Affinity Column Containing the D-Loop Region of mtDNA

The amine-modifed D-loop regions of human mtDNA from 16,022 to 637 (GenBank #NC_001807) were amplified by PCR using the purified human genomic DNA

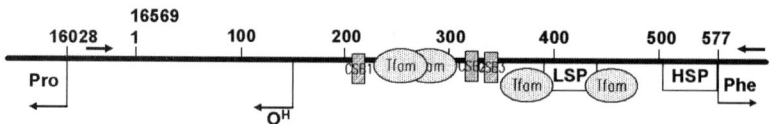

FIGURE 1. PCR synthesis of the D-loop DNA fragment of mtDNA. The D-loop region fragment of mtDNA was amplified with 5′-NH2-modified primers using human blood genomic DNA as a template. The generated DNA fragment was 1.1 kb covering 16060 to 16569 and 1 to 635 region of mtDNA, which contains light- and heavy-strand promoters (LSP and HSP), four Tfam binding sites, and three CSBs. *Arrows* indicate the position of primers.

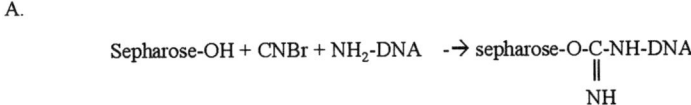

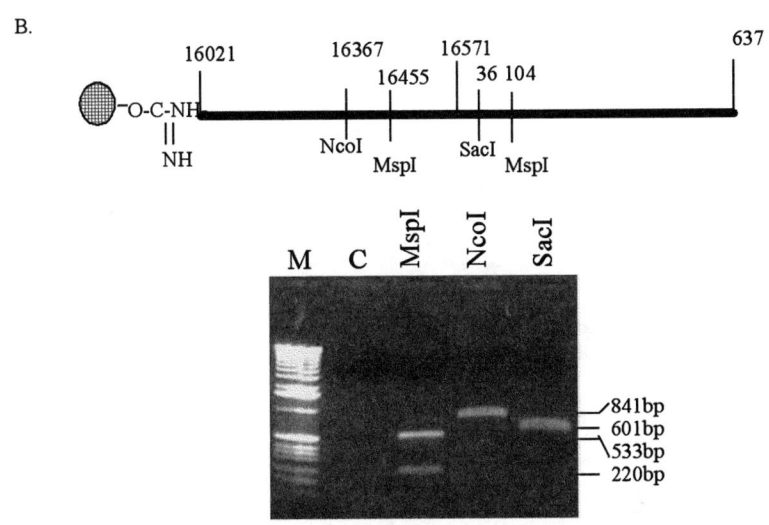

FIGURE 2. Coupling of the D-loop DNA fragment onto CNBr-activated Sepharose beads. (**A**) The CNBr-coupled reaction to link DNA to Sepharose. The CNBr-activated Sepharose was coupled to the amine group of the purified PCR products in 0.1M NaHCO3, 0.5 M NaCl (pH 8.0) at room temperature for 2 h. (**B**) Confirmation of the DNA fragments' conjugation to Sepharose matrix. The coupling reaction of the D-loop region was confirmed by digesting a small aliquot of gel slurry with MspI, NcoI, or SacI, separating the DNA fragment from gel matrix by brief centrifugation and analyzing the supernatant on gel electrophoresis.

(Promega) as a template. FIGURE 1 schematically demonstrates that the D-loop DNA contains LSP, HSP, four Tfam binding sites, and three CSBs. The purified 1.1 kb PCR fragments were cloned into T-vector, and the sequences were verified by automatic DNA sequencing.

The 1.1 kb D-loop DNA fragments were conjugated to CNBr-activated Sepharose via primary amino groups of DNA under mild conditions (FIG. 2A).[15] The conjugation reaction was verified by analyzing the D-loop DNA on the Sepharose beads after digesting a small aliquot of gel slurry with MspI, NcoI, or Sac I. The DNA fragments eluted from the affinity gel beads were separated by 1% agarose gel electrophoresis. As expected, 533 bp and 220 bp fragments were identified with MspI digestion, 841 bp with NcoI, 601 bp with SacI, respectively (FIG. 2B).

D-Loop DNA-Affinity Chromatography of the Mitochondrial Proteins

We isolated the highly enriched mitochondria fraction by ultracentrifugation on linear OptiprepTM gradients media, as summarized in FIGURE 3. The ultracentrifugation step excluded the potential risk of contamination with nuclear proteins as confirmed by Western blot analysis using antibodies against organelle marker proteins. As shown in FIGURE 4, in the pure mitochondria obtained after Optiprep ultracentrifugation, the nuclear marker PARP was not detected whereas hsp60, a mitochondria-matrix protein, was found. The membranous compartment of the cells such as endo-

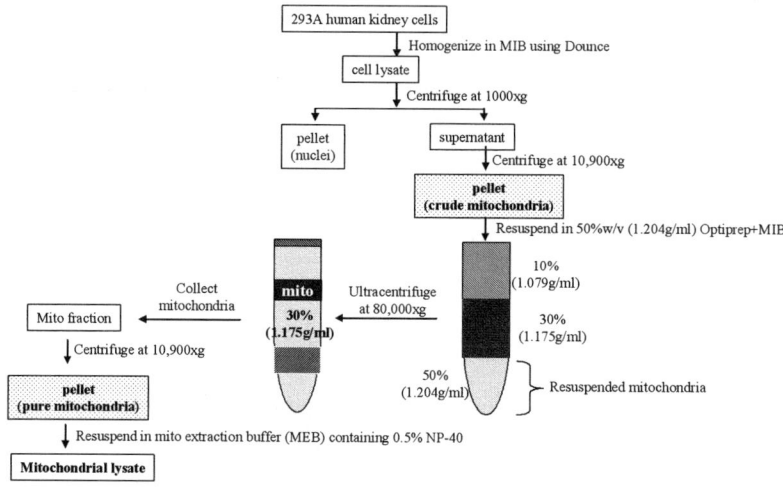

FIGURE 3. Isolation of mitochondria from HEK293A human kidney cells. To isolate the highly enriched mitochondria, the crude mitochondrial pellets, obtained by differential centrifugation, were applied to linear Optiprep™ gradients media. The highly purified mitochondrial fraction pellets were resuspended in mitochondrial isolation buffer (MIB, 25 mM Tris pH 7.4, 0.25 M sucrose, 1mM EDTA) and lysed with 0.5% NP-40.

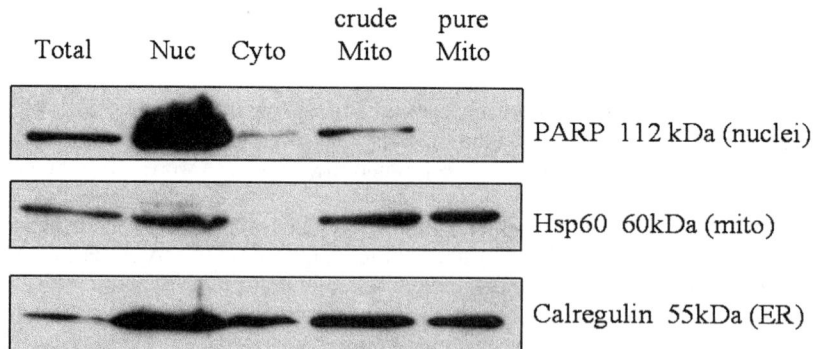

FIGURE 4. Identification of marker proteins in each organelle fraction. Organelle proteins (30 µg) of whole-cell lysates (Total), nuclei pellet (Nuc), cytosol (Cyto), the crude mitochondrial pellets (crude Mito), or the purified mitochondrial fraction by Optiprep media (pure Mito) were analyzed by Western blotting. Antibodies against the marker proteins were utilized; poly (ADP-ribose) polymerase (PARP) for nuclei, heat-shock protein 60 (Hsp60) for mitochondria and calregulin for ER/Golgi. Pure mitochondria fraction used for proteomic analysis lacked nuclear contamination.

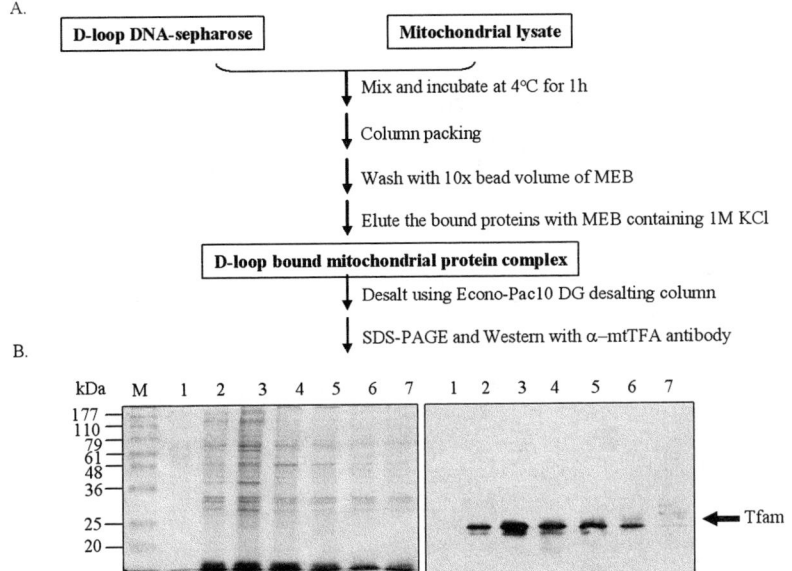

FIGURE 5. The purification scheme of mitochondria protein complex bound to the D-loop region of mtDNA. (**A**) The total mitochondria extracts were incubated directly with D-loop DNA-Sepharose beads and eluted by the buffer containing 1M KCl and desalted. (**B**) The eluted protein fractions contain various sizes of proteins as shown by 12% SDS-PAGE followed by Coomassie staining. Immunoblotting with anti-human Tfam antiserum showed the 25-kDa band corresponding to Tfam, confirming the D-loop DNA affinity of the isolated protein complex. M, size-marker proteins.

plasmic reticulum (ER) or golgi might not be completely separated from mitochondria even after Optiprep step because calregulin, an ER marker, was present. The isolated mitochondria fraction was lysed with 0.5% NP-40, and the total mitochondria extracts were directly incubated to D-loop DNA-conjugated Sepharose beads, as summarized in FIGURE 5. The bound proteins were eluted in high-salt buffer and collected in 0.5 mL fractions. The eluted fraction of D-loop DNA binding proteins contains a lot of proteins with various molecular weights as shown on SDS-PAGE followed by Coomassie staining (FIG. 5). Immunoblotting of the collected fractions with anti-human Tfam serum revealed that 25-kDa band corresponding Tfam were present in the eluted fractions, indicating that the eluted proteins were indeed a protein complex, which binds the D-loop DNA under our affinity-chromatography condition (FIG. 5).

Analysis of the Peptide Sequences of the Proteome by cRPLC/MS/MS

In order to identify other proteins interacting with the mitochondrial D-loop, the affinity-purified protein mixture was separated by 12% SDS-PAGE and viewed by Coomassie blue staining. Each gel band from the Coomassie blue-stained SDS-PAGE was excised, digested, and analyzed by MALDI-TOF. This analysis resulted

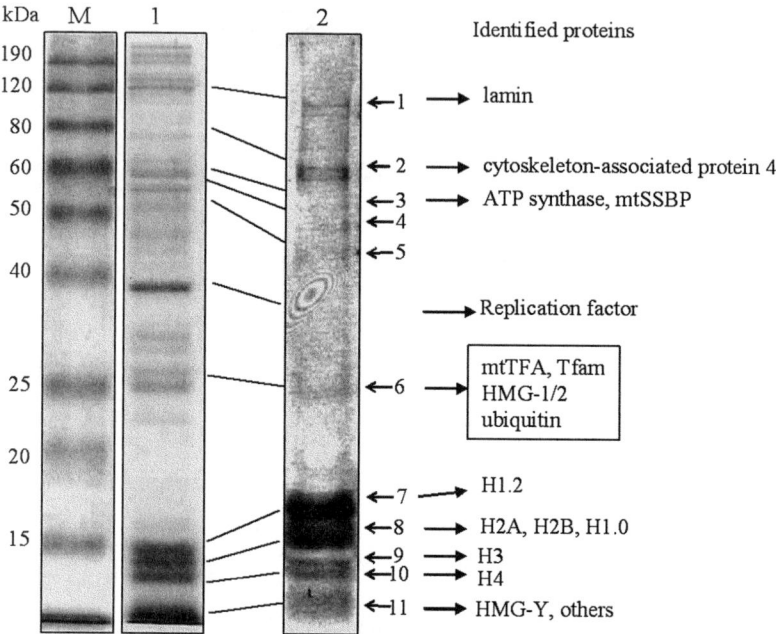

FIGURE 6. MALDI-TOF and LC/Ms/Ms analysis of mitochondrial protein complex bound to the D-loop region on SDS-PAGE. The proteins eluted from the D-loop DNA-Sepharose beads were separated by 15% glycine/SDS-PAGE and stained with 0.1% Coomassie (*lane 1*). The stained gel bands were excised and analyzed by MALDI-TOF. The Tricine/SDS-PAGE gel was transferred onto PVDF, stained with 2% Ponceau S (*lane 2*). The stained bands of the membrane were excised and analyzed by capillary LC/MS/MS. Representative proteins identified from each band were presented on the right. M, size-marker proteins.

in only a few proteins characterized, such as lamin and replication factor, while most of other bands were not confidently identified. The incomplete results of MALDI-TOF analysis may be explained by limited-peptide recovery of these particular proteins after in-gel digestion and insensitive detection of the resultant peptides of these particular proteins by MALDI-TOF. In the next experiment, we employed the capillary RPLC/MS/MS technique to analyze the D-loop DNA binding proteome. The eluted protein fractions were combined and separated on Tricine-based 12% SDS-PAGE. Before staining, the separated proteins on the gel were transferred onto PVDF membrane via semidry blot-transfer method. This method of PVDF membrane transfer was used to increase the peptide recovery.[11] When a conventional transfer buffer containing methanol and glycine for electroblotting or a Tris-based SDS-PAGE was utilized, the peptide recovery from the band was poor, possibly due to less efficient protein transfer to PVDF membrane compared with experiments using Tricine containing buffer. The PVDF membrane electroblotted from the gel was stained with Ponceau S to locate positions of protein bands (FIG. 6). A

TABLE 1. List of representative proteins identified from the bands of SDS-PAGE/PVDF by capillary RPLC/MS/MS

Band	GenBank #	Protein name	No. of amino acids	Peptide sequences	Sequence coverage (%)
2	NP_002030	cytoskeleton-associated protein 4	602	SVGELPSTVESLQKVQEQ VHTLLSQDQAQAAR	5.71
	NP_006816	GRB2-associated binding protein 1	694	QKSSGSGSSVADERV	2.16
3	NP_001677	ATP synthase	529	AIAELGIYPAVDPLDSTSR VLDSGAPIKIPVGPETLGR	7.18
	NP_003134	mitochondrial single-stranded DNA-binding protein (mtSSBP)	148	ESETTTSLVLER QATTIIADNIIFLSDQTK-SLN-RVHLLGR	27.03
6	NP_003192	Tfam, mtTFA	246	KPVSSYLR SAYNVYVAERFQEAKGD-SPQEK	12.20
	NP_002119	HMG-1	215	IKGEHPGLSIGDVAK LKEKYEKDIAAYR	13.02
	NP_002120	HMG-2	209	LKEKYEKDIAAYR SEHPGLSIGDTAK	12.44
	NP_061828	ubiquitin B	229	TITLEVEPSDTIENVK	20.96
7	NP_005310	H1 histone; H1.2	213	ALAAAGYDVEK ALAAAGYDVEKNNSR ASGPPVSELITK ASGPPVSELITKAVAASK GTGASGSFKLNK KASGPPVSELITK PAAPAAAPPAEK PAAPAAAPPAEKAPVK PAAPAAAPPAEKAPVKK SGVSLAALKK SGVSLAALKKALAAAGYDVEK SGVSLAALKKALAAAGYD-VEKNNSR SLVSKGTLVQTK	39.90
	XP_061871	similar to fat3	3522	PSNKIILKVSAKD	0.36
8	NP_066544	H2A histone	128	AGLQFPVGR HLQLAIR HLQLAIRNDEELNK HLQLAIRNDEELNKLLGK HLQLAIRNDEELNKLLGKV-TIAQGGVLPNIQ AVLLPK LLGKVTIAQGGVLPNIQAVLLPK NDEELNKLLGK NDEELNKLLGKVTIAQGGVLPN PNIQAVLLPK SSRAGLQFPVGR VTIAQGGVLPNIQAVLLPK VTIAQGGVLPNIQAVLLPKK	39.37

TABLE 1. (*continued*) **List of representative proteins identified from the bands of SDS-PAGE/PVDF by capillary RPLC/MS/MS**

Band	GenBank #	Protein name	No. of amino acids	Peptide sequences	Sequence coverage (%)
	NP_003510	H2B histone	126	EIQTAVRLLLPGELAK ESYSVYVYKVLKQVHPDTGISSK HAVSEGTK KESYSVYVYK LLLPGELAKHAVSEGTK STITSREIQTAVR VLKQVHPDTGISSK	43.20
	NP_005309	H1 histone; H1.0	194	LVTTGVLKQTK	5.67
9	NP_066403	H3 histone	136	LVREIAQDFK YQKSTELLIR	14.70
10	NP_003486	H4 histone	103	DAVTYTEHAK DNIQGITKPAIR ISGLIYEETRGVLKVFLENVIR TVTAMDVVYALKR VFLENVIRDAVTYTEHAK	58.25
	NP_056249	DKFZP434B168 protein	962	ADAGYGEQELDANSALMELDK	2.19
11	NP_665909	HMG-I/HMG-Y	96	KQPPKEPSEVPTPK	14.59
	NP_003851	Barrier-to-autointegration factor	89	GFDKAYVVLGQFLVLK	17.97
	NP_001773	CD72 antigen	359	SESCRSSLPYICEMTA	4.45
	NP_002903	DNA polymerase zeta catalytic subunit	3,130	RVRGNLQMLEQLDLIGKTSE	0.64

total of 11 bands were observed by Ponceau S staining, and each band was excised for subsequent tryptic digestion and capillary RPLC/MS/MS analysis. Of the 11 protein bands, 8 bands resulted in positive hits by capillary RPLC/MS/MS, and as a whole we identified a total of 20 D-loop DNA binding proteins that include Tfam and mitochondrial single-strand binding protein (mtSSBP), which are D-loop DNA binding proteins. The representative proteins identified from each band are described in FIGURE 6 and TABLE 1. Among others, it is astonishing to note that the distinct low-molecular-weight proteins of the eluted fractions are histones, because mtDNA has been known as a histone-free genome. It is possible that some proteins associated to mitochondrial outer membrane, namely, proteins present outside mitochondria that are enriched by DNA-affinity chromatography. The localization of the identified proteins in/on mitochondria needs to be extensively studied to elucidate their function in/on mitochondria.

DISCUSSION

In the present study, we identified proteins that bound to D-loop DNA using the combined technique of DNA-linked affinity chromatography and 1D-cRPLC/MS/MS. Since we have to usea highly enriched mitochondrial fraction, the purity of the organelle fraction is very important for organelle proteomic analysis. In order to exclude the contamination from nuclei and cytosol, the mitochondria fraction isolated by differential centrifugation was purified again by ultracentrifugation onto linear OptiprepTM gradients media. The application of Optiprep ultracentrifugation successfully removed the nuclei from the mitochondria as determined by Western blot analysis. However, it still did not remove other membranous suborganelles like peroxisome or ER/Golgi, as confirmed by marker enzymes. It is more critical to exclude nuclei than other organelles because we aimed to isolate the proteins with DNA-binding activity.

It is worthy to note that many novel proteins are found to be associated to DNA of the D-loop region, while well-known mtDNA binding proteins such as Tfam and mtSSBP were still observed. For example, the proteins with DNA-binding property such as high-mobility group proteins (HMG-1 and HMG-2) and histone family proteins were also identified; they may function as novel regulators of mitochondrial replication/transcription. Their presence and function in mitochondria need to be confirmed. We note that the proteomic analysis employed in this study, however, is limited in application to identifying the proteins that bind to D-loop DNA in a sequence-specific manner. That is, mitochondrial proteins that have general DNA-binding affinity would be enriched in D-loop DNA-linked affinity chromatography. In this study, we focused on reporting some mitochondrial proteins with general DNA-binding affinity, rather than on verifying whether the identified proteins were sequence-specific or not. Recent studies reported new mitochondrial proteins based on the proteomic analysis.[16,17] and these proteins might be novel and important regulatory factors in mitochondria biogenesis.

From these findings, we can offer new insights into the functional proteomics in binding of the D-loop region and understanding the machinery of mtDNA maintenance and expression. Further investigations, based on the function of the identified proteins, are necessary to explore this understanding. Finally, as the proteomics analysis of this study is limited by the use of 1D-PAGE separation, a shotgun proteomic approach, bypassing the 1D-PAGE separation, might produce a more complete list of D-loop binding proteins. We are currently undertaking global proteomic experiments on the affinity-enriched D-loop binding proteins.

ACKNOWLEDGMENTS

This study was supported by grants (FPR02A6-24-110 and FPR-02-A-5) of 21C Frontier Functional Proteomics Project from the Korean Ministry of Science & Technology, and in part by grants (01-PJ1-PG3-20500-0147 and 02-PJ1-PG1-CH04-0001) of Good Health 21 from the Ministry of Health and Welfare, Korea. S.W.L. and H.K.M. thank the Center for Electro- & Photo-Responsive Molecules at Korea University for supporting the facility.

REFERENCES

1. ANTONETTI, D.A., C. REYNET & C.R. KAHN. 1995. Increased expression of mitochondrial-encoded genes in skeletal muscle of humans with diabetes mellitus. J. Clin. Invest. **95:** 1383–1388.

2. PETERSEN, K.F. et al. 2003. Mitochondrial dysfunction in the elderly: possible role in insulin resistance. Science **300:** 1140–1142.
3. LEE, H.K. et al. 1998. Decreased mitochondrial DNA content in peripheral blood precedes the development of non-insulin-dependent diabetes mellitus. Diabetes Res. Clin. Pract. **42:** 161–167.
4. PARK, K.S. et al. 2001. Depletion of mitochondrial DNA alters glucose metabolism in SK-Hep1 cells. Am. J. Physiol. Endocrinol. Metab. **280:** E1007–E1014.
5. TAANMAN, J.W. 1999. The mitochondrial genome: structure, transcription, translation and replication. Biochim. Biophys. Acta **1410:** 103–123.
6. MICHIKAWA, Y. et al. 1999. Aging-dependent large accumulation of point mutations in the human mtDNA control region for replication. Science **286:** 774–779.
7. COSKUN, P.E., E. RUIZ-PESINI & D.C. WALLACE. 2003. Control region mtDNA variants: longevity, climatic adaptation, and a forensic conundrum. Proc. Natl. Acad. Sci. USA **100:** 2174–2176.
8. CLAYTON, D.A. 1991. Nuclear gadgets in mitochondrial DNA replication and transcription. Trends Biochem. Sci. **16:** 107–111.
9. HARLOW, E. & D. LAND. 1988. Antibodies: A Laboratory Manual. Cold Spring Harbor Laboratory Press. Cold Spring Harbor, NY.
10. MEEUSEN, S. et al. 1999. Mgm101p is a novel component of the mitochondrial nucleoid that binds DNA and is required for the repair of oxidatively damaged mitochondrial DNA. J. Cell Biol. **145:** 291–304.
11. BUNAI, K. et al. 2003. Proteomic analysis of acrylamide gel separated proteins immobilized on polyvinylidene difluoride membranes following proteolytic digestion in the presence of 80% acetonitrile Proteomics **3:** 1738–1749.
12. SCHAGGER, H. & G. VON JAGOW. 1987. Tricine-sodium dodecyl sulfate-polyacrylamide gel electrophoresis for the separation of proteins in the range from 1 to 100 kDa. Anal. Biochem. **166:** 368–379.
13. HIRANO, H. & T. WATANABE. 1990. Microsequencing of proteins electrotransferred onto immobilizing matrices from polyacrylamide gel electrophoresis: application to an insoluble protein. Electrophoresis **11:** 573–580.
14. LEE, Y.H. et al. 2004. Highly informative proteome analysis by combining improved N-terminal sulfonation for de novo peptide sequencing and online capillary reverse-phase liquid chromatography/tandem mass spectrometry. Proteomics **4:** 1684–1694.
15. ALBERTS, B. & G. HERRICK. 1971. DNA-cellulose chromatography. Methods Enzymol. **21:** 198–217.
16. MOOTHA, V.K. et al. 2003. Integrated analysis of protein composition, tissue diversity, and gene regulation in mouse mitochondria. Cell **115:** 629–640.
17. RABILLOUD, T. et al. 2002. Comparative proteomics as a new tool for exploring human mitochondrial tRNA disorders. Biochemistry **41:** 144–150.

Mitochondrial Transcription Factor A in the Maintenance of Mitochondrial DNA

Overview of Its Multiple Roles

DONGCHON KANG AND NAOTAKA HAMASAKI

Department of Clinical Chemistry and Laboratory Medicine, Kyushu University Graduate School of Medical Sciences, Fukuoka 812-8582, Japan

ABSTRACT: Mitochondria have their own genome, which is essential for proper oxidative phosphorylation needed for a large part of ATP production in a cell. Although mitochondrial DNA-less (rho^0) cells can survive under special conditions, the integrity of the mitochondrial genome is critical for survival of multicellular organisms. Mitochondrial transcription factor A (TFAM), originally cloned as transcription factor, is essential for the maintenance of mtDNA. Recently, it has become known that TFAM plays critical roles in multiple aspects to maintain the integrity of mitochondrial DNA: transcription, replication, nucleoid formation, damage sensing, and DNA repair. The effects of TFAM in these aspects are intimately related to each other and to function as a whole for the purpose of maintenance of mtDNA.

KEYWORDS: mitochondria; mitochondrial DNA; mitochondrial transcription factor A; TFAM; reactive oxygen species; ROS; transcription; replication; DNA repair

INTRODUCTION

Mitochondria, which probably evolved from endosymbiotically incorporated organisms, have their own genome. Mitochondria replicate and transcribe their DNA semiautonomously. The circular, approximately 16 kbp of the human mitochondrial genome encodes 13 proteins, 2 rRNAs, and 22 tRNAs. All 13 proteins are considered to be essential subunits of a mitochondrial respiratory chain. The rRNAs and tRNAs, used for constructing mitochondrial translational machineries, are also essential for synthesis of the proteins encoded by mitochondrial DNA (mtDNA).[1] Given that the majority of ATP production depends on the aerobic oxidative phosphorylation executed by the respiratory chain, maintenance of the mitochondrial genome is naturally crucial for individuals to survive normally. Mitochondrial transcription factor A (TFAM), which was cloned as a transcription factor for mtDNA,[2] is known to play multiple roles in addition to the transcription in the maintenance of mtDNA. In this

---

Address for correspondence: Dongchon Kang, Department of Clinical Chemistry and Laboratory Medicine, Kyushu University Graduate School of Medical Sciences, 3-1-1 Maidashi, Higashi-ku, Fukuoka 812-8582, Japan. Voice: +81-92-642-5749; fax: +81-92-642-5772.
kang@mailserver.med.kyushu-u.ac.jp

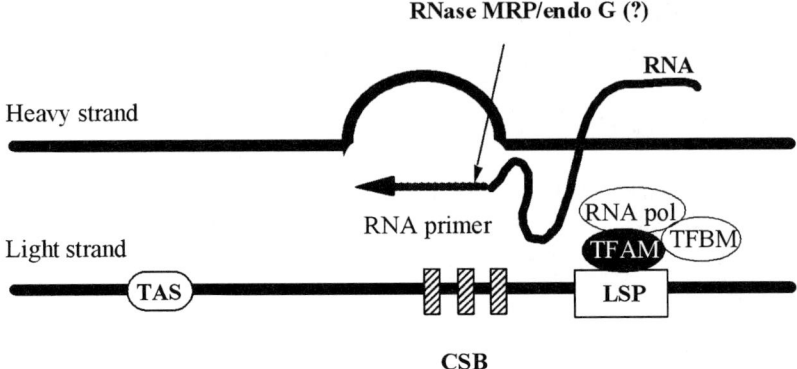

FIGURE 1. Transcription from LSP. TFAM binds to LSP and then recruits mitochondrial RNA polymerase and mitochondrial transcription factor B (TFBM). The C-terminal is responsible for binding to TFBM. TFBM confers the promoter specificity to the TFAM-dependent transcription. Some transcripts originating from LSP make stable RNA-DNA heteroduplexes in the conserved sequence block (CSB). The RNA strand of the heteroduplex is processed by an RNase and serves as a primer for DNA synthesis by DNA polymerase.

review, we focus on the essential roles of TFAM in the maintenance of mtDNA integrity.

TFAM AS A TRANSCRIPTION FACTOR

TFAM was purified from mitochondria of human HeLa cells using transcription activation as a marker[3] and then cloned.[2] TFAM is a member of an HMGB subfamily in high-mobility group (HMG) protein families. TFAM has two HMG boxes and a C-terminal tail composed of 25 amino acids. Each HMG box has DNA-binding activity. Although TFAM has a higher affinity to the two promoters on mtDNA, light- and heavy-strand promoters (LSP and HSP, respectively), it also has nonspecific DNA-binding activity irrespective of DNA sequence. The C-terminal tail is required for the promoter-specific transcription *in vitro*. Abf2p, a counterpart of TFAM in *Saccharomyces cerevisiae*, also has two HMG boxes but not the C-terminal tail.[4] In fact, Abf2p does not activate the transcription *in vitro*[5] or is not required *in vivo* for the expression of mtDNA genes (i.e., transcription). The current general model for the human mitochondrial transcription is outlined in FIGURE 1. TFAM binds to LSP and then recruits mitochondrial RNA polymerase and mitochondrial transcription factor B (TFBM).[6] The C-terminal is responsible for binding to TFBM.[7] Without TFBM, TFAM activates promoter-independent nonspecific transcription in the presence of RNA polymerase. TFBM confers the promoter specificity to the TFAM-dependent transcription. However, it is still unknown how TFBM enhances the transcription from LSP or, in other words, how TFBM recognizes the promoter.

MODULATION OF REPLICATION

Some transcripts originating from LSP make stable RNA-DNA heteroduplexes in the conserved sequence block (CSB).[8,9] The RNA strand of the heteroduplex is processed by an RNase and serves as a primer for DNA synthesis by DNA polymerase. The majority of the nascent DNA-strand synthesis starting from this region terminates prematurely at np 16105, about 50 nucleotides downstream of the terminating association sequence (TAS), resulting in the formation of the triple-stranded structure, D-loop. The D-loop is composed of the parental light and heavy strands, as well as the prematurely terminated nascent DNA strands, called D-loop strands or 7S DNA.[10] These D-loop strands have exactly the same free 5' ends as those of the nascent strands that exceed the termination point and therefore are considered true nascent strands leading to complete replication.[11] In fact, the premature termination is well correlated inversely with the replication rate,[12] suggesting that the DNA synthesis starting from this region may contribute to the replication of mtDNA. Thus, the transcription is coupled with the replication according to the classic strand asymmetric replication model, and TFAM therefore is also essential for the replication of mtDNA. Recently, multiple replication origins for the heavy strand were reported in other than the D-loop region. In this model, called strand-coupled replication, the role of TFAM in the replication is unclear.[13,14]

In addition to the possible straightforward contribution via transcription, TFAM may modulate the mtDNA replication through destabilizing the D-loop structure. As described above, a certain portion of mtDNA molecules steadily take on the D-loop structure. 1-Methyl-4-phenyl-pyridinium ion (MPP^+), a parkinsonism-causing toxin, inhibits selectively the replication of mtDNA at least in part by destabilizing the D-loop.[15–18] Similarly, TFAM at a level normally existing in mitochondria also destabilizes the triple-stranded structure and releases the D-loop strand.[19] Mitochondrial single-stranded DNA-binding protein (SSB) protects the D-loop from this destabilization caused by TFAM. The D-loop structure may be maintained under a balance of the two proteins *in vivo*.[19]

NUCLEOID STRUCTURE OF mtDNA

During the course of examining the D-loop destabilizing effect, we have found that human TFAM is much more abundant than previously thought. TFAM exists about 1,000 molecules per one mtDNA molecule.[19] This amount is almost enough to cover the entire region of mtDNA, given that TFAM occupies about 20 nucleotides. A similar amount of TFAM exists in *Xenopus leavis* oocytes.[20] It has been considered that mtDNA of higher eukaryotes is somewhat naked except for the D-loop region.[21–24] However, in lower eukaryotes, there is good evidence that mtDNA is packaged and forms a large mtDNA-protein complex resembling a nuclear DNA nucleosome.[25,26] Such a nucleosome-like structure of mtDNA is called nucleoid. Recent reports of our and other groups support that mtDNA of higher eukaryotes also takes on the nucleoid structure. Considering that (i) HMG proteins have generally an ability to package DNA, (ii) human TFAM molecules are indeed mostly associated with the mtDNA-proteins complexes, and (iii) human TFAM is abundant, it is very likely that human TFAM makes a major contribution in the formation of the nucleoid structure as a main component.[27,28]

Homozygous gene disruption of Tfam is lethal in mouse and chicken cells.[29,30] The heterozygous disruption decreases the amount of TFAM by about 50%.[29,30] Expression of exogenous TFAM increases the amount of mtDNA in a finely correlated manner to the total amount of TFAM in mouse,[31] chicken,[30] and human cells (unpublished data). When the TFAM expression was suppressed by RNA interference, the amount of mtDNA gradually decreased with the daily decrease in the TFAM amount after the treatment. The decrease in mtDNA was strongly correlated with the decrease in TFAM (unpublished data). These observations of our and other groups suggest that the TFAM amount is a major determinant of the mtDNA amount. Probably, only mtDNA in the nucleoid structure can be stably maintained in mitochondria, and TFAM may be a dose-limiting factor for the number of nucleoids.

DAMAGE AND REPAIR

mtDNA is much more vulnerable than nuclear DNA.[32,33] First, mtDNA is under much stronger oxidative stress than nuclear DNA.[34,35] Mitochondria normally account for ~90% of oxygen consumption in mammalian cells. It is considered that 1–5% of the mitochondrially consumed oxygen is converted to reactive oxygen species (ROS) due to electron leaks from the respiratory chain.[36–38] mtDNA is associated with inner membranes where the ROS-generating respiratory chain exists. Second, mtDNA is subject to chemical damage much more strongly than nuclear DNA.[39] Mitochondria have a membrane potential with matrix-side negative by which ATP synthesis is driven. Because of this membrane potential, lipophilic cations tend to accumulate in mitochondria. Mitochondria can take up lipophilic cations from the cytosol and concentrate these cations inside up to 1000-fold.[40] Unfortunately, many toxic xenobiotics as well as medically beneficial drugs are also lipophilic and have positive charges. Thus, higher mutation rates in mtDNA than in nuclear DNA are expected. In fact, mutation rates of human mtDNA are reported to be several hundredfold higher than nuclear gene mutation rates.[41]

The oxidative damage of mtDNA is evidenced, for example, by a fact that 8-oxoguanine (8-oxoG), an oxidatively modified guanine base, accumulates more and increases more rapidly in mtDNA than in nuclear DNA.[34] TFAM binds more strongly to DNA containing 8-oxoG than to normal DNA. The binding of TFAM to a C:8-oxoG pair is stronger than that of hOGG1,[42] a human functional counterpart of a bacterial MutM that binds to C:8-oxoG and excises the 8-oxoG. Similarly, the binding of TFAM to an A:8-oxoG pair is stronger than that of hMYH,[42] a human homologue of a bacterial MutY that binds to A:8-oxoG and excises the A. Hence, it is naturally expected that TFAM affects the repair of oxidatively damaged DNA by those enzymes.

Human TFAM shows higher affinity to DNA with cisplatin adducts.[42] Cisplatin is an anticancer drug that produces inter- and intrastrand DNA cross-linking. Interestingly, many cisplatin-resistant cell lines overexpress TFAM than do their parental cells,[43] suggesting that TFAM may play a role in protecting mtDNA from modification by cisplatin or in enhancing the repair of cisplatin-modified mtDNA. p53, which is a tumor suppressor and associates preferentially with damaged DNA, is transferred to mitochondria upon death signals.[44,45] p53 physically interacts with TFAM in mitochondria.[43] The interaction enhances the binding of TFAM to cisplatin-modified DNA, but conversely inhibits the binding of TFAM to DNA containing an A:8-oxoG pair.[43] These alterations in properties of TFAM-binding could affect the repair or maintenance of mtDNA and possibly modulate the p53-mediated apoptosis.

BRANCHED STRUCTURE

HMG proteins preferentially bind to branched structures including four-way junctions, typical recombination intermediates.[46] Abf2p promotes or stabilizes Holliday recombination junction intermediates in rho+ mtDNA in *Saccharomyces cerevisiae*.[47] Human TFAM has about a 10-fold higher affinity to a synthesized four-way junction than to a corresponding linear form of DNA.[48] It has long been believed that recombination reactions do not occur, or very rarely occur, in mammalian mitochondria because recombined mtDNA molecules were hardly detected in mitochondria artificially harboring hetroplasmic mtDNAs. However, fairly good recombination activity is detected in rat mitochondrial lysates.[49] Furthermore, prominent recombination intermediates are detected in human heart mtDNA.[50] The presence of intramolecular recombination is suggested based on the fact that partially duplicated mtDNA molecules produced the wild-type mtDNA during cultivation by releasing the partially duplicated parts.[51] Although a physiological role of the intramolecular recombination of mtDNA is unclear at present, the strong binding of TFAM to branched structures may no doubt affect such recombination events. The triple-stranded D-loop in a strand asymmetric replication model and the replication Y-folk suggested in a strand-coupled replication model are also typical branched structures. In this aspect, TFAM should have effects on the proceeding of DNA synthesis.

OXIDATIVE STRESS ON HEART

As described previously, mtDNA is physiologically more vulnerable than in nuclear DNA. In addition, mtDNA is feasibly more damaged by pathological insults as well. For example, when isolated rat cardiomyocytes are treated with 50 μM H_2O_2, the amount of intact 16 kbp mtDNA is decreased by 50% in 15 min.[52] In other words, one-half of mtDNA molecules are rapidly degraded by the 15-min H_2O_2 exposure. In a partial myocardial infarction model of mouse, the non-ischemic region works harder to compensate the loss of function of infarcted myocardium. This working overload causes oxidative stress leading to mitochondrial dysfunction. Under these conditions, the activities of complexes I, III, and IV—all of which contain mtDNA-encoded subunits—all declined, but the activity of complex II which does not contain any mtDNA-encoded subunits was maintained,[53] suggesting that mtDNA is a primary target of ROS. This mitochondrial dysfunction followed by cardiomyopathy was largely lessened by pretreatment with antioxidants. Most importantly, this cardiac dysfunction was almost completely prevented by the overexpression of TFAM, strongly suggesting that TFAM has a protective effect on mtDNA from the oxidative attack and indicating how the integrity of mtDNA is important for survival (unpublished data).

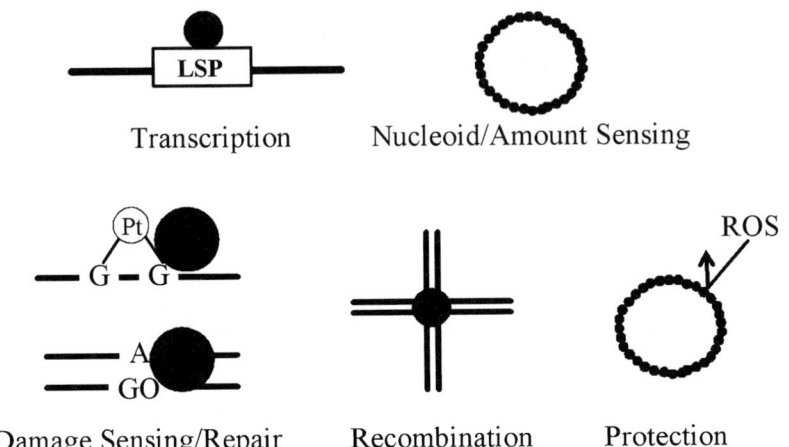

FIGURE 2. Multiple roles of TFAM in maintenance of mtDNA. *Circular ring patterns* represent TFAM. Pt, cisplatin; GO, 8-oxo-guanine; ROS, reactive oxygen species.

In conclusion, mtDNA is essential for individuals to live normally. This crucial genome, however, is very fragile. TFAM plays multiple and critical roles in the maintenance of mtDNA (FIG. 2).

ACKNOWLEDGMENTS

This work was supported, in part, by the Uehara Memorial Foundation, the Naito Foundation, and Grants-in-Aid for Scientific Research from the Ministry of Education, Science, Technology, Sports and Culture of Japan.

REFERENCES

1. KANG, D. *et al.* 1998. Introduction. *In* Mitochondrial DNA Mutations in Aging, Disease and Cancer. K.K. Singh, Ed.: 1–15. Springer-Verlag and R.G. Landes Company. Austin, TX.
2. PARISI, M.A. & D.A. CLAYTON. 1991. Similarity of human mitochondrial transcription factor 1 to high mobility group proteins. Science **252:** 965–969.
3. FISHER, R.P. & D.A. CLAYTON. 1988. Purification and characterization of human mitochondrial transcription factor 1. Mol. Cell. Biol. **8:** 3496–3509.
4. DIFFLEY, J.F. & B. STILLMAN. 1991. A close relative of the nuclear, chromosomal highmobility group protein HMG1 in yeast mitochondria. Proc. Natl. Acad. Sci. USA **88:** 7864–7868.
5. DAIRAGHI, D.J. *et al.* 1995. Addition of a 29 residue carboxyl-terminal tail converts a simple HMG box-containing protein into a transcriptional activator. J. Mol. Biol. **249:** 11–28.
6. FALKENBERG, M. *et al.* 2002. Mitochondrial transcription factors B1 and B2 activate transcription of human mtDNA. Nat. Genet. **31:** 289–294.

7. MCCULLOCH, V. & G.S. SHADEL. 2003. Human mitochondrial transcription factor B1 interacts with the C-terminal activation region of h-mtTFA and stimulates transcription independently of its RNA methyltransferase activity. Mol. Cell. Biol. **23:** 5816–5824.
8. LEE, D.Y. & D.A. CLAYTON. 1996. Properties of a primer RNA-DNA hybrid at the mouse mitochondrial DNA leading-strand origin of replication. J. Biol. Chem. **271:** 24262–24269.
9. OHSATO, T. *et al.* 1999. R-Loop in the replication origin of human mitochondrial DNA is resolved by RecG, a Holliday junction-specific hHelicase. Biochem. Biophys. Res. Commun. **255:** 1–5.
10. SHADEL, G.S. & D.A. CLAYTON. 1997. Mitochondrial DNA maintenance in vertebrates. Annu. Rev. Biochem. **66:** 409–435.
11. KANG, D. *et al.* 1997. In vivo determination of replication origins of human mitochondrial DNA by ligation-mediated polymerase chain reaction. J. Biol. Chem. **272:** 15275–15279.
12. KAI, Y. *et al.* 1999. Mitochondrial DNA replication in human T lymphocytes is regulated primarily at the H-strand termination site. Biochim. Biophys. Acta **1446:** 126–134.
13. BOWMAKER, M. *et al.* 2003. Mammalian mitochondrial DNA replicates bidirectionally from an initiation zone. J. Biol. Chem. **278:** 50961–50969.
14. HOLT, I.J. *et al.* 2000. Coupled leading- and lagging-strand synthesis of mammalian mitochondrial DNA. Cell **100:** 515–524.
15. MIYAKO, K. *et al.* 1999. 1-Methyl-4-phenylpyridinium ion (MPP+) selectively inhibits the replication of mitochondrial DNA. Eur. J. Biochem. **259:** 412–418.
16. MIYAKO, K. *et al.* 1997. The content of intracellular mitochondrial DNA is decreased by 1-methyl-4-phenylpyridinium ion (MPP+). J. Biol. Chem. **272:** 9605–9608.
17. UMEDA, S. *et al.* 2000. The D-loop structure of human mitochondrial DNA is destabilized directly by 1-methyl-4-phenylpyridinium ion (MPP+), a parkinsonism-causing toxin. Eur. J. Biochem. **267:** 200–206.
18. IWAASA, M. *et al.* 2002. 1-Methyl-4-phenylpyridinium ion (MPP+), a toxin that can cause parkinsonism, alters branched structures of DNA. J. Neurochem. **82:** 30–37.
19. TAKAMATSU, C. *et al.* 2002. Regulation of mitochondrial D-loops by transcription factor A and single-stranded DNA-binding protein. EMBO Rep. **3:** 451–456.
20. SHEN, E.L. & D.F. BOGENHAGEN. 2001. Developmentally-regulated packaging of mitochondrial DNA by the HMG-box protein mtTFA during Xenopus oogenesis. Nucleic Acids Res. **29:** 2822–2828.
21. ALBRING, M. *et al.* 1977. Association of a protein structure of probable membrane derivation with HeLa cell mitochondrial DNA near its origin of replication. Proc. Natl. Acad. Sci. USA **74:** 1348–1352.
22. CARON, F. *et al.* 1979. Characterization of a histone-like protein extracted from yeast mitochondria. Proc. Natl. Acad. Sci. USA **76:** 4265–4269.
23. POTTER, D.A. *et al.* 1980. DNA-protein interactions in the *Drosophila melanogaster* mitochondrial genome as deduced from trimethylpsoralen crosslinking patterns. Proc. Natl. Acad. Sci. USA **77:** 4118–4122.
24. DEFRANCESCO, L. & G. ATTARDI. 1981. In situ photochemical crosslinking of HeLa cell mitochondrial DNA by a psoralen derivative reveals a protected region near the origin of replication. Nucleic Acids Res. **9:** 6017–6030.
25. Miyakawa, I. *et al.* 1987. Isolation of morphologically intact mitochondrial nucleoids from the yeast, *Saccharomyces cerevisiae*. J. Cell Sci. **88:** 431–439.
26. SASAKI, N. *et al.* 2003. Glom is a novel mitochondrial DNA packaging protein in Physarum polycephalum and causes intense chromatin condensation without suppressing DNA functions. Mol. Biol. Cell **14:** 4758–4769.
27. KANKI, T. *et al.* 2004. Mitochondrial nucleoid and transcription factor A. Ann. N.Y. Acad. Sci. **1011:** 61–68.
28. ALAM, T.I. *et al.* 2003. Human mitochondrial DNA is packaged with TFAM. Nucleic Acids Res. **31:** 1640–1645.
29. LARSSON, N.G. *et al.* 1998. Mitochondrial transcription factor A is necessary for mtDNA maintenance and embryogenesis in mice. Nat. Genet. **18:** 231–236.

30. MATSUSHIMA, Y. *et al.* 2003. Functional domains of chicken mitochondrial transcription factor A for the maintenance of mitochondrial DNA copy number in lymphoma cell line DT40. J. Biol. Chem. **278:** 31149–31158.
31. EKSTRAND, M.I. *et al.* 2004. Mitochondrial transcription factor A regulates mtDNA copy number in mammals. Hum. Mol. Genet. **13:** 935–944.
32. KANG, D. & N. HAMASAKI. 2003. Mitochondrial oxidative stress and mitochondrial DNA. Clin. Chem. Lab. Med. **41:** 1281–1288.
33. KANG, D. & N. HAMASAKI. 2002. Maintenance of mitochondrial DNA integrity: repair and degradation. Curr. Genet. **41:** 311–322.
34. BECKMAN, K.B. & B.N. AMES. 1996. Detection and quantification of oxidative adducts of mitochondrial DNA. Methods Enzymol. **264:** 442–453.
35. AMES, B.N. *et al.* 1993. Oxidants, antioxidants, and the degenerative diseases of aging. Proc. Natl. Acad. Sci. USA **90:** 7915–7922.
36. KANG, D. *et al.* 1983. Kinetics of superoxide formation by respiratory chain NADH-dehydrogenase of bovine heart mitochondria. J. Biochem. **94:** 1301–1306.
37. ETO, Y. *et al.* 1992. Succinate-dependent lipid peroxidation and its prevention by reduced ubiquinone in beef heart submitochondrial particles. Arch. Biochem. Biophys. **295:** 101–106.
38. PAPA, S. 1996. Mitochondrial oxidative phosphorylation changes in the life span. Molecular aspects and physiopathological implications. Biochim. Biophys. Acta **1276:** 87–105.
39. BANDY, B. & A.J. DAVISON. 1990. Mitochondrial mutations may increase oxidative stress: implication for carcinogenesis and aging? Free Radical Biol. Med. **8:** 523–539.
40. SINGER, T.P. & R.R. RAMSAY. 1990. Mechanism of the neurotoxicity of MPTP. An update. FEBS Lett. **274:** 1–8.
41. KHRAPKO, K. *et al.* 1997. Mitochondrial mutational spectra in human cells and tissues. Proc. Natl. Acad. Sci. USA **94:** 13798–13803.
42. YOSHIDA, Y. *et al.* 2002. Human mitochondrial transcription factor A binds preferentially to oxidatively damaged DNA. Biochem. Biophys. Res. Commun. **295:** 945–951.
43. YOSHIDA, Y. *et al.* 2003. p53 physically interacts with mitochondrial transcription factor A and differentially regulates binding to damaged DNA. Cancer Res. **63:** 3729–3734.
44. MARCHENKO, N.D. *et al.* 2000. Death signal-induced localization of p53 protein to mitochondria. A potential role in apoptotic signaling. J. Biol. Chem. **275:** 16202–16212.
45. SANSOME, C. *et al.* 2001. Hypoxia death stimulus induces translocation of p53 protein to mitochondria. Detection by immunofluorescence on whole cells. FEBS Lett. **488:** 110–115.
46. P-OHLER, J.R. *et al.* 1998. HMG box proteins bind to four-way DNA junctions in their open conformation. EMBO J. **17:** 817–826.
47. MACALPINE, D.M. *et al.* 1998. The high mobility group protein Abf2p influences the level of yeast mitochondrial DNA recombination intermediates in vivo. Proc. Natl. Acad. Sci. USA **95:** 6739–6743.
48. OHNO, T. *et al.* 2000. Binding of human mitochondrial transcription factor A, an HMG box protein, to a four-way DNA junction. Biochem. Biophys. Res. Commun. **271:** 492–498.
49. THYAGARAJAN, B. *et al.* 1996. Mammalian mitochondria possess homologous DNA recombination activity. J. Biol. Chem. **271:** 27536–27543.
50. KAJANDER, O.A. *et al.* 2001. Prominent mitochondrial DNA recombination intermediates in human heart muscle. EMBO Rep. **2:** 1007–1012.
51. TANG, Y. *et al.* 2000. Maintenance of human rearranged mitochondrial DNAs in long-term cultured transmitochondrial cell lines. Mol. Biol. Cell **11:** 2349–2358.
52. SUEMATSU, N. *et al.* 2003. Oxidative stress mediates tumor necrosis factor-alpha-induced mitochondrial DNA damage and dysfunction in cardiac myocytes. Circulation **107:** 1418–1423.
53. IDE, T. *et al.* 2001. Mitochondrial DNA damage and dysfunction associated with oxidative stress in failing hearts following myocardial infarction. Circ. Res. **88:** 529–535.

Mitochondrial Genome Instability and mtDNA Depletion in Human Cancers

HSIN-CHEN LEE,[a] PEN-HUI YIN,[a] JIN-CHING LIN,[b] CHENG-CHUNG WU,[c] CHIH-YI CHEN,[d] CHEW-WUN WU,[e] CHIN-WEN CHI,[a,f] TSENG-NIP TAM,[g] AND YAU-HUEI WEI[h]

Departments of Pharmacology[a] and Biochemistry,[h] School of Medicine, National Yang-Ming University, Taiwan 112, Republic of China

[b]*Department of Radiation Oncology, Section of General Surgery[c] and of Thoracic Surgery,[d] Department of Surgery, Taichung Veterans General Hospital, Taiwan 407, Republic of China*

Departments of Surgery[e] and Medical Research and Education,[f] Taipei Veterans General Hospital, Taiwan 112, Republic of China

[g]*Section of Gastroenterology, Department of Internal Medicine, Cheng Hsin Rehabilitation Medical Center, Taipei, Taiwan 112, Republic of China*

> ABSTRACT: An increase in the rate of glycolysis is one of the metabolic alterations in most cancer cells. However, the role of alterations in mitochondrial function and mitochondrial DNA (mtDNA) in carcinogenesis still remains unclear. In this study, we analyzed the nucleotide sequence of the D-loop and the copy number of mtDNA in 54 hepatocellular carcinomas (HCCs), 31 gastric, 31 lung, and 25 colorectal cancers as well as their corresponding non-tumorous tissues. The results revealed that 42.6% (23/54) of the HCCs, 51.6% (16/31) of the gastric cancers, 22.6% (7/31) of the lung cancers, and 40.0% (10/25) of the colorectal cancers harbored mutation(s) in the D-loop of mtDNA. The mtDNA mutations in 43.5% (10/23) of the HCCs, 62.5% (10/16) of the gastric cancers, 57.1% (4/7) of the lung cancers, and 90.0% (9/10) of the colorectal cancers were changes in the mononucleotide or dinucleotide repeats, deletions, or multiple insertions. Moreover, we found that there is a significant decrease in mtDNA copy number in 57.4% (31/54) of the HCCs, 54.8% (17/31) of the gastric cancers, 22.6% (7/31) of the lung cancers, and 28.0% (7/25) of the colorectal cancers compared with the corresponding non-tumorous tissues. It is noteworthy that the incidence of somatic mutations in the D-loop of mtDNA in the cancers of later stages was higher than that of the early-stage cancers. Taken together, our findings suggest that instability in the D-loop region of mtDNA, together with the decrease in mtDNA copy number, is involved in the carcinogenesis of human cancers.
>
> KEYWORDS: hepatocellular carcinoma; gastric cancer; lung cancer; colorectal cancer; mitochondrial DNA; somatic mutation; copy number

---

Address for correspondence: Dr. Hsin-Chen Lee, Department of Pharmacology, School of Medicine, National Yang-Ming University, Taiwan 112, Republic of China. Voice: +886-2-28267327; fax: +886-2-28264372.
    hclee2@ym.edu.tw

INTRODUCTION

An increase in the capacity of glycolysis accompanied by defective respiration is one of the common alterations in bioenergetic function in many cancer cells. As early as 75 years ago, Warburg hypothesized that cancer cells may have impaired mitochondrial function and that this alteration would result in an elevated rate of glycolysis.[1] In the past few years, it has been demonstrated that glycolysis could be activated by hypoxia and modulated by alteration in the expression of oncogenes or tumor suppressor genes.[2] However, the presumed impairment of the bioenergetic function of mitochondria has never been established in human cancers.

Mitochondria are responsible for the supply of most of the energy needed by the human cell, and they play a key role in the initiation and execution of apoptosis.[3] Moreover, mitochondria are the major intracellular producer of reactive oxygen species (ROS) and are also subject to direct attack by ROS in the organelles of mammalian cells.[3] Impairment in mitochondrial respiratory function not only reduces the supply of energy, which may prevent energy-dependent apoptosis, but also enhances ROS production that may induce mutation and oxidative damage to mitochondrial DNA (mtDNA). It has been proposed that accumulation of mtDNA mutations and alteration in the execution of apoptosis contribute to the onset and progression of various neurodegenerative diseases.[3] Recently, it was reported that the expression level of the β subunit of F_1-ATPase required for mitochondrial ATP synthesis is decreased in human cancers, including liver, gastric, lung, and colorectal cancers.[4,5] These findings suggest that there are common mechanisms by which the bioenergetic function of mitochondria is altered in cancer cells. However, little information is available as to how qualitative and quantitative changes in mtDNA are involved in the alteration of mitochondrial oxidative phosphorylation in cancer tissues.

Human mtDNA is a 16.5-kb circular double-stranded DNA molecule, which is present at a high copy number per cell; the number varies widely with the cell type. Human mtDNA contains genes coding for 13 polypeptides involved in respiration and oxidative phosphorylation, 2 rRNAs, and a set of 22 tRNAs that are essential for the protein synthesis in mitochondria.[6] In addition, mtDNA contains a noncoding region that includes a unique displacement loop (D-loop), which controls replication and transcription of mtDNA.[6] It has been established that the mitochondrial genome is particularly susceptible to oxidative damage and mutation because of the high rate of ROS generation and inefficient DNA repair system in mitochondria.[7,8] Both inherited and somatic mtDNA mutations have been demonstrated to cause severe degenerative disorders.[8,9] Most mutations occur in the tRNA genes and structural genes of mtDNA in the affected tissues of patients with inherited mitochondrial disorders. By contrast, many of the mtDNA mutations found in human cancers are located in the D-loop region, which has been shown to be a "hot spot" for point mutations in many human cancers.[10–14] Because the D-loop is involved in the control of replication and transcription of mtDNA, mutations in this region might cause a decrease in the copy number and/or alteration in gene expression of the mitochondrial genome.[15]

To investigate qualitative and quantitative alterations in mtDNA of human cancer, we analyzed the nucleotide sequence of the D-loop and determined the copy number of mtDNA in the cancerous tissues and corresponding non-cancerous tissues from patients with heptocellular carcinoma (HCC), gastric cancers, lung cancers, and colorectal cancers. The consequence of mtDNA mutations and the mechanisms by which the copy number of mtDNA is changed in human cancers are discussed.

MATERIALS AND METHODS

Collection of Human Cancer Tissues and DNA Extraction

Fifty-four HCCs, 31 lung cancers, and corresponding non-tumorous tissues were obtained with consent from patients and were histologically confirmed at Taichung Veterans General Hospital. Thirty-one gastric cancers and corresponding non-tumorous tissues were obtained with consent from patients at Taipei Veterans General Hospital. In addition, 25 colorectal cancers and corresponding non-tumorous tissues were obtained with consent from patients at Cheng Hsin Rehabilitation Medical Center in Taipei. All the tissues were kept in liquid nitrogen immediately after surgical resection according to a protocol approved by the medical ethics committee for conducting human research at the three hospitals. Total DNA of these tissues was extracted by proteinase K/SDS lysis followed by phenol/chloroform extraction as previously described.[16] The DNA was dissolved in doubly distilled water and frozen at −30°C until use.

Direct Sequencing for Determination of Somatic Mutation in the mtDNA D-Loop

The somatic mutation in the D-loop region of mtDNA was analyzed by direct sequencing of the PCR products. The primer pairs L16190 (nucleotide position [np] 16190–16209, 5′-CCCCATGCTTACAAGCAAGT-3′) and H602 (np 602–583, 5′-GCTTTGAGGAGGTAAGCTAC-3′) were used for the amplification of a 982-bp DNA fragment from the D-loop region of mtDNA. All polymerase chain reaction (PCR) products were purified and sequenced with the AmpliCycle sequencing kit (Perkin-Elmer/Cetus) according to the instructions of the manufacturer.

Determination of mtDNA Copy Number

A competitive PCR method was used for determining the copy number of mtDNA.[16,17] In brief, a known amount of the internal DNA standard (a competitor DNA fragment that contained truncated ND1 and β-actin gene) was added with the DNA sample into the PCR reaction mixture. The PCR was carried out for 25 cycles in a 50-mL reaction mixture containing 200 ng DNA, 200 μM of dNTP, 40 pmol of each primer, 1.0 U of *Taq* DNA polymerase, 50 mM KCl, 1.5 mM $MgCl_2$, 10 mM Tris-HCl (pH 9.0), 0.1% Triton X-100, and 0.01% (w/v) gelatin. The sequences of the primers are β-actin forward primer: 5′-CATGTGCAAGGCCGGCTTCG-3′; β-actin reverse primer: 5′-CTGGGTCATCTTCTCGCGGT-3′; ND1 L-strand primer: 5′-TCTCACCATCGCTCTTCTAC-3′; and ND1 H-strand primer: 5′-TTGGTCTCTGCTAGTGTGGA-3′. The PCR cycles consisted of denaturation at 94°C for 15 s, annealing at 58°C for 15 s, and primer extension at 72°C for 40 s. The PCR products were separated by electrophoresis on a 3% agarose gel at 100 V for 40 min and detected under UV transillumination after ethidium bromide staining. The band intensities of the PCR products of the target and internal standard DNAs were ana-

TABLE 1. Somatic mutations in the D-loop region of mtDNA of human cancers

Nucleotide position	Type of mutation	HCC	Gastric cancer	Lung cancer	Colorectal cancer	Homoplasmy
16260	C→T		1			Yes
16298	C→T	1[a]				Yes
16300	A→G/T	1				heteroplasmy
16390	G→A/G			1		heteroplasmy
16391	G→A/G			1		heteroplasmy
16399	A→A/G		1			heteroplasmy
16438	G→G/A		1[d]			heteroplasmy
16519	C→T	1	1[d]			Yes
70	G→A	1				Yes
72	T→C	4 + 1[b]				Yes
94	G→A	1				Yes
146	T→C	1				Yes
152	T→C	1[a]				Yes
189	A→G	1	1		1[e]	Yes
200	G→A				1	Yes
204	T/C→T/C		1			heteroplasmy
205	G→A		1			Yes
303	7C→8C	1				heteroplasmy
303	8C→7C	2	2			Yes
303	8C→7C		1	1		heteroplasmy
303	8C→9C	1	1		3	Yes
303	8C→9C	1	4	1		heteroplasmy
303	9C→7C	1			1	Yes
303	9C→8C	1			4	Yes
303	9C→8C	1	1 + 1[d]	1		heteroplasmy
303	9C→10C				1[e]	Yes
303	9C→10C			1		heteroplasmy
303	9C→13C	1[b]				heteroplasmy
303	9C→14C	1				heteroplasmy
417	G→A	1[c]				Yes
514	5CA→4CA	1[c]				Yes
298–348/306–356	deletion	1		1		heteroplasmy
311/568	insertions		1			heteroplasmy

a, b, c, d e: The indicated mtDNA mutations were detected in the same tumor samples with identical superscripts.

lyzed. The ratio between the intensities of the two DNA bands was used to calculate the relative DNA content of the target gene. The DNA content of the ND1 gene was normalized with that of the β-actin gene in nuclear DNA to calculate the copy number of mtDNA in each sample.

RESULTS

Somatic Mutation in the D-Loop of mtDNA of Human Cancers

To search for tumor-associated somatic mutation in the D-loop of mtDNA, we directly analyzed the nucleotide sequences in the D-loop region of mtDNA in tumors and in corresponding non-tumorous tissues. The results showed that 42.6% (23/54)

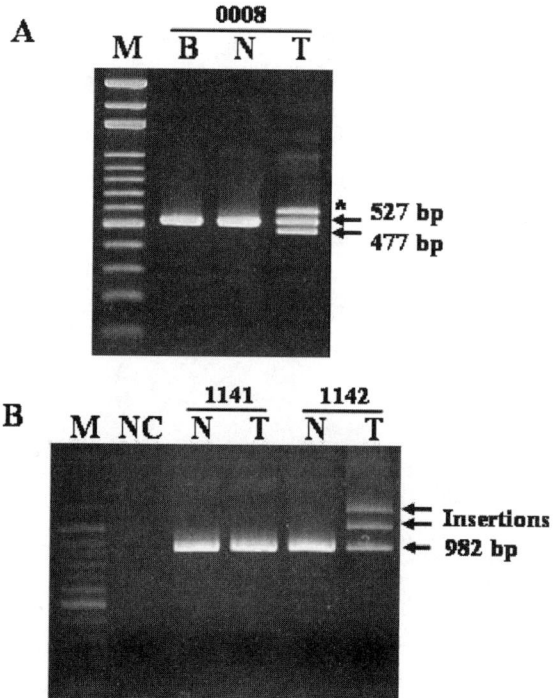

FIGURE 1. The deletion and insertion of mtDNA in human cancers. Human mtDNA with a deletion or insertion in the D-loop was detected by the PCR technique described in MATERIALS AND METHODS. Using the primers L76 (5'-CACGCGAATAGCATTGC-GAGACGCTG-3') and H602, the 527-band 477-bp PCR fragments amplified from the wild-type mtDNA and the 50 bp-deleted mtDNA, respectively, in the tumor part (T) and in the surrounding non-tumor part (N) of lung cancer patient 0008 (**A**) were separated on a 1.5% agarose gel and detected under UV transillumination after ethidium bromide staining. M, 100-bp DNA ladder; B, blood sample; *, one insertion in the D-loop region of mtDNA in the case. Multiple insertions in the D-loop region of mtDNA in the gastric cancer of patient 1142 (**B**) were detected by using the primers L16190 and H602. NC, negative control.

of the HCCs, 51.6% (16/31) of the gastric cancers, 22.6% (7/31) of the lung cancers, and 40.0% (10/25) of the colorectal cancers harbored somatic mutation(s) in the D-loop region of mtDNA (TABLE 1). Moreover, most of the mtDNA mutations in HCCs and colorectal cancers were homoplasmic, but those in the gastric cancers and lung cancers were mostly heteroplasmic. Interestingly, 43.5% (10/23) of the mtDNA mutations detected in the HCCs, 62.5% (10/16) in the gastric cancers, 57.1% (4/7) in the lung cancers, and 90.0% (9/10) in the colorectal cancers were alterations in the mononucleotide repeat located in the polycytidine stretch between np 303–309 of mtDNA (TABLE 1). Furthermore, one HCC contained a change in the dinucleotide repeat (CA)n at np 514; and two cancers (one HCC and one lung cancer) carried a specific 50-bp deletion (FIG. 1A), which was located between np 298/306 and 348/356, and the breakpoints were flanked by a 9-bp direct repeat of 5'-CCAAACCCC-3'.[17] A gastric cancer sample harbored multiple insertions located between np 311 and np 568 of mtDNA (FIG. 1B).

Depletion of mtDNA in Human Cancers

Using a competitive PCR method, we determined the mtDNA copy number of tumors and of corresponding non-tumorous tissues. The results revealed that 57.4% (31/54) of the HCCs, 54.8% (17/31) of the gastric cancers, 22.6% (7/31) of the lung

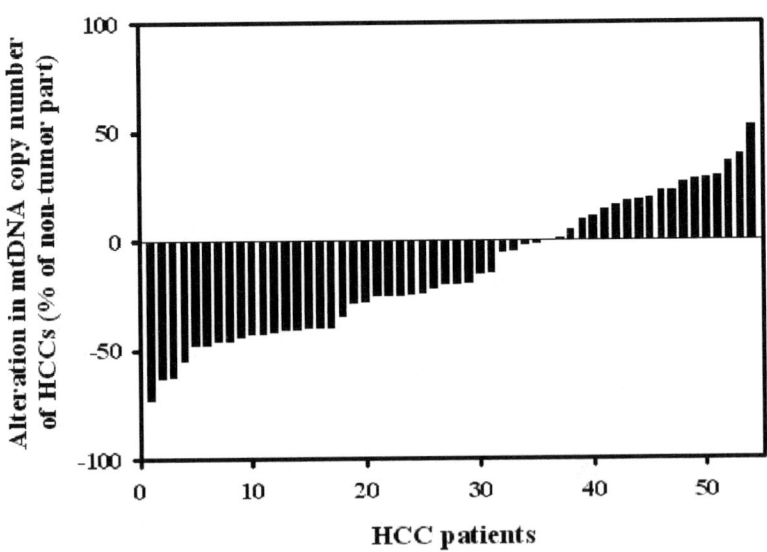

FIGURE 2. Alteration of the copy number of mtDNA in hepatocellular carcinomas. The copy numbers of mtDNA in 54 HCCs and their corresponding non-tumorous tissues were determined by a competitive PCR method described in MATERIALS AND METHODS. The alteration of the copy number of mtDNA in each HCC sample was indicated as the relative percentage of the copy number of mtDNA in the corresponding non-tumorous tissue. A negative value represents a decrease in the copy number of mtDNA in the cancer, while a positive value represents an increase in the mtDNA copy number.

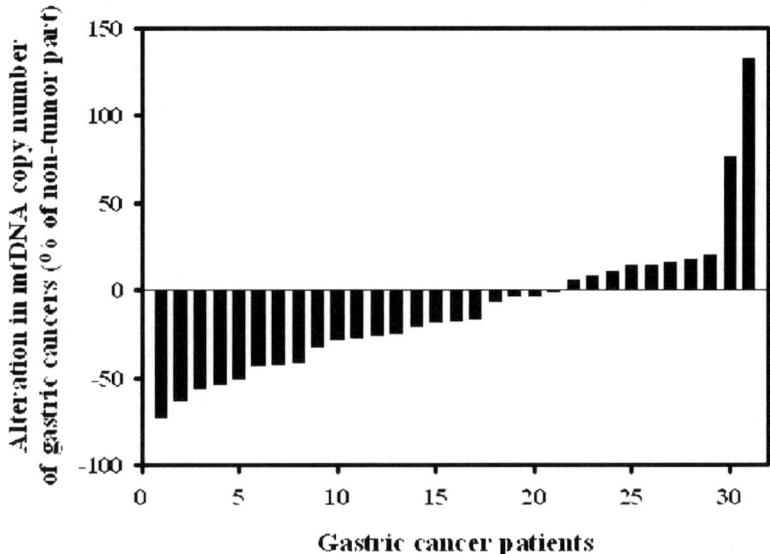

FIGURE 3. Alteration of the copy number of mtDNA in gastric cancers. The copy numbers of mtDNA in 31 gastric carcinomas and their corresponding non-tumorous tissues were determined by a competitive PCR method described in MATERIALS AND METHODS. An alteration of the copy number of mtDNA in each gastric cancer sample was indicated as the relative percentage of the copy number of mtDNA in the corresponding non-tumorous part of the tissue. A negative value represents a decrease in the copy number of mtDNA in the cancer, while a positive value represents an increase in the mtDNA copy number.

cancers, and 28.0% (7/25) of the colorectal cancers had significant decreased mtDNA copy number (below 90%) as compared with the corresponding non-tumorous tissues (FIGS. 2–5). By contrast, 27.8% (15/54) of the HCCs, 22.6% (7/31) of the gastric cancers, 48.4% (15/31) of the lung cancers, and 40.0% (10/25) of the colorectal cancers had significantly higher mtDNA copy numbers (above 110%) as compared with the corresponding non-tumorous tissues (FIGS. 2–5).

Higher Incidence of Mutation in the D-Loop of mtDNA in Cancers at Later Stages

After analysis of the incidence of somatic mutation in the D-loop and depletion of mtDNA in the cancers at various stages, we found that the incidence of somatic mutation of mtDNA in all four kinds of cancers at stages III and IV was higher than those at stages I and II (TABLE 2). In addition, the incidence of mtDNA depletion in colorectal cancers at stages III and IV was higher than those at stages I and II, respectively. However, the increase did not occur in HCCs, gastric cancers, or lung cancers (TABLE 2).

TABLE 2. Incidence of mutation and depletion of mtDNA in human cancers at different tumor stages

TNM stages	Incidence of mtDNA alteration in cancers	
	Somatic mutation in the D-loop (%)	Depletion of mtDNA (%)
HCC		
I + II	9/25 (36)	15/25 (60.0)
III + IV	14/29 (48.3)	16/29 (55.2)
Gastric cancer		
I + II	2/5 (40.0)	3/5 (60.0)
III + IV	14/26 (53.8)	14/26 (53.8)
Lung cancer		
I + II	4/21 (19.0)	5/21 (23.8)
III + IV	3/10 (30.0)	2/10 (20.0)
Colorectal cancer		
I + II	3/11 (27.3)	2/11 (18.2)
III + IV	7/14 (50.0)	5/14 (35.7)

DISCUSSION

Our results revealed that among a total of 141 Taiwanese cancer patients (including 54 HCCs, 31 gastric cancers, 31 lung cancers, and 25 colorectal cancers) examined, 39.7% of the examined cancerous tissues carried mutation(s) in the D-loop of mtDNA. The alterations of mononucleotide repeat located in the polycytidine stretch at np 303–309 were the most frequent ones in these cancers (TABLE 1). The change in the dinucleotide repeat (CA)n at np 514, the specific 50-bp deletion, and multiple insertions were detected in some cancerous tissues. Moreover, the incidence of somatic mutation in the D-loop of mtDNA in these cancers was found to increase with the cancer stages. These findings have supported the notion that the alterations are a result of mtDNA instability in most cancers during the carcinogenesis process.

It is worth noting that the most frequent changes in mtDNA were located in the polycytidine stretch, as well as the specific 50-bp deletion, and multiple insertions were flanked by the sequences containing polycytidine run in the D-loop region. It was recently demonstrated that the D-loop was highly susceptible to oxidative damage as compared with the other regions of mtDNA.[18] The extensive oxidative damage to the polycytidine sequences may result in slipping and/or misincorporation during replication or repair of mtDNA by mitochondrial DNA polymerase. The DNA polymerase γ is the sole DNA polymerase in mitochondria and is responsible for replication and repair of mtDNA. It has been demonstrated that mitochondrial DNA polymerase γ is a target of oxidative damage.[19] Oxidative stress can significantly inhibit DNA polymerase activity and lower its DNA binding efficiency,

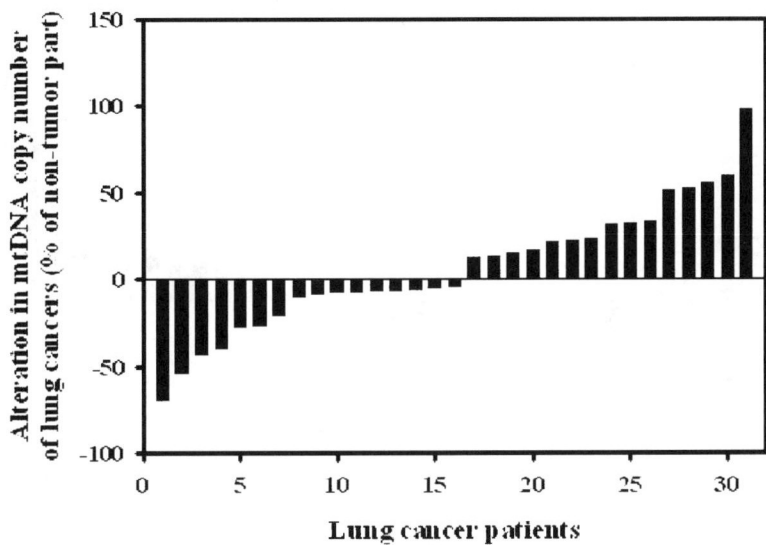

FIGURE 4. Alteration of the copy number of mtDNA in lung cancers. The copy numbers of mtDNA in 31 lung carcinomas and their corresponding non-tumorous tissues were determined by a competitive PCR method described in MATERIALS AND METHODS. An alteration of the copy number of mtDNA in each lung cancer sample was indicated as the relative percentage of the copy number of mtDNA in the corresponding non-tumorous part. The negative value represents the decrease in the copy number of mtDNA in the cancer, while a positive value represents an increase in the copy number of mtDNA.

which may result in less efficient replication and repair of mtDNA.[19] Therefore, oxidative damage to mitochondrial DNA polymerase γ may also contribute to the alteration in the number of polycytidine during replication or repair of mtDNA (FIG. 6).

In addition, we found that 44.0% of the cancers examined in this study had a lower copy number of mtDNA as compared with the corresponding non-tumorous tissues, but 33.3% of the cancers had higher mtDNA content. It was recently reported that point mutation in the D-loop region is associated with a decrease in mtDNA copy number in HCC.[17] Due to the role of the D-loop in controlling both replication and transcription of mtDNA,[15] oxidative damage and/or somatic mutations in the region might interfere with replication and maintenance of mtDNA. Moreover, it was reported that a deregulation in the expression of mitochondrial single-stranded DNA binding protein (mtSSB) is associated with the decrease in mtDNA copy number of HCC.[20] Thus, oxidative damage to the D-loop of mtDNA and/or less efficiency in the DNA replication could be involved in the instability and depletion of mtDNA in the cancerous tissues (FIG. 6).

The mitochondrial genome is more susceptible to DNA damage and consequently acquires mutations at a higher rate than does nuclear DNA because of the lack of histone protection of mtDNA and the high rate of ROS generation and limited DNA repair mechanisms in human mitochondria. It is plausible that when the frequency

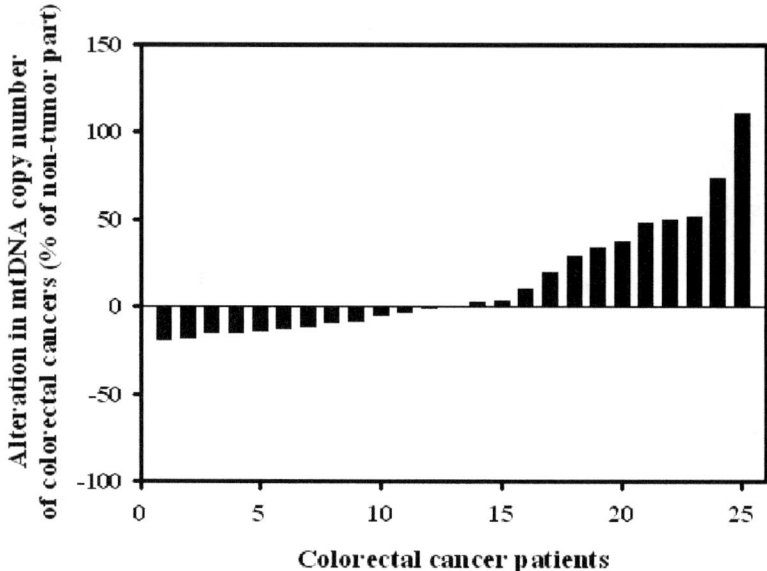

FIGURE 5. Alteration of the copy number of mtDNA in colorectal cancers. The copy numbers of mtDNA in 25 colorectal carcinomas and their corresponding non-tumorous tissues were determined by a competitive PCR method described in MATERIALS AND METHODS. An alteration of the copy number of mtDNA in each colorectal cancer sample was indicated as a relative percentage of the copy number of mtDNA in the corresponding non-tumorous part. A negative value represents the decrease in the copy number of mtDNA in the cancer, while a positive value represents an increase in the mtDNA copy number.

of occurrence and proportion of mtDNA mutations are increased in somatic tissues, nuclear DNA may be subjected to damage or mutation, although at a lower level. Thus, mtDNA mutation and oxidative damage can be used as an index of genome instability in cancer cells. These observations are consistent with the concept that tumor cells are exposed to a higher and persistent oxidative stress compared with the adjacent normal tissues.[21] Chronic oxidative stress may induce DNA damage such as base modifications and strand breaks, which may lead to further mtDNA mutation and chromosomal aberrations that are commonly seen in cancers. Accumulation of oxidative damage and sequence variations in nuclear DNA and mtDNA may ultimately lead to uncontrolled cell growth and malignant transformation of affected tissue cells.

In the present study, a dramatic decrease in the copy number of mtDNA was found in most HCCs (57.4%, FIG. 2) and gastric cancers (54.8%, FIG. 3), but only in 22.6% of lung cancers (FIG. 4), and in 28.0% of colorectal cancers (FIG. 5), respectively. The incidence of the decrease in mtDNA content of these cancers was not increased with their clinical stages. These results indicate that alterations (an increase or decrease) in the copy number of mtDNA occur in the same type of carcinoma, and some tissue-specific factors are involved in reducing the mtDNA copy number of the

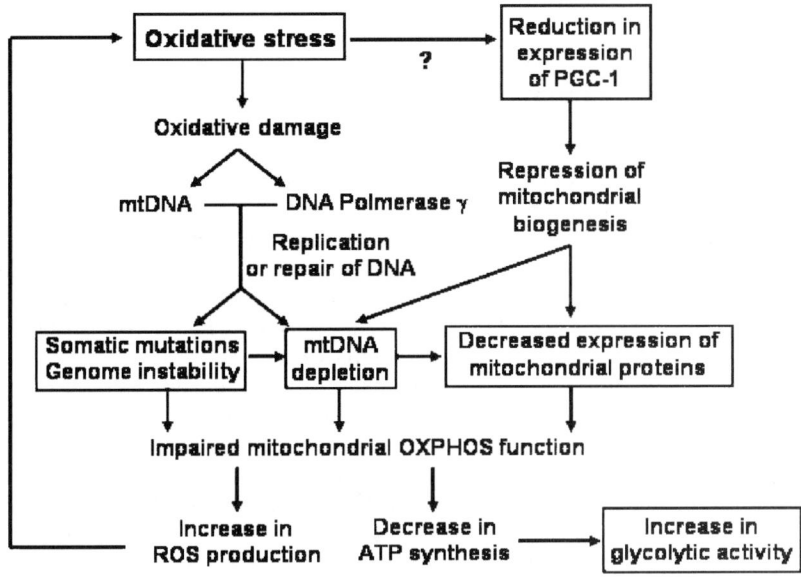

FIGURE 6. Alterations in mitochondria of cancer cells and their possible linkages. Extensive oxidative stress in cancer cells may result in oxidative damages to mitochondrial DNA (mtDNA) and to DNA polymerase γ, which both contribute to somatic mutation and genome instability in mtDNA as well as mtDNA depletion. Moreover, reduced expression of the PPARγ coactivator-1 (PGC-1) could lead to a repression in mitochondrial biogenesis, thereby causing mtDNA depletion and decreased expression of mitochondrial proteins. Mitochondrial genome instability and mtDNA depletion, as well as mitochondrial biogenesis repression, result in the impaired mitochondrial oxidative phosphorylation (OXPHOS) function, which might cause an increase in the production of reactive oxygen species (ROS) and a decrease in the synthesis of ATP. The increased ROS production enhances oxidative stress, and the decreased ATP synthesis stimulates glycolytic activity in cancer cells. The alterations described in the boxes have been observed in several human cancers.

liver and gastric cancers. However, no dominant factors are involved in the alteration of the mtDNA copy number of the lung or colorectal tissues during carcinogenesis.

During the past decade, a number of factors involved in the replication and maintenance of mtDNA, as well as in biogenesis of mitochondria, have been characterized.[22] A transcriptional coactivator, peroxisome proliferator-activated receptor γ coactivator-1 (PGC-1) has been demonstrated to interact with NRF-1 and NRF-2 at the upstream level in the regulation of mitochondrial biogenesis in animal tissues.[23] Recently, the expression of PGC-1 was found to decrease in human HCC, breast, and colon cancers.[20,24,25] It is possible that carcinogenesis of these tissues could be associated with alterations in the quantity and activity of these regulatory factors. The depletion of mtDNA may be the result of the repression of mitochondrial biogenesis (FIG. 6). The observations are consistent with the findings that there is a general downregulation of mitochondrial components and a repression of the program of mitochondrial proliferation in liver cancers.[4,20,26] Our results suggest that in most HCCs and gastric cancers, as well as some other carcinomas (lung and colorectal cancers), carcinogenesis is accompanied by an alteration in mitochondrial biogenesis and a repression of mtDNA replication.[5]

In contrast, we observed an increase in the copy number of mtDNA in 48.4% lung cancers, 40.0% colorectal cancers, 27.8% HCCs, and 22.6% gastric cancers. The increase in mtDNA copy number may result from the compensation for the increased energy need and/or the decreased capacity of oxidative phosphorylation in these cancers. It is possible that the increased mitochondrial biogenesis might be similar to the development of ragged-red fibers as a result of overproliferation of mitochondria in muscle of patients suffering from mitochondrial myopathies.[8,9] Indeed, it was recently reported that the rate of ATP synthesis is decreased in thyroid oncocytoma although the numbers of mitochondria and mtDNA are increased.[27] These are consistent with the previous observations in human kidney and colon cancers[4] that, although there is a selective decrease in the expression of some proteins involved in mitochondrial respiratory function, it may not affect the replication of mtDNA. Thus, the increase in the mtDNA copy number in cancer cells may be a result of the compensation for the mitochondrial dysfunction and energy shortage. Recently, we have demonstrated that oxidative stress can cause an increase in mitochondria and mtDNA content in human cells.[28] Further study is necessary to establish the role of the increase of mtDNA in cancer biology. It is possible that elevated ROS production from impaired mitochondrial respiration might be involved in the upregulation of the copy number of mtDNA in carcinomas.

Cancer cells possess mitochondrial alterations that may contribute to resistance to apoptosis and/or cancer development. It has been known since the 1930s that cancerous cells exhibit enhanced glycolytic ATP generation and decreased oxidative phosphorylation, even under normal oxygen tension—a phenomenon known as the Warburg effect.[1] Moreover, it was reported that many cancer cells have a higher mitochondrial transmembrane potential than do the control cells.[29] Another common feature of cancer cells is their intrinsic oxidative stress.[21] In the present study, the observed mutations in the D-loop region and the alterations in the copy number of mtDNA were often seen in many cancers, which may play a central role in mitochondrial alterations in cancer cells and/or in the multistage process of carcinogenesis (FIG. 6). Cancer cells have increased glycolytic activity, perhaps reflecting a less efficient ATP generation in mitochondria due to respiration injury and/or mtDNA mutation as well as depletion, thus forcing the cancer cells to increase glycolysis to produce ATP required for cellular functions (FIG. 6). These metabolic alterations might be linked to a change that contributes to the resistance and compromised apoptotic potential of cancer cells.[30] Defects in the execution of apoptosis enable neoplastic cells to survive beyond their intended life span, allowing them time to accumulate genetic alterations that deregulate cell proliferation, increase cell motility, and effect other changes that play an important role in tumor pathogenesis. In addition, a lower coupling efficiency of mitochondrial enzymes may lead to more leakage of electrons from the respiratory enzyme complexes to react with oxygen, forming superoxide anions and other ROS (FIG. 6). The increased oxidative stress in cancer cells may lead to further mutagenesis of both mtDNA and nuclear DNA, activation of proto-oncogenes and some transcription factors, genomic instability, chemotherapy resistance, invasion, and metastasis.[21] Therefore, it is possible that alterations in the sequence and copy number of mtDNA in the cancers play some roles in the carcinogenesis and the progression of human cancers as well as resistance to chemotherapy.

In conclusion, we have shown differential alterations in the copy number and mutations of mtDNA that occur in human liver, gastric, lung, and colorectal cancers. Our findings suggest that mtDNA instability, together with the alteration in the copy number of mtDNA, plays an important role in the carcinogenesis of human cancers. The molecular mechanism by which the mtDNA copy number is changed by cancer-associated mutations in the D-loop region awaits further investigation.

ACKNOWLEDGMENTS

This work was supported by grants TCVGH-917106D, TCVGH-927107D, TCVGH-93106D from Taichung Veterans General Hospital, and also by grants NSC 91-2320-B-040-040, NSC 91-2320-B-010-069, NSC 92-2320-B-010-077, NSC 91-2745-B-040-001, NSC 92-2320-B-010-037, and NSC 92-2745-B-040-001 from the National Science Council, Taiwan, Republic of China.

REFERENCES

1. WARBURG, O. 1930. Metabolism of Tumors. Arnold Constable. London.
2. DANG, C.V. & G.L. SEMENZA. 1999. Oncogenic alterations of metabolism. Trends Biochem. Sci. **24:** 68–72.
3. LEE, H.C. & Y.H. WEI. 2000. Mitochondrial role in life and death of the cell. J. Biomed. Sci. **7:** 215.
4. CUEZVA, J.M., M. KRAJEWSKA, M.L. DE HEREDIA, et al. 2002. The bioenergetic signature of cancer: a marker of tumor progression. Cancer Res. **62:** 6674–6681.
5. ISIDORO, A., M. MARTINEZ, P.L. FERNANDEZ, et al. 2004. Alteration of the bioenergetic phenotype of mitochondria is a hallmark of breast, gastric, lung and oesophageal cancer. Biochem. J. **378:** 17–20.
6. ATTARDI, G. & G. SCHATZ. 1988. Biogenesis of mitochondria. Annu. Rev. Cell Biol. **4:** 289–333.
7. CROTEAU, D.L. & V.A. BOHR. 1997. Repair of oxidative damage to nuclear and mitochondrial DNA in mammalian cells. J. Biol. Chem. **272:** 25409–25412.
8. WALLACE, D.C. 1999. Mitochondrial diseases in man and mouse. Science **283:** 1482–1488.
9. WEI, Y.H. & H.C. LEE. 2003. Mitochondrial DNA mutations and oxidative stress in mitochondrial diseases. Adv. Clin. Chem. **37:** 83–128.
10. POLYAK, K., Y. LI, H. ZHU, et al. 1998. Somatic mutations of the mitochondrial genome in human colorectal tumors. Nat. Genet. **20:** 291–293.
11. FLISS, M.S., H. USADEL, O.L. CABALLERO, et al. 2000. Facile detection of mitochondrial DNA mutations in tumors and bodily fluids. Science **287:** 2017–2019.
12. PENTA, J.S., F.M. JOHNSON, J.T. WACHSMAN, et al. 2001. Mitochondrial DNA in human malignancy. Mutat. Res. **488:** 119–133.
13. SANCHEZ-CESPEDES, M., P. PARRELLA, S. NOMOTO, et al. 2001. Identification of a mononucleotide repeat as a major target for mitochondrial DNA alterations in human tumor. Cancer Res. **61:** 7015–7019.
14. NOMOTO, S., K. YAMASHITA, K. KOSHIKAWA, et al. 2002. Mitochondrial D-loop mutations as clonal markers in multicentric hepatocellular carcinoma and plasma. Clin. Cancer Res. **8:** 481–487.
15. SHADEL, G.S. & D.A. CLAYTON. 1997. Mitochondrial DNA maintenance in vertebrates. Annu. Rev. Biochem. **66:** 409–435.

16. LEE, H.C., C.Y. LU, H.J. FAHN, *et al.* 1998. Aging- and smoking-associated alteration in the relative content of mitochondrial DNA in human lung. FEBS Lett. **441:** 292–296.
17. LEE, H.C., S.H. LI, J.C. LIN, *et al.* 2004. Somatic mutations in the D-loop and decrease in the copy number of mitochondrial DNA in human hepatocellular carcinoma. Mutat. Res. **547:** 71–78.
18. MAMBO, E., X. GAO, Y. COHEN, *et al.* 2003. Electrophile and oxidant damage of mitochondrial DNA leading to rapid evolution of homoplasmic mutations. Proc. Natl. Acad. Sci. USA **100:** 1838–1843.
19. GRAZIEWICZ, M.A., B.J. DAY & W.C. COPELAND. 2002. The mitochondrial DNA polymerase as a target of oxidative damage. Nucleic Acids Res. **30:** 2817–2824.
20. YIN, P.H., H.C. LEE, G.Y. CHAU, *et al.* 2004. Alteration of the copy number and deletion of mitochondrial DNA in human hepatocellular carcinoma. Br. J. Cancer **90:** 2390–2396.
21. TOYOKUNI, S., K. OKAMOTO, J. YODOI, *et al.* 1995. Persistent oxidative stress in cancer. FEBS Lett. **358:** 1–3.
22. MORAES, C.T. 2001. What regulates mitochondrial DNA copy number in animal cells? Trends Genet. **17:** 199–205.
23. KNUTTI, D. & A. KRALLI. 2001. PGC-1, a versatile coactivator. Trends Endocrinol. Metab. **12:** 360–365.
24. JIANG, W.G., A. DOUGLAS-JONES & R.E. MANSEL. 2003. Expression of peroxisome-proliferator activated receptor-gamma (PPARg) and the PPARg co-activator, PGC-1, in human breast cancer correlates with clinical outcomes. Int. J. Cancer **106:** 752–757.
25. FEILCHENFELDT, J., M.A. BRUNDLER, C. SORAVIA, *et al.* 2004. Peroxisome proliferators-activated receptors (PPARs) and associated transcription factors in colon cancer: reduced expression of PPARγ-coactivator 1 (PGC-1). Cancer Lett. **203:** 25–33.
26. CUEZVA, J.M., L.K. OSTRONOFF, J. RICART, *et al.* 1997. Mitochondrial biogenesis in the liver during development and oncogenesis. J. Bioenerg. Biomembr. **29:** 365–377.
27. SAVAGNER, F., B. FRANC, S. GUYETANT, *et al.* 2001. Defective mitochondrial ATP synthesis in oxyphilic thyroid tumors. J. Clin. Endocrinol. Metab. **86:** 4920–4925.
28. LEE, H.C., P.H. YIN, C.Y. LU, *et al.* 2000. Increase of mitochondria and mitochondrial DNA in response to oxidative stress in human cells. Biochem. J. **348:** 425–432.
29. CHEN, L.B. 1988. Mitochondrial membrane potential in living cells. Annu. Rev. Cell Biol. **4:** 155–181.
30. CAVALLI, L.R. & B.C. LIANG. 1998. Mutagenesis, tumorigenicity, and apoptosis: are the mitochondria involved? Mutat. Res. **398:** 12–26.

Frequent Occurrence of Mitochondrial Microsatellite Instability in the D-Loop Region of Human Cancers

YUE WANG,[a] VINCENT W. S. LIU,[a] HEXTAN Y. S. NGAN,[a] AND PHILLIP NAGLEY[b]

[a]*Department of Obstetrics & Gynecology, University of Hong Kong, Hong Kong SAR, China*

[b]*Department of Biochemistry and Molecular Biology, Monash University, Clayton, Victoria 3800, Australia*

ABSTRACT: We analyzed the occurrence of mitochondrial microsatellite instability (mtMSI) in 262 pairs of female cancer tissues with the matched normal controls. mtMSI was detected in only 4 of 12 microsatellites found in the mitochondrial genome (3 in the D-loop and 1 in the 12S rRNA gene). Interestingly, 95.6% (87/91) of mtMSI was detected in the D-loop, namely, at nucleotide positions 303–315, 514–523, and 16184–16193. This demonstrates that the D-loop is a hotspot for mtMSI. Different incidences of mtMSI at these three microsatellites were found in the four cancer types (including cervical, endometrial, ovarian, and breast). Together with those mtMSI reported in other studies, the differential occurrence of mtMSI at each of the markers in the D-loop region was observed, indicating that the extent of mtMSI varies from one cancer to another. Although the mechanisms of generation and functional impact of mtMSI are still not clear, the high incidence of mtMSI in the D-loop and its broad distribution in human cancers render it a potential marker for cancer detection.

KEYWORDS: D-loop; mitochondrial DNA; mitochondrial microsatellite instability; cervical carcinoma; endometrial carcinoma; ovarian carcinoma; breast carcinoma

INTRODUCTION

Microsatellites are short stretches of DNA sequence, each containing a motif of nucleotide repeats ranging from one to five DNA bases. The somatic change in length in any such stretch, due to either insertions or deletions of repeating units in a microsatellite, within tumor DNA compared to that of normal tissue is termed microsatellite instability (MSI).[1]

Address for correspondence: Dr. Vincent W.S. Liu, Department of Obstetrics & Gynecology, University of Hong Kong, Room 747, Laboratory Block, Faculty of Medicine, Hong Kong, China. Voice: +852-2819-9367; fax: +852-2816-1947.
vwsliu@hkusua.hku.hk

TABLE 1. Nature and nucleotide positions of mitochondrial microsatellites

Sequence motif	Position[a]	Region
CCCCCCCTCCCCC	303–315	D-loop
$(CA)_5$	514–523	D-loop
CCCCCTCCCC	956–965	12S rRNA
CCCCCCTCCCC	3566–3576	NADH dehydrogenase 1
A_7	4605–4611	NADH dehydrogenase 2
A_7	6692–6698	Cytochrome c oxidase 1
T_7	9478–9484	ATP synthase 6
C_6	12385–12390	NADH dehydrogenase 5
A_8	12418–12425	NADH dehydrogenase 5
$(CCT)_3 (AGC)_3$	12981–12998	NADH dehydrogenase 5
A_7	13231–13237	NADH dehydrogenase 5
CCCCCTCCCC	16184–16193	D-loop

[a]The nucleotide position according to the revised Cambridge reference sequence (http://www.mitomap.org/).

Recently, we and others identified a number of cases of mitochondrial MSI (mtMSI) in mitochondrial DNA (mtDNA) from primary human malignancies.[2–6] To further define the distribution of mtMSI in common female cancers, we extended the mtMSI screening to a series of primary cervical, endometrial, ovarian, and breast cancers. A high incidence of mtMSI in the four cancers was detected. Compared with the incidence of mtMSI in other common human cancers studied, the results emphasize that the prevalence of mtMSI differs among various cancer types, within a distinct microsatellite region. Because of the high frequency of occurrence of mtMSI, they have a good potential for use as markers for cancer detection.

mtMSI ANALYSIS

DNA was isolated from cancer tissues and the matched normal tissues from 262 patients with primary cancer, namely, 71 cervical, 62 endometrial, 78 ovarian, and 51 breast cancers. The mtMSI was classified using polymerase chain reaction (PCR) amplification, followed by polyacrylamide gel electrophoresis, and finally was confirmed by DNA sequencing.[3] In all, 12 microsatellites within the mitochondrial genome were analyzed; the locations were identified by their nucleotide position (np) extremities (TABLE 1).

RESULTS AND DISCUSSION

Four of 12 markers at which mtMSI was detected were located at the following locations: np 303–315, np 514–523, and np 16184–16193 within the D-loop and np

TABLE 2. Incidence of mtMSI starting at np 303, np 514, np 16184, and np 956 in various cancers

Region	Cancer type	Incidence[a] Previous reports[b]	Present data
np 303	Gastric cancer	62.5% (5/8[4])	
	Colorectal cancer	38.6% (7/25[4]; 20/45[11])	
	Gallbladder cancer	38.2% (47/123[12])	
	HNSCC[c]	37.3% (19/51[4])	
	Cervical cancer	35.7% (5/142[2])	23.9% (17/71)
	Hepatocellular cancer	26.1% (10/50[13]; 8/19[14])	
	Prostate cancer	25.0% (0/16[4]; 8/16[15])	
	Bladder cancer	22.6% (4/16[2]; 3/15[4])	
	Breast cancer	20.0% (3/20[2]; 12/64[16]; 4/19[17]; 5/17[4])	13.7% (7/51)
	Endometrial cancer	16.7% (1/6[2])	32.3% (20/62)
	Esophageal cancer	16.0% (1/37[18]; 11/38[6])	
	Lung cancer	16.0% (16/100[4])	
	Glioblastomas	11.8% (2/17[5])	
	Thyroid tumor	11.1% (20/59[19]; 8/166[20]; 5/72[21])	
	Barrett's tumor	10.0% (2/20[22])	
	Astrocytomas	10.0% (1/10[23])	
	Ovarian cancer	0% (0/15[4])	17.9% (14/78)
np 514	Breast cancer	42.5% (17/40[9])	7.8% (4/51)
	Thyroid tumor	25.4% (15/59[19])	
	Gastric cancer	15.6% (5/32[24])	
	Glioblastomas	5.9% (1/17[5])	
	Ovarian cancer		6.4% (5/78)
	Cervical cancer		2.8% (2/71)
	Endometrial cancer		1.6% (1/62)
np 16184	Glioblastomas	23.5% (4/17[5])	
	Astrocytomas	10.0% (1/10[5])	
	Prostate cancer	9.4% (2/16[15]; 1/16[25])	
	Breast cancer	5.3% (1/19[17])	13.7% (7/51)
	Endometrial cancer		14.5% (9/62)
np 956	Endometrial cancer		6.5% (4/62)

[a]The actual number of mtMSI cases relative to the number of cases analyzed is shown in parentheses; literature citation is indicated by the superscript number.
[b]Data ranked by the incidence of mtMSI in previously published literature.
[c]Head and neck squamous cell cancer.

956–965 within the 12S rRNA gene coding region (TABLE 2). In fact, a remarkably high percentage of cases (95.6%; 87 of 91) of mtMSI were detected in the D-loop itself. Interestingly, the other 8 markers are relatively stable and are all located in the polypeptide coding regions (data not shown here). This demonstrates that the D-loop region represents a significant hotspot for mtMSI.

The incidence of mtMSI at np 303 was 32.2% (20 of 62) in endometrial cancer; it is statistically significantly higher than that in breast cancer (13.7%; 7 of 51) ($P = 0.022$) and noticeably higher than that in cervical cancer (23.9%; 17 of 71) and ovarian cancer (17.9%; 14 of 78) (TABLE 2). The incidence of np 303 mtMSI has also been reported (incidence ranging from 10–62.5%) in various human cancers (TABLE 2). The incidence of distinct types of cancer indicates that the extent of mtMSI varies remarkably from one cancer to another.

The wild-type sequence structure of microsatellite at np 303 is C_7TC_5 (according to revised Cambridge sequence, http://www.mitomap.org). This microsatellite includes two consecutive cytosines (C_7 and C_5) separated by a thymine residue at np 310. Interestingly, the alterations were detected only in the poly C upstream to the T residues (FIG. 1A). Why instability only occurred in the upstream poly C stretch is unknown. Both insertion and deletion variants were observed.[3,4]

The microsatellite at np 303 is located at the conserved sequence block II (CSB II) of the heavy-strand replication origin. CSB II contributes to formation of a persistent RNA-DNA hybrid which serves to prime mtDNA replication.[7] The potential impact of the np 303 sequence instability in the process of mtDNA replication is yet to be determined.

MtMSI at np 16184 was the second-most unstable region observed in the present study. We detected 17 mtMSI at np 16184. The incidences were 14.5% (9 of 62) and 15.7% (8 of 51) in endometrial cancer and breast cancer, respectively (TABLE 2). By contrast, mtMSI at this particular locus was not detected in any cases of cervical or ovarian cancer. Previous investigations had found 5.3–23.5% mtMSI at this locus from several types of cancer (TABLE 2).

Microsatellite at np 16184 is located at Hypervariable Region I of the D-loop at the 3′ end of the termination-associated sequence and within the 7S DNA binding site. This region is a remarkable mutational hotspot.[8] The wild-type sequence pattern of np 16184 is C_5TC_4. Although the np 16184 microsatellite has a similar sequence structure pattern to that of the np 303 microsatellite, both being based on a poly C sequence interrupted by a single T, the mtMSI at the former locus had a different pattern of instability. Deletions or insertions were detected in the poly C tract of tumor mtDNA that linked in all cases to a germline T to C transition at np 16189. The elimination of the T generated a long C repeat, generally anticipated to exhibit higher instability and empirically indicating that this position represents a strong hotspot of mtMSI with a $(C)_{7-14}$ pattern of variation (FIG. 1C). These findings indicate that the origin of the mtMSI at np 16184 may be different from that of the one at np 303.

A CA-repeat starting at np 514 is the only dinucleotide microsatellite encountered in mtDNA (TABLE 1). Either insertion or deletion of nucleotides was found to contribute to mtMSI (FIG. 1B). Overall, 12 cases in this study were found to display mtMSI at np 514, of which 2 were cases of cervical cancer, 1 of endometrial cancer, 5 of ovarian cancer, and 4 of breast cancer (TABLE 2). In contrast with previous reports that a very high frequency of mtMSI at np 514 was found in breast cancer,

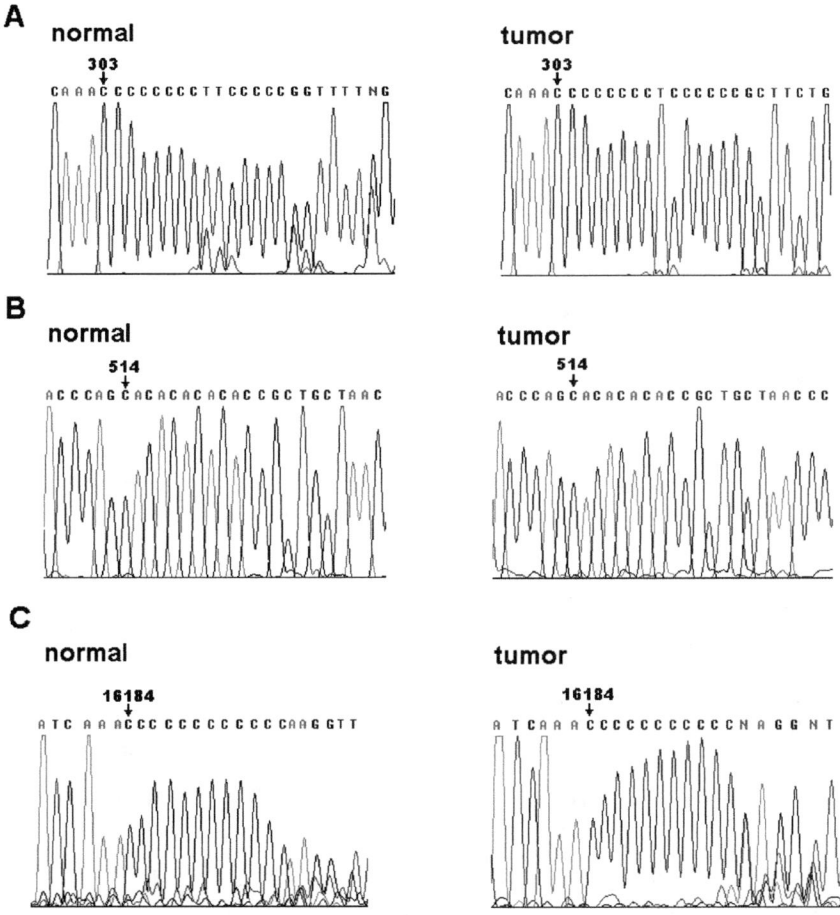

FIGURE 1. Detection of mtMSI by direct sequencing. Numbers above each sequence represent the np of the first base in the microsatellite motif. (**A**) The mtMSI detected at np 303 in a particular endometrial cancer tissue (U120). Both normal and tumor tissue mtDNA exhibited heteroplasmic poly C repeats before the T residue. In normal tissue, the sequence contained mainly C8, C9, and C10 with a small proportion of C7 and gave rise to overlapped signals of the subsequent sequence. In the tumor DNA, the sequence contained mainly C8 with a small proportion of C7 and C9. (**B**) The mtMSI detected at np 514 in a particular ovarian cancer tissue (OV10). The normal tissue sequence of the microsatellite contained five CA repeats and the tumor tissue mtDNA carried four CA repeats. (**C**) The mtMSI detected at np 16184 in a particular endometrial cancer tissue (U37). The T to C transition at np 16189 leads to an extended stretch of C repeats. In normal tissue, sequence contained mainly C12. In tumor tissue, the majority of sequence comprises C11 and C12.

we failed to find high frequency of this mtMSI in any of all four types of cancer that we analyzed (TABLE 2). Thus, this mtMSI generally occurs in common female cancers at relatively lower frequency.

Apparently, mtMSI in the D-loop region represents a common genetic event in human cancers. The broad range of incidence of the four mtMSIs we have described here reveal the potential diverse origin and accumulative process of such mtDNA instability in cancers. Although the mtMSI hotspot has been identified in these cases, it remains a considerable challenge to address the functional impact of the sequence variation. Although previous studies have shown that nuclear microsatellite instability had prognostic significance,[10] such clinical significance for mtMSI remains to be established. However, the high incidence and broad distribution of mtMSI, as described herein, render it a potential marker for cancer detection.

ACKNOWLEDGMENTS

This work was supported by the University of Hong Kong Conference and Research Grants (10203866 and 10205029).

REFERENCES

1. BOLAND, C.R. *et al.* 1998. A National Cancer Institute Workshop on Microsatellite Instability for cancer detection and familial predisposition: development of international criteria for the determination of microsatellite instability in colorectal cancer. Cancer Res. **58:** 5248–5257.
2. PARRELLA, P. *et al.* 2003. Mutations of the D310 mitochondrial mononucleotide repeat in primary tumors and cytological specimens. Cancer Lett. **190:** 73–77.
3. LIU, V.W. *et al.* 2003. High frequency of mitochondrial genome instability in human endometrial carcinomas. Br. J. Cancer **89:** 697–701.
4. SANCHEZ-CESPEDES, M. *et al.* 2001. Identification of a mononucleotide repeat as a major target for mitochondrial DNA alterations in human tumors. Cancer Res. **61:** 7015–7019.
5. KIRCHES, E. *et al.* 2001. High frequency of mitochondrial DNA mutations in glioblastoma multiforme identified by direct sequence comparison to blood samples. Int. J. Cancer **93:** 534–538.
6. KUMIMOTO, H. *et al.* 2004. Frequent somatic mutations of mitochondrial DNA in esophageal squamous cell carcinoma. Int. J. Cancer **108:** 228–231.
7. XU, B. & D.A. CLAYTON. 1995. A persistent RNA-DNA hybrid is formed during transcription at a phylogenetically conserved mitochondrial DNA sequence. Mol. Cell Biol. **15:** 580–589.
8. STONEKING, M. 2000. Hypervariable sites in the mtDNA control region are mutational hotspots. Am. J. Hum. Genet. **67:** 1029–1032.
9. RICHARD, S.M. *et al.* 2000. Nuclear and mitochondrial genome instability in human breast cancer. Cancer Res. **60:** 4231–4237.
10. LAWES, D.A., S. SENGUPTA & P.B. BOULOS. 2003. The clinical importance and prognostic implications of microsatellite instability in sporadic cancer. Eur. J. Surg. Oncol. **29:** 201–212.
11. HABANO, W., S. NAKAMURA & T. SUGAI. 1998. Microsatellite instability in the mitochondrial DNA of colorectal carcinomas: evidence for mismatch repair systems in mitochondrial genome. Oncogene **17:** 1931–1937.
12. TANG, M. *et al.* 2004. Mitochondrial DNA mutation at the D310 (displacement loop) mononucleotide sequence in the pathogenesis of gall bladder carcinoma. Clin. Cancer Res. **10:** 1041–1046.

13. OKOCHI, O. et al. 2002. Detection of mitochondrial DNA alterations in the serum of hepatocellular carcinoma patients. Clin. Cancer Res. **8:** 2875–2878.
14. NOMOTO, S. et al. 2002. Mitochondrial D-loop mutations as clonal markers in multicentric hepatocellular carcinoma and plasma. Clin. Cancer Res. **8:** 481–487.
15. CHEN, J.Z. et al. 2002. Extensive somatic mitochondrial mutations in primary prostate cancer using laser capture microdissection. Cancer Res. **62:** 6470–6474.
16. PARRELLA, P. et al. 2001. Detection of mitochondrial DNA mutations in primary breast cancer and fine-needle aspirates. Cancer Res. **61:** 7623–7626.
17. TAN, D.J., R.K. BAI & L.J. WONG. 2002. Comprehensive scanning of somatic mitochondrial DNA mutations in breast cancer. Cancer Res. **62:** 972–976.
18. HIBI, K. et al. 2001. Mitochondrial DNA alteration in esophageal cancer. Int. J. Cancer **92:** 319–321.
19. MAXIMO, V. et al. 2002. Mitochondrial DNA somatic mutations (point mutations and large deletions) and mitochondrial DNA variants in human thyroid pathology: a study with emphasis on Hurthle cell tumors. Am. J. Pathol. **160:** 1857–1865.
20. LOHRER, H.D., L. HIEBER & H. ZITZELSBERGER. 2002. Differential mutation frequency in mitochondrial DNA from thyroid tumours. Carcinogenesis **23:** 1577–1582.
21. TONG, B.C. et al. 2003. Mitochondrial DNA alterations in thyroid cancer. J. Surg. Oncol. **82:** 170–173.
22. MIYAZONO, F. et al. 2002. Mutations in the mitochondrial DNA D-Loop region occur frequently in adenocarcinoma in Barrett's esophagus. Oncogene **21:** 3780-3783.
23. KIRCHES, E. et al. 2002. Comparison between mitochondrial DNA sequences in low grade astrocytomas and corresponding blood samples. Mol. Pathol. **55**: 204–206.
24. MAXIMO, V. et al. 2001. Microsatellite instability, mitochondrial DNA large deletions, and mitochondrial DNA mutations in gastric carcinoma. Genes Chromosomes Cancer **32:** 136–143.
25. JERONIMO, C. et al. 2001. Mitochondrial mutations in early stage prostate cancer and bodily fluids. Oncogene **20:** 5195–5198.

Analysis of Heteroplasmy in Hypervariable Region II of Mitochondrial DNA in Maternally Related Individuals

MEI-CHEN LO,[a] HORNG-MO LEE,[a] MING-WEI LIN,[b] AND CHIN-YUAN TZEN[c]

[a]*Graduate Institute of Cell & Molecular Biology, Taipei Medical University, Taipei, Taiwan, ROC*

[b]*Faculty of Medicine, School of Medicine, National Yang-Ming University Taipei, Taiwan, ROC*

[c]*Department of Pathology, Mackay Memorial Hospital, Taipei, Taiwan, ROC*

ABSTRACT: Mitochondrial DNA sequences have been widely employed for identity investigation. However, the presence of a heteroplasmic site may complicate sequence analysis for forensic purposes when two samples are compared. To study this potential problem, we analyzed the hypervariable region of the displacement loop in five maternally related individuals, that is, grandmother, mother, one son, and two daughters. The results showed that three of them had a heteroplasmic site at nucleotide position (np) 204, located in the hypervariable region II. By using Bayesian inference to assess the significance of the mother-offspring pairs, a likelihood ratio of 1.78×10^5 was obtained. Therefore, Bayesian inference does not place the prior odds into the context of the two different likelihood ratios derived from the DNA evidence. On the other hand, the chromatogram of the denaturing high-performance liquid chromatography system proved that the single peak in a sequencing chromatogram at np 204 was, in fact, heteroplasmic in nature. This study demonstrated that heteroplasmy is a common occurrence in tissue from normal individuals and should be taken into account in forensic investigation when samples appear to differ at a single nucleotide position by direct sequencing.

KEYWORDS: displacement loop (D-loop); heteroplasmy; likelihood ratio; denaturing high-performance liquid chromatography (dHPLC)

INTRODUCTION

Mitochondrial DNA (mtDNA) is maternally inherited[1–3] and exhibits a high degree of homoplasmy.[4–6] Heteroplasmy has been observed to be an intermediate condition in which new mutations are in the process of segregation to homoplasmy through genetic drift after relatively few generations.[7–9] Heteroplasmic segregation rates in mammals differ widely. Rapid fixation of mutations was observed in a herd

---

Address for correspondence: Chin-Yuan Tzen, M.D., Ph.D., Department of Pathology, Mackay Memorial Hospital, 45, Minsheng Road, Tamshui, Taipei 251, Taiwan. Voice: +8862-2809-4661 ext. 2419; fax: +88622809-3385.

Jeffrey@ms2.mmh.org.tw

of Holstein cattle, where new point mutations resolved to homoplasmy only in two to three generations.[10] Various estimates of the mtDNA mutation rate have been inferred from evolutionary analysis, with a value of 1 per 300 generations.[11]

As with any forensic case, identification of the victims of mass disasters or murders, when bodies are undiscovered for many years, is one of the most challenging fields of forensic medicine. mtDNA analysis can be used effectively when victims and living descendants are separated by many generations. Although mtDNA analysis offered a powerful tool in forensic investigation, it can be problematic due to a single nucleotide "mismatch." In this study, we present a case confounded by mismatch due to heteroplasmy. Both Bayesian inference and denaturing high performance liquid chromatography (dHPLC) could provide a resolution in this situation.

MATERIAL AND METHODS

DNA Extraction and Polymerase Chain Reaction (PCR)

DNA extraction from peripheral leukocytes, buccal cells and urine sediments applied a commercial kit (QIA amp DNA Mini Kit, Germany). The procedures of PCR and DNA sequencing were performed according to the previously described method.[12] Two oligonucleotide primers used to amplify the mtDNA segment from nps 15881 to 725 were as follows: forward primer 5'-TGG GGC CTG TCC TTG TAG TAT-3' (primer 15881A); reverse primer 5'-GGT GAA CTC ACT GG AAC GGG-3' rimer 10B).

Calculation of Likelihood Ratio

The likelihood ratio was calculated as follow:

$$LR = P(E_1/R)/P(E_1/R') \times P(E_2/R)/P(E_2/R')$$
$$= 0.99/0.0044 \times 0.37/0.000468 = 1.78 \times 10^5.$$

where

LR: likelihood ratio;

E_1: sequence match between two individuals;

E_2: the co-occurrence of heteroplasmy at position 204 in two individuals;

R: the group in the same family;

R': the group of an unknown family in which E_1 and E_2 are unrelated;

$P(E_1/R) = E^{-g\mu} = 0.99$;

g: two generational events separating the tested individuals;

μ: the estimation of the displacement loop (D-loop) mutation rate of the mitochondrial DNA = 1/300.

$$P(E_1/R') = 2/455 = 0.0044,$$

where 455 indicates the numbers of unrelated populations.

$$P(E_2/R) = e^{-g\mu} = E^{-2\beta} = 0.37,$$

β = 1/2 (time to homoplasmic fixation = two generational events

$$P(E_2/R') = 2/300 \times 100/1426 = 0.000468,$$

where 300 indicates the numbers of mtDNA mutation rate with a value of 1 per 300 generations and 1426 indicates the numbers of unrelated populations.

dHPLC Analysis of Specific Mutations

Primer pairs to amplify the mtDNA segment from nps 128–553 were as follows: forward primer 5′-CTG TCT TTG ATT CCT GCC TC-3′ (primer 128A); reverse primer 5′-GGT TGG TTC GGG GTA TGG GG-3′ (primer 1AB). dHPLC analysis was performed on an automated dHPLC instrument (varian Helix^{TM} System).[13] Temperatures for successful resolution of heteroduplexes were both calculated by the dHPLC Melt program[14] and experimentally determined for the fragments containing the T204C mutation.

RESULTS AND DISCUSSION

We analyzed the hypervariable region II of the D-loop in five individuals within a family, that is, grandmother, mother, one son, and two daughters. mtDNA was

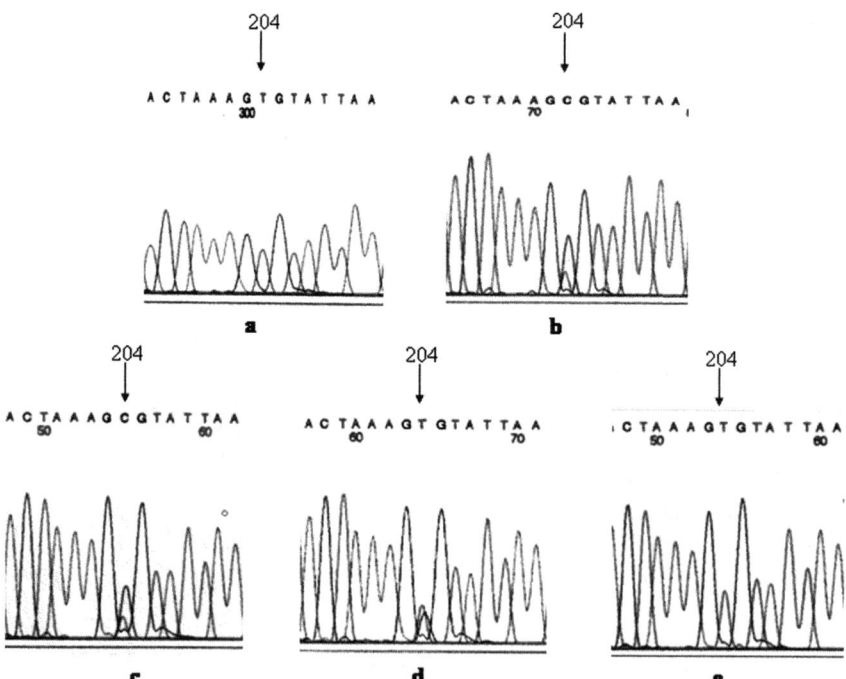

FIGURE 1. Automated sequence chromatographs of peripheral blood showing mtDNA sequence at position 204. Homoplasmy with thymine is shown in grandmother (**a**) and daughter 2 (**e**). Heteroplasmy with cytosine predominating thymine is shown in mother (**b**) and daughter 1 (**c**). Heteroplasmy with thymine predominating cytosine is shown in son (**d**).

TABLE 1. Sequence variation in the mitochondrial DNA D-loop in one family

	HVRII						HVRIII	HVRI		
Anderson sequence:	73	195	204	207	263	298	489	16075	16223	16519
Samples	A	T	T	G	A	C	T	T	C	T
Grandmother										
Blood	G	C	T	A	G	T	C	C	T	C
Buccal cells	G	C	T	A	G	T	C	C	T	C
Urine sediments	G	C	T	A	G	T	C	C	T	C
Mother										
Blood	G	C	C/t	A	G	T	C	C	T	C
Buccal cells	G	C	C/t	A	G	T	C	C	T	C
Urine sediments	G	C	C/c	A	G	T	C	C	T	C
Daughter 1										
Blood	G	C	C/t	A	G	T	C	C	T	C
Buccal cells	G	C	C/t	A	G	T	C	C	T	C
Urine sediments	G	C	T/c	A	G	T	C	C	T	C
Son										
Blood	G	C	T/c	A	G	T	C	C	T	C
Buccal cells	G	C	C/t	A	G	T	C	C	T	C
Urine sediments	G	C	C/t	A	G	T	C	C	T	C
Daughter 2										
Blood	G	C	T	A	G	T	C	C	T	C
Buccal cells	G	C	T	A	G	T	C	C	T	C
Urine sediments	G	C	T	A	G	T	C	C	T	C

[a]In case of multiple signals, the predominant sequence is denoted by a capital letter.

extracted from peripheral leukocytes, buccal cells, and urine sediments, and the sequencing chromatograms showed that the nucleotide at position 204 was either a homoplasmic thymine (grandmother and daughter 2) or a heterolpasmic cytosine/thymine (mother, daughter 1, and son) (FIG. 1 and TABLE 1).

Without knowing that these five individuals were from the same family, one would cast doubt on their maternal relationship because of the concurrent homoplasmy and heteroplasmy at position 204. This is similar to the case of the remains of Tsar Nicholas II's brother, Grand Duke of Russia Georgij Romanov.[15] In 1911, nine sets of skeletal remains excavated from a mass grave near Yekaterinburg, Russia, were believed to include those of Tsar Nicholas II. Although mtDNA sequences from the bone of the presumed Tsar matched those of two living maternal relatives, the bone sample had a mismatched heteroplasmy at position 16169. The presence of heteroplasmy in both the Tsar and Georgij resulted in a likelihood ratio of 1.3×10^8, suggesting that the putative remains were from Tsar Nicholas II.

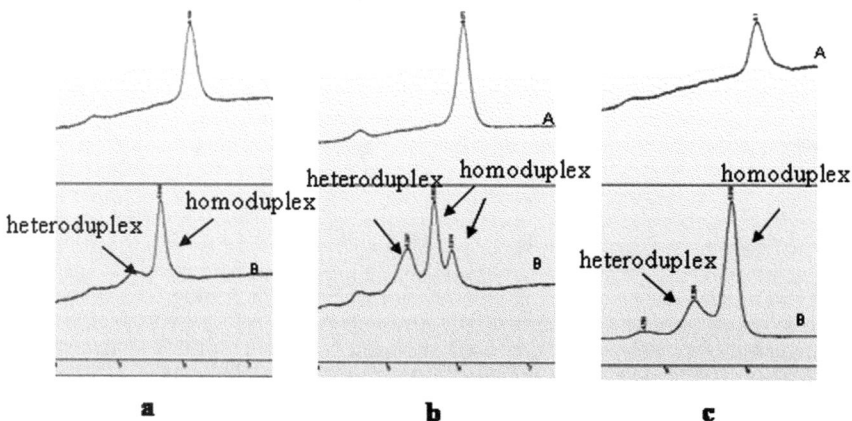

FIGURE 2. dHPLC analysis of np 128–553 for peripheral blood specimens of grandmother (**a**), mother (**b**), and daughter 2 (**c**). *Arrows* indicate peaks of homoduplex and heteroduplex.

Using Bayesian inference in the current study, we estimated that the chance of observing heteroplasmy at the same position in two randomly selected unrelated individuals was 4.68×10^{-4}, resulting in a likelihood ratio of 2.25×10^2. However, when combined with the mtDNA match of the mother-offspring pairs, the likelihood ratio was 1.78×10^5, suggesting that these individuals were maternally related.

As shown in this case, the presence of both homoplasmy and heteroplasmy can cause confusion in forensic casework, although calculation of likelihood ratio may be helpful if a database is available. However, this confusion may be artificial if the "homoplasmy" resulted from false interpretation. Because sequencing chromatograms are only semiquantitative, a single peak does not necessarily indicate homoplasmy in a given nucleotide. To further investigate this potential problem, dHPLC analysis, a sensitive and specific method to detect heteroplasmic mtDNA mutations,[13] was employed. The sensitivity of the dHPLC system was high, with the lowest detection of 0.5 for the mtDNA mutation. The result of dHPLC analysis of nps 128–553 showed that the grandmother and daughter 2 were, in fact, heteroplasmy at position 204 (FIG. 2). This result indicated that all five individuals of this maternal lineage were actually heteroplasmic at position 204. In conclusion, the concurrence of homoplasmy and heteroplasmy in forensic caseworks can be resolved by either calculating the likelihood ratio or examining the homoplasmy by dHPLC.

ACKNOWLEDGMENTS

This work was supported by Grant MMH-E-93002 from Mackay Memorial Hospital to Chin-Yuan Tzen, MD, PhD. The authors thank Tsu-Yen Wu for technical supports and Shu-Huei Wang for secretarial expertise.

REFERENCES

1. CASE, J.T. & D.C. WALLACE. 1981. Maternal inheritance of mitochondrial DNA polymorphisms in cultured human fibroblasts. Somatic Cell Genet. **7:** 103–108.
2. GINTHER, C., L. ISSEL-TARVER & M.C. KING. 1992. Identifying individuals by sequencing mitochondrial DNA from teeth. Nat. Genet. **2:** 135–138.
3. HUTCHISON, C.A., 3rd, *et al.* 1974. Maternal inheritance of mammalian mitochondrial DNA. Nature **251:** 536–538.
4. BODENTEICH, A., L. G. MITCHELL & C. R. MERRIL. 1991. A lifetime of retinal light exposure does not appear to increase mitochondrial mutations. Gene **108:** 305–309.
5. MONNAT, R.J., JR. & L.A. LOEB. 1985. Nucleotide sequence preservation of human mitochondrial DNA. Proc. Natl. Acad. Sci. USA. **82:** 2895–2899.
6. MONNAT, R. J., JR. & D. T. REAY. 1986. Nucleotide sequence identity of mitochondrial DNA from different human tissues. Gene **43:** 205–211.
7. HAUSWIRTH, W. W. & P. J. LAIPIS. 1982. Mitochondrial DNA polymorphism in a maternal lineage of Holstein cows. Proc. Natl. Acad. Sci. USA **79:** 4686–4690.
8. WALLACE, D.C. 1992. Mitochondrial genetics: a paradigm for aging and degenerative diseases? Science **256:** 628–632.
9. WALLACE, D.C. 1992. Diseases of the mitochondrial DNA. Annu. Rev. Biochem. **61:** 1175–1212.
10. ASHLEY, M.V., P.J. LAIPIS & W.W. HAUSWIRTH. 1989. Rapid segregation of heteroplasmic bovine mitochondria. Nucleic Acids Res. **17:** 7325–7331.
11. STONEKING, M. *et al.* 1992. New approaches to dating suggest a recent age for the human mtDNA ancestor. Philos Trans. R. Soc. Lond. B. Biol. Sci. **337:** 167–175.
12. CHEN, M.H., H.M. LEE & C.Y. TZEN. 2002. Polymorphism and heteroplasmy of mitochondrial DNA in the D-loop region in Taiwanese. J. Formos. Med. Assoc. **101:** 268–276.
13. VAN DEN BOSCH, B.J. *et al.* 2000. Mutation analysis of the entire mitochondrial genome using denaturing high performance liquid chromatography. Nucleic Acids Res. **28:** E89.
14. http://insertion.stanford.edu/melt.html.
15. IVANOV, P.L. *et al.* 1996. Mitochondrial DNA sequence heteroplasmy in the Grand Duke of Russia Georgij Romanov establishes the authenticity of the remains of Tsar Nicholas II. Nat. Genet. **12:** 417–420.

Deleted Mitochondrial DNA in Human Luteinized Granulosa Cells

HENG-KIEN AU,[a,b] SHYH-HSIANG LIN,[c] SHIH-YI HUANG,[c] TIEN-SHUN YEH,[d] CHII-RUEY TZENG,[a,b] AND RONG-HONG HSIEH[b,c]

[a]*Department of Obstetrics and Gynecology, Taipei Medical University Hospital, Taipei, Taiwan 112, Republic of China*

[b]*Center for Reproductive Medicine and Sciences, Taipei Medical University, Taipei, Taiwan 112, Republic of China*

[c]*School of Nutrition and Health Sciences, Taipei Medical University, Taipei, Taiwan 112, Republic of China*

[d]*Graduate Institute of Cell and Molecular Biology, Taipei Medical University, Taipei, Taiwan 112, Republic of China*

> ABSTRACT: The rearrangement of mitochondrial DNA in luteinized granulosa cells was determined in order to evaluate the fertilization capacity of oocytes and the development of embryos. Multiple deletions of mtDNA were found in luteinized granulosa cells from *in vitro* fertilization (IVF) patients. The 4977-base pair (bp) deletion was the most frequent deletion found in human granulosa cells. No significant difference was noted between mtDNA deletions of granulosa cells based on the fertilization capacity of oocytes and the development of embryos. To determine the relationship of proportions of mtDNA rearrangements with the aging process, granulosa cells were grouped into three different cohorts according to maternal age: younger than 32 years, between 32 and 37 years, and older than 37 years. No statistical correlation was noted between patient age and the frequency of occurrence of multiple mtDNA deletions. However, an increase in granulosa cell apoptosis was associated with an increase in mtDNA deletions. Accumulation of mtDNA deletions may contribute to mitochondrial dysfunction and impaired ATP production. We concluded that the accumulation of rearranged mtDNA in granulosa cells might not interfere with fertilization of human oocytes and further embryonic development; it was, however, associated with apoptosis processes.
>
> KEYWORDS: deletion; mtDNA; granulosa cell

INTRODUCTION

Mammalian ovaries include several hundreds of thousands of follicles in the primordial and primary stages during ovarian follicular development, among which

---

Address for correspondence: Dr. Rong-Hong Hsieh, School of Nutrition and Health Sciences, Taipei Medical University, Taipei, Taiwan 112, Republic of China. Voice: +886-2-27361661 ext. 6551-128; fax: +886-2-27373112.

hsiehrh@tmu.edu.tw

only limited numbers in each cycle will fully develop and are selected for ovulation, whereas the remaining majority of follicles undergo atresia.[1,2] Granulosa cells play a major role in regulating ovarian physiology, including ovulation and luteal regression.[3] Granulosa cells secrete a wide variety of growth factors that may attenuate gonadotrophin's action in ovaries in paracrine-autocrine processes.[4,5] Most of these factors do not directly affect oocytes but exert their action via granulosa cells. The presence of granulosa cells appears to be beneficial for oocyte maturation and early development.[6] Recent studies have suggested that follicular atresia is associated with apoptosis of granulosa cells.[1,2] Researchers also reported that the incidence of apoptotic bodies in granulosa cells can be used to predict the developmental capacity of oocytes in an IVF program.[7]

In eukaryotic cells, mitochondria are specialized organelles that catalyze the formation of ATP. Two distinct genomes exist in all eukaryotic cells. One is located in the nucleus and is transmitted in a mendelian fashion, whereas the other is located in mitochondria and is transmitted by maternal inheritance. Most human somatic cells contain about 1,000 mitochondria, and each mitochondrion consists of 2 to 10 copies of mtDNA.[8] mtDNA comprises a circular, histone-free molecule composed of 16.6 kb of DNA, present in one or more copies in every mitochondrion. Thirteen protein subunits are required for oxidative phosphorylation of a total of about 80 subunits, the remainder of which are encoded by nuclear genes and imported into the mitochondrion. The mtDNA also contains 2 ribosome subunits and 22 transfer RNA. The oxidative phosphorylation capacity of mitochondria is determined by the interplay between nuclear and mitochondrial genes. mtDNA encodes 13 proteins that are all components of the respiratory chain, whereas nuclear DNA encodes the majority of respiratory chain proteins, which are all proteins that regulate replication and transcription of mtDNA.[9] In recent years, an increasing number of reports have shown that mtDNA deletions are associated with human aging and mitochondrial diseases.[10,11] In this study, the rearrangement of mitochondrial DNA in luteinized granulosa cells was determined in order to evaluae the fertilization capacity of oocytes and the development of embryos.

MATERIAL AND METHODS

Polymerase Chain Reaction

Oocytes and embryos were stored in 20 µL of 1X polymerase chain reaction (PCR) buffer containing 0.05 mg/mL of proteinase K, 20 mM of dithiothreitol (DTT), and 1.7 µM of SDS. After digestion for 1 h at 56°C and 10 min of heat-inactivation of proteinase K at 95°C, the total DNA in the solution was then used as template for the PCR assay. The sequences of the oligonucleotide primers used in this study are listed as follows: H1 (nucleotide position [np] 8285~8304, CTCTA-GAGCCCACTGTAAAG), H2 (np 8781~8800, CGGACTCCTGCCTCACTCAT), H3 (np 9207~9226, ATGACCCACCAATCACATGC), L1 (np 13650~13631, GGG-GAAGCGAGGTTGACCTG), L2 (np 14145~14126, TGTGATTAGGAGTAGGGT-TA), L3 (np 15896~15877, TACAAGGACAGGCCCATTTG), and L4 (np 16410~16391, GAGGATGGTGGTCAAGGGAC).

Semiquantitative RT-PCR

Total RNA extracted from human granulosa cells was used as templates, and cDNA was prepared using the RNA extraction and reverse-transcription polymerase chain reaction (RT-PCR) kit from Ambion (Austin, TX). RT-PCR amplifications were performed with 3 µL of cDNA in a total volume of 50 µL of amplification buffer, 40 pmol of specific primers, and 2.5 units of Taq DNA polymerase (Life Technologies, Grand Island, NY). Sequences of the oligonucleotide primers used in this study are listed as follows: ND2 (forward, np 5101~5120, TAACTACTACCGCATTCCTA; reverse, np 5400-5381, CGTTGTTAGATATGGGGAGT), and GAPDH (forward, CCTTCATTGACCTCAAC; reverse, AGTTGTCATGGATGACC). For semiquantitative amplification, each cycle was carried out at 92°C for 30 s, 58°C for 30 s, and 72°C for 60 s. The reactions were analyzed after 15, 20, 25, 30, 35, and 40 cycles to optimize the linear range of amplification. The PCR reactions were optimized with respect to annealing temperature and numbers of PCR cycles. Each PCR product was run through a 2% agarose gel and was visualized with ethidium bromide staining.

RESULTS

We attempted to determine whether the existence of mtDNA deletions in luteinized granulosa cells affects the fertilization capacity of oocytes and the development of embryos. DNA was extracted from granulosa cells to determine the extent of mtDNA rearrangement by PCR using multiple pairs of primers. Multiple deletions of mtDNA were found in luteinized granulosa cells from IVF patients. The 4977-bp deletion was the most frequent deletion in human granulosa cells. There was no significant difference between mtDNA deletions of granulosa cells with the fertilization capacity of oocytes and the development of embryos. To determine the relationship of proportions of mtDNA rearrangements with the aging process, granulosa cells were grouped into three different cohorts according to maternal age: younger than 32 years, between 32 and 37 years, and older than 37 years. Frequencies of occurrence of mtDNA deletions in human granulosa cells of different age cohorts are listed in TABLE 1. Percentages of 4977-bp rearranged mtDNA were 43.8%, 39.1%, and 47.4% in the three age-stratified cohorts, respectively. There was no statistical correlation between patient age and the frequency of occurrence of the 4977-bp mtDNA deletion or with multiple mtDNA deletions. Whether rearranged mtDNA coexisted

TABLE 1. Frequency of occurrence of mtDNA deletions in human granulosa cells of different age cohorts

Type of rearranged mtDNA	Percentage of rearranged mtDNA (%)		
	<32^a	32–37^a	>37^a
4977 bp	43.8 (7/16)	39.1 (9/23)	47.4 (9/19)
Multiple	56.3 (9/16)	52.2 (12/23)	57.8 (11/19)

^{a}Maternal age (years).

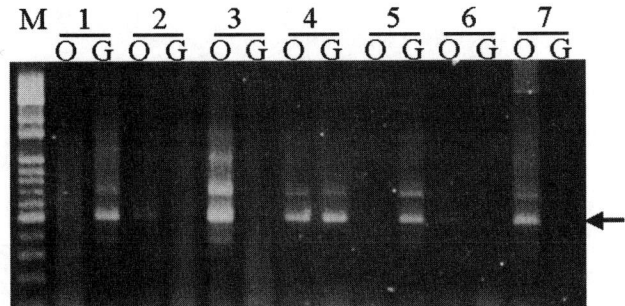

FIGURE 1. Multiple deletions of mtDNA in granulosa cells and oocytes. *Lane 1*, 26 years old; *lane 2*, 30 years old; *lane 3*, 32 years old; *lane 4*, 35 years old; *lane 5*, 36 years old; *lane 6*, 38 years old; and *lane 7*, 40 years old. O, oocyte; G, granulosa cells. *Arrow* indicates a 4977-bp mtDNA deletion.

in oocytes and surrounding granulosa cells was also determined. The data showed that mtDNA deletions were randomly distributed between oocytes and granulosa cells; higher proportions of rearranged mtDNA were found in oocytes (FIG. 1, lanes 2, 3, 6, and 7), and higher frequencies were determined in granulosa cells (FIG. 1, lanes 1 and 5). The independent existence of rearranged mtDNA in oocytes and granulosa cells is shown in FIGURE 1. Expression levels of mitochondrial RNA in granulosa cells of different age groups were determined using a semiquantitative RT-PCR (FIG. 2). Transcripts of mitochondrial NADH dehydrogenase subunit 2 (ND2) were determined and normalized to the GAPDH gene, with no statistically significant differences among different age cohorts.

DISCUSSION

The 4977-bp deletion is the most common mtDNA deletion in human oocytes and embryos.[12,13] Our previous study showed that frequencies of 4977-bp deleted mtDNA were 66.1%, 34.8%, and 21.1% in unfertilized oocytes, arrested embryos, and 3PN embryos, respectively.[13] The 4977-bp mtDNA rearrangement may remove major structural genes containing Fo-F1-ATPase (ATPase 6 and 8), cytochrome oxidase III (CO III), and NADH-CoQ oxidoreductase (ND3, ND4, ND4L, and ND5). This deletion also creates a chimeric gene, which fuses the 5'-portion of ATPase 8 and 3'-portion of the ND5 gene of mtDNA. These mutated genes may cause impaired gene expression by decreasing the expression of the deleted genes or by producing transcripts of fused genes. In this study, the age-independent existence of rearranged mtDNA in granulosa cells indicates that the aging process did not play a major role in the accumulation of mtDNA mutations in granulosa cells.

To evaluate the relation between mitochondrial gene expression of granulosa cells and oocyte fertilization ability, transcripts of mitochondrial NADH dehydrogenase subunit 2 (ND2) were determined. The mtDNA transcripts are polycistronic,[14,15] which means that each gene is separated following precise

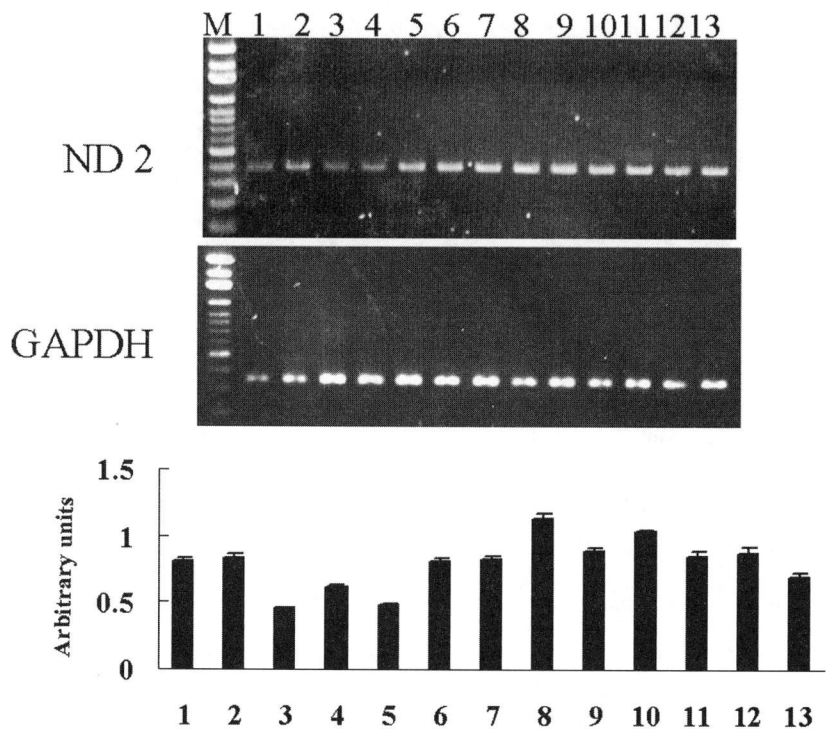

FIGURE 2. Expression levels of mtRNA in granulosa cells of different age groups. *Lanes 1–3*, 25–30 years old; *lanes 4–6*, 30–35 years old; *lanes 7–10*, 35–38 years old; and *lanes 11–13*, 38–42 years old.

endonucleolytic excision of the tRNAs from nascent transcripts. The polycistronic mtDNA transcripts are consistent, with different mtRNA expression levels showing the same pattern in human oocytes.[16] In this study, we first determined the ND2 transcript in order to estimate mitochondrial gene expression in granulosa cells. No statistically significant differences in mtRNA expression patterns were noted among the different age cohorts. The present evidence of the independent existence of rearranged mtDNA in oocytes and granulosa cells supports the accumulation of rearranged mtDNA in granulosa cells possibly not interfering with fertilization of human oocytes and further embryonic development.

ACKNOWLEDGMENTS

This work was supported by Research Grants NSC 91-2320-B-038-026 and NSC 92-2320-B-038-045 from the National Science Council of the Republic of China.

REFERENCES

1. TILLY, J.L. et al. 1991. Involvement of apoptosis in ovarian follicular atresia and postovulatory regression. Endocrinology **129:** 2799–2801.
2. HSUEH, A.J., H. BILLIG & A. TSAFRIRI. 1994. Ovarian follicle atresia: a hormonally controlled apoptotic process. Endocr. Rev. **15:** 707–724.
3. AMSTERDAM, A. & N. SELVARAJ. 1997. Control of differentiation, transformation, and apoptosis in granulosa cells by oncogenes, oncoviruses, and tumor suppressor genes. Endocr. Rev. **18:** 435–461.
4. NAKAMURA, T. et al. 1990. Activin-binding protein from rat ovary is follistatin. Science **247:** 836–838.
5. GRAS, S. et al. 1996. Transient periovulatory expression of pituitary adenylate cyclase activating peptide in rat ovarian cells. Endocrinology **137:** 4779–4785.
6. CANIPARI, R. 2000. Oocyte-granulosa cell interactions. Hum. Reprod. Update **6:** 279–289.
7. NAKAHARA, K. et al. 1997. The incidence of apoptotic bodies in membrane granulosa cells can predict prognoses of ova from patients participating in *in vitro* fertilization programs. Fertil. Steril. **68:** 312–317.
8. GILES, R.E. et al. 1980. Maternal inheritance of human mitochondrial DNA. Proc. Natl. Acad. Sci. USA **77:** 6715–6719.
9. TAANMAN, J.W. 1999. The mitochondrial genome: structure, transcription, translation and replication. Biochim. Biophys. Acta **1410:** 103–123.
10. WEI, Y.H. 1992. Mitochondrial DNA alterations as ageing-associated molecular events. Mutat. Res. **275:** 145–155.
11. WALLACE, D.C. et al. 1999. Mitochondrial DNA variation in human evolution and disease. Gene **238:** 211–230.
12. BRENNER, C.A. et al. 1998. Mitochondrial DNA deletion in human oocytes and embryos. Mol. Hum. Reprod. **4:** 887–892.
13. HSIEH, R.H. et al. 2002. Multiple rearrangements of mitochondrial DNA in unfertilized human oocytes. Fertil. Steril. **77:** 1012–1017.
14. MONTOYA, J. et al. 1982. Identification of initiation sites for heavy-strand and light-strand transcription in human mitochondrial DNA. Proc. Natl. Acad. Sci. USA **79:** 7195–7199.
15. MONTOYA, J., G.L. GAINES & G. ATTARDI. 1983. The pattern of transcription of the human mitochondrial rRNA genes reveals two overlapping transcription units. Cell **34:** 151–159.
16. HSIEH, R.H. et al. 2004. Decreased expression of mitochondrial genes in human unfertilized oocytes and arrested embryos. Fertil. Steril. **81:** 912–918.

Identification of Human-Specific Adaptation Sites of *ATP6*

BEY-LIING MAU,[a] HORNG-MO LEE,[b] AND CHIN-YUAN TZEN[c]

[a]*Division of General Internal Medicine, Sun Yat-Sen Cancer Center, Taipei 251, Taiwan*
[b]*Taipei Medical University, Taipei 251, Taiwan*
[c]*Departments of Pathology, Mackay Memorial Hospital, Taipei 251, Taiwan*

ABSTRACT: Mitochondria play an essential role in forming ATP and generating heat. The proportion of these two depends on the coupling efficiency of electrochemical gradient to synthesize ATP. Therefore, an increased basal metabolic rate caused by partial uncoupling of the mitochondria can be balanced by a high caloric intake provided by a high-fat diet. The recent study by Mishmar *et al.* (Proc. Natl. Acad. Sci. USA 2003; 100: 171–176) suggested that *ATP6* was the most variable gene among human mitochondrial DNAs and probably resulted from the adaptation of *Homo sapiens* to the colder climate during the migration out of Africa. According to this adaptation theory, the *ATP6* of *Homo sapiens* (omnivorous animals consuming fat-containing diet) should be significantly different from that of other primates for permitting human adaptation to the dietary conditions. On the basis of this rationale, we analyzed *ATP6* sequences of 136 unrelated Taiwanese subjects, which then were compared with 1,130 reported sequences. The obtained human consensus from 1,266 individuals was compared with that derived from 42 species of primates other than human. The alignment showed that human *ATP6* harbored 80 variable residues, among which 25 amino acids were conserved in other primates, suggesting that adaptation constraints operating at the amino acid level results in the species-specific difference of *ATP6*. Therefore, these 25 amino acids are probably the human-specific adaptation residues of *ATP6*.

KEYWORDS: *ATP6*; adaptation; evolution

INTRODUCTION

Mitochondria are the central integrators of energy production because of their role in forming ATP and generating heat. The efficiency of coupling electrochemical gradient to synthesize ATP determines how much heat will be produced. This biochemical reaction has long been used to explain the dietary preference consumed by people living in cold areas to generate heat to maintain body temperature. However, the mechanism whereby the uncoupling variation occurs among humans is not known. Recently, Wallace and associates reported that the *ATP6* gene was the most

Address for correspondence: Chin-Yuan Tzen, Department of Pathology, Mackay Memorial Hospital, 45, Minsheng Road, Tamshui, Taipei 251, Taiwan. Voice: +8862-2809-4661 ext. 2491; fax: +8862-2809-3385.
jeffrey@ms2.mmh.org.tw

variable gene among human mitochondrial DNAs (mtDNAs),[1] particularly in the population of the Arctic zone. They explained that *ATP6* variants with reduced coupling might account for the increased basal metabolic rate, and the fast-evolving *ATP6* gene allows rapid adaptation to new climate and dietary conditions. In keeping with this notion, we reported that *ATP6* contains many polymorphisms.[2] This finding was confirmed by Lutz-Bonengel *et al.*[3] and recently was applied by Poetsch *et al.* for forensic analysis.[4]

On the basis of the adaptation theory of *ATP6* evolution, we intended in this study to identify the *ATP6* adaptation sites of *Homo sapiens*. Our rationale is that the human-specific adaptation sites of *ATP6* are variable in humans but not in other primates.

MATERIALS AND METHODS

The inhabitants of Taiwan are a mixed population dominated by Han Chinese. Among them, Minnan Chinese and Hakka Chinese immigrated into Taiwan in the eighteenth century for different reasons, such as trading, famine, and exile. Most of the other mainland Chinese retreated to Taiwan with the government in the last century. These Han Chinese and scattered indigenous peoples of undetermined origin collectively constitute the so-called Taiwanese. In this study, 136 unrelated Taiwanese were subjected to DNA sequencing of the *ATP6*.

Total DNA was extracted from peripheral blood leukocytes (EDTA plasma samples) by using GFX Genomic Blood DNA purification kit (Amersham Pharmacia Biotech UK, Ltd., England) according to the manufacturer's instructions. The final extracts were dissolved in TE buffer and kept in 4°C for later use.

The oligonucleotide primers used to amplify nucleotides 8282 to 9300 included a forward primer 5'-CCC CTC TAG AGC CCA CTG TAA AGC-3' and a backward primer 5'-CTA GGC CGG AGG TCA TTA GG-3', both of which were designed according to the Anderson mitochondrial sequence.[5] PCR was performed in a DNA thermal cycler (GeneAmp PCR System 9600; Perkin-Elmer, Foster City, CA) according to the previously described procedures.[2]

The PCR products were sequenced using the ABI PRISM BigDye Terminator Cycle Sequencing Ready Reaction Kit with AmpliTaq DNA polymerase FS (Perkin-Elmer). Fluorescent dye–labeled DNA fragments then were analyzed on a semi-automated DNA sequencer and ABI Prism 377 Genetic Analyzer (PE Applied Biosystems, Foster City, CA). Each DNA fragment was sequenced in both sense and antisense directions.

RESULTS AND DISCUSSION

Characterization of Homo sapiens **ATP6**

The DNA sequences from position 8527 to 9207, which encodes ATP6, were obtained from 136 unrelated Taiwanese individuals. Analysis of these stretches of 681 bases revealed 41 haplotypes, from which 20 variants of *ATP6* could be deduced from the haplotypes containing synonymous alterations (TABLE 1). The remaining

TABLE 1. Analysis of *ATP6* peptide sequences among 136 unrelated Taiwanese individuals

Variant	Polymorphisms					Frequency (n)
1	A11T	A20T	T112A			0.74% (1)
2	G16S	T112A				0.74% (1)
3	A20T	T53I	T59A	I77T	T112A	0.74% (1)
4	I24T	T59A	T112A			0.74% (1)
5	F26S	T112A				0.74% (1)
6	Y36H	T53A	T112A			0.74% (1)
7	T59A	T112A	N185D			0.74% (1)
8	T59A	T112A	I201T	V213I		0.74% (1)
9	G111S	T112A				0.74% (1)
10	T112A	I192S				0.74% (1)
11	T112A	I209T				0.74% (1)
12	T13A	T112A				1.47% (2)
13	T59A	T112A	N185S			1.47% (2)
14	T13A	H90Y	T112A			3.68% (5)
15	A20T	T59A	T112A			3.68% (5)
16	A20T	T112A				5.89% (8)
17	T112A	S176N				6.62% (9)
18	A20T	T53I	T59A	T112A		7.35% (10)
19	T112A					24.26% (33)
20	T59A	T112A				37.50% (51)

haplotypes contained nonsynonymous single-nucleotide polymorphism. After comparing the amino acid sequences of these 20 variants aligned with that deduced from "Cambridge Reference Sequences" or the so-called Anderson sequence, we found that one amino acid difference was seen in 33 individuals, two differences in 75, three differences in 16, four differences in 11, and five differences in 1 (TABLE 1). The most common variant was T59A/T112A ($n = 51$), followed by T112A ($n = 33$). Because every examined individual had an A to G transition at position 8860, causing the amino acid substitution of alanine for threonine at codon 112, the Anderson sequence appears to be atypical at this site. There were 208 conserved residues (labeled H1 in FIG. 1) among 136 Taiwanese. Phylogenetic analysis using the neighbor-joining method showed no statistically well-supported monophyletic clade and represented a star-like evolution (data not shown).[2]

Recently, Lutz-Bonengel *et al.* reported 14 variants of *ATP6* among 109 individuals.[6] Comparing the H1 consensus with those of Lutz-Bonengel (labeled H2 in FIG. 1), we identified 194 conserved residues among these 245 individuals. Conserved residues were similarly obtained after comparing amino acids deduced from 241 DNA sequences of NCBI[7] (labeled H3 in FIG. 1) and from 34 DNA sequences of Pfam[8] (labeled H4 in FIG. 1). From these analyzed individuals ($n = 520$), a total of 78 variants of *ATP6* were identified. There are an additional 20 polymorphic res-

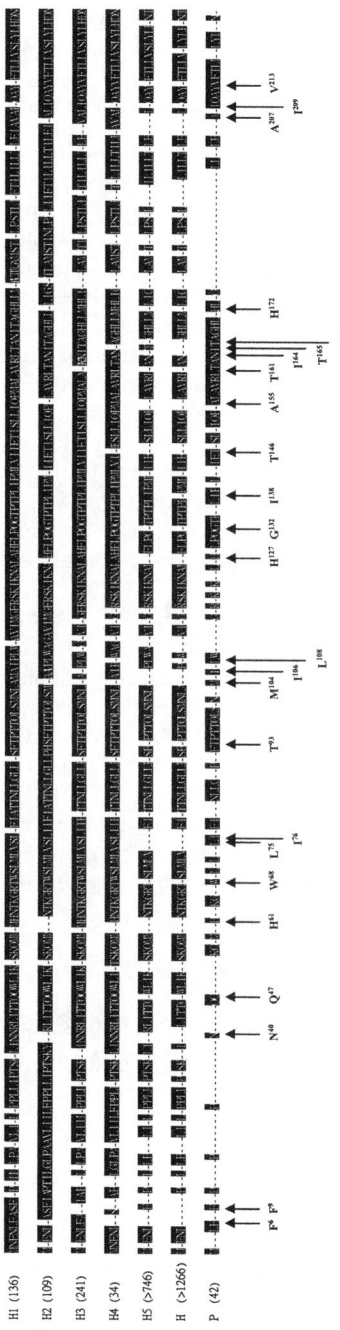

FIGURE 1. Conserved and variable amino acids of *ATP6* in humans and primates. Conserved amino acids are presented in *filled boxes*, whereas the variable residues are labeled by *dashes*. H_1, H_2, H_3, H_4, and H_5 as indicated to the *left* of the sequences represent the consensus identified in Taiwan ($n = 136$), German[6] ($n = 109$), NCBI[7] ($n = 241$), Pfam[8] ($n = 34$), and two public databases[9,10] ($n > 746$), respectively. H and P represent consensus in human ($n > 1,266$) and primate (42 species, excluding *Homo sapiens*), respectively. The *arrows* point to the human-specific variable amino acids.

idues reported in public databases,[9,10] of which the data were collected from more than 746 individuals. Taken together, there were 146 conserved amino acids and 80 variable residues (labeled H in FIG. 1) in *Homo sapiens ATP6*.

ATP6 *Comparison between* Homo sapiens *and Other Primates*

De Giorgi *et al.*[11] compared *ATP6* sequences and found that the amino acid identity of *ATP6* peptides between sea urchin and sea star (separating 560 million years earlier) is approximately 67%, whereas the identify is approximately 37% between echinoderms and mammals (vertebrates and invertebrates separating 600 million years earlier), suggesting that functional constraints operating at the amino acid level werephylum- to species-specific rather than time dependent. This is consistent with the concept that amino acids recover the expected topology more reliably than nucleotide sequences.[12] This might raise the question of whether there are amino acid residues that are varied in humans but conserved in other primates.

To identify primate consensus of *ATP6*, we randomly selected 42 species of primates from public databases for alignment. These included 29 species in Platyrrhini, 1 species in Tarsii, 2 species in Strepsirhini, and 10 species in Catarrhini other than *Homo sapiens*. From these primates excluding *Homo sapiens*, we found that 45% (102/226) of amino acids were conserved (labeled P in FIG. 1). Among 80 variable residues of human *ATP6*, 25 amino acids were conserved in other primates. These residues were F^6, F^9, N^{40}, Q^{47}, H^{61}, W^{68}, L^{75}, I^{76}, T^{93}, M^{104}, I^{106}, L^{108}, H^{127}, G^{132}, I^{138}, T^{146}, A^{155}, T^{161}, I^{164}, T^{165}, A^{166}, H^{172}, A^{207}, I^{209}, and V^{213} (arrows in FIG. 1).

The substitution rate of *ATP6* is relatively high (1.52×10^{-8}).[13] Wallace and associates[1] found that the *ATP6* gene was the most variable gene among human mtDNAs, particularly variable in the Arctic zones. From their point of view, the variation in *ATP6* would permit humans to adapt the changes in diet and climate. Based on this rationale, the 25 amino acids that are variable in humans but conserved in primates appear to be human-specific adaptation sites of *ATP6*. If *ATP6* variants have reduced coupling efficiency to synthesize ATP, the increased heat production is advantageous for survival in cold areas. As to the increased basal metabolic rate, it could be met by a higher caloric intake, such as consuming a high-fat diet.

ACKNOWLEDGMENTS

This work was supported by grant MMH-E-93002 from Mackay Memorial Hospital to C.-Y.T. The authors thank Ms. Tsu-Yen Wu for technical support and Ms. Shu-Huei Wang for secretarial expertise.

REFERENCES

1. MISHMAR, D. *et al.* 2003. Natural selection shaped regional mtDNA variation in humans. Proc. Natl. Acad. Sci. USA **100:** 171–176.
2. TZEN, C.Y., T.Y. WU & H.F. LIU. 2001. Sequence polymorphism in the coding region of mitochondrial genome enzcompassing position 8389-8865. Forensic Sci. Int. **120:** 204–209.
3. LUTZ-BONENGEL, S. *et al.* 2003. Sequence polymorphisms within the human mitochondrial genes MTATP6, MTATP8 and MTND4. Int. J. Legal Med. **117:** 133–142.

4. POETSCH, M. *et al.* 2003. The impact of mtDNA analysis between positions nt8306 and nt9021 for forensic casework. Mitochondrion **3:** 133–137.
5. ANDERSON, S. *et al.* 1981. Sequence and organization of the human mitochondrial genome. Nature **290:** 457–465.
6. LUTZ-BONENGEL, S. *et al.* 2003. Sequence polymorphisms within the human mitochondrial genes MTATP6, MTATP8 and MTND4. Int. J. Legal Med. **117:** 133–142.
7. NATIONAL CENTER FOR BIOTECHNOLOGY INFORMATION. http://www.ncbi.nlm.nih.gov/entrez/guery.fcgi; accessed March 12, 2004.
8. WASHINGTON UNIVERSITY IN ST. LOUIS, MO. http://pfam.wustl.edu/cgi-bin/getdesc?name=ATP-synt_A; accessed March 2, 2004.
9. MITOMAP. http://www.mitomap.org/; accessed February 15, 2004.
10. Uppsala University. http://www.genpat.uu.se/mtDB/polysites.html; accessed March 16, 2004.
11. DE GIORGI, C. *et al.* 1997. Lineage-specific evolution of echinoderm mitochondrial ATP synthase subunit 8. J. Bioenerg. Biomembr. **29:** 233–239.
12. RUSSO, C.A., N. TAKEZAKI & M. NEI. 1996. Efficiencies of different genes and different tree-building methods in recovering a known vertebrate phylogeny. Mol. Biol. Evol. **13:** 525–536.
13. INGMAN, M. *et al.* 2000. Mitochondrial genome variation and the origin of modern humans. Nature **408:** 708–713.

Repeated Ovarian Stimulations Induce Oxidative Damage and Mitochondrial DNA Mutations in Mouse Ovaries

HSIANG-TAI CHAO,[a] SHU-YU LEE,[b] HORNG-MO LEE,[c] TIEN-LING LIAO,[b] YAU-HUEI WEI,[d] AND SHU-HUEI KAO[b]

[a]*Department of Obstetrics and Gynecology, Taipei Veterans General Hospital, Taipei, Taiwan*

[b]*Graduate Institute of Biomedical Technology, Taipei Medical University, Taipei, Taiwan*

[c]*Graduate Institute of Cellular and Molecular Biology, Taipei Medical University, Taipei, Taiwan*

[d]*Department of Biochemistry, National Yang-Ming University, Taipei, Taiwan*

ABSTRACT: Superovulation by injection of exogenous gonadotropin is the elementary method to produce *in vivo*–derived embryos for embryo transfer in women. Increased oocyte aneuploidy, embryo mortality, fetal growth retardation, and congenital abnormalities have been studied at higher-dose stimulations. Ovarian and oocyte biological aging possibly may have adverse implications for human oocyte competence with repeated hyperstimulation. In this study, we found that reduced competence for the human oocyte has been associated with degenerative embryo upsurge during embryo culture and failure to develop into the blastocyst stage in the three, four, five, and six stimulation cycles. On the other hand, the numbers of ovulated oocytes were decreased in the groups with more ovarian stimulation. More aggregated mitochondria were found in the cytoplasm of the repetitively stimulated embryos. Higher amounts of oxidative damage including 8-OH-dG, lipoperoxides, and carbonyl proteins were also revealed in the ovaries with more cycle numbers of ovarian stimulation. Higher proportions of mtDNA mutations were also found. The detected molecular size of the mutated band was approximately 675 bp. Increased amounts of carbonyl proteins were also revealed after repeated stimulation. An understanding of the relationship between oocyte competence and ovarian responses to stimulation in the mouse may provide insights into the origin of oocyte defects and the biology of ooplasmic aging that could be of clinical relevance in the diagnosis and treatment of human infertility.

KEYWORDS: oocyte; aging; ovarian stimulation; oxidative damage; mtDNA mutations

Address for correspondence: Shu-Huei Kao, Graduate Institute of Biomedical Technology, Taipei Medical University, Taipei, Taiwan 110. Voice: +886-2-27361661-3312; fax: +886-2-27324510.
kaosh@tmu.edu.tw

INTRODUCTION

Despite the recent development of the technology of *in vitro* production of embryos, superovulation by injection of exogenous gonadotropin is still the fundamental method to produce *in vivo*–derived embryos for embryo transfer in women.[1] Ovulation stimulation permits the growth and development of supernumerary dominant follicles and the ability to time the initiation of preovulatory oocyte maturation. However, the yield and quality of embryo raised after ovarian hyperstimulation for *in vitro* fertilization (IVF) are variable and unpredictable owing to variations in ovarian response, fertilization rate, and embryo development.[1] In mammals such as mice, rats, and hamsters, reduced fertility and preimplantation and postimplantation mortality have been indicated as consequences of a single round of ovarian stimulation using standard doses of gonadotrophins.[2,3] At higher doses, increased frequencies of oocyte aneuploidy,[4] embryo mortality, fetal growth retardation, and congenital abnormalities have been reported.[5] Reduced viability with ovarian stimulation is often attributed to adverse "maternal" factors such as inadequate uterine synchrony or receptivity.[5] Comparing natural and stimulated ovarian cycles in animals have shown detrimental effects of gonadotropin stimulation, possibly caused by a failure in the coordination of folliculogenesis and oogenesis.

It is common for women who have undergone several cycles of ovarian stimulation before pregnancy is achieved to have declined ovarian competence.[6] Maternal age-related depletion of ovarian reserve that normally begins in women after about age 30 to 33 years is reported.[7] Declined ovarian competence is age related or occurs prematurely in women with early-age menopause.[8] The decreased competence is found in the second-round *in vitro* fertilization cycle in women of various ages.[9,10] The number of stimulation cycles may represent an important factor related to the probability of pregnancy.

In this study, we proposed that declined oocyte quality and developmental competence were produced from the aged ovaries after repeated gonadotropin stimulations. We also searched what contributed to adverse oocyte competence after repeated ovarian hyperstimulation. Mitochondrial DNA (mtDNA) mutation is generally demonstrated as the common biomarkers of human aging and degenerative disease.[11,12] In this study, we examined the occurrence of mtDNA mutation and oxidative damage in the repeat stimulated ovaries. Understanding of the relationship between oocyte competence and ovarian responses to stimulation in the mouse may provide insights into the origin of oocyte defects that could be of clinical relevance in the diagnosis and treatment of human infertility.

MATERIALS AND METHODS

Ovarian Stimulation Schedule

Cycles of ovarian stimulation were initiated in 5–6-week-old B6/57J mice according to the standard protocol of 5 IU of pregnant mare serum gonadotropin (PMSG; Sigma Chemical Co., St. Louis, MO) administered and followed after 48 h by 5 IU of human chorionic gonadotropin (HCG; Serono Singapore Pte Ltd., Taiwan Branch). Six cycles of ovarian stimulation were performed in the same week to prevent the aging effect. Ovulated oocytes and embryos were collected from the oviductal ampulla at 14 and 38 h after HCG injection, respectively. Cumulus masses were disassociated with hyaluronidase (Sigma Chemical Co.), and adherent coronal cells were removed by repeated passage of oocytes through a glass micropipette. Ovaries were also removed for light microscopic histological analysis between stimulations at each interval. The quality and development competence of collected oocytes were examined.

Determination of the mtDNA Mutations in Human or Mouse Tissue

Total DNA Extraction

At the end of ovulation induction, the mouse tissues were collected including ovaries and liver. All the samples were checked by microscopy for morphological changes and stored at $-196°C$ until analysis. All tissue samples were minced into small pieces and incubated at 56°C for 2 h in 50 µL lysis buffer containing 2% SDS and 50 mM Tris-HCl (pH 8.3) and then followed by the phenol-chloroform extraction method. All the DNA samples were finally conserved in 200 µL of 10 mM Tris-HCl, pH 8.3.

Synthesis of Oligonucleotide Primers

Oligonucleotide primers used for the amplification of the target sequences of mtDNA were chemically synthesized by Protec, Inc. (Taipei, Taiwan). The nucleotide sequences and sizes of the PCR products obtained from these primer pairs are L8265 (5′-AATTACAGGCTTCCGACACA-3′) and H13375 (5′-TTTAGGCTTAG-GTTGAGAGA-3′).

Detection of Mutated mtDNA by PCR

The desired target sequence of mtDNA were amplified from 15 to 20 ng of each DNA sample in a 50-µL reaction mixture containing 200 µM of each dNTP, 0.4 µM of each primer, 1 unit of Ampli-Taq DNA polymerase (Perkin-Elmer/Cetus, Roche Molecular System, Inc., Branchburg, NJ), 50 mM KCl, 1.5 mM $MgCl_2$, and 10 mM Tris-HCl, pH 8.3. PCR was conducted for 35 cycles in a DNA thermal cycler (Perkin-Elmer/Cetus) using the thermal profile of denaturation at 94°C for 40 s, annealing at 55°C for 40 s, and primer extension at 72°C for 40 s. Long-range PCR procedures were followed as the thermal profile of denaturation at 94°C for 2 min, annealing at 68°C for 1 min, and primer extension at 72°C for 2 min.

Analysis of Oxidative Damage in Cellular Molecules

Determination of 8-OH-dG

An aliquot of 100 µg of DNA dissolved in 100 µL of 10 mM Tris-HCl (pH 7.4)/ 0.1 mM DFAM was digested by incubation with 1 µL of DNase I (20 units/µl) and 11 µL of 0.1 M $MgCl_2$ solution at 37°C for 30 min. After adjusting the pH to 5.0 by adding 4.8 µL of 1 M sodium acetate (pH 5.3) and 1.2 µL of 0.1 M $ZnSO_4$, we digested the DNA sample with 5 µL of nuclease P1 (1 unit/3 µL in 20 mM sodium acetate, pH 5.3) at 65°C for 10 min. The DNA molecules were hydrolyzed to the corresponding nucleosides by incubation with 5 µL of 1 U/µL alkaline phosphatase for 30 min at 37°C. Processed DNA samples are separated by C-18 column (particle size 5 µm, 200 × 4.6 mm; JT Baker, Inc., Phillipsburg, NJ) on an HPLC system (Jasco, Inc., Easton, MD) connected in series with an ECD detector (Bioanalytical Systems, Inc., West Lafayette, IN) and a UV detector (at 254 nm). Elution was performed at the flow rate of 0.8 mL/min for 40 min with a mobile phase consisting of 12.5 mM citric acid, 25 mM sodium acetate, and 10 mM acetic acid containing 6% of methanol (pH 5.8).

Determination of Lipid Peroxide

A 50-µL aliquot of each sample was pipetted into a test tube containing 0.6 mL of 0.44 M phosphoric acid. After mixing, 0.2 mL of 42 mM thiobarbituric acid (TBA) solution was added to a final concentration of 7 mM and then placed in a 95°C dry bath for 1 h. The samples then were cooled and neutralized with 1 N NaOH in methanol before HPLC analysis. An aliquot of 20 µL supernatant obtained above is injected into a narrow-pore C_{18} column (4.6 × 250 mm, particle size 5 µm) using a Jasco PU-980 pump (Tokyo, Japan) with a solvent system made of methanol and 50 mM phosphate buffer (pH 6.8; 4:6, v/v) at the flow rate of 1 mL/min. The eluent is monitored with a fluorescence detector (excitation at 525 nm; emission at 550 nm).

Determination of Carbonyl Protein

For the assessment of protein carbonyl content, plasma proteins were derivatized with DNP before SDS–polyacrylamide gel electrophoresis on 10% gradient gels, followed by Western transfer to nitrocellulose filters. Protein equivalent to 0.12 µL of plasma was loaded onto each lane for electrophoresis. DNP-derivatized tissue protein carbonyl groups were sequentially reacted with rabbit anti–DNP and goat anti–rabbit immunoglobulin G (IgG) antibodies (OxyBlot Oxidized Protein Detection Kit; Oncor, Gaithersburg, MD, now supplied by Intergen Company, Purchase, NY) followed by chemiluminescence detection with ECL Western Blotting Detection Reagents (Amersham Pharmacia Biotech, Piscataway, NJ).

Confocal Images of Mitochondrial Membrane Potential of the Repeated Stimulated Oocytes

The collected oocytes were gently stained for mitochondria in 0.4 mL PBS containing 0.5 µmol JC-1 (Molecular Probes, Eugene, OR) for 30 min at room temperature. The cells then were imaged with a Olympus OLS 3000 laser scanning confocal microscope equipped with a krypton/argon laser (excitation 488 nm and emission 610 nm). Fluorescence was detected with a 585-nm–long pass filter.

RESULTS

Declined Development Competence of Embryos

In this study, we detected the development competence of embryos from differentially repeated stimulation (TABLE 1). Large populations of embryos were degenerative and failed to develop into blastocyst stage in the three, four, five, and six

TABLE 1. Characterization of the developmental competencies of mouse embryos

Stimulation cycles	Day 1			Day 2			Day 3			Day 4			Day 5		
	A	1c	2c	A	2c	4c	A	4c	8c	A	8c	M	A	M	B
1	4	16	8	5	4	15	4	4	11	5	3	7	3	3	4
2	6	12	4	8	6	2	5	2	1	3	0	0	0	0	0
3	8	10	2	5	4	3	4	3	0	0	0	0	0	0	0
4	10	6	2	5	2	1	3	0	0	0	0	0	0	0	0
5	9	6	1	6	1	0	1	0	0	0	0	0	0	0	0
6	5	3	0	3	0	0	0	0	0	0	0	0	0	0	0

A, apoptosis; 1c, fertilized one-cell stage; 2c, two-cell stage; 4c, four-cell stage; 8c, eight-cell stage; M, morula stage of embryo; B, blastocyst stage.

TABLE 2. Declined oocyte qualities revealed with repeated ovarian stimulation

Stimulation cycle	No. of mice	Collected oocytes	Degenerative oocytes	GV stage	MII stage
1	3	48	6	6	36
3	4	59	20	16	23
6	5	71	64	2	5

GV, geminal vesicle; MII, metaphase II.

stimulation cycles (TABLE 1). The fertilization rates were seen to be similar in all tested groups. On the other hand, the declined qualities of ovulated oocytes were found in the groups with repeated stimulation (TABLE 2). A higher proportion of degenerative oocytes and immature oocytes (GV stage) were collected from the mice with six stimulation cycles (TABLE 2). After staining the oocytes with JC-1 (FIG. 1), we observed the embryos with a confocal microscope. More aggregated mitochondria that were found in aged embryos were located in the cytoplasm of embryos with repeated stimulation cycles.[13,14]

Oxidative Damage in the Ovaries and Livers

We detected the oxidative damage including 8-OH-dG and lipoperoxide contents (malondialdehyde) and carbonyl protein in the ovaries and livers with differential stimulation. It was clear that higher amounts of oxidative damage existed in the ovaries with more cycles of ovarian stimulation (TABLE 3). There was a positive correlation between oxidative damage and cycles of repeated ovarian stimulation. Strongly immunoreacted proteins for the carbonyl groups were detected in the high stimulation groups (FIG. 2).

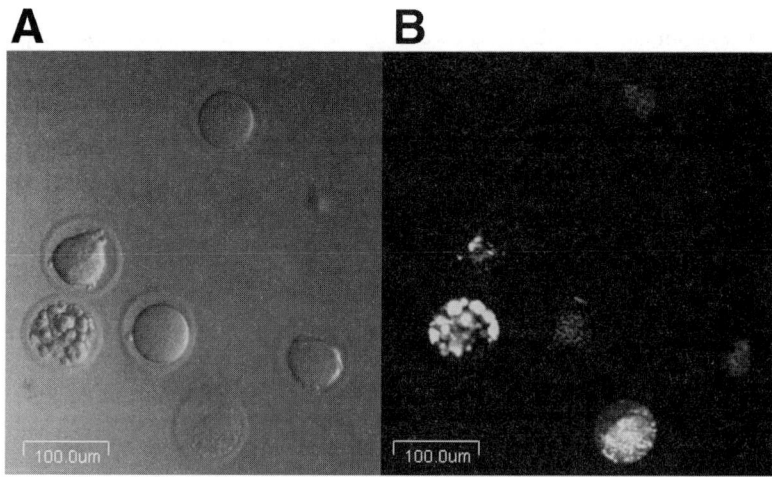

FIGURE 1. Morphological change in the oocytes with six stimulation cycles. All collected oocytes were stained with JC-1. **A** was visualized under phase-contrast microscopy. **B** was visualized under confocal microscopy.

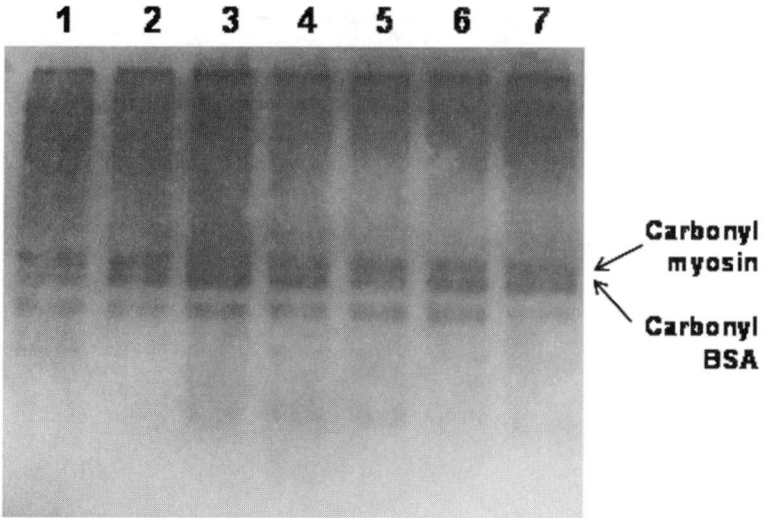

FIGURE 2. Immunochemical detection of protein carbonyls in mouse ovaries with differentially repeated stimulation. Proteins (10 mg) were electrophoresed on by DNPH-SDS-PAGE under reducing conditions and transferred to a nylon membrane. Oxidized proteins were detected immunochemically as described in the text. *Lane 1*, young group (4 weeks); *lane 2*, one stimulation cycle; *lane 3*, two cycles; *lane 4*, three cycles; *lane 5*, four cycles; *lane 6*, five cycles; and *lane 7*, six cycles of stimulation.

TABLE 3. Content of lipid peroxides and 8-OH-dG of mouse ovaries characterized

No. of cycles	Lipid peroxides (pmol/mg protein)	8-OH-dG (8-OH-dG/dG, ×10⁻³%)
1	2.156 ± 0.024	0.086 ± 0.014
2	2.756 ± 0.376	3.487 ± 0.658
3	2.987 ± 0.746	5.438 ± 1.633
4	3.255 ± 0.926	6.225 ± 1.677
5	3.651 ± 1.004*	8.462 ± 2.889
6	4.579 ± 1.278*	10.941 ± 2.546*

$*P < 0.05$.

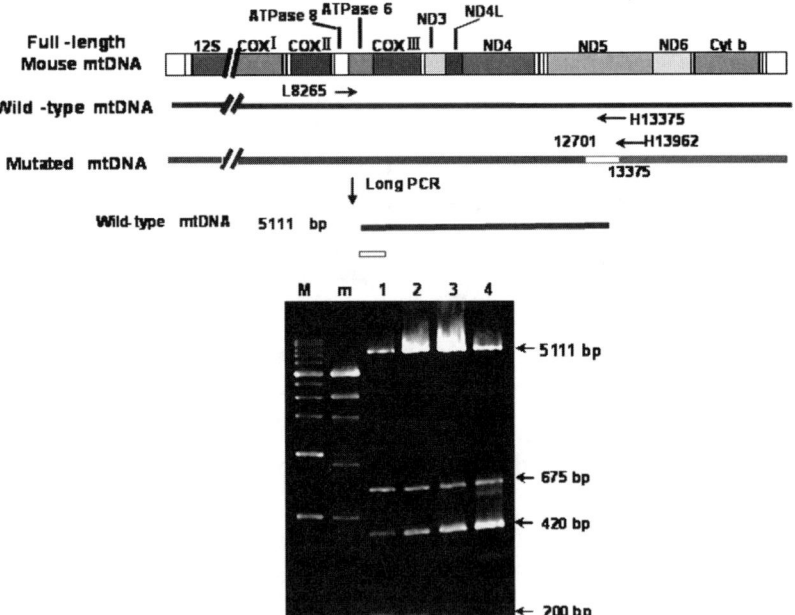

FIGURE 3. Demonstration of large-scale deletions of mtDNA in mouse ovaries with differentially repeated stimulation by long-range PCR. Using the primer sets L8265-H13375, we detected large-scale deletions of mtDNA in repeated stimulated ovaries. The 5,111-bp band was amplified from the wild-type mtDNA. There were three mutated bands generated; the molecular sizes were approximately 675, 420, and 200 bp. *Lane M*, 100-bp DNA ladder; *lane m*, 1-Kb DNA ladder; *lane 1*, one stimulation cycle; *lane 2*, three repeated cycles; *lane 3*, five cycles; and *lane 4*, six cycles of stimulation.

Accumulation of Large-scale Deletion of Mouse mtDNA with Repeated Stimulation

Higher proportions of mtDNA large-scale deletions were also shown when there were more cycles of ovarian stimulation. Using the primer sets L8265-H13375, we detected large-scale deletions of mtDNA in repeated stimulated ovaries. The 5,111-bp band was amplified from the wild-type mtDNA. There were three mutated bands generated; the molecular sizes were approximately 675, 420, and 200 bp (FIG. 3).

DISCUSSION

Hormonal stimulation increases the probability of pregnancy by increasing the number of oocytes retrieved, which again might provide us with more embryos to undergo embryo transfer and implantation. The maturation of follicles and oocytes are dependent on the hormonal environment. The differential hormone environment is found between the natural cycle and stimulation cycle. Ovarian stimulation was shown to have an impact on apoptosis in the granulose cells and ovary bioaging.[15–17] Hormone environment and culture conditions contribute to the developmental capacity of an embryo. Recently, Van Blerkom and Davis[10] demonstrated the effect of repeated ovarian stimulation in mice, resulting in a significant increase in the frequency of spindle defects resulting in chromosomal errors with each series of ovarian stimulation.[18]

In this study, reduced competence for the human embryo has been associated with an upsurge of degenerative embryos during embryo culture and failure to develop into blastocyst stage in the three, four, five, and six stimulation cycles. On the other hand, the numbers of ovulated oocytes were decreased in the groups with ovarian stimulation. More immature or degenerative oocytes were collected in the ovaries with repeated stimulation. More aggregated mitochondria were found in the cytoplasm of the repetitively stimulated embryos. The aggregated pattern was revealed in the aged oocytes.[18] Declined ovarian competence is suggested to be age related. The detrimental factors must be generated in the environment of the repeated stimulated ovary.

In this study, the multiple mtDNA rearrangements and oxidative damage were detected in the ovaries with repeated stimulation. Large amounts of oxidative damage including 8-OH-dG, lipoperoxide contents (malondialdehyde), and carbonyl proteins were also revealed in the ovaries with more cycle numbers of ovarian stimulation. According to these data, the increased oxidative stress is raised in the environment of a stimulated ovary. However, an understanding of the relationship between oocyte competence and ovarian responses to stimulation in the mouse may provide insights into the origin of oocyte defects and the biology of ooplasmic aging that could be of clinical relevance in the diagnosis and treatment of human infertility.

ACKNOWLEDGMENTS

This work was supported by research grant NSC90-2320-B-038-048 from the National Science Council of the Republic of China and grant VGH93-299 from the Taipei Veterans Hospital.

REFERENCES

1. BOLAND, M.P., D. GOULDING & J.F. ROCHE. 1991. Alternative gonadotropins for superovulation in cattle. Theoriogenology **35:** 5–17.
2. MCKIERNAN, S. & B. BAVISTER. 1998. Gonadotropin stimulation of donor females decreases post-implantation viability of cultured one-cell hamster embryos. Hum. Reprod. **13:** 724–729.
3. VOGEL, R. & H. SPIELMANN. 1992. Genotoxic and embryotoxic effects of gonadotropin-hyperstimulated ovulation of murine oocytes and preimplantation embryos, and term fetuses. Reprod. Toxicol. **6:** 329–333.
4. MAILHES, J., F. MARCHETTI & D. YOUNG. 1995. Synergism between gonadotropins and vinblastine relative to the frequencies of metaphase I, diploid and aneuploid mouse oocytes. Mutagenesis **10:** 185–188.
5. MA, S., D. KALOUSEK, B. YUEN & Y. MOON. 1997. Investigation of effects of pregnant mare serum gonadotropin (PMSG) on the chromosomal complement of CD-1 mouse embryos. J. Assist. Reprod. Genet. **14:** 162–169.
6. DE BOER, P., F. VAN DER HOEVEN, E. WOLTERS & J. MATTHEIJ. 1991. Embryo loss, blastomere development and chromosomal constitution after human chorionic gonadotropin-induced ovulation in mice and rats with regular cycles. Gynecol. Obstet. Invest. **32:** 200–205.
7. NAVOT, D., P.A. BERGH, M.A. WILLIAMS, *et al.* 1991. Poor oocyte quality rather than implantation failure as a cause of age-related decline in female fertility. Lancet **337:** 1375–1377.
8. KLINE, J., A. KINNEY, B. LEVIN & D. WARBURTON. 2000. Trisomic pregnancy and earlier age menopause. Am. J. Hum. Genet. **67:** 395–404.
9. COPPERMAN, A.B., C.E. SELICK, L. GRUNFELD, *et al.* 1995. Cumulative number and morphological score of embryos resulting in success: realistic expectations from *in vitro* fertilization-embryo transfer. Fertil. Steril. **64:** 88–92.
10. VAN BLERKOM, J. & P. DAVIS. 2001. Differential effects of repeated ovarian stimulation on cytoplasmic and spindle organization in metaphase II mouse oocytes matured *in vivo* and *in vitro*. Hum. Reprod. **16:** 757–764.
11. KEEFE, D.L., T. NIVEN-FAIRCHILD, S. POWELL & S. BURADAGUNTA. 1995. Mitochondrial deoxyribonucleic acid deletions in oocytes and reproductive aging in women. Fertil. Steril. **64:** 577–583.
12. WAN, Q.H., H. WU, T. FUJIHARA & S.G. FANG. 2004. Which genetic marker for which conservation genetics issue? Electrophoresis **25:** 2165–2176.
13. SZOLTYS, M., J. GALAS, A. JABLONKA & Z. TABAROWSKI. 1994. Some morphological and hormonal aspects of ovulation and superovulation in the rat. J. Endocrinol. **141:** 91–100.
14. ZIEBE, S., S. BANGSBOLL, K.L.T. SCHMIDT, *et al.* 2004. Embryo quality in natural versus stimulated IVF cycles. Hum. Reprod. **19:** 1457–1460.
15. COMBELLES, C.M.H. & D.F. ALBERTINI. 2003. Assessment of oocyte quality following repeated gonadotropin stimulation in the mouse. Biol. Reprod. **68:** 812–821.
16. WHELAN, J.G., III & N.F. VLAHOS. 2000. The ovarian hyperstimulation syndrome. Fertil. Steril. **73:** 883–896.
17. TARÍN, J.J., S. PÉREZ-ALBALÁ & A. CANO. 2001. Cellular and morphological traits of ooytes retrieved from aging mice after exogenous ovarian stimulation. Biol. Reprod. **65:** 141–150.
18. WILDING, M., B. DALE, M. MARINO, *et al.* 2001. Mitochondrial aggregation patterns and activity in human oocytes and preimplantation embryos. Hum. Reprod. **16:** 909–917.

Calcium Stimulates Mitochondrial Biogenesis in Human Granulosa Cells

TIEN-SHUN YEH,[a] JAU-DER HO,[b] VIVIAN WEI-CHUNG YANG,[c] CHII-RUEY TZENG,[d] AND RONG-HONG HSIEH[e]

[a]*Graduate Institute of Cell and Molecular Biology, Taipei Medical University, Taipei, Taiwan 112, Republic of China*

[b]*Department of Ophthalmology, Taipei Medical University Hospital, Taipei, Taiwan 112, Republic of China*

[c]*Graduate Institute of Biomedical Materials, Taipei Medical University, Taipei, Taiwan 112, Republic of China*

[d]*Department of Obstetrics and Gynecology, Taipei Medical University Hospital, Taipei, Taiwan 112, Republic of China*

[e]*Department of Nutrition and Health Sciences, Taipei Medical University, Taipei, Taiwan 112, Republic of China*

ABSTRACT: Ovarian granulosa cells are known to play a key role in regulating ovarian physiology. Age increases apoptosis in follicular granulosa cells and subsequently decreases ovarian fecundity. The aging ovary also contains fewer follicles that possess fewer granulosa cells. The viability of follicular granulosa cells may be essential for development of the oocyte. Calcium ion plays an important role in a variety of biological processes, including gene expression, cell cycle regulation, and cell death. To study the ability of mitochondrial biogenesis in human granulosa cells, we determined the mitochondrial marker proteins, including the nuclear-encoded NADH-ubiquinone oxidoreductase alpha subunit 9 (NDUFA9) and mitochondrial-encoded COX I, after treatment of the cells with the calcium ionophore A23187. We showed that the expression of these mitochondrial marker proteins in human granulosa cells increased with changes in cytosolic Ca^{2+} using the ionophore A23187. Treatment of granulosa cells with 0.5 μM of A23187 for 120 h increased the levels of NDUFA 9 and COX I subunit by up to 2.6- and 2.4-fold, respectively. Raising Ca^{2+} by exposing granulosa cells to 1 μM of A23187 for 48 h significantly increased mitochondrial transcription factor (mtTFA) gene expression by up to 2.9-fold. Our results indicate that the adaptive responses of granulosa cells to increased Ca^{2+} may include upregulation of mitochondrial proteins and that mtTFA may be involved in such a mitochondrial biogenesis pathway.

KEYWORDS: calcium; granulosa cell; mitochondria

Address for correspondence: Dr. Rong-Hong Hsieh, School of Nutrition and Health Sciences, Taipei Medical University, Taipei, Taiwan 112, Republic of China. Voice: +886-2-27361661 ext. 6551-128; fax: +886-2-27373112.
 hsiehrh@tmu.edu.tw

INTRODUCTION

Ovarian granulosa cells play a major role in regulating ovarian physiology, including ovulation and luteal regression.[1] Granulosa cells secrete a wide variety of growth factors that may attenuate the action of gonadotrophin in the ovary in paracrine-autocrine processes.[2,3] Most of these factors do not directly affect the oocyte but exert their action via granulosa cells. The presence of granulosa cells appears to be beneficial for oocyte maturation and early development.[4] Nevertheless, granulosa cells might also have a negative effect on the oocyte. It has been demonstrated that the increased apoptotic potential in oocytes of aged mice is caused by the presence of granulosa cells.[5] The decline in reproductive ability with age in women is associated with a loss of follicles and a decrease in oocyte quality. Aging-associated apoptosis increases in follicular granulosa cells and consequently decreases ovarian fecundity.[6,7]

In eukaryotic cells, mitochondria are specialized organelles that catalyze the formation of ATP. Two distinct genomes exist in all eukaryotic cells. One is located in the nucleus and is transmitted in a Mendelian fashion, whereas the other is located in mitochondria and is transmitted by maternal inheritance. The mitochondria in an oocyte must have produced and stored all the energy required for the resumption of meiosis II, fertilization, and development of the embryo.[8,9] Deficiency in mitochondrial ATP production may be associated with impairment of oocyte fertilization.[10,11] Normal function of mitochondria in follicular granulosa cells may be needed for growth factor production and subsequent paracrine effects in oocyte development. Calcium ion plays an important role in a variety of biological processes, including gene expression, cell cycle regulation, and cell death. In this study, the ability of mitochondrial biogenesis in human granulosa cells was determined by incubation of the cells with calcium ion.

MATERIALS AND METHODS

Collection of Human Granulosa Cells

We collected human granulosa cells from patients undergoing *in vitro* fertilization by gonadotrophin-stimulated cycles. This study was approved by the Institutional Review Board of Taipei Medical University Hospital. Follicular fluids aspirated from each patient were pooled and were centrifuged for 10 min at $800 \times g$ at room temperature.

Cell Cultures and Drug Treatment

A23187 was purchased from Sigma (St. Louis, MO). It was prepared as a stock solution in Me_2SO at concentrations of 0.25, 0.5, and 1 µM. Granulosa cells were cultured with HTF medium containing 2% human plasma in an incubator at 37°C. The dose-dependent effects of A23187, from 0.25 to 1 µM, on mitochondrial respiratory chain subunits were evaluated. Total RNA extracted from harvested cells was used as templates, and cDNA was prepared using the RNA extraction kit and reverse-transcription polymerase chain reaction kit from Ambion (Austin, TX).

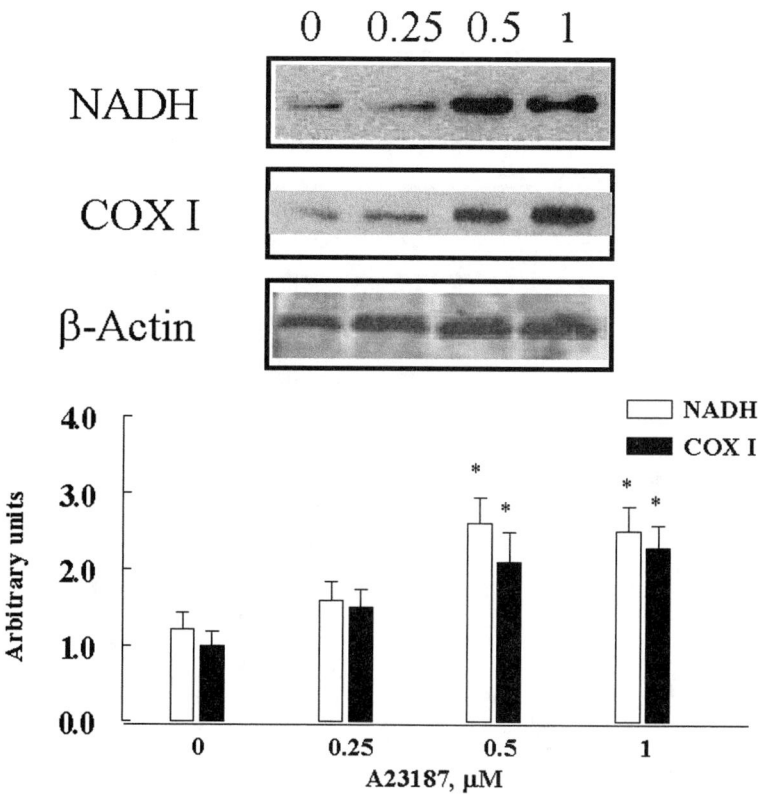

FIGURE 1. Western blots of nuclear- and mtDNA-encoded proteins in human granulosa cells treated for 96 h with A23187 at concentrations ranging from 0.25 to 1 μM.

RESULTS

To study the effects of calcium ion on mitochondrial biogenesis, the mitochondrial marker proteins were determined. Western blot analysis was used to determine the effect of A23187 treatment on the nuclear-encoded NADH-ubiquinone oxidoreductase alpha subunit 9 (NDUFA9) and mitochondrial-encoded COX I levels. A23187 caused a dose-dependent increase in NDUFA9 and COX I levels of up to 2.5- and 2.3-fold ($P < 0.05$; FIG. 1). The effect of A23187 on both the nuclear- and mitochondrial-encoded subunits was also time dependent (FIG. 2). A23187 exposure for 120 h increased NDUFA9 and COX I levels by 2.6- and 2.4-fold, respectively ($P < 0.05$). The possible role of mitochondrial transcription factor A (mtTFA) in mitochondrial biogenesis also was examined. A23187 exposure for 24 and 48 h at 0.5 and 1 μM resulted in an increase in mtTFA mRNA levels by 2.5- (0.5 μM, 24 h), 2.9- (0.5 μM, 48 h), 3.1- (1 μM, 24 h), and 2.9-fold increases (1 μM, 48 h). Taken together, these data show calcium-dependent increases in both nuclear-encoded NDUFA9

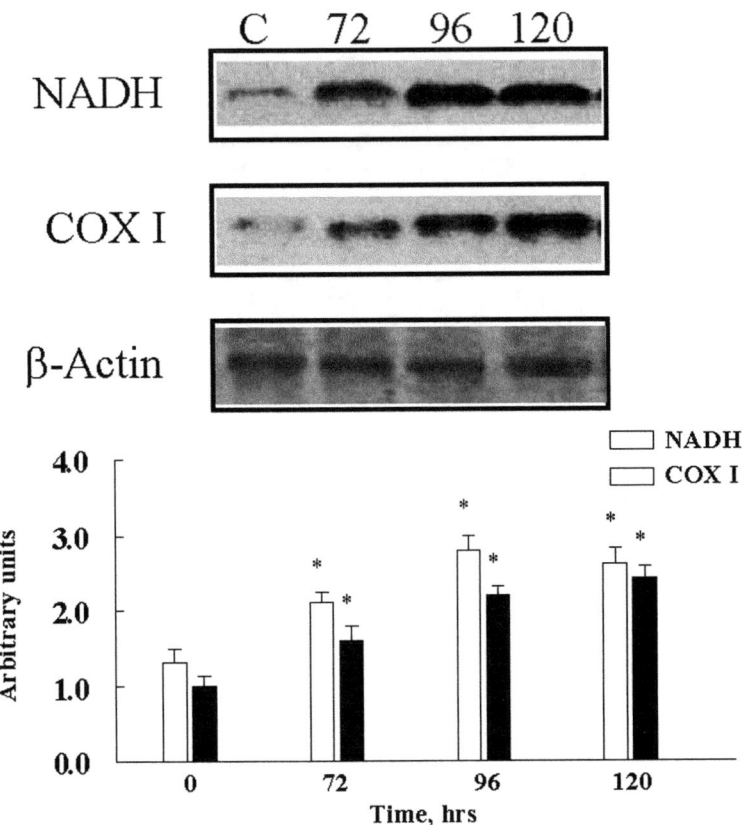

FIGURE 2. Western blots of nuclear- and mtDNA-encoded proteins in human granulosa cells treated with 0.5 μM of A23187 for 72 to 120 h.

and mitochondrial-encoded COX I expression levels and suggest that mtTFA may be involved in this process.

DISCUSSION

Mitochondrial biogenesis requires the symphonious expression of mtDNA and nuclear genes that both encode mitochondrial proteins and their regulatory factors. One of these factors is mtTFA, which activates mtDNA transcription and replication.[12] Cellular responses to environmental changes including energy demands should be reflected in changes in the physiological state of mitochondria. Our data indicate increased expression levels of mtTFA mRNA in human granulosa cells after Ca^{2+} stimulation. This event was accompanied by changes in mitochondrial biogenesis, reflected by an increase in NDUFA9 and COX I levels. These findings are similar to an increase in mitochondrial biogenesis induced by Ca^{2+} in muscle,[13] and by mitogens in human lymphocytes[14] and murine splenocytes.[15]

Ca^{2+} acts as a second messenger in a variety of biological processes.[16,17] In this study, the effects of Ca^{2+} on both nuclear- and mitochondrial-encoded subunits were dose and time dependent. Changes in expression levels of mtTFA were also dependent on Ca^{2+}. Previous studies have provided evidence that cytosolic Ca^{2+} concentration is involved in regulating mitochondrial biogenesis in skeletal muscle.[13,18] The adaption of mitochondria in this process involves induction of mtTFA.[13] This experimental evidence indicates that Ca^{2+} upregulates mtTFA expression, leading to mitochondrial biogenesis. Results from this study are in accordance with the contention that Ca^{2+} is one of the signals that mediate mitochondrial biogenesis in differentiated granulosa cells.

ACKNOWLEDGMENTS

This work was supported by research grants NSC 91-2320-B-038-026 and NSC 92-2320-B-038-045 from the National Science Council of the Republic of China.

REFERENCES

1. AMSTERDAM, A. & N. SELVARAJ. 1997. Control of differentiation, transformation, and apoptosis in granulosa cells by oncogenes, oncoviruses, and tumor suppressor genes. Endocrinol. Rev. **18:** 435–461.
2. NAKAMURA, T. *et al.* 1990. Activin-binding protein from rat ovary is follistatin. Science **247:** 836–838.
3. GRAS, S. *et al.* 1996. Transient periovulatory expression of pituitary adenylate cyclase activating peptide in rat ovarian cells. Endocrinology **137:** 4779–4785.
4. CANIPARI, R. 2000. Oocyte-granulosa cell interactions. Hum. Reprod. Update **6:** 279–289.
5. PEREZ, G.I. & J.L. TILLY. 1997. Cumulus cells are required for the increased apoptotic potential in oocytes of aged mice. Hum. Reprod. **12:** 2781–2783.
6. SEIFER, D.B. & F. NAFTOLIN. 1998. Moving toward an earlier and better understanding of perimenopause. Fertil. Steril. **69:** 387–388.
7. BOPP, B.L. & D.B. SEIFER. 1998. Oocyte loss and the perimenopause. Clin. Obstet. Gynecol. **41:** 898–911.
8. PIKO, L. & K.D. TAYLOR. 1987. Amounts of mitochondrial DNA and abundance of some mitochondrial gene transcripts in early mouse embryos. Dev. Biol. **123:** 364–374.
9. EBERT, K.M., H. LIEM & N.B. HECHT. 1988. Mitochondrial DNA in the mouse preimplantation embryo. J. Reprod. Fertil. **82:** 145–149.
10. HSIEH, R.H. *et al.* 2002. Multiple rearrangements of mitochondrial DNA in unfertilized human oocytes. Fertil. Steril. **77:** 1012–1017.
11. HSIEH, R.H. *et al.* 2004. Decreased expression of mitochondrial genes in human unfertilized oocytes and arrested embryos. Fertil. Steril. **81:** 912–918.
12. VIRBASIUS, J.V. & R.C. SCARPULLA. 1994. Activation of the human mitochondrial transcription factor A gene by nuclear respiratory factors: a potential regulatory link between nuclear and mitochondrial gene expression in organelle biogenesis. Proc. Natl. Acad. Sci. USA **91:** 1309–1313.
13. OJUKA, E.O. *et al.* 2003. Raising Ca^{2+} in L6 myotubes mimics effects of exercise on mitochondrial biogenesis in muscle. FASEB J. **17:** 675–681.
14. KAIN, K.H., V.L. POPOV & N.K. HERZOG. 2000. Alterations in mitochondria and mtTFA in response to LPS-induced differentiation of B-cells. Biochim. Biophys. Acta **1494:** 91–103.

15. RATINAUD, M.H. & R. JULIEN. 1986. Variation of cellular mitochondrial activity in culture: analysis by flow cytometry. Biol. Cell **58:** 169–172.
16. HARDINGHAM, G.E. *et al.* 1997. Distinct functions of nuclear and cytoplasmic calcium in the control of gene expression. Nature **385:** 260–265.
17. MEANS, A.R. 1994. Calcium, calmodulin and cell cycle regulation. FEBS Lett. **347:** 1–4.
18. LAWRENCE, J.C. JR. & W.J. SALSGIVER. 1983. Levels of enzymes of energy metabolism are controlled by activity of cultured rat myotubes. Am. J. Physiol. **244:** C348–C355.

Dynamics of Mitochondria and Mitochondrial Ca^{2+} near the Plasma Membrane of PC12 Cells

A Study by Multimode Microscopy

DE-MING YANG,[a,f] CHUNG-CHIH LIN,[b,f,g] HSIA YU LIN,[c] CHIEN-CHANG HUANG,[d] DIN PING TSAI,[c] CHIN-WEN CHI,[a] AND LUNG-SEN KAO[d,e,h]

[a]*Department of Medical Research and Education, Taipei Veterans General Hospital, Taipei, Republic of China*

[b]*Department of Life Sciences, Chung Shan Medical University, Taichung, Republic of China*

[c]*Department of Physics, National Taiwan University, Taipei, Republic of China*

[d]*Institute of Biochemistry, National Yang-Ming University, Taipei, Republic of China*

[e]*Department of Life Sciences, National Yang-Ming University, Taipei, Republic of China*

> ABSTRACT: The goal of this study is to examine whether there is a difference in the regulation of Ca^{2+} between mitochondria near the cell surface and mitochondria in the cytosol. Total internal reflection fluorescence and epifluorescence microscopy were used to monitor changes in the mitochondrial Ca^{2+} ([Ca^{2+}]$_{mt}$) between the mitochondria near the plasma membrane and those in the cytosol. The results show that [Ca^{2+}]$_{mt}$ near the plasma membrane increased earlier and decayed slower after high K$^+$ stimulation than average mitochondria in the cytosol. In addition, the changes in [Ca^{2+}]$_{mt}$ in the mitochondria near the cell surface after a second stimulation were larger than those induced by the first stimulation. The results provide direct evidence to support the hypothesis that mitochondria in different subcellular localization show differential responses to the influx of extracellular Ca^{2+}.
>
> KEYWORDS: total internal reflection fluorescence microscopy; mitochondria; mitochondrial Ca^{2+}; PC12 cells

INTRODUCTION

Mitochondria play multiple roles and these include generation of ATP, control of apoptosis,[1] and buffering of cytosolic Ca^{2+} ([Ca^{2+}]$_i$).[2] The spatial localization of mi-

---

[f]De-Ming Yang and Chung-Chih Lin contributed equally to this work.

[g]Current address: Department of Life Sciences, National Yang-Ming University, Taipei, Taiwan, R.O.C.

[h]Also affiliated with Brain Research Center, University System of Taiwan, Taiwan, R.O.C.

Address for correspondence: Lung-Sen Kao, Ph.D., Department of Life Sciences, National Yang-Ming University, Taipei, Taiwan, R.O.C. Voice: +886-2-2826-7268; fax: +886-2-2823-4898.

lskao@ym.edu.tw

tochondria near endoplasmic reticulum or the plasma membrane has been proposed to be tightly associated with the homeostasis of $[Ca^{2+}]_i$.[3,4] However, there is very limited information available regarding differential changes in mitochondria located at different subcellular locations. In this study, we used total internal reflection (TIR) fluorescence microscopy,[5,6] which detects only fluorescence near the plasma membrane; we combined this with conventional epifluorescence (EPI) microscopy, which detects fluorescence changes of total mitochondria population in a thicker focal plane, to monitor Ca^{2+} changes in mitochondria. By comparing the images of the same cell obtained alternately by TIR and EPI fluorescence microscopy, we were able to provide direct evidence that supports the hypothesis that mitochondria located near the cell surface respond differently from the total population of mitochondria to an influx of extracellular Ca^{2+}.

MATERIALS AND METHODS

Cell Culture

Rat adrenal pheochromocytoma (PC12) cells were cultured as described previously.[7,8]

Labeling of Fluorescence Indicators

MitoTracker Green and rhod-2 were obtained from Molecular Probes (Eugene, OR). PC12 cells were incubated with 5 µM of the acetoxymethyl esters form of MitoTracker Green for 15 min or 4.4 µM rhod-2 for 45 min at 37°C in the loading buffer (for composition see Yang *et al.*[7]). PC12 cells were stimulated by high K^+ buffer as described previously.[7]

Multimode Fluorescence Microscopy

A dual condenser for TIR/EPI illumination (T.I.L.L. Photonics, Munich BioRegio, Germany) was added to a confocal laser scanning (CLS) microscopic system (FV300 upgrade; Olympus, Tokyo, Japan). A high numerical aperture objective (×60, PlanApo, N.A. = 1.45, Olympus) with the TIR condenser was used to achieve the TIR illumination.[7] For CLS microscopic system, an Argon 488-nm laser (10 mW) was used to excite MitoTracker Green, and a He-Ne 543-nm laser (1 mW) was used to excite rhod-2. The focal plane is controlled and locked by a Z-step motor on the CLS microscopic system. The confocal and differential interference contrast (DIC) images were acquired by a photomultiplier tube. Both TIR and EPI images were acquired alternately using a fast-cooled CCD (IMAGO QE, T.I.L.L. Photonics) and controlled by commercial software (TILLvisION 4.0, T.I.L.L. Photonics).

RESULTS AND DISCUSSION

Morphology of Mitochondria

The mitochondria of PC12 cells were stained by a specific fluorescence marker, MitoTracker Green (FIG. 1). Images of mitochondria of the same cell under CLS,

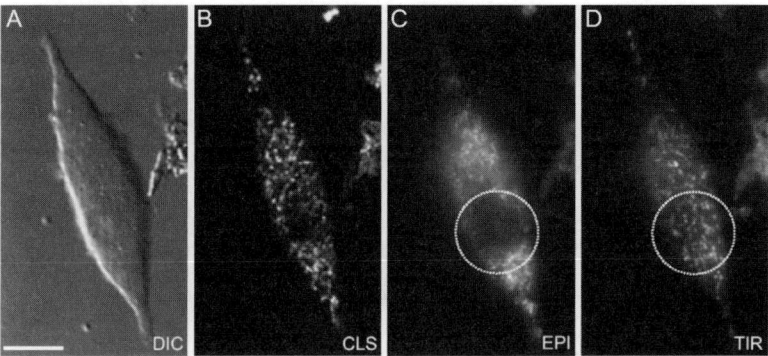

FIGURE 1. Images of mitochondria under a multimode microscope. PC12 cells were stained by MitoTracker Green. Images of the same cell under confocal laser scanning (CLS) mode (**A**), differential interference contrast (DIC) (**B**), epifluorescence (**C**), and total internal reflection (**D**) illumination are shown. Scale bar = 10 μm.

EPI, and TIR fluorescence microscopes were compared (FIG. 1). The morphology of mitochondria showed as dots and rod structures under the confocal and EPI modes, and more mitochondria showed a tubular form near the plasma membrane under the TIR mode. A better and clearer image of each mitochondrion could be obtained under the TIR and confocal modes. Mitochondria in some areas could not be detected (circled regions, FIG. 1C vs. 1D) under the EPI mode even though the optical section is thicker than that obtained under the TIR mode.

Dynamics of Mitochondrial Ca^{2+} and Mitochondria near the Cell Surface

Mitochondrial Ca^{2+} was monitored by a Ca^{2+} indicator rhod-2 that is known to be retained in the mitochondria. The confocal images of rhod-2 and MitoTracker Green were completely overlapped and are shown in FIGURE 2A. The results show that changes in rhod-2 fluorescence represent the changes in Ca^{2+} within the mitochondria. The microscope system allows time-dependent rhod-2 fluorescence changes to be alternately monitored by the TIR and EPI fluorescence modes and this approach was used. Changes in rhod-2 fluorescence of individual mitochondria under TIR illumination show that, after high K^+ stimulation, the rhod-2 fluorescence in the cytosol, but not that in the nucleus, increased immediately and remained high during the stimulation (FIG. 2C). To compare the mitochondria near the plasma membrane with those deep in the cytosol, we alternately recorded time-dependent changes of $[Ca^{2+}]_{mt}$ in the same cells under TIR and EPI illuminations (FIG. 2D). Rhod-2 fluorescence signals under TIR illumination were lower than those obtained under EPI illumination because the fluorescence collected from the evanescent field by TIR illumination is thinner than that from the focal plane by EPI illumination. The results show that TIR signals, which reflect changes in $[Ca^{2+}]_{mt}$ near the plasma membrane, increased earlier and decayed more slowly upon high K^+ stimulation than the EPI signals that represent the $[Ca^{2+}]_{mt}$ of the total mitochondria in a thicker focal plane. In addition, the changes in TIR fluorescence signal on a second stimu-

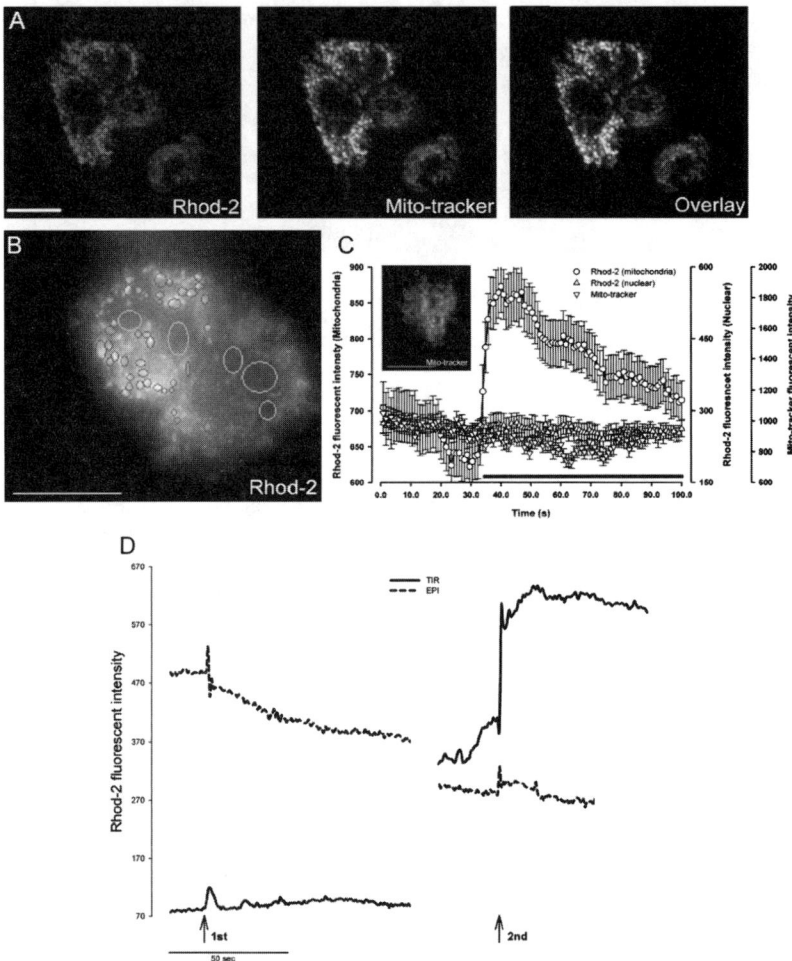

FIGURE 2. Time-dependent changes of $[Ca^{2+}]_{mt}$ in different subcellular location. (**A**) Confocal images of rhod-2, MitoTracker Green, and the overlaid images of PC12 cell. (**B**) TIR fluorescence image of rhod-2–stained PC12 cells. Mitochondria (57 *filled circles*) and regions in nuclei (5 *open circles*) were selected for analysis. As a negative control, a cell stained with MitoTracker Green was also analyzed similarly (19 *circles*). (**C**) Time-dependent changes of rhod-2 fluorescence in mitochondrial and nuclear regions. The *black bar* indicates the point of high K^+ stimulation. Data are mean ± SEM of the fluorescence intensity of the areas selected in **B**. (**D**) Time-dependent changes of TIR (*solid lines*) and EPI fluorescence (*dashed lines*) signals. The cell received two stimulations. The second stimulation (2nd) was applied 5 min after the first one (1st). *Arrows* indicate where a 0.5-s high K^+ stimulation occurred. Scale bars = 10 μm.

lation is larger and remained at a higher level for a longer time than those that occurred on the first stimulation. In contrast, the EPI fluorescence signals show that the fluorescence changes induced by the two stimulations were similar. It is possible that after the first stimulation, mitochondria near the plasma membrane continue to accumulate Ca^{2+} which leads to a higher $[Ca^{2+}]_{mt}$ level and saturates the pathways involved in maintaining the mitochondrial Ca^{2+} homeostasis; therefore, the subsequent stimulation induced a higher $[Ca^{2+}]_{mt}$ increase and remained at a plateau for a longer time. As a negative control, no changes in the fluorescence of MitoTracker were found upon high K^+ stimulation. This shows that the results indicate that the increase in rhod-2 fluorescence is not caused by the movement of mitochondria toward plasma membrane (FIG. 2C).

The results demonstrate that the response of mitochondria near the plasma membrane to the influx of extracellular Ca^{2+} was faster than the response of those located deep in the cytosol. It provides direct evidence to support the hypothesis that mitochondria near the plasma membrane are more active in taking up cytosolic Ca^{2+} than those deep in the cytosol. Finally, the detailed fluorescence images obtained using TIR illumination revealed the dynamics and heterogeneity of mitochondria in living cells.

ACKNOWLEDGMENTS

This research is supported by the grants from National Science Council (NSC-90-2112-M-002-047, NSC-90-2311-B-040-006), Taipei Veterans General Hospital (VGH92-385), Ministry of Education (Program for Promoting Academic Excellence of Universities, Grant 89-6-FA22-4), Brain Research Center, University System of Taiwan and Chung Shan Medical University (CSMU91-OM-B-014).

REFERENCES

1. ADRAIN, C. & S.J. MARTIN. 2001. The mitochondrial apoptosome: a killer unleashed by the cytochrome seas. Trends Biochem. Sci. **26:** 390–397.
2. CARAFOLI, E. 2003. Historical review: mitochondria and calcium: ups and downs of an unusual relationship. Trends Biochem. Sci. **28:** 175–181.
3. RIZZUTO, R. et al. 1998. Close contacts with endoplasmic reticulum as determinants of mitochondrial Ca^{2+} responses. Science **280:** 1763–1766.
4. YANG, D.-M. & L.-S. KAO. 2001. Relative contribution of the Na^+/Ca^{2+} exchanger, mitochondria and endoplasmic reticulum in the regulation of cytosolic Ca^{2+} and catecholamine secretion of bovine adrenal chromaffin cells. J. Neurochem. **76:** 210–216.
5. AXELROD, D. 1989. Total internal reflection fluorescence microscopy. Methods Cell Biol. **30:** 245–270.
6. STEPHENS, D.J. & V.J. ALLAN. 2003. Light microscopy techniques for living cell imaging. Science **300:** 82–86.
7. YANG, D.-M. et al. 2003. Tracking of secretory vesicles of PC12 cells using total internal reflection fluorescence microscopy. J. Microsc. **209:** 223–227.
8. HUANG, C.-M. & L.-S. KAO. 1996. Nerve growth factor, epidermal growth factor, and insulin differentially potentiate ATP-induced $[Ca^{2+}]_i$ rise and dopamine secretion in PC12 cells. J. Neurochem. **66:** 124–130.

Propofol Specifically Inhibits Mitochondrial Membrane Potential but Not Complex I NADH Dehydrogenase Activity, Thus Reducing Cellular ATP Biosynthesis and Migration of Macrophages

GONG-JHE WU,[a,d,f] YU-TING TAI,[b,f] TA-LIANG CHEN,[c] LI-LING LIN,[d] YUNE-FANG UENG,[e] AND RUEI-MING CHEN[b,d]

[a]*Department of Anesthesiology, Shin Kong Wu Ho-Su Memorial Hospital, Taipei, Taiwan*

[b]*Department of Anesthesiology, Wan-Fang Hospital, College of Medicine, Taipei Medical University, Taipei, Taiwan*

[c]*Taipei City Hospital and Taipei Medical University, Taipei, Taiwan*

[d]*Graduate Institute of Medical Sciences, College of Medicine, Taipei Medical University, Taipei, Taiwan*

[e]*National Research Institute of Chinese Medicine, Taipei, Taiwan*

ABSTRACT: Propofol is a widely used intravenous anesthetic agent. Our previous study showed that a therapeutic concentration of propofol can modulate macrophage functions. Mitochondria play critical roles in the maintenance of macrophage activities. This study attempted to evaluate further the effects of mitochondria on the propofol-induced suppression of macrophage functions using mouse macrophage-like Raw 264.7 cells as the experimental model. Macrophages were exposed to a clinically relevant concentration of propofol for 1, 6, and 24 h. Analysis by the Trypan blue exclusion method revealed that propofol was not cytotoxic to macrophages. Exposure of macrophages to propofol did not affect mitochondrial NADH dehydrogenase activity of complex I. However, analysis of flow cytometry showed that propofol significantly decreased the mitochondrial membrane potential of macrophages. Cellular levels of ATP in macrophages were significantly reduced after propofol administration. In parallel with the dysfunction of mitochondria, the chemotactic analysis showed that exposure to propofol significantly inhibited the migration of macrophages. This study shows that a therapeutic concentration of propofol can specifically reduce the mitochondrial membrane potential, but there is no such effect on complex I NADH dehydrogenase activity. Modulation of the mitochondrial membrane potential may decrease the biosynthesis of cellular ATP and thus reduce the chemotactic activity of macrophages. This study provides *in vitro* data

[f]Gong-Jhe Wu and Yu-Ting Tai contributed equally to this paper.
  Address for correspondence: Ruei-Ming Chen, Ph.D., Graduate Institute of Medical Sciences, College of Medicine, Taipei Medical University. No. 250, Wu-Hsing St. Taipei 110, Taiwan. Voice: +886-2-29307930 ext. 2159; fax: +886-2-86621119.
  rmchen@tmu.edu.tw; rmchen@wanfang.gov.tw

to validate mitochondrial dysfunction as a possible critical cause for propofol-induced immunosuppression of macrophage functions.

KEYWORDS: propofol; macrophages; mitochondrial membrane potential; ATP; cell migration

INTRODUCTION

Propofol is an intravenous anesthetic agent used for induction and maintenance of anesthesia in surgical procedures.[1] Previous studies have demonstrated that propofol has immunomodulatory effects on activities of human neutrophils and leukocytes.[2,3] Macrophages play important roles in cellular host defense against infection and tissue injury.[4] Dysfunction of macrophages can decrease a host's nonspecific cell-mediated immunity.[5] An *ex vivo* study showed that anesthesia with propofol or isofluorene intraoperatively decreased phagocytotic and microbicidal activities of alveolar macrophages.[6] Our previous study further showed that propofol has immunosuppressive effects on macrophage functions via suppression of cell migration, phagocytotic activity, oxidative ability, and cytokine production.[7]

Macrophages are activated in response to stimulation. Adenosine triphosphate (ATP), synthesized by the mitochondrial respiration and oxidative phosphorylation system, is required for maintenance of macrophage activities. In murine polymicrobial sepsis, a decrease in cellular ATP levels is positively correlated with the marked suppression of lymphocyte and macrophage functions.[8] Therefore, mitochondria, important energy-producing organelles, may participate in macrophage activation.[9] The integrities of the mitochondrial membrane potential and NADH dehydrogenase activity of complex I are two typical factors involved in ATP biosynthesis. In this study, we attempted to evaluate the role of mitochondria in propofol-induced macrophage dysfunction from the viewpoints of cell viability, NADH dehydrogenase activity, mitochondrial membrane potential, ATP synthesis, and cell migration ability.

MATERIALS AND METHODS

Cell Culture and Drug Treatment

Murine macrophage-like Raw 264.7 cells were used in this study as the experimental model. Macrophages were cultured in RPMI-1640 medium (Gibco-BRL, Grand Island, NY) supplemented with 10% fetal calf serum, L-glutamine, penicillin (100 IU/mL), and streptomycin (100 μg/mL) in 75-cm^2 flasks at 37°C in a humidified atmosphere of 5% CO_2. Cells were grown to confluence before propofol administration. Propofol, donated by Zeneca Limited (Macclesfield, Cheshire, UK), was stored under nitrogen, protected from light, and freshly prepared by dissolving it in dimethyl sufoxide (DMSO) for each independent experiment. DMSO in the medium was kept to less than 0.1% to avoid the toxicity of this solvent to macrophages. Propofol at 50 μM, which corresponds to a clinical plasma concentration, was chosen to be the dosage used in this study.[10]

Assay of Cell Viability

For evaluating the cytotoxicity of propofol to macrophages, cell viability was determined by the Trypan blue exclusion method. In brief, macrophages (2×10^5) were cultured in 24-well tissue culture plates. After propofol administration, macrophages were trypsinized with 0.1% trypsin-EDTA (Gibco-BRL). After centrifugation and washing, macrophages were suspended in phosphate-buffered saline (0.14 M NaCl, 2.6 mM KCl, 8 mM Na_2HPO_4, and 1.5 mM KH_2PO_4) and then stained with an equal volume of Trypan blue dye (Sigma, St. Louis, MO). Dead cells, blue in appearance, were counted under a reverse-phase microscope.

Assay of Mitochondrial NADH Dehydrogenase Activity of Complex I

Activity of NADH dehydrogenase of complex I was determined by a colorimetric 3-(4,5-dimethylthiazol-2-yl)-2,5-diphenyltetrazolium bromide assay.[11] In brief, macrophages (2×10^4) were seeded in 96-well tissue culture clusters overnight. After drug treatment, cells were renewed with medium containing 0.5 mg/mL 3-(4,5-dimethylthiazol-2-yl)-2,5-diphenyltetrazolium bromide for another 3 h. The blue formazan product in cells was dissolved in DMSO and measured spectrophotometrically at a wavelength of 570 nm.

Determination of the Mitochondrial Membrane Potential

The mitochondrial membrane potential in macrophages was determined according to the method of Chen.[12] In brief, macrophages (5×10^5) were seeded in 12-well tissue culture plates overnight and then treated with propofol for different time intervals. After propofol administration, macrophages were harvested and incubated with 3,3'-dihexyloxacarbocyanine ($DiOC_6(3)$), a positively charged dye, at 37°C for 30 min in a humidified atmosphere of 5% CO_2. After washing and centrifugation, cell pellets were suspended in phosphate-buffered saline. The intracellular fluorescence intensities were quantified using a flow cytometer (FACS Calibur, Becton Dickinson, San Jose, CA).

Quantification of Cellular Adenosine Triphosphate Levels

Cellular ATP levels in macrophages were determined by a bioluminescence assay as described previously.[7] This assay was based on luciferase's requirement for ATP in producing emission light according to the protocol provided in the Molecular Probes ATP determination kit (Molecular Probes, Eugene, OR). A wavelength (560 nm) emitted after a luciferase-mediated reaction of ATP with luciferin was detected using a WALLAC VICTOR 2 1420 multilabel counter (Welch Allyn, Turku, Finland).

Assay of Chemotactic Activity

The migrating capacity of macrophages was determined using the Costar Transwell cell culture chamber inserts, pore size 8 µm, according to the application guide provided by Corning Costar (Cambridge, MA). Rich RPMI-1640 medium (1.5 mL) was first added to 12-well tissue cluster plates (Corning Costar), and the transwells

TABLE 1. Effects of propofol on cell viability and complex I NADH dehydrogenase activity

Time (h)	Cell viability (cell number × 1,000)	NADH dehydrogenase activity (OD values at 570 nm)
0	521 ± 58	0.814 ± 0.125
1	542 ± 67	0.798 ± 0.132
6	497 ± 53	0.825 ± 0.201
24	516 ± 84	0.783 ± 0.097

Macrophages were exposed to 50 µM propofol for 1, 6, and 24 h. Cell viability was determined by the Trypan exclusion assay as described in MATERIALS AND METHODS. Mitochondrial complex I NADH dehydrogenase activity was assayed by a colorimetric method. Each value represents the mean ± SEM for $n = 6$.

were inserted into the plates. Macrophages (1×10^5) suspended with propofol in 0.5 mL of rich medium were added to the inside of transwells and cultured at 37°C for 1, 6, and 24 h in an atmosphere of 5% CO_2. Macrophages that migrated to the bottom surface of the polycarbonate filters were counted in each field and averaged for three fields with the aid of a crosshair micrometer (Nikon, Tokyo, Japan).

Statistical Analysis

Statistical differences between groups were considered significant when the P value of the Duncan's multiple range test was less than 0.05. Statistical analysis between groups over time was conducted by two-way ANOVA.

RESULTS

The toxicity of propofol to macrophages was determined by morphological observation and a Trypan blue exclusion method. Administration of 50 µM propofol in macrophages for 1, 6, and 24 h did not alter cell morphologies (data not shown). Analysis by the Trypan blue exclusion test showed that exposure to 50 µM propofol for 1, 6, and 24 h was noncytotoxic to macrophages (TABLE 1). After administration of macrophages with propofol for 1, 6, and 24 h, mitochondrial NADH dehydrogenase activity of complex I was not affected (TABLE 1).

A flow cytometric method was performed to quantify the mitochondrial membrane potential of macrophages. In untreated macrophages, a potential peak was detected (FIG. 1A). After propofol administration for 1 h, the potential peak had obviously shifted to the left. In 6-h–treated macrophages, propofol also had a suppressive effect on the mitochondrial membrane potential. In 24-h–treated groups, propofol showed no effect on the potential peak (FIG. 1A). Peaks analyzed by flow cytometry were quantified, and data are shown in FIGURE 1B. The basal fluorescence intensities of the mitochondrial membrane potential in untreated macrophages were 132 ± 15 arbitrary units (FIG. 1B). After propofol administration for 1 h, the mitochondrial membrane potential was reduced by 38%. In 6-h–treated macrophages, propofol caused a significant 30% decrease in the mitochondrial membrane poten-

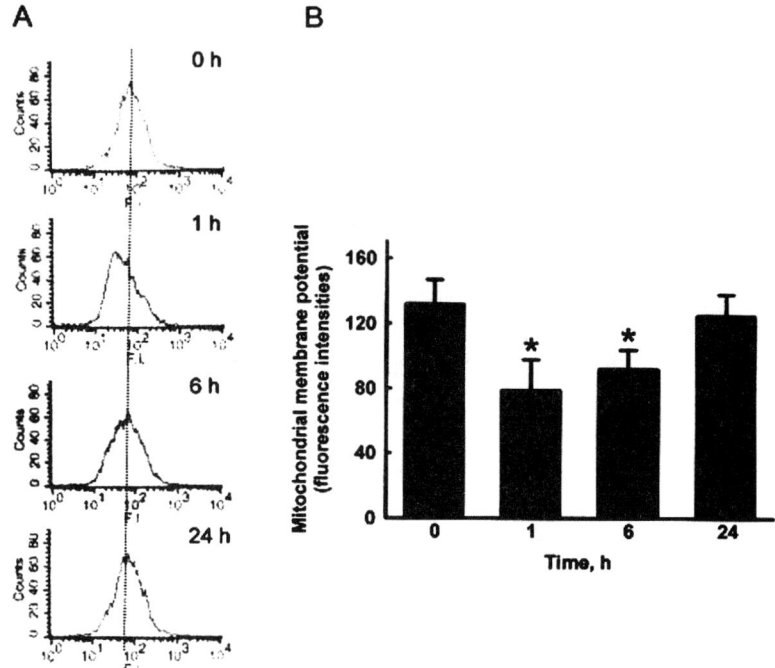

FIGURE 1. Effects of propofol on the mitochondrial membrane potential. Macrophages were exposed to 50 μM propofol for 1, 6, and 24 h. The mitochondrial membrane potential of macrophages was quantified by a flow cytometric method (**A**). The area of each peak was calculated and statistically analyzed (**B**). Each value represents the mean ± SEM for $n = 6$. *Values significantly differ from the respective control, $P < 0.05$.

tial. However, administration of propofol in macrophages for 24 h did not change the mitochondrial membrane potential (FIG. 1B).

To evaluate the effects of propofol on ATP synthesis, levels of intracellular ATP were determined by a bioluminescence assay. In untreated macrophages, average levels of cellular ATP were 35 ± 11 pmol (FIG. 2). After administration of propofol for 1 h, cellular ATP levels were significantly decreased by 43%. In 6-h-treated-macrophages, propofol decreased levels of macrophage ATP by 31%. Levels of cellular ATP were not affected after propofol administration for 24 h.

Chemotactic activities of macrophages were determined by a transwell test to validate effects of propofol on cell migration. In untreated macrophages, cells could migrate to the bottom layer of the transwell (FIG. 3A). After propofol administration for 1 h, macrophages migrating to the bottom layer were decreased. In 6-h-treated macrophages, propofol reduced the migrating capacity of macrophages to the bottom layer. Administration of propofol in macrophages for 24 h did not influence cell migration (FIG. 3A). Fractions of macrophages that migrated to the bottom layers of transwells were counted, and data are shown in FIGURE 3B. In 1-h-treated macrophages, propofol caused a significant 48% decrease in chemotactic activity. After propofol administration for 6 h, the chemotactic activity of macrophages was reduced

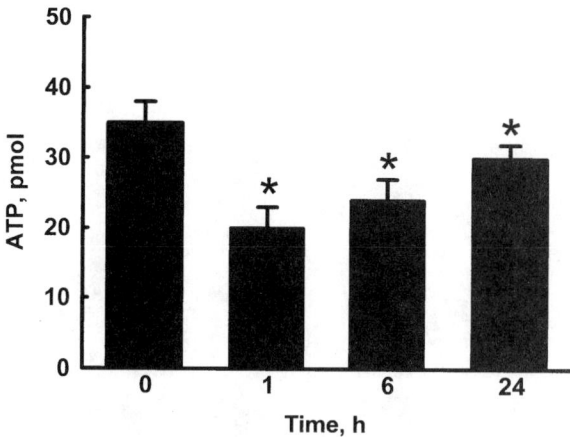

FIGURE 2. Effects of propofol on ATP biosynthesis. Macrophages were exposed to 50 μM propofol for 1, 6, and 24 h. Levels of cellular ATP were determined by a bioluminescence assay. Each value represents the mean ± SEM for $n = 6$. *Values significantly differ from the respective control, $P < 0.05$.

by 40% (FIG. 3B). However, when macrophages were exposed to propofol for 24 h, cell migration was not affected.

DISCUSSION

This study shows that propofol significantly decreases the mitochondrial membrane potential of macrophages. In parallel with modulation of the mitochondrial membrane potential, propofol reduced macrophage ATP levels. However, the activity of mitochondrial NADH dehydrogenase of complex I was not affected after propofol administration. The concentration of propofol used in this study was 50 μM which is within the range of clinical plasma concentrations.[10] Thus, this study provides *in vitro* data to validate that a therapeutic concentration of propofol can decrease ATP biosynthesis in macrophages through a reduction in the mitochondrial membrane potential, but not via modulation of NADH dehydrogenase activity of complex I.

In parallel with suppression of the mitochondrial membrane potential and ATP production, propofol significantly decreased the chemotactic activity of macrophages. Previous studies reported that a decrease in cellular ATP levels can modulate lymphocyte and macrophage functions.[8,13] Cellular ATP contributes to the regulation of cell migration.[14,15] Thus, one of the critical reasons that explains the propofol-induced suppression of macrophage chemotactic activity is the reduction of ATP biosynthesis. Data from the Trypan blue exclusion assay revealed that a therapeutic concentration of propofol did not affect the integrity of the macrophage plasma membranes. Determination of the integrity of plasma membranes can validate whether cells have experienced an insult. According to the current data, we suggest

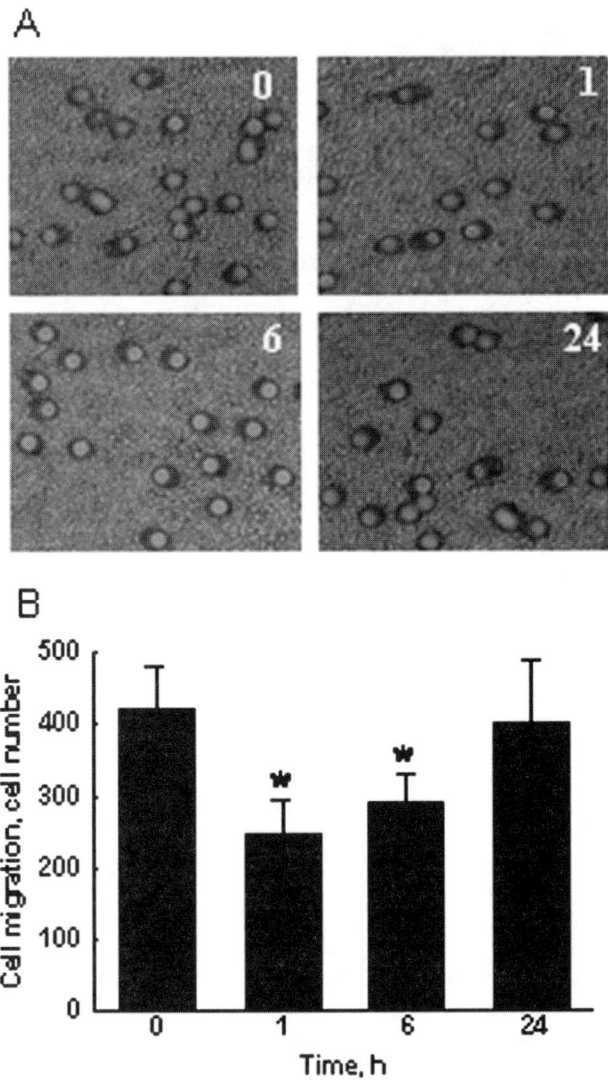

FIGURE 3. Effects of propofol on chemotactic activities. Macrophages were exposed to 50 μM propofol for 1, 6, and 24 h. Chemotactic activity was determined by the transwell method (**A**). The number of cells that migrated to the bottom layer was counted and statistically analyzed (**B**). Each value represents the mean ± SEM for $n = 6$. *Values significantly differ from the respective control, $P < 0.05$.

that the mechanism of propofol-induced suppression of macrophage functions does not occur via the death mechanism.

Balancing the mitochondrial membrane potential is important for ATP synthesis.[16] Depolarization of the transmembrane potential will lead to disruption of the respiratory chain function and ATP synthesis. A previous study showed that depolarization of the mitochondrial membrane potential increases the release of apoptotic factors, including cytochrome c and reactive oxygen species from mitochondria to cytoplasm, which induces cell apoptosis.[17] This study reveals that propofol can reduce the mitochondrial membrane potential. Therefore, modulation of the mitochondrial membrane potential is one of critical factors involved in the propofol-induced suppression of ATP synthesis and macrophage functions. Propofol is a hydrophobic anesthetic agent. One possible reason that may explain the suppressive effects of propofol on the mitochondrial membrane potential is the lipophilic characteristic of this anesthetic agent. Propofol may accumulate in the mitochondrial membrane and depolarize the membrane potential through interference with the integrity of mitochondrial membranes.

Propofol significantly reduced the mitochondrial membrane potential, ATP synthesis, and cell migration. However, these suppressive effects decreased with time. In 1-h–treated macrophages, propofol had the greatest inhibitory effects. After administration of propofol in macrophages for 6 and 24 h, these suppressive effects partially or completely disappeared. The major explanation for the time-dependent decrease in the inhibitory effects of propofol is that this anesthetic agent is progressively decomposed after exposure to visible light and aerobic conditions.[1] Both cytochrome P450–dependent monooxygenases and uridine diphosphate glucuronosyltransfease contribute to the metabolism of propofol into 4-hydroypropofol.[18] These endogenous enzymes are detectable in Raw 264.7 cells.[19,20] Therefore, the metabolism of propofol in macrophages by these endogenous enzymes may be another reason to explain the decreasing effects of this anesthetic agent on the mitochondrial membrane potential, ATP synthesis, and cell migration.

In conclusion, this study has shown that at therapeutic concentrations propofol can specifically reduce the mitochondrial membrane potential without affecting NADH dehydrogenase activity of complex I. Modulation of the mitochondrial membrane potential leads to a decrease in cellular ATP levels. In parallel with mitochondrial dysfunction, propofol significantly inhibits macrophage migration. Therefore, the mechanism of propofol-induced suppression of chemotactic activity is through modulation of the mitochondrial membrane potential and ATP synthesis, but not through the death effect.

ACKNOWLEDGMENTS

This study was supported by the Shin Kong Wu Ho-Su Memorial Hospital (SKH-TMU-92-10) and the Wan-Fang Hospital (93TMU-WAH-11), Taipei, Taiwan.

REFERENCES

1. SEBEL, P.S. & J.D. LOWDON. 1989. Propofol: a new intravenous anesthetic. Anesthesiology **71:** 260–277.

2. MIKAWA, K., H. AKAMATSU, K. NISHINA, et al. 1998. Propofol inhibits human neutrophil functions. Anesth. Analg. **87:** 695–700.
3. HOFBAUER, R., M. FRASS, H. SALFINGER, et al. 1999. Propofol reduces the migration of human leukocytes through endothelial cell monolayers. Crit. Care Med. **27:** 1843–1847.
4. NATHAN, C.F. 1987. Neutrophil activation on biological surfaces. Massive secretion of hydrogen peroxide in response to products of macrophages and lymphocytes. J. Clin. Invest. **80:** 1550–1560.
5. LANDER, H.M. 1997. An essential role for free radicals and derived species in signal transduction. FASEB J. **11:** 118–124.
6. KOTANI, N., H. HASHIMOTO, D.I. SESSLER, et al. 1998. Intraoperative modulation of alveolar macrophage function during isoflurane and propofol anesthesia. Anesthesiology **89:** 1125–1132.
7. CHEN, R.M., C.H. WU, H.C. CHANG, et al. 2003. Propofol suppresses macrophage functions and modulates mitochondrial membrane potential and cellular adenosine triphosphate levels. Anesthesiology **98:** 1178–1185.
8. AYALA, A. & I.H. CHAUDRY. 1996. Immune dysfunction in murine polymicrobial sepsis: mediators, macrophages, lymphocytes and apoptosis. Shock **6:** S27–S38.
9. DIEHL, A.M. & J.B. HOEK. 1999. Mitochondrial uncoupling: role of uncoupling protein anion carriers and relationship to thermogenesis and weight control "the benefits of losing control." J. Bioenerg. Biomembr. **31:** 493–506.
10. GEPTS, E., F. CAMU, I.D. COCKSHOTT, et al. 1987. Disposition of propofol administered as constant rate intravenous infusions in humans. Anesth. Analg. **66:** 1256–1263.
11. WU, C.H., T.L. CHEN, T.G. CHEN, et al. 2002. Nitric oxide modulates pro- and anti-inflammatory cytokines in lipopolysaccharide-activated macrophages. J. Trauma **55:** 540–545.
12. CHEN, L.B. 1988. Mitochondria membrane potential in living cells. Annu. Rev. Cell Biol. **4:** 155–181.
13. OSHIMI, Y., S. MIYAZAKI & S. ODA. 1999. ATP-induced Ca^{2+} response mediated by P2U and P2Y purinoceptors in human macrophages: signalling from dying cells to macrophages. Immunology **98:** 220–227.
14. FREDHOLM, B.B. 1997. Purines and neutrophil leukocytes. Gen. Pharmacol. **28:** 345–350.
15. DI VIRGILIO, F., P. CHIOZZI, D. FERRARI, et al. 2001. Nucleotide receptors: an emerging family of regulatory molecules in blood cells. Blood **97:** 587–600.
16. LEE, I., E. BENDER, S. ARNOLD, et al. 2001. New control of mitochondrial membrane potential and ROS formation—a hypothesis. Biol. Chem. **382:** 1629–1636.
17. LY, J.D., D.R. GRUBB & A. LAWEN. 2003. The mitochondrial membrane potential in apoptosis; an update. Apoptosis **8:** 115–128.
18. SIMONS, P.J., I.D. COCKSHOTT, E.J. DOUGLAS, et al. 1988. Disposition in male volunteers of a subanaesthetic intravenous dose of an oil in water emulsion of ^{14}C-propofol. Xenobiotica **18:** 429–440.
19. NAKAMURA, M., S. IMAOKA, F. AMANO, et al. 1998. P450 isoforms in a murine macrophage cell line, RAW264.7, and changes in the levels of P450 isoforms by treatment of cells with lipopolysaccharide and interferon-γ. Biochem. Biophys. Acta **1385:** 101–106.
20. GANOUSIS, L.G., D. GOON, T. ZYGLEWSKA, et al. 1992. Cell-specific metabolism in mouse bone marrow stroma: studies of activation and detoxification of benzene metabolites. Mol. Pharmacol. **42:** 1118–1125.

Abnormal Mitochondrial Structure in Human Unfertilized Oocytes and Arrested Embryos

HENG-KIEN AU,[a,b] TIEN-SHUN YEH,[c] SHU-HUEI KAO,[d] CHII-RUEY TZENG,[a,b] AND RONG-HONG HSIEH[b,e]

[a]*Department of Obstetrics and Gynecology, Taipei Medical University Hospital, Taipei, Taiwan*

[b]*Center for Reproductive Medicine and Sciences, Tairpei Medical University, Taipei, Taiwan*

[c]*Graduate Institute of Cell and Molecular Biology, Taipei Medical University, Taipei, Taiwan*

[d]*Graduate Institute of Biomedical Technology, School of Medicine, Taipei Medical University, Taipei, Taiwan*

[e]*School of Nutrition and Health Sciences, Taipei Medical University, Taipei, Taiwan*

ABSTRACT: To clarify the relationship between mitochondria and embryo development, we collected human unfertilized oocytes, early embryos, and arrested embryos. Unfertilized oocytes and poor-quality embryos were collected, and the ultrastructure of mitochondria was determined by transmission electron micrography. Four criteria for determining the mitochondrial state were mitochondrial morphology, cristae shape, location, and number of mitochondria. In mature oocytes, mitochondria were rounded with arched cristae and a dense matrix and were distributed evenly in the ooplasm. In pronuclear zygotes, the size and shape of mitochondria were similar to those in mature oocytes; however, mitochondria appeared to migrate and concentrate around pronuclei. In this study, 67% of examined unfertilized oocytes had fewer mitochondria in the cytoplasm. A decreased number of mitochondria located near the nucleus was also demonstrated in 60% of arrested embryos. Fewer differentiated cristae were determined in all three arrested blastocyst stages of embryos. The relative expressions of oxidative phosphorylation genes in oocytes and embryos were also determined. These data imply that inadequate redistribution of mitochondria, unsuccessful mitochondrial differentiation, or decreased mitochondrial transcription may result in poor oocyte fertilization and compromised embryo development.

KEYWORDS: embryo; mitochondria; oocyte

Address for correspondence: Dr. Rong-Hong Hsieh, School of Nutrition and Health Sciences, Taipei Medical University, Taipei 110, Taiwan, Republic of China. Voice: +886-2-27361661, ext. 6551-128; fax: +886-2-27373112.
   hsiehrh@tmu.edu.tw

INTRODUCTION

Mature oocytes contain approximately 10^5 mitochondria, but these are structurally undifferentiated compared with those of later embryo stages.[1,2] Throughout oogenesis and early embryogenesis, mitochondria in germ cells differ in appearance from those of somatic cells. Mitochondria in female germ cells assume a unique spherical profile, and an elongated mitochondrial morphology can be observed after implantation.[3,4] There are significant differences in net ATP content between oocytes, and low concentrations of ATP are generated in oocytes and early embryos.[5] Mitochondrial function can affect the physiology of embryos in many ways. This organelle has been recognized as the "powerhouse" of the cell because of its role in oxidative metabolism. The electron transfer chain consists of four respiratory enzyme complexes arranged on the mitochondrial inner membrane.

In recent years, an increasing number of reports have shown that mtDNA mutations are associated with human aging and mitochondrial diseases.[6–8] Declining mitochondrial function in older women may contribute to declining fertility.[9,10] Male subfertility and sperm dysfunction are also associated with defective mitochondrial function.[11,12] The loss of mitochondrial activity in oocytes obtained from aging couples therefore may contribute to lower embryo development and pregnancy rates.[13] To determine the relationships of mitochondrial structure and location with the ability for embryo development, we compared the ultrastructure of mitochondria through oocytes to early embryos by electron microscopy.

MATERIALS AND METHODS

Human Oocytes and Embryo Collection

This study was approved by the institutional review board of Taipei Medical University Hospital. Unfertilized oocytes were donated to our laboratory for research from patients enrolled in an *in vitro* fertilization program. In addition, embryos that were abnormally arrested and tripronucleus zygotes unsuitable for embryonic replacement or cryopreservation were also donated and used for the following experiments. Fresh human oocytes were obtained after informed consent in cases in which the donation of these oocytes to the research program would have little effect on the outcome of an *in vitro* fertilization cycle.

Electron Microscopy

Human oocytes and early embryos were fixed for 2 h in 2% paraformaldehyde and 2.5% glutaraldehyde in 0.2 M cacodylate buffer, washed in 0.1 M cacodylate buffer containing 0.2 M sucrose three times, and postfixed for 2 h in 1% osmium tetroxide. Dehydration was achieved by a graded series of 35, 50, 75, 95, and 100% ethanol, respectively. Samples then were infiltrated in a mixture of ethanol and spurr (Electron Microscopy Sciences, Fort Washington, PA) and were embedded in spurr. Ultrathin sections were cut on a Leica AG ultramicrotome, placed on 200-mesh copper grids, stained with uranyl acetate and lead citrate, and photographed on a Hitachi T-600 electron microscope.

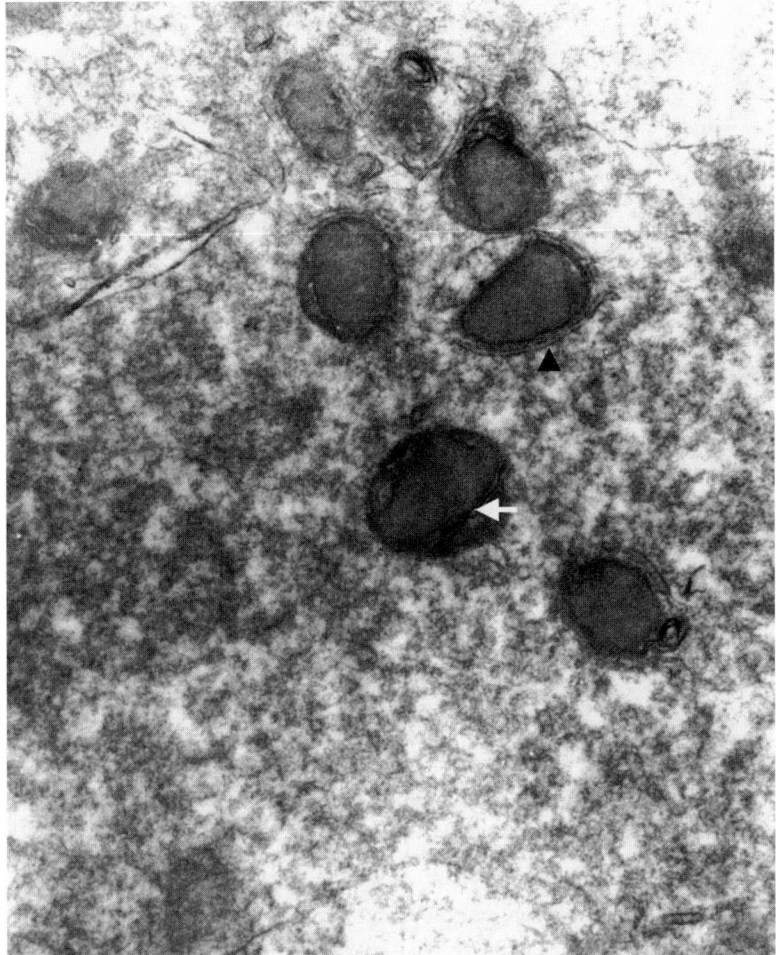

FIGURE 1. Ultrastructure of mitochondria in mature oocytes. The arched cristae were determined by electron microscopy. Mitochondrial cristae had an arched shape and were located in the mitochondria periphery (*arrow*). The smooth endoplasmic reticulum was also present together with mitochondria (*arrowhead*). Original magnification ×25,000.

RESULTS

To study the ultrastructure of mitochondria in human oocytes and early embryos, we examined normal mitochondrial structure and location by electron microphotography. In mature human oocytes, mitochondria are the prominent organelle. Mitochondria were rounded and possessed a dense matrix. Mitochondrial cristae had an arched shape and were located in the mitochondria periphery (FIG. 1). The smooth endoplasmic reticulum was also present together with mitochondria (FIG. 1). Some complexes existed in mature oocytes, which consisted of mitochondria aligned

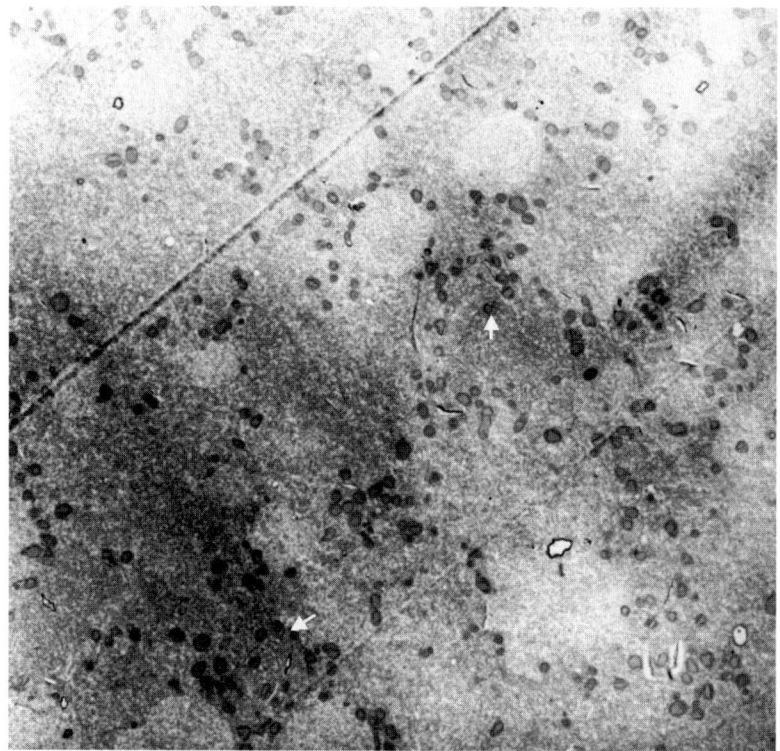

FIGURE 2. Multivesicular complexes distributed in mature oocytes. Mitochondria are arranged around vesicules and form multiple complexes scattered evenly throughout the ooplasm. Multivesicular complexes are indicated (*arrows*). Original magnification ×3,000.

around a vesicule. Multivesicular complexes were randomly distributed in mature oocytes (FIG. 2). Pronuclear zygotes had a similar size and shape compared with mature oocytes. The multivesicular complexes were also still observed at this stage. The mitochondria migrated and were concentrated around the pronuclei. In the eight-cell stage of embryos, mitochondria that were more elongated were seen together with rounded elements, and the cristae were more differentiated (FIG. 3). Some of the mitochondria began to form transverse cristae in the blastocysts (FIG. 4). A decrease in the number of mitochondria was also observed.

To determine whether the number and differentiation patterns of mitochondria affect the ability of oocyte fertilization and embryo development, we also collected 12 unfertilized oocytes and 15 arrested embryos to study their mitochondrial structure and location. There were multiple vacuoles and fewer mitochondria in unfertilized oocytes compared with functional mature oocytes. Eight unfertilized oocytes with less than 100 mitochondria were examined. Significantly decreased numbers of mitochondria located near the nucleus were observed in arrested embryos. Nine of 15 arrested embryos were characterized by this phenomenon. Peripheral arched cristae that were insufficiently differentiated to transverse cristae also were determined in

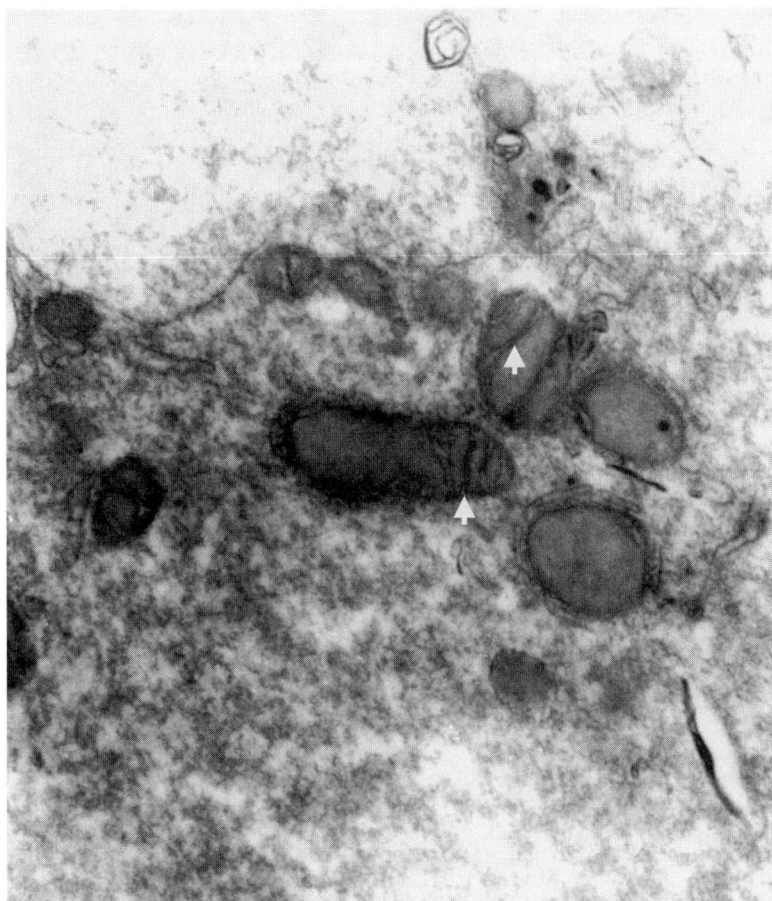

FIGURE 3. Electron microscopic examination of the eight-cell stage of an embryo. There are both round and elongated mitochondria in the embryo. Mitochondria are differentiated with few transverse cristae (*arrows*). Original magnification ×25,000.

all three arrested blastocysts examined. These data indicate that a reduced number of mitochondria may affect the fertilization potential of oocytes. Arrested embryos may occur because of a lack of redistribution of mitochondria or successful mitochondrial differentiation.

DISCUSSION

In our studies, mitochondria present peripherally arched to transverse cristae from mature oocytes to the blastocyst stage. The dynamic nature of cristae may be caused by proteins, which mediate electron transport and oxidative phosphorylation, being bound to the inner mitochondrial membrane. The varied crista structures repre-

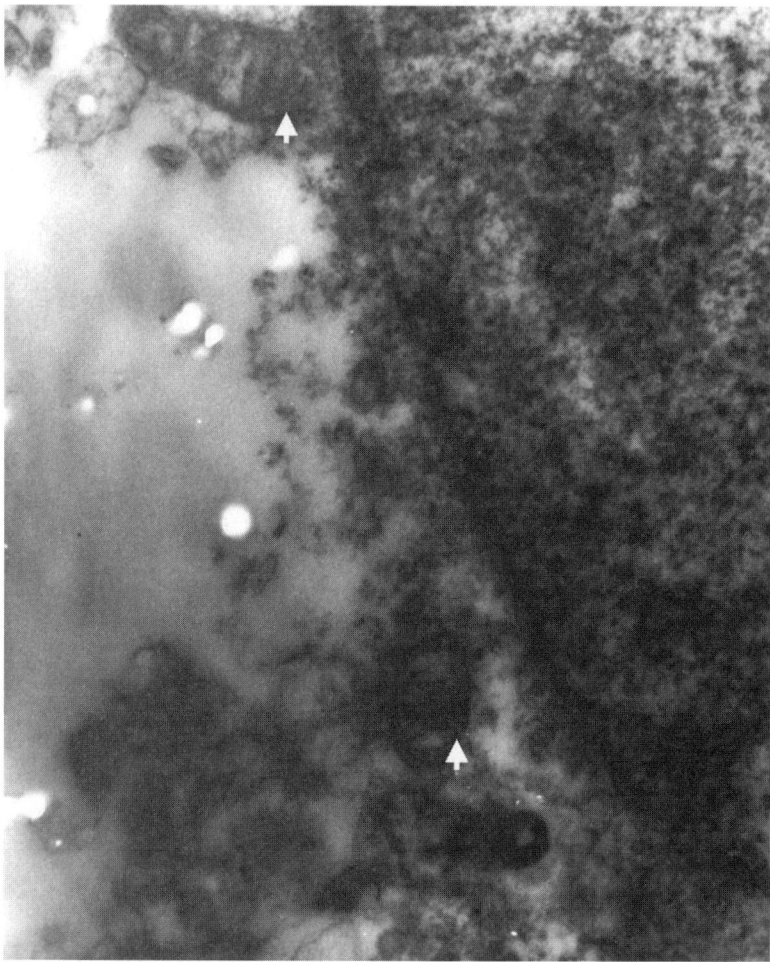

FIGURE 4. Electron microscopic examination of a blastocyst. Some of the mitochondria are differentiated with fully transverse cristae (*arrows*). Original magnification ×25,000.

sent mitochondria that progress from the arrested to the active state with embryo development. The relationship between energy production and cristae area has been shown in other studies. Proportional increases in respiratory chain enzymes and cristae surface areas have been observed.[14] The high energy demand of cells is met by an increase in the surface area of cristae.[15] Cristae differentiation may provide an efficient energy power supply for embryo development. Mitochondrial cristae change from a tubulovesicular pattern to a sparse, lamellar configuration in primordial germ cells during differentiation into oogonia.[16] Throughout oogenesis to early embryogenesis, despite cristae changes, mitochondria are also differentiated into various shapes to fit the energy requirements of different developmental stages. Mitochondria vary considerably in size and structure depending on their source and metabolic state. Mitochondria in mature oocytes assume a unique spherical profile. The arrested state of round mitochondria in ovulatory oocytes was also reported by other groups.[4] Postfertilization changes in mitochondria are characterized by a gradual transition from round or oval mitochondria with a dense matrix and few arched cristae to forms that are more elongated, possessing a lighter matrix and more numerous cristae oriented transverse to the long axis of the mitochondria.[16] Increased mitochondrial metabolism appears to coincide with a decrease in density of the mitochondrial matrix and an increase in the number of cristae.

In mature oocytes, mitochondria are the prominent organelles and are evenly distributed in the cytoplasm. After fertilization, mitochondria become concentrated in the center of the oocyte, around the developing pronuclei (FIG. 3). The mitochondria are persistently located around the nucleus from fertilization to the early developmental stage. Pronuclear formation and fusion presumably require energy. Mitochondria were reported to move close to the nucleus along microtubules to satisfy this energy requirement.[17,18] The observation that mitochondrial DNA replication in somatic cells is preferentially located close to the nucleus,[19] with human pachytene oocytes giving the appearance of a necklace of mitochondria around the nucleus,[20,21] implies that mitochondria migrate close to the nucleus when replication is required in both germ cells and somatic cells. In immature oocytes, mitochondrial aggregation is granular and clumped. Maturation of oocytes to metaphase I or II leads to the appearance of evenly distributed mitochondria.[13] Mitochondria evenly distributed in the cytoplasm are translocated to the perinucleus area as embryos develop.

There is a decrease in the number of mitochondria in normal blastocysts compared with mature oocytes. This may result from the original mitochondria segregating into the blastomeres without biogenesis of mitochondria from fertilization to the blastocyst stage. With oocyte maturity at ovulation, mitochondrial amplification[22] and mtDNA replication cease.[23] The gap between oogenesis and resumption of new mtDNA synthesis means that mitochondria are diluted and partitioned into multiplying daughter blastomeres. At ovulation, each oocyte contains around 10^5 mitochondria.[5] The mtDNA does not replicate until gastrulation in diverse species.[22–24] In arrested embryos, we also observed that fewer mitochondria existed in the cytoplasm. There were not enough mitochondria to supply energy for embryo development because of less-functional mitochondria or defective mitochondria in aging oocytes.

The average expression proportions of the eight studied genes were 4.4, 5.8, and 12.9 in unfertilized oocytes, arrested embryos, and tripronucleus zygotes, respectively. Higher expression levels in tripronucleus zygotes compared with unfertilized oocytes and arrested embryos were determined.[5] In this study, the arrested embryos collected at the two- to four-cell stage and tripronucleus zygotes collected at around the eight-cell stage had normal growth rates. In previous studies, Piko and Taylor reported that mouse mtDNA does not replicate during preimplantation development but is transcribed actively from the two-cell stage.[26] There is an approximately 30-fold increase during cleavage through the blastocyst stage.[22] Embryos with normal growth rates are assumed to have more than two times the expression level compared with unfertilized oocytes. However, there were no significant differences in expression levels between unfertilized oocytes and arrested embryos. Reduced mitochondrial transcription may affect the development of embryos. There was a three-fold greater expression level in 3PN compared with unfertilized oocytes. Mitochondrial RNA expression does not seem to be modified in embryos developing with abnormal tripronucleus. The expression of the ATPase 6 gene in unfertilized oocytes decreases compared with that in early cleavage–stage embryos.[27] We previously determined multiple deletions of mtDNA in unfertilized oocytes and arrested embryos, as well as significant increases in the proportion of deleted mtDNA in unfertilized oocytes.[28] It is probable that there is a minimum requirement for ATP content for normal embryo development including chromosomal segregation, normal mitosis, and physiological events. Fully differentiated mitochondria, successful translocation, an optimal amount of mitochondria, and sufficient transcripts may be the minimum requirements for embryo development. Our study results provide some criteria for selecting adequately developed oocytes.

ACKNOWLEDGMENTS

This work was supported by research grants NSC 91-2320-B-038-026 and NSC 92-2320-B-038-045 from the National Science Council of the Republic of China.

REFERENCES

1. SATHANANTHAN, A.H. 1997. Ultrastructure of the human egg. Hum. Cell **10**: 21–38.
2. VAN BLERKOM, J. 2000. Intrafollicular influences on human oocyte developmental competence: perifollicular vascularity, oocyte metabolism and mitochondrial function. Hum. Reprod. **15**: 173–188.
3. SHEPARD, T.H., L.A. MUFFLEY & L.T. SMITH. 2000. Mitochondrial ultrastructure in embryos after implantation. Hum. Reprod. **15**: 218–228.
4. SMITH, L.C. & A.A. ALCIVAR. 1993. Cytoplasmic inheritance and its effects on development and performance. J. Reprod. Fertil. **48**: 31–43.
5. VAN BLERKOM, J., P.W. DAVIS & J. LEE. 1995. ATP content of human oocytes and developmental potential and outcome after in-vitro fertilization and embryo transfer. Hum. Reprod. **10**: 415–424.
6. HSIEH, R.H. et al. 2001. A novel mutation in the mitochondrial 16S rRNA gene in a patient with MELAS syndrome, diabetes mellitus, hyperthyroidism and cardiomyopathy. J. Biomed. Sci. **8**: 328–335.
7. LEE, H.C. & Y.H. WEI. 2000. Mitochondrial role in life and death of the cell. J. Biomed. Sci. **7**: 2–15.
8. WALLACE, D.C. 1999. Mitochondrial diseases in man and mouse. Science **283**: 1482–1488.
9. BRENNER, C.A. et al. 1998. Mitochondrial DNA deletion in human oocytes and embryos. Mol. Hum. Reprod. **4**: 887–892.
10. KEEFE, D.L. et al. 1995. Mitochondrial deoxyribonucleic acid deletions in oocytes and reproductive aging in women. Fertil. Steril. **64**: 577–583.
11. KAO, S.H., H.T. CHAO & Y.H. WEI. 1998. Multiple deletions of mitochondrial DNA are associated with the decline of motility and fertility of human spermatozoa. Mol. Hum. Reprod. **4**: 657–666.
12. ST. JOHN, J.C., I.D. COOKE & C.L. BARRATT. 1997. Mitochondrial mutations and male infertility. Nat. Med. **3**: 124–125.
13. WILDING, M., et al. 2001. Mitochondrial aggregation patterns and activity in human oocytes and preimplantation embryos. Hum. Reprod. **16**: 909–917.
14. NICHOLLS, D.G. & S.L. BUDD. 2000. Mitochondria and neuronal survival. Physiol. Rev. **80**: 315–360.

15. GOSSLAU, A. *et al.* 2001. Cytological effects of platelet-derived growth factor on mitochondrial ultrastructure in fibroblasts. Comp. Biochem. Physiol. A Mol. Integr. Physiol. **128:** 241–249.
16. JANSEN, R.P. & K. DE BOER. 1998. The bottleneck: mitochondrial imperatives in oogenesis and ovarian follicular fate. Mol. Cell. Endocrinol. **145:** 81–88.
17. BARNETT, D.K., J. KIMURA & B.D. BAVISTER. 1996. Translocation of active mitochondria during hamster preimplantation embryo development studied by confocal laser scanning microscopy. Dev. Dynamics **205:** 64–72.
18. VAN BLERKOM, J., J. SINCLAIR & P. DAVIS. 1998. Mitochondrial transfer between oocytes: potential applications of mitochondrial donation and the issue of heteroplasmy. Hum. Reprod. **13:** 2857–2868.
19. DAVIS, A.F. & D.A. CLAYTON. 1996. In situ localization of mitochondrial DNA replication in intact mammalian cells. J. Cell. Biol. **135:** 883–893.
20. BAKER, T.G. & L.L. FRANCHI. 1967. The fine structure of oogonia and oocytes in human ovaries. J. Cell. Sci. **2:** 213–224.
21. GONDOS, B. 1987. Comparative studies of normal and neoplastic ovarian germ cells: ultrastructure and pathogenesis of dysgerminoma. Int. J. Gynecol. Pathol. **6:** 124–131.
22. TAYLOR, K.D. & L. PIKO. 1995. Mitochondrial biogenesis in early mouse embryos: expression of the mRNAs for subunits IV, Vb, and VIIc of cytochrome *c* oxidase and subunit 9 of H^+-ATP synthase. Mol. Reprod. Dev. **40:** 29–35.
23. LARSSON, N.G. *et al.* 1998. Mitochondrial transcription factor A is necessary for mtDNA maintenance and embryogenesis in mice. Nat. Genet. **18:** 231–236.
24. EL MEZIANE, A., J.C. CALLEN & J.C. MOUNOLOU. 1989. Mitochondrial gene expression during *Xenopus laevis* development: a molecular study. EMBO J. **8:** 1649–1655.
25. HSIEH, R.H. *et al.* 2004. Decreased expression of mitochondrial genes in human unfertilized oocytes and arrested embryos. Fertil. Steril. **81:** 912–918.
26. PIKO, L. & K.D. TAYLOR. 1987. Amounts of mitochondrial DNA and abundance of some mitochondrial gene transcripts in early mouse embryos. Dev. Biol. **123:** 364–374.
27. LEE, S.H. *et al.* 2000. Mitochondrial ATPase 6 gene expression in unfertilized oocytes and cleavage-stage embryos. Fertil. Steril. **73:** 1001–1005.
28. HSIEH, R.H. *et al.* 2002. Multiple rearrangements of mitochondrial DNA in unfertilized human oocytes. Fertil. Steril. **77:** 1012–1017.

Oxidative Damage and Mitochondrial DNA Mutations with Endometriosis

SHU-HUEI KAO,[a] HSIENG-CHIANG HUANG,[a] RONG-HONG HSIEH,[b]
SU-CHEE CHEN,[c] MING-CHUAN TSAI,[a] AND CHII-REUY TZENG[d]

[a]*Graduate Institute of Biomedical Technology, School of Medicine,
Taipei Medical University, Taipei, Taiwan*

[b]*School of Nutrition and Health Sciences, Taipei, Taiwan*

[c]*Department of Obstetrics and Gynecology, Cathay General Hospital, Taipei, Taiwan*

[d]*Department of Obstetrics and Gynecology, Taipei Medical University Hospital,
Taipei, Taiwan*

ABSTRACT: Endometriosis, a frequently encountered disease in gynecology, is a considerable threat to the physical, psychological, and social integrity of women. Moreover, up to 50% of infertile patients have this disease. The etiology and pathogenesis of this important disease are poorly understood; it is defined as an ectopic location for endometrium-like glandular epithelium and stroma outside of the uterine cavity. It still remains an open question as to what extent the peritoneal environment influences the establishment and/or progression of endometriosis. As a result of such stress, a sterile, inflammatory reaction with the secretion of growth factors, cytokines, and chemokines is generated, which is especially deleterious to successful reproduction. Significantly higher amounts of oxidative damage were detected in endometriotic lesions than in controlled normal endometrium, including mitochondrial DNA (mtDNA) rearrangement, 8-OH-deoxyguanosine (8-OH-dG), and lipoperoxide contents. There were approximately sixfold increases in 8-OH-dG and lipoperoxides in chocolate cysts compared with normal endometrial tissues. A novel 5,335-bp deletion of mtDNA was identified in endometriotic tissue. According to these results, we propose that oxidative stress and mtDNA mutations might be anticipated in the initiation or progression of endometriosis. Only by understanding the mechanisms involved in the pathogenesis of endometriosis can we develop a basis for new diagnostic and therapeutic approaches.

INTRODUCTION

Endometriosis is an invasive but benign gynecological disease that is histologically characterized by the presence of endometrium-like glands and stroma outside of the uterus. It is one of the most frequently encountered diseases in gynecology, affecting 15 to 50% of women in their reproductive life span.[1] Clinical observations

Address for correspondence: Prof. Chii-Reuy Tzeng, Department of Obstetrics and Gynecology, Taipei Medical University, 250, Wu-Hsing Street, Taipei 110, Taiwan. Voice: +886-2-27372181 ext. 1996; fax: +886-2-27358406,
kaosh@tmu.edu.tw

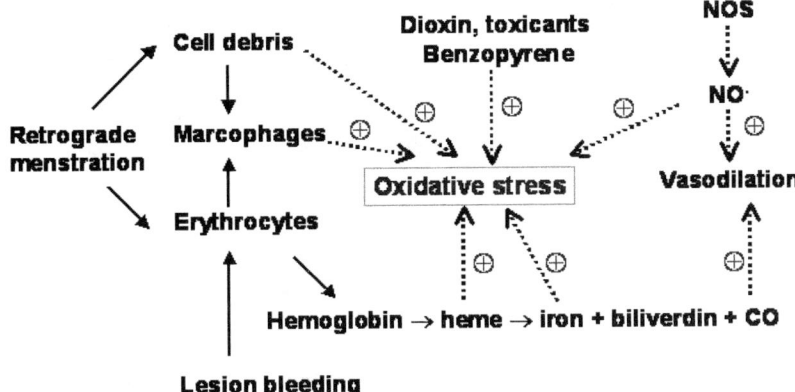

FIGURE 1. Hypothesis explaining oxidative stress in the peritoneal cavity of women with endometriosis.[4] Boldface type indicates factors that have specifically been studied in relation to pelvic endometriosis. CO, carbon dioxide; NO, nitric oxide; NOS, nitric oxide synthase.

and *in vitro* experiments imply that endometriotic cells are invasive and able to metastasize. The endometriotic tissue often undergoes cyclic proliferation and breakdown similar to eutopic endometrium, resulting in local inflammatory reactions. These processes cause the cyclical character of endometriosis with dysmenorrhea, dyspareunia, pelvic pain, catamenial hematuria, and other symptoms derived from the affected organ. Moreover, up to 50% of infertile patients have this disease.[2]

The pathophysiology of this disease still remains elusive. Recent studies have suggested that menstrual effluent contains factors that induce alterations in the morphology of the peritoneal mesothelium,[3] which may create adhesion sites for endometrial cells. Endometriosis is a multifactorial disease associated with a general inflammatory response in the peritoneal cavity. Oxidative stress has been proposed as a potential factor involved in the pathogenesis of the disease[4] and may be responsible for local destruction of the peritoneal mesothelium, thereby creating adhesion sites for ectopic endometrial cells. Several hypotheses have been proposed to explain why oxidative stress is induced in cases of pelvic endometriosis (FIG. 1). Oxidatively damaged erythrocytes,[5] apoptotic endometrial cells, or undigested endometrial tissue[6] may become transplanted into the peritoneal cavity and signal the recruitment and activation of mononuclear phagocytes. Women with endometriosis are prone to react to this stimulus with an inadequate macrophage scavenger receptor response. Activated macrophages in the peritoneal cavity generate oxidative stress, which consists of lipid peroxides, their degradation products, and products formed from their interaction with low-density lipoprotein (LDL), apoprotein,[7] and other proteins.[8] Moreover, autoantibodies to malondialdehyde-modified low-density lipoprotein, oxidized low-density lipoprotein, and lipid peroxide-modified serum albumin markers of oxidative stress are significantly increased in women with endometriosis.[9] As a result of such a stress, a sterile, inflammatory reaction with secretion of growth factors, cytokines, and chemokines is generated, which is deleterious especially to successful reproduction.

Production of reactive oxygen species appears to be increased in women in whom endometriosis is developing and progressing. Most reports have discussed the potential consequences of increased oxidative stress in endometriosis relative to decreased fertility. We propose that oxidative stress might be anticipated in the initiation or progression of endometriosis. In this study, we attempted to elucidate relationships among oxidative stress, mitochondrial DNA (mtDNA) mutations, and endometriosis.

MATERIALS AND METHODS

Analysis of Oxidative Damage in Cellular Molecules

Determination of 8-OH-Deoxyguonosine (8-OH-dG)

Tissue samples were scraped using a cell lifter in 1.5 mL TE buffer, and 75 µL of 10% SDS, 30 mL of 200 mM butylated hydroxytoluene, and 15 mL of RNase A stock solution (10 µg/mL) were added; then the lysate was incubated at 37°C for 1 h and incubated with proteinase K (100 µg/mL) at 55°C for 12 h. The lysate was extracted by a phenol/chloroform method. An aliquot of 100 µg of DNA dissolved in 100 µL of 10 mM Tris-HCl (pH 7.4)/0.1 mM DFAM was digested by incubation with 1 µL of DNase I (20 U/µL) and 11 µL of a 0.1 M $MgCl_2$ solution at 37°C for 30 min. After adjusting the pH to 5.0 by adding 4.8 µL of 1 M sodium acetate (pH 5.3) and 1.2 µL of 0.1 M $ZnSO_4$, the DNA sample was digested with 5 µL of nuclease P1 (1 U/3 µL in 20 mM sodium acetate, pH 5.3) at 65°C for 10 min. The DNA molecules were hydrolyzed to the corresponding nucleosides by incubation with 5 mL of 1 U/µL alkaline phosphatase for 30 min at 37°C. Processed DNA samples were separated on a C-18 column (particle size 5 mm, 200 × 4.6 mm; JT Baker, Phillipsburg, NJ) on an HPLC system (Jasco, Easton, MD) connected in series with an ECD detector (Bioanalytical Systems, West Lafayette, IN) and a UV detector (at 254 nm; Jasco). Elution was performed at a flow rate of 0.8 mL/min for 40 min with a mobile phase which consisted of 12.5 mM citric acid, 25 mM sodium acetate, and 10 mM acetic acid containing 6% methanol (pH 5.8).

Determination of Lipid Peroxide

In each analytical run, a reagent blank, 1,1,3,3-tetraethoxypropane standard working solutions, and the sample were assayed in duplicate. An aliquot of 50 µL of each sample was pipetted into a test tube containing 0.6 mL of 0.44 M phosphoric acid. After mixing, 0.2 mL of a 42 mM thiobarbituric acid solution was added to a final concentration of 7 mM and then placed in a 95°C dry bath for 1 h. The samples were then cooled and neutralized with 1 N NaOH in methanol before the HPLC analysis. An aliquot of 20 µL of supernatant obtained above was injected into a narrow-pore C18 column (4.6 × 250 mm, particle size 5 mm) using a Jasco PU-980 pump with a solvent system made of methanol and 50 mM phosphate buffer (pH 6.8; 4:6, v/v) at a flow rate of 1 mL/min. The eluent was monitored using a Jasco fluorescence detector using excitation at 525 nm and emission at 550 nm.

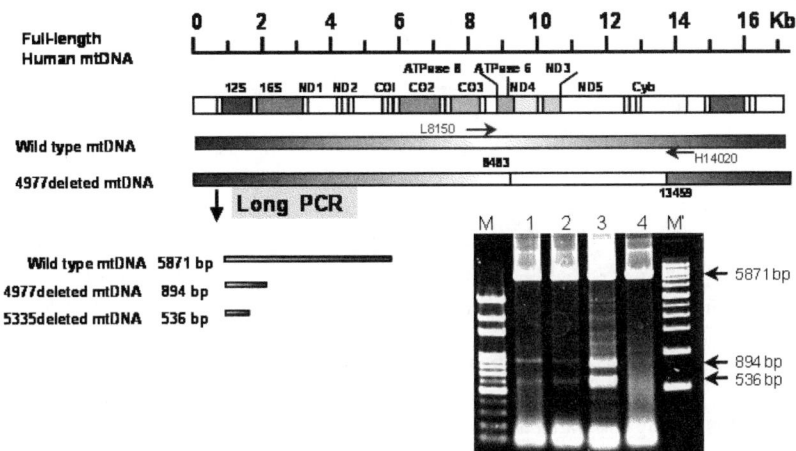

FIGURE 2. (**Top panel**) A scheme illustrating the strategy for the determination of multiple mtDNA deletions of the endometrium by the long-range PCR technique. Using the primer pair L8150-H14020, we generated three types of PCR products. The 5,871-bp fragment was produced from wild-type mtDNA, an 894-bp fragment was from 4,977-bp deleted mtDNA, and a 536-bp fragment was from a 5,335-bp mtDNA deletion. (**Bottom panel**) Electrophorectogram of PCR products amplified from mtDNA with specific deletions in tissues from women with or without endometriosis. Lanes 1 to 3 were examined from ovaries, myometrium, and endometrium from three individuals with endometriosis, respectively. *Lane M* is a 100-bp DNA ladder and *lane M'* is a 10-kb DNA ladder.

Determination of mtDNA Mutations in Human Tissue

Total DNA Extraction

All samples then were checked by microscopy for morphological changes and stored at −196°C until analysis. All tissues were minced into small pieces and incubated at 56°C for 2 h in 50 µL lysis buffer containing 2% SDS and 50 mM Tris-HCl (pH 8.3), followed by a phenol-chloroform extraction method. All DNA samples were finally preserved in 200 µL of 10 mM Tris-HCl (pH 8.3).

Synthesis of Oligonucleotide Primers

Oligonucleotide primers used for amplification of the target sequences of mtDNA and genomic DNA were chemically synthesized by Protec (Taipei, Taiwan). The nucleotide sequences of used primer pairs were L8150 (5'-CCGGGGGTATACTACG-GTCA-3') and H14020 (5'-ATAGCTTTTCTAGTCAGGTT-3'). Sizes of the PCR products obtained from these primer pairs are shown in FIGURE 2.

Detection of mtDNA Mutations by PCR

The desired target sequence of mtDNA was amplified from 15~20 ng of each DNA sample in a 50-µL reaction mixture containing. 200 µM of each dNTP, 0.4 µM of each primer, 1 unit of Ampli-Taq DNA polymerase (Perkin-Elmer/Cetus, Roche

TABLE 1. Contents of 8-OH-dG and lipid peroxides examined in this study

Type of tissue	8-OH-dG/dG ($\times 10^{-3}$%)	Lipid peroxide content (pmol/μg protein)
Normal endometrium	0.17 ± 0.07 ($n = 5$)	1.50 ± 0.25 ($n = 10$)
Chocolate cyst	1.21 ± 0.10 ($n = 10$)	4.48 ± 0.44 ($n = 10$)
Endometriotic endometrium	0.72 ± 0.17 ($n = 4$)	2.61 ± 0.66 ($n = 7$)
Myometrium	0.55 ± 0.18 ($n = 5$)	1.85 ± 0.11 ($n = 6$)
Myoma	0.66 ± 0.30 ($n = 9$)	2.80 ± 0.72 ($n = 12$)
Ovary	0.62 ± 0.05 ($n = 9$)	2.68 ± 0.44 ($n = 10$)
Peritoneal fluid	0.73 ($n = 1$)	3.34 ($n = 1$)

Molecular System, Branchburg, NJ), 50 mM KCl, 1.5 mM $MgCl_2$, and 10 mM Tris-HCl (pH 8.3). PCR was performed for 30 cycles in a DNA thermal cycler (Perkin-Elmer/Cetus) using the thermal profile of denaturation at 94°C for 40 s, annealing at 55°C for 40 s, and primer extension at 72°C for 40 s. Long PCR proceeded with the thermal profile of denaturation at 94°C for 2 min, annealing at 68°C for 1 min, and primer extension at 72°C for 2 min.

Detection of mtDNA Mutations by PCR

The desired target sequence of mtDNA was amplified from 500 fmol of each DNA sample in a 20-μL reaction mixture containing A-dye terminator, T-dye terminator, C-dye terminator, G-dye terminator, dITP, dATP, dCTP, dTTP, 3.2 pmol primer, 10 units of AmpliTaq DNA polymerase (Epicentric Technologies, Oldendorf, Germany), and 1 × sequencing buffer. PCR was performed for 30 cycles of denaturation at 96°C for 30 s, 50°C for 15 s, and primer extension at 60°C for 4 min. PCR products were then separated by electrophoresis on 6% polyacrylamide gels containing 8 M urea at 65 W for 4 h.

RESULTS

Higher Content of Lipid Peroxides in Endometriotic Tissues

In this study, we detected lipid peroxides (as malondialdehyde) in the endometrium, myoma, adenoma, ovary, and chocolate cyst from women with endometriosis and from normal individuals (TABLE 1). The content of lipoperoxides was 1.50 ± 0.25 ($n = 10$), 4.48 ± 0.44 ($n = 10$), 2.61 ± 0.66 ($n = 7$), 1.85 ± 0.11 ($n = 6$), 2.80 ± 0.72 ($n = 12$), 2.68 ± 0.44 ($n = 10$), and 3.34 ($n = 1$) in the normal endometrium, chocolate cyst, endometriotic endometrium, myometrium, myoma, ovary, and peritoneal fluid, respectively. Higher contents of lipoperoxides were detected in tissues from eutopic and ectopic endometriosis. There was approximately 6.2-fold higher lipid peroxides in the chocolate cyst than normal endometrial tissue

Higher Amount of 8-OH-dG in Endometriotic Tissues

8-OH-dG was detected using HPLC-ECD in the endometrium, myoma, adenoma, ovary, and chocolate cyst from women with endometriosis and from normal individu-

TABLE 2. MtDNA mutation with the 4,977-bp deletion and 5,335-bp deletion were examined in each sample collected from 46 women with or without endometriosis

Type of tisssue	Endometriosis (endometriosis/tissue no.)	4,977-bp mtDNA deletion	5,335-bp mtDNA deletion
Adenomyoma	8/8	2/8	2/8
Chocolate cyst	12/12	4/12	7/12
Endometrium	4/8	2/8	1/8
Myometrium	3/8	1/8	0/8
Myoma	11/17	4/17	4/17
Ovary	2/3	1/3	0/3
Peritoneal fluid	1/1	1/1	0/1

als. The content of 8-OH-dG was 0.17 ± 0.07 ($\times 10^{-3}\%$, $n = 5$), 1.21 ± 0.10 ($\times 10^{-3}\%$, $n = 10$), 0.72 ± 0.17 ($\times 10^{-3}\%$, $n = 4$), 0.55 ± 0.18 ($\times 10^{-3}\%$, $n = 5$), 0.66 ± 0.30 ($\times 10^{-3}\%$, $n = 9$), 0.62 ± 0.05 ($\times 10^{-3}\%$, $n = 9$), and 0.73 ($\times 10^{-3}\%$, $n = 1$) in the normal endometrium, chocolate cyst, endometriotic endometrium, myometrium, myoma, ovary, and peritoneal fluid, respectively (TABLE 1). Higher contents of 8-OH-dG were detected in tissues from eutopic and ectopic endometriosis. There was approximately 7.2-fold higher 8-OH-dG in the chocolate cyst than in the normal endometrium.

Accumulation of Large-scale Deletions and DNA Rearrangement of mtDNA in Endometriotic Tissues

Accumulations of mtDNA rearrangements have been shown in aged tissues, degenerated diseases, and several types of cancer in humans. In our study, we detected mtDNA mutations in such tissues as the endometrium, myoma, adenoma, ovary, and chocolate cyst from women with endometriosis and from normal individuals (TABLE 2). Using the primer pair, L8150-H14020, we generated three types of PCR products. A 5,871-bp fragment was produced from the wild-type mtDNA, an 894-bp DNA fragment was from the 4,977-bp deleted mtDNA, and a nearly 536-bp fragment was from a 5,335-bp mtDNA deletion. A scheme illustrates the strategy for determination of multiple mtDNA deletions in various human tissues by the long-range PCR techniques (FIG. 2). In the bottom panel, lanes 1 to 3 are from ovaries, myometrium, and endometrium from three individuals with endometriosis, respectively. Lane M is a 100-bp DNA ladder, and lane M' is a 10-kb DNA ladder in FIGURE 2.

A 4,977-bp Deletion and a Novel Deletion of mtDNA in Endometriotic Tissues

We applied primer-shift PCR to ensure the existence of a 4,977-bp deletion and DNA sequencing to identify the novel 5,335-bp mtDNA deletion found in endometriotic tissue. Furthermore, we sequenced the generated PCR products. In FIGURE 3, a schematic illustration is given of the nucleotide sequence flanking the junction sites at the 5' end of the novel 5,335-bp deletion on the heavy strand of mtDNA in the endometriotic tissue. It shows a 10-nucleotide indirect repeat (5'-CCTAT-

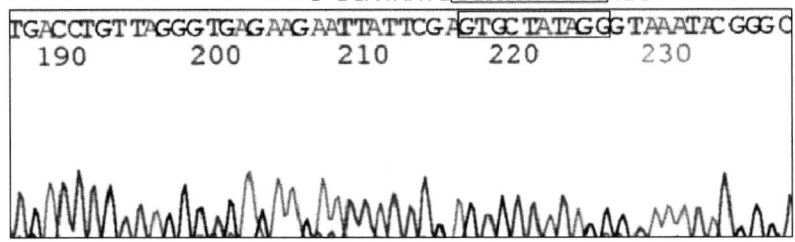

FIGURE 3. Schematic illustration of the nucleotide sequence flanking the junction sites at the 5′ end of the novel 5,335-bp deletion on the light strand of mtDNA in the endometrium. It reveals a three-nucleotide indirect repeat (5′-CCTATAGCAC-3′) located at the junction site at nucleotide position (np) 8263~8272 or np 13598~13607 (5′ to 3′) on the heavy strand of mtDNA.

AGCAC-3′) located at the junction sites at nucleotide position (np) 8263–8272 or np 13598–13607 (5′ to 3′) on the heavy strand of mtDNA.

DISCUSSION

Endometriosis is a frequent disorder that commonly presents with infertility and pelvic pain and affects younger women of childbearing age. However, despite a growing number of reports on endometriosis, the pathophysiology of this disease remains poorly understood. Although the precise etiology of endometriosis is unclear, it is generally considered to involve multiple genetic, environmental, immunological, angiogenic and endocrine processes.

Recent studies have suggested that menstrual effluent contains factors that induce alterations in the morphology of the peritoneal mesothelium,[3] which may create adhesion sites for endometrial cells. Attachment of endometrial cells appears to be enhanced by induction of adhesion molecules[10] and overexpression of matrix metalloproteinases[11] and plasminogen activators,[12] which ensure local destruction of the extracellular matrix in endometriosis. After adhesion, endometrial cells proliferate and gradually invade the peritoneal tissue. Some factors induce vascularization of endometriotic implants, allowing their further development. Cytokines[13–15] and growth factors,[16] such as transforming growth factor-β, interleukin-8, interleukin-1, tumor necrosis factor, interferon-γ,[17] and vascular growth factor,[18] have been implicated as inducers of attachment, proliferation, and neovascularization.

Oxidative stress has been proposed as a potential factor involved in the pathogenesis of the disease.[8,19] This disease is characterized by the increased presence of activated macrophages, erythrocyte destruction, iron deposition,[20] and associated increases in growth-promoting activities and the production of inflammatory cytokines. In this study, significantly higher amounts of oxidative damage were detected in endometriotic lesions than in controlled normal endometrium such as the mtDNA rearrangement, 8-OH-dG, and lipoperoxide contents. In the future, we will explore the identification of the molecular pathway and factors of reactive oxygen species generation and eradication. A better understanding of the mechanisms of reactive oxygen species detoxification and further investigation of their effect on the peritoneal environment are essential to obtaining new insights into this disease and eventually developing new diagnostic and therapeutic strategies.

ACKNOWLEDGMENTS

This work was supported by research grant NSC91-2314-B-038-013 from the National Science Council of the R.O.C.

REFERENCES

1. STARZINSKI-POWITZ, A., H. HANDROW-METZMACHER & S. KOTZIAN. 1998. The putative role of cell adhesion molecules in endometriosis: can we learn from tumor metastasis? Mol. Med. Today **5:** 304–309.
2. BATTISTA, G. 1991. Mild endometriosis and infertility: a clinical review of epidemiological data, diagnostic pitfalls, and classification limits. Obstet. Gynecol. Surv. **46:** 374–379.
3. KOKS, C.A., A.Y. DEMIR-WEUSTEN, P.G. GROOTHUIS, et al. 2000. Menstruum induces changes in mesothelial cell morphology. Gynecol. Obstet. Invest. **50:** 13–18.
4. LANGENDONCKT, A.V., F. CASANAS-ROUX & J. DONNEZ. 2002. Oxidative stress and peritoneal endometriosis. Fertil. Steril. **77:** 861–870.
5. ARUMUGAM, K. & Y.C. YIP. 1995. De novo formation of adhesion in endometriosis. The role of iron and free radical reactions. Fertil. Steril. **64:** 62–64.
6. MURPHY, A.A. , N. SANTANAM & S. PARTHASARATHY. 1998. Endometriosis: a disease of oxidative stress? Semin. Reprod. Endocrinol. **16:** 263–273.
7. OTA, H., S. IGARASHI, J. HATAZAWA & T. TANAKA. 1999. Endometriosis and free radicals. Gynecol. Obstet. Invest. **48:** 29–35.
8. LANGEDONCKT, A.V., F. CASANAS-ROUX & J. DONNEZ. 2002. Oxidative stress and peritoneal endometriosis. Fertil. Steril. **77:** 861–870.
9. SHANTI, A., N. SANTANAM, A.J. MORALES, et al. 1999. Autoantibodies to markers of oxidative stress are elevated in women with endometriosis. Fertil. Steril. **71:** 1115–1118.
10. BELIARD, A., J. DONNEZ, M. NILSOLLE & J.M. FOIDART. 1997. Localization of laminin, fibronectin, E-cadherin and integrins in endometrium and endometriosis. Fertil. Steril. **67:** 266–272.
11. KOKORINE, I., M. NISOLLE, J. DONNEZ, et al. 1997. Expression of interstitial collagenase (matrix metalloproteinase-1) is related to the activity of human endometriotic lesions. Fertil. Steril. **68:** 246–251.
12. SILLEM, M., S. PRIFTI, M. NEHER & B. RUNNEBAUM. 1992. Extracellular matrix remodeling in the endometrium and its possible relevance to the pathogenesis of endometriosis. Hum. Reprod. Update **4:** 730–735.

13. MURPHY, A.A., N. SANTANAM, A.J. MORALES & S. PARTHASARATHY. 1998. Lysophosphatidyl choline, a chemotactic factor for monocytes/T-lymphocytes is elevated in endometriosis. J. Clin. Endocrinol. Metab. **83:** 2110–2113.
14. SAWASTRI, S., N. DESAI, J.A. ROCK & N. SIDELL. 2000. Retinoic acid suppresses interleukin-6 production in human endometrial cells. Fertil. Steril. **73:** 1012–1019.
15. BEDAIWY, M.A., T. FALCONE, R.K. SHARMA, *et al.* 2002. Prediction of endometriosis with serum and peritoneal fluid markers: a prospective controlled trial. Hum. Reprod. **17:** 426–431.
16. GIUDICE, L.C., B.A. DSUPIN, S.E. GARGOSKY, *et al.* 1994. The insulin-like growth factor system in human peritoneal fluid: its effects on endometrial stromal cells and its potential relevance to endometriosis. J. Clin. Endocrinol. Metab. **79:** 1284–1293.
17. VIANTIER, D., M. COSSON & P. DUFOUR. 2000. Is endometriosis an endometrial disease? Eur. J. Obstet. Gynecol. Reprod. Biol. **91:** 113–125.
18. DONNEZ, J., P. SMOES, S. GILLEROT, *et al.* 1998. Vascular endothelial growth factor (VEGF) in endometriosis. Hum. Reprod. **13:** 1686–1690.
19. HALME, J., S. BECKER & R. WING. 1984. Accentuated cyclic activation of peritoneal macrophages in patients with endometriosis. Am. J. Obstet. Gynecol. **148:** 85–90.
20. VAN LANGENDONCKT, A., F. CASANAS-ROUX, J. EGGERMONT & J. DONNEZ. 2004. Characterization of iron deposition in endometriotic lesions induced in the nude mouse model. Hum. Reprod. **19:** 1265–1271.

Neuroprotective Role of Coenzyme Q10 against Dysfunction of Mitochondrial Respiratory Chain at Rostral Ventrolateral Medulla during Fatal Mevinphos Intoxication in the Rat

F.C.H. LI,[a,b] H.P. TSENG,[a,b] AND A.Y.W. CHANG[a,b]

[a]*Department of Biological Science, National Sun Yat-sen University, Kaohsiung, Taiwan, Republic of China*

[b]*Center for Neuroscience, National Sun Yat-sen University, Kaohsiung, Taiwan, Republic of China*

ABSTRACT: We evaluated the functional changes in the mitochondrial respiratory chain at the rostral ventrolateral medulla (RVLM), the medullary origin of sympathetic vasomotor tone, in an experimental model of fatal organophosphate poisoning using the insecticide mevinphos (Mev). We also investigated the neuroprotective role of coenzyme Q10 (CoQ10) in this process. Intravenous administration of Mev (1 mg/kg) in Sprague–Dawley rats maintained with propofol elicited an initial hypertension followed by hypotension, accompanied by bradycardia, with death ensuing within 10 min. Enzyme assay revealed a significant depression of the activity of nicotinamide adenine dinucleotide cytochrome *c* reductase, succinate cytochrome *c* reductase, and cytochrome *c* oxidase in the RVLM during this fatal Mev intoxication. ATP production also underwent a significant decrease. Pretreatment by microinjection bilaterally of CoQ10 (4 μg) into the RVLM significantly prevented mortality, antagonized the cardiovascular suppression, and reversed the depressed mitochondrial respiratory enzyme activities, or reduced ATP production in the RVLM induced during Mev intoxication. Our results indicated that dysfunction of mitochondrial respiratory chain and energy production at the RVLM takes place during fatal Mev intoxication. We further demonstrated thdat CoQ10 provides neuroprotection against Mev-induced cardiovascular depression and fatality through maintenance of activity of the key mitochondrial respiratory enzymes in the RVLM.

KEYWORDS: mevinphos intoxication; dysfunction of mitochondrial respiratory chain; bioenergetic production; rats

Address for correspondence: Alice Y.W. Chang, Ph.D., Center for Neuroscience, National Sun Yat-sen University, Kaohsiung 80424, Taiwan, ROC. Voice: +886-7-5253629; fax: +886-7-5255801.
achang@mail.nsysu.edu.tw

INTRODUCTION

Brain death is associated with the permanent termination of essentially all brain functions, particularly the autonomic cardiovascular regulatory mechanisms in the brainstem. Thus, the current legal definition of death in Taiwan, similar to that in most developed countries, is brainstem death.[1] In our search for the mechanisms that underlie brain death, our group demonstrated previously that a common denominator exists among patients who succumbed to systemic inflammatory response syndrome,[2] severe brain injury,[3] or organophosphate poisoning.[4] We found that death is invariably preceded by a dramatic reduction or loss of a life-and-death signal that we detected from the systemic arterial pressure (SAP) signals. Our group further established that this life-and-death signal is related to the functional integrity of the brainstem.[3] It is also associated with the sympathetic neurogenic vasomotor tone[5] that takes origin from the premotor sympathetic neurons[6] at the rostral ventrolateral medulla (RVLM).[7]

The primary organelle for cellular energy generation and oxygen consumption is the mitochondrion.[8] As the major machinery for cellular ATP production, the mitochondrion presents itself as an important target for cellular insults. It follows that mitochondrial dysfunction, leading to bioenergetic failure in the RVLM, may be associated with brain death. This study evaluated this hypothesis, using a rat model of fatal mevinphos (Mev) intoxication that closely resembles the clinical conditions of organophosphate poisoning.

As an organophosphate of the P = O type, Mev (3-[dimethoxyphosphinyl-oxyl]-2-butenoic acid methyl ester) exerts direct inhibition on acetylcholinesterase in the brain.[9] Previous studies from our laboratory[10–12] revealed that a crucial brain site on which Mev acts to elicit cardiovascular toxicity is the RVLM. We therefore evaluated the functional changes in mitochondrial respiratory chain at the RVLM during fatal Mev intoxication, and the neuroprotective role of coenzyme Q10 (CoQ10), a mobile electron carrier in the mitochondrial respiratory chain[13] that also acts as an antioxidant[14–16] in this process.

MATERIALS AND METHODS

Animals

Adult, male, specific pathogen–free Sprague–Dawley rats purchased from the Experimental Animal Center, National Science Council, Taiwan, Republic of China, were used. All experimental procedures were approved by our institutional animal care committee.

General Preparation

SAP was recorded via the right femoral artery, and heart rate (HR) was estimated instantaneously from the digitized SAP signals.[10–12,18] During the experiment, animals received continuous intravenous infusion of propofol (Zeneca, Macclesfield, UK) at 30 mg/kg/h, which provided satisfactory anesthetic maintenance while preserving the capacity of central cardiovascular regulation.[17] Animals were allowed to breathe spontaneously with room air, and body temperature was maintained at 37°C via a heating pad.

Microinjection of CoQ10

Microinjection bilaterally of CoQ10 into the RVLM, at a volume of 50 nL, was carried out using a 27-gauge needle that was connected to a 0.5-μL (Hamilton microsyringe, Reno, NV). The coordinates used were as follows: 4.5–5 mm posterior to the lambda, 1.8–2.1 mm lateral to the midline, and 8.1–8.4 mm below the dorsal surface of the cerebellum.[10–12,18] Microinjection of artificial cerebrospinal fluid (aCSF) served as the vehicle and volume control.

Assay for Activity of Mitochondrial Respiratory Enzymes or ATP Production

Tissues on both sides of the ventrolateral part of medulla oblongata from each animal, at the level of RVLM (0.5–2.5 mm rostral to the obex),[12,18] were collected 5–10 min after administration of Mev or saline. Baseline control samples of RVLM were collected from rats that received only preparatory surgery under anesthesia. Isolation of mitochondria from tissue samples was carried out at 4°C and completed within 2 h according to procedures reported previously.[18] As a routine, samples pooled from 7–8 animals in each experimental group were used for the analysis of mitochondrial enzyme activities or ATP production. All enzyme assays were performed using a thermostatically regulated (37°C) UV/visible spectrophotometer (Amersham, Uppsala, Sweden), with reagents purchased from Sigma (St. Louis, MO). Determination of nicotinamide adenine dinucleotide (NADH) cytochrome c reductase (NCCR; marker for Complexes I + III), succinate cytochrome c reductase (SCCR; marker for Complexes II + III), or cytochrome c oxidase (CCO; marker for Complex IV) activity was carried out as reported previously.[18] ATP concentration was measured using an ATP bioluminescence assay kit (Roche, Mannheim, Germany). At least quadruplicate determination was carried out for each tissue sample in all enzyme assays. Total protein in the mitochondrial suspension was estimated using a protein assay kit (Bio-Rad, Hercules, CA).

Statistical Analysis

All values are expressed as mean ± standard error of the mean. The averaged value of MSAP or HR calculated every 20 min after microinjection of test agents, or the activity of mitochondrial respiratory enzymes or ATP production, was used for statistical analysis. One-way or two-way analysis of variance with repeated measures was used, as appropriate, to assess group means, followed by the Scheffé multiple-range test for post hoc assessment of individual means. Mortality was defined as death within 10 min after administration of Mev and was assessed by the Fisher exact test. A P value less than 0.05 was considered to be statistically significant.

RESULTS AND DISCUSSION

Intravenous administration of a lethal dose of Mev (1 mg/kg) in Sprague–Dawley rats maintained with propofol resulted in 100% mortality within 10 min (TABLE 1).

TABLE 1. Effect of pretreatment with microinjection bilaterally of aCSF or CoQ10 into RVLM on mortality rate in rats that received Mev (1 mg/kg, i.v.) or saline

	aCSF Saline	aCSF Mev	CoQ10 Saline	CoQ10 Mev
Survived	7	0	12	8
Died	0	12*	0	0+
Mortality rate (%)	0	100*	0	0+

*$P < 0.05$ vs. aCSF + saline group and +$P < 0.05$ vs. aCSF + Mev group in the Fisher exact test.

There was a significant bradycardia, accompanied by an initial hypertension that was followed by an abrupt hypotension (FIG. 1). We interpret these results to reflect the vagomimetic action of Mev via accumulation of acetylcholine in the heart. We also reason that it is conceivable that heightened sympathetic vasomotor outflows from the RVLM (data not shown), triggered as a baroreceptor reflex response to bradycardia, are responsible for the initial hypertension. More importantly, it is most likely that the subsequent hypotension reflects a progressive reduction in the activity of the premotor sympathetic neurons in the RVLM, which, together with bradycardia, leads to fatality.

One potential mechanism for the progressive reduction in RVLM neuronal activity is bioenergetic failure. In the process of oxidative phosphorylation,[8] electrons are passed along a transport chain that consists of four respiratory enzyme complexes arranged in a specific orientation in the mitochondrial inner membrane. These electrons are generated from the oxidation of fuel molecules by oxygen, and their transport through the respiratory chain leads to the generation of ATP. It is therefore intriguing that enzyme assay revealed a significant depression of the activity of NCCR, SCCR, and CCO in the RVLM during this fatal Mev intoxication. ATP production also underwent a decrease after the dysfunction of mitochondrial enzyme activity. Pretreatment by microinjection bilaterally of CoQ10 (4 μg) into the RVLM significantly prevented mortality (TABLE 1) and antagonized the Mev-induced bradycardia and hypotension (FIG. 1). CoQ10 pretreatment also reversed the significantly depressed activity of all three mitochondrial respiratory chain enzymes and the reduction of ATP production in the RVLM seen during fatal Mev intoxication (FIG. 2).

During oxidative phosphorylation,[8] electrons from the reducing agent, NADH, move from Complex I through CoQ10 to Complex III, and then through cytochrome c to Complex IV. Electrons from succinate enter the respiratory chain through flavin adenine dinucleotide (FAD), which is covalently linked to Complex II, and move to Complexes III and IV by way of CoQ10. The passage of electrons between these complexes releases energy that is stored in the form of a proton gradient across the membrane by Complexes I, III, and IV, and is used by ATPase to produce ATP from ADP. We observed in this study that the activity of NCCR, SCCR, and CCO in the RVLM underwent a decrease during fatal Mev intoxication, along with reduced ATP production. It follows that the fatal cardiovascular effects of Mev should arise from a dysfunction of Complexes I, II, III, and IV in the mitochondrial electron transport chain at the RVLM. Furthermore, it is likely that pathways linked to both NADH and FAD are engaged in the speculated bioenergetic failure.

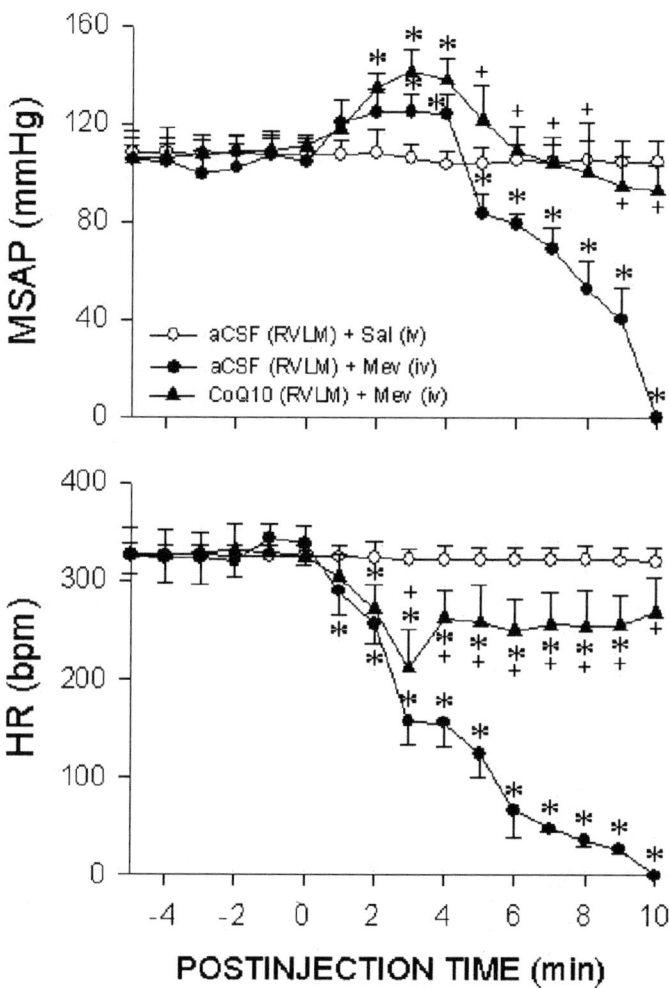

FIGURE 1. Time-course changes in mean systemic arterial pressure (MSAP) and heart rate (HR) in rats that were pretreated with microinjection bilaterally of CoQ10 or aCSF into the RVLM, followed by intravenous mevinphos (Mev, 1 mg/kg) or saline (Sal). Values are presented as mean ± SEM, n = 7–8 animals per experimental group. *$P < 0.05$ versus aCSF plus Sal group; $^+P < 0.05$ versus aCSF plus Mev group at corresponding time points in the Scheffé multiple-range analysis.

That CoQ10 confers neuroprotection against Mev-induced cardiovascular depression and fatality through maintenance of activity of the key mitochondrial respiratory enzymes in the RVLM provided further credence to the above notion. CoQ10 is a quinone derivative with a long isoprenoid tail that renders the molecule highly nonpolar and enables it to diffuse rapidly in the hydrocarbon phase of the inner mitochondrial membrane.[13] CoQ10 is also the only electron carrier in the respiratory

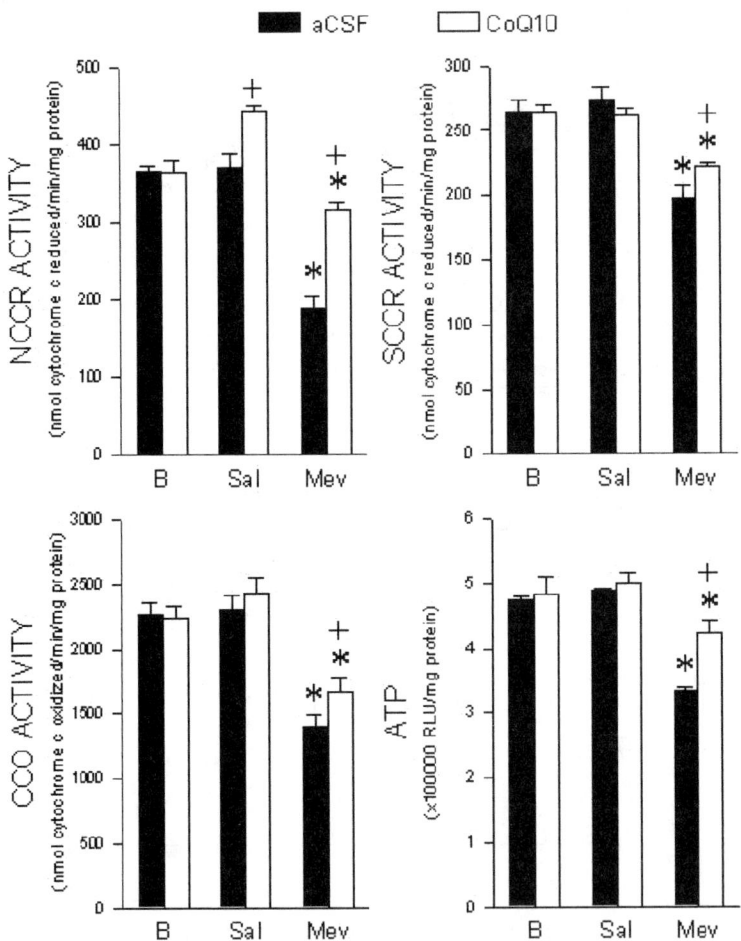

FIGURE 2. Enzyme assay for activity of NADH cytochrome *c* reductase (NCCR; marker for Complex I + III), succinate cytochrome *c* reductase (SCCR; marker for Complex II + III), cytochrome *c* oxidase (CCO; marker for Complex IV), or ATP production in mitochondria isolated from the ventrolateral medulla of rats that were pretreated with microinjection bilaterally of CoQ10 or aCSF into the RVLM, followed by i.v. mevinphos (Mev, 1 mg/kg) or saline (Sal). Values are presented as mean ± SEM of quadruplicate analyses from samples pooled from $n = 7-8$ animals per experimental group. *$P < 0.05$ versus baseline control (B) or corresponding Sal group; $^+P < 0.05$ versus aCSF-pretreated group in the Scheffé multiple-range analysis.

chain that is not tightly bound or covalently attached to a protein. In addition to serving as a highly mobile carrier of electrons between the flavoproteins and the cytochromes in the mitochondrial respiratory chain,[13] the fully reduced form (ubiquinol) of CoQ10 reportedly acts as a potent antioxidant.[14] Endogenous ubiquinol inhibits lipid peroxidation in biologic membranes and protects mitochondrial membrane proteins and DNA from oxidative damage.[13,15] In addition, it was suggested previously that CoQ10 can serve as a neuroprotective agent,[16] which brings about clinical and biochemical improvements in muscular dystrophies or neurogenic atrophies, mitochondrial encephalomyopathies, or neurodegenerative disorders that include Parkinson's disease, Huntington's disease, and Friedreich's ataxia.[13,19,20] This study further indicated that application of CoQ10 to the RVLM also provided neuroprotection against fatal Mev intoxication.

In conclusion, this study revealed that dysfunction of mitochondrial respiratory chain and energy production at the RVLM takes place during fatal Mev intoxication. We further demonstrated that CoQ10 provides neuroprotection against Mev-induced cardiovascular depression and fatality through maintenance of activity of the key mitochondrial respiratory enzymes in the RVLM.

REFERENCES

1. HUNG, T.P. & S.T. CHEN. 1995. Prognosis of deeply comatose patients on ventilators. J. Neurol. Neurosurg. Psychiatry **58:** 75–80.
2. YIEN, H.W., S.S. HSEU, L.C. LEE, et al. 1997. Spectral analysis of systemic arterial pressure and heart rate signals as a prognostic tool for the prediction of patient outcome in intensive care unit. Crit. Care Med. **25:** 258–266.
3. KUO, T.B., H.W. YIEN, S.S. HSEU, et al. 1997. Diminished vasomotor component of systemic arterial pressure signals and baroreflex in brain death. Am. J. Physiol. **273:** H1291–H1298.
4. YEN, D.H., H.W. YIEN, L.M. WANG, et al. 2000. Spectral analysis of systemic arterial pressure and heart rate signals in patients with acute respiratory failure induced by severe organophosphate poisoning. Crit. Care Med. **28:** 2805–2811.
5. YANG, M.W., T.B. KUO, S.M. LIN, et al. 1995. Continuous, on-line, real-time spectral analysis of systemic arterial pressure signals during cardiopulmonary bypass. Am. J. Physiol. **268:** H2329–H2335.
6. ROSS, C.A., D.A. RUGGIERO, D.H. PARK, et al. 1984. Tonic vasomotor control by the rostral ventrolateral medulla: effect of electrical or chemical stimulation of the area containing C_1 adrenaline neurons on arterial pressure, heart rate, and plasma catecholamines and vasopressin. J. Neurosci. **4:** 474–494.
7. KUO, T.B., C.C. YANG & S.H. CHAN. 1997. Selective activation of vasomotor component of SAP spectrum by nucleus reticularis ventrolateralis in rats. Am. J. Physiol. **272:** H485–H492.
8. HATEFI, Y. 1985. The mitochondrial electron transport and oxidative phosphorylation system. Annu. Rev. Biochem. **54:** 1015–1069.
9. TAKAHASHI, H., T. KOJIMA, T. IKEDA, et al. 1991. Differences in the mode of lethality produced through intravenous and oral administration of organophosphorus insecticides in rats. Fundam. Appl. Toxicol. **16:** 459–468.
10. YEN, D.H., J.C. YEN, W.B. LEN, et al. 2001. Spectral changes in systemic arterial pressure signals during acute mevinphos intoxication in the rat. Shock **15:** 35–41.
11. CHANG, A.Y., J.Y. CHAN, F.J. KAO, et al. 2001. Engagement of inducible nitric oxide synthase at rostral ventrolateral medulla during mevinphos intoxication in the rat. J. Biomed. Sci. **8:** 475–483.
12. CHAN, J.Y., S.H. CHAN & A.Y. CHANG. 2004. Differential contribution of NOS isoforms in the rostral ventrolateral medulla to cardiovascular responses associated with mevinphos intoxication in the rat. Neuropharmacology **46:** 1184–1194.

13. ERNSTER, L. & G. DALLNER. 1995. Biochemical, physiological and medical aspects of ubiquinone function. Biochim. Biophys. Acta **1271:** 195–204.
14. FREI, B., M.C. KIM & B.N. AMES. 1990. Ubiquinol-10 is an effective lipid-soluble antioxidant at physiological concentrations. Proc. Natl. Acad. Sci. USA **87:** 4879–4883.
15. FORSMARK-ANDRÉE, P. & L. ERNSTER. 1994. Evidence for a protective effect of endogenous ubiquinol against oxidative damage to mitochondrial protein and DNA during lipid peroxidation. Mol. Aspects Med. **15:** S73–S81.
16. MATTHEWS, R.T., L. YANG, S. BROWNE, *et al.* 1998. Coenzyme Q_{10} administration increases brain mitochondrial concentrations and exerts neuroprotective effects. Proc. Natl. Acad. Sci. USA **95:** 8892–8897.
17. YANG, C.H., M.H. SHYR, T.B. KUO, *et al.* 1995. Effects of propofol on nociceptive response and power spectra of electroencephalographic and systemic arterial pressure signals in the rat: correlation with plasma concentration. J. Pharmacol. Exp. Ther. **275:** 1568–1574.
18. CHUANG, Y.C., J.L. TSAI, A.Y. CHANG, *et al.* 2002. Dysfunction of the mitochondrial respiratory chain in the rostral ventrolateral medulla during experimental endotoxemia in the rat. J. Biomed. Sci. **9:** 542–548.
19. PEPPING, J. 1999. Coenzyme Q_{10}. Am. J. Health Syst. Pharm. **56:** 519–521.
20. SHULTS, C.W. & A.H. SCHAPIRA. 2001. A cue to queue for CoQ? Neurology **57:** 375–376.

Mitochondrial Dysfunction and Oxidative Stress as Determinants of Cell Death/Survival in Stroke

PAK H. CHAN

Department of Neurosurgery, Department of Neurology and Neurological Sciences, and Program in Neurosciences, Stanford University School of Medicine, Stanford, California 94305-5487, USA

ABSTRACT: Mitochondria are the powerhouse of the cell. Their primary physiological function is to generate ATP through oxidative phosphorylation via the electron transport chain. Reactive oxygen radicals generated from mitochondria have been implicated in acute brain injuries, like stroke and neurodegeneration. Recent studies have shown that mitochondrially formed oxidants are mediators of molecular signaling and have implicated mitochondria-dependent apoptosis involving pro- and antiapoptotic protein binding, the release of cytochrome *c* and Smac, the activation of downstream caspase-9 and -3, and the fragmentation of DNA. Oxidative stress and the redox state are also implicated in the survival signaling pathway that involves phosphatidylinositol 3-kinase (PI3-K)/Akt and downstream signaling molecular bindings like Bad/Bcl-X_L and phosphorylated Bad/14-3-3. Genetically modified mice (SOD1, SOD2) or rats that overexpress or are deficient in superoxide dismutase have provided strong evidence in support of the role of mitochondrial dysfunction and oxidative stress as determinants of neuronal death/survival after stroke and neurodegeneration.

KEYWORDS: mitochondrion; oxidative stress; apoptosis; PI3-kinase; Akt/PKB; Bad; cerebral ischemia

MITOCHONDRIA-DEPENDENT PATHWAY OF APOPTOSIS

The cell death signaling pathway in mitochondria has recently been demonstrated in the ischemic brain with the release of mitochondrial cytochrome *c*, a water-soluble peripheral-membrane protein of mitochondria and an essential component of the mitochondrial respiratory chain (FIG. 1).[1] Cytochrome *c* is translocated from the mitochondria to the cytosolic compartment after transient focal cerebral ischemia in rats,[2] in brain slices that are subjected to hypoxia-ischemia, and in vulnerable hippocampal CA_1 neurons after transient global cerebral ischemia.[3] Mitochondria are involved in both the necrosis and apoptosis pathways, which depend on the severity

---

Address for correspondence: Pak H. Chan, Ph.D., Neurosurgical Laboratories, Stanford University, 1201 Welch Rd., MSLS #P314, Stanford, CA 94305-5487. Voice: 650-498-4457; fax: 650-498-4550.
phchan@stanford.edu

of the insult or the nature of the signaling pathways.[4,5] In most instances, severe cerebral ischemia renders the mitochondria completely dysfunctional for adenosine triphosphate (ATP) production, which ensures necrotic cell death. *In vitro* studies demonstrate that various cellular or biochemical signaling pathways involve mitochondria in apoptosis by releasing cytochrome *c* to the cytoplasm. Cytochrome *c* interacts with the CED-4 homologue, Apaf-1, and deoxyadenosine triphosphate, forming the apoptosome and leading to activation of caspase-9.[6–8] Caspase-9, which is presumably an initiator of the cytochrome *c*–dependent caspase cascade, then activates caspase-3, followed by activation of caspase-2, -6, -8, and -10 downstream.[9] Caspase-3 also activates caspase-activated DNase and leads to DNA damage. In cerebral ischemia studies, caspase-3 and -9 have also been shown to play a key role in neuronal death after ischemia.[10] The downstream caspases cleave many substrate proteins including poly(ADP-ribose) polymerase (PARP). Substrate cleavage causes DNA injury and subsequently leads to apoptotic cell death, but excessive activation of PARP causes depletion of nicotinamide-adenine dinucleotide and ATP, which ultimately leads to cellular energy failure and death. There are proteins that can prevent caspase activation in the cytosol. The inhibitor-of-apoptosis protein (IAP) family suppresses apoptosis by preventing activation of procaspases and also by inhibiting the enzymatic activity of active caspases.[11,12] The second mitochondria-derived activator of caspase (Smac) is also released by apoptotic stimuli and binds IAPs, thereby promoting activation of caspase-3.[13] A recent study showed that mitochondrial release of

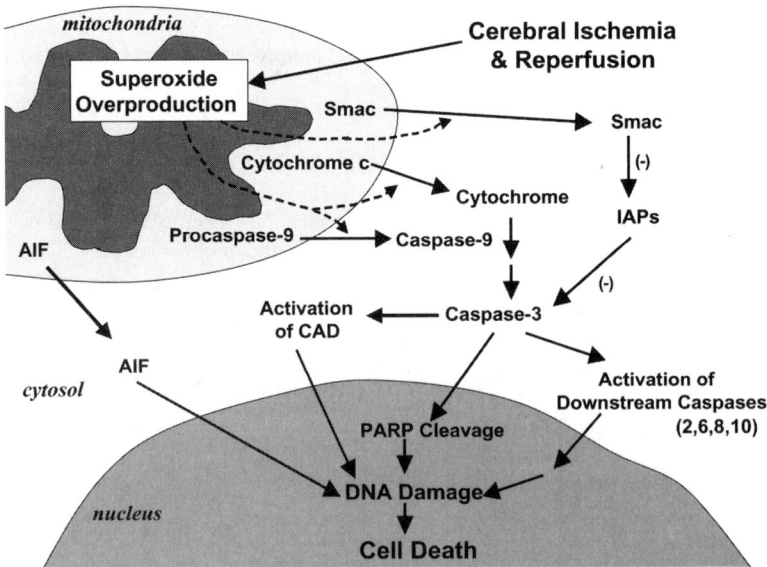

FIGURE 1. Intrinsic mitochondria-dependent pathway of apoptosis in cerebral ischemia and reperfusion.

cytochrome c and Smac preceded caspase activation after global ischemia, suggesting the importance of IAP inhibition as well as caspase activation.[10]

It has been well established that the processes involving mitochondrial release of cytochrome c, the formation of the apoptosome, and subsequent activation of caspase-3 and DNA fragmentation after transient cerebral ischemia are exacerbated in mice that are deficient in manganese superoxide dismutase (SOD2), one of the major mitochondrial antioxidant enzymes responsible for scavenging superoxide radicals formed in mitochondria (FIG. 1).[14,15] In addition, increased mitochondrial superoxide radical formation, cytochrome c and Smac release, and apoptosis are significantly reduced in transgenic (Tg) mice or rats that overexpress cytosolic copper/zinc-superoxide dismutase (SOD1) after transient focal or global cerebral ischemia, suggesting that oxidative stress is an upstream event of mitochondria-dependent apoptosis.[3,4,10,16–18]

BCL-2 FAMILY PROTEINS AS MOLECULAR TARGETS IN CELL DEATH/SURVIVAL SIGNALING

The Bcl-2 family proteins have one or more Bcl-2 homology domains and play a crucial role in intracellular apoptotic signal transduction by regulating permeability of the mitochondrial membrane.[19] Although still controversial, many researchers believe that mitochondrial cytochrome c is released through the permeability transition pore (PTP), which is part of the voltage-dependent anion channel, and that Bcl-2 family proteins directly regulate the PTP.[20] Among these proteins, Bax, Bcl-X_S, Bak, Bid, and Bad are proapoptotic. They eliminate the mitochondrial membrane potential by affecting the PTP and facilitating the release of cytochrome c. Conversely, Bcl-2 and Bcl-X_L function to conserve the membrane potential and block the release of cytochrome c. As expected, after focal cerebral ischemia, a decreased infarct was observed in Tg mice that overexpress Bcl-2[21] and in Bid knockout (KO) animals, whereas Bcl-2 KO mice showed an increased infarct. These findings, especially in the studies using proapoptotic/antiapoptotic protein-Tg/KO animals, suggest the importance of mitochondrial permeability regulation and Bcl-2 family proteins in ischemic cerebral injury.

BAD AS A TARGET OF CELL SURVIVAL SIGNALING

Bad is an important proapoptotic member of the Bcl-2 family that links the upstream cell survival signaling pathway and downstream pathway to inactivate antiapoptotic Bcl-2 family proteins.[22] *In vitro* studies show that Bad resides in an inactive complex with the molecular chaperone 14-3-3 via the phosphorylation of four serine residues (Ser-112, -136, -155, and -170).[23] With apoptotic stimuli, Bad is dephosphorylated, dissociated from 14-3-3, and translocated to the outer membrane of the mitochondria, where it subsequently dimerizes with Bcl-X_L and promotes mitochondrial cytochrome c release.[23] Ser-155 residue is important for direct interaction between Bad and Bcl-X_L, and its phosphorylation is regulated by several upstream signaling pathways. After cerebral ischemia, dephosphorylation and translocation of Bad from the cytosol to the mitochondria are observed, and dimerization of Bad progresses with

Bcl-X_L in the early stages after middle cerebral artery occlusion (MCAO).[24] These results suggest the pivotal function of Bad in ischemic cell death.

There are several pathways to the inhibition of the proapoptotic function of Bad. Ras is considered to play a central role in signaling for growth factor–mediated resistance to apoptosis. Ras can directly activate phosphatidylinositol 3-kinase (PI3-K), an upstream effector for activation of Akt/PKB. Akt/PKB is an initiator of the downstream pathways that inhibit the apoptotic pathways. Akt phosphorylates Bad

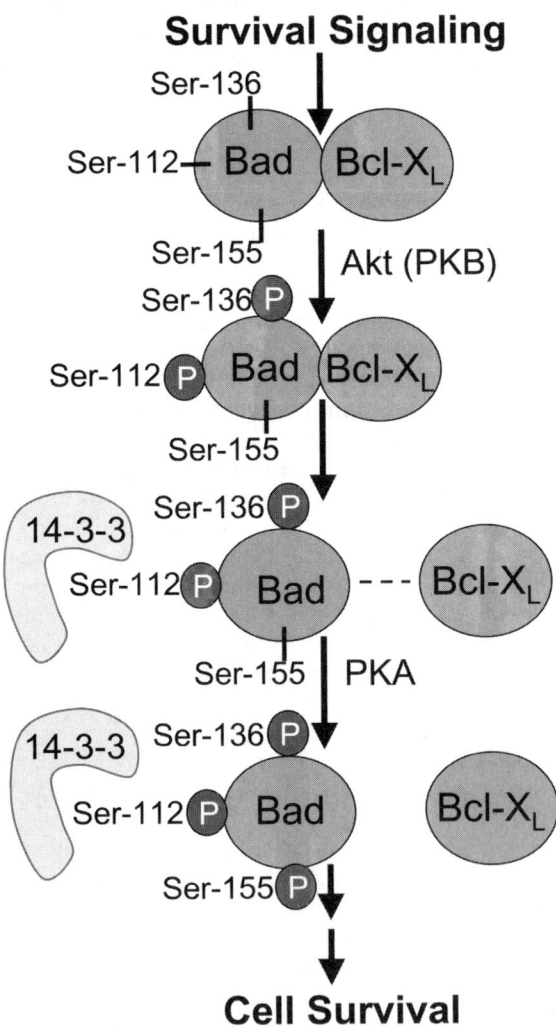

FIGURE 2. The survival signaling pathway involving PI3-K/Akt and PKA in Bad phosphorylation and its binding to 14-3-3 in surviving neurons after cerebral ischemia and reperfusion.

and obviates its inhibitory effects on Bcl-X_L, ultimately inhibiting the release of cytochrome c by blocking channel formation on the mitochondrial membrane by Bax. Akt also inhibits proteolytic activity of caspase-9 by phosphorylating it on Ser-196.[25] In addition, Akt can translocate into the nuclei and inactivate a proapoptotic member of the Forkhead family of transcription factors by phosphorylation, thereby inhibiting activation of the Fas pathway of apoptosis.[26] One mitogen-activated protein kinase (MAPK) family member, extracellular signal-regulated kinase (ERK), has two isoforms (ERK 1/2), which are constitutively expressed in the normal brain[27] and are activated by MAPK/ERK kinase 1/2. Active ERK 1/2 inactivates Bad through phosphorylation of 90-kDa ribosomal S6 kinases. Transforming growth factor-β1 has been shown to suppress Bad activity by phosphorylation of Bad at the Ser-112 site via activation of the ERK pathway in both *in vivo* cerebral ischemia models and *in vitro* studies.[28] Evidence from our laboratory has shown that phosphorylation of ERK 1/2 is involved in apoptosis and cell death after transient MCAO.[29] Phosphorylation of the Ser-155 residue in Bad is regulated by protein kinase A (PKA) in studies *in vitro*.[30] In rodent focal cerebral ischemia models, intra-

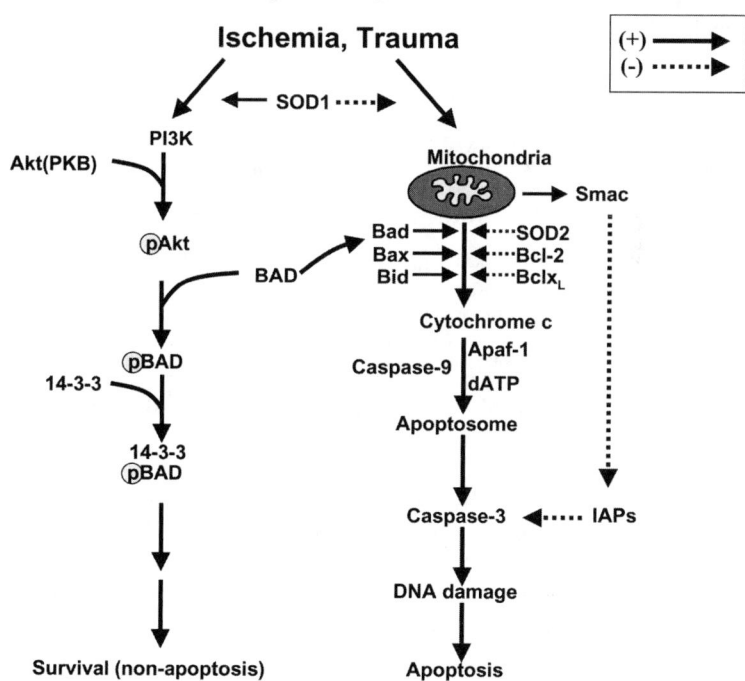

FIGURE 3. Life and death signaling involving mitochondria and oxidative stress after cerebral ischemia and reperfusion. Note that a proapoptotic protein is the missing link (a molecular switch?) between mitochondria-dependent cell death signaling and the PI3-K/Akt survival signaling pathway.

ventricular injection of a PKA inhibitor, H89, effectively suppressed PKA activity[31] and dimerization of Bad/Bcl-X_L and subsequent apoptotic cell death.[24] This cumulative evidence suggests that Akt and PKA pathways inhibit Bad function as cell survival signaling pathways after cerebral ischemia (FIG. 2).

It is important to point out that the survival signaling pathway PI3-K/Akt and Bad phosphorylation are upregulated by the redox state of the cell and that this signaling pathway is upregulated in ischemic animals that overexpress SOD1.[24] Since Bad dephosphorylation is known to be involved in mitochondria-dependent cell death, whereas Bad phosphorylation by Akt and PKA is involved in cell survival signaling after ischemia and reperfusion, we have proposed that Bad is a molecular switch and a target for cell death/survival after cerebral ischemia and reperfusion. We also propose that the redox state and the copper/zinc-superoxide dismutase level play a major role in turning this switch on and off in ischemic neurons (FIG. 3).

ACKNOWLEDGMENTS

This work was supported by National Institutes of Health grants P50 NS14543, RO1 NS25372, RO1 NS36147, and RO1 NS38653 and by an American Heart Association Bugher Foundation Award. I thank Cheryl Christensen for editorial assistance and Elizabeth Hoyte for figure preparation.

REFERENCES

1. BOYER, P.D. *et al.* 1977. Oxidative phosphorylation and photophosphorylation. Ann. Rev. Biochem. **46:** 955–026.
2. FUJIMURA, M. *et al.* 1998. Cytosolic redistribution of cytochrome c after transient focal cerebral ischemia in rats. J. Cereb. Blood Flow Metab. **18:** 1239–1247.
3. SUGAWARA, T. *et al.* 1999. Mitochondrial release of cytochrome *c* corresponds to the selective vulnerability of hippocampal CA1 neurons in rats after transient global cerebral ischemia. J. Neurosci. **19:** 1–6.
4. FUJIMURA, M. *et al.* 2000. The cytosolic antioxidant copper/zinc-superoxide dismutase prevents the early release of mitochondrial cytochrome c in ischemic brain after transient focal cerebral ischemia in mice. J. Neurosci. **20:** 2817–2824.
5. GREEN, D.R. & J.C. REED. 1998. Mitochondria and apoptosis. Science **281:** 1309–1312.
6. KUIDA, K. *et al.* 1998. Reduced apoptosis and cytochrome c-mediated caspase activation in mice lacking caspase 9. Cell **94:** 325–337.
7. LI, P. *et al.* 1997. Cytochrome c and dATP-dependent formation of Apaf-1/caspase-9 complex initiates an apoptotic protease cascade. Cell **91:** 479–489.
8. YOSHIDA, H. *et al.* 1998. Apaf1 is required for mitochondrial pathways of apoptosis and brain development. Cell **94:** 739–750.
9. SLEE, E.A. *et al.* 1999. Ordering the cytochrome c-initiated caspase cascade: hierarchical activation of caspases-2, -3, -6, -7, -8, and -10 in a caspase-9-dependent manner. J. Cell Biol. **144:** 281–292.
10. SUGAWARA, T. *et al.* 2002. Overexpression of copper/zinc superoxide dismutase in transgenic rats protects vulnerable neurons against ischemic damage by blocking the mitochondrial pathway of caspase activation. J. Neurosci. **22:** 209–217.
11. DEVERAUX, Q.L. & J.C. REED. 1999. IAP family proteins—suppressors of apoptosis. Genes Dev. **13:** 239–252.
12. MILLER, L.K. 1999. An exegesis of IAPs: salvation and surprises from BIR motifs. Trends Cell Biol. **9:** 323–328.

13. CHAI, J. et al. 2000. Structural and biochemical basis of apoptotic activation by Smac/DIABLO. Nature **406:** 855–862.
14. NOSHITA, N. et al. 2001. Manganese superoxide dismutase affects cytochrome c release and caspase-9 activation after transient focal cerebral ischemia in mice. J. Cereb. Blood Flow Metab. **21:** 557–567.
15. KIM, G.W. et al. 2002. Manganese superoxide dismutase deficiency exacerbates cerebral infarction after focal cerebral ischemia/reperfusion in mice. Implications for the production and role of superoxide radicals. Stroke **33:** 809–815.
16. SUGAWARA, T. et al. 2002. Overexpression of SOD1 protects vulnerable motor neurons after spinal cord injury by attenuating mitochondrial cytochrome c release. FASEB J. express article 10.1096/fj.02-0251fje; summary. FASEB J. **16:** 1997–1999.
17. FUJIMURA, M. et al. 1999. Copper-zinc superoxide dismutase prevents the early decrease of apurinic/apyrimidinic endonuclease and subsequent DNA fragmentation after transient focal cerebral ischemia in mice. Stroke **30:** 2408–2415.
18. SAITO, A. et al. 2004. Oxidative stress is associated with XIAP and Smac/DIABLO signaling pathways in mouse brains after transient focal cerebral ischemia. Stroke **35:** 1443–1448.
19. YUAN, J. & B.A. YANKNER. 2000. Apoptosis in the nervous system. Nature **407:** 802–809.
20. SHI, Y. 2001. A structural view of mitochondria-mediated apoptosis. Nat. Struct. Biol. **8:** 394–401.
21. MARTINOU, J.-C. et al. 1994. Overexpression of BCL-2 in transgenic mice protects neurons from naturally occurring cell death and experimental ischemia. Neuron **13:** 1017–1030.
22. YANG, E. et al. 1995. Bad, a heterodimeric partner for Bcl-XL and Bcl-2, displaces Bax and promotes cell death. Cell **80:** 285–291.
23. ZHA, J. et al. 1996. Serine phosphorylation of death agonist BAD in response to survival factor results in binding to 14-3-3 not BCL-X(L). Cell **87:** 619–628.
24. SAITO, A. et al. 2003. Overexpression of copper/zinc superoxide dismutase in transgenic mice protects against neuronal cell death after transient focal ischemia by blocking activation of the Bad cell death signaling pathway. J. Neurosci. **23:** 1710–1718.
25. CARDONE, M.H. et al. 1998. Regulation of cell death protease caspase-9 by phosphorylation. Science **282:** 1318–1321.
26. BRUNET, A. et al. 1999. Akt promotes cell survival by phosphorylating and inhibiting a Forkhead transcription factor. Cell **96:** 857–868.
27. BOULTON, T.G. et al. 1991. ERKs: a family of protein-serine/threonine kinases that are activated and tyrosine phosphorylated in response to insulin and NGF. Cell **65:** 663–675.
28. ZHU, Y. et al. 2002. Transforming growth factor-β1 increases Bad phosphorylation and protects neurons against damage. J. Neurosci. **22:** 3898–3909.
29. NOSHITA, N. et al. 2002. Copper/zinc superoxide dismutase attenuates neuronal cell death by preventing extracellular signal-regulated kinase activation after transient focal cerebral ischemia in mice. J. Neurosci. **22:** 7923–7930.
30. LIZCANO, J.M., N. MORRICE & P. COHEN. 2000. Regulation of BAD by cAMP-dependent protein kinase is mediated via phosphorylation of a novel site, Ser^{155}. Biochem. J. **349:** 547–557.
31. KIMURA, S. et al. 1998. cAMP-dependent long-term potentiation of nitric oxide release from cerebellar parallel fibers in rats. J. Neurosci. **18:** 8551–8558.

Oxidative Damage in Mitochondrial DNA Is Not Extensive

KOK SEONG LIM, KANDIAH JEYASEELAN, MATTHEW WHITEMAN, ANDREW JENNER, AND BARRY HALLIWELL

Department of Biochemistry, National University of Singapore, 8 Medical Drive, Singapore 117597

ABSTRACT: Since 1988 several research groups have reported greater levels of oxidative damage in mitochondrial DNA than in nuclear DNA, while others have suggested that the greater damage in mtDNA might be due to artifactual oxidation. The popular theory that mtDNA is more heavily damaged *in vivo* than nDNA does not stand on firm ground. Using an improved GC-MS method and pure mtDNA, our analyses revealed that the damage level in mtDNA is not higher, and may be somewhat lower, than that in nDNA.

KEYWORDS: mtDNA; nDNA; oxidative damage; GC-MS; 8-OH guanine

INTRODUCTION

In 1988, Richter *et al.*[1] claimed that mitochondrial DNA (mtDNA) suffers greater oxidative damage than does nuclear DNA (nDNA). They reported a level of 8-hydroxy-2′-deoxyguanosine (8-OHdG) in mtDNA 16 times higher than that in nDNA.[1] This report received much attention in the scientific community because it was the first observation that revealed a higher level of oxidative damage end products in mtDNA than in nDNA, although it had long been known that mtDNA, in contrast to nDNA, is located in close proximity to free radical generation sites of the electron transport chain in mitochondria and is not protected by histones. Several arguments supported the observation. First, mitochondria are thought to be the most important intracellular source of reactive oxygen species. Second, some data suggested that DNA damage is less efficiently repaired in mitochondria than in nuclei. Third, the age-dependent increase in mtDNA mutations, especially deletions, appeared to strengthen the hypothesis that mtDNA is more susceptible to oxidative damage.[2,3] Analyses of mtDNA deletions in different brain areas[4] showed that the highest frequencies of mtDNA deletions were found in those areas with the highest metabolic rates. This, together with the fact that the mtDNA genome contains few,

---

Address for correspondence: Professor Barry Halliwell, Department of Biochemistry, National University of Singapore, 8 Medical Drive, Singapore 117597. Voice: +65-6874-3247; fax: +65-6779-1453.
bchbh@nus.edu.sg

if any, noncoding bases, has led some to believe that one of the direct consequences of free radical damage to mtDNA is deletion. However, it has been argued that these specific deletions are unlikely to be produced by the random attack of reactive oxygen species such as hydroxyl radical (OH·).[5]

Since the report of Richter et al.,[1] several groups have carried out comparative studies of oxidative damage between mtDNA and nDNA (e.g., Refs. 6 and 7), while others studied the oxidative damage in either nuclei or mitochondria alone. The results consistently showed greater damage in mtDNA, supporting the existing notion. However, when Beckman and Ames[8] compared all reported values of oxidative base damage products in mtDNA and nDNA, they found that the range of measured values spans over five orders of magnitude. When the lowest values generated using high-performance liquid chromatography–electrochemical detection (HPLC-ECD) were compared, the level of 8-OHdG in mtDNA was no higher than that of nDNA. Indeed, over the past few years it is increasingly apparent that the bulk of the variation between published values of 8-OHdG in DNA may be due to the methods used, rather than to endogenous levels, and arguments rage over the validity of the various methods used, with no clear answer in sight.[5,8–11] Another possibility causing an artifact is cross contamination of nDNA and mtDNA. mtDNA constitutes only 1% of the total cellular DNA. If mtDNA is truly more oxidized and if contamination of mtDNA by nDNA occurs, the measurement of damage levels in mtDNA will be prone to underestimation. It is therefore essential to obtain mtDNA of high purity in order to make a comparison of oxidative damage. At present, the theory that mtDNA is more heavily oxidatively damaged than nDNA does not stand on firm ground.[8]

We embarked on this study of oxidative damage in mitochondria and nuclei to address these two methodological issues. Analysis of oxidative damage in highly purified mtDNA and nDNA was performed using gas chromatography–mass spectrometry (GC-MS) using ethanethiol to prevent the artifactual generation of oxidized DNA bases.[12] Heavy isotopes were also used as internal standards for some of the base damage products.

MATERIALS AND METHODS

Chemicals

Ethylenediaminetetraacetic acid (EDTA), ethidium bromide, boric acid, glycerol, and ethylene glycol-bis(2-aminoethylether)-N,N,N',N'-tetraacetic acid (EGTA) were obtained from Sigma (St. Louis, MO); DNAzol from Invitrogen (Carlsbad, CA); sucrose from Bio-Rad (Hercules, CA); RNase A, RNase T1, and proteinase K from Roche (Mannheim, Germany); sodium chloride (NaCl) and ethanol from Merck (Darnstadt, Germany); Tris base from Fisher (Fairlawn, NJ); DYNAzyme II DNA polymerase from Finnzymes (Espoo, Finland); agarose from BioWhittaker Molecular Applications (East Rutherford, NJ); dNTP mixture and DNA markers from Promega (Madison,WI); ethanethiol from Fluka (Buchs SG, Switzerland); formic acid from Riedel-de Haen (Seelze, Germany), heavy isotopes (internal standards for GC-MS) from Cambridge Isotopes Laboratories (Andover, MA); N,O-bis(trimethylsilyl)trifluoroacetamide (BSTFA) with 1% trimethylchlorosilane (TMCS) and acetonitrile from Pierce (Rockford, IL).

Animals

Outbred male Sprague-Dawley rats were used in this study. They were maintained on a 12/12-h light-dark cycle with lights on at 7 A.M. and food and water available *ad libitum*. Animals were sacrificed by cervical dislocation. The tissues were immediately removed and used for isolation of nuclei and mitochondria.

Isolation of DNA

Five volumes of liver tissue were minced and homogenized in 95 volumes of TES buffer (100 mM Tris pH 7.4, 10 mM EDTA, 250 mM Sucrose) using a Teflon-glass homogenizer. The homogenate was first centrifuged at $959 \times g$ for 20 min to pellet unbroken cells and nuclei, which were then used to isolate nDNA. The supernatant was then centrifuged at $5,524 \times g$ for 20 min to remove any contaminating nuclei. The supernatant, which contained the mitochondria, was centrifuged twice more at the same speed. The final supernatant was centrifuged at $22,095 \times g$ for 20 min. The nuclear and mitochondrial pellets were washed once by resuspension in TES buffer followed by centrifugation at $959 \times g$ and $22,095 \times g$, respectively, for 20 min. These pellets were used for DNA isolation using DNAzol according the manufacturer's instruction with the following modifications. Mitochondrial and nuclear pellets were resuspended in approximately 8 volumes of DNAzol and homogenized. The homogenates were incubated with 100 µg/mL RNase A and 100 units/mL RNase T1 at 37°C for 30 min and then with 100 µg/mL proteinase K at 25°C for 1 h. The supernatant was collected by centrifugation at $14,000 \times g$ for 10 min, and DNA was precipitated using a half-volume of ethanol. DNA was washed once with DNAzol/ethanol [70:30 (v/v)] and then twice with 80% (v/v) ethanol before being redissolved in sterile water and incubated at 37°C for 15 min to aid dissolution. DNA was then again incubated with 100 µg/mL RNase A and 100 units/mL RNase T1 for 60 min at 37°C. Protein was removed by addition of SDS to 1% (w/v) followed by precipitation using 1/4 volume of saturated NaCl. The protein precipitate was removed by centrifugation at $14,000 \times g$ for 10 min and DNA precipitated using 2.5 volumes of ethanol.

Sample Pooling

Sample pooling was necessary because the yield of pure mtDNA was low (<5 µg/rat). The DNA samples from sets of 5–8 rats were pooled to give 11 independent DNA samples, each containing approximately 35 µg DNA, which were later used for verification of purity, quantitation, and GC-MS analysis.

Quantitation of DNA

Absorbance spectra between 220 nm–320 nm were obtained for all samples, and the DNA concentrations were determined using the absorbance at 260 nm with background correction at 320 nm.

Verification of Purity

The purity of isolated nDNA and mtDNA was determined by checking for the presence of the cytochrome *b* gene and the β-actin gene in the DNA preparations using the polymerase chain reaction (PCR). PCR reactions were carried out in a Perkin Elmer GeneAmp PCR system. The PCR profile for cytochrome b was as follows: an initial denaturation at 94°C for 1 min, followed by 27 cycles of denaturation at 94°C for 30 s, annealing at 65°C for 30 s, extension at 72°C for 1 min, and a final extension at 72°C for 10 min. The cytochrome b gene, which is coded by mtDNA, is identified as a 467-bp fragment, resulting from amplification with primers 5'-ATCATCAAC-CACTCCTTTATCGACC-3' and 5'-AAGCCTCCTCAGATTCATTCGACTA-3'. The PCR profile for β-actin was as follows: an initial denaturation at 94°C for 1 min, followed by 27 cycles of denaturation at 94°C for 30 s, annealing at 69°C for 30 s, extension at 72°C for 1 min and a final extension at 72°C for 10 min. β-Actin gene, which is coded by nDNA, is identified as a 395-bp fragment, resulting from amplification with primers 5'-CCTTCAACACCCCAGCCATGTACG-3' and 5'-GCATCGGAACCGCTCATTGCC-3'.

Restriction Digestion of DNA

DNA (0.05–0.1 µg) was digested with 20 units of each restriction enzyme in 10 µL of appropriate buffer at 37°C for 3 h.

Agarose Gel Electrophoresis

DNA was mixed with 1/5 volume of gel-loading buffer [1.86% (w/v) EDTA, 0.5% (w/v) bromophenol blue, 50% (v/v) glycerol], and then loaded into a well of a 0.5% (w/v) agarose gel (for DNA) or 1.5% (w/v) agarose gel (for PCR product) in TBE buffer (89 mM Tris base, 89 mM boric acid, 2 mM EDTA, pH 8.3). Electrophoresis was performed in TBE for 60 min at 80 V.[13] After electrophoresis, the gel was stained with ethidium bromide (0.5 µg/mL in TBE) and photographed. A HindIII digest of a DNA or 1-kb DNA ladder was used as size marker for DNA, and 100 bp or 123 bp DNA ladder for PCR amplification products.

Gas Chromatography–Mass Spectrometric Analysis of DNA

Hydrolysis, derivatization, and analysis of DNA were performed as described previously,[12,14] with minor modifications. Approximately 35 µg of the pooled mtDNA and nDNA samples were freeze-dried overnight. Hydrolysis was carried out by addition of 0.5 mL of 60% (v/v) formic acid and heating at 140°C for 45 min in an evacuated, sealed hydrolysis tube. The internal standards were added to the cooled hydrolyzed samples and freeze-dried. These include 6-azathymine (Sigma), 2,6-diamino-4-hydroxy-5-formamidopyrimidine-formyl-^{13}C-4-amino-5-amido-^{15}N$_2$ (FAPy guanine), 4,6-diamino-5-formamidopyrimidine-formyl-^{13}C-diamino-^{15}N$_2$ (FAPy adenine), 5-(hydroxymethyl)uracil-4,5-^{13}C$_2$-5'5'-d$_2$ (5-OH, Me uracil), 5-hydroxycytosine-1,3-^{15}N$_2$-2-^{13}C (5-OH cytosine), 8-hydroxyadenine-8-^{13}C-6,9-diamino-^{15}N$_2$ (8-OH adenine), xanthine-1,3-^{15}N$_2$, 8-hydroxyguanine-8-^{13}C-7,9-^{15}N$_2$ (8-OH guanine), thymine glycol-α,α,α,6-d$_4$, 5-(hydroxymethyl)hydantoin-1,3-^{15}N$_2$-2,4-^{13}C$_2$ (5-OH, Me hydantoin), and 5-hydroxyhydantoin-1,3-^{15}N$_2$-2,4-^{13}C$_2$ (5-OH hydantoin) (approximately 0.5 nmol of each). Derivatization was carried out for 2 h at 25°C using BSTFA (+1% TMCS)/acetonitrile /ethanethiol [16:3:1(v/v)] mixture.[12]

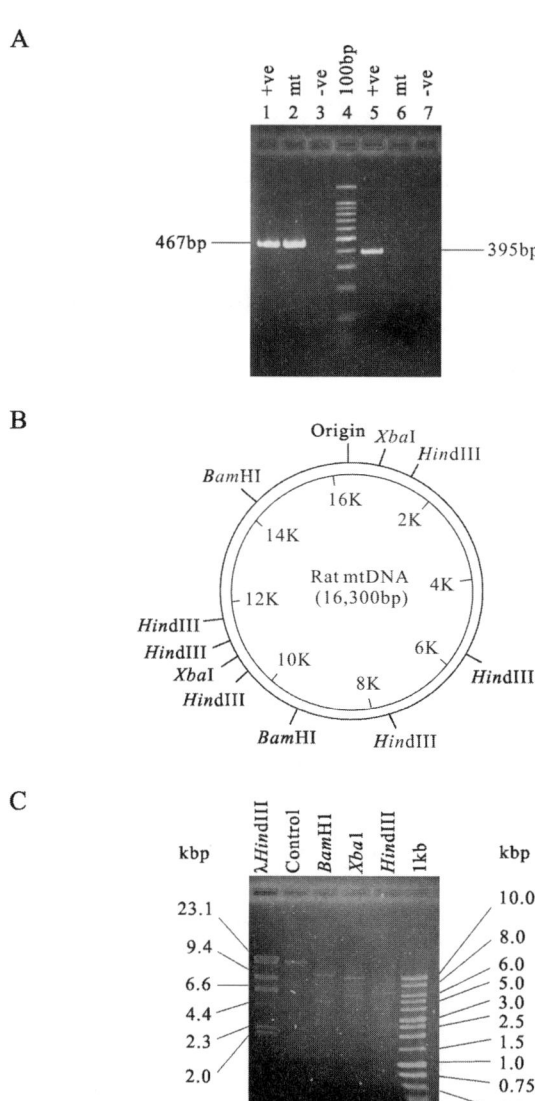

FIGURE 1. (**A**) Mitochondria from rat liver were isolated by differential centrifugation. mtDNA isolated using DNAzol method was used as template for primers specific to the cytochrome *b* gene (*lanes 1–3*) and the β-actin gene (*lanes 5–7*). Lanes *1* and *5* contain characterized mtDNA and nDNA (positive control) as templates, respectively; *lanes 2* and *6* contain mtDNA samples used in our study as templates; *lanes 3* and *7* contain no DNA (negative control). (**B**) Restriction map of rat mtDNA. Restriction site locations are BamH1 9361 and 14436; XbaI 612 and 10891; HindIII 1477, 5608, 7685, 10255, 11065, and 11230. (**C**) mtDNA restriction fragments on 0.5% (w/v) agarose gel stained with ethidium bromide. mtDNA was digested using the restriction enzymes indicated in appropriate buffer at 37°C for 3 h.

RESULTS

Isolation of Pure Mitochondrial DNA

The method of mitochondrial isolation used in our study resulted in mtDNA samples of desired purity for comparison of damage between nDNA and mtDNA. As shown in FIGURE 1A, our mtDNA sample contains the cytochrome b gene, but the β-actin gene was not detectable.

Characterization of Mitochondrial DNA

Restriction digestion of the mtDNA was also done to confirm its characteristics and purity. The results of restriction digest confirmed that the 16,300-bp mtDNA was intact and free of RNA, and was cut into two fragments by restriction enzyme BamH1 (5075-bp and 11,225-bp fragments) or XbaI (6,021-bp and 10,279-bp fragments), and cut into six fragments by HindIII [165 bp, 810 bp, 2,077 bp, 2,570 bp, 4,131 bp, and 6,547 bp; the first two fragments were too small to be visualized in our 0.5%(w/v) agarose gel] (FIG. 1C). Restriction digestion of nDNA by BamH1 generates DNA smears which spread from the top to the bottom of the gel, with two distinct bands still appearing at molecular sizes corresponding to the two fragments generated from restriction digestion of mtDNA using BamH1 (not shown).

Comparison between Phenol and DNAzol Method

This study used DNAzol to prepare DNA for analysis by GC-MS. Salmon testis DNA reextracted using the DNAzol method gave similar levels of all base damage products to DNA prepared using a phenol method (data not shown).

Comparison between Mitochondrial and Nuclear DNA

Analysis of the pure mtDNA and nDNA extracted using DNAzol revealed that the damage in mtDNA is not higher than that in nDNA (TABLE 1). In contrast to previous

TABLE 1. Oxidative base damage products in mtDNA and nDNA isolated from rat liver tissue using DNAzol method

	mtDNA (nmol/mg DNA)	nDNA (nmol/mg DNA)	Ratio n/mt	P Value
5-OH, Me hydantoin	0.044 ± 0.016	0.058 ± 0.038	1.32	0.296
5-OH uracil	0.028 ± 0.017	0.023 ± 0.011	0.82	0.450
5-OH, Me uracil	0.105 ± 0.130	0.142 ± 0.122	1.35	0.502
5-OH cytosine	0.009 ± 0.009	0.122 ± 0.119	13.60	0.005
FAPy adenine	0.044 ± 0.018	0.064 ± 0.007	1.45	0.044
8-OH adenine	0.090 ± 0.074	0.036 ± 0.016	0.40	0.031
FAPy guanine	0.043 ± 0.014	0.097 ± 0.025	2.26	0.003
8-OH guanine	0.022 ± 0.012	0.137 ± 0.062	6.23	$1.85e^{-5}$

NOTE: Each data point represents the mean ± SEM for 11 samples from two independent experiments. 5-Formyl uracil and 2-OH adenine were present at levels too low for detection.

studies, 8-OH guanine level is higher in nDNA than in mtDNA. Other base oxidation products, e.g., Fapy guanine, Fapy adenine, and 5-OH cytosine were also higher in nDNA.

Pure mtDNA in our experiments was isolated from the $22,095 \times g$ mitochondrial pellet collected after a $5,524 \times g$ centrifugation. This latter speed, intended to remove remaining nuclei, might however have removed most of the heavy mitochondria, leaving only light mitochondria; therefore, it is possible that the pure mtDNA obtained for the analysis represented, to a large extent, the DNA from light mitochondria. If oxidative adducts mostly lie in the heavy mitochondria, we might have eliminated most of them. To investigate this possibility, we also collected $5,524 \times g$ pellets as crude mitochondrial preparations for analysis. The presence of mtDNA in these preparations was confirmed by agarose gel electrophoresis (not shown). Analysis of the DNA from crude mitochondrial preparations (cmDNA) showed similar results to that with mtDNA—damage in cmDNA was not higher than that in nDNA. Therefore, the comparison of these three types of DNA—nDNA, cmDNA, and pure mtDNA— revealed that the most heavily damaged base products were in nDNA and not mtDNA or cmDNA.

DISCUSSION

Our failure to replicate the findings of more extensive damage in mtDNA is interesting. In contrast to most previous reports, oxidative base damage products in nDNA were found to be generally higher in our experiments, despite the fact that two independent experiments were carried out on two separate occasions involving different GC columns, batches of chemicals, and a total number of 80 rats.

TABLE 2 summarizes the oxidative damage levels in mtDNA and nDNA reported in the literature. While the levels of nDNA damage reported in most studies differ by only one order of magnitude, the claimed damage levels in mtDNA reveal a large variation. In particular, one study[7] reported levels of mtDNA damage which differed largely from others, and all the damage levels were at least fivefold higher than those in other studies, reaching the largest difference of 125-fold in the case of 5-OH, Me hydantoin (1738 vs. 14 lesions/10^6 bases) when compared to the current study. Such a huge variation occurred despite the fact that the same analytical method was used to study oxidative damage in both Zastawny et al.[7] and the current study (TABLE 2). Although RNA content was reported to be < 5% in that report,[7] it was not clear how much overestimation any RNA oxidation products would contribute to the levels of damage observed; therefore, it was difficult to estimate the actual levels of damage in mtDNA. Further, the degree of purity of the mtDNA used in the analysis was also not clear from that report. Hamilton et al.,[6] on the other hand, reported very low levels of damage in both mtDNA and nDNA. The level of 8-OHdG in mtDNA in that study was in fact close to the lowest value reported so far[15] when all studies were compared, including those that studied mtDNA damage alone.[8]

Our data are consistent with the report by Anson et al.,[16] claiming no elevated levels of oxidative base damage in mtDNA (TABLE 2). While the levels were similar in both mtDNA and nDNA for most base damage products (including 8-OH guanine), three lesions—5-OH uracil, 8-OH adenine, and 5-OH cytosine—were found to be significantly higher in nDNA than in mtDNA, and four other lesions were

TABLE 2. Comparison of results from several studies on oxidative DNA damage

	Richter et al.[1] 1988	Zastawny et al.[7] 1998	Anson et al.[16] 1999	Hamilton et al.[6] 2001	Current study 2004
Mitochondrial DNA (lesions/10^6 bases)					
5-OH hydantoin		1137	121		
5-OH, Me hydantoin		1738	49		14
5-OH uracil			7		9
5-OH, Me uracil		200			33
5-OH cytosine		223	15		3
FAPy adenine			8		14
8-OH adenine			43		28
FAPy guanine			75		18
8-OH guanine	130	758	135	0.43	7
Nuclear DNA (lesions/10^6 bases)					
5-OH hydantoin		27	155		
5-OH, Me hydantoin		36	33		18
5-OH uracil			12		7
5-OH, Me uracil		10	14		45
5-OH cytosine		34	56		39
FAPy adenine			24		20
8-OH adenine			133		12
FAPy guanine			55		31
8-OH guanine	8	171	263	0.07	43

present at lower levels in mtDNA than in nDNA, although not achieving statistical significance. It should be noted that while the level published in this report[16] for 8-OH guanine in mtDNA was 135 lesions/10^6 bases, it was later reported by the same author[17] that such mtDNA damage level might have been overestimated by threefold.

In contrast, direct comparisons between nDNA and mtDNA in several studies have seemingly revealed greater oxidative damage to mtDNA.[1,6,7] Three possibilities have been suggested: first, mtDNA suffers greater oxidative damage *in vivo* [due to lack of histones and close proximity to the electron transport chain (ETC)]; second, mtDNA damage is less efficiently repaired than nDNA; third, mtDNA is more prone to *ex vivo* artifactual oxidation. However, recent studies have challenged the concept of inefficient repair of mtDNA. Evidence to suggest that mitochondria lacked DNA repair mechanisms came originally from the observation that ultraviolet damage is not repaired in mitochondria,[18,19] while it is efficiently removed in the nuclei by nucleotide excision repair (NER). In addition, damage caused by cisplatin and nitrogen mustard is inefficiently repaired in mitochondria.[19] From these studies, it had been widely assumed that mitochondria have little DNA-repair capacity. It was after the discovery of the base excision repair (BER) mechanism in mitochondria that this view began to change. BER was shown to efficiently remove from mtDNA damage induced by alkylating agents, alloxan, streptozotocin, and acridine orange.[20–22] Following the subsequent reports on the purification of enzymes involved in the removal of 8-OHdG,[23] it is now evident that mitochondria repair oxidative DNA damage. Strand breaks and 8-OHdG are efficiently repaired in mtDNA, as fast as or even faster than in nDNA.[20,24] Another possible source of error is that mtDNA can be damaged by free radical generation from the ETC during isolation of mitochondria. Generation of reactive oxygen species by mitochondria occurs at several sites in the ETC, although the rates and relative importance of each site depend on the tissue and species origin of the mitochondria.[25,26] Consistent with this idea was a report which showed, using enzymatic/Southern blot assay, that the levels of 8-OHdG were approximately threefold higher when measured in mtDNA purified from isolated mitochondria than when measured without prior mitochondrial isolation.[17] However, the isolated rat liver mitochondria used in our studies did not produce hydrogen peroxide *in vitro*, consistent with the observation by St-Pierre *et al.*,[27] whereas rat heart mitochondria (used as a control) did. We also did not observe an increase in oxidative damage when mitochondria or tissue homogenates were incubated at room temperature for up to 3 h, measured as protein carbonyls, DNA base damage products, and malondialdehyde.

Other kinds of artifacts have also been observed. Beckman *et al.*[28] claimed an inverse correlation between the ratio of 8-OHdG/dG and the amount of DNA analyzed. A severalfold increase in the measured value was observed when the total amount of DNA hydrolyzed drops below 10 µg before analysis by HPLC-ECD. The reason for this is unknown, but it was suggested that comparisons between samples should be performed on equivalent amounts of DNA and meticulous attention be paid to system hygiene in view of the sensitivity of ECD. In the current study, we used 40 µg DNA in the first experiment and 20–30 µg DNA in the second experiment to measure oxidative damage in both mtDNA and nDNA.

While the two arguments that used to support the hypothesis that mtDNA damage is extensive—inefficient oxidative damage repair and *ex vivo* oxidation of mtDNA by free radicals, are now challenged—the question of whether *in vivo* mtDNA suffers more oxidative damage remains a difficult one to answer. The close proximity of DNA to the ETC and lack of histones are still relevant. However, Thorslund *et al.*[24] discussed the possibility that mtDNA, which is free of histones, may allow greater accessibility of repair enzymes to the lesions in mtDNA than in highly condensed nDNA.

mtDNA mutates at a much faster rate than nDNA,[29] a phenomenon often quoted as evidence that oxidative damage to mtDNA must be extensive. Mutations, especially deletions, are observed in various tissues.[30–34] Deletions often involve large portions of the mitochondrial genome, located between the two origins of replication, and are thought to underlie certain rare diseases of oxidative phosphorylation, including myopathies and encephalomyopathies. The levels of these deletions, measured with PCR techniques, increase markedly with age in heart[30] and brain.[4,35,36] Correlations have also been reported between deletion levels and levels of 8-OHdG in heart mtDNA,[37] thus suggesting that oxidative damage to mtDNA may be an initiating event. However, the most important consequences of unrepaired oxidative DNA damage are point mutations.[38] The mutation that would be expected to result from 8-OH guanine is a G-T transversion,[39] but mutational consequences specific to G-T transversion are rarely reported in mitochondrial DNA.[40] This suggests that factors other than 8-OH guanine play a large role in the mutation, such as damage to mtDNA polymerase or repair systems, possibly by reactive oxygen species.[5]

In conclusion, our data show that mtDNA damage is not always higher than nDNA damage and may even be lower for some base lesions. Thus, while future work in our and other laboratories will continue to refine our knowledge of the oxidative damage in DNA, the literature claim that mtDNA damage is more extensive than nDNA needs reconsideration.

ACKNOWLEDGMENTS

We are grateful to the National Medical Research Council of Singapore for its generous research support.

REFERENCES

1. RICHTER, C., J.W. PARK & B.N. AMES. 1988. Normal oxidative damage to mitochondrial and nuclear DNA is extensive. Proc. Natl. Acad. Sci. USA **85:** 6465–6467.
2. ARNHEIM, N. & G. CORTOPASSI. 1992. Deleterious mitochondrial DNA mutations accumulate in aging human tissues. Mutat. Res. **275:** 157–167.
3. CORTOPASSI, G.A. & N. ARNHEIM. 1990. Detection of a specific mitochondrial DNA deletion in tissues of older humans. Nucleic Acids Res. **18:** 6927–6933.
4. CORRAL-DEBRINSKI, M. et al. 1992. Mitochondrial DNA deletions in human brain: regional variability and increase with advanced age. Nat. Genet. **2:** 324–329.
5. HALLIWELL, B. 1999. Oxygen and nitrogen are pro-carcinogens. Damage to DNA by reactive oxygen, chlorine and nitrogen species: measurement, mechanism and the effects of nutrition. Mutat. Res. **443:** 37–52.
6. HAMILTON, M.L. et al. 2001. A reliable assessment of 8-oxo-2-deoxyguanosine levels in nuclear and mitochondrial DNA using the sodium iodide method to isolate DNA. Nucleic Acids Res. **29:** 2117–2126.
7. ZASTAWNY, T.H. et al. 1998. Comparison of oxidative base damage in mitochondrial and nuclear DNA. Free Radical Biol. Med. **24:** 722–725.
8. BECKMAN, K.B. & B.N. AMES. 1999. Endogenous oxidative damage of mtDNA. Mutat. Res. **424:** 51–58.
9. HELBOCK, H.J. et al. 1998. DNA oxidation matters: the HPLC-electrochemical detection assay of 8-oxo-deoxyguanosine and 8-oxo-guanine. Proc. Natl. Acad. Sci. USA **95:** 288–293.
10. SENTURKER, S. & M. DIZDAROGLU. 1999. The effect of experimental conditions on the levels of oxidatively modified bases in DNA as measured by gas chromatography-mass spectrometry: how many modified bases are involved? Prepurification or not? Free Radic. Biol. Med. **27:** 370–380.
11. COLLINS, A.R. et al. 2004. Are we sure we know how to measure 8-oxo-7,8-dihydroguanine in DNA from human cells? Arch. Biochem. Biophys. **423:** 57–65.
12. JENNER, A. et al. 1998. Measurement of oxidative DNA damage by gas chromatography-mass spectrometry: ethanethiol prevents artifactual generation of oxidized DNA bases. Biochem. J. **331:** 365–369.
13. SAMBROOK, J. & D.W. RUSSELL. 2001. Molecular Cloning : a Laboratory Manual. Cold Spring Harbor Laboratory Press. Cold Spring Harbor, NY.
14. ENGLAND, T.G. et al. 1998. Determination of oxidative DNA base damage by gas chromatography-mass spectrometry. Effect of derivatization conditions on artifactual formation of certain base oxidation products. Free Radic. Res. **29:** 321–330.
15. HIGUCHI, Y. & S. LINN. 1995. Purification of all forms of HeLa cell mitochondrial DNA and assessment of damage to it caused by hydrogen peroxide treatment of mitochondria or cells. J. Biol. Chem. **270:** 7950–7956.

16. ANSON, R.M. *et al.* 1999. Measurement of oxidatively induced base lesions in liver from Wistar rats of different ages. Free Radic. Biol. Med. **27:** 456–462.
17. ANSON, R.M., E. HUDSON & V.A. BOHR. 2000. Mitochondrial endogenous oxidative damage has been overestimated. FASEB J. **14:** 355–360.
18. CLAYTON, D.A., J.N. DODA & E.C. FRIEDBERG. 1974. The absence of a pyrimidine dimer repair mechanism in mammalian mitochondria. Proc. Natl. Acad. Sci. USA **71:** 2777–2781.
19. LEDOUX, S.P. *et al.* 1992. Repair of mitochondrial DNA after various types of DNA damage in Chinese hamster ovary cells. Carcinogenesis **13:** 1967–1973.
20. TAFFE, B.G. *et al.* 1996. Gene-specific nuclear and mitochondrial repair of formamidopyrimidine DNA glycosylase-sensitive sites in Chinese hamster ovary cells. Mutat. Res. **364:** 183–192.
21. DRIGGERS, W.J., S.P. LEDOUX & G.L. WILSON. 1993. Repair of oxidative damage within the mitochondrial DNA of RINr 38 cells. J. Biol. Chem. **268:** 22042–22045.
22. LEDOUX, S.P. *et al.* 1993. Repair of N-methylpurines in the mitochondrial DNA of xeroderma pigmentosum complementation group D cells. Carcinogenesis **14:** 913–917.
23. CROTEAU, D.L. *et al.* 1997. An oxidative damage-specific endonuclease from rat liver mitochondria. J. Biol. Chem. **272:** 27338–27344.
24. THORSLUND, T. *et al.* 2002. Repair of 8-oxoG is slower in endogenous nuclear genes than in mitochondrial DNA and is without strand bias. DNA Repair **1:** 261–273.
25. BARJA, G. 1999. Mitochondrial oxygen radical generation and leak: sites of production in states 4 and 3, organ specificity, and relation to aging and longevity. J. Bioenerg. Biomembr. **31:** 347–366.
26. TURRENS, J.F. 2003. Mitochondrial formation of reactive oxygen species. J. Physiol. **552:** 335–344.
27. ST-PIERRE, J. *et al.* 2002. Topology of superoxide production from different sites in the mitochondrial electron transport chain. J. Biol. Chem. **277:** 44784–44790.
28. BECKMAN, K.B. & B.N. AMES. 1996. Detection and quantification of oxidative adducts of mitochondrial DNA. Methods Enzymol. **264:** 442–453.
29. WALLACE, D.C. *et al.* 1987. Sequence analysis of cDNAs for the human and bovine ATP synthase beta subunit: mitochondrial DNA genes sustain seventeen times more mutations. Curr. Genet. **12:** 81–90.
30. CORTOPASSI, G.A. *et al.* 1992. A pattern of accumulation of a somatic deletion of mitochondrial DNA in aging human tissues. Proc. Natl. Acad. Sci. USA **89:** 7370–7374.
31. EDRIS, W. *et al.* 1994. Detection and quantitation by competitive PCR of an age-associated increase in a 4.8-kb deletion in rat mitochondrial DNA. Mutat. Res. **316:** 69–78.
32. GADALETA, M.N. *et al.* 1992. Mitochondrial DNA copy number and mitochondrial DNA deletion in adult and senescent rats. Mutat. Res. **275:** 181–193.
33. MELOV, S. *et al.* 1994. Detection of deletions in the mitochondrial genome of *Caenorhabditis elegans*. Nucleic Acids Res. **22:** 1075–1078.
34. SIMONETTI, S. *et al.* 1992. Accumulation of deletions in human mitochondrial DNA during normal aging: analysis by quantitative PCR. Biochim. Biophys. Acta **1180:** 113–122.
35. SOONG, N.W. *et al.* 1992. Mosaicism for a specific somatic mitochondrial DNA mutation in adult human brain. Nat. Genet. **2:** 318–323.
36. FILBURN, C.R. *et al.* 1996. Mitochondrial electron transport chain activities and DNA deletions in regions of the rat brain. Mech. Ageing Dev. **87:** 35–46.
37. HAYAKAWA, M. *et al.* 1992. Age-associated oxygen damage and mutations in mitochondrial DNA in human hearts. Biochem. Biophys. Res. Commun. **189:** 979–985.
38. WALLACE, S.S. 2002. Biological consequences of free radical-damaged DNA bases. Free Radic. Biol. Med. **33:** 1–14.
39. KOUCHAKDJIAN, M. *et al.* 1991. NMR structural studies of the ionizing radiation adduct 7-hydro-8-oxodeoxyguanosine (8-oxo-7H-dG) opposite deoxyadenosine in a DNA duplex. 8-Oxo-7H-dG(syn).dA(anti) alignment at lesion site. Biochemistry **30:** 1403–1412.
40. HORAI, S. & K. HAYASAKA. 1990. Intraspecific nucleotide sequence differences in the major noncoding region of human mitochondrial DNA. Am. J. Hum. Genet. **46:** 828–842.

Enhanced Generation of Mitochondrial Reactive Oxygen Species in Cybrids Containing 4977-bp Mitochondrial DNA Deletion

MEI-JIE JOU,[a] TSUNG-I PENG,[b] HONG-YUEH WU,[a] AND YAU-HUEI WEI[c]

[a]*Department of Physiology and Pharmacology, Chang Gung University, Tao-Yuan, Taiwan*

[b]*Department of Neurology, Lin-Ko Medical Center, Chang Gung Memorial Hospital, Tao-Yuan, Taiwan*

[c]*Department of Biochemistry, National Yang-Ming University, Taipei, Taiwan*

ABSTRACT: The poor bioenergetic state in mitochondria containing mtDNA with the 4977-bp deletion has been well documented. However, information on mitochondrial reactive oxygen species (ROS) generation at rest or under intense oxidative stress in mitochondria lacking the 4977-bp mtDNA fragment inside intact living cells was insufficient. We used cybrids containing truncated mtDNA lacking the 4977-bp fragment and measured ROS levels inside cybrids by fluorescence probe, 2′,7′-dichlorodihydrofluorescein (DCF), and confocal microscopy. Mitochondrial ROS at resting state was slightly higher in cybrids containing 4977-bp deletion mtDNA as compared to cybrids without mtDNA defects. For intense oxidative stress treatment, cybrids were treated with 5 mM H_2O_2 for 10 min. Consecutive DCF images were acquired after H_2O_2 had been washed away. Progressive increase of DCF signals, especially in the mitochondrial area, was observed in cybrids containing 4977-bp deletion mtDNA, even long after the brief, intense H_2O_2 treatment. This result suggests that a feed-forward, self-accelerating vicious cycle of mitochondrial ROS production could be initiated in cybrids containing 4977-bp deletion fragment mitochondria after brief, intense H_2O_2 treatment. This mechanism may play an important role in the pathophysiology of the disease process caused by mitochondria containing mtDNA with the 4977-bp deletion.

KEYWORDS: mitochondrial DNA deletion; cybrids; oxidative stress; hydrogen peroxide; reactive oxygen species; fluorescent probe; confocal microscopy

Address for correspondence: Tsung-I Peng, M.D., Ph.D., Department of Neurology, Chang Gung Memorial Hospital, Lin-Ko Medical Center, No. 5 Fu-Shin Street, Gwei-Shan, Tao-Yuan, 333 Taiwan. Voice: +886-3-328-1200 ext. 8347; fax: +886-3-328-8849.
tipeng@adm.cgmh.org.tw

INTRODUCTION

Mitochondrial DNA (mtDNA) encodes 13 polypeptides of the electron transport chain (ETC) and contains 22 tRNA genes for mitochondrial protein synthesis. Mutations or deletions of mtDNA result in serious mitochondrial dysfunction and have been suggested to be the basis of many human disease processes. A specific mtDNA deletion lacking the 4977-bp fragment occurs between two 13-bp direct repeats in the mtDNA sequence,[1] at nucleotide positions 13447–13459 and 8470–8482. This 4977-bp deletion affects genes encoding seven polypeptide components of the mitochondrial respiratory chain and 5 of the 22 tRNA genes for mitochondrial protein synthesis. The seven disrupted polypeptides comprise four polypeptides for complex I, one polypeptide for complex IV, and two polypeptides for complex V.

Cybrids containing various amounts of the truncated mtDNA (Δ-mtDNA) lacking the 4977-bp fragment had been constructed by Wei and Murphy[2] by fusing a human osteosarcoma cell line depleted of mtDNA (ρ^0) with enucleated skin fibroblasts from a patient with chronic progressive external ophthalmoplegia (CPEO). In the resulting cybrids that contained Δ-mtDNA exceeding a threshold level of about 50%, the mitochondrial membrane potential ($\Delta\Psi m$), rate of ATP synthesis, and cellular ATP/ADP ratio decreased.[2] Respiratory-chain enzyme activities measured in skeletal muscle specimens of 20 patients harboring large-scale mtDNA deletions were inversely related to the degree of heteroplasmy of mtDNA deletions.[3] In this study, however, even low degrees of heteroplasmy of mtDNA deletions were found to result in biochemical abnormalities, indicating the absence of any well-defined mtDNA deletion threshold in skeletal muscle.

Mitochondrial ETC has been recognized as the major source of reactive oxygen species (ROS) generation. Electrons passing through mitochondrial ETC leak out from complex I or complex III and interact with molecular oxygen (O_2) to form superoxide anion (O^{2-}), which is quickly dismutated by the mitochondrial Mn-superoxide dismutase (Mn-SOD) into H_2O_2. Impaired respiratory-chain function may lead to enhanced production of ROS via larger electron leakage. This elevated ROS production from defective ETC may result in further oxidative stress to mitochondria. Still, such oxidative stress may not be an obligate mediator of disease provoked by mtDNA mutations.[4] Cytoprotective responses, including increases in the levels of Bcl-2 and Bfl-1, pro-survival proteins that inhibit apoptosis,[4] and overexpression of the mitochondrial Mn-SOD,[5] have been reported.

Optimal conditions for ROS production in rat brain mitochondria require either a hyperpolarized membrane potential ($\Delta\Psi m$) or a substantial level of complex I inhibition.[6] This mode of ROS generation is very sensitive to depolarization of $\Delta\Psi m$, and even the depolarization associated with ATP generation was sufficient to inhibit ROS production.[6] It is not clear yet whether defective mitochondria with truncated mtDNA (Δ-mtDNA) lacking the 4977-bp fragment have enhanced production of ROS. In this study, fluorescence images of ROS in cybrids containing 4977-bp mtDNA deletion were acquired during resting state and under oxidative stress. Even though mitochondrial membrane potentials of cybrids containing 4977-bp mtDNA deletion were low, there were enhanced productions of mitochondrial ROS in these cybrids at rest and under oxidative stress.

MATERIALS AND METHODS

Cell Preparation—Establishment and Culture of Cybrids

Cybrid clones containing various proportions of Δ-mtDNA were established according to the procedures developed by King and Attardi.[7,8] The mtDNA-free cells (ρ^0) were from human 143B osteosarcoma cells treated with ethidium bromide. Human skin fibroblasts established from a patient with CPEO were enucleated and then fused with ρ^0 cells to create a range of cybrid clones. The amount of Δ-mtDNA in each clone was quantitated by Southern hybridization using a 1577-bp mtDNA probe specific for the D-loop region (nucleotide positions 16455–16462) wild-type mtDNA.[2] Cells were cultured in medium consisting of Dulbecco's modified Eagle's medium (Life Technologies, Grand Island, NY) supplemented with 10% (v/v) fetal bovine serum and with uridine (50 mg/L) and pyruvate (100 mg/L) to stabilize the proportion of Δ-mtDNA in each cybrid clone during culture. Cells were plated onto No. 1 glass coverslips for fluorescent measurement experiments (Model No. 1, VWR Scientific, San Francisco, CA).

Chemicals and Fluorescent Probes

All fluorescent probes were purchased from Molecular Probes Inc. (Eugene, OR), and chemicals were obtained from Sigma (St. Louis, MO). Intracellular ROS was detected using an intracellular ROS dye, 2′,7′-dichlorodihydrofluorescein (DCF). The nonfluorescent DCF is oxidized by intracellular ROS to form the highly fluorescent DCF. Loading concentrations of fluorescent probes used were DCF 1 mM. Fluorescent probes were all loaded at room temperature for 30 min of incubation to allow intracellular deacetylation of the ester form of dyes. After loading, cells were rinsed three times with HEPES buffered saline (140 mM NaCl/5 mM KCl/1 mM $MgCl_2$/2 mM $CaCl_2$/10 mM glucose/5 mM HEPES, pH 7.4).

Conventional and Confocal Imaging Microscopy

Conventional fluorescence images were obtained using a Zeiss inverted microscope (AxioVert 200M) equipped with a mercury lamp (HBO 103), a cool CCD camera (coolsnap fx), and Zeiss objectives (Plan-NeoFluar 100X, N.A. 1.3 oil). The filter set used for detecting DCF was No. 10 (Excitation: BP 450–490 nm; Emission: BP 515–565 nm). Confocal fluorescence images were obtained using a Leica SP2 MP (Leica-Microsystems, Wetzlar, Germany) fiber-coupling system equipped with an argon-laser system. During fluorescence imaging, the illumination light was reduced to the minimal level by using a neutral density filter (3%) to prevent the photosensitizing effect from the interaction of light with fluorescent probes. All images were processed and analyzed using a MetaMorph software (Universal Imaging Corp., West Chester, PA).

RESULTS AND DISCUSSION

Two clones of cybrid were used in this study. Southern blot hybridization analysis of these two lines of cybrids revealed the proportion of 4977-bp deletion Δ-mtDNA in

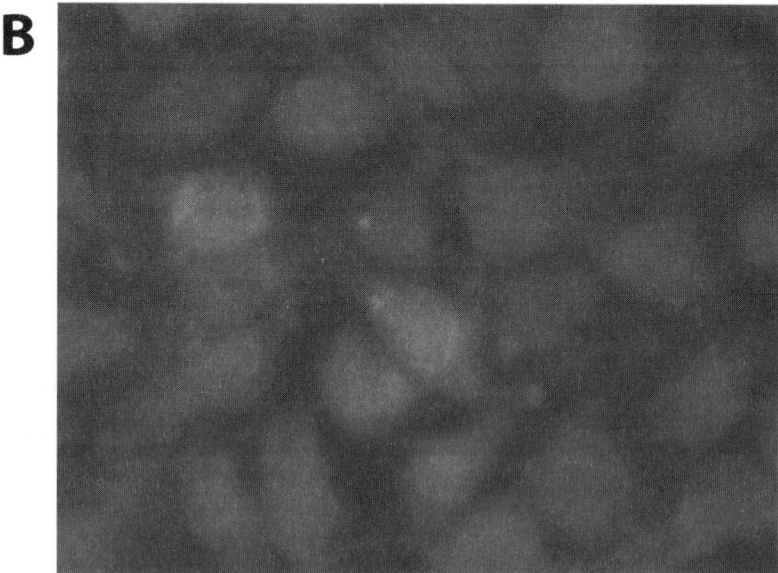

FIGURE 1. DCF fluorescence images of cybrids without 4977-bp deletion Δ-mtDNA before (**A**) and after (**B**) treatment with 5 mM H_2O_2 for 10 min. (**B**) DCF image was taken 35 min after H_2O_2 had been washed away.

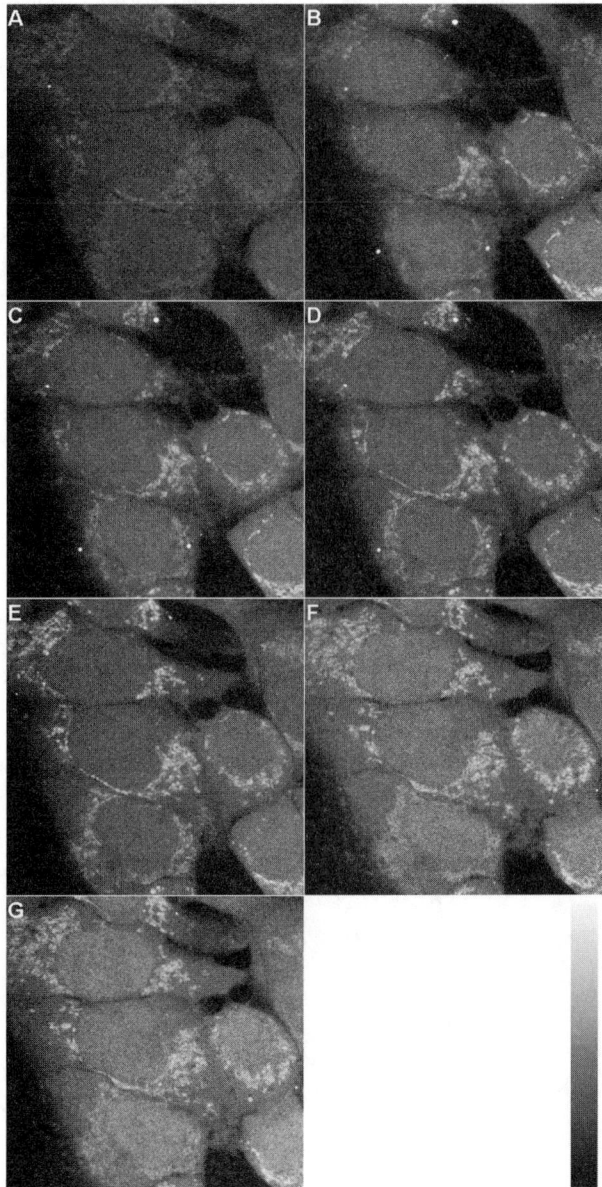

FIGURE 2. DCF fluorescence images of cybrids with 4977-bp deletion Δ-mtDNA before (**A**) and after (**B–G**) treatment with 5 mM H_2O_2 for 10 min. (**B–G**) DCF image was taken at 5-min intervals after H_2O_2 had been washed away.

one line to be 0% (wild type) and about 80% in another line (deletion type). Representative confocal DCF fluorescence images of intracellular ROS of these two lines of cybrids are shown in FIGURES 1 and 2.

At resting state, DCF fluorescence signals in wild-type (FIG. 1A) and deletion-type (FIG. 2A) cybrids were both low. However, DCF signal intensities in deletion-type cybrids were higher than those in wild-type cybrids. In addition, DCF signals in deletion-type cybrids showed obvious punctate distribution, while this dot/thread pattern was not evident in DCF images of wild-type cybrids. These punctate high-signal areas colocalized well with areas stained with mitochondria marker, Mito-Red (data not shown), a finding similar to our previous reports.[9–11] These findings indicate that cybrids containing 4977-bp deletion Δ-mtDNA have higher basal mitochondrial ROS production, which may well be a result of defective respiratory-chain function caused by Δ-mtDNA deletion.

Compensatory overexpression of the mitochondrial Mn-SOD[5] has been suggested to play a role in reducing oxidative damage associated with large-scale mtDNA deletions. This mechanism may also exist in the Δ-mtDNA cybrids used in this study and may help to reduce basal mitochondrial ROS levels in these cybrids. Without this mechanism, mitochondrial DCF intensities in Δ-mtDNA cybrids would have been even higher. However, this notion awaits confirmation.

While increased ROS production can result from mitochondrial dysfunction, mitochondria can be further damaged by ROS generated by them. Hence, a feed-forward, self-accelerating vicious cycle[12] may ensue. To reach this status, however, the concentration of ROS is critical. It has been suggested that short-term incubation of rat brain mitochondria with low micromolar range H_2O_2 induced increased H_2O_2 production at complex I of electron transport chain.[12] Exogenous H_2O_2 treatment has been widely used to cause potent oxidative stress[12,13] and has been demonstrated to be a potent apoptosis-inducing agent in many kinds of cell preparations.[14–16] For apoptosis induction in these studies, H_2O_2 in micromolar ranges were used; yet, the durations were rather long, ranging from hours to days. For intense oxidative stress treatment to in-

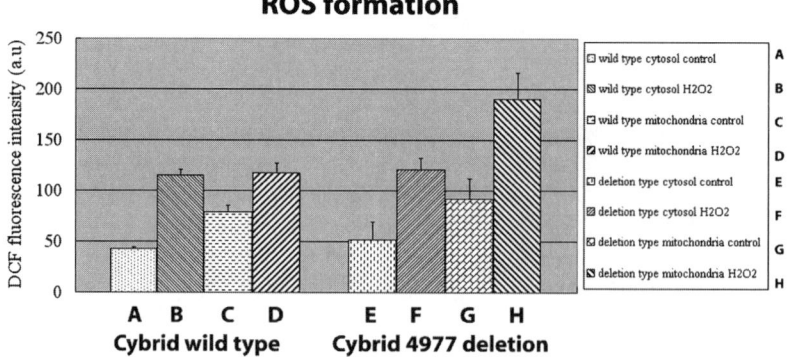

FIGURE 3. Quantitative analysis of DCF signal intensities in cytosol or mitochondrial areas, in wild-type or deletion-type cybrids, before or after H_2O_2 treatment (5 mM H_2O_2 for 10 min).

duce ROS production, low micromolar ranges of H_2O_2 were used for short durations, usually minutes. To challenge the cybrids with intense oxidative stress, we treated cybrids with 5 mM H_2O_2 for 10 min, washed away H_2O_2 before subsequent ROS imaging, and monitored the changes of ROS afterwards. (FIGS. 1B and 2B–G).

As shown in FIGURE 2B–G, DCF images of Δ-mtDNA cybrids were taken every 5 min after H_2O_2 had been washed away. Long after the brief, intense H_2O_2 treatment, progressive increase of DCF signals, especially at the mitochondrial area, could clearly be seen in Δ-mtDNA cybrids. In contrast to this observation, much weaker DCF signals and fewer mitochondrial bright-up areas could be observed 35 min after the same brief intense H_2O_2 treatment in cybrids without Δ-mtDNA (FIG. 1B). Quantitative analysis of DCF signal intensities in cytosol or mitochondrial areas, in wild-type or deletion-type cybrids, before or after H_2O_2 treatment are illustrated in FIGURE 3. Increases of DCF signals were most dramatic in the mitochondrial area in Δ-mtDNA deletion cybrids after the short and intense H_2O_2 treatment (FIG. 3H).

Apoptotic cell death could be induced by this brief, intense H_2O_2 treatment. However, the vulnerability was different in cybrids with or without 4977-bp deletion Δ-mtDNA. In FIGURE 4C, phase-contrast images of cybrids without 4977-bp deletion Δ-mtDNA were taken 25 h after brief, intense H_2O_2 treatment. Most of the cells shrank

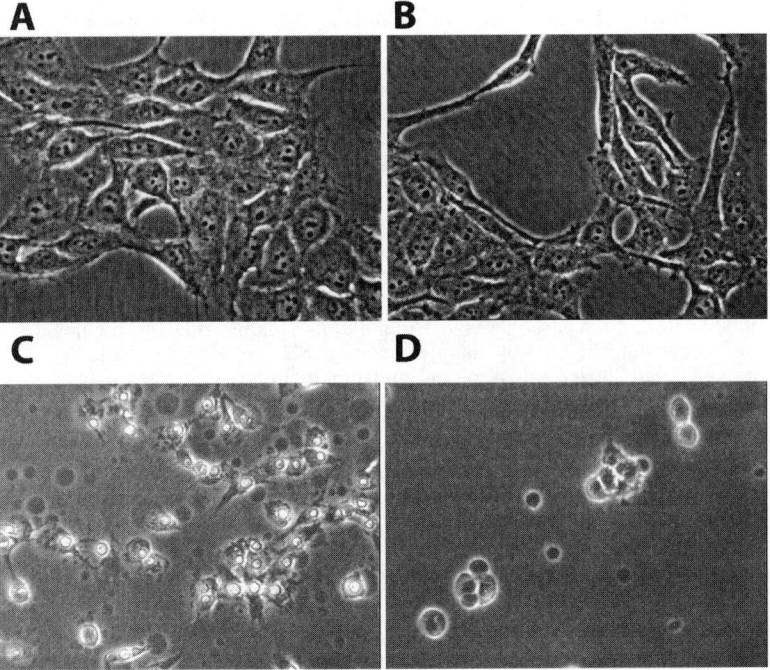

FIGURE 4. Apoptosis induced by brief, intense hydrogen peroxide treatment (5 mM H_2O_2 for 10 min). Phase-contrast images of cybrids (**A**) without Δ-mtDNA before H_2O_2 treatment; (**B**) with 4977-bp Δ-mtDNA deletion before H_2O_2 treatment; (**C**) 25 h after H_2O_2 treatment of the same cells in **A**; and (**D**) 150 min after H_2O_2 treatment of the same cells in **B**.

and could be stained with PI (data not shown). In FIGURE 4D, phase-contrast images of cybrids with 4977-bp deletion Δ-mtDNA were taken 150 min after brief, intense H_2O_2 treatment. Many cells had been lost and all the remaining cells were swollen.

In summary, feed-forward, self-intensified production of mitochondrial ROS could be initiated in cybrids containing 4977-bp deletion fragment mitochondria after brief, intense H_2O_2 treatment. A strong oxidative stress may activate further damage in mitochondria already in faulty states.

ACKNOWLEDGMENTS

This research was supported by Grants CMRP 930, CMRPG 33011 (to T.-IP.), CMRPD 32054 (to M.-J.J.) from the Chang Gung Research Foundation, and Grant NSC 91-2320-B-182-056 (to M.-J.J.) from the National Science Council.

REFERENCES

1. SCHON, E.A. et al. 1989. A direct repeat is a hotspot for large-scale deletion of human mitochondrial DNA. Science **244:** 346–349.
2. PORTEOUS, W.K. et al. 1998. Bioenergetic consequences of accumulating the common 4977-bp mitochondrial DNA deletion. Eur. J. Biochem. **257:** 192–201.
3. SCHRODER, R. et al. 2000. New insights into the metabolic consequences of large-scale mtDNA deletions: a quantitative analysis of biochemical, morphological, and genetic findings in human skeletal muscle. J. Neuropathol. Exp. Neurol. **59:** 353–360.
4. MOTT, J.L. et al. 2001. Oxidative stress is not an obligate mediator of disease provoked by mitochondrial DNA mutations. Mutat. Res. **474:** 35–45.
5. LU, C.Y. et al. 1999. Oxidative damage elicited by imbalance of free radical scavenging enzymes is associated with large-scale mtDNA deletions in aging human skin. Mutat. Res. **423:** 11–21.
6. VOTYAKOVA, T.V. & I.J. REYNOLDS. 2001. DeltaPsi(m)-dependent and -independent production of reactive oxygen species by rat brain mitochondria. J. Neurochem. **79:** 266–277.
7. KING, M.P. & G. ATTARDI. 1988. Injection of mitochondria into human cells leads to a rapid replacement of the endogenous mitochondrial DNA. Cell **52:** 811–819.
8. KING, M.P. & G. ATTARDI. 1989. Human cells lacking mtDNA: repopulation with exogenous mitochondria by complementation. Science **246:** 500–503.
9. JOU, M.J. et al. 2002. Critical role of mitochondrial reactive oxygen species formation in visible laser irradiation-induced apoptosis in rat brain astrocytes (RBA-1). J. Biomed. Sci. **9:** 1–16.
10. JOU, M.J. et al. 2004. Mitochondrial reactive oxygen species generation and calcium increase induced by visible light in astrocytes. Ann. N.Y. Acad. Sci. **1011:** 45–56.
11. PENG, T.I. & M.J. JOU. 2004. Mitochondrial swelling and generation of reactive oxygen species induced by photoirradiation are heterogeneously distributed. Ann. N.Y. Acad. Sci. **1011:** 112–122.
12. KUDIN, A.P. et al. 2004. Characterization of superoxide-producing sites in isolated brain mitochondria. J. Biol. Chem. **279:** 4127–4135.
13. CHEN, Q. et al. 2003. Production of reactive oxygen species by mitochondria: central role of complex III. J. Biol. Chem. **278:** 36027–36031.
14. LIU, L., J.R. TRIMARCHI & D.L. KEEFE. 2000. Involvement of mitochondria in oxidative stress-induced cell death in mouse zygotes. Biol. Reprod. **62:** 1745–1753.
15. TAKEYAMA, N. et al. 2002. Role of the mitochondrial permeability transition and cytochrome c release in hydrogen peroxide-induced apoptosis. Exp. Cell. Res. **274:** 16–24.
16. AKAO, M. et al. 2003. Mechanistically distinct steps in the mitochondrial death pathway triggered by oxidative stress in cardiac myocytes. Circ. Res. **92:** 186–194.

S-Nitrosoglutathione and Hypoxia-Inducible Factor–1 Confer Chemoresistance against Carbamoylating Cytotoxicity of BCNU in Rat C6 Glioma Cells

DING-I YANG,[a] SHANG-DER CHEN,[b] JIU-HAW YIN,[c] AND CHUNG Y. HSU[d,e]

[a]*Institute of Neuroscience, Tzu Chi University, Hualien, Taiwan*

[b]*Department of Neurology, Chang Gung Memorial Hospital, Kaohsiung, Taiwan*

[c]*Department of Neurology, Armed Forces Tao-Yuan General Hospital, Tao-Yuan, Taiwan*

[d]*Department of Neurology, Washington University School of Medicine, St. Louis, Missouri, USA*

[e]*Taipei Medical University, Taipei, Taiwan*

> ABSTRACT: BCNU (1,3-bis[2-chloroethyl]-1-nitrosourea) is the mainstay in glioblastoma multiform chemotherapy with only minimal effects. BCNU may kill tumor cells via carbamoylating cytotoxicity, which irreversibly inhibits glutathione reductase with resultant accumulation of oxidized form of glutathione causing oxidative stress. S-nitrosoglutathione (GSNO) is a product of glutathione and nitric oxide interaction. We report that GSNO formation may underlie carbamoylating chemoresistance mediated by activation of inducible nitric oxide synthase. Transactivation of hypoxia-inducible factor–1 (HIF-1)–responsive genes reduces oxidative stress caused by glutathione depletion. We also noted that preconditioning of C6 glioma cells to induce HIF-1 and its downstream genes confers chemoresistance against carbamoylating cytotoxicity of BCNU.
>
> KEYWORDS: alkylation; carbamoylation; chemoresistance; chemotherapy; hypoxia; iNOS; nitric oxide

S-NITROSOGLUTATHIONE FORMATION UNDERLYING NITRIC OXIDE–DEPENDENT CARBAMOYLATING CHEMORESISTANCE

Glioblastoma multiform (GBM) is the most common type of primary brain tumor accounting for more than 40% of neoplasm in the central nervous system.[1] A combination of surgery, radiotherapy, and chemotherapy results in survival of approximately 14 months.[2] 1,3-Bis(2-chloroethyl)-1-nitrosourea (BCNU) is the mainstay in chemotherapy of GBM, in part because of its lipophilic character that allows better

---

Address for correspondence: Chung Y. Hsu, M.D., Ph.D., Taipei Medical University, No. 250, Wu-Hsing Street, Taipei 110, Taiwan. Voice: +886-2-27361661-2016; fax: +886-2-23787795.
hsuc@tmu.edu.tw

Ann. N.Y. Acad. Sci. 1042: 229–234 (2005). © 2005 New York Academy of Sciences.
doi: 10.1196/annals.1338.025

passage across the blood–brain barrier.[3] Unfortunately, BCNU does not appear to substantially prolong median survival. The mechanism of GBM resistance to BCNU chemotherapy remains to be fully delineated. Variations in multidrug resistance genes,[4] DNA repair activity such as O6-methylguanine–DNA methyltransferase,[5] glutathione S-transferase, and intracellular glutathione contents[6] all have been speculated to cause BCNU chemoresistance.

Nitric oxide (NO) is a free radical gas mediating several physiological functions including modulation of cell viability. NO is synthesized from arginine and oxygen catalyzed by nitric oxide synthases (NOS). Expression of inducible NOS (iNOS) has been demonstrated in human glioma[7,8] and in a variety of brain tumors or peritumor areas, with its mRNA levels higher in malignant glioma than normal brain tissues[9] or meningioma.[10] In addition to its well-established cytotoxicity at higher concentrations, NO may contribute to antioxidant action via its interaction with glutathione to form S-nitrosoglutathione (GSNO), an antioxidant that is two orders of magnitude more potent than the reduced form of glutathione (GSH).[11,12] At micromolar concentrations, GSNO is capable of neutralizing oxidative stress exerted by peroxynitrite,[13] a highly reactive species derived from interaction of NO with superoxide anions.[14]

Using a panel of different compounds each carrying alkylating, carbamoylating, or both tumoricidal activities, we have reported previously that overexpression of iNOS conferred chemoresistance against carbamoylating agents, including BCNU, in rat C6 glioma cells.[15,16] Suppression of iNOS expression by an antisense strategy or inclusion of L-NAME, a NOS inhibitor, attenuated BCNU chemoresistance in C6 cells.[15] To further characterize the molecular mechanism underlying this novel *in vitro* effect, we have explored the potential involvement of GSNO mediating the observed iNOS effects. Our findings suggest that GSNO likely plays an important role in this iNOS-induced chemoresistance against carbamoylating agents.[17] Several lines of evidence support this contention. First, among the three NO donors tested, only GSNO conferred BCNU chemoresistance. Exogenous GSNO also enhanced chemoresistance against all the carbamoylating agents tested including cyclohexyl isocyanate and 2-chloroethyl isocyanate, the respective carbamoylating moiety of 1-(2-chloroethyl)-3-cyclohexyl-1-nitrosourea (CCNU) and BCNU, but not alkylating agents such as temozolomide. Second, experimental manipulations expected to increase or decrease cellular GSNO contents correspondingly affected carbamoylating chemoresistance. Specifically, chemoresistant C6 glioma cells, such as those exposed to cytokines for iNOS induction or overexpressing iNOS by a gene transfer strategy, contain significantly higher levels of GSNO as detected by immunocytochemistry using a GSNO-specific antibody and by HPLC analysis of medium GSNO contents. Third, copper ions have been shown to modulate GSNO action.[18,19] We found that neocuproine (a Cu^+ chelator), but not cuprizone (a Cu^{2+} chelator), was effective in blocking the development of chemoresistance against carbamoylating agents induced by iNOS overexpression as well as that induced by exogenous GSNO.

Cellular proteins undergoing carbamoylation may lose their biological functions. Carbamoylation may render those enzymes critically involved in maintaining cellular redox homeostasis, such as glutathione reductase, irreversibly nonfunctional, thereby leading to accumulation of GSSG.[20–22] In this respect, carbamoylation may be considered as a chemical-induced oxidative stress that can be antagonized by antioxidants. Recently, GSNO has been reported to act as an antioxidant two orders of magnitude more potent than GSH, capable of protecting brain dopaminergic neurons against iron-induced oxidative stress.[13,23] Heightened iNOS expression has been detected in malignant glioblastomas as compared with normal brain tissues[9] or meningiomas.[10] Surgical procedures or radiation therapy may also result in inflammatory responses leading to enhanced iNOS expression. Formation of GSNO therefore may occur in malignant brain tumors before the initiation of BCNU therapy, thereby causing chemoresistance against carbamoylating agents. Interestingly, we have observed that *S*-nitroso-*N*-acetyl-D,L-penicillamine (SNAP), another NO-releasing nitrosothiol, was effective in increasing GSNO formation and neutralizing BCNU cytotoxicity under a pretreatment condition, but not in a cotreatment paradigm. Similarly, overexpression of iNOS also has to occur before application of BCNU to develop NO-mediated chemoresistance.[15,17] Further investigation is required to examine whether GSNO metabolism in malignant brain tumors is distinct from normal brain tissues or meningiomas.

PUTATIVE ROLES OF HYPOXIA-INDUCIBLE FACTOR–1 IN GBM CHEMORESISTANCE

Hypoxia-inducible factor–1 (HIF-1), a heterodimeric protein complex consisting of alpha (HIF-1α) and beta (HIF-1β or ARNT; aryl hydrocarbon receptor nuclear translocator) subunits, is a key regulator of mammalian oxygen homeostasis. HIF-1α expression is tightly regulated by the cellular oxygen tension,[24,25] whereas the expression of HIF-1β is oxygen independent. The activity of this basic helix-loop-helix transcription factor is increased in most cells in response to low oxygen tension.[26,27] In addition to tissue hypoxia, several reagents including cobalt chloride and iron chelator desferrioxamine (DFO) are also known to induce HIF-1.[28] Recently, a growing body of evidence has also shown HIF-1 activation in response to hypoxia in tumors.[29,30] HIF-1 appears to play a key role in cancer growth by transactivating genes such as erythropoietin (EPO),[31] vascular endothelial growth factor (VEGF),[32] and iNOS[33] that may confer cytoprotective as well as angiogenic effects. Indeed, differential regulation of VEGF, HIF-1α, and angiopoietin-1, -2, and -4 by hypoxia and ionizing radiation was observed in human glioblastoma.[34] Clinical studies revealed strong nuclear expression of HIF-1α protein in the majority of glioblastomas and anaplastic astrocytomas, particularly surrounding areas of necrosis in glioblastomas.[35] Upregulation of HIF-1α mRNA was also detected with a significant increase in glioblastomas compared with lower grade tumors.[35]

We have recently demonstrated that preconditioning of C6 glioma cells with reagents mimicking hypoxia and hence capable of HIF-1 induction, namely, cobalt chloride and DFO, enhanced carbamoylating but not alkylating chemoresistance against BCNU.[33] Expression of HIF-1α protein and HIF-1 DNA binding activity were induced by cobalt chloride pretreatment based on Western blotting and electrophoretic mobility shift assay, respectively.[33] Downregulation of cobalt-mediated HIF-1 activation, either by coincubation with cadmium ions or transfection with HIF-specific oligodeoxynucleotide (ODN) decoy or an antisense phosphorothioate ODN against HIF-1α, abolished at least in part the carbamoylating chemoresistance associated with cobalt preconditioning, suggesting a putative role of HIF-1 implicated in the observed chemoresistance.

Hypoxia has been shown to increase chemoresistance against BCNU in human glioma cell lines.[36] The expression of the drug resistance genes was, however, unchanged under this condition, suggesting alternative mechanisms that may exist in hypoxia-induced chemoresistance. We have provided experimental evidence supporting the contention that HIF-1 induction under hypoxia may contribute to acquired chemoresistance against BCNU through inhibition of its carbamoylating cytotoxicity. The molecular mechanisms underlying HIF-1–mediated chemoresistance against carbamoylating cytotoxicity of chloroethylnitrosoureas remain unclear but may involve the transcriptional activation of genes downstream of HIF-1. Genes that are up-regulated by microenvironmental hypoxia through activation of HIF include glucose transporters, glycolytic enzymes, and angiogenic growth factors such as VEGF and EPO.[28,37,38] Transactivation of these genes may contribute to HIF-1–dependent protection against oxidative stress caused by glutathione depletion in primary cortical neurons.[39] Thus, carbamoylating action constitutes a chemical-induced oxidative stress that may be neutralized in a hypoxic microenvironment with resultant HIF-1 activation. We have provided direct experimental evidence supporting an important role of HIF-1 in the observed preconditioning effects.

CONCLUSIONS

Results from these studies suggest that GSNO formation as a result of iNOS expression as well as induction of HIF-1 and its target genes may represent important mechanisms underlying the development of chemoresistance, at least *in vitro*, against carbamoylating agents in glioma cells that are independent of the well-known angiogenesis actions of NO and HIF-1.[40] Such effects are also distinct from other known mechanisms of resistance to chemotherapeutic agents, such as the induction of O6-alkylguanine-DNA alkyltransferase[41] and DNA mismatch repair,[42] which are more likely to render GBM resistant to the alkylating action of chloroethylnitrosoureas. Pharmacological modulation of GSNO formation and/or HIF-1 activation may open a novel avenue to reduce chemoresistance against BCNU in GBM.

ACKNOWLEDGMENTS

This work was supported by a Tzu Chi University Research Grant (TCMRC93119A-01) and a National Science Council grant (NSC93-2314-B-320-002) to D.-I.Y. and by Washington University School of Medicine-Pharmacia Biomedical Research Support Program, Lowell B. Miller Memorial Research Grant (National Brain Tumor Foundation), and a National Science Council grant (NSC92-2314-B-038-030) to C.Y.H.

REFERENCES

1. KLEIHUES, P. *et al.* 1995. Histopathology, classification and grading of gliomas. Glia **15:** 211–221.
2. RAJKUMAR, S.V. *et al.* 1999. Phase I evaluation of pre-irradiation chemotherapy with carmustine and cisplatin and accelerated radiation therapy in patients with high-grade gliomas. Neurosurgery **44:** 67–73.

3. PAOLETTI, P. 1984. Therapeutic strategy for central nervous system tumors: present status, criticism and potential. J. Neurosurg. Sci. **28:** 51–60.
4. NUTT, C.L. *et al.* 2000. Differential expression of drug resistance genes and chemosensitivity in glial cell lineages correlate with differential response of oligodendrogliomas and astrocytomas to chemotherapy. Cancer Res. **60:** 4812–4818.
5. ROLHION, C. *et al.* 1999. O(6)-methylguanine-DNA methyltransferase gene (MGMT) expression in human glioblastomas in relation to patient characteristics and p53 accumulation. Int. J. Cancer **84:** 416–420.
6. ALI-OSMAN, F. *et al.* 1990. Glutathione content and glutathione-S-transferase expression in 1,3-bis(2-chloroethyl)-1-nitrosourea-resistant human malignant astrocytoma cell lines. Cancer Res. **50:** 6976–6980.
7. FUJISAWA, H. *et al.* 1995. Inducible nitric oxide synthase in a human glioblastoma cell line. J. Neurochem. **64:** 85–91.
8. KATO, S. *et al.* 2003. Immunohistochemical expression of inducible nitric oxide synthase (iNOS) in human brain tumors: relationships of iNOS to superoxide dismutase (SOD) proteins (SOD1 and SOD2), Ki-67 antigen (MIB-1) and p53 protein. Acta Neuropathol. (Berl.) **105:** 333–340.
9. HARA, E. *et al.* 1996. Expression of heme oxygenase and inducible nitric oxide synthase mRNA in human brain tumors. Biochem. Biophys. Res. Commun. **224:** 153–158.
10. ELLIE, E. *et al.* 1995. Differential expression of inducible nitric oxide synthase mRNA in human brain tumours. Neuroreport **7:** 294–296.
11. CHIUEH, C.C. & P. RAUHALA. 1999. The redox pathway of S-nitrosoglutathione, glutathione and nitric oxide in cell to neuron communications. Free Radic. Res. **31:** 641–650.
12. JU, T.C. *et al.* 2004. Protective effects of S-nitrosoglutathione against neurotoxicity of 3-nitropropionic acid in rat. Neurosci. Lett. **362:** 226–231.
13. RAUHALA, P. *et al.* 1998. Neuroprotection by S-nitrosoglutathione of brain dopamine neurons from oxidative stress. FASEB J. **12:** 165–173.
14. BECKMAN, J.S. *et al.* 1990. Apparent hydroxyl radical production by peroxynitrite: implications for endothelial injury from nitric oxide and superoxide. Proc. Natl. Acad. Sci. USA **87:** 1620–1624.
15. YIN, J.H. *et al.* 2001. Inducible nitric oxide synthase neutralizes carbamoylating potential of 1,3-bis(2-chloroethyl)-1-nitrosourea in C6 glioma cells. J. Pharmacol. Exp. Ther. **297:** 308–315.
16. YANG, D.I. *et al.* 2002. NO-mediated chemoresistance in C6 glioma cells. Ann. N.Y. Acad. Sci. **962:** 8–17.
17. YANG, D.I. *et al.* 2004. Nitric oxide and BCNU chemoresistance in C6 glioma cells: role of S-nitrosoglutathione. Free Radic. Biol. Med. **36:** 1317–1328.
18. GORDGE, M.P. *et al.* 1995. Copper chelation-induced reduction of the biological activity of S-nitrosoglutathione. Br. J. Pharmacol. **114:** 1083–1089.
19. AL-SA'DONI, H.H. *et al.* 1997. Neocuproine, a selective Cu(I) chelator, and the relaxation of rat vascular smooth muscle by S-nitrosothiols. Br. J. Pharmacol. **121:** 1047–1050.
20. TEW, K.D. *et al.* 1985. Carbamoylation of glutathione reductase and changes in cellular and chromosome morphology in a rat cell line resistant to nitrogen mustards but collaterally sensitive to nitrosoureas. Cancer Res. **45:** 2326–2333.
21. JOCHHEIM, C.M. & T.A. BAILLIE. 1994. Selective and irreversible inhibition of glutathione reductase in vitro by carbamate thioester conjugates of methyl isocyanate. Biochem. Pharmacol. **47:** 1197–1206.
22. VANHOEFER, U. *et al.* 1997. Carbamoylation of glutathione reductase by *N,N*-bis(2-chloroethyl)-*N*-nitrosourea associated with inhibition of multidrug resistance protein (MRP) function. Biochem. Pharmacol. **53:** 801–809.
23. RAUHALA, P. *et al.* 1996. S-nitrosothiols and nitric oxide, but not sodium nitroprusside, protect nigrostriatal dopamine neurons against iron-induced oxidative stress in vivo. Synapse **23:** 58–60.
24. WANG, G.L. *et al.* 1995. Hypoxia-inducible factor 1 is a basic-helix-loop-helix-PAS heterodimer regulated by cellular O2 tension. Proc. Natl. Acad. Sci. USA **92:** 5510–5514.

25. JIANG, B.H. *et al.* 1996. Hypoxia-inducible factor 1 levels vary exponentially over a physiologically relevant range of O2 tension. Am. J. Physiol. **271:** C1172–C1180.
26. SEMENZA, G.L. & G.L. WANG. 1992. A nuclear factor induced by hypoxia via de novo protein synthesis binds to the human erythropoietin gene enhancer at a site required for transcriptional activation. Mol. Cell. Biol. **12:** 5447–5454.
27. WANG, G.L. & G.L. SEMENZA. 1993. General involvement of hypoxia-inducible factor 1 in transcriptional response to hypoxia. Proc. Natl. Acad. Sci. USA **90:** 4304–4308.
28. SEMENZA, G.L. *et al.* 1994. Transcriptional regulation of genes encoding glycolytic enzymes by hypoxia-inducible factor 1. J. Biol. Chem. **269:** 23757–23763.
29. SEMENZA, G.L. 1998. Hypoxia-inducible factor 1: master regulator of O2 homeostasis. Curr. Opin. Genet. Dev. **8:** 588–594.
30. SEMENZA, G.L. 2000. Expression of hypoxia-inducible factor 1: mechanisms and consequences. Biochem. Pharmacol. **59:** 47–53.
31. WANG, G.L. & G.L. SEMENZA. 1993. Desferrioxamine induces erythropoietin gene expression and hypoxia-inducible factor 1 DNA-binding activity: implications for models of hypoxia signal transduction. Blood **82:** 3610–3615.
32. KIMURA, H. *et al.* 2000. Hypoxia response element of the human vascular endothelial growth factor gene mediates transcriptional regulation by nitric oxide: control of hypoxia-inducible factor-1 activity by nitric oxide. Blood **95:** 189–197.
33. YANG, D.I. *et al.* 2004. Carbamoylating chemoresistance induced by cobalt pretreatment in C6 glioma cells: putative roles of hypoxia-inducible factor-1. Br. J. Pharmacol. **141:** 988–996.
34. LUND, E.L. *et al.* 2004. Differential regulation of VEGF, HIF1alpha and angiopoietin-1, -2 and -4 by hypoxia and ionizing radiation in human glioblastoma. Int. J. Cancer **108:** 833–838.
35. SONDERGAARD, K.L. *et al.* 2002. Expression of hypoxia-inducible factor 1alpha in tumours of patients with glioblastoma. Neuropathol. Appl. Neurobiol. **28:** 210–217.
36. LIANG, B.C. 1996. Effects of hypoxia on drug resistance phenotype and genotype in human glioma cell lines. J. Neurooncol. **29:** 149–155.
37. EBERT, B.L. *et al.* 1995. Hypoxia and mitochondrial inhibitors regulate expression of glucose transporter-1 via distinct cis-acting sequences. J. Biol. Chem. **270:** 29083–29089.
38. SHWEIKI, D. *et al.* 1992. Vascular endothelial growth factor induced by hypoxia may mediate hypoxia-initiated angiogenesis. Nature **359:** 843–845.
39. ZAMAN, K. *et al.* 1999. Protection from oxidative stress-induced apoptosis in cortical neuronal cultures by iron chelators is associated with enhanced DNA binding of hypoxia-inducible factor-1 and ATF-1/CREB and increased expression of glycolytic enzymes, p21(waf1/cip1), and erythropoietin. J. Neurosci. **19:** 9821–9830.
40. SWAROOP, G.R. *et al.* 1998. Effects of nitric oxide modulation on tumour blood flow and microvascular permeability in C6 glioma. Neuroreport **9:** 2577–2581.
41. FRIEDMAN, H.S. *et al.* 1998. DNA mismatch repair and O6-alkylguanine-DNA alkyltransferase analysis and response to Temodal in newly diagnosed malignant glioma. J. Clin. Oncol. **16:** 3851–3857.
42. SANDERSON, R.J. & D.W. Mosbaugh. 1998. Fidelity and mutational specificity of uracil-initiated base excision DNA repair synthesis in human glioblastoma cell extracts. J. Biol. Chem. **273:** 24822–24831.

Celecoxib Induces Heme-Oxygenase Expression in Glomerular Mesangial Cells

CHUN-CHENG HOU,[a] SU-LI HUNG,[b] SHU-HUEI KAO,[b] TSO HSIAO CHEN,[a,c] AND HORNG-MO LEE[d,e]

[a]*Department of Internal Medicine, Taipei Medical University–Wang Fang Hospital, Taipei, Taiwan*

[b]*Graduate Institute of Biomedical Technology, Taipei, Taiwan*

[c]*Graduate Institute of Medical Sciences, Taipei, Taiwan*

[d]*Graduate Institute of Cell and Molecular Biology, Taipei Medical University, Taipei, Taiwan*

[e]*Department of Laboratory Medicine, Taipei Medical University–Wang Fang Hospital, Taipei, Taiwan*

ABSTRACT: Nonsteroidal anti-inflammatory drugs (NSAIDs) are frequently used as analgesics. They inhibit cyclooxygenases (COX), preventing the formation of prostaglandins, including prostacyclin and thromboxane. A serious side effect of COX-1 and COX-2 inhibitors is renal damage. To investigate the molecular basis of the renal injury, we evaluated the expression of the stress marker, heme oxygenase–1 (HO-1), in celecoxib-stimulated mesangial cells. We report here that a COX-2 selective NSAID, celecoxib, induced a concentration- and time-dependent increase of HO-1 expression in glomerular mesangial cells. Celecoxib-induced HO-1 protein expression was inhibited by actinomycin D and cycloheximide, suggesting that *de novo* transcription and translation are required in this process. N-acetylcysteine, a free radical scavenger, strongly decreased HO-1 expression, suggesting the involvement of reactive oxygen species (ROS). Celecoxib-induced HO-1 expression was attenuated by pretreatment of the cells with SP 600125 (a specific JNK inhibitor), but not SB 203580 (a specific p38 MAPK inhibitor), or PD 98059 (a specific MEK inhibitor). Consistently, celecoxib activated c-Jun N-terminal kinase (JNK) as demonstrated by kinase assays and by increasing phosphorylation of this kinase. N-acetylcysteine reduced the stimulatory effect of celecoxib on stress kinase activities, suggesting an involvement of JNK in HO-1 expression. On the other hand, LY 294002, a phosphatidylinositol 3-kinase (PI-3K)–specific inhibitor, prevented the enhancement of HO-1 expression. This effect was correlated with inhibition of the phosphorylation of the PDK-1 downstream substrate Akt/protein kinase B (PKB). In conclusion, our data suggest that celecoxib-induced HO-1 expression in glomerular mesangial cells may be mediated by ROS via the JNK–PI-3K cascade.

KEYWORDS: celecoxib; glomerular mesangial cells; heme oxygenase

Address for correspondence: Horng-Mo Lee, Ph.D., Institute of Cell and Molecular Biology, Taipei Medical University, 250 Wu-Hsing Street, Taipei, Taiwan. Voice: +8862-2736-1661; fax: +8862-2732-4510.
  leehorng@tmu.edu.tw

Ann. N.Y. Acad. Sci. 1042: 235–245 (2005). © 2005 New York Academy of Sciences.
doi: 10.1196/annals.1338.026

INTRODUCTION

Nonsteroidal anti-inflammatory drugs (NSAIDs) are frequently used in the treatment of rheumatisms and other chronic inflammatory diseases.[1] Most NSAIDs in clinical use are cyclooxygenase (COX) inhibitors. COX catalyzes the synthesis of eicosanoids from arachidonic acid.[2] There are two isoforms of COX: a constitutively expressed COX-1 and an inducible COX-2.[3] Both isoforms are expressed in the adult mammalian kidney. Prostaglandins modulate renal microvascular hemodynamics, renin release, and tubular salt and water reabsorption.[4] Depletion of prostaglandins in kidney may result in renal damages reflected by a reduction in glomerular filtration rate, renal blood flow, and diminished sodium and potassium excretion.[5] Inhibition of COX activity in the kidney by NSAIDs has relatively mild consequences in healthy individuals but in vulnerable subjects can lead to serious adverse events in kidney due to depletion of prostaglandins. The decline in renal function is especially pronounced in the elderly and in patients with pre-existing renal disease.[6]

Heme oxygenase-1 (HO-1) catalyzes the rate-limiting step in heme degradation, resulting in the formation of iron, carbon monoxide, and biliverdin, which is subsequently converted to bilirubin by biliverdin reductase. Recent attention has focused on the biological effects of the reaction product(s), which exert important antioxidant, anti-inflammatory, and cytoprotective functions. There are three heme oxygenase isoforms: an inducible isoform, HO-1, and two constitutively expressed isoforms, HO-2 and HO-3. Induction of HO-1 occurs as an adaptive and beneficial response to several injury signaling processes and has been implicated in many clinically relevant disease states including acute renal injury. Increased HO-1 expression protects kidneys from oxidative injuries,[7] rhabdomyolysis,[8] cisplatin nephrotoxicity,[9] acute renal failure,[10] and ischemic/reperfusion-mediated tissue injury.[11]

The cellular signaling mechanisms that mediate HO-1 expression in glomerular mesangial cells remain unclear. It has been shown that HO-1 gene expression is mediated through activation of the JNK and p38 MAPK pathways in primary cultures of rat hepatocytes.[12] In the present study, we examined the effects of a selective COX-2 inhibitor, celecoxib (Celebrex), on the induction of HO-1 expression in glomerular mesangial cells. This study also investigated the mechanisms by which celecoxib induced HO-1 expression. The data clearly show that the treatment of glomerular mesangial cells with celecoxib results in increased free radical generation, which triggers a signal transduction cascade involving JNK and PI-3K activation. These stress signals ultimately induce HO-1 expression in glomerular mesangial cells.

MATERIALS AND METHODS

PD 98059, SB 203580, and SP 600125 were purchased from Calbiochem (San Diego, CA). Dubecco's modified Eagle medium (DMEM), fetal calf serum (FCS), penicillin, and streptomycin were purchased from Life Technologies (Gaithersburg, MD). Affinity-purified polyclonal antibodies to c-Jun phosphorylated ATF-2 and phosphorylated c-Jun, phosphorylated Akt (Thr^{473}), and phosphorylated p38 MAPK were obtained from Transduction Laboratory (Lexington, KY). Antibodies specific for HO-1 and β-actin were purchased from Santa Cruz Biochemicals (Santa Cruz, CA). Horseradish peroxidase–conjugated anti–rabbit IgG antibody was purchased from

BioRad (Hercules, CA). 5-Bromo-4-chloro-3-indolyl-phosphate/4-nitro blue tetrazolium substrate was purchased from Kirkegaard and Perry Laboratories (Gaithersburg, MD). Protease inhibitor cocktail tablets were purchased from Boehringer Mannheim (Mannheim, Germany). The p38 MAPK assay kit was purchased from New England Biolabs (Beverly, MA). Celecoxib is a gift from Pharmacia Co. (Northumberland, England). All other chemicals were purchased from Sigma (St. Louis, MO).

Cell Culture and Preparation of Cell Lysates

Glomerular mesangial cells from an SV40 transgenic mouse (ATCC CRL-1927) were cultured in DMEM supplemented with 13.1 mM $NaHCO_3$, 13 mM glucose, 2 mM glutamine, 10% of heat inactive fetal calf serum (FCS), and 10% horse serum and penicillin (100 U/mL)/streptomycin (100 mg/mL). Cells were attached to a petri dish after a 24-h incubation. Cells were plated at a concentration of 1×10^5 cells/mL and used for experiments when they reached 80% confluency. All reagents were added directly to the culture at a volume of 100 μL/10 M medium. Cultures were maintained in a humidified incubator under 5% CO_2 at 37°C. After reaching 80% confluence, cells were treated with various concentrations of celecoxib for indicated time intervals and incubated in a humidified incubator under 5% CO_2 at 37°C. In some experiments, cells were pretreated with specific inhibitors as indicated followed by celecoxib and incubated for more 16 h. After incubation, cells were harvested, chilled on ice, and washed three times with ice-cold phosphate-buffered saline (PBS). Cells were lysed by adding lysis buffer containing 10 mM Tris HCl (pH 7.5), 1 mM EGTA, 1 mM $MgCl_2$, 1 mM sodium orthovanadate, 1 mM DTT, 0.1% mercaptoethanol, 0.5% Triton X-100, and the protease inhibitor cocktails (final concentrations: 0.2 mM PMSF, 0.1% aprotinin, 50 μg/mL leupeptin). Cells adhering to the plates were scraped off using a rubber policeman and stored at −70°C for further measurements.

Polyacrylamide Gel Electrophoresis and Western Blotting

Electrophoresis was ordinarily conducted at different percentages of sodium dodecyl sulfate–polyacrylamide electrophoresis (SDS-PAGE). After electrophoresis, proteins on the gel were electrotransferred onto a polyvinyldifluoride (PVDF) membrane. After transfer, the PVDF paper was washed once with PBS and twice with PBS plus 0.1% Tween 20. The PVDF membrane then was blocked with blocking solution containing 3% bovine serum albumin in PBS containing 0.1% Tween 20 for 1 h at room temperature. The PVDF membrane was incubated with a solution containing primary antibodies in the blocking buffer for 2 h. Finally, the PVDF papers were incubated with enzyme-linked secondary antibodies for 1 h and then visualized by incubating with colorigenic substrate (nitroblue tetrazolium and 5-bromo-4-chloro-3-indolyl-phosphate; Sigma) or developed using a chemiluminescence kit (Amersham, Buckinghamshire, UK), and the immunoreactive bands were visualized by autoradiography.

Measurement of Intracellular Reactive Oxygen Species Generation

Intracellular reactive oxygen species (ROS) generation was assessed using 2′,7′-dichlorofluorescein diacetate (DCFH-DA; Molecular Probes, Leiden, the

Netherlands) as described by Chandel et al.[13] In brief, mesangial cells were cultured in petri dishes and incubated with 10 µM DCFH-DA for 30 min. ROS in the cells caused oxidation of DCFH, which produces a fluorescent product. The intracellular fluorescence then was measured using flow cytometry.

Statistical Analysis

Results are expressed as means ± SEM from the number of independent experiments performed. One-way analysis of variance (ANOVA) was used to assess the difference of means among groups, and the Student's two-tailed t-test was used to determine the difference of means between two groups. A P value less than 0.05 was considered significant.

RESULTS

Celecoxib Induction of HO-1 Expression

Exposure of mesangial cells to celecoxib induced the expression of a 32-kDa band corresponding to HO-1 in a dose-dependent (FIG. 1A) and time-dependent (FIG. 1B) manner. HO-1 expression was maximum at approximately 10 µM of Celebrex. Celecoxib-induced HO-1 expression was apparent at 4 h and reached maximum at 16 h. The induction of HO-1 by celecoxib could be blocked with actinomycin D (1 µM; FIG. 1C) or cycloheximide (10 µg/mL) pretreatment (FIG. 1D), suggesting that celecoxib-induced HO-1 expression required *de novo* transcription and translation.

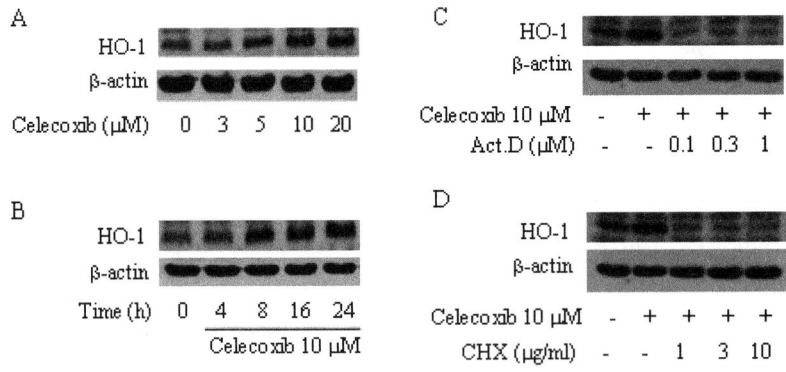

FIGURE 1. Celecoxib induced a time- and dose-dependent increase in HO-1 expression in mesangial cells. (**A**) Cells were incubated with various concentrations (3, 5, 10, 20 µM) of celecoxib at 37°C for 24 h; (**B**) cells were incubated with celecoxib (10 µM) at 37°C for various time periods (4, 8, 16, 24 h). In (**C**) and (**D**), cells were pretreated with actinomycin D (0.1, 0.3, 1 µM) (**C**) and cycloheximide (1, 3, 10 µg/mL) (**D**) for 30 min before being incubated with celecoxib (10 µM) for 16 h. Cell lysates were electrophoresed and probed by Western blot with HO-1–specific antibodies. Equal loading in each lane was demonstrated by the similar intensities of β-actin.

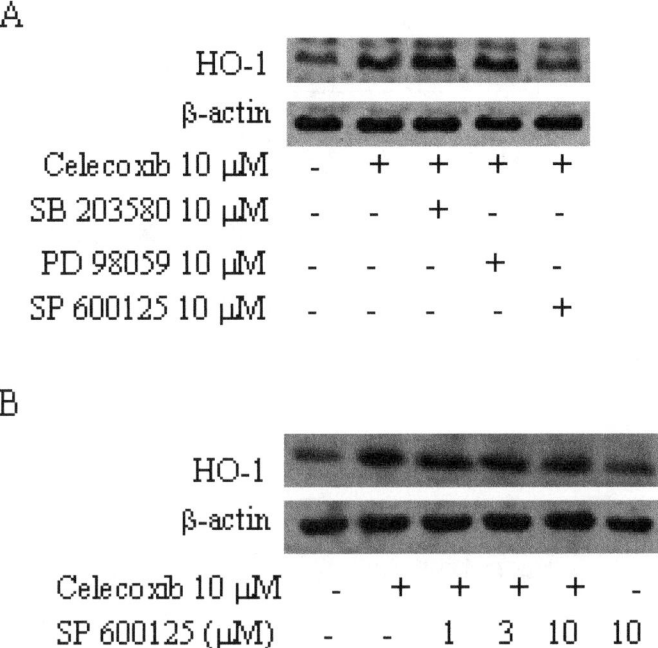

FIGURE 2. Roles of MAPKs in celecoxib-induced HO-1 expression in mesangial cells. **(A)** Cell were pretreated with various MAPK inhibitors, including SB 203580 (10 μM), PD 98059 (10 μM), and SP 600125 (10 μM) for 30 min before being incubated with celecoxib (10 μM) for 16 h. **(B)** Cells were pretreated with various concentrations SP 600125 (1, 3, 10 μM) for 30 min before being incubated with celecoxib (10 μM) for 16 h and lysed. Cell lysates were electrophoresed and probed by Western blot with HO-1–specific antibodies. Equal loading in each lane was demonstrated by the similar intensities of β-actin.

Involvement of the MAPK Signaling Pathway in Celecoxib-Induced HO-1 Expression

It has been demonstrated that mitogen-activated protein kinases (MAPKs) play important roles in celecoxib-induced HO-1 expression in macrophages.[12] To elucidate the mechanisms responsible for changes in HO-1 protein expression, we pretreated mesangial cells with MEK inhibitor (PD 98059), or JNK inhibitor (SP 600125), or p38 MAPK inhibitor (SB 203580), before incubation with celecoxib for 16 h. Pretreatment for 30 min with SP 600125 (10 μM) attenuated celecoxib-induced HO-1 expression. However, pretreatment with PD 98059 (10 μM) or SB 203580 (10 μM) did not alter celecoxib-induced HO-1 protein expression (FIG. 2). These findings suggest that JNK and p38 MAPK, but not MEK, are involved in the induction of HO-1 expression by celecoxib.

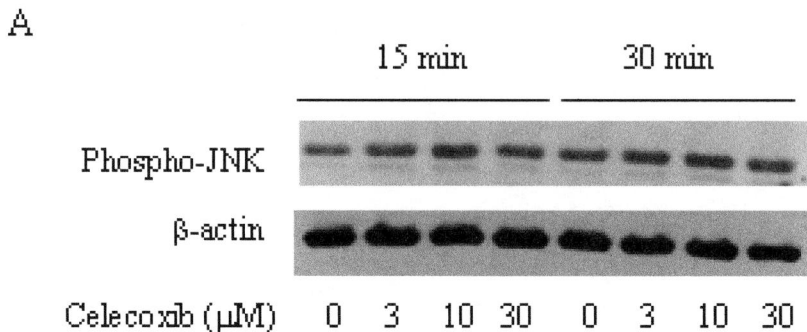

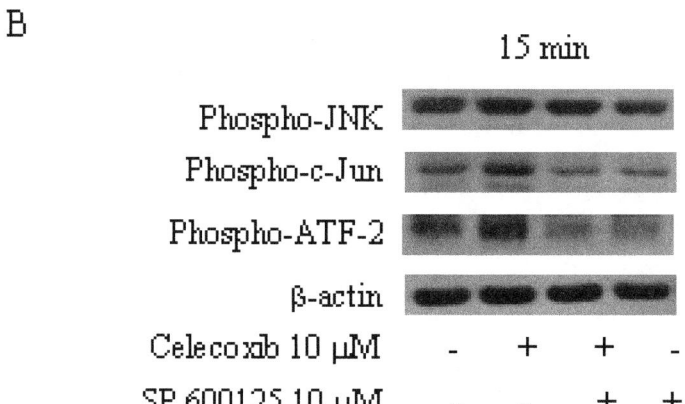

FIGURE 3. Celecoxib activates JNK in mesangial cells. (**A**) Cells were pretreated with various concentrations of celecoxib (3, 10, 30 µM) for 15 or 30 min and lysed. Cell lysates were electrophoresed and probed by Western blot with phospho-JNK–specific antibodies. (**B**) Cells were pretreated with 10 µM SP 600125 for 30 min before being incubated with 10 µM celecoxib for 15 min and lysed. Cell lysates were electrophoresed and probed by Western blot with phospho-JNK, phospho-c-Jun, and phospho-ATF-2–specific antibodies. Equal loading in each lane was demonstrated by the similar intensities of β-actin.

Activation of JNK by Celecoxib

To determine whether JNK was activated by celecoxib, we measured the activity of this kinase after celecoxib stimulation. As shown in FIGURE 3A, addition of celecoxib increased the phosphorylation of JNK; the maximum response was seen at 10 µM. Furthermore, celecoxib (10 µM) increased the phosphorylation of c-Jun and ATF-2; both are the downstream substrates of JNK. This effect of celecoxib could be blocked by pretreatment with SP 600125, a JNK inhibitor (FIG. 3B).

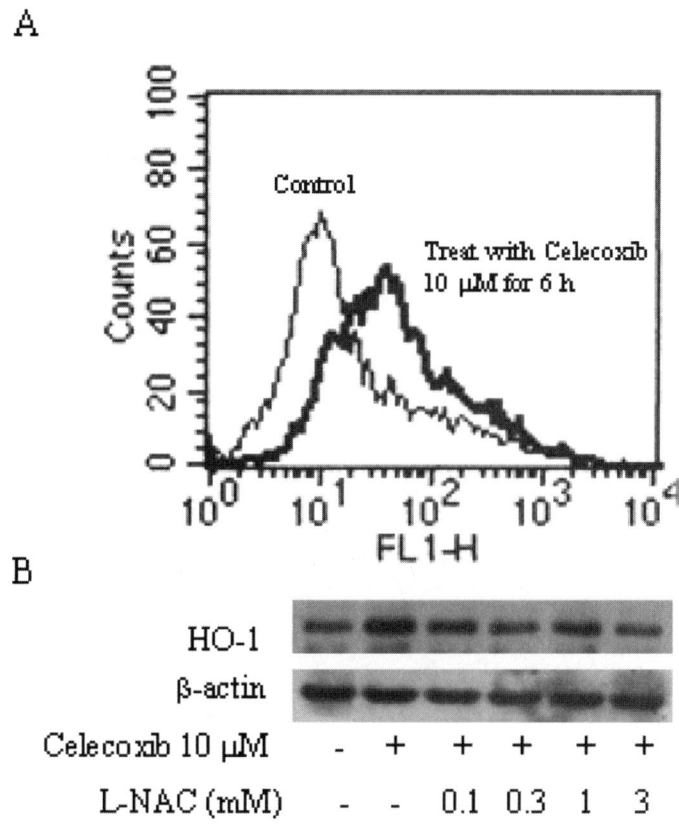

FIGURE 4. Roles of reactive oxygen species in celecoxib-induced HO-1 expression in mesangial cells. (**A**) Cells were incubated with DCFH-DA (10 μM) for 6 h in the presence of celecoxib (10 μM) as described in MATERIALS AND METHODS. (**B**) Cells were pretreated with various concentrations of L-NAC (0.1, 0.3, 1, 3 mM) before being incubated with celecoxib (10 μM) for 16 h and lysed. Cell lysates were electrophoresed and probed by Western blot with HO-1–specific antibodies. Equal loading in each lane was demonstrated by the similar intensities of β-actin.

Roles of ROS Generation in Celecoxib-Induced HO-1 Expression

The PI-3K pathway has been implicated in the stress-induced increase in ROS. We therefore examined whether celecoxib stimulated ROS generation using 2′,7′-dichlorofluorescein diacetate (DCFH-DA). The fluorescent probe can be oxidized by ROS and generates a fluorescent compound. As shown in FIGURE 4A, celecoxib increased ROS production. *l*-NAC (1 mM), an antioxidant, abolished celecoxib-induced HO-1 protein expression, suggesting that ROS generation caused by celecoxib is involved in this signaling process (FIG. 4B). Taken together, these data suggest that celecoxib may increase ROS generation, leading to activation of JNK, which, in turn, regulates HO-1 expression in mesangial cells.

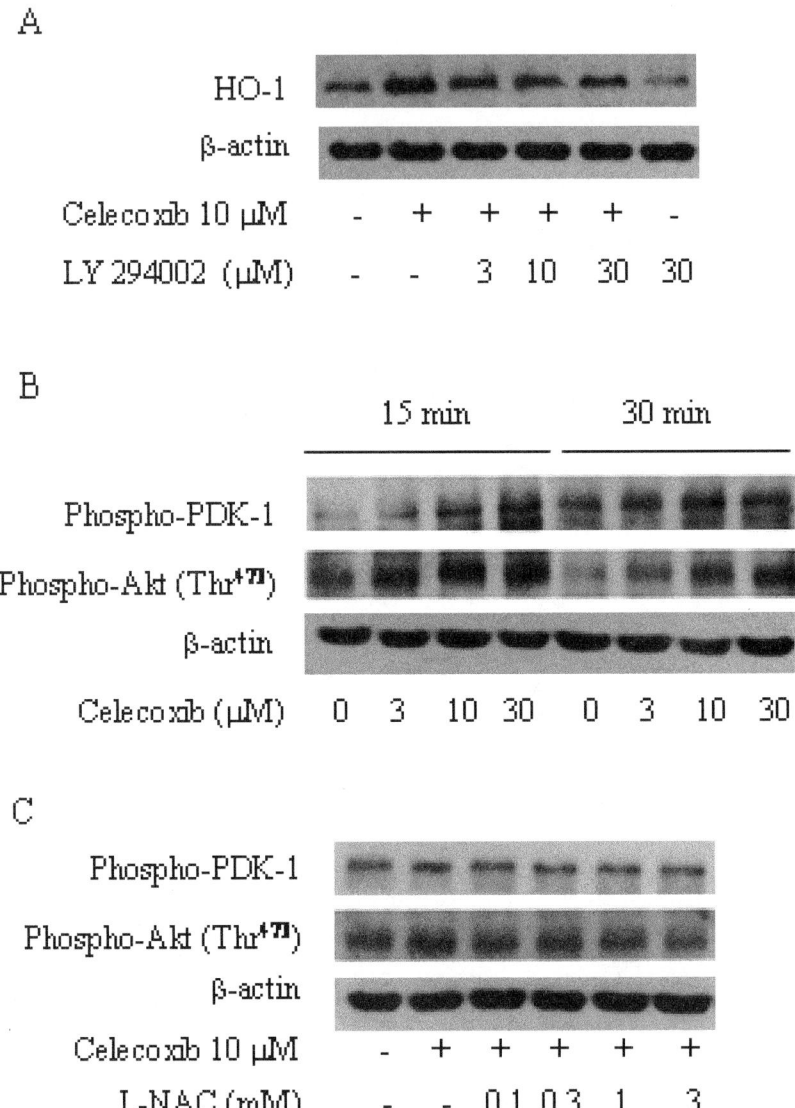

FIGURE 5. Roles of PI-3K signaling pathway in celecoxib-induced HO-1 expression in mesangial cells. (**A**) Cells were pretreated with various concentrations LY 294002 (3, 10, 30 μM) for 30 min before being incubated with celecoxib (10 μM) for 16 h. Cell lysates were electrophoresed and probed by Western blot with HO-1– and β-actin–specific antibodies. (**B**) Cells were treated with various concentrations of celecoxib (3, 10, 30 μM) for 15 or 30 min and lysed. Cell lysates were electrophoresed and probed by Western blot with phospho-PDK-1 and phospho-Akt (Thr^{473})–specific antibodies. (**C**) Cells were pretreated with various concentrations of L-NAC (0.1, 0.3, 1, 3 mM), and cell lysates were electrophoresed and probed by Western blot with phospho-PDK-1 and phospho-Akt (Thr^{473})–specific antibodies. Equal loading in each lane was demonstrated by the similar intensities of β-actin.

Roles of PI-3K in Celecoxib-Induced HO-1 Expression in Glomerular Mesangial Cells

The ROS-dependent HO-1 expression has been shown to be mediated through the phosphatidylinositol 3-kinase (PI-3K)/Akt pathway. We next investigated whether the PI-3K/Akt signaling pathway is involved in celecoxib-induced HO-1 expression in glomerular mesangial cells. As shown in FIGURE 5A, pretreatment with a specific inhibitor of PI-3K (LY 294002) reduced HO-1 expression. FIGURE 5B shows that celecoxib (10 µM) increased the phosphorylated form of its downstream effectors, Akt/PKB in mesangial cells, suggesting celecoxib can activate the PI-3K–PKB/Akt signaling pathway. *l*-NAC, an antioxidant, also inhibited celecoxib-stimulated PDK-1 and Akt phosphorylation (FIG. 5C).

DISCUSSION

In the present study, we demonstrate that celecoxib induced HO-1 expression in glomerular mesangial cells. HO-1 is a stress enzyme which can be induced under various experimental paradigms, including exposure to heavy metals, UV radiation, and other oxidative stresses. Given that NSAIDs have been implicated in causing renal damage, HO-1 expression may be considered as a marker of cellular stress. On the contrary, induction of HO-1 in macrophages has been shown to be essential for the anti-inflammatory effects of 15-deoxy-Delta 12,14-prostaglandin J2.[14] HO-1 induction may mediate its anti-inflammatory response through production of carbon monoxide, a product of heme catabolism catalyzed by heme oxygenase. Carbon monoxide, in turn, mediates the anti-inflammatory response by activation of the MAPK pathway.[15] The findings that celecoxib and another COX-2 inhibitor, SC58125, upregulate HO-1 induction in RAW 264.7 macrophages[16] suggest that these NSAIDs may exert their anti-inflammatory effect via induction of HO-1 expression.

Transcriptional regulation of HO-1 gene expression involves the activation of the JNK and p38 MAPK pathways in primary cultures of rat hepatocytes[12] and in mammary epithelial cells.[17] We demonstrated that inhibition of JNK/SAPK but not MEK or p38 MAPK by specific pharmacological inhibitors blocked celecoxib-induced HO-1 expression in glomerular mesangial cells. We further confirmed that celecoxib may activate JNK and increase the phosphorylation of its downstream substrates in mesangial cells. The finding that *N*-acetylcysteine, a free radical scavenger, substantially reduced HO-1 expression induced by celecoxib suggests ROS is involved in this process. We also provide direct evidence that celecoxib increased ROS generation, which is likely responsible for the signaling event leading to JNK activation. The stress signal ultimately induces HO-1 expression in glomerular mesangial cells.

Furthermore, we demonstrated that celecoxib activated the PI-3K pathway as demonstrated by increasing phosphorylation of its downstream effectors PKB/Akt. These results support the notion that celecoxib may exert its protective effect by activating a PI-3K–HO-1 cascade. However, our results differ from those found in cancer cell lines. In human prostate cancer cells[18] and in human colon cancer HT-29 cells,[19] celecoxib inhibits the PI-3K signaling pathway which leads to cancer cell apoptosis. In cultured mesangial cells, inhibition of PI-3K by a specific PI-3K inhibitor inhibited the phosphorylation of the downstream substrate PDK-1 and PKB/Akt and

HO-1 expression without inducing cell apoptosis. Note that the concentrations used in our study (10 μM) are slightly lower than those used in prostate cancer cells (25–50 μM) and in HT-29 cells (25–100 μM). Whether celecoxib induces mesangial cell apoptosis at higher concentrations remains to be determined.

In conclusion, this study demonstrates that celecoxib stimulates ROS generation, activating a signal transduction cascade involving JNK and PI-3K. These stress signals ultimately induce HO-1 expression in glomerular mesangial cells.

ACKNOWLEDGMENTS

We are grateful to Pharmacia Co. for providing celecoxib and to Shu-Ting Tsai for her skilled technical assistance. This study was supported by a National Science Council grant (NSC92-2314-B-038-039) to Horng-Mo Lee.

REFERENCES

1. INSEL, P.A. 1996. Analgesic-antipyretic and antiinflammatory agents and drugs employed in the treatment of gout. In Goodman & Gilman's The Pharmacological Basis of Therapeutics. 9th edit. J.G. Hardman, A.G. Gilman & L.E. Limbird, Eds.: 617–657. McGraw Hill. New York.
2. LEVY, G.N. 1997. Prostaglandin H syntheses, nonsteroidal anti-inflammatory drugs, and colon cancer. FASEB J. **11:** 234–237.
3. SMITH, W.L. et al. 1996. Prostaglandin endoperoxide H synthases (cyclooxygenases)-1 and -2. J. Biol. Chem. **271:** 33157–33160.
4. HARRIS, R.C. 2000. Cyclooxygenase-2 in the kidney. J. Am. Soc. Nephrol. **11:** 2387–2394.
5. SWAN, S.K. et al. 2000. Effect of cyclooxygenase-2 inhibition on renal function in elderly persons receiving a low-salt diet. Ann. Int. Med. **133:** 1–9.
6. DUNN, M. 2000. Are COX-2 selective inhibitors nephrotoxic? Am. J. Kidney Dis. **35:** 976–977.
7. MORIMOTO, K. et al. 2001. Cytoprotective role of heme oxygenase (HO)-1 in human kidney with various renal diseases. Kidney Int. **60:** 1858–1866.
8. NATH, K.A. et al. 1992. Induction of heme oxygenase is a rapid, protective response in rhabdomyolysis in the rat. J. Clin. Invest. **90:** 267–270.
9. AGARWAL, A. et al. 1995. Induction of heme oxygenase in toxic renal injury: a protective role in cisplatin nephrotoxicity in the rat. Kidney Int. **48:** 1298–1307.
10. SHIMIZU, H. et al. 2000. Protective effect of heme oxygenase induction in ischemic acute renal failure. Crit. Care Med. **28:** 809–817.
11. OGAWA, T. et al. 2001. Contribution of nitric oxide to the protective effects of ischemic preconditioning in ischemia-reperfused rat kidneys. J. Lab. Clin. Med. **138:** 50–58.
12. KIETZMANN, T. et al. 2003. Transcriptional regulation of heme oxygenase-1 gene expression by MAP kinases of the JNK and p38 pathways in primary cultures of rat hepatocytes. J. Biol. Chem. **278:** 17927–17936.
13. CHANDEL, N.S. et al. 2000. Reactive oxygen species generated at mitochondrial complex III stabilize hypoxia-inducible factor-1a during hypoxia. J. Biol. Chem. **275:** 25130–25138.
14. LEE, T.S. et al. 2003. Induction of heme oxygenase-1 expression in murine macrophages is essential for the anti-inflammatory effect of low dose 15-deoxy-Delta 12,14-prostaglandin J2. J. Biol. Chem. **278:** 19325–19330.
15. OTTERBEIN, L.E. et al. 2000. Carbon monoxide has anti-inflammatory effects involving the mitogen-activated protein kinase pathway. Nat. Med. **6:** 422–428.
16. ALCARAZ, M.J. et al. 2001. Heme oxygenase-1 induction by nitric oxide in RAW 264.7 macrophages is upregulated by a cyclo-oxygenase-2 inhibitor. Biochim. Biophys. Acta **1526:** 13–16.

17. ALAM, J. et al. 2000. Mechanism of heme oxygenase-1 gene activation by cadmium in MCF-7 mammary epithelial cells. Role of p38 kinase and Nrf2 transcription factor. J. Biol. Chem. **275:** 27694–27702.
18. HSU, A.L. et al. 2000. The cyclooxygenase-2 inhibitor celecoxib induces apoptosis by blocking Akt activation in human prostate cancer cells independently of Bcl-2. J. Biol. Chem. **275:** 11397–11403.
19. ARICO, S. et al. 2002. Celecoxib induces apoptosis by inhibiting 3-phosphoinositide-dependent protein kinase-1 activity in the human colon cancer HT-29 cell line. J. Biol. Chem. **277:** 27613–27621.

Oxidative Stress–Induced Depolymerization of Microtubules and Alteration of Mitochondrial Mass in Human Cells

CHENG-FENG LEE,[a] CHUN-YI LIU,[a] RONG-HONG HSIEH,[b] AND YAU-HUEI WEI[a]

[a]*Department of Biochemistry and Molecular Biology, National Yang-Ming University, Taipei 112, Taiwan*

[b]*School of Nutrition and Health Sciences, Taipei Medical University, Taipei 110, Taiwan*

ABSTRACT: Mitochondrial biogenesis is a biological process that has been intensively studied over the past few years. However, the detailed molecular mechanism underlying this increase in mitochondria remains unclear. To investigate the mechanism of such a mitochondrial proliferation, we examined alterations in mitochondria of human osteosarcoma 143B cells that had been treated with 100 to 500 µM hydrogen peroxide (H_2O_2) for 48 h. The results showed that mitochondrial mass of the cell was increased with the increase of the concentration of H_2O_2. On the other hand, by using real-time PCR techniques, we observed the changes of mitochondrial DNA (mtDNA) content in the cells exposed to oxidative stress. The copy number of mtDNA was increased by treatment with a low dose of H_2O_2 but was drastically decreased after treatment with H_2O_2 higher than 300 µM. Transmission electron microscopic images revealed that mitochondria were abnormally proliferated in cells exposed to oxidative stress. Moreover, we found that the percentage of 143B cells arrested at the G_2/M phase increased upon treatment with H_2O_2. Immunostaining and microtubule fractionation assay revealed that microtubules were depolymerized in the cells that had been treated with H_2O_2. To understand the effect of microtubules depolymerization on the mitochondrial mass, we treated the cells with several kinds of microtubule-active drugs, which arrest cultured cells at the G_2/M phase. The results showed that mitochondrial mass and mtDNA copy number all were increased after such treatments. Taking these findings together, we suggest that oxidative stress–induced microtubule derangement is one of the molecular events involved in the increase of mitochondrial mass upon treatment of human cells with H_2O_2.

KEYWORDS: oxidative stress; microtubule; mitochondrial mass; osteosarcoma

INTRODUCTION

It has been well documented that abnormal proliferation of mitochondria occurs frequently in affected tissues of patients with mitochondrial myopathy and muscle of aged individuals. This is one of the most prominent clinical hallmarks in mito-

Address for correspondence: Professor Yau-Huei Wei, Department of Biochemistry and Molecular Biology, National Yang-Ming University, Taipei 112, Taiwan. Voice: +886-2-2826-7118; fax: +886-2-2826-4843.
joeman@ym.edu.tw

chondrial diseases such as chronic progressive external ophthalmoplegia (CPEO) and myoclonic epilepsy and ragged-red fiber (MERRF) syndromes. Morphologic features of giant mitochondria were observed in nerve and muscle tissues of elderly subjects and patients with mitochondrial disorders.[1,2] Moreover, giant mitochondria also appeared in mtDNA-depleted ρ^0 cells,[3] and mtDNA depletion caused a reduction in the amount of mitochondrial inner membranes.[4] Therefore, structure and distribution of mitochondria are closely related to the oxidative status of mammalian cells.

Mitochondrial biogenesis has been a subject of intensive study in the past few years. Many factors, such as oxidative stress, deficiency in ATP production, cell cycle stage, and even the structure of microtubules, have been shown to affect the mitochondrial mass of mammalian cells.[5–8] Mitochondria proliferate independently throughout the cell cycle and are inherited by daughter cells upon cell division. They actively move along cytoskeletal tracks and frequently change their shape and size through fission and fusion of the organelles.

Microtubules seem to be the major components of cytoskeletal systems that are involved in the regulation of the distribution of mitochondria in the cells. It has been reported that microtubules in interphase are involved in mitochondrial biogenesis aside from participation in the regulation of mitochondrial distribution in mammalian cells.

In this study, we treated human osteosarcoma 143B cells with sublethal doses of H_2O_2 to increase reactive oxygen species (ROS) in the cells, and we investigated the role that microtubules may play in oxidative stress–induced alteration of mitochondrial mass.

MATERIALS AND METHODS

Cell Culture

Human osteosarcoma 143B TK⁻ cells were grown at 37°C in a humidified atmosphere with 5% CO_2/ 95% air in Dulbecco's modified Eagle's medium (Life Technologies, Grand Island, NY) containing 100 µg/mL pyruvate and 50 µg/mL uridine and supplemented with 5% fetal bovine serum.

Determination of Mitochondrial Mass

The fluorescent dye 10-*n*-nonyl-acridine orange (NAO; Molecular Probes, Eugene, OR), which binds specifically to cardiolipin at the inner mitochondrial membrane independently of membrane potential ($\Delta\Psi_m$), was used to monitor the mitochondrial mass.[9] Cells at subconfluent stage were trypsinized and resuspended in 0.5 mL of PBS containing 2.5 µM of NAO. After incubation for 10 min at 25°C in the dark, cells were immediately transferred to a tube for analysis with an EPICS XL-MCL flow cytometer (Beckman-Coulter, Miami, FL).

Determination of the Relative Content of mtDNA

A LightCycler-FastStar DNA Master SYBR Green I kit (Roche Diagnostics, Mannheim, Germany) was used to perform quantitative PCR on the LightCycler

PCR machine (Roche Diagnostics). The relative content of mtDNA was determined by amplification of a DNA fragment each of the ND1 gene and β-actin gene (as internal standard) with specific primer pairs (5′-GGAGTAATCCAGGTCGGT-3′ and 5′-TGGGTACAATGAGGAGTAGG-3′; 5′-CATGTGCAAGGCCGGCTTC-3′ and 5′-CTGGGTCATCTTCTCGCGGT-3′), respectively. PCR was usually performed under the following conditions: initial 6-min denaturation at 95°C followed by 31 cycles of 5 s at 95°C, 10 s at 62°C, and 20 s at 72°C. The relative content of mtDNA in the cell was then calculated by the RelQuant software.

Transmission Electron Microscope

Human osteosarcoma 143B TK⁻ cells were fixed for 2 h in 2% paraformaldehyde and 2.5% glutaraldehyde in 0.2 M cacodylate buffer, washed in 0.1 M cacodylate buffer containing 0.2 M sucrose three times, and postfixed for 2 h in 1% osmium tetroxide. Dehydration was achieved by 35, 50, 75, 95, and 100% ethanol, respectively. Samples then were infiltrated in a mixture of ethanol and spurr (Electron Microscopy Sciences, Fort Washington, PA) and were embedded in spurr. Ultrathin sections were cut on a Leica AG ultramicrotome, placed on 200-mesh copper grids, stained with uranyl acetate and lead citrate, and photographed on a Hitachi T-600 electron microscope.

Cell Cycle Analysis

Cells were trypsinized and fixed with 70% ethanol. After washing twice with PBS buffer, cells were resuspended in 1 mL of cell cycle assay buffer (1% Triton X-100, 0.1 mg/mL RNase A and 4 μg/mL propidium iodide in PBS buffer) and further incubated for 30 min. Samples were stored in the dark at 4°C until cell cycle analysis, which was performed with a flow cytometer (Beckman-Coulter, Miami, FL).

Microtubule Fractionation and Quantitative Immunoblotting

Polymerized and monomeric fractions of tubulin were isolated by the method described by Banan et al.[10] Cells were pelleted and subsequently resuspended in microtubule stabilization buffer containing 0.1 M PIPES, pH 6.9, 30% glycerol, 5% dimethyl sulfoxide, 1 mM $MgSO_4$, 10 μg/mL anti-protease cocktail (Roche Diagnostics, Mannheim, Germany), 1 mM EGTA, and 1% Triton X-100 at room temperature for 20 min. Cell lysate was centrifuged at 105,000 × g for 45 min at 4°C. The supernatant containing the soluble monomeric tubulin was gently removed. The polymerized tubulin fraction then was resuspended in calcium-containing microtubule depolymerization buffer containing 0.1 M PIPES, pH 6.9, 1 mM $MgSO_4$, 10 μg/mL anti-protease cocktail (Roche Diagnostics), and 10 mM $CaCl_2$, at 4°C for 1 h. Subsequently, the samples were centrifuged at 48,000 × g for 15 min at 4°C, and the supernatant was removed. The remaining pellet was treated with the calcium-containing microtubule depolymerization buffer twice by resuspension and centrifugation. Polymerized and monomeric fractions of tubulin were recovered by incubating with 10 μM Taxol and 1 mM GTP at 37°C for 1 h to promote polymerization. Microtubules were recovered by centrifugation and resuspended in the microtubule stabilization buffer. An aliquot of 1 μg of protein was subjected to electrophoresis on a 10% polyacrylamide gel and blotted onto a piece of Hybond-P⁺ membrane

TABLE 1. Cell biological features of 143B cells after treatment with H_2O_2 for 48 h

H_2O_2 (μM)	Viability of the cells (%) ($N = 3$)	Depolarization of mitochondrial membrane	Relative amount of mitochondrial mass (%) ($N = 5$)	Relative amount of mitochondrial DNA (%) ($N = 4$)	Percentage of the cells at the indicated cell cycle (%)		
					G_0/G_1	S	G_2/M
0	89.1 ± 1.8	8.1 ± 3.2	100.0 ± 2.9	100.0 ± 4.3	55.0 ± 3.7	20.3 ± 2.2	24.5 ± 5.2
100	89.2 ± 1.5	3.3 ± 1.7	131.7 ± 7.3**	138.3 ± 20.5*	38.5 ± 13.0	19.2 ± 1.7	48.1 ± 2.5**
200	87.3 ± 0.9	6.7 ± 2.8	140.9 ± 7.5**	175.9 ± 33.5**	31.5 ± 2.6**	21.0 ± 1.8	45.5 ± 2.8**
300	86.6 ± 2.4	11.4 ± 6.0	144.1 ± 7.2**	136.7 ± 17.4*	31.6 ± 2.8**	20.1 ± 4.5	45.0 ± 5.6**
400	85.3 ± 2.8	11.7 ± 2.0	154.9 ± 10.9**	86.2 ± 13.0	17.9 ± 6.5**	14.9 ± 3.6	65.6 ± 8.8**
500	81.8 ± 3.9*	9.2 ± 4.5	145.8 ± 14.0**	30.4 ± 7.3**	9.2 ± 3.2**	19.1 ± 3.8	70.5 ± 2.5**

*$P < 0.05$ vs. control; **$P < 0.01$ vs. control.

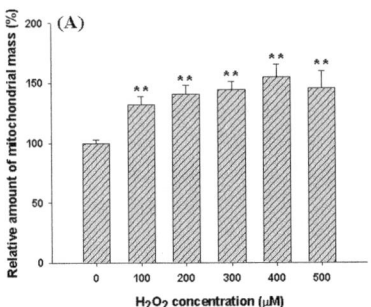
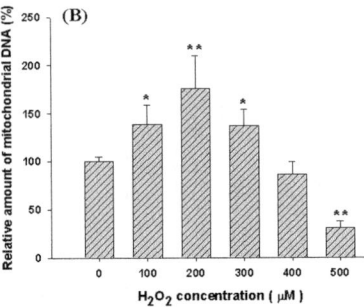

FIGURE 1. Hydrogen peroxide–induced changes of mitochondrial mass and mtDNA content in 143B cells. (**A**) The relative NAO intensity of 143B cells increased after H_2O_2 treatment for 48 h ($n = 5$; **$P < 0.01$ vs. control). (**B**) Relative content of mtDNA was determined by Q-PCR after the cells had been treated with H_2O_2 for 48 h. The results showed that mtDNA content increased in 143B cells that had been treated with 100–300 μM H_2O_2, but decreased in the cells that had been treated with 500 μM H_2O_2 ($n = 4$; *$P < 0.05$ vs. control; **$P < 0.01$ vs. control).

(Amersham Biosciences, Uppsala, Sweden). Western blotting was performed at room temperature with anti–α-tubulin antibody as described previously.[11] To quantify the relative levels of tubulin, we measured the optical density of the bands with a laser densitometer.

RESULTS

We first demonstrated that mitochondrial mass was increased in 143B cells that had been treated with 100 to 500 μM of H_2O_2 for 48 h (FIG. 1A). After treatment with 100–300 μM H_2O_2, the copy number of mtDNA was increased in 143B cells but decreased in the cells that had been treated with 500 μM H_2O_2 (FIG. 1B). This dual-phase response to oxidative stress was not caused by the possibility that cells were undergoing apoptosis or necrosis (TABLE 1). Moreover, we showed that H_2O_2 treatment may alter the morphology of mitochondria; mitochondria were enlarged and their cristae were disorganized after treatment with varying concentrations of H_2O_2 for 48 h (FIG. 2). Moreover, H_2O_2 treatment cause cell cycle arrest at G_2/M (FIG. 3) and disruption of microtubules in 143B cells (FIG. 4). Quantitative immunoblotting demonstrated that polymerized tubulin was decreased by approximately 30% after 400 μM H_2O_2 treatment; there was no significant decrease in cells treated with 200 μM H_2O_2. On the other hand, the monomeric tubulin was increased by approximately 15% and 25% after treatment with 200 and 400 μM H_2O_2, respectively. These results suggest that H_2O_2 induced depolymerization of microtubules and resulted in the alteration of mitochondrial mass in human cells. Furthermore, after treatment of 143B cells with Taxol, colchicine, and nocodazole, we found that mitochondrial mass and mtDNA copy number of the cells were all increased (FIG. 5A and B). This finding was consistent with our contention that H_2O_2 may disrupt the organization of microtubules and thus alter the mitochondrial mass in human cells.

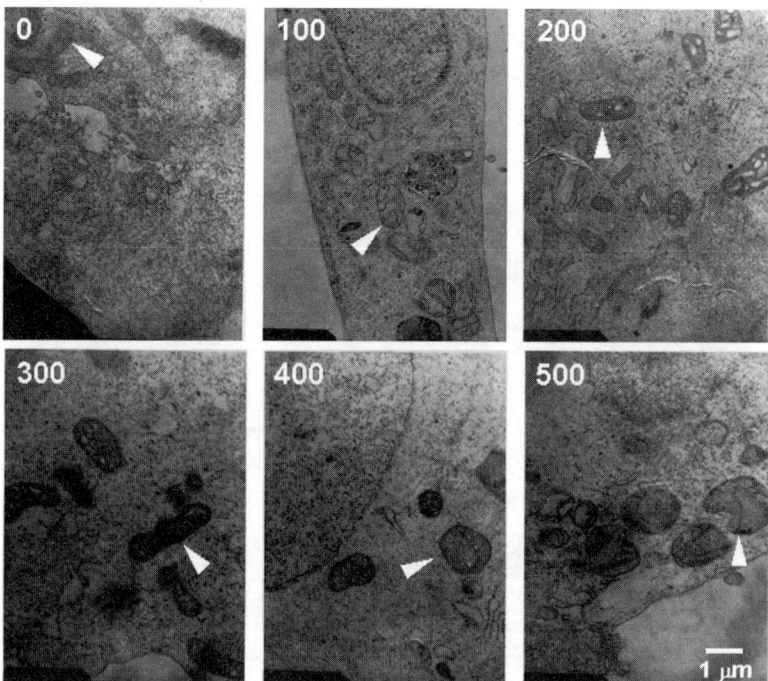

FIGURE 2. The ultrastructure of mitochondria in 143B cells treated with H_2O_2. Under examination by TEM, we found that mitochondria were enlarged and their cristae were disorganized after treatment of 143B cells with varying concentrations (0–500 μM) of H_2O_2 for 48 h. *White arrowheads* indicate the single mitochondrion inside the cell.

DISCUSSION

Human cells regulate their energy production by mitochondrial oxidative phosphorylation (OXPHOS) according to their needs. Depending on the challenge, mitochondria respond to the energy demands by subtle change in the activity of respiratory enzymes, by changing the expression of constituent enzyme subunits, or by increasing the number and size of the organelles. It is known that mitochondrial proliferation occurs in skeletal muscles in response to increased contractile activity, in adipose tissues in response to adaptive thermogenesis, and in cardiac myocytes in response to electrical stimulation or hypothermia. In clinical settings, the proliferation of abnormal mitochondria is generally observed in skeletal muscles of patients with a defective respiratory chain of mitochondria caused by mutation or depletion of mtDNA. It is thought that ROS, generated continuously by the respiratory chain, might play a role in mitochondrial biogenesis. However, solid evidence in support of this hypothesis has not yet been available.

The coordination between mitochondrial and cytoplasmic events leading to the growth and division of mammalian cells is largely unknown. It was reported that the mitochondrial mass was closely related to the cell cycle.[8] Our results confirmed the

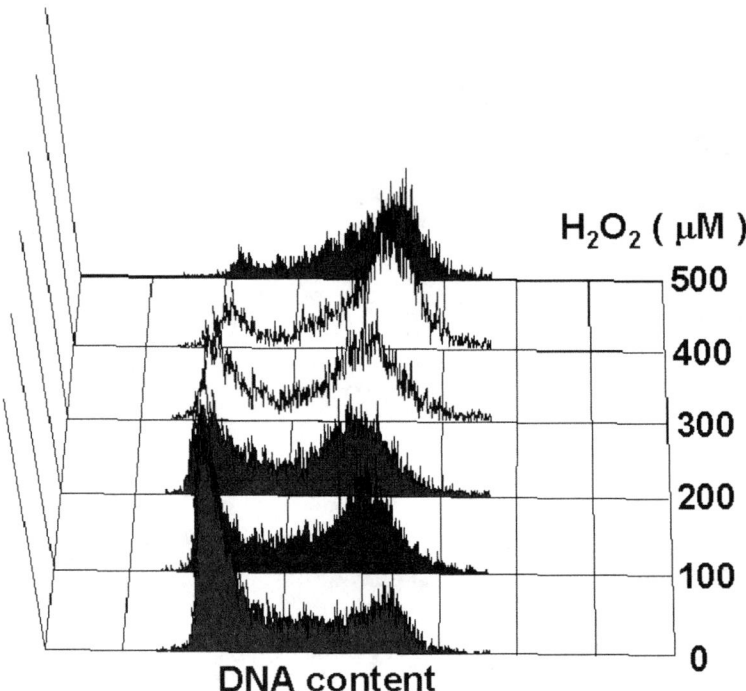

FIGURE 3. Hydrogen peroxide–induced cell cycle arrest at G_2/M in 143B cells. The percentage of 143B cells arrested at G_2/M phase increased with the increase of H_2O_2 concentration. Cells were fixed and stained with propidium iodide. The DNA content of the cell was estimated by use of a flow cytometer.

notion that stress-induced mitochondrial mass increase was correlated to the arrest of the cells at G_2/M phase.

The key elements that regulate cell cycle in mammalian cells are microtubules. Microtubules are the major components of cytoskeletal systems that are responsible for the regulation of the mitochondrial distribution in the cell. Several studies demonstrated that oxidants disassemble the microtubules in human cells.[10] By using immunostaining and microtubule fractionation assay, we showed that microtubules were depolymerized in the cells after H_2O_2 treatment (FIG. 4). This finding suggests that H_2O_2 induces depolymerization of microtubules and results in the alteration of mitochondrial mass in human cells. To substantiate this notion, we treated human osteosarcoma cells and hepatocellular carcinoma cells with microtubule-active drugs, such as Taxol, colchicine, and nocodazole, which stabilize or destabilize microtubules and arrest cultured cells at the G_2/M phase. The results showed that mitochondrial mass and mtDNA copy number of 143B cells were all increased after treatment with these drugs.

The copy number of mtDNA is an important index of mitochondrial biogenesis. Our results showed that mtDNA copy number was increased by treatment with a low concentration of H_2O_2. However, the mtDNA content was drastically decreased after

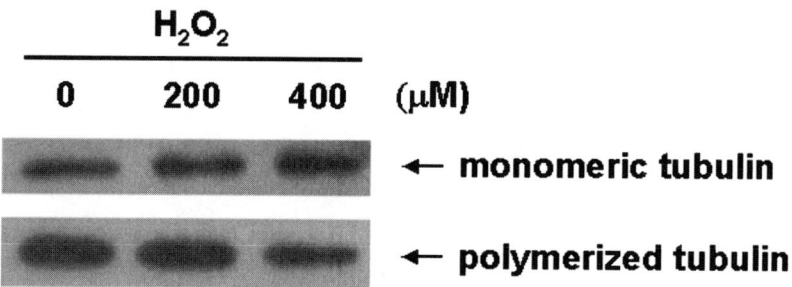

FIGURE 4. Microtubule fractionation and immunoblotting for the assessment of tubulin assembly and disassembly in 143B cells after H_2O_2 treatment. The results showed that the content of monomeric tubulins increased and that of polymerized tubulins was decreased, respectively, in the cells that had been treated with 200 and 400 µM H_2O_2 for 48 h.

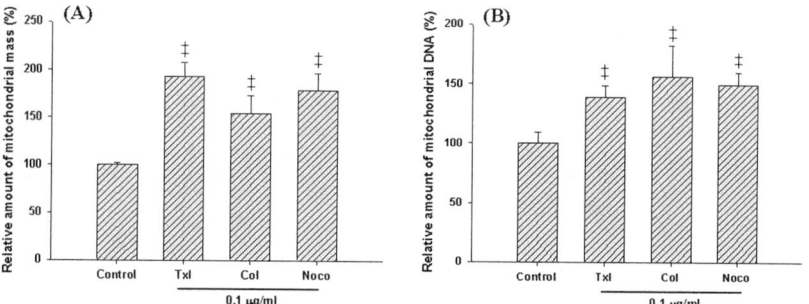

FIGURE 5. Increase of mitochondrial mass and mtDNA content in 143B cells after treatment with microtubule-active drugs. After treatment with 0.1 µg/mL of Taxol, colchicine, and nocodazole, respectively, the 143B cells were arrested at the G_2/M phase. The results showed that mitochondrial mass (**A**) and mtDNA content (**B**) increased after such treatments ($n = 4$; $\ddagger P < 0.005$ vs. control).

treatment with H_2O_2 at a concentration higher than 300 µM. This phenomenon may be caused by the low fidelity of the mtDNA template with high oxidative damage or caused by low activity of impaired mitochondrial DNA polymerase γ.[12] This dual-phase change of mtDNA copy number is consistent with our previous findings in the cybrids harboring different proportions of mtDNA with a 4,977-bp deletion and in leukocytes from healthy subjects of different ages.[9,13]

It has been reported that Taxol, but not colchicine or nocodazole, promotes the biogenesis of mitochondria in mammalian cells.[5] However, in this study, we demonstrated that the alteration in the stability of microtubules might control the biogenesis of mitochondria and that a H_2O_2-induced mitochondrial mass increase might be a result of the depolymerization of microtubules. It is plausible that H_2O_2 disrupts the organization of microtubules and alters the mitochondrial mass. Further studies are warranted to clarify the mechanism of action of H_2O_2 on the depolymerization of microtubules and on mitochondrial biogenesis in human cells.

ACKNOWLEDGMENTS

This work was supported by a grant (NSC92-2321-B-010-011-YC) from the National Science Council, Executive Yuan, Taiwan.

REFERENCES

1. LEDDA, M., C. MARTINELLI & E. PANNESE. 2001. Quantitative changes in mitochondria of spinal ganglion neurons in aged rabbits. Brain Res. Bull. **54:** 455–459.
2. NAKANO, T., K. IMANAKA, H. UCHIDA, *et al.* 1987. Myocardial ultrastructure in Kearns-Sayre syndrome. Angiology **38:** 28–35.
3. NAKADA, K., T. ONO & J.I. HAYASHI. 2002. A novel defense system of mitochondria in mice and human subjects for preventing expression of mitochondrial dysfunction by pathogenic mutant mtDNAs. Mitochondrion **2:** 59–70.
4. GILKERSON, R.W., D.H. MARGINEANTU, R.A. CAPALDI, *et al.* 2000. Mitochondrial DNA depletion causes morphological changes in the mitochondrial reticulum of cultured human cells. FEBS Lett. **474:** 1–4.
5. KARBOWSKI, M., J.H. SPODNIK, M. TERANISHI, *et al.* 2001. Opposite effects of microtubule-stabilizing and microtubule-destabilizing drugs on biogenesis of mitochondria in mammalian cells. J. Cell Sci. **114:** 281–291.
6. LEE, H.C., P.H. YIN, C.Y. LU, *et al.* 2000. Increase of mitochondria and mitochondrial DNA in response to oxidative stress in human cells. Biochem. J. **348:** 425–432.
7. SWEET, S. & G. SINGH. 1995. Accumulation of human promyelocytic leukemic (HL-60) cells at two energetic cell cycle checkpoints. Cancer Res. **55:** 5164–5167.
8. SWEET, S. & G. SINGH. 1999. Changes in mitochondrial mass, membrane potential, and cellular adenosine triphosphate content during the cell cycle of human leukemic (HL-60) cells. J. Cell. Physiol. **180:** 91–96.
9. WEI, Y.H., C.F. LEE, H.C. LEE, *et al.* 2001. Increases of mitochondrial mass and mitochondrial genome in association with enhanced oxidative stress in human cells harboring 4,977 bp-deleted mitochondrial DNA. Ann. N.Y. Acad. Sci. **928:** 97–112.
10. BANAN, A., S. CHOUDHARY, Y. ZHANG, *et al.* 2000. Oxidant-induced intestinal barrier disruption and its prevention by growth factors in a human colonic cell line: role of the microtubule cytoskeleton. Free Radic. Biol. Med. **28:** 727–738.
11. LIU, C.Y., C.F. LEE, C.H. HONG, *et al.* 2004. Mitochondrial DNA mutation and depletion increase the susceptibility of human cells to apoptosis. Ann. N.Y. Acad. Sci. **1011:** 133–145.
12. GRAZIEWICZ, M.A., B.J. DAY & W.C. COPELAND. 2002. The mitochondrial DNA polymerase as a target of oxidative damage. Nucleic Acids Res. **30:** 2817–2824.
13. LIU, C.S., C.S. TSAI, C.L. KUO, *et al.* 2003. Oxidative stress-related alteration of the copy number of mitochondrial DNA in human leukocytes. Free Radic. Res. **37:** 1307–1317.

Blood Lipid Peroxides and Muscle Damage Increased following Intensive Resistance Training of Female Weightlifters

JEN-FANG LIU,[a] WEI-YIN CHANG,[a] KUEI-HUI CHAN,[b] WEN-YEE TSAI,[c] CHEN-LI LIN,[d] AND MEI-CHIEH HSU[c]

[a]*Graduate Institute of Nutrition and Health Science, Taipei Medical University, Taipei, Taiwan, ROC*

[b]*Department of Sports Business Management, Da Yeh University, Changhua, Taiwan, ROC*

[c]*Graduate Institute of Sports Science, National College of Physical Education and Sports, Taoyuan, Taiwan, ROC*

[d]*Department of Athletic Training and Health, National College of Physical Education and Sports, Taoyuan, Taiwan, ROC*

ABSTRACT: The aim of this study was to examine changes in muscle cell injury and antioxidant capacity of weightlifters following a 1-week intensive resistance-training regimen. Thirty-six female subjects participated in this study, and their ages ranged from 18 to 25 years. The sample group included 19 elite weightlifters with more than 3 years of weightlifting training experience, while the control group comprised 17 non-athletic individuals. Compared with non-athletes, weightlifters had significantly lower glutathione peroxidase activity and plasma vitamin C concentrations. Weightlifters also had significantly higher malondialdehyde + 4-hydroxy 2-(E)-nonenal (MDA+4-HNE) and thiobarbituric acid–reactive substance (TBARS) levels and creatine kinase (CK) activity. For weightlifters, the plasma vitamin E level and the activity of superoxide dismutase (SOD) decreased, and CK activity increased significantly ($P < 0.05$) after a 1-week intensive resistance-training regimen. Both the TBARS levels and CK activity returned to values of pre-intensive training after a 2-day rest. The MDA+4-HNE level strongly correlated with CK activity in weightlifters ($P < 0.05$). In conclusion, both long-term exercise training and 1 week of intensive resistance training resulted in increased oxidative stress and cell injury in female weightlifters. Furthermore, proper rest after intensive training was found to be important for recovery.

KEYWORDS: weightlifters; resistance training; lipid peroxidation; cellular damage

Address for correspondence: Mei-Chieh Hsu, Graduate Institute of Sports Science, National College of Physical Education and Sports, Taoyuan, Taiwan, ROC. Voice: +886-3-3283201 ext. 2619; fax: +886-3-3311843.
meichich@mail.ncpes.edu.tw

INTRODUCTION

There is no doubt that exercise has many health benefits. Exercise is recommended for the prevention and management of many chronic diseases and for the maintenance of optimal health. However, performing strenuous physical activity can increase oxygen consumption by up to 10- to 15-fold over resting levels in order to meet energy demands. Exercise increases metabolism, and metabolic leakage from mitochondrial electron transport chains is now considered to be the most important source of oxygen-derived radicals. Increased oxygen uptake during exercise is accompanied by an elevation in reactive oxygen species (ROS), which is considered to be oxidative stress.[1] Oxidative stress has been linked to fatigue of skeletal and respiratory muscles, impaired performance and recovery, and the etiology of various diseases.[2–4]

Different athletes follow different training programs. Resistance training, also called strength training or weight training, is reported to have many benefits, such as weight control, prevention of osteoporosis, improvement of cardiovascular risk factors, and injury prevention.[5,6] Resistance training is a regular and important training program for weightlifters and power athletes, not only for improving the strength of muscles but also for maintaining a proper body composition.[7,8] However, an improper resistance-training program may increase cellular damage and oxidative stress in athletes. During resistance exercise, ischemia-reperfusion of muscles and production of free radicals via oxidative bursts from neutrophils occur and are considered serious.[9] The purpose of this study was to investigate cellular damage, the degree of oxidative stress, and antioxidative status both with habitual training and after a 1-week intensive resistance-training program.

MATERIALS AND METHODS

Human Subjects

The study was conducted at the National College of Physical Education and Sports, Taoyuan, Taiwan. Nineteen female elite weightlifters and 17 age-matched female non-athletes were recruited for this experiment, which was approved by the Taipei Medical University Subjects Committee. All weightlifters were members of the varsity team and participated regularly in national competitions. All weightlifters had undergone exercise training for more than 3 years. All subjects were free-living, and asked to maintain their usual food intake. The purpose and possible discomfort of the study were fully explained to each subject. All subjects signed informed consent forms. In a pre-study interview, each subject was asked to stop taking any vitamin and nutritional supplements 2 weeks before the study began.

Experimental Design

The resistance-training program was designed by a nationally certified senior weightlifting coach for each athlete, and all weightlifters were asked to record their training program. All athletes had similar training regimens. The high-intensity resistance-training lasted for 1 week, including 6 days, for two sessions per day, for 2–3 h per session. Each subject performed adequate warm-up sets. After the warm-up sets were completed, the resistance-training program was designed to increase the strength of all major muscle groups. A session consisted of the following: 3 sets of 3 repetitions on the shoulder press; 3 sets of 5 repetitions on the back squat; 1 set of 2 repetitions on abdominal curls; 3 sets of 5 repetitions on the dead lift; 1 set of 2 repetitions on the snatch; and 4 sets of 5 repetitions on the good morning. Resistance was set at 80% of 1 repetition maximum (1-RM). To control for aerobic exercise effects, weightlifters were required to rest for 30 to 60 s between each set and for 2 to 3 min between each exercise. A 2-day rest followed the 1-week high-intensity resistance training.

Sample Collection and Analysis

The total body fat and fat-free mass (FFM) of subjects were measured using a Biodynamics Body Composition Analyzer (Biodynamics Corp., Seattle, WA) according to a bioelectrical impedance analysis (BIA). Ten milliliters of blood were collected from each subject after 12 h of fasting, prior to and after the 1-week training, as well as from all weightlifters, after the 2-day rest that followed. Red blood cell portions of whole-blood samples were separated out for subsequent superoxide dismutase (SOD) and glutathione peroxidase (GSH-Px) analyses using commercial kits (Randox Laboratories Ltd., Crumlin, Antrum, UK). Both plasma vitamin C and vitamin E were analyzed using reverse-phase high-performance liquid chromatography (HPLC) with a UV detector at 254 nm, and a fluorescence detector with excitation at 292 nm and emission at 340 nm, respectively. The 4×250 mm and 4×125 mm Lichrosphere 100RP-18 column (Merck, Darmstadt, Germany), containing 5-mm particles protected by a guard column, were used for vitamin C and vitamin E analysis, respectively. The mobile phase for vitamin C contained 0.5 mM PICB (1-Pentane sulfuric acid sodium salt) adjusted to pH 3.1 with glacial acetic acid. Pure methanol was used as the mobile phase for vitamin E analysis. Plasma leptin was measured with an immunoradioassay using a commercial kit. The measurement of total antioxidant status (TAS) was based on the method of Miller et al.[10] using commercial kits. Plasma creatine kinase (CK) activity was determined in duplicate using a colorimetric assay method with a Johnson & Johnson DT-60 at 680 nm. Plasma lipid peroxidation was assessed by a modified spectrofluorometric measurement of thiobarbituric acid–reactive substances (TBARS). Plasma glutathione (GSH), malondialdehyde (MDA), and 4-hydroxy-2(E)-nonenal (4-HNE) were also directly determined using a commercial kit (Calibiochem-Novabiochem Co., San Diego, CA).

Statistical Analysis

Values are expressed as the mean ± SD. Differences between weightlifters and non-athletes were compared using the independent-sample t-test. Data obtained before and after the 1-week training, and after the 2-day rest, were compared using Student's paired t-test with the Statistical Analysis System (SAS) program. Differences of $P < 0.05$ were considered significant.

TABLE 1. Characteristics of weightlifters and non-athletes[a]

Characteristics	Weightlifters ($n = 19$)	Non-athletes ($n = 17$)
Age (years)	20.7 ± 1.6	21.9 ± 0.9
Height (cm)	156.8 ± 6.9	160.8 ± 5.2
Weight (kg)	67.9 ± 7.1*	52.7 ± 7.0
Body mass index (kg/m^2)	27.6 ± 8.1*	20.3 ± 5.1
Body fat (%)	23.6 ± 3.4	22.0 ± 3.8
(kg)	14.9 ± 3.7*	11.7 ± 2.2
Fat-free mass (%)	76.4 ± 3.4	77.3 ± 4.3
(kg)	47.6 ± 7.1*	41.4 ± 6.3
Arm circumference (cm)	12.2 ± 1.2*	10.6 ± 1.0
Weightlifting history (years)	5.3 ± 1.8	—

[a]Values are the mean ± SD.
*Significantly different from the non-athlete group by independent-sample t-test ($P < 0.05$).

RESULTS AND DISCUSSION

The characteristics of subjects are shown in TABLE 1. Both groups were similar in age and height. The body weight, body mass index (BMI), and arm circumference of weightlifters were higher than those of non-athletes. There were no significant differences in body fat (%) or FFM (%) between the two groups; however, weightlifters had a higher amount of both body fat (kg) and FFM (kg) than did non-athletes. Because most weightlifters in this study had more than 3 years of weightlifting history and regular exercise training and weightlifting experience, they had a different body composition from that of the non-athletes. The data reported are similar to those of previous studies.[11,12]

The biochemical parameters of weightlifters and non-athletes are presented in TABLE 2. Compared with non-athletes, weightlifters had higher plasma levels of vitamin E and GSH, and lower vitamin C and GSH-Px activity than did non-athletes. However, the creatine kinase (CK) activity and TBARS levels in weightlifters were significantly higher than those of the control group. Long-term regular training fostered a worse antioxidative status and more serious cellular damage in weightlifters compared with non-athletes.

TABLE 3 presents changes in biochemical parameters after the 1 week of intensive training and after a 2-day rest. The level of plasma vitamin E showed a significant 51% decrease after 1 week of high-intensity training, but increased after the 2-day rest. The same was true for SOD, which showed a significant 14% activity decrease after 1 week of high-intensity training. Levels of vitamin C and TAS and the activity of GSH-Px decreased after 1 week of high-intensity training; however, there were no significant differences between pre-training levels and those seen after the 2-day rest, respectively.

As seen in TABLE 4, after weight training, CK activity significantly increased, -by 95.4%, after 1 week of high-intensity training, but significantly decreased to a level similar to that of pre-training after a 2-day rest. After 1 week of high-intensity train-

TABLE 2. Levels of plasma biochemical parameters of weightlifters and non-athletes[a]

Plasma Biochemical Parameters	Weightlifters ($n = 19$)	Non-athletes ($n = 17$)
Leptin (ng/mL)	11.70 ± 4.85	10.8 ± 2.94
Vitamin C (mol/L)	38.20 ± 17.02*	49.15 ± 13.81
Vitamin E (mol/L)	23.93 ± 10.27*	20.85 ± 8.26
Glutathione (μmol/L)	287.34 ± 81.34*	198.34 ± 46.32
TAS (mmol/L)	1.54 ± 0.29	1.15 ± 0.21
SOD (U/mg Hb)	18.20 ± 2.59	17.46 ± 2.65
GSH-Px (U/g Hb)	26.25 ± 11.57*	38.07 ± 6.02
Creatine kinase (U/L)	117.01 ± 45.72*	80.85 ± 14.14
MDA+4-HNE (μmol/L)	78.87 ± 17.24	72.42 ± 17.11
TBARS (μmol/L)	5.82 ± 1.96*	3.89 ± 0.78

[a]Values are the mean ± SD.
*Significantly different from the non-athlete group by independent-sample t-test ($P < 0.05$).
ABBREVIATIONS: TAS, total antioxidant status; SOD, superoxide dismutase; GSH-Px, glutathione peroxidase; MDA+4-HNE, malondialdehyde + 4-hydroxy 2-(E)-nonenal; TBARS, thiobarbituric acid–reactive substances.

TABLE 3. Levels of blood antioxidant substances and activities of antioxidant enzymes of weightlifters before and after 1 week of resistance training and after a 2-day rest[a]

	Pre-training ($n = 19$)	Post-training ($n = 19$)	After a 2-day rest ($n = 17$)
Vitamin C (μmol/L)	38.20 ± 17.02	33.30 ± 10.32	36.14 ± 7.25
Vitamin E (μmol/L)	23.93 ± 1027	11.35 ± 5.29*	17.11 ± 6.18
Glutathione (μmol/L)	287.34 ± 81.34	301.85 ± 65.8	283.35 ± 42.64
TAS (mmol/L)	1.54 ± 0.29	1.28 ± 0.57	1.29 ± 0.47
SOD (U/mg Hb)	18.20 ± 2.59	15.64 ± 2.89*	16.09 ± 2.47
GSH-Px (U/g Hb)	26.25 ± 11.57	20.54 ± 7.42	25.29 ± 8.96

[a]Values are the mean ± SD.
*Significantly different from pre-training by Student's paired t-test ($P < 0.05$).
ABBREVIATIONS: TAS, total antioxidant status; SOD, superoxide dismutase; GSH-Px, glutathione peroxidase.

TABLE 4. Activities of creatine kinase and levels of lipid peroxidation of weightlifters before and after 1 week of resistance training, and after a 2-day rest[a]

	Pre-training ($n = 19$)	Post-training ($n = 19$)	After a 2-day rest ($n = 17$)
Creatine kinase (U/L)	117.01 ± 45.72	220.60 ± 109.52*	118.84 ± 42.25[#]
MDA+4-HNE (μmol/L)	78.87 ± 17.24	101.75 ± 20.29*	100.38 ± 20.58
TBARS (μmol/L)	5.82 ± 1.96	7.23 ± 1.90*	5.09 ± 2.14[#]

[a]Values are the mean ± SD.
*Significantly different from pre-training by Student's paired t-test ($P < 0.05$).
[#]Significantly different from post-training by Student's paired t-test ($P < 0.05$).

ing, there were significant increases in the indices of lipid peroxidation of 29% for MDA+-4-HNE, and 21% for TBARS. Similar to CK activity, the levels of TBARS significantly decreased after the 2-day rest. A significant positive correlation between plasma CK activity and MDA+4-HNE levels was found ($r = 0.38$, $P < 0.05$). Both MDA+4-HNE and TBARS are usually used as quantitative markers for free radical interactions with cell membranes.[13]

In this study, the increased activity of the cytoplasmic enzyme CK indicated that tissue damage had occurred, especially in muscles.[4] The CK, MDA+4-HNE, and TBARS responses agreed with those of some previous investigations involving aerobic-type exercise.[14] It is unclear, however, why both resistance exercise and aerobic-type exercise result in similar increases in free radical production. It might be expected that the mechanisms for the increase in free radical formation during resistance exercise and aerobic-type exercise would differ.

Resistance training consists of repetitive, static muscle actions. These include concentric and eccentric muscle actions, which are considered to be, respectively, a low- or high-intensity resistance exercise protocol. Based on previous investigations, it was determined that the intensity of the exercise protocol used is a primary factor in creating a physiological environment for increased free radical production.[15,16] Plasma MDA was also found to be elevated after heavy resistance exercise involving upper and lower body muscles in recreational weight-training exercises.[13]

As previously mentioned, hyperoxia at the site of mitochondria may be a more prominent mechanism for free radical formation during aerobic-type exercise.[17,18] In contrast, intense muscle contractions associated with resistance exercise may result in ischemia-reperfusion at the site of the active muscles. The free radicals act as a mediator of ischemia-reperfusion injury to skeletal muscles and result in muscle injury accompanied by increased amounts of CK. Kanter *et al.*[19,20] have also shown that plasma MDA measurements correlate with CK activity during exercise. It is now clear that high-intensity whole-body resistance exercise can result in the formation of free radicals. These free radicals may play a role in how muscle tissues adapt to the physiological stress caused by resistance exercise. Ischemia-reperfusion during resistance exercise at the site of the muscle, and post-exercise production of free radicals via oxidative bursts from neutrophils, are realistic concepts that must be seriously considered.[9]

In conclusion, this study demonstrates that both regular long-term exercise training and 1-week intensive resistance training can result in increased oxidative stress and cell injury in female weightlifters. Further, proper rest after intensive training is necessary in order to reduce exercise-induced oxidative damage and allow the athletes to recover. The results obtained from this study may provide coaches and sports specialists with a foundation and reference information from which to design proper training programs for weightlifters, power athletes, and muscle builders in the future.

ACKNOWLEDGMENTS

We wish to express our deep appreciation to all the subjects for their dedication to this study. This work was supported by a grant (NSC88-2314-B038-002) from the National Science Council, Taiwan, R.O.C.

REFERENCES

1. AROMA, O.I. 1994. Free radicals and antioxidant strategies in sports. J. Nutr. Biochem. **5:** 370–381.
2. BARCLAY, J.K. & M. HANSEL. 1991. Free radicals may contribute to oxidative skeletal muscle fatigue. Can. J. Physiol. Pharmacol. **69:** 279–284.
3. SCHWARTZ, C.J. et al. 1991. The pathogenesis of atherosclerosis: an overview. Clin. Cardiol. **14:** I1–I6.
4. VINCENT, H.K. & K.R. VINCENT. 1997. The effect of training status on the serum creatine kinase response, soreness and muscle function following resistance exercise. Int. J. Sports Med. **18:** 431–437.
5. DINUBILE, N.A. 1991. Strength training. Clin. Sports Med. **10:** 33–62.
6. VERRILL, D.E. & P.M. RIBISL. 1996. Resistance exercise training in cardiac rehabilitation. An update. Sports Med. **21:** 347–383.
7. KRAEMER, W.J., N.A. RATAMESS & D.N. FRENCH. 2002. Resistance training for health and performance. Curr. Sports Med. Rep. **1:** 165–171.
8. DESCHENES, M.R. & W.J. KRAEMER. 2002. Performance and physiologic adaptations to resistance training. Am. J. Phys. Med. Rehab. **81:** S3–S16.
9. PYNE, D.B. 1994. Regulation of neutrophil function during exercise. Sports Med. **17:** 245–258.
10. MILLER, N.J. et al. 1993. A novel method for measuring antioxidant capacity and its application to monitoring the antioxidant status in premature neonates. Clin. Sci. **84:** 407–412.
11. CHARETTE, S.L. et al. 1991. Muscle hypertrophy response to resistance training in older women. J. Appl. Physiol. **70:** 1912–1916.
12. CULLINEN, K. & M. CALDWELL. 1998. Weight training increases fat-free mass and strength in untrained young women. J. Am. Diet. Assoc. **98:** 414–418.
13. MCBRIDE, J.M. et al. 1998. Effect of resistance exercise on free radical production. Med. Sci. Sports Exerc. **30:** 67–72.
14. SUMIDA, S. et al. 1989. Exercise-induced lipid peroxidation and leakage of enzymes before and after vitamin E supplementation. Int. J. Biochem. **21:** 835–838.
15. SAHLIN, K. et al. 1992. Repetitive static muscle contractions in humans—a trigger of metabolic and oxidative stress? Eur. J. Appl. Physiol. Occup. Physiol. **64:** 228–236.
16. SAXTON, J.M., A.E. DONNELLY & H.P. ROPER. 1994. Indices of free-radical-mediated damage following maximum voluntary eccentric and concentric muscular work. Eur. J. Appl. Physiol. **68:** 189–193.
17. SEYAMA, A. 1993. The role of oxygen-derived free radicals and the effect of free radical scavengers on skeletal muscle and ischemia/reperfusion injury. Surg. Today **23:** 1060–1067.
18. WALKER, P.M. 1991. Ischemia/reperfusion injury in skeletal muscle. Ann. Vasc. Surg. **5:** 399–402.
19. KANTER, M.M., G.R. LESUMES & L.A. KAMINSKY. 1988. Serum creatine kinase and lactate dehydrogenase changes following an eighty kilometer race. Relationship to lipid peroxidation. Eur. J. Appl. Physiol. **57:** 60–63.
20. KANTER, M.M., L.A. NOLTE & J.O. HOLLOSZY. 1993. Effects of an antioxidant vitamin mixture on lipid peroxidation at rest and postexercise. J. Appl. Physiol. **74:** 965–969.

Anti-Inflammatory and Antioxidative Effects of Propofol on Lipopolysaccharide-Activated Macrophages

RUEI-MING CHEN,[a,b,e] TYNG-GUEY CHEN,[b,e] TA-LIANG CHEN,[c] LI-LING LIN,[a] CHIA-CHEN CHANG,[a] HWA-CHIA CHANG,[b] AND CHIH-HSIUNG WU[d]

[a]*Graduate Institute of Medical Sciences, College of Medicine, Taipei Medical University, Taipei, Taiwan*

[b]*Department of Anesthesiology, Wan-Fang Hospital, College of Medicine, Taipei Medical University, Taipei, Taiwan*

[c]*Taipei City Hospital and Taipei Medical University, Taipei, Taiwan*

[d]*Department of Surgery, Taipei Medical University Hospital, College of Medicine, Taipei Medical University, Taipei, Taiwan*

> ABSTRACT: Sepsis is a serious and life-threatening syndrome that often occurs in intensive care unit (ICU) patients. During sepsis, inflammatory cytokines and nitric oxide (NO) can be overproduced, causing tissue and cell injury. Propofol is an intravenous agent used for sedation of ICU patients. Our previous study showed that propofol has immunosuppressive effects on macrophage functions. This study was designed to evaluate the anti-inflammatory and antioxidative effects of propofol on the biosyntheses of tumor necrosis factor α (TNF-α), interleukin 1β (IL-1β), IL-6, and NO in lipopolysaccharide (LPS)-activated macrophages. Exposure to a therapeutic concentration of propofol (50 μM), LPS (1 ng/mL), or a combination of these two drugs for 1, 6, and 24 h was not cytotoxic to the macrophages. ELISA revealed that LPS increased macrophage TNF-α, IL-1β, and IL-6 protein levels in a time-dependent manner, whereas propofol significantly reduced the levels of LPS-enhanced TNF-α, IL-1β, and IL-6 proteins. Data from RT-PCR showed that LPS induced TNF-α, IL-1β, and IL-6 mRNA, but propofol inhibited these effects. LPS also increased NO production and inducible nitric oxide synthase (iNOS) expression in macrophages. Exposure of macrophages to propofol significantly inhibited the LPS-induced NO biosynthesis. The present study shows that propofol, at a therapeutic concentration, has anti-inflammatory and antioxidative effects on the biosyntheses of TNF-α, IL-1β, IL-6, and NO in LPS-activated macro-phages and that the suppressive effects are exerted at the pretranslational level.
>
> KEYWORDS: propofol; macrophages; anti-inflammation; TNF-α; IL-1β; IL-6; antioxidation; nitric oxide

---

[e] R.-M.C. and T-G.C. contributed equally to this paper.

Address for correspondence: Chih-Hsiung Wu, Department of Surgery, Taipei Medical University Hospital, or Ruei-Ming Chen, Graduate Institute of Medical Sciences, College of Medicine, Taipei Medical University, No. 250, Wu-Hsing St., Taipei 110, Taiwan. Voice: +886-2-29307930 ext. 2161; fax: +886-2-86621119.

chwu@tmu.edu.tw or rmchen@tmu.edu.tw

INTRODUCTION

Sepsis can cause multiple system organ failure, which often leads to high mortality for intensive care unit (ICU) patients.[1,2] Lipopolysaccharide (LPS), an outer-membrane component of gram-negative bacteria, is one of the critical factors contributing to the pathogenesis of sepsis.[3] *In vivo* studies demonstrated that exposure of animals and human volunteers to LPS can induce inflammatory responses seen in sepsis.[3] Macrophages play important roles in infection-associated tissue injury.[4,5] During inflammation, macrophages can produce and release large amounts of inflammatory cytokines and oxidants into the general circulation to exert the systemic effects that occur in the sepsis syndrome.[6,7] Modulating these inflammatory factors may influence the LPS-induced septic pathogenesis.

Propofol (2,6-diisopropylphenol) is one of the widely used intravenous anesthetic agents for induction and maintenance of anesthesia for surgical procedures.[8] In the ICU, propofol can be used as an effective sedative agent. Propofol has the advantages of rapid onset, short duration of action, and rapid elimination.[9] Studies using human neutrophils and leukocytes demonstrated that propofol might have immunomodulating effects.[10,11] In response to LPS stimuli, tumor necrosis factor α (TNF-α), interleukin 1β (IL-1β), IL-6, and nitric oxide (NO) are typical inflammatory factors that are produced by activated macrophages and are involved in immune responses and host defense.[12] Our previous study showed that propofol can modulate macrophage functions via the suppression of migration, phagocytosis, and oxidative ability.[13] Data from another study revealed that therapeutic concentrations of propofol can protect macrophages from NO-induced insults to the cell.[14] The present study was designed to evaluate the anti-inflammatory and antioxidative effects of propofol on the biosyntheses of TNF-α, IL-1β, IL-6, and NO in LPS-activated macrophages.

MATERIALS AND METHODS

Cell Culture and Drug Treatment

Mouse macrophage-like Raw 264.7 cells were purchased from American Type Culture Collection (Rockville, MD). Macrophages were maintained in RPMI 1640 medium (Gibco-BRL, Grand Island, NY) supplemented with 10% fetal calf serum, 100 IU/mL penicillin, and 100 μg/mL streptomycin in 75-cm^2 flasks at 37°C in a humidified atmosphere of 5% CO_2. These cells were grown to confluence before drug treatment.

Propofol donated by Zeneca Limited (Macclesfield, Cheshire, UK) was stored under nitrogen, protected from light, and freshly prepared by dissolving it in dimethyl sulfoxide (DMSO) for each independent experiment. DMSO in the medium was kept to <0.1% to avoid the toxicity of this solvent to macrophages. According to the clinical application, propofol at 50 μM, which corresponds to a clinical plasma concentration,[15] was chosen as the dosage administered in this study. Control macrophages were treated with DMSO only.

Assay of Cell Viability

Macrophage viability was analyzed using a variation of the colorimetric 3-(4,5-dimethylthiazol-2-yl)-2,5-diphenyltetrazolium bromide (MTT) assay, which is based on the ability of live cells to reduce MTT to a blue formazan product.[16] Macrophages (1×10^4) were seeded in 96-well tissue culture plates overnight. After drug treatment, cells were renewed with medium containing 0.5 mg/ml MTT and then cultured for another 3 h. The blue formazan product in cells was dissolved in DMSO and measured spectrophotometrically at a wavelength of 570 nm.

Determination of Nitrite

After drug treatment, the amounts of nitrite, an oxidative product of NO, in the culture medium of macrophages was detected according to the technical bulletin of Promega's Griess reagent system (Promega, Madison, WI).

Enzyme-Linked Immunosorbent Assay

The amounts of TNF-α, IL-1β, and IL-6 in the culture medium of macrophages exposed to LPS, propofol, and a combination of propofol and LPS were determined following the standard protocols of the ELISA kits purchased from Endogen (Woburn, MA).

Immunoblotting Analysis

After pretreatment with the drugs for 24 h, macrophages were washed with phosphate-buffered saline (PBS) (0.14 M NaCl, 2.6 mM KCl, 8 mM Na_2HPO_4, and 1.5 mM KH_2PO_4), and cell lysates were collected after dissolving the cells in 50 µL of ice-cold radioimmunoprecipitation assay (RIPA) buffer (25 mM Tris-HCl, pH 7.2; 0.1% sodium dodecylsulfate; 1% Triton X-100; 1% sodium deoxycholate; 0.15 M NaCl; and 1 mM EDTA). Protein concentrations were quantified by a bicinchonic acid (BCA) protein assay kit (Pierce, Rockford, IL). Cytosolic proteins (100 µg) were resolved on a 12% polyacrylamide gel and electrophoretically blotted onto a piece of nitrocellulose membrane. Cellular iNOS protein was immunodetected using a mouse monoclonal antibody against the mouse iNOS protein (Transduction Laboratories, Lexington, KY). Cellular β-actin was immunodetected using a mouse monoclonal antibody against mouse β-actin (Sigma, St. Louis, MO) as an internal standard. Intensities of immunoreactive bands were determined using a UVIDOCMW version 99.03 digital-imaging system (Uvtec, Cambridge, UK).

RT-PCR Assay

Messenger RNA from macrophages exposed to LPS, propofol, and a combination of LPS and propofol were prepared for the RT-PCR analyses of TNF-α, IL-1β, IL-6, iNOS, and β-actin following manufacturer's instructions of the ExpressDirect™ mRNA Capture and RT System for the RT-PCR kit (Pierce, Rockford, IL). Oligonucleotides for PCR analyses of TNF-α, IL-1β, IL-6, iNOS, and β-actin were designed and synthesized by Clontech Laboratories (Palo Alto, CA). The oligonucleotide sequences of these primers were as follows:

TNF-α: 5'-ATGAGCACAGAAAGCATGATCCGC-3' and
3'-CTCAGGCCCGTCCAGATGAAACC-5'

IL-1β: 5'-ATGGCAACTGTTCCTGAACTCAACT-3' and
3'-TTTCCTTTCTTAGATATGGACAGGAC-5'

IL-6: 5'-ATGAAGTTCCTCTCTGCAAGAGACT-3' and
3'-CACTAGGTTTGCCGAGTAGATCTC-5'

iNOS: 5'-CCCTTCCGAAGTTTCTGGCAGCAGC-3' and
5'-CGACTCCTTTTCCGCTTCCTGAG-3'

β-actin: 5'-GTGGGCCGCTCTAGGCACCAA-3' and
3'-CTTTAGCACGCACTGTAGTTTCTC-5'.

The PCR reaction was carried out using 35 cycles, including 94°C for 45 s, 60°C for 45 s, and 72°C for 2 min. The PCR products were loaded onto a 1.8% agarose gel containing 0.1 μg/mL ethidium bromide, and were electrophoretically separated. DNA bands were visualized and photographed under UV-light exposure. The intensities of the DNA bands in the agarose gel were quantified with the aid of the UVI-DOCMW digital-imaging system.

Statistical Analysis

The statistical significance of differences between the control and drug-treated groups was evaluated by Student's t-test; the difference was considered statistically significant at P values of <0.05. The statistical difference between LPS and a combination of the propofol- and LPS-treated groups was considered significant when the P value of the Duncan multiple range test was <0.05.

RESULTS

Exposure to 50 μM propofol for 1, 6, and 24 h was not cytotoxic to macrophages (TABLE 1). Survival of macrophages was not influenced by LPS administration. Co-treatment with propofol and LPS for 1, 6, and 24 h did not affect macrophage viability.

ELISA analyses revealed that levels of TNF-α, IL-1β, and IL-6 proteins were low but detectable in untreated macrophages (FIG. 1). In macrophages, administration of LPS for 1 h caused significant 12-, 7-, and 6-fold increases of TNF-α, IL-1β, and IL-6 levels, respectively (FIG. 1A). Propofol did not change the amounts of TNF-α, IL-1β, and IL-6 in macrophages. However, co-treatment with propofol and LPS significantly reduced levels of LPS-enhanced TNF-α, IL-1β, and IL-6 by 45, 77, and 83%, respectively. In macrophages treated for 6 h, LPS increased the levels of TNF-α, IL-1β, and IL-6 by 10-, 12-, and 5-fold, respectively (FIG. 1B). Propofol did not change the basal levels of these inflammatory cytokines. Meanwhile, propofol significantly

TABLE 1. Effects of propofol and LPS on macrophage viability[a]

Treatment	Cell viability, OD values at 570 nm		
	1 h	6 h	24 h
Control	0.623 ± 0.063	0.668 ± 0.072	0.769 ± 0.082
Propofol	0.661 ± 0.059	0.683 ± 0.051	0.743 ± 0.081
LPS	0.642 ± 0.076	0.660 ± 0.089	0.785 ± 0.075
Propofol plus LPS	0.617 ± 0.068	0.652 ± 0.074	0.767 ± 0.075

[a]Macrophages were exposed to 50 mM propofol, 1 ng/mL LPS and a combination of propofol and LPS for 1, 6, and 24 h. Cell viability was determined by the MTT assay as described in MATERIALS AND METHODS. Each value represents the mean ± SEM for $n = 6$.

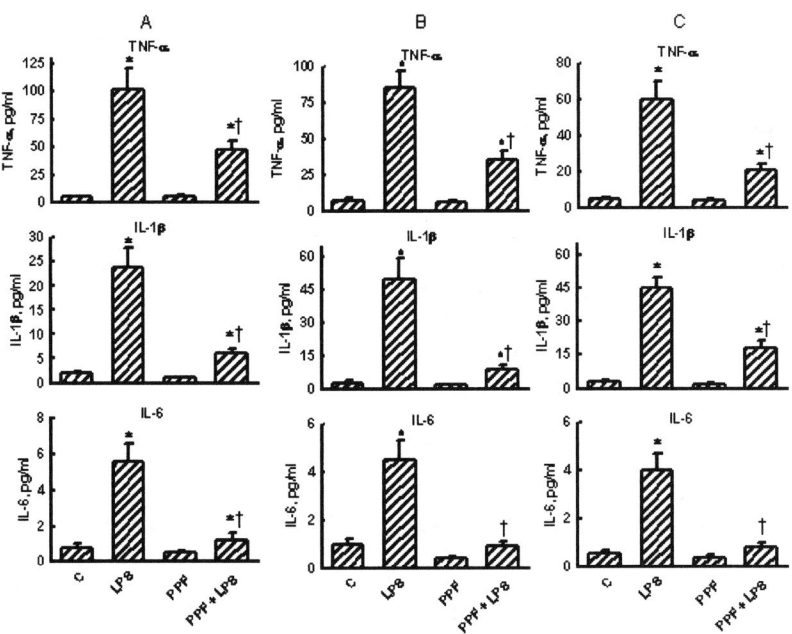

FIGURE 1. Effects of propofol on LPS-enhanced TNF-α, IL-1β, and IL-6 protein levels. Macrophages were exposed to 50 µM propofol (PPF), 1 ng/mL lLPS, and a combination of PPF and LPS for 1 h (**A**), 6 h (**B**), and 24 h (**C**). The amounts of TNF-α, IL-1β, and IL-6 in the culture medium were analyzed by the ELISA method. Each value represents the mean ± SEM for $n = 9$. *Values are significantly different from the respective control, $P < 0.05$. †Values are significantly different between the combined PPF and LPS and LPS treatment groups.

TABLE 2. Effects of propofol and LPS on TNF-α, IL-1β, and IL-6 mRNA production by macrophages

Treatment	1-h-treated macrophages		
	TNF-α	IL-1β	IL-6
Control	n.d.	n.d.	n.d.
LPS	5673 ± 872	1310 ± 672	682 ± 79
Propofol	n.d.	n.d.	n.d.
Propofol + LPS	1522 ± 256†	533 ± 87†	299 ± 55†
Treatment	6-h-treated macrophages		
	TNF-α	IL-1β	IL-6
Control	876 ± 491	n.d.	n.d.
LPS	3712 ± 546	7124 ± 917	1565 ± 226
Propofol	665 ± 569	n.d.	n.d.
Propofol + LPS	1279 ± 330†	723 ± 177†	257 ± 59†
Treatment	24-h-treated macrophages		
	TNF-α	IL-1β	IL-6
Control	n.d.	n.d.	n.d.
LPS	4316 ± 774	5819 ± 748	827 ± 116
Propofol	513 ± 491	n.d.	n.d.
Propofol + LPS	1198 ± 311†	477 ± 88†	n.d.

[a]Macrophages were exposed to 50 μM propofol, 1 ng/mL lipopolysaccharide (LPS), and a combination of propofol and LPS for 1, 6, and 24 h. RT-PCR analyses of TNF-α, IL-1β, and IL-6 mRNA were carried out, and DNA bands were quantified as described in MATERIALS AND METHODS. Each value represents the mean ± SEM for $n = 6$. †Values significantly differ between the combined propofol and LPS treatment and the LPS treatment groups, $P < 0.05$. n.d. not detected.

reduced LPS-enhanced levels of TNF-α, IL-1β, and IL-6 proteins by 62, 84, and 76%, respectively. In macrophages treated for 24 h, administration of LPS resulted in 3-, 3-, and 4-fold increases in TNF-α, IL-1β, and IL-6, respectively (FIG. 1C). Propofol did not influence macrophage TNF-α, IL-1β, and IL-6 levels. Exposure to propofol significantly reduced LPS-increased levels of TNF-α, IL-1β, and IL-6 proteins by 67, 67, and 80%, respectively.

RT-PCR analyses showed that administration of LPS in macrophages for 1, 6, and 24 h induced TNF-α, IL-1β, and IL-6 mRNA (FIG. 2). Co-treatment of macrophages with propofol and LPS for 1, 6, and 24 h inhibited LPS-induced TNF-α, IL-1β, and IL-6 mRNA. The DNA bands of RT-PCR products were quantified using a digital imaging system (TABLE 2). In macrophages treated for 1 h, propofol significantly reduced LPS-induced levels of TNF-α, IL-1β, and IL-6 mRNA by 83, 61, and 56%, respectively (TABLE 2). In groups treated for 6 h, the levels of LPS-induced TNF-α, IL-1β, and IL-6 mRNA were significantly decreased by 66, 90, and 84%, respectively, following propofol administration. In macrophages treated for 24 h, administration of propofol suppressed 72% and 92% of the LPS-induced TNF-α and IL-1β mRNA, respectively. Propofol completely inhibited LPS-induced IL-6 mRNA (TABLE 2).

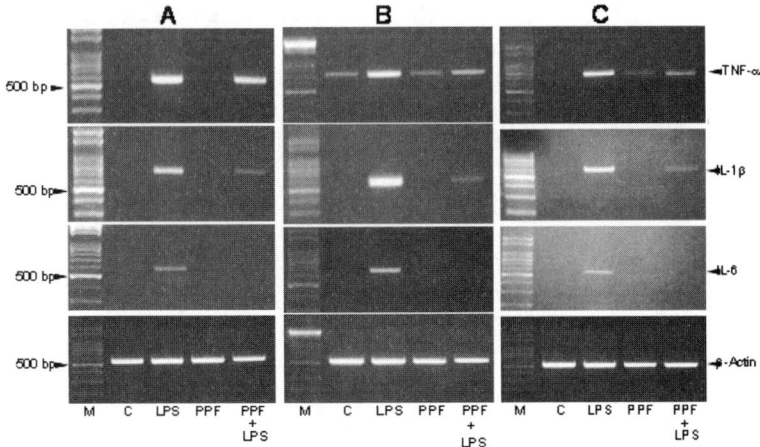

FIGURE 2. Effects of propofol on LPS-induced TNF-α, IL-1β, and IL-6 mRNA levels. Messenger RNA from macrophages exposed to 50 μM propofol (PPF), 1 ng/mL lipopolysaccharide (LPS), and a combination of PPF and LPS for 1 h (**A**), 6 h (**B**), and 24 h (**C**) was prepared for RT-PCR analyses of TNF-α, IL-1β, and IL-6 mRNA. The PCR products were loaded onto a 1.8% agarose gel containing 0.1 μg/mL ethidium bromide and electrophoretically separated. DNA bands were visualized and photographed under UV light exposure. RT-PCR analysis of β-actin was used as the internal control. M, DNA marker; C, control.

TABLE 3. Effects of propofol and LPS on nitrite production and iNOS expression by macrophages[a]

Treatment	Nitrite (μM)	iNOS protein (arbitrary units)	iNOS mRNA (arbitrary units)
Control	5 ± 1	n.d.	n.d.
Propofol	4 ± 1	n.d.	n.d.
LPS	32 ± 7*	679 ± 137	3217 ± 624
Propofol plus LPS	11 ± 3*,†	198 ± 58*,†	882 ± 177*,†

[a]Macrophages were exposed to 50 μM propofol, 1 ng/mL LPS, and a combination of propofol and LPS for 24 h. The levels of nitrite produced by macrophages were determined by the Griess reaction method. Analyses by immunoblotting and RT-PCR were carried out to determine the levels of iNOS protein and mRNA. Each value represents the mean ± SEM for $n = 3$. *Asterisk* (*), values are significantly different from the respective control ($P < 0.05$); *dagger* (†), values are significantly different between the combined propofol and LPS treatment and the LPS treatment groups ($P < 0.05$); n.d., not detected.

Nitrite was detectable in untreated macrophages, and propofol did not affect nitrite production. Administration of LPS caused a 6-fold increase of nitrite in the medium with macrophages (TABLE 3). Propofol significantly reduced LPS-augmented nitrite levels in macrophages by 66%. In untreated macrophages, iNOS protein and mRNA were undetectable (TABLE 3). Propofol did not influence iNOS protein or mRNA levels, but LPS increased the protein and mRNA levels of iNOS in macrophages (TABLE 3). Co-treatment with propofol and LPS significantly inhibited LPS-induced levels of iNOS protein and mRNA by 71% and 73%, respectively.

DISCUSSION

Macrophages play critical roles in cellular host defense against infection or tissue injury.[4,5] This study showed that in response to LPS stimulation, macrophages produced large amounts of TNF-α, IL-1β, IL-6, and NO. Induction of TNF-α, IL-1β, and IL-6 plays a major role in mediating endotoxin-induced pathogenesis of septic shock, such as fever, metabolic acidosis, diarrhea, hypotension, and disseminated intravascular coagulation.[7,12,17] Overproduction of NO can induce cell apoptosis.[14,18] Slowing the release of these inflammatory factors from macrophages may retard the inflammatory responses to LPS stimulation. One of our previous studies showed that propofol has immunosuppressive effects on macrophage functions.[13] The present study further shows that propofol significantly suppressed LPS-induced TNF-α, IL-1β, IL-6, and NO production by macrophages. The concentration of propofol used in this study was 50 µM, which corresponds to a clinical plasma concentration.[15] Therefore, a therapeutic concentration of propofol has anti-inflammatory and antioxidative effects on the biosyntheses of TNF-α, IL-1β, IL-6, and NO in LPS-activated macrophages.

LPS increased the levels of TNF-α, IL-1β, and IL-6 proteins in macrophages. Following LPS administration, TNF-α, IL-1β, and IL-6 mRNA were induced. Thus LPS pretranslationally induces the expression of TNF-α, IL-1β, and IL-6. Propofol reduced LPS-enhanced levels of TNF-α, IL-1β, and IL-6 proteins. RT-PCR analyses revealed that propofol could inhibit LPS-induced TNF-α, IL-1β, and IL-6 mRNA production. Data from the viability test showed that administration of macrophages with propofol, LPS, or a combination of these two drugs was not cytotoxic to macrophages. Therefore, propofol-induced suppression of TNF-α, IL-1β, and IL-6 production in macrophages occurred because of a pretranslational mechanism but not because of the cytotoxic effect of this sedative agent. Induction of inflammatory cytokines by LPS in macrophages has been reported to be via the toll-like receptor 4–mediated signal transduction pathway.[19] Toll-like receptor 4 is a type I transmembrane protein.[20] In response to LPS stimulation, propofol is so lipophilic that it might disrupt the plasma membrane and modulate the toll-like receptor 4–mediated induction of inflammatory cytokines in macrophages.

Exposure of macrophages to LPS significantly increased the levels of nitrite in the culture medium containing macrophages. Nitrite is an oxidative product of NO. The elevation of nitrite corresponds to the increase in NO by macrophages. LPS induced iNOS at the protein and mRNA levels. Co-treatment with propofol and LPS significantly decreased nitrite levels and iNOS expression. Thus the suppressive effects of propofol on LPS-induced NO biosynthesis are exerted by a pretranslational event. Our previous study showed that propofol could protect macrophages from exogenous NO-induced cell apoptosis.[14] The present study has further demonstrated that propofol inhibits LPS-induced endogenous NO biosynthesis in macrophages. LPS can increase endogenous NO through CD14-dependent induction of iNOS.[21] CD14 is a membrane protein in macrophages. Propofol could modulate the interaction of CD14 and LPS, which leads to the inhibition of iNOS expression and a decrease in NO production by macrophages.

LPS is recognized as one of the best-characterized signal effectors for triggering macrophages.[3] Concentrations of LPS used in in vitro and in vivo studies are about 1 μg/mL and 1 mg/kg body weight, respectively.[5,22] Bysani et al.[23] reported that the plasma concentration of LPS in a patient with fatal *Klebsiella pneumoniae* sepsis was 25 ng/mL In this study, we used ELISA and RT-PCR analyses to demonstrate that LPS at a low concentration (1 ng/mL) could apparently induce TNF-α, IL-1β, IL-6, and iNOS at both the protein and RNA levels. Our previous study showed that the inhibition of LPS-induced NO production by *N*-monomethyl argentine, one of the L-arginine analogues, can transcriptionally reduce TNF-α and IL-1β at the protein and mRNA levels.[16] The propofol-induced inhibition of NO synthesis might contribute to the suppression of LPS-induced TNF-α and IL-1β production.

In conclusion, this study has shown that LPS induces production of TNF-α, IL-1β, IL-6, and NO. In LPS-activated macrophages, propofol suppresses the biosyntheses of TNF-α, IL-1β, IL-6, and iNOS at the protein and mRNA levels. Thus, the modulating mechanism of propofol on LPS-induced production of TNF-α, IL-1β, IL-6, and NO occurs at a pretranslational level. Taken together, the results of this study show that at the therapeutic concentration, propofol has anti-inflammatory and antioxidative effects on the biosyntheses of TNF-α, IL-1β, IL-6, and NO in LPS-activated macrophages.

ACKNOWLEDGMENTS

The authors thank Ms.Wan-Lu Lee for technical support and data collection for the experiment. This study was supported by Grant NSC92-2314-B-038-010 from the National Science Council, Taiwan, ROC.

REFERENCES

1. LYNN, W.A. & J. COHEN. 1995. Management of septic shock. J. Infect. **30:** 207–212.
2. TABBUTT, S. 2001. Heart failure in pediatric septic shock: utilizing inotropic support. Crit. Care Med. **29:** S231–S236.
3. RAETZ, C.R., R.J. ULEVITCH & S.D. WRIGHT. 1991. Gram-negative endotoxin: an extraordinary lipid with profound effects on eukaryotic signal transduction. FASEB J. **5:** 2652–2660.
4. MOLLOY, R.G., J.A. MANNICK & M.L. RODRICK. 1993. Cytokines, sepsis and immunomodulation. Br. J. Surg. **80:** 289–297.
5. WEST, M.A., M.H. LI, S.C. SEATTER, et al. 1994. Pre-exposure to hypoxia or septic stimuli differentially regulates endotoxin release of tumor necrosis factor, interleukin-6, interleukin-1, prostaglandin E2, nitric oxide, and superoxide by macrophages. J. Trauma **37:** 82–89.
6. STOUT, R.D. 1993. Macrophage activation by T cells: cognate and non-cognate signals. Curr. Opin. Immunol. **5:** 398–403.
7. WEST, M.A., S.C. SEATTER, J. BELLINGHAM, et al. 1995. Mechanisms of reprogrammed macrophage endotoxin signal transduction after lipopolysaccharide pretreatment. Surgery **118:** 220–228.
8. SEBEL, P.S. & J.D. LOWDON. 1989. Propofol: a new intravenous anesthetic. Anesthesiology **71:** 260–277.

9. BRYSON, H.M., B.R. FULTON & D. FAULDS. 1995. Propofol. An update of its use in anaesthesia and conscious sedation. Drugs **50:** 513–559.
10. DAVIDSON, J.A., S.J. BOOM, F.J. PEARSALL, et al. 1995. Comparison of the effects of four i.v. anaesthetic agents on polymorphonuclear leucocyte function. Br. J. Anaesth. **74:** 315–318.
11. HOFBAUER, R., M. FRASS, H. SALFINGER, et al. 1999. Propofol reduces the migration of human leukocytes through endothelial cell monolayers. Crit. Care Med. **27:** 1843–1847.
12. CARSWELL, E.A., L.J. OLD & R.L. KASSEL. 1975. An endotoxin-induced serum factor that causes necrosis of tumors. Proc. Natl. Acad. Sci. USA **72:** 3666–3670.
13. CHEN, R.M., C.H. WU, H.C. CHANG, et al. 2003. Propofol suppresses macrophage functions and modulates mitochondrial membrane potential and cellular adenosine triphosphate levels. Anesthesiology **98:** 1178–1185.
14. CHANG, H., S.Y. TSAI, T.L. CHEN, et al. 2002. Therapeutic concentrations of propofol protect mouse macrophages from nitric oxide-induced cell death and apoptosis. Can. J. Anesth. **49:** 477–480.
15. GEPTS, E., F. CAMU & I.D. COCKSHOTT. 1987. Disposition of propofol administered as constant rate intravenous infusions in humans. Anesth. Analg. **66:** 1256–1263.
16. WU, C.H., T.L. CHEN, T.G. CHEN, et al. 2002. Nitric oxide modulates pro- and anti-inflammatory cytokines in lipopolysaccharide-activated macrophages. J. Trauma **55:** 540–545.
17. BEUTLER, B. & A. CERAMI. 1986. Cachectin and tumor necrosis factor as two sides of the same biological coin. Nature **320:** 584–588.
18. CHEN, R.M., H.C. LIU, Y.L. LIN, et al. 2002. Nitric oxide induces osteoblast apoptosis through the de novo synthesis of Bax protein. J. Orthop. Res. **20:** 295–302.
19. HOSHINO, K., O. TAKEUCHI, T. KAWAI, et al. 1999. Toll-like receptor 4 (TLR 4)-deficient mice are hyporesponsive to lipopolysaccharide: evidence for TLR 4 as the *lps* gene product. J. Immunol. **162:** 3749–3752.
20. MEDZHITOV, R., P. PRESTON-HURLBURT, E. KOPP, et al. 1993. MyD88 is an adaptor protein in the hToll/IL-1 receptor family signaling pathways. Mol. Cell **2:** 253–258.
21. KIRKLAND, T.N., F. FINLEY, D. LETURCO, et al. 1993. Analysis of lipopolysaccharide binding by CD14. J. Biol. Chem. **268:** 24818–24824.
22. MILLAR, C.G. & C. THIEMERMANN. 1997. Intrarenal haemodynamics and renal dysfunction in endotoxaemia: effects of nitric oxide synthase inhibition. Br. J. Pharmacol. **121:** 1824–1830.
23. BYSANI, G.K., J.L. SHENEP, W.K. HILDNER, et al. 1990. Detoxification of plasma containing lipopolysaccharide by adsorption. Crit. Care Med. **18:** 67–71.

Protective Effect of α-Keto-β-Methyl-n-Valeric Acid on BV-2 Microglia under Hypoxia or Oxidative Stress

HSUEH-MEEI HUANG,[a] HSIO-CHUNG OU, HUAN-LIAN CHEN,[a] ROLIS CHIEN-WEI HOU, AND KEE CHING G. JENG

Department of Education and Research, Taichung Veterans General Hospital, Taichung 40705, Taiwan

ABSTRACT: The α-ketoglutarate dehydrogenase complex (KGDHC) is a mitochondrial enzyme in the TCA cycle. Inhibition of KGDHC activity by α-keto-β-methyl-n-valeric acid (KMV) is associated with neuron death. However, the effect of KMV in microglia is unclear. Therefore, we investigated the effect of KMV on BV-2 microglial cells exposed to hypoxia or oxidative stress. The results showed that KMV (1–20 mM) enhanced the cell viability under hypoxia. KMV dose-dependently reduced ROS and LDH releases from hypoxic BV-2 cells. KMV also reduced ROS production and enhanced the cell viability under H_2O_2 but failed to reduce the SIN-1 and sodium nitroprusside (SNP) toxicity. KMV also reduced caspase-3 and -9 activation under stress. These results suggest that KMV protects BV-2 cells from stress and acts by reducing ROS production through inhibition of KDGHC.

KEYWORDS: hypoxia; oxidative stress; reactive oxygen species; cell viability; KMV; caspase-3; caspase-9

INTRODUCTION

Mitochondrial dysfunction has been implicated in several neurodegenerative diseases.[1–2] The α-ketoglutarate dehydrogenase complex (KGDHC) is a mitochondrial enzyme complex that oxidatively decarboxylates α-ketoglutarate to succinyl-CoA in the tricarboxylic acid (TCA) cycle. Reduction of KGDHC activities occurs in neurodegenerative disorders.[3–6] Moreover, diminished KGDHC activity also occurs in the fibroblasts of patients with Alzheimer's disease (AD).[7] Thus, diminished KGDHC activity may have important implications in neurodegenerative processes.

Address for correspondence: K.C.G. Jeng, Department of Education and Research, Taichung Veterans General Hospital, Taichung 40705, Taiwan. Voice: +8864-23592525-4038; fax: +8864-2359-2705.

kcjengmr@vghtc.gov.tw

[a]Present address: Dementia Research, Burke Medical Research Institute, White Plains, NY 11605.

α-Keto-β-methyl-n-valeric acid (KMV) is a substrate for KGDHC. KMV inhibits KGDHC activities in a dose- and time-dependent manner as assessed with a histochemical assay.[8–9] Inhibition of KGDHC by KMV promotes translocation of mitochondrial cytochrome c to the cytosol, activates caspase-3, and increases LDH release in PC12 cells.[10] These findings are consistent with the contention that reducing KGDHC activity may lead to mitochondrial dysfunction and neuronal degeneration.

Microglial activation has been shown to contribute to neurodegenerative processes. Whether microglia respond to reduced KGDHC activities in a manner similar to that of neurons has not been previously examined. In the present study, we studied the effects of KMV inhibition of KGDHC in a microglial cell line, BV-2. One of the major consequences of mitochondrial dysfuction is the excessive generation of reactive oxygen species (ROS), which has also been linked to a number of neurodegenerative processes including those caused by cerebral ischemia/reperfusion.[11] We applied two experimental paradigms of oxidative stress, namely, hypoxia and exposure to exogenous oxidants (H_2O_2, 3-morpholinosydnonimine [SIN-1] or sodium nitroprusside [SNP]) to examine the effect of KMV inhibition of KGDHC activity in BV-2 cells.

MATERIALS AND METHODS

Dulbecco's modified Eagle's medium (DMEM; GIBCO, Grand Island, NY), fetal calf serum (FCS; Hyclone, Logan, UT), and antibiotics (Sigma, St. Louis, MO) were purchased from each respective company. α-Keto-β-methyl-n-valeric acid (KMV), 3-morpholinosydnonimine (SIN-1), sodium nitroprusside (SNP), and 3-(4,5-dimethyl-thiazol-2-yl)-2,5-diphenyl tetrazolium bromide (MTT) were purchased from Sigma (St. Louis, MO) and 2′,7′-dichlorodihydrofluorescein diacetate (H_2DCF-DA) from Molecular Probes (Eugene, OR). The LDH diagnostic kit was from Boehringer Mannheim (Mannheim, Germany). Rabbit anti–mouse caspase-3, caspase- 9, and BCL-2 (Calbiochem, San Diego, CA) and streptavidin–horseradish peroxidase-conjugated goat anti–rabbit IgG (Jackson, West Grove, PA) were obtained from the sources indicated.

BV-2 cells were maintained with DMEM and supplemented with 10% FCS, 100 U/mL penicillin, and 100 µg/mL streptomycin, at 37°C in a humidified incubator with 5% CO_2. For the experimental paradigms conducted in the present study, BV2 cells were seeded at density 2×10^4 per well on poly-D-lysine–precoated 96-well plates containing the same medium. Prior to the experiments, BV-2 cell cultures were washed twice with serum-free DMEM (without glucose, phenol red, or L-glutamine). BV-2 cells were then exposed to hypoxia (with glucose deprivation) in a humidified temperature-controlled hypoxia chamber (Bugbox, Ruskinn, England), which was purged with a 85% N_2/10% H_2/ 5% CO_2 atmosphere, at 37°C. KMV was added immediately before hypoxia or exposure to H_2O_2, SIN-1, or SNP. For ROS generation, BV-2 cells on poly-D-lysine–precoated 96-well plate were loaded with DCF-DA (10 µM) for 1 h at 37°C and rinsed with serum-free medium prior to addition of KMV and hypoxia or exposure to oxidants. The DCF fluorescence was read at excitation 485 nm and emission 538 nm by a spectrophotometer (LabSystem, Fluoroskan Ascent, Helsinki, Finland).

MTT Assay

The cell viability was assessed based on the content of metabolized blue formazan from 3-(4,5-dimethyl-thiazol-2-yl)-2,5-diphenyl tetrazolium bromide (MTT) converted by mitochondrial dehydrogenases in live cells.

LDH Assay

Cytotoxicity was measured as the release from damaged cells of the cytosolic enzyme lactic dehydrogenase (LDH), using a LDH diagnostic kit. The absorbance values were read at 490/630 nm on an automated SpectraMAX 340 (Molecular Devices, Sunnyvale, CA) microtiter plate reader.

Western Blotting

Cells were centrifuged and resuspended in lysis buffer and sonicated. Protein extracts were separated on 12% sodium dodecyl sulfate-polyacrylamide gels and transferred to immobilon polyvinylidene difluoride membranes (Millipore, Bedford, MA). The membranes were blocked for non-specific binding and then incubated with rabbit anti-caspases (1:2000) and anti-BCL-2 (1:1000). Subsequently, the membranes were incubated with conjugated antibody. The caspase-3, caspase-9, and BCL-2 bands were detected by chemiluminescence according to the manufacturer's instruction (ECL, Amersham, Berkshire, UK). The band intensity was quantified with a densitometric scanner (PDI, Huntington Station, NY).

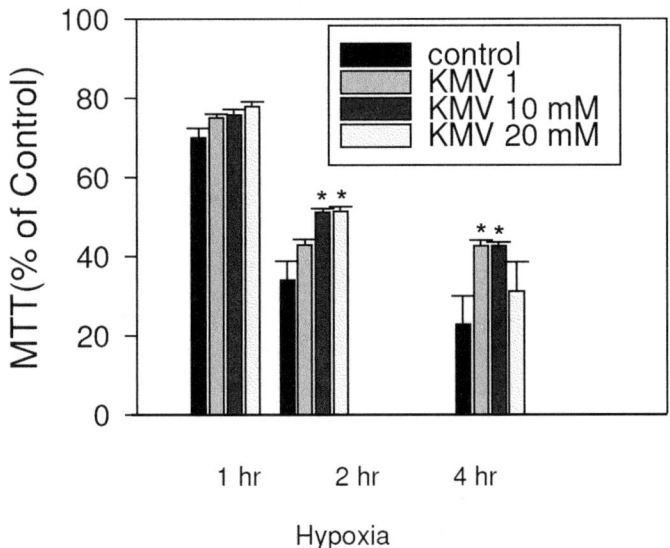

FIGURE 1. Effect of KMV on hypoxia-induced cytotoxicity in BV-2 cells. The cells in glucose-free medium were treated with hypoxia. KMV (10 and 20 mM) enhanced cell viability under hypoxia; *$P < 0.01$.

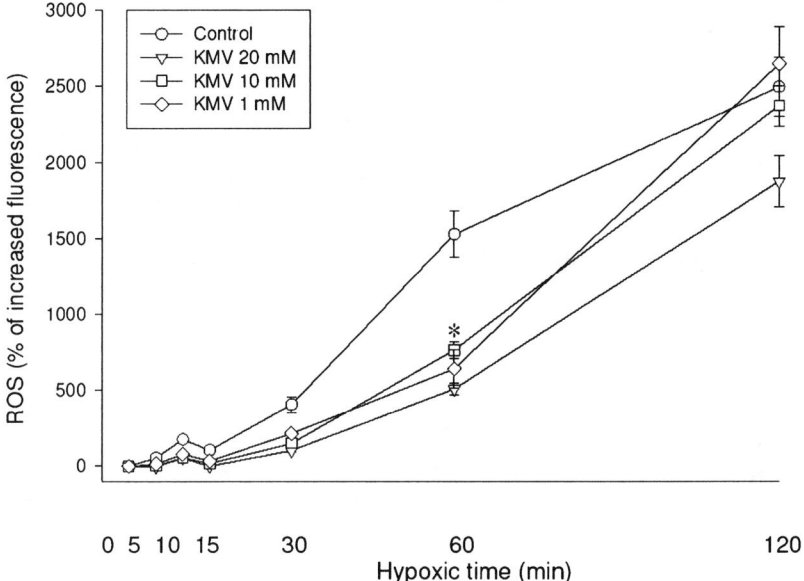

FIGURE 2. KMV reduced hypoxia-induced ROS production in BV-2 cells. The increase levels in DCF-ROS during hypoxia were reduced by KMV significantly; $*P < 0.01$.

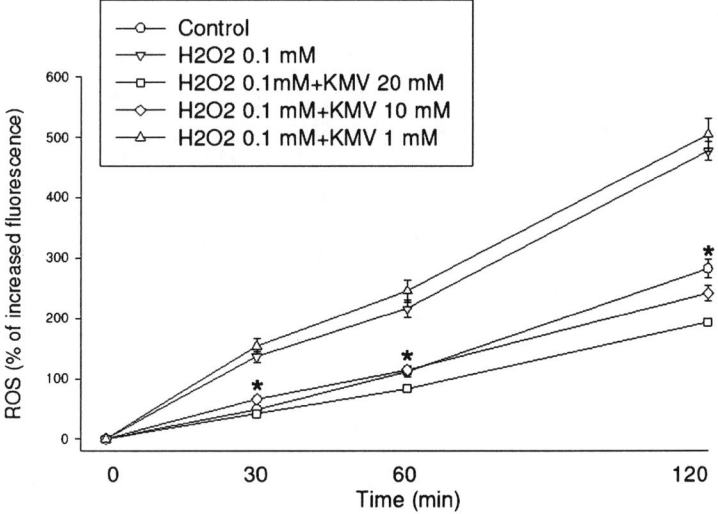

FIGURE 3. KMV reduced oxidative stress–induced ROS production. H_2O_2 (0.1 mM) increased the ROS production in BV-2 cells and KMV reduced ROS; $*P < 0.01$.

RESULTS

KMV at concentrations of 1–20 mM enhanced cell viability (6% and 14%, respectively; FIG. 1). Consistent with this finding, we also noted that KMV reduced the extent of LDH released into the medium under hypoxia by 15–20% (data not shown). Hypoxia-induced ROS generation in BV-2 cells based on the DCF-DA assay was dose-dependently reduced by KMV (FIG. 2). Hypoxia-induced activation of caspases was examined by Western blotting. The results showed that phosphorylated caspase-3 and -9 proteins were activated by stresses, but not BCL-2. KMV significantly ($P < 0.05$) reduced hypoxic-activated caspase-9 and marginally reduced caspase-3, 39% and 9%, respectively, from the hypoxic control (FIG. 4A).

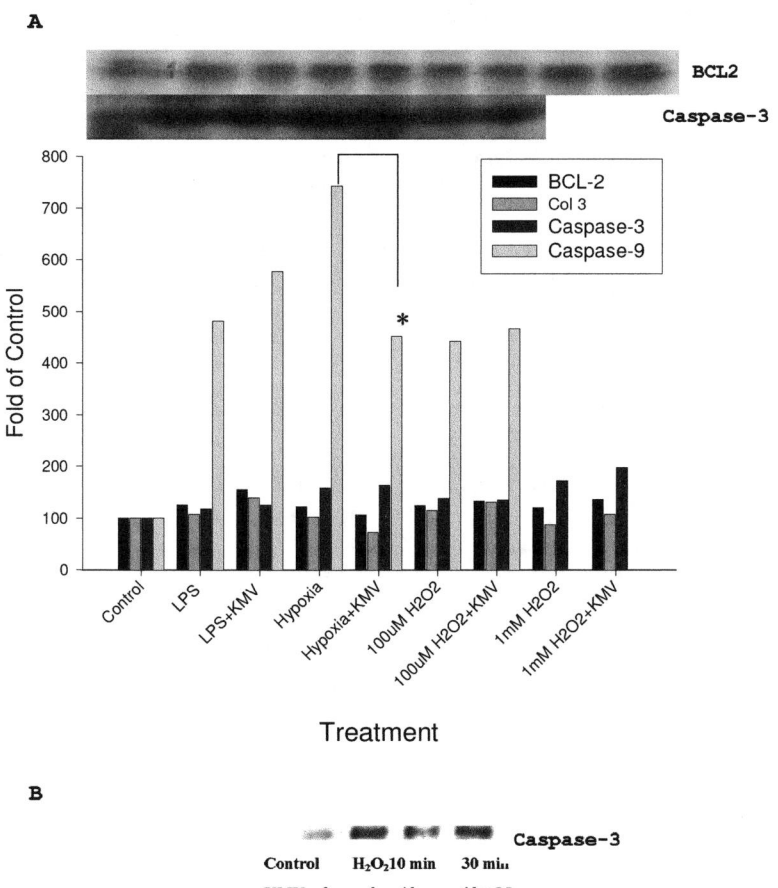

FIGURE 4. Hypoxia-induced apoptosis. Western blot results showed that phosphorylated caspase-3 proteins increased with stress. KMV reduced caspase-3 under (**A**) hypoxia and (**B**) H_2O_2 (0.1 mM); *$P < 0.05$.

H_2O_2 (1 and 10 mM) caused cell death in BV-2 cells to a much greater extent than did SIN-1 (0.100 mM) or SNP (1 mM). At the concentrations tested, SIN-1 and SNP, two donors of nitric oxide (NO), had little cytotoxic effect on BV-2 cells Despite KMV's differential effects on BV-2 cell viability under oxidative stress exerted by H_2O_2, SNP, and SIN-1, it dose-dependently reduced ROS generation, including that caused by H_2O_2 and SIN-1 (FIG. 3). Caspase-3 activation in BV-2 cells secondary to H_2O_2 exposure was reduced by KMV based on Western blot analysis (−39% from the control, FIG. 4B).

DISCUSSION

Mitochondria produce ATP to drive various physiological functions, and they regulate cellular ion homeostasis. In the process, ROS are generated. In response to death signaling, mitochondrial dysfunction may occur, resulting in the release of factors that cause apoptotic cell death.[10–13] KGDHC is an mitochondrial enzyme complex for ATP synthesis via the TCA cycle. A reduction in KGDHC activity may reduce energy output that is needed to maintain cell function and viability. Thus, it is expected that KMV inhibition of KGDHC results in mitochondrial dysfunction characterized by excessive generation of ROS and activation of apoptotic cascade; this has been shown in neuronal cells.[9,10] Thus, KMV exposure may activate cellular events similar to those in neurodegenrative processes. Microglia, a cell type in the central nervous system, have been shown to play an active role in facilitating neurodegenerative processes. The KMV effects on this cell type have not been previously examined. We found that, in contrast to neuronal cells, KMV enhances BV-2 cell viability and ROS generation induced by hypoxia and H_2O_2. This unique action of KMV in BV-2 cells suggests that microglia may have a different signaling process in response to mitochondrial dysfunction secondary to KGDHC inhibition. Results derived from the present study also indicate that the BV-2 cell responded differently under oxidative stress exerted by H_2O_2 versus NO donors, such as SIN-1 or SNP. Further studies are needed to explore the underlying mechanisms that cause these differential effects of KMV.

REFERENCES

1. BEAL, M.F. 1998. Mitochondrial dysfunction in neurodegenerative diseases. Biochim. Biophys. Acta **1366**: 211–223.
2. SCHAPIRA, A.H. 1998. Mitochondrial dysfunction in neurodegenerative disorders. Biochim. Biophys. Acta **1366**: 225–233.
3. GIBSON, G.E., K.F. SHEU & J.P. BLASS. 1998. Abnormalities of mitochondrial enzymes in Alzheimer disease. J. Neural Transm. **105**: 855–870.
4. GIBSON, G.E. et al. 2000. The α-ketoglutarate dehydrogenase complex in neurodegeneration. Neurochem. Int. **36**: 97–112.
5. GIBSON, G.E. et al. 2003. Deficits in a tricarboxylic acid cycle enzyme in brains from patients with Parkinson's disease. Neurochem. Int. **43**: 129–135.
6. MIZUNO, Y. et al. 1995. Role of mitochondria in the etiology and pathogenesis of Parkinson's disease. Biochim. Biophys. Acta **1271**: 265–274.
7. SHEU, K.F. et al. 1994. Abnormality of the α-ketoglutarate dehydrogenase complex in fibroblasts from familial Alzheimer's disease. Ann. Neurol. **35**: 312–318.

8. PARK, L.C. *et al.* 1999. Metabolic impairment induces oxidative stress, compromises inflammatory responses, and inactivates a key mitochondrial enzyme in microglia. J. Neurochem. **72:** 1948–1958.
9. HUANG, H.M. *et al.* 2003. Inhibition of the α-ketoglutarate dehydrogenase complex alters mitochondrial function and cellular calcium regulation. Biochim. Biophys. Acta **1637:** 119–126.
10. HUANG, H.M. *et al.* 2003. Inhibition of alpha-ketoglutarate dehydrogenase complex promotes cytochrome c release from mitochondria, caspase-3 activation, and necrotic cell death. J. Neurosci. Res. **74:** 309–317.
11. SEATON, T.A., J.M. COOPER & A.H. SCHAPIRA. 1997. Free radical scavengers protect dopaminergic cell lines from apoptosis induced by complex I inhibitors. Brain Res. **777:** 110–118.
12. VISWANATH, V. *et al.* 2001. Caspase-9 activation results in downstream caspase-8 activation and bid cleavage in 1-methyl-4-phenyl-1,2,3,6-tetrahydropyridine–induced Parkinson's disease. J. Neurosci. **21:** 9519–9528.
13. JONES, P.A. *et al.* 2004. Apoptosis is not an invariable component of in vitro models of cortical cerebral ischaemia. Cell Res. **14:** 241–250.

Oxidative Toxicity in BV-2 Microglia Cells: Sesamolin Neuroprotection of H_2O_2 Injury Involving Activation of p38 Mitogen-Activated Protein Kinase

ROLIS CHIEN-WEI HOU,[a] CHIA-CHUAN WU,[b] JING-RONG HUANG,[b] YUH-SHUEN CHEN,[c] AND KEE-CHING G. JENG[b]

[a]*Jen-Teh Junior College of Medical and Nursing Management, Miaoli, Taiwan*

[b]*Department of Education and Research, Taichung Veterans General Hospital, Taichung, Taiwan*

[c]*Department of Food Nutrition, Hungkuang University, Taichung, Taiwan*

ABSTRACT: Reactive oxygen species (ROS) has been proposed to play a pathogenic role in neuronal injury. Sesame antioxidants that inhibit lipid peroxidation and regulate cytokine production may suppress ROS generation. In this study, we focused on the effect of sesamolin on H_2O_2-induced neurotoxicity and ROS production in the murine microglial cell line BV-2. Results indicate that the H_2O_2 elicited BV-2 cell death in a concentration- and time-dependent manner. ROS generation in BV-2 cells was time-dependently increased by the H_2O_2 treatment. Sesamolin reduced ROS generation in BV-2 cells. p38 mitogen-activated protein kinase (MAPK) and caspase-3 were also activated in BV-2 cells under H_2O_2 stress. Sesamolin was able to inhibit H_2O_2-induced p38 MAPK and caspase-3 activation and cell death. In addition, sesamolin preserved superoxide dismutase and catalase activities in BV-2 cells under H_2O_2 stress. In conclusion, sesamolin protects microglia against H_2O_2-induced cell injury and this protective effect was accompanied by its inhibition of p38 MAPK and caspase-3 activation and ROS production.

KEYWORDS: sesamolin; H_2O_2; ROS; SOD; p38 MAPK

INTRODUCTION

Microglia play an active role in host defense and repair in the central nervous system. Excessive formation of reactive oxygen species (ROS) causes lipid peroxidation and has been implicated as a major mechanism that causes tissue injury following ischemia.[1] Antioxidants may protect brain cells from ischemic injury mediated by ROS.[2–4]

Address for correspondence: K.C. Jeng, Department of Education and Research, Taichung Veterans General Hospital, Taichung 40705, Taiwan. Fax: +886-4-2359-2705.
kcjengmr@vghtc.gov.tw

ROS are involved in the activation of multiple signaling pathways, including the mitogen-activated protein kinases (MAPKs) cascade that may contribute to cell death. p38 MAPK has been shown to be involved in cell death caused by trauma, cellular stresses, infection, and ischemia.[5]

Sesamolin is one of the most abundant lignans in sesame seeds.[6] Sesame lignans inhibit Δ5-desaturase and protect hepatic and neuronal cells against toxic chemicals and hypoxic insults, respectively.[3] Caspases are a family of specific cysteine proteases that play important roles in the execution of apoptotic death. Caspase-3 is a major effector in the apoptotic cascade induced by a variety of stimuli.[7] In this study, we examined the effect of sesamolin on the H_2O_2-induced injury in BV-2 cells and the associated signaling events.

MATERIALS AND METHODS

Sesamolin was purified from sesame oil as previously described.[3] SB203580 (p38 MAPK inhibitor) was obtained from Biomol (Plymouth Meeting, PA), 3-(4,5-dimethyl-thiazol-2-yl)-2,5-diphenyl tetrazolium bromide (MTT) and 2′,7′-dichlorodihydrofluorescein diacetate (H_2DCF-DA) were purchased from Sigma (St. Louis, MO) and Molecular Probes (Eugene, OR), respectively. LDH diagnostic kit was from Boehringer Mannheim (Mannheim, Germany). Rabbit anti-mouse caspase 3 (Calbiochem, San Diego, CA), mouse anti-human HSP70 (Calbiochem) anti-phospho MAPKs (Promega, Madison, WI) and streptavidin–horseradish peroxidase-conjugated goat anti-rabbit IgG (Jackson ImmunoResearch Laboratories, West Grove, PA) were obtained from each source.

BV-2 cells, from a microglial cell line, were maintained in DMEM supplemented with 10% fetal bovine serum, 100 U/mL penicillin and 100 μg/mL streptomycin at 37°C under 5% CO_2. Cells were pretreated with sesamolin or vehicle before H_2O_2 exposure. Sesamolin was dissolved in dimethylsulfoxide (DMSO, final concentration <0.1%).[4] The vehicle contained the identical concentration of DMSO without sesamolin.

MTT Assay

The cell viability was assessed based on the content of metabolized blue formazan from 3-(4,5-dimethyl-thiazol-2-yl)-2,5-diphenyl tetrazolium bromide (MTT) converted by mitochondrial dehydrogenases in live cells as previously described.[3]

LDH Assay

Cytotoxicity was measured as the release from damaged cells of the cytosolic enzyme lactic dehydrogenase (LDH), using a LDH diagnostic kit. BV-2 cells with or without various concentrations of sesamolin were treated with 0.1 mM H_2O_2 for 1 h, and the supernatant was used for assay of LDH activity. The absorbance values were read at 490/630 nm on an automated SpectraMAX 340 (Molecular Devices, Sunnyvale, CA) microtiter plate reader. Data were expressed as the mean percent of the H_2O_2 control value.[3]

Western Blotting

Cells were centrifuged at 200 × g for 10 min at 4°C. The cell pellets were resuspended in lysis buffer (20 mM Tris-HCl, pH 7.5, 137 mM NaCl, 1 mM phenylmethylsulfonylfluoride, 10 µg/mL aprotinin, 10 µg/mL leupeptin, and 5 µg/mL pepstain A), and sonicated. Protein concentration was determined by Bradford assay (Bio-Rad, Hemel, Hempstead, UK). Cell homogenates containing 50 µg protein were separated on 12% sodium dodecyl sulfate-polyacrylamide gels and transferred to immobilon polyvinylidene difluoride membranes (Millipore, Bedford, MA). The membranes were incubated for 1 h with 5% dry skim milk in TBST buffer (0.1 M Tris-HCl, pH 7.4, 0.9% NaCl, 0.1% Tween-20) to block non-specific binding, then incubated with rabbit anti–caspase-3 (1:2000), and anti-phospho MAPKs (1:1000). Subsequently, the membranes were incubated with secondary antibody streptavidin–horseradish peroxidase-conjugated goat anti–rabbit IgG. Caspase-3 and phosphorylated-MAPK bands were detected by chemiluminescence according to the manufacturer's instruction (ECL, Amersham, Berkshire, UK). The band intensity was quantified with a densitometric scanner (PDI, Huntington Station, NY).

ROS Generation

ROS was determined with the 2′,7′-dichlorodihydrofluorescein diacetate (H_2DCF-DA) assay. The cells were incubated with 10 µM H_2DCF-DA for 60 min at 37°C, followed by exposure to 0.1 mM H_2O_2 for 1 or 2 h and washed with Hanks balanced salt solution. Fluorescence was monitored on a Fluoroskan Ascent fluorometer (Labsystems Oy, Helsinki, Finland) at excitation/emission wavelengths of 485/538 nm.

SOD and Catalase Assays

Catalase activity was assayed by the disappearance of substrate H_2O_2 as measured spectrophotometrically at 240 nm. Total superoxide dismutase (SOD) activity was determined by the inhibition of nitrite formation from hydroxylammonium in the presence of $O_2^{-\cdot}$ generators. One unit of SOD is defined as the amount required for 50% of inhibition from the initial of nitrite formation.[3]

RESULTS

Sesamolin and p38 MAPK inhibitor SB203580 protected BV-2 cells from H_2O_2-injury. After 1 h of 0.1 mM H_2O_2-stress, sesamolin and SB203580 increased BV-2 cell viability by 15–50% based on the MTT assay. Sesamolin dose-dependently reduced cell damage 30–35% as compared to the LDH release of H_2O_2-control ($P < 0.01$; FIG. 1).

Sesamolin diminished DCF-sensitive ROS production in BV-2 cells following H_2O_2 exposure (FIG. 2). The ROS scavenging effect was tested on SOD and catalase activities in BV-2 cells under 0.1 mM H_2O_2-stress for 1 or 2 h. The results showed that sesamolin maintained SOD and catalase activities, with SOD activity preserved to a greater extent (FIG. 3). The effects of sesamolin on MAPKs caspase-3 activation in BV-2 cells were further examined following H_2O_2 exposure. Western blotting

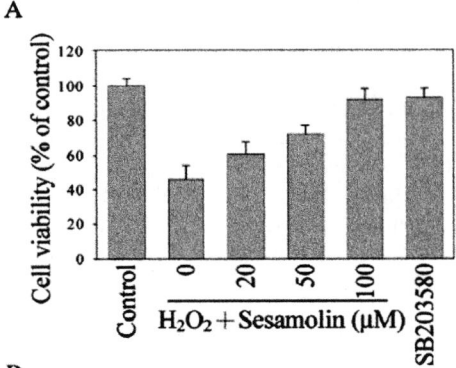

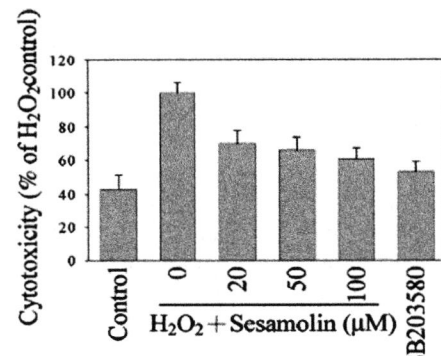

FIGURE 1. Effects of sesamolin on H_2O_2-induced changes in BV-2 cell viability and on the extent of cell death. Sesamolin dose-dependently improved cell viability based on the MMT assay (**A**) and reduced the extent of cell death based on the amount of LDH released (**B**). SB203580 also protected BV-2 cells against H_2O_2-induced death. Data are expressed as mean ± SEM from five separate experiments. * $P \leq 0.01$.

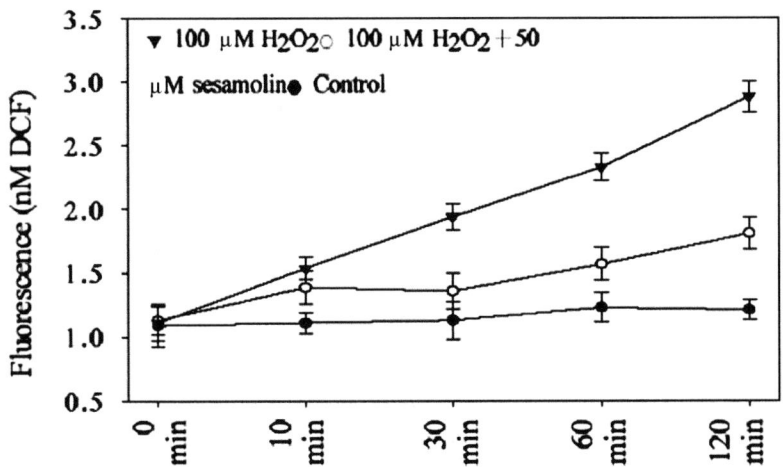

FIGURE 2. Effects of sesamolin on H_2O_2-induced ROS generation. BV-2 cells were treated with 0.1 mM H_2O_2 followed by the addition of 50 µM sesamolin for 1 h.

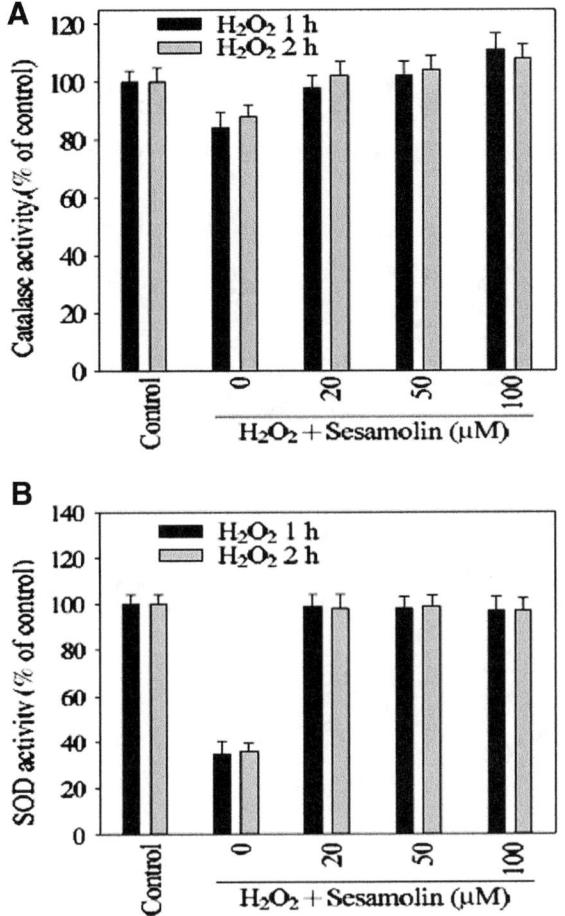

FIGURE 3. Effects of sesamolin on antioxidant enzyme activitiess in H_2O_2-stressed BV-2 microglia. Sesamolin maintained (**A**) catalase and (**B**) SOD activities ($P < 0.01$) in BV-2 cells under H_2O_2-stressed treatment for 1 h and 2 h.

showed that the phosphorylation of MAPKs and caspase-3 increased with time in BV-2 cells under H_2O_2 stress. Sesamolin, at a concentration of 50 μM, (FIG. 4A) and SB203580 at a concentration of 20 μM (FIG. 4B) significantly suppressed H_2O_2-induced expression of phospho-p38 MAPK, and the active form of caspase-3.

DISCUSSION

In the present study, sesamolin dose-dependently protected BV-2 cells against H_2O_2-induced death. The protective mechanism of sesamolin in H_2O_2-induced BV-

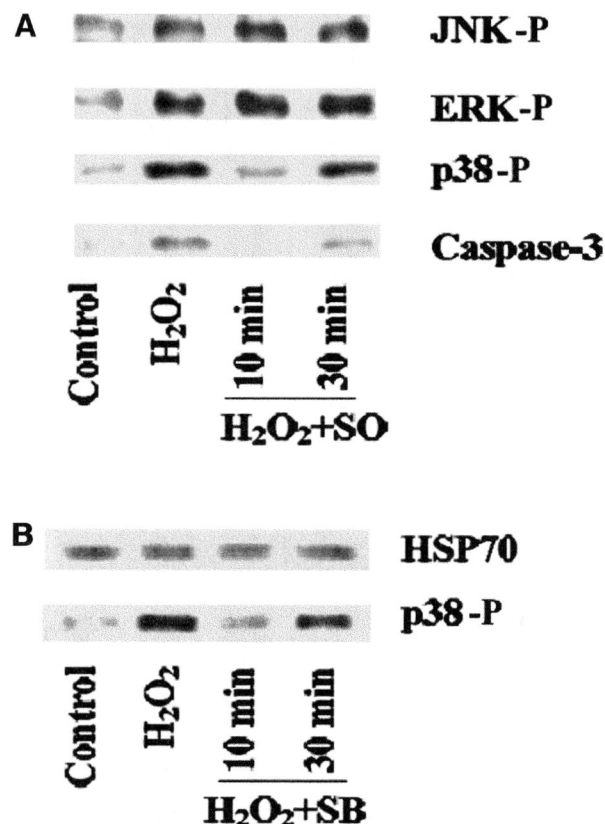

FIGURE 4. Sesamolin inhibited H_2O_2-induced phospho-MAP kinases and caspase-3. Western blots showed that (**A**) sesamolin (SO, 50 μM) or (**B**) SB203580 (SB) decreased the H_2O_2-induced phospho-p38 MAPK (p38-P), and caspase-3. Lane 1, BV-2 control; lane 2, H_2O_2 10 min; lane 3, SO or SB for 10 min, $P < 0.01$; lane 4, H_2O_2 plus SO or SB for 30 min.

2 cell death remained to be established. The findings that seamolin reduced ROS generation and preserved SOD and catalase activities, confirm the antioxidant potency of seamolin.[8–13]

H_2O_2 activated p38 MAPK and the downstream apoptotic cascade was reflected by an increase in the expression of the active form of caspase-3. Results showing the seasmolin was able to inhibit H_2O_2-induced activation of p38 MAPK and caspase-3 suggest that this cascade may be pivotal in H_2O_2-induced BV-2 cell death. The observation that SB203580 (20 mM) blocked p38 MAPK activation and BV-2 cell death is in agreement with the contention that the p38 MAP kinase–caspase-3 cascade[14–17] is involved in the BV-2 cell death signaling processes which can be blocked by seamolin, probably through its antioxidant action.

In summary, H_2O_2-induced BV-2 cell death can be partially prevented by sesamolin. The protection is probably provided by scavenging ROS during H_2O_2 stress. Furthermore, sesamolin protection against H_2O_2-induced BV-2 cell death was accompanied by its inhibition of the activation of p38 MAPK and caspase-3.

REFERENCES

1. PAHLMARK, K., J. FOLBERGROVA, M.L. SMITH & B. K. SIESJO. 1993. Effects of dimethylthiourea on selective neuronal vulnerability in forebrain ischemia in rats. Stroke **24:** 731–737.
2. WANG, M.J., W.W. LIN, H.L. CHEN, et al. 2002. Silymarin protects dopaminergic neurons against lipopolysaccharide-induced neurotoxicity by inhibiting microglia activation. Eur. J. Neurosci. **16:** 2103–2112.
3. HOU, R.C.W., H.M. HUANG, J.T. TZEN & K. C. JENG. 2003. Protective effects of sesamin and sesamolin on hypoxic neuronal and PC12 cells. J. Neurosci. Res. **74:** 123–133.
4. HOU, R.C., H.L. CHEN, J.T. TZEN & K.C. JENG. 2003. Effect of sesame antioxidants on LPS-induced NO production by BV2 microglial cells. Neuroreport **14:** 1815–1819.
5. OBATA, T., G.E. BROWN & M.B. YAFFE. 2000. MAP kinase pathways activated by stress: the p38 MAPK pathway. Crit. Care Med. **28:** N67–N77.
6. FUKUDA, Y., M. NAGATA, T. OSAWA & M. NAMIKI. 1986. Contribution of lignan analogues to antioxidant activity of refined unroasted sesame seed oil. J. Am. Oil Chem. Soc. **63:** 1027–1031.
7. COHEN, G.M. 1997. Caspases: the executioners of apoptosis. Biochem. J. **326:** 1–16.
8. KOPPENOL, W.H., J.J. MORENO, W.A. PRYOOR, et al. 1992. Peroxynitrite, a cloaked oxidant formed by nitric oxide and superoxide. Chem. Res. Toxicol. **5:** 834–842.
9. FUJIMURA, M., Y. MORITA-FUJIMURA, N. NOSHITA, et al. 2000. The cytosolic antioxidant copper/zinc-superoxide dismutase prevents the early release of mitochondrial cytochrome c in ischemic brain after transient focal cerebral ischemia in mice. J. Neurosci. **20:** 2817–2824.
10. CHAN, P. H., M. KAWASE, K. MURAKAMI, et al. 1998. Overexpression of SOD1 in transgenic rats protects vunerable neurons against ischemic damage after global cerebral ischemia and reperfusion. J. Neurosci. **18:** 8292–8299.
11. GHADGE, G.D., J.P. LEE, V.P. BINDOKAS, et al. 1997. Mutant superoxide dismutase-1-linked familial amyotrophic lateral sclerosis: molecular mechanisms of neuronal death and protection. J. Neurosci. **17:** 8756–8766.
12. TROY, C.M. & M.L. SHELANSKI. 1994. Down-regulation of copper/zinc superoxide dismutase causes apoptotic death in PC12 neuronal cells. Proc. Natl. Acad. Sci. USA **91:** 6384–6387.
13. TANAKA, M., A. SOTOMATSU, T. YOSHIDA, et al. 1994. Detection of superoxide production by activated microglia using a sensitive and specific chemiluminescence assay and microglia-mediated PC12h cell death. J. Neurochem. **63:** 266–270.
14. JUNN, E. & M.M. MOURADIAN. 2001. Apoptotic signaling in dopamine-induced cell death: the role of oxidative stress, p38 mitogen-activated protein kinase, cytochrome c and caspases. J. Neurochem. **78:** 374–383.
15. HORSTMANN, S., P.J. KAHLE & G.D. BORASIO. 1998. Inhibitors of p38 mitogen-activated protein kinase promote neuronal survival in vitro. J. Neurosci. Res. **52:** 483–490.
16. BARONE, F.C., E.A. IRVING, A.M. RAY, et al. 2001. Inhibition of p38 mitogen-activated protein kinase provides neuroprotection in cerebral focal ischemia. Med. Res. Rev. **21:** 129–145.
17. SARKER, K.P., M. NAKATA, I. KITAJIMA, et al. 2000. Inhibition of caspase-3 activation by SB 203580, p38 mitogen-activated protein kinase inhibitor in nitric oxide-induced apoptosis of PC-12 cells. J. Mol. Neurosci. **15:** 243–250.

Angiotensin II Stimulates Hypoxia-Inducible Factor 1α Accumulation in Glomerular Mesangial Cells

TSO-HSIAO CHEN,[a,b] JIN-FONG WANG,[c] PAUL CHAN,[a,b] AND HORNG-MO LEE[d,e]

[a]*Department of Internal Medicine, Taipei Medical University—Wan-Fang Hospital, Taipei, Taiwan*

[b]*Graduate Institute of Medical Sciences, Taipei Medical University, Taipei, Taiwan*

[c]*Graduate Institute of Biomedical Technology, Taipei Medical University, Taipei, Taiwan*

[d]*Graduate Institute of Cell and Molecular Biology, Taipei Medical University, Taipei, Taiwan*

[e]*Department of Laboratory Medicine, Taipei Medical University—Wan-Fang Hospital, Taipei, Taiwan*

ABSTRACT: Hypoxia increases hypoxia-inducible factor 1α (HIF-1α) protein levels by inhibiting ubiquitination and degradation of HIF-1α, which regulates the transcription of many genes. Recent studies have revealed that many ligands can stimulate HIF-1α accumulation under nonhypoxic conditions. In this study, we show that angiotensin II (Ang II) increased HIF-1α protein levels in a time- and dose-dependent manner under normoxic conditions. Treatment of mesangial cells with Ang II (100 nM) increased production of reactive oxygen species (ROS). Ang II (100 nM) increased the phosphorylation of PDK-1 and Akt/PKB in glomerular mesangial cells. Ang II–stimulated HIF-1α accumulation was blocked by the phosphatidylinositol 3-kinase (PI-3K) inhibitors, Ly 294001, and wortmannin, suggesting that PI-3K was involved. Because increased ROS generation by Ang II may activate the PI-3K–PKB/Akt signaling pathway, these results suggest that Ang II may stimulate a ROS-dependent activation of the PI-3K–PKB/Akt pathway, which leads to HIF-1α accumulation.

KEYWORDS: angiotensin II; hypoxia-inducible factor 1; phosphatidylinositol 3-kinase; Akt; reactive oxygen species

INTRODUCTION

Angiotensin II (Ang II) plays a central role in the pathophysiology of renal diseases. In addition to its hemodynamic actions, Ang II exerts several nonhemodynamic effects. Ang II causes mesangial cell proliferation and regulates the expression of

Address for correspondence: Horng-Mo Lee, Ph.D., Institute of Cell and Molecular Biology, Taipei Medical University, 250 Wu-Hsing St., Taipei, Taiwan. Voice: +886-2-2736-1661 ext. 3310; fax: +886-2-2732-4510.
 leehorng@tmu.edu.tw

several genes that are involved in intracellular signaling cascades in glomerular mesangial cells.[1] Recently, it has been shown that Ang II is involved in the process of tissue destruction in chronic renal diseases[2] and that angiotensin-converting enzyme inhibitors slow the progress of renal diseases.[3] One of the Ang II–induced pathological effects is mediated through expression of vascular endothelial growth factor (VEGF). Ang II–induced VEGF expression has been attributed to the pathogenesis of diabetic nephropathies.[4] VEGF is involved in angiogenesis, wound healing, and inflammation, and it plays a major role in a chronic cyclosporine nephrotoxicity and diabetes-associated microvasculopathies and glomerulosclerosis.[5,6]

VEGF is transcriptionally regulated by hypoxia-inducible factor 1 (HIF-1).[6] HIF-1 is a heterodimer composed of the basic helix–loop–helix proteins HIF-1α and the aryl hydrocarbon nuclear translocator that is also known as HIF-1β.[7] An active HIF-1 heterodimer binds to the HIF-1 binding site within the hypoxia response element and enhances transcription of hypoxia-inducible genes involved in glucose/energy metabolism, cell proliferation and viability, erythropoiesis, iron metabolism, vascular development, or remodeling. Whereas HIF-1β is found in most cells, HIF-1α is undetectable in normoxic conditions. The availability of HIF-1α is determined predominantly by stability regulation of HIF-1α via proline hydroxylation.[8] HIF-1α is degraded under normoxic conditions by von Hippel–Lindau protein (pVHL). In addition to hypoxia, mitochondrial generation of reactive oxygen species (ROS), including superoxide and H_2O_2, may also cause HIF-1 accumulation and subsequent expression of genes inducible by HIF-1 activation under normoxic conditions.[9]

Several signaling pathways, including phosphatidylinositol 3-kinase (PI-3K), serine/threonine kinases (e.g., protein kinase C), and mitogen-activated protein kinase (MAPK) have been implicated as the signal transduction pathways in the regulation of HIF-1α accumulation in a cell-specific manner.[10–12] In this study, we demonstrated that Ang II stimulated ROS production and activated the PI-3K–PDK-1–PKB/Akt pathway, leading to increased HIF-1α accumulation in glomerular mesangial cells.

MATERIALS AND METHODS

Materials

Dulbecco's modified Eagle medium (DMEM), fetal calf serum (FCS), and other reagents used in cell cultures were purchased from Life Technologies (Gaithersburg, MD); antibodies specific for α-tubulin, p38 MAPK, phospho-PDK-1, and phospho-Akt/protein kinase B (PKB) antibodies from Transduction Laboratory (Lexington, KY); Ly 294002, PD 98059, SB 203580, and wortmannin from Calbiochem (San Diego, CA); horseradish peroxidase–conjugated anti–rabbit immunoglobulin G (IgG) antibody from Bio-Rad (Hercules, CA); 5-bromo-4-chloro-3-indolyl-phosphate/4-nitro blue tetrazolium substrate from Kirkegaard and Perry Laboratories (Gaithersburg, MD); chemiluminescence kits from Amersham (Buckinghamshire, UK); 2′,7′-dichlorofluorescein diacetate (DCFH-DA) from Molecular Probes (Eugene, OR); protease inhibitor cocktail tablets from Boehringer Mannheim (Mannheim, Germany); and Ang II and all other chemicals from Sigma (St. Louis, MO).

Cell Culture and Preparation of Cell Lysates

Simian virus 40–transformed mesangial cells were cultured in Leibovitz's L-15 medium supplemented with 13.1 mM NaHCO$_3$, 13 mM glucose, 2 mM glutamine, 10% heat-inactivated fetal bovine serum, and penicillin (100 U/mL)–streptomycin (100 mg/mL). Cells were plated at a concentration of 10^5 cells/mL and maintained in a humidified incubator under a 5% CO$_2$ atmosphere at 37°C. Cells were used for experiment when they reached approximately 80% confluence. All reagents were added directly to the culture at a volume of 100 µL/10 mL of medium. For preparation of cell lysates, cells were harvested, chilled on ice, and washed three times with ice-cold phosphate-buffered saline (PBS). Later procedures were carried out on ice unless otherwise specified. Cells were lysed by adding lysis buffer containing 10 mM Tris–HCl (pH 7.5), 1 mM EGTA, 1 mM MgCl$_2$, 1 mM sodium orthovanadate, 1 mM dithiothreitol, 0.1% mercaptoethanol, 0.5% Triton X-100, and the protease inhibitor cocktails (final concentrations: 0.2 mM phenylmethylsulfonyl fluoride, 0.1% aprotinin, 50 µg/mL leupeptin). Cell lysates were stored at –70°C for further measurements.

Polyacrylamide Gel Electrophoresis and Western Blotting

Electrophoresis was carried out with sodium dodecyl sulfate–polyacrylamide gel electrophoresis (SDS–PAGE) (7.5%). Following electrophoresis, proteins on the gel were electrotransferred onto a polyvinylidene difluoride (PVDF) membrane. After transfer, the PVDF paper was washed once with PBS and twice with PBS plus 0.1% Tween 20. The PVDF membrane was then blocked with blocking solution containing 3% bovine serum albumin in PBS containing 0.1% Tween 20 for 1 h at room temperature. The PVDF membrane was incubated with a solution containing primary antibodies in the blocking buffer. Finally, the PVDF paper was incubated with peroxidase-linked anti–mouse IgG antibodies for 1 h and then developed using a commercially available chemiluminescence kit.

Measurement of Intracellular ROS Generation

Intracellular ROS generation was assessed using 2',7'-dichlorofluorescein diacetate (DCFH-DA) as described by Chandel et al.[10] In brief, mesangial cells were cultured in petri dishes and incubated with 10 µM DCFH-DA for 30 min. Cells were washed and incubated with Ang II (100 nM) for various periods of time. ROS in the cells caused oxidation of DCFH, generating a fluorescent product (DCF). The intracellular fluorescence of DCF was then measured using flow cytometry. Data were normalized to fluorescence intensities obtained from untreated control subjects.

RESULTS

Ang II Effect on HIF-1α Protein Levels

In glomerular mesangial cells deprived of FCS for 16 h, the protein level of HIF-1α was barely detected under normoxic conditions. Exposure of serum-deprived mesangial cells to Ang II or CoCl$_2$ rapidly increased HIF-1α cellular protein levels. The

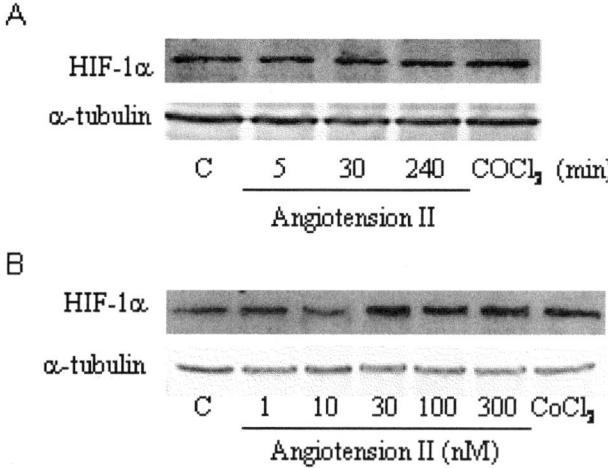

FIGURE 1. HIF-1α protein levels in glomerular mesangial cells. (**A**) Cells were incubated with 100 nM of Ang II for different periods. After incubation, the cells were lysed, and protein levels of HIF-1α were immunodetected with HIF-1α–specific antibody. (**B**) Glomerular mesangial cells were incubated with various concentrations of Ang II for 4 h, and protein levels of HIF-1α were immunodetected with HIF-1α–specific antibody using Western blot analysis. Equal loading in each lane was demonstrated by the similar intensities of α-tubulin.

maximum response was seen 4 h after Ang II exposure (FIG. 1A). Ang II (30–300 nM) increased HIF-1α cellular protein levels in a dose-dependent manner. The maximal effect was seen at a concentration of 100 nM (FIG. 1B).

Ang II Effect on Intracellular ROS Generation

Ang II has been shown to modulate the NAD(P)H oxidase system and increase intracellular ROS levels in several cell types. We next investigated whether Ang II increased the intracellular ROS levels using DCFH-DA. As shown in FIGURE 2, treatment of mesangial cells with Ang II increased the intracellular ROS production. The increase in DCF fluorescence induced by Ang II became evident at 30 min and reached the maximum at 4 h.

Ang II Activation of PI-3K Pathway

We then examined whether Ang II stimulates the PI-3K pathway. Activation of PI-3K may activate and phosphorylate PDK-1, which in turn phosphorylates the downstream PKB/Akt kinase at specific phosphorylation sites on Ser^{473}. As depicted in FIGURE 3, incubation of mesangial cells with Ang II increased PDK-1 and PKB/Akt phosphorylation in a dose-dependent manner. Ang II did not increase the protein level of PI-3K.

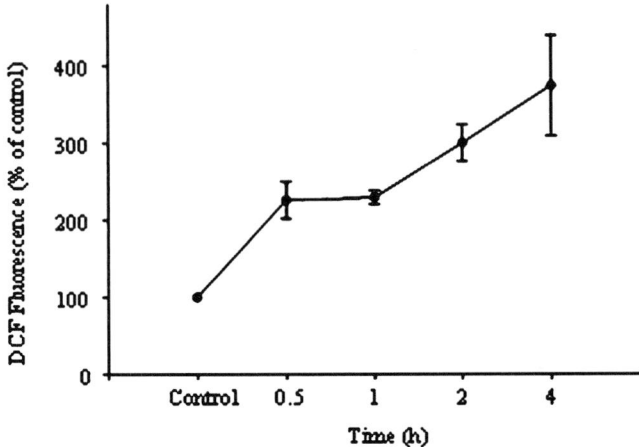

FIGURE 2. Ang II increases intracellular ROS generation in glomerular mesangial cells. Cells were incubated with DCFH-DA (10 µM) for 6 h in the presence of Ang II (100 nM) for various periods. Results are expressed relative to untreated control and are expressed as the mean ± SEM of three independent experiments performed in triplicate. *$P < 0.05$ compared with untreated control.

PI-3K–dependent Stimulation of HIF-1α Accumulation by Ang II

To investigate the link between Ang II–stimulated PI-3K activation and increases in HIF-1α protein levels, we pretreated mesangial cells with specific pharmacological inhibitors of PI-3K, Ly 294002, and wortmannin. As shown in FIGURE 4, Ang II–stimulated PKB/Akt phosphorylation was inhibited with Ly 294002 or wortmannin pretreatment. In accordance with this observation, Ang II–stimulated HIF-1α accumulation was also blocked by Ly 294002 (50 µM), suggesting that PI-3K plays an important role in mediating HIF-1α accumulation.

DISCUSSION

Ang II plays a key role in the regulation of fluid and electrolyte balance and has been implicated in the pathogenesis of renal diseases. HIF-1 controls the expression of several genes under hypoxic conditions and has been linked to Ang II activity in renal diseases. However, the roles of HIF-1 and the cellular signaling mechanisms that regulate the increase of HIF-1α protein level by Ang II in mesangial cells have not been elucidated. In this study, we demonstrate that Ang II increased HIF-1α protein levels in cultured mesangial cells under normoxic conditions. We present data showing that an increase in the HIF-1α protein level was accompanied by an increase in ROS generation and the activation of the PI-3K pathway. Thus, Ang II may bypass hypoxic condition to increase HIF-1α protein levels, which in turn lead to induction of HIF-1α–responsive genes such as VEGF, resulting in several pathological conditions.

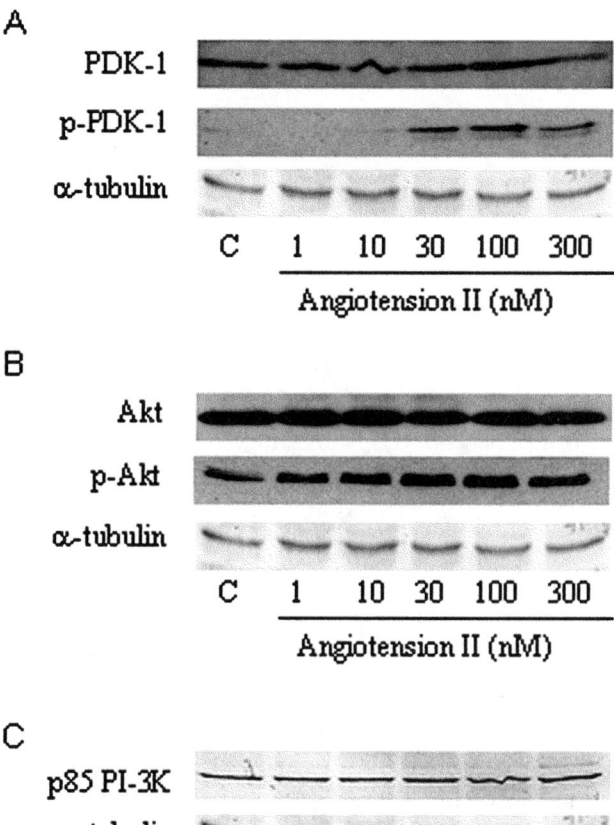

FIGURE 3. Ang II stimulates PI-3K signaling pathway in mesangial cells. (**A**) Cells were incubated with various concentrations of Ang II for 30 min. After incubation, the cells were lysed and then immunodetected with PDK1 and p-PDK1–specific antibody (**A**) or with Akt and p-Akt473–specific antibody (**B**), or p85 PI-3K–specific antibody (**C**). Equal loading in each lane was demonstrated by the similar intensities of α–tubulin.

Free radicals have been shown to play a crucial role in mediating several signaling pathways. We demonstrated that Ang II increased the ROS production in glomerular mesangial cells. In vascular smooth muscle cells, Ang II increases the HIF-1α protein level by a ROS-dependent activation of the PI-3K pathway.[11] In glomerular mesangial cells, increased generation of ROS by Ang II may activate the PI-3K pathway as well. Activation of the PI-3K pathway may modulate the activation of eukaryotic translation initiation factor 4F (eIF-4F) and/or the ribosomal S6 protein by

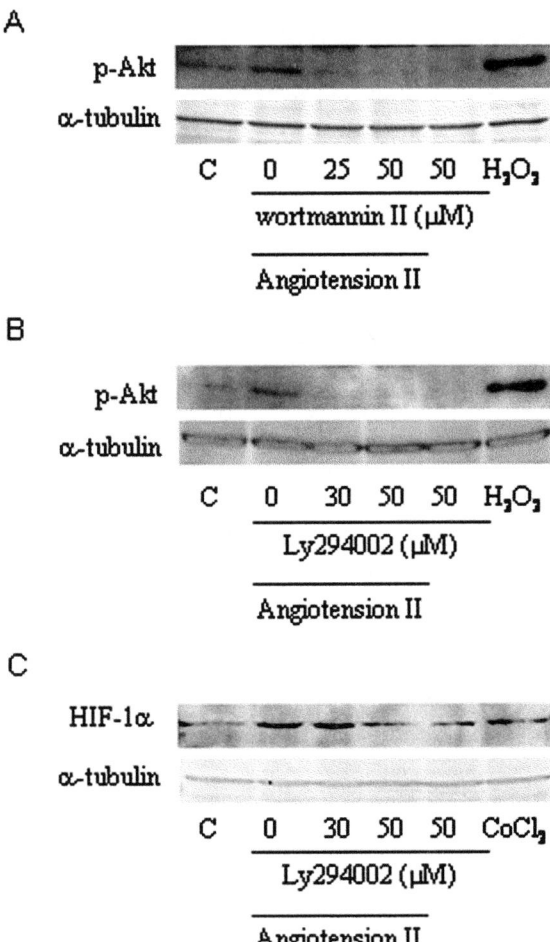

FIGURE 4. Effects of wortmannin on Ang II–stimulated increase in p-Akt473 activity and HIF-1α accumulation in rat mesangial cells. In (**A**) and (**B**), mesangial cells were pretreated with wortmannin (0–50 μM) or Ly 294002 (0–50 μM) for 30 min before the addition of 100 nM Ang II for 30 min. After incubation, the cells were lysed, and the p-Akt473 activity was determined with an immunodetected p-Akt473–specific antibody as described in MATERIALS AND METHODS. Equal loading in each lane was demonstrated by the similar intensities of α-tubulin. In (**C**), mesangial cells were pretreated with Ly 294002 (0–50 μM) for 30 min before the addition of 100 nM Ang II for 4 h. After incubation, the cells were lysed, and the HIF-1α expression was determined with an HIF-1α–specific antibody as described in MATERIALS AND METHODS. Equal loading in each lane was demonstrated by the similar intensities of α-tubulin.

the PI-3K/p70S6K/mTOR pathway and subsequently increase HIF-1α protein level.[11]

Ang II activates many signaling pathways. PI-3K/Akt is a common signaling pathway activated by a variety of ligands and is involved in transcriptional regulation of VEGF in the kidney.[11] We demonstrate that incubation of mesangial cells with specific inhibitors of PI-3K potently blocked the increase in HIF-1α protein levels by Ang II, suggesting that PI-3K is involved in Ang II–induced HIF-1α accumulation. We also show that treatment of mesangial cells with Ang II stimulated the phosphorylation of PDK-1 and PKB/Akt, both downstream effectors in the PI-3K pathway. These data agree with those found in vascular smooth muscle cells showing that the PI-3K pathway can be activated by Ang II.[11,12]

In conclusion, our data suggest that Ang II increases ROS production and activates the PI-3K signaling pathway, leading to an increase in HIF-1α accumulation. Given the role of HIF-1α in controlling the expression of several genes that are involved in the pathophysiology of renal diseases, Ang II may exert its pathogenetic role via the regulation of HIF-1α accumulation.

REFERENCES

1. KIM, S. & H. IWAO. 2000. Molecular and cellular mechanisms of angiotensin II-mediated cardiovascular and renal diseases. Pharmacol. Rev. **52:** 11–34.
2. ZATZ, R. & C.K. FUJIHARA. 2002. Mechanisms of progressive renal disease: role of angiotensin II, cyclooxygenase products and nitric oxide. J. Hypertens. **20:** S37–S44.
3. WOLF, G. 1998. Angiotensin II is involved in the progression of renal disease: implication of non-hemodynamic mechanisms. Nephrologie **19:** 451–456.
4. WILLIAMS, B. 1998. A potential role for angiotensin II-induced vascular endothelial growth factor expression in the pathogenesis of diabetic nephropathy. Miner. Electrolyte Metab. **24:** 400–405.
5. AIELLO, L.P. & J.S. WONG. 2000. Role of vascular endothelial growth factor in diabetic vascular complications. Kidney Int. **77:** S113–S119.
6. SHIHAB, F.S., W.M. BENNETT, J. ISAAC, et al. 2002. Angiotensin II regulation of vascular endothelial growth factor and receptors Flt-1 and KDR/Flk-1 in cyclosporine nephrotoxicity. Kidney Int. **62:** 422–433.
7. RISAU, W. 1997. Mechanisms of angiogenesis. Nature **386:** 671–674.
8. HUANG, L.E., E.A. PETE, M. SCHAU, et al. 2002. Leu-574 of HIF-1α is essential for the von Hippel-Lindau (VHL)-mediated degradation pathway. J. Biol. Chem. **277:** 41750–41755.
9. ENOMOTO, N., N. KOSHIKAWA, M. GASSMANN, et al. 2002. Hypoxic induction of hypoxia-inducible factor-1α and oxygen-regulated gene expression in mitochondrial DNA-depleted HeLa cells. Biochem. Biophys. Res. Commun. **297:** 346–352.
10. CHANDEL, N.S., C.S. MCCLINTOCK, C.E. FELICIANO, et al. 2000. Reactive oxygen species generated at mitochondrial complex III stabilize hypoxia-inducible factor-1α during hypoxia. J. Biol. Chem. **275:** 25130–25138.
11. PAGE, E.L., G.A. ROBITAILLE, J. POUYSSEGUR & D.E. RICHARD. 2002. Induction of hypoxia-inducible factor-1α by transcriptional and translational mechanisms. J. Biol. Chem. **277:** 48403–48409.
12. USHIO-FUKAI, M., R.W. ALEXANDER, M. AKERS, et al. 1999. Reactive oxygen species mediate the activation of Akt/protein kinase B by angiotensin II in vascular smooth muscle cells. J. Biol. Chem. **274:** 22699–22704.

Effects of Glucose and α-Tocopherol on Low-Density Lipoprotein Oxidation and Glycation

CHUN-JEN CHANG,[a] RONG-HONG HSIEH,[b] HUI-FANG WANG,[b] MEI-YUN CHIN,[b] AND SHIH-YI HUANG[b]

[a]*Department of Metabolism and Endocrinology, Wan-Fang Hospital, Taipei, Taiwan*

[b]*School of Nutrition and Health Sciences, Taipei Medical University, Taipei, Taiwan*

ABSTRACT: Glycation of blood proteins is considered to be a major contributor to hyperglycemic complications in diabetes mellitus patients. In this study, we demonstrate the efficacy of α-tocopherol in reducing low-density lipoprotein (LDL) oxidation and glycation *in vitro*. Native LDL isolated from healthy subjects was exposed to various concentrations of glucose and malondialdehyde (MDA) with or without α-tocopherol enrichment for 7 days in sealed vacuum ampoules. The degree of glycation, copper-induced lag time, content of thiobarbituric acid–reactive substances (TBARS), and α-tocopherol levels in LDL were then assessed. LDL lag time was significantly reduced with high levels of glucose and MDA. α-Tocopherol enrichment dramatically inhibited the oxidation of LDL in the lag-time assay. However, the length of incubation time was inversely related to the LDL lag time. Longer incubation time resulted in shorter LDL lag time, with or without α-tocopherol enrichment. The level of TBARS associated with LDL oxidation was highest in native, MDA-supplemented, and high-glucose samples. The α-tocopherol levels were inversely related to glucose levels and incubation times. In conclusion, high-glucose concentrations heightened the oxidative susceptibility of LDL. α-Tocopherol enrichment reduced this trend and prevented LDL from undergoing architectural modification.

KEYWORDS: α-tocopherol; glucose; LDL; oxidation; glycation

INTRODUCTION

The primary therapeutic goal in treating diabetic patients is to maintain an optimal blood glucose level. Recently, therapies of diabetes have been extended to prevention of complications of diabetes, including macrovascular disorders secondary to accelerated atherogenesis in large vessels,[1–3] as well as microvascular disturbances such as foot ulcers and retinopathy.[4] The rationale for placing a high priority on controlling blood glucose levels in diabetic patients is to prevent aberrant and irreversible sugar–protein alterations. Poorly controlled blood glucose levels with enhanced oxidative capability of glucose modification of proteins have been proposed

Address correspondence to: Shih-Yi Huang, School of Nutrition and Health Sciences, Taipei Medical University, 250 Wu-Hsing St., Taipei 110, Taiwan. Voice: +886-2-27361661 ext. 6550; fax: +886-2-27373112.
sihuang@tmu.edu.tw

Ann. N.Y. Acad. Sci. 1042: 294–302 (2005). © 2005 New York Academy of Sciences.
doi: 10.1196/annals.1338.052

to be instrumental in the development of vascular complications in diabetes. These pathological processes may involve lipid peroxidation and irreversible oxidative modifications of proteins (i.e., collagen and low-density lipoprotein [LDL]).[5,6] However, the precise mechanisms that facilitate these complications of diabetes are poorly understood. High blood-glucose levels favor nonenzymatic reactions, resulting in the generation of low-molecular-weight aldehydes such as methylglyoxal and malondialdehyde.[7,8] Proteins' interaction with reducing sugars (i.e., glucose) and lipids in nonenzymatic manners may also generate Amadori products, leading to the formation of advanced glycation end products (AGEs). Amadori products as well as reactive oxygen species (ROS), such as superoxide and hydrogen peroxide,[9–11] are likely to be involved in glycation-induced enhancement of the oxidative modifications of LDL. Nonenzymatic modification of proteins can contribute to the dysfunction of functional proteins such as lipoprotein, collagen, and membrane components.[12,13]

In vitro and *in vivo* studies have shown that hyperglycemia leads to increased oxidative stress and endothelial dysfunction. In addition, many studies suggest that patients with diabetes appear to have a decreased antioxidant capability reflected by lower levels of vitamin E, glutathione, and reduced activities of antioxidant enzymes, such as catalase, superoxide dismutase, or glutathione peroxidase. α-Tocopherol is considered to be a strong antioxidative reagent that may be salutary to slow down progression of diseases that involve oxidative stresses. However, the role of α-tocopherol in preventing LDL oxidation and glycation has not been fully explored. In this study, we sought to demonstrate the efficacy of α-tocopherol in delaying or reducing LDL oxidation caused by high glucose levels under heightened oxidative states.

MATERIALS AND METHODS

Materials

All solutions were prepared with deionized distilled water (Milli Q system; Millipore-Waters, Billerica, MA) prior to use. Glucose, α-tocopherol, EDTA, $CuCl_2$, and malondialdehyde (MDA) were purchased from Sigma (St. Louis, MO). The analytical columns (Poly-Prep Chromatography Column; Bio-Rad, Hercules, CA) for LDL glycation were obtained from Pierce Chemical Company (Rockford, IL). All other chemicals were of analytical grade and all solvents were of high-performance liquid chromatography (HPLC) grade from TEDIA (Fairfield, OH) with the highest purity and laboratory standards.

LDL Preparation

Native LDL

Ten healthy subjects were recruited from clinics at the Municipal Wan-Fang Hospital, Taipei, Taiwan. Blood (30 mL) was drawn and centrifuged to collect native LDL. LDL (density, 1.019–1.063 g/mL) was prepared by density gradient ultracentrifugation as described previously.[14] After isolation, LDL was dialyzed with an appropriate volume (corresponding to 1000-fold dilution of the LDL fraction) of phosphate-buffered saline (PBS; 8 g/L of NaCl, 0.2 g/L of KCl, 1.44 g/L of Na_2HPO_4, and 0.21 g/L of NaH_2PO_4, pH 7.4) for 4 h at 4°C twice. Dialyzed LDL fractions were pooled and dialyzed with 2000-fold volume of PBS for 16 h at 4°C. The redialyzed LDL fractions (native LDL) were pooled for glycation and oxidation assays. The protein content was analyzed using a commercial kit from Bio-Rad.

Vitamin E–Enriched LDL

Partial plasma was fortified with a predetermined volume (10 µL of dimethyl sulfoxide [DMSO]/mL of serum) of a prepared vitamin E solution (250 µmol α-tocopherol acetate/L of DMSO) and incubated at 37°C under shaking, in a germ- and oxygen-free atmosphere for 4 h, as described elsewhere.[15] Preparation of the vitamin E–enriched plasma followed the previously described procedures of LDL preparations.[16] The final LDL fraction was named vitamin E–enriched LDL.

Assays of LDL Glycation and Oxidation

Native and vitamin E–enriched LDL fractions were incubated at 37°C with different levels of glucose and MDA with shaking, in germ- and oxygen-free sealed ampoules for 7 days. Samples were assayed at different time intervals (on days 0, 1, 3, and 7). Assays to determine the degree of glycation, copper-induced lag time, content of thiobarbituric acid–reactive substances (TBARS), and α-tocopherol levels were conducted as reported previously.[17]

Incubated samples were separated into glycated (fraction A) and nonglycated (fraction B) LDL via *m*-aminophenyl boronate affinity chromatography. The partially modified step-by-step procedures followed those described elsewhere.[18] In brief, fractions were measured at 414 nm, and absorbance (optical density [OD]) was calculated to derive the degree of LDL glycation. The degree of LDL glycation (%) was expressed as [OD of fraction A × 2/(OD of fraction A × 2 + OD of fraction B × 5.9)] × 100%.

To measure the oxidative stress of LDL, we determined the level of MDA, a product of lipid peroxidation that reacts with TBARS, based on the MDA–TBARS complex at 532 nm with a spectrophotometer (Hitachi U2000; Hitachi, Tokyo, Japan); the detailed preparation procedures for measuring the MDA–TBA complex are described elsewhere.[17] Quantities of MDA presented in the results were adjusted based on the LDL protein content.

In vitro oxidizability of LDL was evaluated by a spectrophotometric technique described by Esterbauer *et al*.[15] Glucose-treated LDL samples were adjusted to a protein concentration of 200 µg/mL of PBS in the presence of 5 mM of Cu^{2+}. The oxidation kinetics was monitored by the change in absorbance at 234 nm with a Hitachi spectrophotometer (Hitachi U2000) at room temperature for 180 min. The oxidative resistance of LDL was assessed in terms of the period when no oxidation occurred (lag phase) and was calculated as the intercept of extrapolations of the parts of the curve representing the lag and propagation phases. The propagation rate was measured from the slope of the absorbance curve during the propagation phase based on the levels of conjugated dienes.

Vitamin E levels in serum were measured by HPLC (Hitachi L-7100 with an L-7420 UV/VIS detector) with a reversed-phase C_{18} column (Vydac 201TP54 C_{18}; 4.6 × 250 mm; 5 µm) with a 100% methanol mobile phase, 1 mL/min flow rate, and detection at 292 nm.[19]

Statistical Analysis

Data were analyzed using the Statistical Package for Social Sciences, version 9.0 (SPSS, Chicago, IL). Results are expressed as mean ± standard deviation. The difference between the means of samples incubated under different conditions was analyzed statistically using analysis of variance and Fisher's test. Statistical calculations were conducted using the Student's t test for paired observations. A P value less than .05 was considered statistically significant.

RESULTS

The glycation level of LDL increased significantly with 25 and 50 mM glucose on day 7 ($P < 0.05$); with the addition of MDA, the glycation level was significantly elevated with 25 mM and 50 mM glucose on day 3 ($P < 0.05$) and day 7 ($P < 0.01$). The lag time dramatically decreased with high glucose concentrations (25 and 50 mM) and longer incubation times (days 3 and 7) (TABLES 1 and 2). The TBARS formation associated with LDL oxidation was highest in native, MDA-supplemented, and high-glucose level samples. α-Tocopherol levels were inversely correlated with the glucose level (25 mM, $P < 0.01$; 50 mM, $P < 0.001$) and incubation time (day 3, $P < 0.01$; day 7, $P < 0.001$) (TABLE 2). α-Tocopherol enrichment dramatically inhibited the oxidation of LDL as reflected by lag-time prolongation (25 mM, $P < 0.01$; 50 mM, $P < 0.001$). However, the incubation time was inversely related to the LDL lag time, with longer incubation time resulting in shorter LDL lag time with or without α-tocopherol enrichment.

DISCUSSION

In this study, the volunteers were healthy and were not known to be suffering from hyperlipidemia, hyperglycemia, or other metabolic diseases. The normal biochemical characteristics of these blood samples were ascertained based on fasting blood sugar, hemoglobin A_{1c} (HbA_{1c}), total cholesterol, LDL, HDL, and total triglyceride levels that were within reference ranges. Among key blood proteins, LDL appeared to be more susceptible to glucose modification than Hb in streptozotocin-induced hyperglycemic rats in a glucose concentration–dependent response.[20] This finding suggests that the extent of LDL glycation reflects the oxidation status better than HbA_{1c} in a short-term study. In our study, subjects were healthy, with stable HbA_{1c} levels (3.78 ± 0.79%) that were within the reference range; the pooled LDL representing native LDL may have excluded many confounding factors such as difference in glycation level and cholesterol content among subjects. Variation among subjects could have affected the subsequent *in vitro* glycation studies.

TABLE 1. Glycation levels and vitamin E contents of LDL under various levels of glucose incubation[1,2]

Treatment	Glucose conc. (mM)	Glycation (%)				Vitamin E concentration (nmol/mg protein)			
		Day 0	Day 1	Day 3	Day 7	Day 0	Day 1	Day 3	Day 7
Native LDL	0	3.21 ± 0.29	3.54 ± 0.42	3.93 ± 0.26	3.83 ± 0.25	16.7 ± 0.5^a	8.6 ± 0.2^b	7.6 ± 0.2^c	3.8 ± 0.3^d
	5	3.46 ± 0.12	3.60 ± 0.17	4.26 ± 0.48	4.25 ± 0.33	11.5 ± 1.0^{a*}	7.7 ± 0.2^{b*}	6.9 ± 0.3^{c*}	3.4 ± 0.2^{d*}
	25	3.36 ± 0.23^a	3.96 ± 0.62^a	4.00 ± 0.31^a	5.34 ± 0.44^{b*}	9.3 ± 0.6^{a*}	7.1 ± 0.3^{b*}	6.8 ± 0.1^{c*}	2.8 ± 0.1^{d*}
	50	3.91 ± 0.23^a	3.98 ± 0.60^a	4.68 ± 0.12^a	5.47 ± 0.34^{b*}	8.6 ± 0.4^{a*}	6.6 ± 0.1^{b*}	6.2 ± 0.1^{b*}	2.6 ± 0.1^{c*}
Native LDL with MDA	0	3.95 ± 0.19	3.85 ± 0.53	3.98 ± 0.11	3.93 ± 0.26	9.8 ± 0.8^a	6.8 ± 0.2^b	6.8 ± 0.2^b	3.0 ± 0.2^c
	5	3.84 ± 0.37	4.10 ± 0.80	4.44 ± 0.25	4.77 ± 0.60	8.7 ± 0.5^a	6.5 ± 0.2^b	6.2 ± 0.2^{b*}	2.4 ± 0.1^{c*}
	25	3.79 ± 0.69^a	4.34 ± 0.14^a	4.97 ± 0.25^{b*}	7.58 ± 0.27^{c*}	7.7 ± 0.4^{a*}	6.4 ± 0.2^b	6.3 ± 0.1^{b*}	1.7 ± 0.2^{c*}
	50	3.84 ± 0.48^a	4.58 ± 0.16^a	5.08 ± 0.13^a	8.90 ± 0.29^{c*}	6.6 ± 0.3^{a*}	6.3 ± 0.1^b	3.9 ± 0.0^c	1.7 ± 0.1^{d*}
Vit E-enriched LDL	0	3.65 ± 0.22	3.88 ± 0.50	3.80 ± 0.24	3.67 ± 0.19	68.7 ± 0.6^a	47.8 ± 0.4^b	25.6 ± 1.1^c	20.7 ± 0.7^d
	5	3.51 ± 0.19	3.87 ± 0.48	4.03 ± 0.13	4.06 ± 0.17	25.7 ± 1.4^{a*}	20.4 ± 0.4^{b*}	16.9 ± 1.4^{c*}	14.2 ± 0.3^{c*}
	25	3.89 ± 0.13	3.89 ± 0.29	4.18 ± 0.67	4.78 ± 0.16*	22.9 ± 1.2^{a*}	18.4 ± 0.6^{b*}	11.3 ± 0.2^{c*}	10.8 ± 0.5^{c*}
	50	3.69 ± 0.26^a	3.97 ± 0.36^a	4.37 ± 0.50^a	5.13 ± 0.18^{b*}	23.0 ± 0.8^{a*}	18.2 ± 0.3^{b*}	10.1 ± 0.5^{c*}	7.9 ± 0.4^{d*}
Vit E-enriched LDL with MDA	0	3.44 ± 0.91	3.87 ± 0.82	4.07 ± 0.30	4.28 ± 0.18	66.1 ± 0.9^a	36.1 ± 1.6^b	21.3 ± 0.5^b	13.1 ± 0.4^d
	5	3.61 ± 0.08^a	3.85 ± 0.23^a	4.12 ± 0.41^a	4.31 ± 0.19^b	19.5 ± 0.4^{a*}	11.3 ± 0.7^{b*}	10.0 ± 0.6^{b*}	9.5 ± 0.2^{c*}
	25	3.98 ± 0.16^a	4.16 ± 0.26^a	4.54 ± 0.34^a	5.61 ± 0.57^{b*}	16.5 ± 0.2^{a*}	10.0 ± 0.2^{b*}	9.1 ± 0.5^{c*}	8.9 ± 0.8^{c*}
	50	4.15 ± 0.16^a	4.31 ± 0.44^a	5.07 ± 0.28^{b*}	6.04 ± 0.06^{c*}	16.0 ± 0.2^{a*}	8.2 ± 0.4^{b*}	8.0 ± 0.3^{b*}	7.1 ± 0.2^{c*}

[1] Superscripts in rows with different letters indicate a significant difference from the day 0 value ($P < 0.05$).
[2] Superscripts in columns with an asterisk "*" indicate a significant difference from the glucose-free status ($P < 0.05$).

TABLE 2. Oxidative status of LDL under various levels of glucose incubation[1,2]

Treatments	Glucose conc. (mM)	LDL lagtime (min)				TBARs (nmol MDA/mg protein)			
		Day 0	Day 1	Day 3	Day 7	Day 0	Day 1	Day 3	Day 7
Native LDL	0	41.8 ± 4.1	40.6 ± 3.2	41.6 ± 2.4	37.2 ± 4.0	0.84 ± 0.08^a	0.93 ± 0.04^a	2.56 ± 0.19^b	3.00 ± 0.27^c
	5	41.9 ± 2.4^a	40.0 ± 3.3^a	41.0 ± 3.3^a	33.7 ± 2.4^b	0.94 ± 0.09^a	0.98 ± 0.12^a	2.73 ± 0.20^b	6.93 ± 0.83^{c*}
	25	40.1 ± 3.2^a	36.1 ± 1.6^a	38.0 ± 2.4^a	31.3 ± 1.8^b	1.01 ± 0.18^a	1.10 ± 0.19^a	2.76 ± 0.22^b	7.78 ± 0.88^{c*}
	50	41.3 ± 2.4^a	34.6 ± 2.1^b	30.0 ± 2.5^{b*}	28.4 ± 2.0^{c*}	1.08 ± 0.16^a	1.08 ± 0.19^a	3.31 ± 0.21^{b*}	9.11 ± 0.68^{c*}
Native LDL with MDA	0	41.1 ± 1.4^a	39.7 ± 2.8^a	35.8 ± 3.2^a	33.5 ± 3.2^b	0.94 ± 0.18^a	1.25 ± 0.21^a	5.12 ± 0.24^b	9.65 ± 0.65^c
	5	39.7 ± 4.0^a	35.4 ± 2.5^a	34.6 ± 3.1^a	30.3 ± 2.9^b	1.06 ± 0.30^a	1.27 ± 0.25^a	5.36 ± 0.29^b	9.99 ± 0.64^c
	25	38.9 ± 1.9^a	31.2 ± 1.8^{b*}	27.6 ± 3.1^{b*}	25.7 ± 1.6^{b*}	1.19 ± 0.09^a	1.32 ± 0.28^a	5.82 ± 0.12^b	11.84 ± 1.35^{c*}
	50	38.5 ± 2.3^a	28.0 ± 2.5^{b*}	22.8 ± 1.8^{c*}	21.0 ± 2.2^{c*}	1.18 ± 0.21^a	1.42 ± 0.26^a	7.44 ± 0.79^{b*}	11.58 ± 0.61^{c*}
Vit E-enriched LDL	0	58.1 ± 3.9	50.2 ± 2.8	50.2 ± 3.0	49.8 ± 2.7	0.77 ± 0.07^a	0.82 ± 0.11^a	1.83 ± 0.09^b	2.15 ± 0.13^c
	5	55.2 ± 3.4^a	49.8 ± 2.2^a	49.0 ± 1.6^a	47.4 ± 3.2^b	0.78 ± 0.08^a	0.87 ± 0.00^a	1.78 ± 0.07^b	2.28 ± 0.14^c
	25	53.9 ± 3.5^a	48.0 ± 2.8^a	46.1 ± 2.4^b	41.7 ± 2.2^{c*}	1.00 ± 0.12^a	0.89 ± 0.11^a	1.93 ± 0.14^b	2.47 ± 0.25^c
	50	49.8 ± 2.7^a	45.7 ± 2.7^a	43.3 ± 3.1^a	40.6 ± 5.3^{b*}	1.13 ± 0.17^{a*}	1.19 ± 0.17^{a*}	1.97 ± 0.13^b	2.95 ± 0.17^{c*}
Vit E-enriched LDL with MDA	0	57.5 ± 4.3^a	51.4 ± 3.6^a	49.8 ± 3.0^a	44.9 ± 3.6^b	0.84 ± 0.08^a	1.09 ± 0.14^a	1.83 ± 0.29^b	2.89 ± 0.38^c
	5	53.2 ± 2.5	49.0 ± 3.2	47.7 ± 2.6	45.2 ± 4.2	0.79 ± 0.02^a	1.08 ± 0.27^a	1.93 ± 0.29^b	2.48 ± 0.21^c
	25	49.7 ± 3.0	48.7 ± 2.5	44.0 ± 5.1	38.5 ± 1.8^a	0.83 ± 0.14^a	1.13 ± 0.24^a	1.95 ± 0.20^b	3.02 ± 0.26^c
	50	48.8 ± 2.8^{a*}	45.5 ± 2.2^a	43.4 ± 3.6^a	36.3 ± 2.6^{b*}	1.08 ± 0.06^a	1.24 ± 0.19^a	2.18 ± 0.05^b	3.41 ± 0.30^{c*}

[1] Superscripts in rows with different letters indicate a significant difference from the day 0 value ($P < 0.05$).
[2] Superscripts in columns with an asterisk "*" indicate a significant difference from the glucose-free status ($P < 0.05$).

When proteins react with reducing sugars (i.e., glucose) and lipids in a series of nonenzymatic reactions, they generate Amadori products, ultimately leading to the formation of advanced glycation end products (AGEs). Amadori products as well as ROS, such as superoxide and hydrogen peroxide,[9–11,21] are likely to be involved in glycation-induced enhancement of the oxidization modification of LDL. In our study, we applied 5 (90 mg/dL), 25 (450 mg/dL), and 50 (900 mg/dL) mM glucose to simulate normoglycemic and hyperglycemic status and to evaluate glucose concentration–dependent modifications of LDL. Not surprisingly, formation of glycated LDL increased with increasing glucose level and incubation time. The incubation time and the level of glucose are the two major contributing factors to the glycation of heterogeneous compounds. Much has been shown to suggest that the complications of diabetes are associated with oxidative stress induced by the generation of free radicals.[9,11] Glycated LDL, resembling modified LDL, is highly prone to oxidation. Glycation and oxidation processes are closely linked and occur simultaneously in diabetes, accelerating LDL oxidization. Studies have reported that some stable chronic patients such as those with cardiovascular disease and diabetes not only had a shortened oxidative lag phase during copper-catalyzed oxidation of LDL but also showed elevated TBARS levels in the blood.[23,24] Vitamin E exerts antioxidant actions in chronic diseases associated with heightened oxidative stress.[17] In this study, with or without α-tocopherol enrichment, a significant increase in MDA production by LDL and a decrease in the LDL lag time at various glucose levels and incubation times were observed. The results suggest that higher glucose levels and longer incubation times contributed to the irreversible architectural alterations of LDL. Some clinical studies[25,26] showed that diminished vitamin E levels were accompanied by increased LDL susceptibility to chemical modification in type 2 diabetes mellitus patients with or without macrovascular complications. Free radicals (i.e., superoxide) and their byproduct, MDA, another ROS, may both contribute to the initial step of glycation, chemically modifying fat-enriched albumin to accelerate lipid oxidation.[27] Chemical modification of proteins by ROS results in the formation of more susceptive carbonyl residues and MDA, which contribute to enzyme and lipoprotein dysfunction.

In conclusion, high glucose levels heightened the oxidative susceptibility of LDL. α-Tocopherol enrichment reduced the oxidative susceptibility of LDL and reduced LDL architectural modification.

ACKNOWLEDGMENTS

This study was supported by the National Science Council of Taiwan (NSC-89-2320- B-038-056) and a Taipei Medical University–Wan-Fang Hospital Research Grant (93TMU-WFH-13).

We thank Chia-Hui Liao for her skillful assistance.

REFERENCES

1. KANNEL, W.B. & D.L. MCGEE. 1979. Diabetes and cardiovascular disease. The Framingham study. JAMA **241:** 2035–2038.

2. PYORALA, K., M. LAAKSO. & M. UUSITUPA. 1987. Diabetes and atherosclerosis: an epidemiologic view. Diabetes **3**: 463–524.
3. BIERMAN, E.L. 1992. Atherogenesis in diabetes. Arterioscler. Thromb. **12**: 647–656.
4. CHAKRABARTI, S., M. CUKIERNIK, D. HILEETO, et al. 2000. Role of vasoactive factors in the pathogenesis of early changes in diabetic retinopathy. Diabetes Metab. Res. Rev. **16**: 393–407.
5. LYONS, T.J. 1993. Glycation and oxidation: a role in the pathogenesis of atherosclerosis. Am. J. Cardiol. **71**: 26B–31B.
6. BROWNLEE, M. 1996. Advanced glycation endproducts in diabetic complications. Curr. Opin. Endocrinol. Diabetes **3**: 291–297.
7. JAKUŠ, V. & N. RIETBROCK. 2004. Advanced glycation end-products and the progress of diabetic vascular complications. Physiol. Res. **53**: 131–142.
8. LYONS, T. & A.J. JENKINS. 1997. Glycation, oxidation and lipoxidation in the development of the complications of diabetes mellitus: a carbonyl stress hypothesis. Diabetes Rev. **5**: 365–391.
9. OOKAWARA, T., N. KAWAMURA, Y. KITAGAWA & N. TANIGUCHI. 1992. Site-specific and random fragmentation of Cu,Zn-superoxide dismutase by glycation reaction. J. Biol. Chem. **267**: 18505–18510.
10. NAGAI, R., N. KAWAMURA, Y. KITAGAWA & N. TANIGUCHI. 1997. Hydroxyl radical mediates N^ε-(carboxymethyl)lysine formation from Amadori products. Biochem. Biophys. Res. Commun. **234**: 167–172.
11. KAWAMURA, M., N. HEINECKE & A. CHAIT. 1994. Pathophysiological concentrations of glucose promote oxidative modification of low density lipoprotein by a superoxide-dependent pathway. J. Clin. Invest. **94**: 771–778.
12. VLASSARA, H., M. BROWNLEE. & A. CERAMI. 1986. Nonenzymatic glycosylation: role in the pathogenesis of diabetic complications. Clin. Chem. **32**: B37–B41.
13. WITZTUM, J.L. & D. STEINBERG. 1991. Role of ox-LDL in atherogenesis. J. Clin. Invest. **88**: 1785–1792.
14. VIANA, M., C. BARBAS, B. BONET, et al. 1996. In vitro effects of a flavonoid-rich extract on LDL oxidation. Atherosclerosis **123**: 83–91.
15. ESTERBAUER, H., M. DIEBER-ROTHENEDER, G. STRIEGL & G. WAEG. 1991. Role of vitamin E in preventing the oxidation of low-density lipoprotein. Am. J. Clin. Nutr. **53**: 3145–3215.
16. JAIN, S.K. & M. PALMER. 1997. The effect of oxygen radical metabolities and vitamin E on glycosylation of proteins. Free Radic. Biol. Med. **22**: 593–596.
17. JIALAL, I., C.J. FULLER & B.A. HUET. 1995. The effect of α-tocopherol supplementation on LDL oxidation. A dose-response study. Atheroscler. Thromb. Vasc. Biol. **15**: 190–198.
18. MANI, I., J. PATEL & U.V. MANI. 1987. Measurement of serum glycosylated proteins and glycosylated low density lipoprotein fraction in diabetes mellitus. Biochem. Med. Metab. Biol. **37**: 184–189.
19. ARNAUD, J., I. FORTIS, S. BLACHIER, et al. 1991. Simultaneous determination of retinol, α-tocopherol and β-carotene in serum by isocratic high-performance liquid chromatography. J. Chromatogr. **572**: 103–116.
20. YOUNG, I.S., J.J. TORNEY & E.R. TRIMBLE. 1992. The effect of ascorbate supplementation on oxidative stress in the streptozotocin diabetic rat. Free Radic. Biol. Med. **13**: 41–46.
21. OHKUWA, T., Y. SATO & M. NAOI. 1995. Hydroxyl radical formation in diabetic rats induced by streptozotocin. Life Sci. **56**: 1789–1798.
22. HUNT, J.V., M.A. BOTTOMS, K. CLARE, et al. 1994. Glucose oxidation and low-density lipoprotein-induced macrophage ceroid accumulation: possible implications for diabetic atherosclerosis. Biochem. J. **300**: 243–249.
23. MORO, E., P. ALESSANDRINI, C. ZAMBON, et al. 1999. Is glycation of low density lipoproteins in patients with type 2 diabetes mellitus a LDL pre-oxidative condition? Diabet. Med. **16**: 663–669.
24. AVOGARO, P., G. CAZZOLATO & G. BITTOLO-BON. 1991. Some questions concerning a small, more electronegative LDL circulating in human plasma. Atherosclerosis **91**: 163–171.

25. HAIDARI, M., E. JAVADI, M. KADKHODAEE & A. SANATI. 2001. Enhanced susceptibility to oxidation and diminished vitamin E content of LDL from patients with stable coronary artery disease. Clin. Chem. **47:** 1234–1240.
26. PRONAI, L., K. HIRAMATSU, Y. SAIGUSA & H. NAKAZAWA. 1991. Low superoxide scavenging activity associated with enhanced superoxide generation by monocytes from male hypertriglyceridemia with and without diabetes. Atherosclerosis **90:** 39–47.
27. MULLARKEY, C.J., D. EDELSTEIN & M. BROWN. 1990. Free radical generation by early glycation products: a mechanism for accelerated atherogenesis in diabetes. Biochem. Biophys. Res. Commun. **173:** 932–939.

Identification of Three Mutations in the Cu,Zn-Superoxide Dismutase (Cu,Zn-SOD) Gene with Familial Amyotrophic Lateral Sclerosis

Transduction of Human Cu,Zn-SOD into PC12 Cells by HIV-1 TAT Protein Basic Domain

CHIH-MING CHOU,[a] CHANG-JEN HUANG,[b] CHWEN-MING SHIH,[a] YI-PING CHEN,[a] TSANG-PAI LIU,[c] AND CHIEN-TSU CHEN[a]

[a]*Department of Biochemistry, Taipei Medical University, Taipei, Taiwan*

[b]*Institute of Biological Chemistry, Academia Sinica, Taipei, Taiwan*

[c]*Division of General Surgery, Mackay Memorial Hospital, Taipei, Taiwan*

> ABSTRACT: The most frequent genetic causes of amyotrophic lateral sclerosis (ALS) determined so far are mutations occurring in the gene coding for copper/zinc superoxide dismutase (Cu,Zn-SOD). The mechanism may involve the formation of hydroxyl radicals or malfunctioning of the SOD protein. Wild-type SOD1 was constructed into a transcription–translation expression vector to examine the SOD1 production *in vitro*. Wild-type SOD1 was highly expressed in *Escherichia coli*. Active SOD1 was expressed in a metal-dependent manner. To investigate the possible roles of genetic causes of ALS, a human Cu,Zn-SOD gene was fused with a gene fragment encoding the nine amino acid transactivator of transcription (Tat) protein transduction domain (RKKRRQRRR) of human immunodeficiency virus type 1 in a bacterial expression vector to produce a genetic in-frame Tat–SOD1 fusion protein. The expressed and purified Tat–SOD1 fusion proteins in *E. coli* can enter PC12 neural cells to observe the cellular consequences. Denatured Tat–SOD1 was successfully transduced into PC12 cells and retained its activity via protein refolding. Three point mutations, E21K, D90V, and D101G, were cloned by site-directed mutagenesis and showed lower SOD1 activity. In undifferentiated PC12 cells, wild-type Tat–SOD1 could prevent DNA fragmentation due to superoxide anion attacks generated by 35 mM paraquat, whereas mutant Tat–D101G enhanced cell death. Our results demonstrate that exogenous human Cu,Zn-SOD fused with Tat protein can be directly transduced into cells, and the delivered enzymatically active Tat–SOD exhibits a cellular protective function against oxidative stress.
>
> KEYWORDS: amyotrophic lateral sclerosis (ALS); copper/zinc superoxide dismutase (Cu, Zn-SOD); transactivator of transcription (Tat) protein

---

Address correspondence to: Chien-Tsu Chen, Department of Biochemistry, Taipei Medical University, No. 250, Wu-Hsing St., Taipei 110, Taiwan. Voice: +886-2-2736-1661 ext. 2500; fax: +886-2-2738-6527.
 chenctsu@tmu.edu.tw

Ann. N.Y. Acad. Sci. 1042: 303–313 (2005). © 2005 New York Academy of Sciences.
doi: 10.1196/annals.1338.053

INTRODUCTION

Amyotrophic lateral sclerosis (ALS) is a progressive and irreversible disease characterized by degeneration of the motor neurons. This disease leads to weakness of the muscles, causing difficulties with walking, speech, and swallowing. The precise cause of the disease remains unclear. Previously, mutations of the Cu,Zn-superoxide dismutase gene (*SOD1*) were identified by Rosen et al. in a rare subgroup of patients, and most of the patients were found to have a family history of ALS.[1,2] Most ALS cases are sporadic, although more than 10% are familial cases. Most of the familial ALS (FALS) cases have autosomal-dominant traits, and about 25% of the FALS cases are reported to have a mutation in the *SOD1* gene.[1,3] SOD1 catalyzes the dismutation of the superoxide radical to hydrogen peroxide and oxygen to provide cellular defense against oxidative damage.[4] More than 90 mutations, involving 59 of 153 amino acid residues, have been described previously.[5] Identification of these mutations revealed that free radicals may play a critical role in the pathogenesis of the disease.[3] The mechanism of pathogenesis—whether SOD mutation produced motor neuron death because of the loss of SOD enzymatic activity or gain of an adverse function—needs to be further elucidated. In other studies involving the transduction of mutant SOD genes into nonneuronal cells and transgenic mice, the data demonstrated that there is no correlation between SOD activities and the frequency or severity of the disease.[7,8] These data suggest that mutant SOD does not cause FALS due to the lack or decrease of SOD activity.[6–8] In addition, in studies of SOD-knockout mice, the observed viability of motor neurons also supports this contention.[9] Wong et al. recently proposed that the FALS-linked mutant SOD failed to bind or shield copper (Cu^{2+}) as effectively as wild-type enzyme, and this change led to the enhancement of Cu^{2+}-catalyzed oxidative reactions.[10]

Currently, gene therapy has been widely exploited and is considered a promising method to introduce therapeutic proteins into cells.[11] There are also reports indicating that the basic domain of human immunodeficiency virus type 1 (HIV-1) transactivator of transcription (Tat) protein can traverse biological membranes efficiently in a process termed protein transduction.[12–15] Although the mechanism is still unclear, it is widely accepted that the transduction occurs in a receptor- and transporter-independent fashion that appears to target the lipid bilayer directly.[15,16] Furthermore, Tat proteins have recently been shown to serve as carriers to direct uptake of heterologous proteins, including ovalbumin, L-galactosidase, and horseradish peroxidase, into the cells *in vitro* and *in vivo*. These data suggest that HIV-1 Tat proteins have tremendous potential to deliver large compounds into the cells.[17,18]

In this study, we report an effective approach for studying the mechanisms underlying neuronal death induced by the mutant SOD. Our results suggested that the wild-type and mutant Tat–SOD1 fusion protein effectively transduced into differentiated rat pheochromocytoma PC12 cells, and the function of the transduced Tat–SOD1 was found to possess a protective effect against oxidative stress in these cells.

MATERIALS AND METHODS

Materials

Restriction endonucleases and T_4 DNA ligase were purchased from Promega (Madison, WI). Oligonucleotides were synthesized from Gibco BRL (Gaithersburg, MD) custom primers. Isopropyl-L-D-thiogalactopyranoside (IPTG) was obtained from Promega. Plasmid pQE30 and Ni^{2+}–nitrilotriacetic acid (NTA) Sepharose superflow was purchased from Qiagen (Valencia, CA). *Escherichia coli* strain JM109 (DE3) was obtained from Stratagene (La Jolla, CA). Human Cu,Zn-SOD1 cDNA fragment was isolated using the polymerase chain reaction (PCR) technique using the human liver cDNA library. Polyclonal antibodies raised against human Cu,Zn-SOD were produced in our laboratory.

Isolation of the Full-Length Human Cu,Zn-SOD1

Total RNA was isolated from human liver using the RNAzol reagent (Tel-Test, Friendswood, TX) according to the instructions of the manufacturer. To isolate the cDNA covering the complete open reading frame (ORF) of human Cu,Zn-SOD1 according to the sequences of one human expressed sequence tag accession number K00065, Genbank), the full-length human Cu,Zn-SOD1 was amplified with a Cu,Zn-SOD sense primer, CuZn-SOD-F: 5'-ATG GCG ACG AAG GCC GTG TGC GTG CTG-3', and a Cu,Zn-SOD antisense primer, CuZn-SOD-R: 5'- TTA TTG GGC GAT CCC AAT TAC ACC ACA-3'. PCR amplification was performed in a 50-mL reaction mixture containing 2 mL of first-strand cDNA, 0.5 mg of primers, 1.5 mM $MgCl_2$, 0.2 mM dNTP, and 2.5 U of HiFi-DNA polymerase (Yeastern Biotech, Taipei, Taiwan). Samples were incubated in a thermal cycler (Hybaid MultiBlock System; Hybaid Limited, Franklin, MA). The cDNA from brain as template was used for PCR amplification using the program of 94°C for 3 min; 40 cycles of 94°C for 30 s, 55°C for 30 s, and 72°C for 60 s; and a final extension step at 72°C for 15 min. All RACE products were ligated into pGEM-T easy vector (Promega) and subjected to sequence analysis.

Construction of Expression Clones

The expression plasmid pQE–SOD1 was constructed by inserting the full-length human Cu,Zn-*SOD1* cDNA into pQE30 at the *Sph*I and *Pst*I sites, which allows generation of the Cu,Zn-SOD1 protein with in-framed His-tag at the N-terminal end.

The pQE-Tat-SOD1 expression vector was constructed to express the basic domain (amino acids 49–57) of HIV-1 Tat as a fusion with Cu,Zn-SOD1 as follows. In brief, two oligonucleotides were synthesized and annealed to generate a double-stranded oligonucleotide encoding nine amino acids from the basic domain of HIV-1 Tat. Their sequences are (*Bam*HI-Tat-F) 5'-GAT CCG GAA GAA GCG GAG ACA GCG ACG AAG ACG-3' and (*Bam*HI-Tat-R) 5'-GAT CCG TCT TCG TCG CTG TCT CCG CTT CTT CCG-3'. The double-stranded oligonucleotide was directly ligated into the *Bam*HI-digested pQE–SOD1 to generate the His–Tat-SOD1 expression plasmid pQE–Tat-SOD1. Similarly, other expression constructs with three point mutations, such as pQE–Tat-SOD1-E21K, pQE–Tat-SOD1-D90V, and pQE–Tat-

SOD1-D101G, respectively, were also generated by using site-directed mutagenesis. The mutated sequences were confirmed by sequence analysis.

Expression and Purification of Recombinant Proteins

Five expression constructs mentioned above, for example, pQE–SOD1, pQE–Tat-SOD1, pQE–Tat-SOD1-E21K, pQE–Tat-SOD1-D90V, and pQE–Tat-SOD1-D101G, were expressed in *E. coli* JM109. A colony of *E. coli* cells was separately inoculated into Luria-Bertani broth in the presence of 100 mg/ml ampicillin, and the culture was grown overnight at 37°C until its optical density at 600 nm reached 1.2. To induce expression of these recombinant proteins, IPTG was added to a final concentration of 1.0 mM, and the incubation was continued for 4 h. All recombinant proteins were collected and purified by Ni^{2+}–NTA Sepharose metal affinity resins according to the manufacturer's instructions (Qiagen). The purity of these proteins was examined by 10% SDS-PAGE.

Transduction of Wild-Type and a Variety of Mutant Forms of Tat–SOD Proteins into PC12 Cells

Rat pheochromocytoma PC12 cells were cultured in Dulbecco's modified Eagle's medium (DMEM), supplemented with 5% fetal bovine serum (FBS; HyClone, Logan, UT), 10% horse serum (HS), penicillin G (100 U/mL), streptomycin (100 µg/mL), and L-glutamine (2 mM) in a humidified atmosphere of 5% CO_2 at 37°C. Various culture reagents used were purchased from HyClone.

For the transduction of Tat–SOD1 and mutant SODs, PC12 cells were grown to confluence on a six-well plate. The culture medium was then replaced with 1 mL of fresh DMEM without FBS. After PC12 cells were treated with various concentrations of Tat–SOD for 1 h, the cells were harvested with trypsin–EDTA and washed with phosphate-buffered saline (PBS). The cells were then lysed, and cell extracts were subjected to SOD enzyme assay and Western blot analysis. The dismutation activity of SOD in cell extracts was measured with the pyrogallol reaction.[19]

For Western blotting, cell lysates were separated on a 10% SDS-PAGE , and the proteins were then electrophoretically transferred to a polyvinylidene difluoride (PVDF) membrane according to the method of Towbin.[20] The membranes were incubated with 5% skim milk in PBS, and then with a polyclonal anti–human Cu,Zn-SOD antibody (hSOD) for 16 h at 4°C. The PVDF membranes were extensively washed and incubated with a horseradish peroxidase–conjugated goat anti–rabbit immunoglobulin G antibody (diluted 1:3000; Jackson ImmunoResearch Labs, West Grove, PA) for 1 h at room temperature. Immunoreactive bands were visualized with the enhanced chemiluminescence substrate kit (NEN, Boston, MA) according to the manufacturer's protocol. The intracellular stability of transduced Tat–SOD1 and mutant forms of SOD1 was estimated as follows: PC12 cells were treated with 0.5 µg/mL denatured Tat–SOD1 for 0, 1, 2, 4 and 6 h; the cells were washed; and cell extracts for a SOD enzyme assay and Western blot analysis were prepared. The biological activity of transduced Tat–SOD1 and mutant forms of SOD1 was assessed by the cell viability of PC12 cells treated with paraquat (methyl viologen), which is well known as an intracellular superoxide anion generator. Cells (2×10^5) were incubated in six-well plate at 70% confluence and were allowed to attach. The cells were first treated with 1.5 μM denatured Tat–SOD1 and mutant forms of SOD1 for 4 h, respectively, followed by the addition of 35 mM paraquat for 2 h. Cell viability was estimated using the dye exclusion method.[26]

Determination of Superoxide Dismutase Activity

Superoxide dismutase (SOD) activity was measured according to the method of Marklund.[19] Superoxide anion radical is involved in the auto oxidation of pyrogallol. At alkaline pH, SOD dismutates superoxide, thereby inhibiting the auto-oxidation of pyrogallol. The stock pyrogallol solution contained 1.0 mL of 0.5 M HCl and was stored at 30°C. The assay mixture, containing 500 μL of buffer (50 mM Tris–cacodylic acid buffer, pH 8.4, 1 mM diethylenetriamine pentaactic acid, 1 mM potassium phosphate buffer, pH7.0), 50 μL of sample, and 400 μl of deionized water, was put into a cuvette. The control sample contained 500 μL of assay buffer and 450 μL of water. The optical density of each sample was measured at 420 nm before the addition of pyrogallol. The increase in absorbance was measured at 10-s intervals and lasted for 3 min. SOD specific activity was expressed as units per milligram.

RESULTS AND DISCUSSION

The Cu,Zn-SOD protein plays an important role to maintain a healthy balance between oxidants and antioxidants and is one of the cell's self-defense systems against oxygen-derived free radicals. Recently, studies revealed that the point mutations of Cu,Zn-SOD have been linked to FALS.[25] Using the transgenic mouse model that expresses a mutant Cu,Zn-SOD, previous work has shown the mice to develop symptoms of ALS.[18,21,22]

To develop a simple but efficient system to express and purify the cell-permeable SOD protein, we constructed the Tat–SOD1 expression vector (pQE–Tat-SOD1), which contains *SOD1* cDNA sequences encoding the human Cu,Zn-SOD, Tat protein transduction domain (Tat 49–57), and His-tagged sequence at the amino terminus (FIG. 1A). We prepared recombinant pQE–Tat with wild-type and mutant SODs as vectors for gene delivery into PC12 cells (FIG. 1B). The three point mutations of E21K, D90V, and D101G were identified from FALS-related patients in Taiwan. The presence of the mutation in the SOD cDNA was confirmed by PCR and DNA sequencing (data not shown). The Tat–wt, Tat–E21K, Tat–D90V, and Tat–D101G Tat fusion proteins were constructed. We also constructed the SOD expression vector (pQE–SOD1) to produce control SOD protein without an HIV-1 Tat protein transduction domain (FIG. 1B).

Transfected bacterial cells induced with IPTG were lysed in PBS buffer. The recombinant proteins were purified by Ni^{2+}–NTA Sepharose affinity chromatography and then subjected to SDS–10% PAGE. FIGURE 2 shows the protein bands visualized by Western blot analysis. The recombinant SOD1 and Tat–SOD1 proteins have an apparent molecular mass of 17 and 18 kDa, respectively (FIG. 2). However, the recombinant fusion proteins migrated to higher-molecular-weight positions than those of the expected sizes on the SDS–PAGE (as shown in FIG. 2), which is consistent with the previous reports.[23,24]

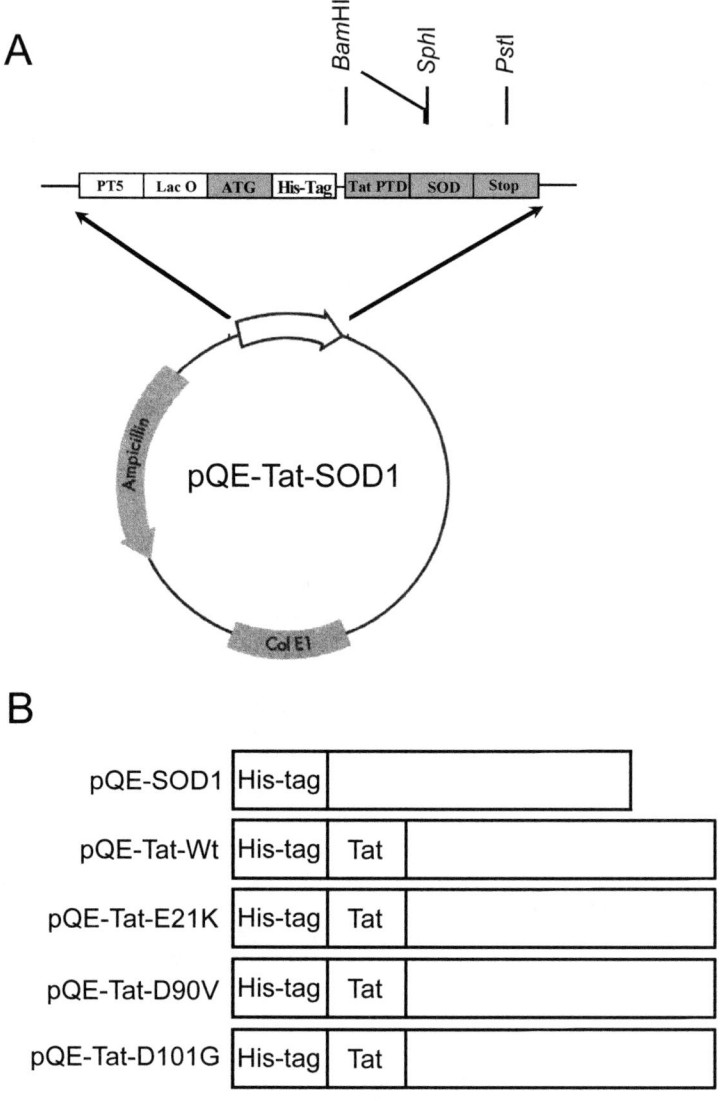

FIGURE 1. Contrustion of SOD1 and Tat–SOD1 expression system (pQE–Tat-SOD1) based in the vector pQE30. The synthetic Tat oligomer was cloned into the *Bam*HI site, and human Cu,Zn *SOD* cDNA was cloned into *Sph*I and *Pst*I sites of pQE (**A**). Expressed Tat–wt, Tat–E21K, Tat–D90V, Tat–D101G, and control SOD fusion proteins (**B**).

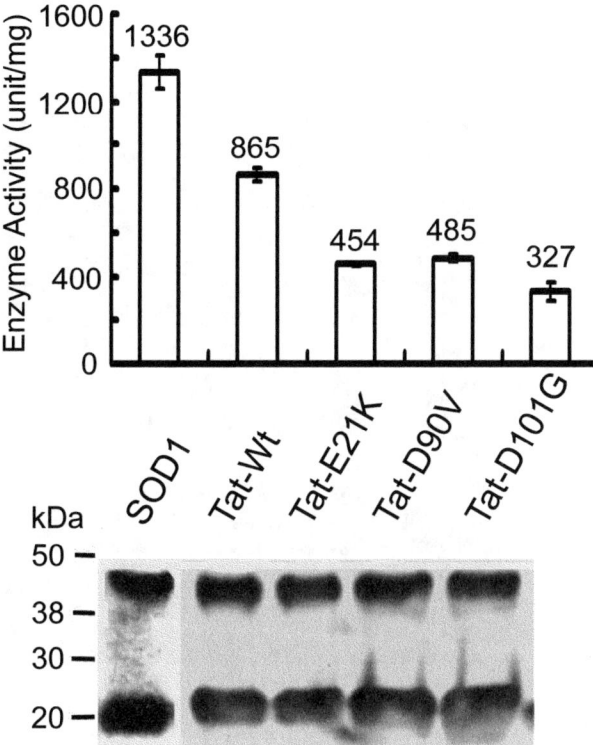

FIGURE 2. Active human wild-type SOD1 and Tat–SOD fusion protein expressed in *E. coli* indicates two forms, 23 kDa (monomer) and 46 kDa (dimer) in Western blot. Rather than expression in a mammalian cell, we constructed the *SOD1* gene into pQE30 vector and transformed it into *E. coli*. Active protein was produced with an activity of 1335 SOD U/mg; the other Tat fusion proteins, Tat–wt, Tat–E21K, Tat–D90V, and Tat–D101G, were produced with an activity of 865, 485, 454, and 327 SOD U/mg, respectively. The purification protein were subjected to 10%SDS–PAGE, transferred to PVDF, and detected with anti–human Cu,Zn-SOD antiserum. The protein molecular mass markers (kDa) are indicated on the left. The result illustrates that two protein forms, 23 and 46 kDa, representing the monomer and dimer, respectively, of SOD1.

The recombinant protein pQE–SOD1 has been produced with a specific activity of 1335 SOD U/mg (FIG. 2), whereas the other Tat fusion proteins, Tat–wt, Tat–E21K, Tat–D90V, and Tat–D101G, produced less specific activity of 865, 485, 454, and 327 SOD U/mg , respectively (FIG. 2).

To further evaluate the transduction ability of Tat–SOD1, 1.5 µM of Tat–SOD1 proteins purified under denaturing conditions was added to the culture media of PC12 cells for 1, 2, 4, and 6 h and then were analyzed by the immunofluorescence staining technique. As shown in FIGURE 3A, the time-dependent manner of transduction indicated that Tat–SOD1 was rapidly transduced into cells and that the levels of transduced proteins in the cultured PC12 cells were increased during the treatments.

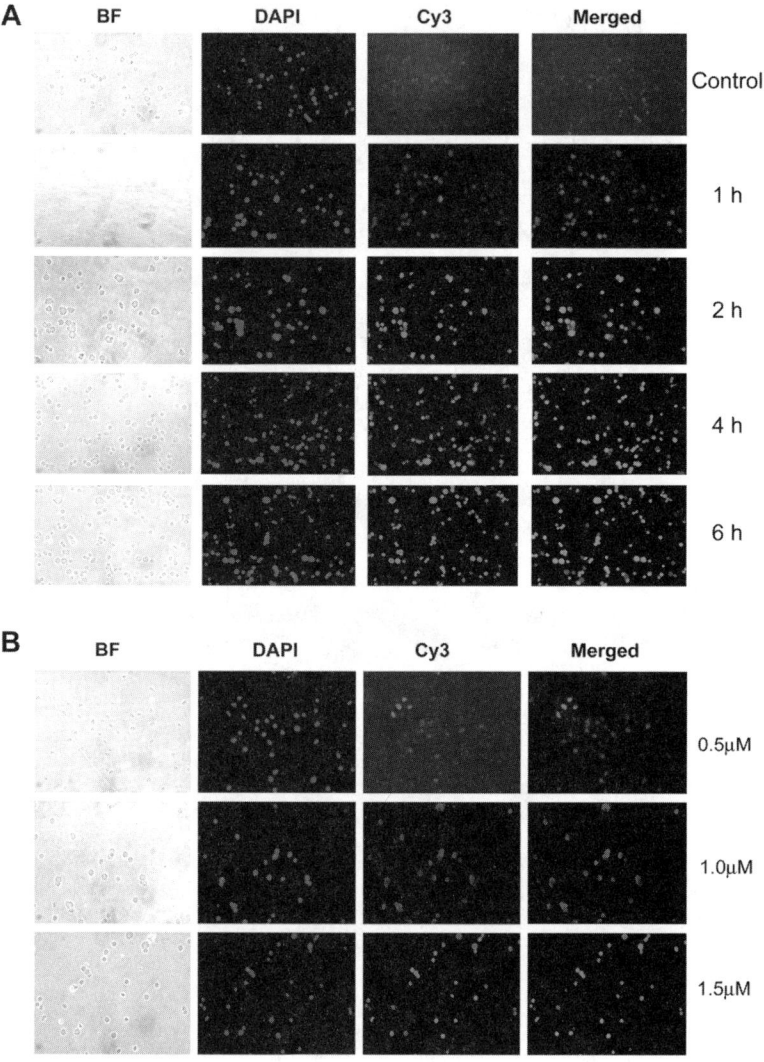

FIGURE 3. Visualization of Tat–SOD transduced into undifferentiated PC12 cells by immunofluorescence assay. PC12 cells plated in a six-well plate were treated with 1.5 μM of Tat–SOD to culture media and incubated for various time intervals (**A**). PC12 cells were treated with Tat–SOD at concentrations of 0.5–1.5 μM for 1 h (**B**). Then, the transduced fusion proteins into the cells were analyzed by immunofluorescence microscopy. DAPI, nuclear staining; Cy3, SOD1 staining.

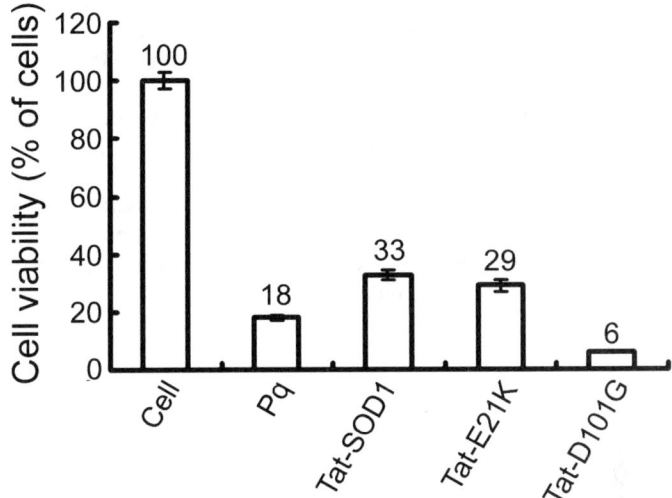

FIGURE 4. Tat–SOD protects PC12 cells from attack of the superoxide anion generated by 35 mM paraquat, but Tat–D101G enhanced cell death. Tat-tagged wild-type and mutant SOD were transduced into PC12 undifferentiated cells as shown. Cell viability experiment was performed in presence of paraquat (35 mM) and revealed that Tat–SOD protects PC12 cells from superoxide anion attack. After the cells were exposed to 35 mM paraquat without Tat–SOD, only 18% of the PC12 cells were viable; the viability was significantly decreased when pretreated with Tat–D101G. The cell viability of PC12 cells pretreated with 1.5 μM Tat–SOD, Tat–E21K, and Tat–D101G, was 33%, 29%, and 6%, respectively.

It was reported that Tat–β-galactosidase fusion protein was transduced rapidly into HepG2 cells, reaching near-maximum intracellular concentrations in less than 15 min.[18] The difference in the transduction period may be due to the properties of transduced Tat fusion protein, such as the degree of unfolding, polarity, and the molecular shape of the protein. The dose dependency of the transduction of denatured Tat–SOD1 fusion proteins was further analyzed. Various concentrations (0.5, 1.0, and 1.5 μM) of denatured Tat–SOD1 proteins were added to the culture media of PC12 cells for 1 h, and the levels of transduced proteins were measured by immunofluorescence staining. As shown in FIGURE 3B, the PC12 cells were treated with 1.5 μM Tat–SOD protein; more than 90% of the cells were transduced in 1 h.

To determine whether transduced Tat–SOD protein could play its biological role in the cells, we tested the Tat–SOD and mutant SODs on cell viability under oxidative stress. After the cells were exposed to 35 mM paraquat without Tat–SOD for 2 h, approximately 18% of the PC12 cells were viable, and the viability was significantly decreased when pretreated with Tat–D101G (FIG. 4). When the cells were pretreated with 1.5 μM Tat–SOD, Tat–E21K, and Tat–D101G, the viability of PC12 cells was 33%, 29%, and 6%, respectively (FIG. 4).

We further analyzed SOD1 activity of three point mutations, E21K, D90V, and D101G, which were cloned via site-directed mutagenesis in our laboratory. All three point mutations were found to exhibit lower SOD1 activity. In undifferentiated PC12 cells, wild-type Tat–SOD1 escaped DNA fragmentation due to superoxide anion attacks generated by 35 mM paraquat; however, mutant Tat–D101G enhanced cell death.

In conclusion, these results demonstrate that exogenous human Cu,Zn-SOD fused with Tat protein can be directly transduced into the cells, and the delivered enzymatically active Tat–SOD fusion exhibits a cellular protective function against oxidative stress. Therefore, this transduction may allow the therapeutic delivery of Cu,Zn-SOD for the various disorders related to this antioxidant enzyme. However, whether mutant Tat-D101G–enhanced cell death may be due to increased production of hydroxyl radicals needs to be elucidated.

ACKNOWLEDGMENTS

We thank Dr. M.T. Lee for critically reading the manuscript and for helpful discussion. This research was supported by grants from the National Science Council (NSC91-2113-M-038-003, NSC92-2113-M-038-005), Taiwan, Republic of China.

REFERENCES

1. ROSEN, D.R., T. SIDDIQUE, D. PATTERSON, et al. 1993. Mutations in Cu/Zn superoxide dismutase gene are associated with familial amyotrophic lateral sclerosis. Nature **362:** 59–62.
2. DE BELLEROCHE, J., R.W. ORRELL & L. VIRGO. 1996. Amyotrophic lateral sclerosis: recent advances in understanding disease mechanisms. J. Neuropathol. Exp. Neurol. **55:** 749–759.
3. DENG, H.X., A. HENTATI, J.A. TAINER, et al. 1993. Amyotrophic lateral sclerosis and structural defects in Cu,Zn superoxide dismutase. Science **261:** 1047–1051.
4. FRIDOVICH, I. 1995. Superoxide radical and superoxide dismutases. Annu. Rev. Biochem. **64:** 97–112.
5. SATO, T., Y. YAMAMOTO, T. NAKANISHIB, et al. 2004. Identification of two novel mutations in the Cu/Zn superoxide dismutase gene with familial amyotrophic lateral sclerosis: mass spectrometric and genomic analyses. J. Neurol. Sci. **218:** 79–83.
6. BORCHELT, D.R., M.K. LEE, H.S. SLUNT, et al. 1994. Superoxide dismutase 1 with mutations linked to familial amyotrophic lateral sclerosis possesses significant activity. Proc. Natl. Acad. Sci. USA **91:** 8292–8296.
7. GURNEY, M.E., H. PU, A.Y. CHIU, et al. 1993. Motor neuron degeneration in mice that express a human Cu,Zn superoxide dismutase mutation. Science **264:** 1772–1775.
8. RIPPS, M.E., G.W. HUNTLEY, P.R. HOF, et al. 1995. Transgenic mice expressing an altered murine superoxide dismutase gene provide an animal model of amyotrophic lateral sclerosis. Proc. Natl. Acad. Sci. USA **92:** 689–693.
9. REAUME, A.G., J.L. ELLIOTT, E.K. HOFFMAN, et al. 1996. Motor neurons in Cu/Zn superoxide dismutase-deficient mice develop normally but exhibit enhanced cell death after axonal injury. Nat. Genet. **13:** 43–47.
10. WONG, P.C., C.A. PARDO, D.R. BORCHELT, et al. 1995. An adverse property of a familial ALS-linked SOD1 mutation causes motor neuron disease characterized by vacuolar degeneration of mitochondria. Neuron **14:** 1105–1116.
11. VERMA, I.M. & N. SOMIA. 1997. Gene therapy—promises, problems and prospects. Nature **389:** 239–242.
12. FRANKEL, A.D. & C.O. PABO. 1988. Cellular uptake of the Tat protein from human immunodeficiency virus. Cell **55:** 1189–1193.

13. GREEN, M. & P.M. LOEWENSTEIN. 1988. Autonomous functional domains of chemically synthesized human immunodeficiency virus tat trans-activator protein. Cell **55:** 1179–1188.
14. MA, M. & A. NATH. 1997. Molecular determinants for cellular uptake of Tat protein of human immunodeficiency virus type 1 in brain cells. J. Virol. **71:** 2495–2499.
15. VIVÈS, E., P. BRODIN & B. LEBLEU. 1997. A truncated HIV-1 Tat protein basic domain rapidly translocates through the plasma membrane and accumulates in the cell nucleus. J. Biol. Chem. **272:** 16010–16017.
16. DEROSSI, D., S. CALVET, A. TREMBLEAU, et al. 1996. Cell internalization of the third helix of the Antennapedia homeodomain is receptor-independent. J. Biol. Chem. **271:** 18188–18193.
17. FAWELL, S., J. SEERY, Y. DAIKH, et al. 1994. Tat-mediated delivery of heterologous proteins into cells. Proc. Natl. Acad. Sci. USA **91:** 664–668.
18. SCHWARTZE, S.R., A. HO, A. VOCERO-AKBANI, et al. 1999. In vivo protein transduction: delivery of a biologically active protein into the mouse. Science **285:** 1569–1572.
19. MARKLUND, S. & G. MARKLUND. 1974. Involvement of the superoxide anion radical in the autoxidation of pyragallol and a convenient assay for superoxide dismutase. Eur. J. Biochem. **47:** 469–474.
20. TOWBIN, H., T. STAEHELIN & J. GORDON. 1979. Electrophoretic transfer of proteins from poly-acrylamide gels to nitrocellular sheet: procedure and some applications. Proc. Natl. Acad. Sci. USA **76:** 4350–4354.
21. YIM, H.S., J.H. KANG, P.B. CHOCK, et al. 1997. A familial amyotrophic lateral sclerosis- associated A4V Cu, Zn-superoxide dismutase mutant has a lower K_m for hydrogen peroxide. Correlation between clinical severity and the K_m value. J. Biol. Chem. **272:** 8861–8863.
22. YIM, M.B., J.H. KANG, H.S. YIM, et al. 1996. A gain-of-function of an amyotrophic lateral sclerosis-associated Cu,Zn-superoxide dismutase mutant: an enhancement of free radical formation due to a decrease in K_m for hydrogen peroxide. Proc. Natl. Acad. Sci. U.S.A. **93:** 5709–5714.
23. CHOI, S.Y., H.Y. KWON, O.B. KWON, et al. 1999. Hydrogen peroxide-mediated Cu,Zn-superoxide dismutase fragmentation: protection by carnosine, homocarnosine and anserine. Biochim. Biophys. Acta **1472:** 651–657.
24. OOKAWARA, T., N. KAWAMURA, Y. KITAGAWA, et al. 1992. Site-specific and random fragmentation of Cu,Zn-superoxide dismutase by glycation reaction. J. Biol. Chem. **267:** 18505–18510.
25. GHADGE, G.D., J.P. LEE, V.P. BINDOKAS, et al. 1997. Mutant superoxide dismutase-1-linked familial amyotrophic lateral sclerosis: molecular mechanisms of neuronal death and protection. J. Neurosci. **17:** 8756–8766.
26. MCATEER, J.A. & J. DAVIS. 1994. Microscopy of the living cell. In Basic Cell Culture: A Practical Approach. J.M. Davis, Ed.: 128–130. IRL Press. Oxford.

Kainic Acid–Induced Oxidative Injury Is Attenuated by Hypoxic Preconditioning

CHENG-HAO WANG,[a] ANYU CHANG,[b] MAY-JYWAN TSAI,[c]
HENRICH CHENG,[c,d,e] LI-PING LIAO,[f] AND ANYA MAAN-YUH LIN[a,b,f]

[a]*Department of Physiology, National Yang-Ming University, Taipei, Taiwan*

[b]*Institute of Biopharmaceutical Sciences, National Yang-Ming University, Taipei, Taiwan*

[c]*Institute of Pharmacology, National Yang-Ming University, Taipei, Taiwan*

[d]*Department of Surgery, National Yang-Ming University, Taipei, Taiwan*

[e]*Neural Regeneration Laboratory, Department of Neurosurgery, Taipei Veterans General Hospital, Taipei, Taiwan*

[f]*Department of Medical Research and Education, Taipei Veterans General Hospital, Taipei, Taiwan*

> ABSTRACT: Female Wistar rats were subjected to 380 mmHg in an altitude chamber for 15 h/day for 28 days. Hypoxic preconditioning attenuated kainic acid (KA)–induced oxidative injury, including KA-elevated lipid peroxidation and neuronal loss in rat hippocampus. Furthermore, KA-induced translocation of cytochrome *c* and apoptosis-inducing factor from mitochondria to cytosol was attenuated in the hypoxic rats. In addition, hypoxic preconditioning attenuated the KA-induced reduction in glutathione content and superoxide dismutase as well as KA-induced increase in glutathione peroxidase. Although local infusion of KA increased hippocampal NF-κB binding activity in the normoxic rat, hypoxia further enhanced KA-elevated NF-κB binding activity. Moreover, hypoxic preconditioning potentiated the KA-induced increase in Bcl-2 level in the lesioned hippocampus. Our data suggest that hypoxic preconditioning exerts its neuroprotection of KA-induced oxidative injury via enhancing NF-κB activation, upregulating the antioxidative defense system, and attenuating the apoptotic process.
>
> KEYWORDS: hypoxic preconditioning; kainate; apoptosis; NF-κB; Bcl-2

INTRODUCTION

Preconditioning stress by a sublethal insult including hypoxia[1,2] and serum deprivation[3,4] is reportedly protective against the deleterious effects of subsequent prolonged lethal injury. For example, *in vitro* studies have demonstrated that

---

Address for correspondence: Anya M.Y. Lin, Ph.D., Department of Medical Research and Education, Veterans General Hospital–Taipei, Taipei, Taiwan. Voice: +886-2-28712121 ext. 2688; fax: +886-2-28751562.
myalin@vghtpe.gov.tw

glutamate-induced neurotoxicity of cerebellar granule cells was attenuated by hypoxic preconditioning.[5] Furthermore, preconditioning with serum deprivation prevented 1-methyl-4-phenyl-pyridinium (MPP$^+$)-induced apoptosis in SH-SY5Y cells.[3] *In vivo* studies showed that hypoxic pretreatment inhibited the development of arrhythmia and myocardial infarction,[6,7] reduced secondary insults during renal ischemia/reperfusion,[1] and attenuated brain damage by systemic administration of kainic acid (KA)[8,9] and intranigral infusion of iron.[2]

Several studies have focused on the mechanisms underlying the preconditioning protection. Preconditioning with serum deprivation or ischemia has been found to increase the activity of transcription factors, including AP-1[4] and NF-κB.[10] Furthermore, preconditioning stress reportedly increased the expression of antioxidant-related proteins, including thioredoxins and NOS1,[4] and suppressed enzymes for free radical generation, that is, microsomal cytochrome P450.[11] These data indicate that preconditioning treatment may create a cellular antioxidative condition.

KA is commonly used as a tool to induce neurotoxicity for studying neurodegenerative diseases of the central nervous system. Systemic administration of KA elicits status epileticus[12,13] and neuronal death in rat brain.[8,12–16] One of the mechanisms underlying KA-induced neurotoxicity is oxidative stress, including free radical[13,14,16–18] and 8-hydroxy-2-deoxyguanosine formation,[12,14] lipid peroxidation, and protein oxidation.[13,19,20] Furthermore, KA-induced neuronal damage was attenuated by antioxidants, including melatonin and estrogen.[18,21] In this study, the effect of hypoxic preconditioning on KA-induced neurotoxicity was investigated *in vivo*. In addition, the mechanisms involved in the hypoxic preconditioning were investigated.

MATERIALS AND METHODS

Adult female Wistar–Kyoto rats were randomly divided into two groups. The hypoxic group was subjected to 380 mmHg in an altitude chamber for 15 h/day for 28 days, and the other was maintained in the normoxic condition. Both normoxic and hypoxic rats were anesthetized with chloral hydrate (450 mg/kg of body weight for normoxic and 360 mg/kg for hypoxic rats). KA (0.2 μg/5 μL of filtered Kreb's Ringer solution) was infused stereotaxically into one hippocampus with the following coordinates: 5.3 mm posterior, 4.8 mm lateral to the bregma zero and 6.8 mm below the cortical surface. The contralateral hippocampus in each animal served as the control.

Measurement of Hematocrit Levels

Blood was withdrawn and hematocrit levels measured using heparinized microhematocrit capillary tubes (Oxford Labware, St. Louis, MO). Capillaries were centrifuged, and hematocrit was calculated as percentage of red blood cell sediments in whole blood.

Fluorescence Assay of Lipid Peroxidation in Hippocampus

Seven days after intrahippocampal infusion of KA, rats were killed by decapitation, and lipid peroxidation in the hippocampus was measured. Hippocampi were homogenized in chilled chloroform and methanol. An aliquot of the chloroform and methanol layer was scanned using a spectrofluorometer (excitation $\lambda = 356$ nm; emission $\lambda = 426$ nm).[22]

Measurement of Glutathione, SOD, Catalase, and GPx in Hippocampus

Aliquots of the hippocampal supernatants were added to 0.6 mM DTNB, 0.2 mg/mL NADPH, and 20 U/mL GSH reductase. GSH was determined by measuring the formation of 5-thio-2-nitrobenzoic acid at 412 nm.[23]

Aliquots of the hippocampal supernatants were added to 0.2 mM xanthine and 0.3 mM epinephrine. Xanthine oxidase was diluted so that an assay mixture without the SOD source yielded 0.03 OD units/min at 320 nm. SOD activity was determined when reaction was started by the addition of xanthine oxidase, and absorbance was measured at 320 nm continuously for 3 min.[24]

Aliquots of the hippocampal supernatants were added to 10 mM H_2O_2 to initiate the assay. The catalase activity was determined from the slope of the absorbance (240 nm) for the 3 min of the reaction.[25]

Aliquots of the hippocampal supernatants were added to 1 mM NaN_3, 1 mM GSH, 0.2 mM NADPH, and 0.24 U of GSH reductase. GPx activity was determined when the reaction was started by adding 100 µL of 205 mM H_2O_2 and was measured at 340 nm for 5 min.[26]

Assessment of CA3 Cell Counts

Rats were perfused transcardially using 0.9% saline followed by fixative consisting of 4% paraformaldehyde in 0.1 M phosphate-buffered saline (PBS). Brains were removed and placed in the 30% sucrose buffer solution overnight and frozen-sectioned coronally at 30 µm using a cryostat. Sections were stained with 1% Cresyl violet. Assessment of KA-induced CA3 cell counts was performed by an observer blinded to treatment using grid morphometric techniques under low- and high-power magnification fields (×40 and ×400 magnification, respectively). Surviving neurons exhibiting normal morphology within the anatomically defined field of interest were included in counts. The viable neurons in the KA-infused hippocampus were normalized to the number of neurons on the contralateral intact side and expressed as percentage of the control.

Western Blot Analysis of Cytochrome c, Bcl-2

Hippocampal tissue was homogenized in mitochondrial isolation buffer (70 mM sucrose, 210 mM mannitol, 5 mM Tris–HCl, 1 mM EDTA, 40 µM fluorocitrate, pH 7.4), and the cytosolic fraction was used for cytochrome *c* immunoblotting. For Bcl-2, hippocampal tissue was homogenized in lysis buffer (1% proteinase K in PBS). Protein samples were run on 12% sodium dodecyl sulfate–polyacrylamide gel electrophoresis (SDS-PAGE) and then transferred onto a nitrocellulose membrane (Bio-Rad, Richmond, CA). Blots were probed with a mouse monoclonal antibody (7H8.2C12; BD Pharmingen, San Diego, CA) against the denatured cytochrome *c* or Bcl-2 antibody (BD Transduction Laboratory, San Diego, CA). Afterward, the membrane was incubated with horseradish peroxidase–conjugated goat anti–mouse immunoglobulin G (Chemicon, Temecula, CA). The immunoreaction was visualized using Amersham enhanced chemiluminescence (Amersham Pharmacia Biotech,

Buckinghamshire, UK). After this detection, the bound primary and secondary antibodies were stripped. The membrane was reprobed with β-actin (Chemicon).

Immunofluorescence for Detecting Apoptosis-Inducing Factor

Brain sections (16 µm) were incubated in 0.3% Triton X-100, blocked with 3% goat serum, and then incubated with mouse anti–apoptosis-inducing factor (AIF; Santa Cruz Biotech, Santa Cruz, CA). The sections were incubated in fluorescein isothiocyanate–conjugated goat anti–mouse immunoglobulin G, (Jackson ImmunoRes., West Grove, PA) and mounted in 50% glycerol and visualized under a fluorescence microscope.

In Situ Labeling of DNA Fragmentation

Apoptotic cells were detected using a terminal deoxytransferase-mediated dUTP nick-end labeling (TUNEL) *in situ* detection kit (Clontech, Palo Alto, CA). Sections were penetrated with proteinase K at room temperature and then were incubated in TUNEL incubation buffer for 1 h at 37°C. After washing, the sections were incubated with propidium iodide (10 ng/mL) and were examined by a fluorescence microscope.

Electrophoretic Mobility Shift Assay (EMSA)

The nuclear protein was extracted with nuclear protein extraction buffer (20 mM HEPES, 25% glycerol, 0.4% NaCl, 1.5 mM $MgCl_2$, 0.2 mM EDTA, proteinase inhibitor, pH 7.9). The binding reaction was performed with a biotin end-labeled NF-κB oligonucleotide (5'-AGT TGA GGG GAC TTT CCC AGG-3') using Lightshift Chemiluminescent EMSA kit (Pierce Chemical Co., Rockford, IL). Biotin-labeled NF-κB oligonucleotide was incubated with the extracted nuclear protein. DNA–protein complexes were resolved with 5% PAGE and then transferred to nylon membrane via electroblotting. The biotin end-labeled DNA was detected using streptavidin–horseradish peroxidase conjugated with a chemiluminescent substrate according to the manufacturer's instructions. For the antibody supershift experiment, reaction of NF-κB p65 antibody with nuclear fraction was performed before adding the probe.

Statistical Analysis

Data were expressed as the mean ± standard error of the mean. Statistical comparisons were made using the Student's *t* test or one-way analysis of variance (ANOVA) followed by a Bonferroni multiple-comparisons procedure as a post-hoc analysis. Differences among means were considered statistically significant when the *P* value was less than .05.

RESULTS AND DISCUSSION

Hypoxic Treatment Increased Hematocrit Level

One of the mechanisms for systemic adaptation to reduced oxygen partial pressure involves elevated hematocrit levels.[27] In this study, hypoxic treatment significantly increased the hematocrit level. The hematocrit of normoxic rats averaged

$48 \pm 7\%$ ($n = 10$), whereas that of hypoxic rats averaged $67 \pm 1\%$ ($n = 10$, $P < 0.05$ by the Student's t test). In our preliminary data, hematocrit was increased 1 week after hypoxic treatment, and elevated levels were maintained throughout the hypoxic preconditioning period. However, neuroprotection by hypoxic preconditioning of iron-induced oxidative injury in the nigrostriatal dopaminergic system[2] or cortical infarction by transient middle cerebral artery occlusion[28] required at least 2 weeks of hypoxic preconditioning. It appears that the hypoxia-induced changes in hematocrit preceded other neuroprotection-related events.

Hypoxic Preconditioning Reduced KA-Induced Oxidative Injury

Intrahippocampal infusion of KA dose-dependently increased lipid peroxidation, a biological marker for free radicals that attack cells in the normoxic rats (FIG. 1A). Furthermore, our previous data showed that local infusion of KA increased the level of 2,3-dihydroxylbenzoic acid, an adduct of salicylate and hydroxyl radicals, indicating that KA can generate hydroxyl radicals.[17,18] At the same time, the number of CA3 neurons in the KA-infused hippocampus was reduced in a concentration-dependent manner (FIG. 2). These data support previous findings that KA may increase the formation of reactive oxygen species,[13,14,16–18] increase oxidative

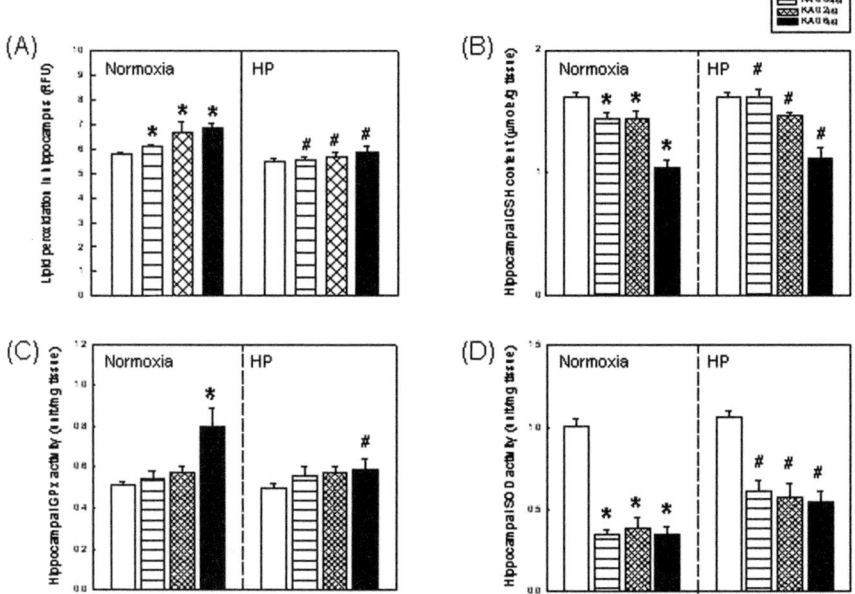

FIGURE 1. Effects of hypoxic pretreatment on lipid peroxidation, GSH content, GPx, and SOD activity in rat hippocampus. Seven days after KA infusion, lipid peroxidation (A; reported as relative fluorescence units [RFU]), GSH (B), GPx (C), and SOD (D) were measured in the hippocampus. Values shown are mean ± SEM. ($n = 7$–9). *$P < 0.05$ in the KA-infused group compared with those in the intact control in the normoxic rats; #$P < 0.05$ in the KA-infused group in hypoxia-preconditioned (HP) rats compared with those in the normoxic rats by a one-way ANOVA followed by a Bonferroni test as a post-hoc analysis.

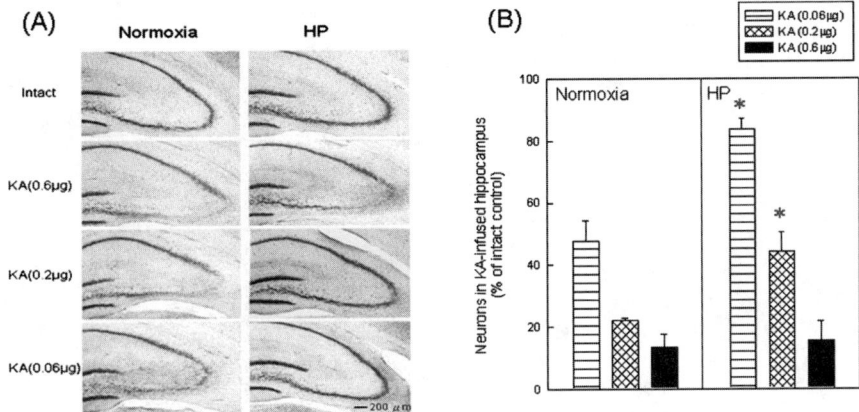

FIGURE 2. KA-induced hippocampal CA3 neuronal death was attenuated by hypoxic preconditioning. Seven days after KA infusion, a dose-dependent loss of CA3 neurons was observed in the KA-infused hippocampus compared with the intact hippocampus in the normoxic rat. However, the KA-induced neuronal loss was attenuated in the infused hippocampus in hypoxia-pretreated rats. *$P < 0.05$ in the hypoxic mice compared with those in the normoxic rats (**A**). Statistical results of protection by hypoxic preconditioning of KA-induced neuronal loss in rat hippocampus (**B**). Values shown are mean ± SEM. ($n = 7$–9).

stress,[12–14,16,17,19] and result in neuronal loss.[8,12–14] Whereas hypoxic preconditioning had no effect on the basal lipid peroxidation and CA3 neuronal survival in the intact hippocampus, hypoxic pretreatment attenuated KA-induced elevation in lipid peroxidation (FIG. 1A) and neuronal loss in the hippocampal CA3 area (FIG. 2). Compared with previous studies that reported that hypoxic treatment reduced KA-induced excess synaptic excitation[8] and brain edema,[9] our study clearly demonstrates that hypoxic preconditioning reduced not only oxidative stress but also neuronal loss by intrahippocampal infusion of KA.

Hypoxic Preconditioning Attenuated KA-Induced Alterations in Antioxidative Defense Systems

Consistent with previous studies,[18,19] GSH content was reduced and GPx was increased in the infused hippocampus of normoxic rats 7 days after KA infusion (FIG. 1). Furthermore, hypoxic preconditioning attenuated KA-induced reduction in GSH content in the infused hippocampus (FIG. 1B). However, hypoxic preconditioning differentially affected antioxidative defense enzymes. Hypoxic treatment attenuated the KA-induced increase in GPx activity (FIG. 1C) and reduction in SOD (FIG. 1D) but did not alter catalase activity (data not shown) in the infused hippocampus. The involvement of antioxidative defense mechanisms in preconditioning-induced protection is still controversial. Elevation in Cu,Zn-SOD and Mn-SOD levels has been reported in intermittent hypoxia or serum deprivation.[4,11,29] However, GPx was not simultaneously increased after intermittent hypoxia.[11]

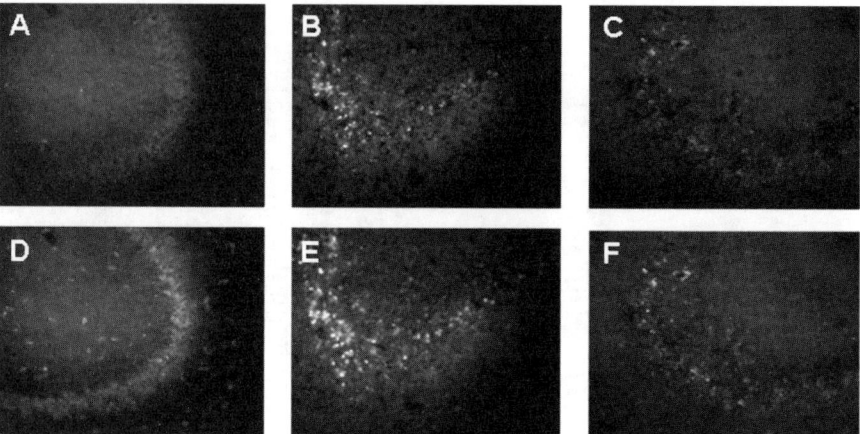

FIGURE 3. Attenuation by hypoxic preconditioning of KA-induced TUNEL-positive cells in the rat hippocampus. Representative photomicrographs from the intact hippocampus (**A**, TUNEL staining; **D**, superimposition of TUNEL and PI staining) and KA-infused hippocampus (**B**, TUNEL staining; **E**, superimposition of TUNEL and PI staining) of normoxic rat. The number of TUNEL-positive cells was reduced in KA-infused hippocampus in the hypoxic rat (**C**, TUNEL staining; **F**, superimposition of TUNEL and PI staining). Similar results were replicated.

Hypoxic Preconditioning Decreased KA-Induced Apoptosis

Three days after KA infusion, TUNEL-positive cells appeared in the infused hippocampus (FIG. 3), whereas no TUNEL-positive cells were observed in the intact hippocampus of the same normoxic rat (FIG. 3). Hypoxic pretreatment reduced the number of TUNEL-positive cells in the KA-infused hippocampus (FIG. 3). The mechanism underlying KA-induced apoptosis was investigated. Intrahippocampal infusion of KA elevated the cytosolic content of cytochrome c in the infused hippocampus (FIG. 4A). Although hypoxic preconditioning had no effect on cytosolic cytochrome c levels in the contralateral hippocampus without KA exposure, hypoxic pretreatment attenuated KA-induced increases in cytosolic cytochrome c level in the ipsilateral hippocampus (FIG. 4A). The distribution of AIF, an indicator of caspase-independent apoptosis,[30] was also examined. When nonlesioned contralateral hippocampus of the same rat was used as a control (FIG. 4B), neurons in the section prepared from the KA-infused hippocampus had a strong, homogenous fluorescent staining of AIF in the cytoplasm (FIG. 4B). Whereas *in vitro* and *in vivo* studies have shown an involvement of apoptosis in KA-induced neurotoxicity in the central nervous system,[4,31] our study provides further evidence that both caspase-dependent and caspase-independent pathways were activated in KA-induced neurotoxicity. Moreover, our data show that hypoxic preconditioning attenuated apoptosis via attenuating both caspase-dependent and caspase-independent pathways.

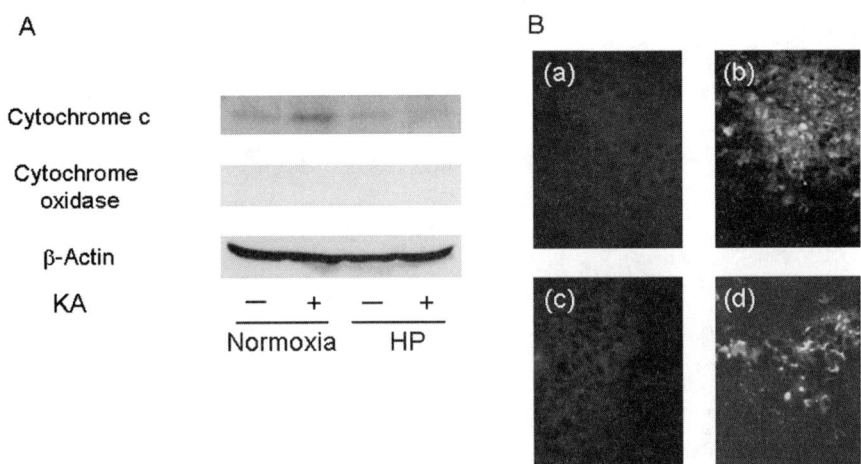

FIGURE 4. Hypoxic preconditioning attenuated KA-induced elevations in cytosolic cytochrome c and apoptosis-inducing factor (AIF) in the infused hippocampus. (**A**) Three hours after intrahippocampal infusion of KA, cytosolic cytochrome c was elevated in the infused hippocampus using western blot. Hypoxic preconditioning attenuated KA-induced elevation in cytosolic cytochrome c level in the hippocampus. Similar results were observed in duplicates. (**B**) A strong and homogenous immunofluorescent staining of AIF was revealed in the KA-infused hippocampus (b) compared with the intact hippocampus (a) of the same normoxic rat 3 h after an intrahippocampal infusion of KA. Hypoxic preconditioning attenuated KA-elevated immunofluorescence of AIF in the infused hippocampus (d) compared with that of intact hippocampus of the same hypoxic rat (c). Results were repeated in independent experiments.

Hypoxic Preconditioning Enhanced KA-elevated NF-κB Activity and Bcl-2 Expression

Oxidative stress and elevation in calcium levels are important inducers of NF-κB activation.[10,15,31–33] Moreover, several anti-apoptosis–related gene products, including Bcl-2 and thioredoxin, were reportedly elevated by KA.[10,32] Indeed, our study demonstrated KA-induced elevation in NF-κB activation and Bcl-2 expression in the lesioned hippocampus (FIG. 5). In addition, our data show that hypoxic preconditioning not only increased NF-κB activity in the contralateral hippocampus but also augmented KA-induced elevations in NF-κB activity and Bcl-2 expression in the ipsilateral hippocampus exposed to KA (FIG. 5A). NF-κB, a transcription factor, is ubiquitously expressed as an inducible regulator of many genes and plays a pivotal role in cell death or survival pathways.[31–33] The mechanisms underlying positive and negative modulation of NF-κB activity in neurons subjected to oxidative stress have not yet been determined. In this study, elevations in NF-κB activation and Bcl-2 expression may be responsible for neuroprotection by hypoxic preconditioning of KA-induced oxidative injury in rat brain.

Taken together, our present and previous studies suggest that hypoxic preconditioning is protective against the neurotoxicity induced by intrahippocampal infusion of KA, intranigral infusion of iron,[2] and transient middle cerebral artery occlusion.[28]

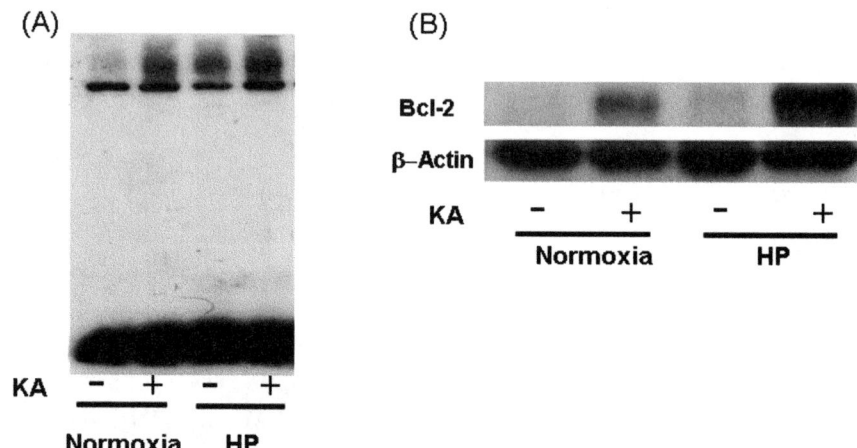

FIGURE 5. Hypoxic preconditioning enhanced KA-induced elevations in NF-κB activation and Bcl-2 expression in rat hippocampus. (**A**) Three hours after KA infusion, NF-κB activity was increased in the infused hippocampus measured by EMSA. Hypoxic preconditioning further increased KA-induced elevation in NF-κB activity in the infused hippocampus. Similar results were observed in triplicate experiments. (**B**) Four hours after an intrahippocampal infusion of KA, Bcl-2 expression was elevated in the infused hippocampus by Western blot. Hypoxic preconditioning augmented the KA-induced elevation in Bcl-2 expression in the infused hippocampus. Protein (50 μg) was loaded in each lane in all experiments. Similar results were observed in duplicates.

Although other studies suggest that new protein synthesis and gene regulation contribute to the preconditioning protection,[3,4,10,11] our studies demonstrate that transcription factor activation, antioxidative protein synthesis, and anti-apoptosis are involved in the neuroprotection of hypoxic preconditioning.

ACKNOWLEDGMENTS

This study was supported by grants from the National Science Council (NSC93-2320-B-075-008) and the Taipei Veterans General Hospital (VGH94-370), Taipei, Taiwan.

The authors express their gratitude to Dr. C.Y. Chai at the Institute of Biomedical Sciences, Academia Sinica, for his encouragement and support. Special thanks are due to Dr. C.F. Chen at National Taiwan University, Taipei, Taiwan for his kind supplying of the hypoxic rats and Dr. R.K. Freund at the University of Colorado, Health Sciences Center, Denver, Colorado, for editing this paper.

REFERENCES

1. CHIEN, C.T., C.F. CHEN, S.M. HSU, *et al.* 1999. Protective mechanism of preconditioning hypoxia attenuates apoptosis formation during renal ischemia/reperfusion phase. Transplant. Proc. **31:** 2012–2013.

2. LIN, A.M., C.F. CHEN & L.T. HO. 2002. Neuroprotective effect of intermittent hypoxia on iron-induced oxidative injury in rat brain. Exp. Neurol. **176:** 328–335.
3. ANDOH, T., S.Y. LEE, B. CHOCK, *et al.* 2000. Preconditioning regulation of Bcl-2 and p66^{shc} by human NOS1 enhances tolerance to oxidative stress. FASEB J. **14:** 2144–2146.
4. ANDOH, T., P.B. CHOCK & C.C. CHIUEH, 2002. Role of thioredoxin in protection against oxidative stress-induced apoptosis in SH-SY5Y cells. J. Biol. Chem. **277:** 9655–9660.
5. WICK, A., W. WICK, J. WALTENBERGER, *et al.* 2002. Neuroprotection by hypoxic preconditioning requires sequential activation of vascular endothelial growth factor receptor and Akt. J. Neurosci. **22:** 6401–6407.
6. VOVC, E. 1998. Arrhythmic effect of adaptation to intermittent hypoxia. Folia Medica **40:** 51–54.
7. ZHUANG, J. & Z. ZHOU. 1999. Protective effect of intermittent hypoxic adaptation on myocardium and its mechanisms. Biol. Signals Recept. **8:** 316–322.
8. POHLE, W. & C. RAUCA. 1994. Hypoxia protects against the neurotoxicity of kainic acid. Brain Res. **644:** 27–304.
9. EMERSON, M.R., F.E. SAMSON & T.L. PAZDERNIK. 1999. Hypoxia preconditioning attenuates brain edema associated with kainic acid-induced status epilepticus in rats. Brain Res. **825:** 189–193.
10. BLONDEAU, N., C. WIDMANN, M. LAZDUNSKI, *et al.* 2001. Activation of the nuclear factor-κB is a key event in brain tolerance. J. Neurosci. **21:** 4668–4677.
11. GULYAEVA, N.V. & E.N. TKATCHOUK, 1998. Antioxidant effects of interval hypoxia training in rat brain [abstract]. 12th European Society of Neurochemistry Meeting, S31C.
12. LAN, J., D.C. HENSHALL, R.P. SIMON, *et al.* 2000. Formation of the base modification 8-OH-2-deoxyguanosine and DNA fragmentation following seizures induced by systemic kainic acid in the rat. J. Neurochem. **74:** 302–309.
13. UEDA, Y., H. YOKOYAMA, R. NIWA, *et al.* 1997. Generation of lipid radicals in the hippocampal extracellular space during kainic acid-induced seizure in rats. Epilepsy Res. **26:** 329–333.
14. LIANG, L.P., Y.S. HO & M. PATEL. 2000. Mitochondrial superoxide induction in kainate-induced hippocampal damage. Neuroscience **101:** 563–570.
15. NAKAI, M., Z.H. QIN, J.F. CHEN, *et al.* 2000. Kainic acid-induced apoptosis in rat striatum is associated with NF-κB activation. J. Neurochem. **74:** 647–658.
16. BOLDYREV, A.A., D.O. CARPENTER, M.J. HUENTELMAN, *et al.* 1999. Sources of reactive oxygen species production in excitotoxin-stimulated cerebellar granule cells. Biochem. Biophys. Res. Commun. **256:** 320–324.
17. CHENG, H., Y.S. FU & J.W. GUO. 2004. Ability of GDNF to diminish free radical production leads to protection against kainate-induced excitotoxicity in hippocampus. Hippocampus **14:** 77–86.
18. DABBENI-SALA, F., M. FLOREANI, D. FRANCESCHINI, *et al.* 2001. Kainic acid induced selective mitochondrial oxidative phosphorylation enzyme dysfunction in cerebellar granule neurons: protective effects of melatonin and GSH ethylester. FASEB J. **15:** 1786–1788.
19. GILBERTI, E.A. & L.D. TROMBETTA. 2000. The relationship between stress protein induction and the oxidative defense system in the rat hippocampus following kainic acid administration. Toxicol. Lett. **116:** 17–26.
20. KIM, H.C., G. BING, W.K. JHOO, *et al.* 2002. Oxidative damage causes formation of lipofuscin-like substances in the hippocampus of the senescence-accelerated mouse after kainate treatment. Behav. Brain Res. **131:** 211–220.
21. ORTIZ, G.G., M.Y. SANCHEZ-RUIZ, D.X. TAN, *et al.* 2001. Melatonin and vitamin E reduce damage induced by kainic acid in the hippocampus: potassium-stimulated GABA release. J. Pineal Res. **31:** 62–67.
22. KIKUGAWA, K., T. KATO, M. BEPPU, *et al.* 1989. Fluorescent and crosslinked protein formed by free radical and aldehyde species generated during lipid peroxidation. Adv. Exp. Med. Biol. **266:** 345–356.

23. GRIFFITH, O.W. 1980. Determination of glutathione and glutathione disulfide using glutathione reductase and 2-vinylpyridine. Anal. Biochem. **106:** 207–212.
24. MISRA, H.P. & I. FRIDOVICH. 1989. The role of superoxide anion in the autoxidation of epinephrine and a simple assay for superoxide dismutase. J. Biol. Chem. **264:** 7761–7764.
25. AEBI, H. 1984. Catalase *in vitro*. Methods Enzymol. **105:** 121–126.
26. FLOHE, L. & W. A. GUNZLER. 1984. Assay of glutathione peroxidase. Methods Enzymol. **105:** 114–121.
27. HEINICKE, K., N. PROMMER, J. CAJIGAL, *et al.*, 2003. Long term exposure to intermittent hypoxia results in increased hemoglobin mass, reduced plasma volume, and elevated erythropoietin plasma levels in man. Eur. J. Appl. Physiol. **88:** 535–543.
28. LIN, A.M., S.W. DOONG & C.F. CHEN, 2003. Hypoxic preconditioning prevents cortical infarction by transient focal ischemia-reperfusion. Ann. N.Y. Acad. Sci. **993:** 168–178.
29. DUAN, C.L., F.S. YAN, X.Y. SONG, *et al.* 1999. Changes of superoxide dismutase, glutathione peroxidase and lipid peroxide in the brain of mice preconditioned by hypoxia. Biol. Signals Recept. **8:** 256–260.
30. LU, C.X., T.J. FAN, G.B. HU, *et al.* 2003. Apoptosis-inducing factor and apoptosis. Acta Biochim. Biophys. Sin. (Shanghai) **35:** 881–885.
31. JIANG, X., D. ZHU, P. OKAGAKI, *et al.* 2003. *N*-methyl-D-aspartate and TrkB receptor activation in cerebellar cells: an *in vitro* model of preconditioning to stimulate intrinsic survival pathways in neurons. Ann. N.Y. Acad. Sci. **993:** 134–145.
32. YALCIN, A., L. KANIT & E.Y. SOZMEN. 2004. Altered gene expressions in rat hippocampus after kainate injection with or without melatonin pre-treatment. Neuroscience **359:** 65–68.
33. MATTSON, M.P., C. CULMSEE, Z.F. YU, *et al.* 2000. Roles of NF-κB in neuronal survival and plasticity. J. Neurochem. **74:** 443–456.

Antioxidant N-Acetylcysteine Blocks Nerve Growth Factor–Induced H_2O_2/ERK Signaling in PC12 Cells

LIANG-YO YANG,[a] WUN-CHANG KO,[b] CHUN-MAO LIN,[d] JIA-WEI LIN,[c] JEN-CHINE WU,[d,e] CHIEN-JU LIN,[d,e] HUEY-HWA CHENG,[d] AND CHWEN-MING SHIH[d]

[a]*Department of Physiology, Taipei Medical University, Taipei 110, Taiwan*

[b]*Graduate Institute of Pharmacology, Taipei Medical University, Taipei 110, Taiwan*

[c]*Department of Neurosurgery, Taipei Medical University–Affiliated Taipei Municipal Wan-Fang Hospital, Taipei 116, Taiwan*

[d]*Department of Biochemistry, Taipei Medical University, Taipei 110, Taiwan*

[e]*Graduate Institute of Medical Science, Taipei Medical University, Taipei 110, Taiwan*

> ABSTRACT: We investigated whether H_2O_2, superoxide, and ERK participate in nerve growth factor (NGF)–induced signaling cascades and whether antioxidant N-acetylcysteine (NAC) regulates these NGF-induced responses. PC12 cells were cultured in medium containing NGF or vehicle with or without NAC pretreatment, and the intracellular H_2O_2 and superoxide levels and the amount of phosphorylated ERK were evaluated by flow cytometry and Western blotting, respectively. We found that NGF increased intracellular H_2O_2 concentration and activated ERK but failed to affect intracellular superoxide level. Moreover, NAC counteracted these NGF-induced responses. These findings demonstrate that NAC blocks the NGF-induced H_2O_2/ERK signaling in PC12 cells.
>
> KEYWORDS: antioxidant; N-acetylcysteine; PC12 cells; ERK; NGF; hydrogen peroxide; reactive oxygen species

INTRODUCTION

The cellular redox reaction plays a pivotal role in a variety of normal physiological functions (e.g., normal cardiovascular function,[1] normal sperm function,[2] and nerve growth factor [NGF] signaling[3,4]) and pathological processes.[5,6] Recent findings point out that reactive oxygen species (ROS) mediate many effects of angio-

---

Address for correspondence: Chwen-Ming Shih, Department of Biochemistry, Taipei Medical University, 250 Wu Hsing St., Taipei 110, Taiwan. Voice: +86-2-27361661 ext. 3151; fax: +886-2-8642-1158.
cmshih@tmu.edu.tw

tensin II, which are vital for normal cardiovascular function.[1] Other evidence also indicates that the spermatozoa produce ROS, which promote sperm capacitation.[2] Moreover, ROS mediate NGF-induced neuronal differentiation[3] and phosphorylation of cAMP-responsive element-binding protein (CREB).[4] In contrast, overproduction of ROS leads to vascular injury associated with angiotensin II–dependent hypertension.[5] Accumulating evidence also suggests that oxidative stress contributes significantly to the development of neurodegenerative diseases.[6–8]

Hydrogen peroxide and superoxide are two common ROS that are generated in the cellular redox reaction and play an essential role in various signaling pathways, regulatory processes, or pathophysiological conditions.[9–12] Application of platelet-derived growth factor (PDGF) to vascular smooth muscle cells (VSMCs) induces activation of tyrosine phosphorylase and mitogen-activated protein kinase (MAPK), DNA production, and a brief increase of intracellular hydrogen peroxide levels.[9] Inhibition of hydrogen peroxide production blocks the PDGF-induced responses in the VSMCs mentioned above, suggesting that hydrogen peroxide mediates these PDGF effects on VSMCs.[9] Evidence indicates that hydrogen peroxide controls neuronal plasticity by altering the action of some special calcium-dependent phosphatases[12] and acts as a key second messenger to mediate the effects of metabolic oxidative stress.[11] Application of angiotensin II leads to elevated mean arterial blood pressure and increased production of superoxide anion in rats, which are decreased and counteracted by treatment of liposome-entrapped superoxide dismutase (SOD), suggesting that superoxide mediates, at least in part, the angiotensin II–induced hypertension in rats.[10]

NGF exerts a variety of effects on sensory neurons, sympathetic neurons, and pheochromocytoma (PC12) cells, including cell survival, neurite outgrowth, and cell differentiation, by binding to NGF receptors and activation of NGF signaling pathways.[13–16] Recent evidence indicates that NGF results in a momentary increase of intracellular ROS in PC12 cells and that antioxidant N-acetylcysteine (NAC) blocks this NGF-induced elevation of intracellular ROS.[3] Catalase counteracts the production of intracellular ROS, neurite growth, and activation of tyrosine phosphorylase induced by NGF, and therefore hydrogen peroxide is suggested to mediate these NGF-induced responses.[3] Nonetheless, the role of ROS in NGF-induced signaling and the downstream signaling cascades of ROS induced by NGF in PC12 cells remain poorly understood.

In this study, we tested the hypotheses that hydrogen peroxide, superoxide, and ERK activation play an important role in NGF-induced signaling in PC12 cells and that antioxidant NAC counteracts these NGF-induced responses in PC12 cells. Our results indicated that NGF had no effect on the intracellular concentration of superoxide in PC12 cells. Nonetheless, NGF increased the intracellular hydrogen peroxide level and the phosphorylation of ERK in PC12 cells, and these NGF-induced responses were blocked by NAC pretreatment. Moreover, a selective MAPK inhibitor, PD098059, suppressed phosphorylation of ERK. Taken together, these findings demonstrate that the production of hydrogen peroxide and activation of ERK are two important steps of NGF-induced MAPK signaling in PC12 cells and strongly suggest that hydrogen peroxide mediates the NGF-induced MAPK/ERK signaling.

MATERIALS AND METHODS

Cell Culture and Chemicals

In a humidified incubator with continuous aeration of 5% CO_2 and 95% air at 37°C, PC12 cells were incubated and maintained in complete Dulbecco's Modified Eagle's Medium (DMEM) (HyClone, Logan, UT) containing penicillin G (100 U/mL), streptomycin (100 µg/mL), amphotericin B (0.25 µg/mL), L-glutamine (2 mM), nonessential amino acids (1 mM), fetal bovine serum (FBS; 5%), and horse serum (10%). At the beginning of all experiments, unless otherwise stated, PC12 cells were plated onto poly-L-lysine–coated tissue culture dishes (10^6 cells/100-mm plate) including a low-serum DMEM medium with 1% FBS; the medium was replenished every 48 h.

Measurement of Intracellular Hydrogen Peroxide and Superoxide

For the measurement of hydrogen peroxide, following starvation in low-serum DMEM with 1% FBS for 24 h, PC12 cells were treated with 10 µM 2′,7′-dichlorodihydrofluorescein diacetate (DCFH-DA; Molecular Probes, Eugene, OR) for 20 min, and then NGF (100 ng/mL) (catalog no. 556-NG-100; R & D Systems, Minneapolis, MN) or vehicle was added to PC12 cells, which were collected 1, 3, 5, 10, and 20 min after NGF or vehicle application and then trypsinized for immediate detection of intracellular hydrogen peroxide. The dichlorodihydrofluorescein (DCFH), the deacetylated product of DCFH-DA by intracellular esterases, reacts with H_2O_2 to form dichlorofluorescein (DCF), which is an oxidized fluorescent compound. The amount of intracellular hydrogen peroxide can be quantified by detection of DCF using a flow cytometer with excitation and emission wavelengths set at 488 nm and 525–550 nm (FL1-H), respectively.[17]

In a separate experiment, we investigated whether antioxidant NAC has any effect on the NGF-induced elevation of intracellular hydrogen peroxide in PC12 cells because NGF treatment did increase the production of intracellular hydrogen peroxide in PC12 cells in the preceding experiment. In addition to the control group and the NGF treatment group, we added an additional group with NAC (5 mM) pretreatment (Sigma Chemical Co., St. Louis, MO) 4 h prior to addition of NGF (100 ng/mL). To determine whether NAC can clear the hydrogen peroxide entering PC12 cells, we also administered NAC to PC12 cells 4 h before addition of H_2O_2 (200 µM) (Merck & Co., Whitehouse Station, NJ) to PC12 cells besides the control group and the H_2O_2 treatment group. The intracellular concentration of hydrogen peroxide in PC12 cells was measured 1 min after NGF or H_2O_2 treatment by the same procedures described earlier.

For detection of superoxide, following starvation in low-serum DMEM with 1% FBS for 24 h, 5 µM dihydroethidium (HEt) was added to PC12 cells 5 min before addition of NGF (100 ng/mL) or vehicle. PC12 cells were collected 1, 3, 5, 10, and 20 min after NGF or vehicle treatment and then trypsinized for immediate evaluation of intracellular superoxide by using a flow cytometer. When HEt enters the cells, it will react with superoxide to form ethidium, which will be incorporated into the nuclear DNA[18] and can be detected by a flow cytometer with excitation and emission wavelengths set at 488 and 637 nm (FL2-H), respectively.

Western Blot Analysis

PC12 cells were incubated in a low-serum DMEM with 1% FBS for 24 h, and NGF (100 ng/mL) was added to the medium for 5, 10, 20, 40, 60, 120, and 180 min; PC12 cells without NGF treatment served as the negative control group. PC12 cells were scraped, washed, and lysed at a density of 10^6 cells/50 μL of lysis buffer (25 mM HEPES, 5 mM EDTA, 0.1 mM sodium deoxycholate, 1.5% Triton X-100, 0.1% sodium dodecyl sulfate [SDS], 0.5 M NaCl)[19] (Merck & Co., Inc.), including a protease inhibitor cocktail (Roche Molecular Biochemicals, Mannheim, Germany). PC12 cells were then incubated on ice for 20 min. Following centrifugation of the lysates at 15,000 × g and 4°C for 15 min, the amount of proteins in the supernatant was quantified using Bio-Rad Protein Assay Dye Reagent (Bio-Rad Laboratories, Hercules, CA). After addition of sample buffer (60 mM Tris–HCl, pH 6.8, 2% SDS, 10% glycerol, 140 mM β-mercaptoethanol, 0.002% bromophenol blue), each lysate was boiled for 5 min and loaded for electrophoresis in an SDS–polyacrylamide gel (30 μg of protein/lane). Proteins on the gel were electrotransferred onto polyvinylidene difluoride (PVDF) membranes, and the PVDF membranes were then incubated with anti-ERK2 antibody (1:2000) (catalog no. sc-154; Santa Cruz Biotechnology, Santa Cruz, CA) or anti–phospho-ERK1/2 antibody (1:2000) (catalog no. 9101S; New England Biolabs, Beverly, MA). After several rinses, the PVDF membranes were incubated with appropriate horseradish peroxidase–conjugated secondary antibodies (1:10,000 dilution) and enhanced chemiluminescence reagent (Amersham Pharmacia Biotech, Piscataway, NJ). The amount of total ERK1/2 (t-ERK1/2) or phosphorylated ERK1/2 (p-ERK1/2) was quantified by a densitometer (Gel Doc 2000; Bio-Rad Laboratories). For the relative ERK activity, the p-ERK1/2 over the t-ERK1/2 in the negative control group was calculated, set to 1.00, and used as the unit to express the ratio of p-ERK1/2 over the t-ERK1/2 for the remaining groups.

In an additional experiment, we examined whether antioxidant NAC blocks the NGF effect on the activation of ERK in PC12 cells, because NGF treatment activated ERK in PC12 cells in the previous experiment. In addition to the negative control group and the NGF treatment groups with or without dimethyl sulfoxide (DMSO) (Sigma Chemical Co.), we added additional groups with pretreatment of PC12 cells with different doses of NAC (5, 10, or 20 mM) 4 h prior to addition of NGF to PC12 cells. To determine whether ERK activation is a downstream event of MAPK in NGF-treated PC12 cells, we also applied different doses of a specific MAPK inhibitor PD098059 (10, 20, or 30 μM in DMSO solution) or DMSO 1 h prior to addition of NGF to PC12 cells. PC12 cells were collected 5 min after NGF treatment, and the amounts of p-ERK1/2 and t-ERK1/2 were measured by the same procedures described previously; the relative ERK activity was calculated and expressed in the same way described earlier.

RESULTS

Application of NGF (100 ng/mL) caused an increase of intracellular hydrogen peroxide but not superoxide levels in PC12 cells, and antioxidant NAC counteracted the NGF effect on the production of intracellular hydrogen peroxide. NGF treatment led

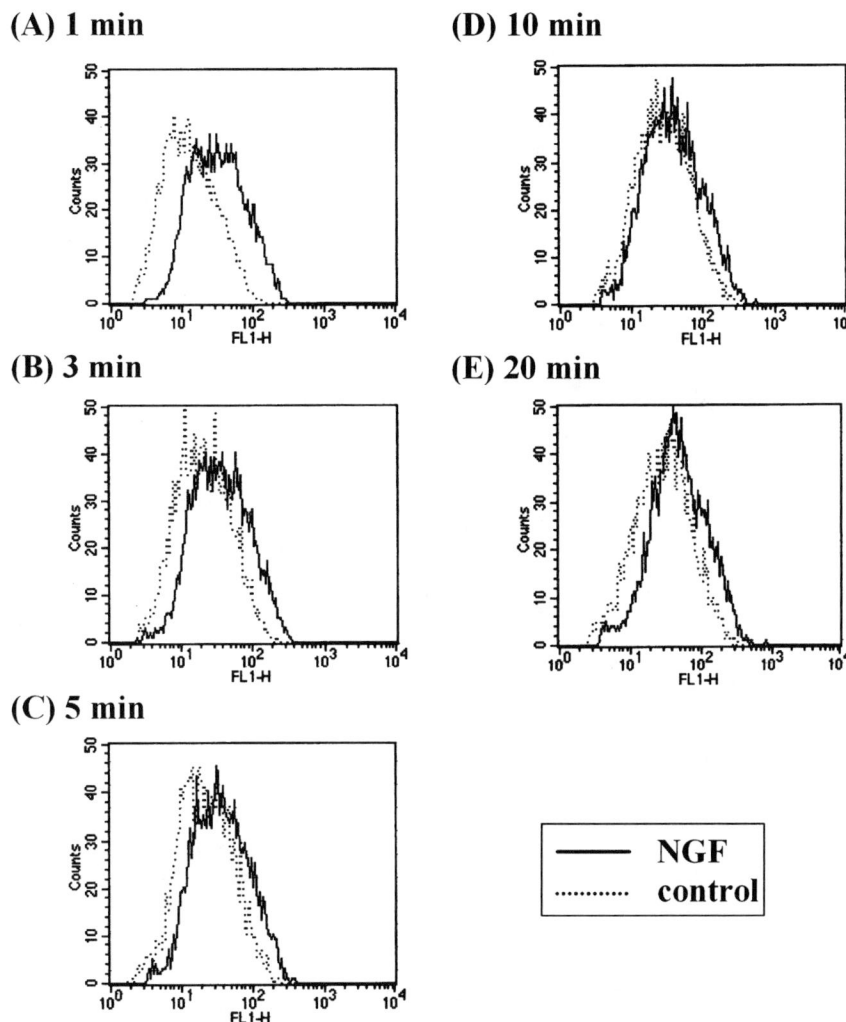

FIGURE 1. NGF induces generation of intracellular hydrogen peroxide in PC12 cells. PC12 cells were treated with 10 μM 2′,7′-dichlorodihydrofluorescein diacetate (DCFH-DA) for 20 min before addition of NGF (100 ng/mL) or vehicle. Intracellular level of hydrogen peroxide was measured by detection of dichlorofluorescein (DCF) in the FL1-H level by flow cytometry at (**A**) 1, (**B**) 3, (**C**) 5, (**D**) 10, and (**E**) 20 min after NGF or vehicle application. NGF caused a 3-, 1.8-, and 1.4-fold increase in intracellular hydrogen peroxide levels 1, 3, and 5 min, respectively, after application. Concentration of intracellular hydrogen peroxide returned to the control level 10 min after NGF application.

to an elevation of intracellular hydrogen peroxide measured by flow cytometry 1, 3, and 5 min after treatment when compared with the control group (FIG. 1A–C). The intracellular concentration of hydrogen peroxide induced by NGF reached the highest level 1 min after application (3 times the control value, FIG. 1A), decreased gradually between 3 min (1.8 times the control value, FIG. 1B) and 5 min (1.4 times the control value, FIG. 1C) after NGF administration, and returned to the control level 10 min after treatment (FIG. 1D). NAC pretreatment (5 mM) counteracted the NGF-induced twofold rise of intracellular hydrogen peroxide 1 min after NGF treatment in a separate experiment (FIG. 2A). Application of hydrogen peroxide (200 µM) to PC12 cells

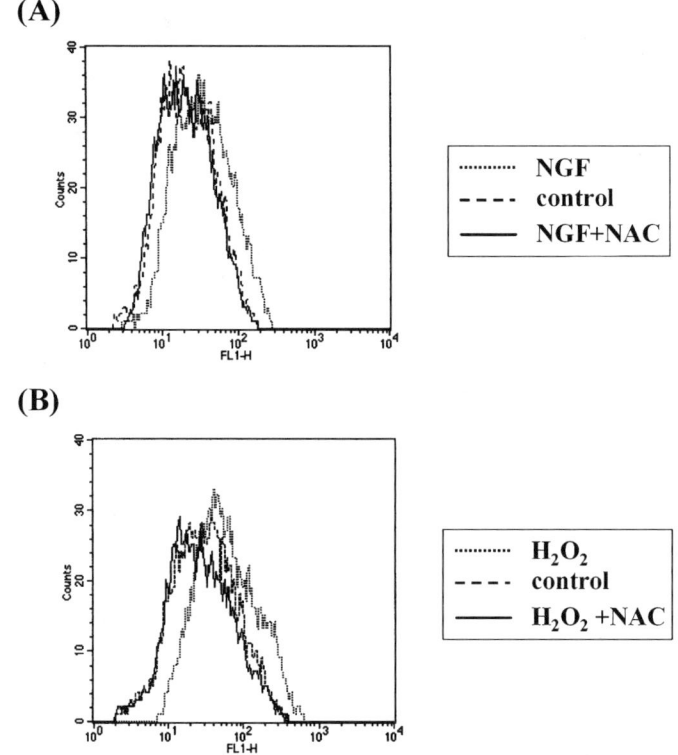

FIGURE 2. Antioxidant N-acetylcysteine (NAC) counteracts the (**A**) NGF-induced and (**B**) H_2O_2-induced production of hydrogen peroxide in PC12 cells. PC12 cells were pretreated with 5 mM NAC 4 h before NGF or H_2O_2 application. We applied 10 µM $2'7'$-dichlorodihydrofluorescein diacetate (DCFH-DA) to PC12 cells 20 min before NGF (100 ng/mL), H_2O_2 (200 µM), or vehicle was added to the culture medium. Amount of dichlorofluorescein (DCF) in the FL1-H level that reflected the intracellular level of hydrogen peroxide was measured by flow cytometry at 1 min after NGF, H_2O_2, or vehicle application. (**A**) NGF increased the intracellular level of hydrogen peroxide in PC12 cells, and NAC pretreatment abolished this NGF effect. (**B**) Hydrogen peroxide caused an increase in intracellular hydrogen peroxide level in PC12 cells, and NAC pretreatment counteracted this hydrogen peroxide effect.

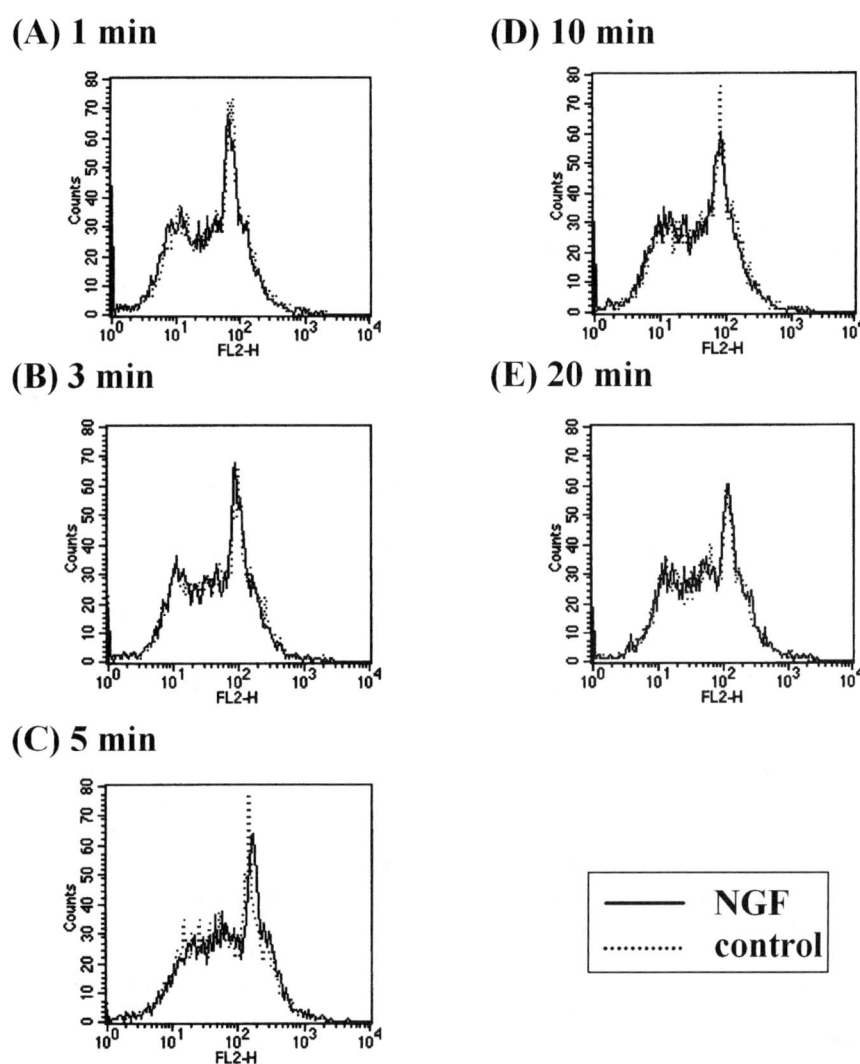

FIGURE 3. NGF has no effect on the generation of intracellular superoxide in PC12 cells. PC12 cells were treated with 5 μM dihydroethidium (HEt) for 5 min before addition of NGF (100 ng/mL) or vehicle. The intracellular level of superoxide was measured by detection of ethidium in the FL2-H level by flow cytometry at (**A**) 1, (**B**) 3, (**C**) 5, (**D**) 10, and (**E**) 20 min after NGF or vehicle application. Our data showed that NGF failed to affect the production of superoxide at all time points examined.

resulted in a 4.2-fold increase of intracellular hydrogen peroxide (FIG. 2B), and NAC pretreatment prevented this hydrogen peroxide effect (FIG. 2B). In contrast, NGF application failed to affect the level of intracellular superoxide in PC12 cells when compared with the control group at all time points examined (FIG. 3).

Western blot data showed that application of NGF (100 ng/mL) to PC12 cells induced a 2- to 8.7-fold increase in the relative ERK activity (p-ERK1/2 over t-ERK1/2) from 5–180 min after treatment (FIG. 4). The relative ERK activity reached 8.7 times the control value at 5 min after NGF application, dropped to ~ 4–5 times the control value 10–40 min after NGF treatment, and stayed at around 2 times the control value 1–3 h after treatment (FIG. 4B).

Antioxidant NAC or a selective MAPK inhibitor PD098059 pretreatment blocked the NGF-induced ERK activation in PC12 cells. NGF treatment alone (lane 2 of FIG. 5) caused a significant increase of ERK activation in PC12 cells when compared with the negative control group without NGF treatment (lane 1 of FIG. 5). Pretreatment of PC12 cells with antioxidant NAC 4 h before NGF application suppressed the NGF-induced activation of ERK at all three doses (5, 10, and 20 mM for lanes 7, 8, and 9 of FIG. 5, respectively) when compared with the positive control groups (lanes 2 and 3 of FIG. 5: lane 2, NGF only; lane 3, NGF plus DMSO) 5 min after NGF application. NGF plus DMSO (the solvent for PD098059, lane 3 of FIG. 5) caused a slight increase of ERK activity in PC12 cells when compared with the group receiving NGF alone (lane 2 of FIG. 5). Pretreatment of PC12 cells with a selective MAPK inhibitor PD098059 1 h prior to NGF treatment inhibited the NGF-induced ERK phosphorylation only at two higher doses (20 and 30 μM for lanes 5 and 6 of FIG. 5, respectively).

DISCUSSION

NGF induces generation of intracellular hydrogen peroxide in PC12 cells, but the time for NGF-induced H_2O_2 production in this study differs from that reported in the earlier study.[3] Recent evidence shows that application of NGF stimulates a brief increase of intracellular ROS in PC12 cells 10 min after treatment, and this ROS is suggested to be hydrogen peroxide because NGF fails to increase the ROS in PC12 cells transfected with PS3CAT carrying human catalase.[3] In this study, our results showed that application of NGF caused a threefold increase of intracellular hydrogen peroxide in PC12 cells 1 min after treatment (FIG. 1A). This NGF-induced elevation of intracellular hydrogen peroxide decreased gradually and returned to the control level 10 min after NGF application (FIG. 1B–D). The difference in the time for NGF-induced H_2O_2 production between our study and the earlier study may result from the 10-min difference in the DCFH-DA incubation time prior to NGF application between these two studies (20 min for this study and 10 min for the earlier study). Our findings confirm that NGF stimulates the production of hydrogen peroxide in PC12 cells and support the notion that ROS in nontoxic amounts act as second messengers in a variety of signaling pathways.[20]

NGF has no effect on the intracellular level of superoxide in PC12 cells. Accumulating evidence indicates that various cytokines (e.g., interleukin 1 [IL-1] and tumor necrosis factor α [TNF-α]) and peptide growth factors (e.g., epidermal growth factor [EGF]) lead to generation of superoxide in different types of cells.[21,22] IL-1 or TNF-α stimulates release of ROS in human fibroblasts, and the released ROS has

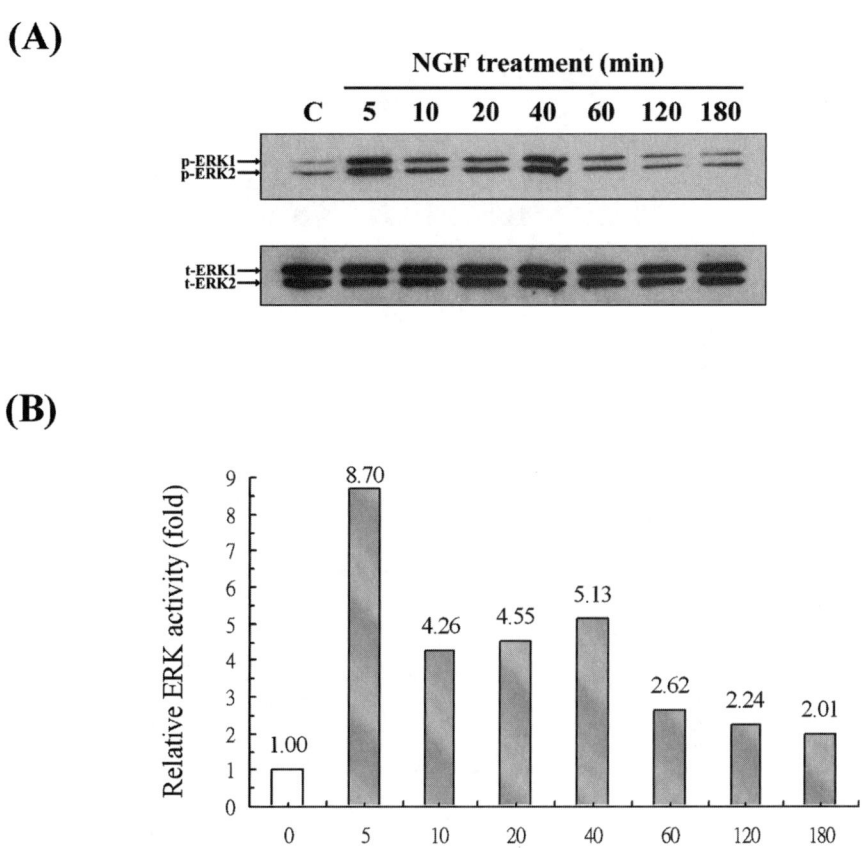

FIGURE 4. NGF induces phosphorylation of ERK in PC12 cells. (**A**) NGF (100 ng/mL) was added to PC12 cells, which were collected 5, 10, 20, 40, 60, 120, and 180 min after NGF application along with control PC12 cells. PC12 cells were then processed for detection of phosphorylated ERK1/2 (p-ERK1/2) and total ERK1/2 (t-ERK1/2) by Western blotting. NGF treatment significantly increased the amount of p-ERK1/2 with a peak at 5 min after application, and the increased p-ERK1/2 lasted for at least 3 h, while keeping the t-ERK1/2 unchanged. (**B**) Amounts of p-ERK1/2 and t-ERK1/2 were quantified by a densitometer, and relative ERK activity of all treatments was calculated as described in MATERIALS AND METHODS. NGF induced a 2- to 8.7-fold increase of ERK activity in PC12 cells within 3 h after NGF application. The NGF-induced ERK activity peaked with an 8.7-fold increase at 5 min after NGF treatment, decreased to ~ 4–5 times the control value between 10 and 40 min after application, and remained at about 2 times the control value even 3 h after NGF treatment.

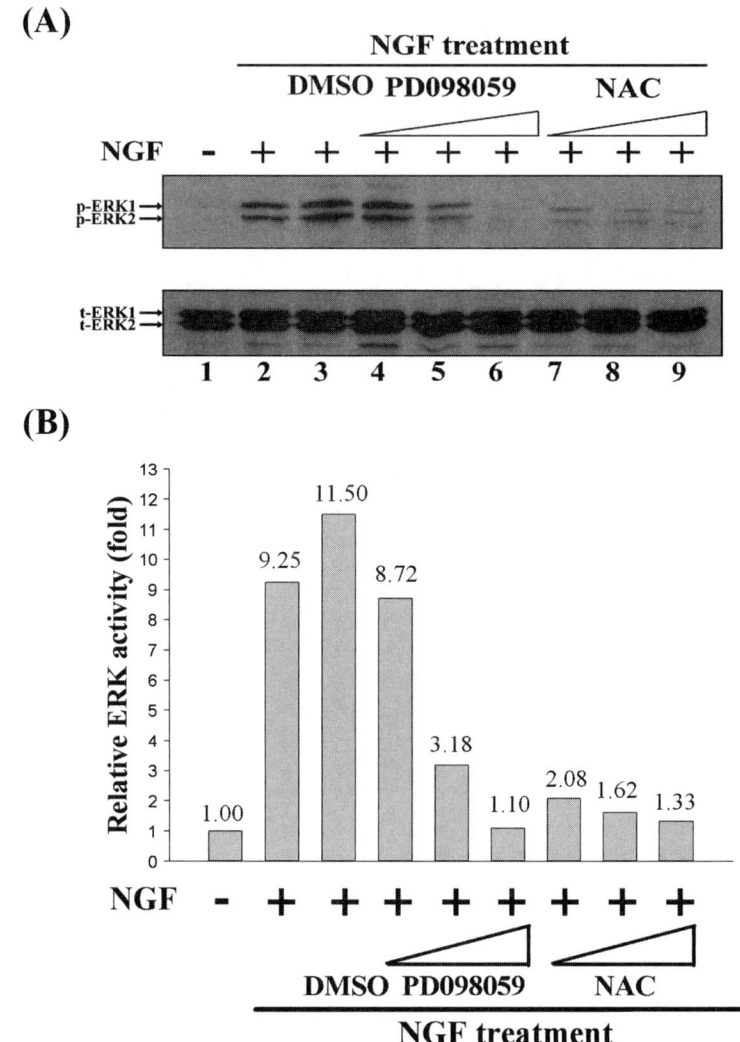

FIGURE 5. Both antioxidant N-acetylcysteine (NAC) and a selective MAPK inhibitor, PD098059, prevent the NGF-increased ERK phosphorylation in PC12 cells. PC12 cells were pretreated with NAC (5, 10, and 20 mM for lanes 7, 8, and 9, respectively), PD098059 (10, 20, and 30 µM for lanes 4, 5, and 6, respectively), DMSO (lane 3), or vehicle (lanes 1 and 2) for 4, 1, 1, and 1 h, respectively. Afterward, NGF (100 ng/mL) or vehicle was applied to PC12 cells, which were collected 5 min after NGF or vehicle application and processed for detection of phosphorylated ERK1/2 (p-ERK1/2) and total ERK1/2 (t-ERK1/2) by Western blotting. (**A**) NGF treatment significantly increased the amount of p-ERK1/2. All three doses of NAC (lanes 7, 8, and 9) and two higher doses of PD098059 (lanes 5 and 6) suppressed the NGF-induced activation of ERK1/2. (**B**) The amounts of p-ERK1/2 and t-ERK1/2 were quantified by densitometer, and the relative ERK activity for all groups was calculated as described in MATERIALS AND METHODS. All three doses of NAC and two higher doses of PD098059 dramatically inhibited the NGF-induced phosphorylation of ERK.

been proven to be the superoxide.[21] Application of EGF to PC12 cells induces increased production of ROS, and pretreatment of PC12 cells with lipoxygenase inhibitor NDGA (inhibitor of superoxide-producing enzyme) abolishes the EGF effect on ROS production, suggesting that the ROS induced by EGF in PC12 cells is likely to be the superoxide.[22] NGF induces ROS in PC12 cells,[3] and NGF increases the generation of superoxide in neutrophils stimulated by phorbol 12-myristate 13-acetate (PMA).[23] In this study, therefore, we also examined whether NGF resulted in production of superoxide in PC12 cells. Our results indicated that NGF treatment was unable to affect the level of intracellular superoxide in PC12 cells at all time points examined (FIG. 3). Our findings fail to support the idea that the superoxide participates in NGF signaling in PC12 cells.

Application of NGF leads to a significant elevation of ERK activity in PC12 cells, and NGF-increased ERK activation is abolished by pretreatment of PC12 cells with a selective MAPK inhibitor PD098059. NGF signaling pathways include at least the Ras–Raf–MEK–MAPK pathway, PI-3 kinase pathway, and phospholipase C pathway,[13–16,24–26] and the Ras–Raf–MEK–MAPK pathway is the most well studied. Nonetheless, the detailed mechanisms of this pathway are still poorly understood. In this study, we found that NGF treatment increased the relative ERK activity in PC12 cells, which is consistent with earlier findings in sympathetic neurons and in PC12 cells.[26,27] Pretreatment with a specific MAPK inhibitor PD098059 1 h prior to NGF treatment prevented the NGF-induced ERK activation in PC12 cells, which replicates the earlier findings reported in sympathetic neurons.[26] Our findings demonstrate that ERK activation is the downstream cascade of NGF-induced MAPK signaling in PC12 cells.

Antioxidant and glutathione precursor NAC pretreatment counteracts the effects of NGF on the increase of intracellular hydrogen peroxide and the activation of ERK in PC12 cells. NAC can protect cells from oxidative damage by its direct scavenging ROS ability and by its indirect ability to increase synthesis of antioxidant glutathione.[28–30] Evidence shows that NAC inhibits the *c-fos* expression and AP-1 activation induced by NGF in PC12 cells.[31] NAC also blocks the NGF-induced MAPK activation and reduces the level of phosphorylated MAPKs in PC12 cells, which is independent of glutathione production.[31] Moreover, NAC prevents the elevated intracellular ROS, neurite extension, activation of AP-1, and phosphorylation of tyrosine in PC12 cells induced by NGF.[3] In this study, H_2O_2 application caused an increase of intracellular hydrogen peroxide, and NAC pretreatment counteracted the H_2O_2-induced production of hydrogen peroxide in PC12 cells. Most importantly, pretreatment of PC12 cells with antioxidant NAC blocked the NGF-increased generation of hydrogen peroxide and activation of ERK in PC12 cells. NAC inhibition of NGF-induced H_2O_2 production and ERK activation and suppression of NGF-induced ERK activity by a specific MAPK inhibitor PD098059 together strongly suggest that NAC inhibits the NGF-induced MAPK/ERK signaling cascades by depleting NGF-induced intracellular hydrogen peroxide.

ACKNOWLEDGMENTS

This research was supported by National Science Council grants NSC 90-2320-B-038-052 (L.Y.Y.), NSC 91-2320-B-038-010 (L.Y.Y.), and NSC 93-3112-B-038-005 (L.Y.Y.), and NSC 92-2320-B-038-055 (C.M.S.).

REFERENCES

1. ZIMMERMAN, M.C. & R.L. DAVISSON. 2004. Redox signaling in central neural regulation of cardiovascular function. Prog. Biophys. Mol. Biol. **84:** 125–149.
2. BAKER, M.A. & R.J. AITKEN. 2004. The importance of redox regulated pathways in sperm cell biology. Mol. Cell. Endocrinol. **216:** 47–54.
3. SUZUKAWA, K., K. MIURA, J. MITSUSHITS, et al. 2000. Nerve growth factor-induced neuronal differentiation requires generation of Rac1-regulated reactive oxygen species. J. Biol. Chem. **275:** 13175–13178.
4. BEDOGNI, B., G. PANI, R. COLAVITTI, et al. 2003. Redox regulation of cAMP-responsive element-binding protein and induction of manganous superoxide dismutase in nerve growth factor-dependent cell survival. J. Biol. Chem. **278:** 16510–16519.
5. TOUYZ, R.M., F. TABET & E.L. SCHIFFRIN. 2003. Redox-dependent signalling by angiotensin II and vascular remodelling in hypertension. Clin. Exp. Pharmacol. Physiol. **30:** 860–866.
6. EMERIT, J., M. EDEAS & F. BRICAIRE. 2004. Neurodegenerative diseases and oxidative stress. Biomed. Pharmacother. **58:** 39–46.
7. OLANOW, C.W. 1993. A radical hypothesis for neurodegeneration. Trends Neurosci. **16:** 439–444.
8. MARKESBERY, W.R. 1997. Oxidative stress hypothesis in Alzheimer's disease. Free Radic. Biol. Med. **23:** 134–147.
9. SUNDARESAN, M., Z.X. YU, V.J. FERRANS, et al. 1995. Requirement for generation of H_2O_2 for platelet-derived growth factor signal transduction. Science **270:** 296–299.
10. LAURSEN, J.B., S. RAJAGOPALAN, Z. GALIS, et al. 1997. Role of superoxide in angiotensin II-induced but not catecholamine-induced hypertension. Circulation **95:** 588–593.
11. SONG, J.J. & Y.J. LEE. 2003. Catalase, but not MnSOD, inhibits glucose deprivation-activated ASK1-MEK-MAPK signal transduction pathway and prevents relocalization of Daxx: hydrogen peroxide as a major second messenger of metabolic oxidative stress. J. Cell Biochem. **90:** 304–314.
12. KAMSLER, A. & M. SEGAL. 2004. Hydrogen peroxide as a diffusible signal molecule in synaptic plasticity. Mol. Neurobiol. **29:** 167–178.
13. LEWIN, G.R. & Y.A. BARDE. 1996. Physiology of the neurotrophins. Annu. Rev. Neurosci. **19:** 289–317.
14. FACCHINETTI, F., S. FUREGATO, T. TERRAZZINO & A. LEON. 2002. H_2O_2 induces upregulation of Fas and Fas ligand expression in NGF-differentiated PC12 cells: modulation by cAMP. J. Neurosci. Res. **69:** 178–188.
15. DAS, K.P., T.M. FREUDENRICH & W.R. MUNDY. 2004. Assessment of PC12 cell differentiation and neurite growth: a comparison of morphological and neurochemical measures. Neurotoxicol. Teratol. **26:** 397–406.
16. PIERCHALA, B.A., R.C. AHRENS, A.J. PADEN & E.M. JOHNSON, JR. 2004. Nerve growth factor promotes the survival of sympathetic neurons through the cooperative function of the protein kinase C and phosphatidylinositol 3-kinase pathways. J. Biol. Chem. **279:** 27986–27993.
17. BEHL, C., J.B. DAVIS, R. LESLEY & D. SCHUBERT. 1994. Hydrogen peroxide mediates amyloid beta protein toxicity. Cell **77:** 817–827.
18. BINDOKAS, V.P., J. JORDAN, C.C. LEE & R.J. MILLER. 1996. Superoxide production in rat hippocampal neurons: selective imaging with hydroethidine. J. Neurosci. **16:** 1324–1336.
19. SIMIZU, S., M. TAKADA, K. UMEZAWA & M. IMOTO. 1998. Requirement of caspase-3(-like) protease-mediated hydrogen peroxide production for apoptosis induced by various anticancer drugs. J. Biol. Chem. **273:** 26900–26907.
20. SCHRECK, R. & P.A. BAEUERLE. 1991. A role for oxygen radicals as second messengers. Trends Cell Biol. **1:** 39–42.
21. MEIER, B., H.H. RADEKE, S. SELLE, et al. 1989. Human fibroblasts release reactive oxygen species in response to interleukin-1 or tumour necrosis factor-alpha. Biochem. J. **263:** 539–545.

22. MILLS, E.M., K. TAKEDA, Z.X. YU, et al. 1998. Nerve growth factor treatment prevents the increase in superoxide produced by epidermal growth factor in PC12 cells. J. Biol. Chem. **273:** 22165–22168.
23. KANNAN, Y., H. USHIO, H. KOYAMA, et al. 1991. 2.5S nerve growth factor enhances survival, phagocytosis, and superoxide production of murine neutrophils. Blood **77:** 1320–1325.
24. NUSSER, N., E. GOSMANOVA, Y. ZHENG & G. TIGYI. 2002. Nerve growth factor signals through TrkA, phosphatidylinositol 3-kinase, and Rac1 to inactivate RhoA during the initiation of neuronal differentiation of PC12 cells. J. Biol. Chem. **277:** 35840–35846.
25. OPPENHEIM, R.W. & J.E. JOHNSON. 2003. Programmed cell death and neurotrophic factors. *In* Fundamental Neuroscience. L.R. Squire, F.E. Bloom, S.K. McConnell, et al., Eds.: 499–532. Academic Press. New York.
26. KIM, I.J., K.M. DRAHUSHUK, W.Y. KIM, et al. 2004. Extracellular signal-regulated kinases regulate dendritic growth in rat sympathetic neurons. J. Neurosci. **24:** 3304–3312.
27. PARRAN, D.K., S. BARONE, JR & W.R. MUNDY. 2004. Methylmercury inhibits TrkA signaling through the ERK1/2 cascade after NGF stimulation of PC12 cells. Brain Res. Dev. Brain Res. **149:** 53–61.
28. ARUOMA, O.I., B. HALLIWELL, B.M. HOEY & J. BUTLER. 1989. The antioxidant action of *N*-acetylcysteine: its reaction with hydrogen peroxide, hydroxyl radical, superoxide, and hypochlorous acid. Free Radic. Biol. Med. **6:** 593–597.
29. STEENVOORDEN, D.P., D.M. HASSELBAINK & G.M. BEIJERSBERGEN VAN HENEGOUWEN. 1998. Protection against UV-induced reactive intermediates in human cells and mouse skin by glutathione precursors: a comparison of *N*-acetylcysteine and glutathione ethylester. Photochem. Photobiol. **67:** 651–656.
30. SEVILLANO, S., A.M. DE LA MANO, M.A. MANSO, et al. 2003. *N*-acetylcysteine prevents intra-acinar oxygen free radical production in pancreatic duct obstruction-induced acute pancreatitis. Biochim. Biophys. Acta **1639:** 177–184.
31. KAMATA, H., C. TANAKA, H. YAGISAWA, et al. 1996. Suppression of nerve growth factor-induced neuronal differentiation of PC12 cells. *N*-acetylcysteine uncouples the signal transduction from ras to the mitogen-activated protein kinase cascade. J. Biol. Chem. **271:** 33018–33025.

Effect of Enhanced Prostacyclin Synthesis by Adenovirus-Mediated Transfer on Lipopolysaccharide Stimulation in Neuron-Glia Cultures

MAY-JYWAN TSAI,[a] SONG-KUN SHYUE,[b] CHING-FENG WENG,[c] YING CHUNG,[a] DANN-YING LIOU,[a] CHI-TING HUANG,[a] HUAI-SHENG KUO,[a] MENG-JEN LEE,[a,d] PEI-TEH CHANG,[a] MING-CHAO HUANG,[a] WEN-CHENG HUANG,[a] K.D. LIOU,[a] AND HENRICH CHENG[a,e,f,g]

[a]*Neural Regeneration Laboratory, Department of Neurosurgery, Neurological Institute, Taipei Veterans General Hospital, Taipei, Taiwan*

[b]*Institute of Biomedical Sciences, Academia Sinica, Taipei, Taiwan*

[c]*Department of Biotechnology, National Dong-Hwa University, Hualien, Taiwan*

[d]*Department of Biotechnology, Fooyin University, Kaoshiang, Taiwan*

[e]*Department of Surgery, National Yang-Ming University, Taipei, Taiwan*

[f]*Center for Neural Regeneration, Taipei Veterans General Hospital, Taipei, Taiwan*

[g]*Institute of Pharmacology, National Yang-Ming University, Taipei, Taiwan*

ABSTRACT: Prostacyclin (PGI_2) is known as a short-lived, potent vasodilator and platelet anti-aggregatory eicosanoid. This work attempts to selectively augment PGI_2 synthesis in neuron-glia cultures by adenoviral (Ad) gene transfer of PGI synthase (PGIS) or bicistronic cyclooxygenase 1 (COX-1)/PGIS and examines whether PGI_2 confers protection against lipopolysaccharide (LPS) stimulation. Cultures released low levels of eicosanoids. Upon Ad-PGIS or Ad-COX-1/PGIS infection, cultures selectively increased prostacyclin release. Both PGIS- and COX-1/PGIS–overexpressed cultures contained fewer microglial numbers. Further, they significantly attenuated LPS-induced iNOS expression and lactate, nitric oxide, and TNF-α production. Taken together, enhanced prostacyclin synthesis in neuron-glial cultures reduced microglia number and suppressed LPS stimulation.

KEYWORDS: prostaglandin; inflammation; neuroprotection; gene transfer; cyclooxygenase; prostacyclin synthase

Address for correspondence: Henrich Cheng, Neural Regeneration Laboratory, Department of Neurosurgery, Neurological Institute, Taipei Veterans General Hospital, No. 322, Shih-pai Road, Sec. 2, Taipei, Taiwan 11217. Voice: +886-2-28712121 ext. 7718; fax: +886-2-28757588.
hc_cheng@vghtpe.gov.tw

Ann. N.Y. Acad. Sci. 1042: 338–348 (2005). © 2005 New York Academy of Sciences.
doi: 10.1196/annals.1338.031

INTRODUCTION

Inflammation is an important contributor to neuronal damage in neurodegenerative diseases. Microglia, the resident immune cells in the brain, play a pivotal role in inflammatory reaction. Several lines of evidence have shown that lipopolysaccharide (LPS) treatment elicits neurotoxicity through microglial activation.[1–3] Activated microglia produce a wide array of factors, including cytokine, nitric oxide, and eicosanoids.[4,5] The accumulated impact of these factors on neurons can lead to neuronal death.[3]

Prostacyclin (prostaglandin I_2; PGI_2) is synthesized by the sequential action of cyclooxygenase (COX) 1 or 2 and prostacyclin synthase (PGIS). It is rapidly produced following tissue injury or inflammation[6] and is a paracrine mediator with two main effects: vasodilation and inhibition of platelet aggregation.[7] PGI_2 is cytoprotective at the blood–brain barrier and in cerebral circulation.[8,9] Because of its instability *in vivo,* several PGI_2 analogs have been developed and demonstrated to reduce ischemic brain damage.[8,10,11] In contrast, PGE_2 and PGD_2 have been reported as functional important products of COX in the brain, involving induction of astrogliosis.[12–14] PGE_2 is the most abundant prostaglandin in the brain. Despite its classic role as a proinflammatory molecule, much *in vivo* and *in vitro* evidence has demonstrated the anti-inflammatory effects of PGE_2.[15]

We constructed recombinant adenovirus (Ad) vectors to encode for green fluorescence protein (GFP), COX-1, COX-2, PGIS, and bicistronic COX-1/PGIS, allowing for stable delivery of these genes to infected cells. By the combined infection of Ad-COX-1 or -2 and Ad-PGIS, we selectively enhanced PGI_2 production in endothelial cells.[16,17] This work attempts to selectively augment PGI_2 synthesis in primary neuron-glia culture by direct adenoviral gene transfer of PGIS or bicistronic COX-1/PGIS and examines whether it confers protection against damage induced by LPS. LPS, a cell wall component of gram-negative bacteria, is a powerful immune challenge associated with an increase in the circulatory levels of numerous cytokines.

MATERIALS AND METHODS

Reagents

LPS (*E. coli* 0111:B4) was purchased from Sigma-Aldrich Co. (St. Louis, MO). Unless stated otherwise, all other chemicals were purchased from Sigma-Aldrich Co. [1-^{14}C] arachidonic acid and radioisotopic prostanoids were purchased from Amersham Biosciences (Buckinghamshire, UK). Acetonitrile and other organic solvents were obtained from Merck (Darmstadt, Germany).

Recombinant Adenovirus Production

Replication-defective adenoviruses containing phosphoglycerate kinase (PGK) (containing only the promoter but not the encoding genes), GFP (expressing the green fluorescence protein of jelly fish), COX-1, PGIS, and bicistronic COX-1/PGIS were produced as previously described.[16] Ad-GFP, Ad-PGIS, and Ad-COX-1/PGIS used PGK as a driving promoter. The viral titers of the purified Ads were determined by a plaque-forming assay.

Neuron-Glial Mixed Culture

Mixed neuron-glia cultures were prepared from cerebrocortical regions of embryonic Sprague-Dawley rat fetus at gestation day 15, as described by Hung et al.[18] Briefly, cells were dissociated with mixtures of papain/protease/deoxyribonuclease I (0.1%:0.1%:0.03%) and plated onto polylysine-coated multiwell plates. Cells were maintained in Dulbecco modified Eagle medium (DMEM) supplemented either with 10% fetal bovine serum (FBS) as serum-containing medium or with N2 (Gibco BRL, Grand Island, NY) as serum-free medium. During LPS challenge, the culture medium was changed to DMEM supplemented with 2% FBS and 2% horse serum.

Tranducing Cells with Ads

Cells were fed with maintaining medium. Ad was added to each well with a multiplicity of infection (MOI, pfu/cell) of 50. Ad-GFP was used for optimizing the infection, and Ad-PGK was used as vector (mock) control. Two days after Ad-GFP infection, cells were processed for immunohistochemical staining. Ad-GFP–infected cells in serum-containing medium were double-immunostained with cell markers and GFP. Several cell markers were used: MAP-2 or βIII tubulin (Chemicon, Temecula, CA) for neurons, glial fibrillary acidic protein (GFAP; Chemicon) for astrocytes, *Griffonia simplicifolia* lectin isolectin B4 (GSA) or ED-1 (Serotec, Oxford, UK) for microglia, and Reca-1 (Serotec) for endothelial cells. Ad-PGK, Ad-PGIS, or Ad-COX-1/PGIS, each at MOI 50, was added to cultured cells in serum-containing medium (DMEM plus 10% FBS). Two days later, cells were treated with LPS at a dose of 100 ng/mL for 2 days. The culture medium was then assayed for lactate dehydrogenase (LDH) release and for lactate, nitric oxide (NO), and tumor-necrosis factor a (TNF-α) measures, while cells were processed for immunocytochemistry.

Chemical Assays

LDH activity in the medium was measured spectrophotometrically as the oxidation of NADH in the presence of pyruvate at 340 nm.[19] Lactate release in the medium was assayed spectrophotometrically at 340 nm as NAD^+ reduction via a sequential reaction of LDH and alanine aminotransferase.[20] The production of NO was assayed as accumulation of nitrite in medium using colorimetric reaction with Griess reagent.[21] Briefly, after 2 days of LPS treatment, the culture supernatants (150 mL) were collected and mixed with 50 mL of Griess reagent (1% sulfanilaminde/0.1% naphthyl ethylene diamine dihydrochloride/2% phosphoric acid) and incubated at room temperature for 10 min. The absorbance was measured at 540 nm. Sodium nitrite ($NaNO_2$) was used as the standard to calculate nitrogen dioxide (NO_2^-) concentrations. TNF-α production in medium was measured by using an ELISA kit from R&D Systems (Minneapolis, MN) according to the manufacturer's instruction.

Extraction and Analysis of Arachidonic Acid Metabolites

Cultured cells after Ad infection were incubated in DMEM (serum-free) containing 10 mM [1-^{14}C] arachidonic acid (AA) at 37°C for 10 min. The cells were saved for Western blot analysis, and the media, containing released eicosanoids, were extracted by Sep-Pak C_{18} Cartridge (Waters Associates, Milford, MA) as previously described.[22] In brief, extracted ^{14}C-labeled AA metabolites were analyzed by reverse-phase HPLC column chromatography equipped with an on-line radioisotopic detector (Packard 150-TP). The stationary phase was Inertsil 7 ODS-3 (4.6 × 150 mm; Vercopak, Taiwan). The mobile phase consisted of gradient elution between solvent A (acetonitrile) and solvent B (0.1% acetic acid, pH 3.7) under the following conditions at flow rate of 1 mL/min: 34% B for 10 min, 34–40% B within 4 min, 40–50% B within 1 min, 50% B for 5 min, 50–75% B within 10 min, 75–100% B within 10 min, and 100% B for 10 min. The eicosanoids were identified by their retention times with the authentic radioisotopic standards.

Western Blot Analysis

Cultured neurons were washed twice with phosphate-buffered saline (PBS) and solubilized in 350 mL lysis buffer containing PBS, 1% NP-40, and a protease inhibitor kit (BM, Germany). The supernatant was collected after centrifugation. Protein was quantified by BCA assay (Pierce, Rockford, IL); protein (5μg) of the cell lysate was analyzed by Western blot (SDS-PAGE), using 8% or 12% gels, as previously described.[16] Many primary antibodies were used, including anti-COX-1 (Cayman Chemical, Ann Arbor, MI), anti-COX-2 (Chemicon), anti-iNOS (BD Biosciences, Caribbean, CA), anti-PGIS (customer-ordered synthesis), and anti-β-actin (Santa Cruz Biotech, Santa Cruz, CA).

RESULTS AND DISCUSSION

Adenovirus Preferentially Infected Non-Neuronal Cells

To optimize the infection condition and characterize the cell-type infective pattern, mixed neuron-glial cultures maintained in serum-free or serum-containing medium were infected with 50 MOI Ad-GFP. Two days after infection, cells were fixed and processed for immunohistochemistry. As shown in FIGURE 1A and B, Ad-GFP predominantly transfected non-neuronal cells and neurons were sparsely infected. No neurons were found to express GFP in serum-containing medium (FIG. 1B), whereas few neurons (<5% of MAP-2–positive cells) possessed GFP(+) in serum-free medium (FIG. 1A). This may be due to the presence of serum proteins, such as serum albumin, which attenuate adenoviral infectivity by nonspecific binding to adenovirus.[23,24] We performed double-immunostaining of GFP with various cell markers in serum-containing Ad-GFP–infected cultures. The results showed that among GFP(+) cells, about 30% of non-neuronal cells were GFAP(+) astrocytes (FIG. 1C), 50% of cells were ED-1(+) microglia (FIG. 1D), and <5% cells each were Reca-1(+) endothelial cells (data not shown).

Overexpression of PGIS or COX-1/PGIS–Enhanced Prostacyclin Synthesis

Mixed neuron-glia cultures were each infected with mock control (Ad-PGK), Ad-PGIS, and Ad-COX-1/PGIS at 50 MOI. Two days later, cells either were treated with LPS or were harvested to measure AA metabolic activity and protein expression. FIGURE 2A demonstrates that mixed neuron-glia cultures produced low basal levels

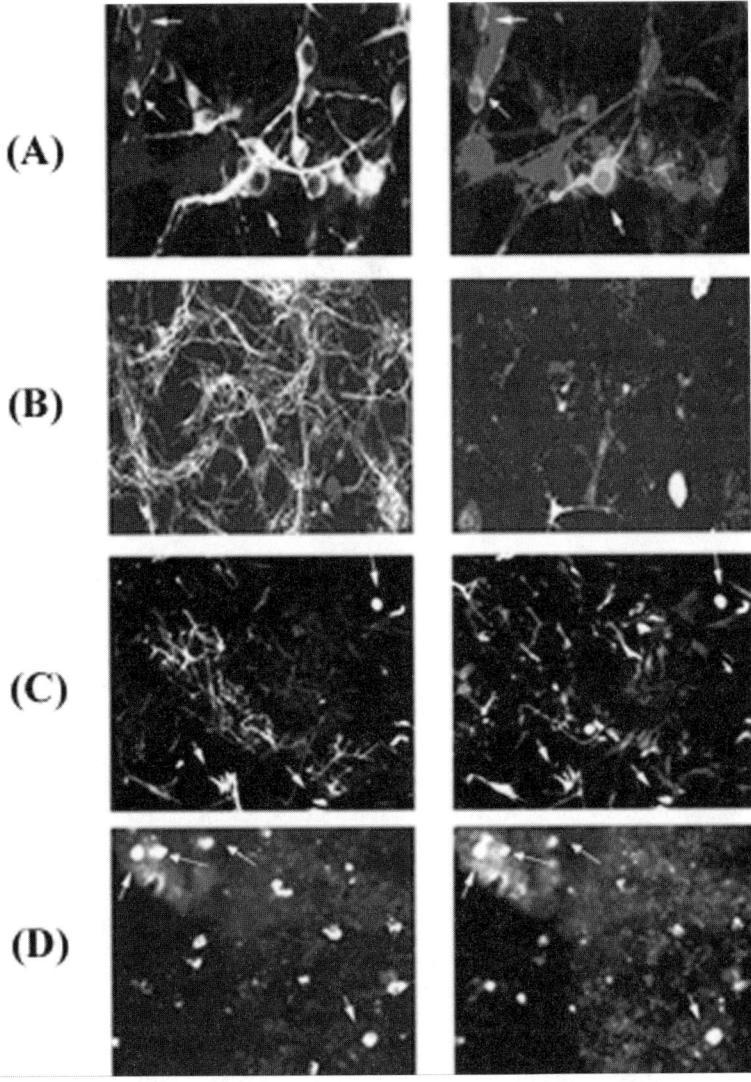

FIGURE 1. Low neuronal infection efficiency with Ad-GFP in neuron-glia cultures. Cultures were infected by Ad-GFP (at 50 MOI) for 2 days and processed for morphological analysis. (**A**) Cells were maintained in serum-free medium during Ad-GFP infection. *Left*, MAP-2 positive; *right*, GFP positive. (**B–D**) Cells were maintained in serum-supplemented medium during Ad-GFP infection. *Left panels* in **B, C,** and **D** are MAP-2, GFAP, and ED-1 positive, respectively; *right panels* in **B–D** are GFP positive. Arrow indicates colocalization of two staining antibodies. Images presented are from one experiment and are representative of at least three independent experiments.

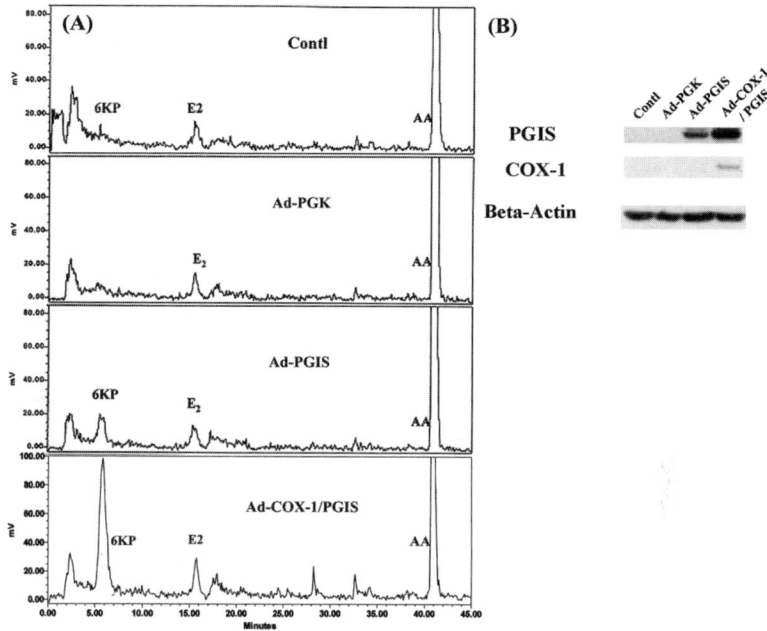

FIGURE 2. Adenovirus-mediated overexpression of COX-1 and PGIS and the resultant augmentation of 6KP synthesis in neuron-glia cultures. (**A**) Analysis of eicosanoids generated by Ad-infected cultures in response to [1-^{14}C] AA treatment. 6KP is a degradation product of prostacyclin. Peaks of the first 5-min fractions are nonspecific. Each prostanoid peak was verified by co-elution with an authentic radiolabeled prostanoid. (**B**) Western blot analysis of COX-1 and PGIS protein levels in cultured neurons, each infected with 50 MOI of Ad. Both figures are representative of two experiments with similar results.

of eicosanoids. Upon Ad-PGIS or Ad-COX1/PGIS infection, the level of ^{14}C-labeled 6-keto PGF$_{1a}$ (6KP), a degradation product of PGI$_2$, was enhanced. Ad-PGIS–infected cells produced more PGI$_2$ than that of control or mock control (Ad-PGK only). Furthermore, ^{14}C-labeled AA metabolites were efficiently shunted through ^{14}C-labeled 6KP synthesis in bicistronic Ad-COX-1/PGIS–infected cells. The overexpressed COX-1 actively converted ^{14}C-labeled AA to PGH$_2$, which was hydrolyzed to PGE$_2$ and was further converted to ^{14}C-labeled 6KP by the overexpressed PGIS in the same cell. This is consistent with the results of our previous study, which found that the optimal ratios for achieving large PGI$_2$ augmentation without overproduction of other prostanoids were 1 or 2 parts of Ad-COX-1 to 1 part of Ad-PGIS.[17] Western blot analysis of these Ad-infected cells are shown in FIGURE 2B. Significantly enhanced PGIS and/or COX-1 proteins were detected in Ad-PGIS– and Ad-COX-1/PGIS–infected cells, respectively. This indicates that the transduced enzymes are functionally active in producing prostacyclin from AA. Because Ad-GFP predominantly infected microglia or astroglia, Ad-PGIS– or Ad-COX-1/PGIS–infected cells would increase synthesis of prostacyclin, which was then released and

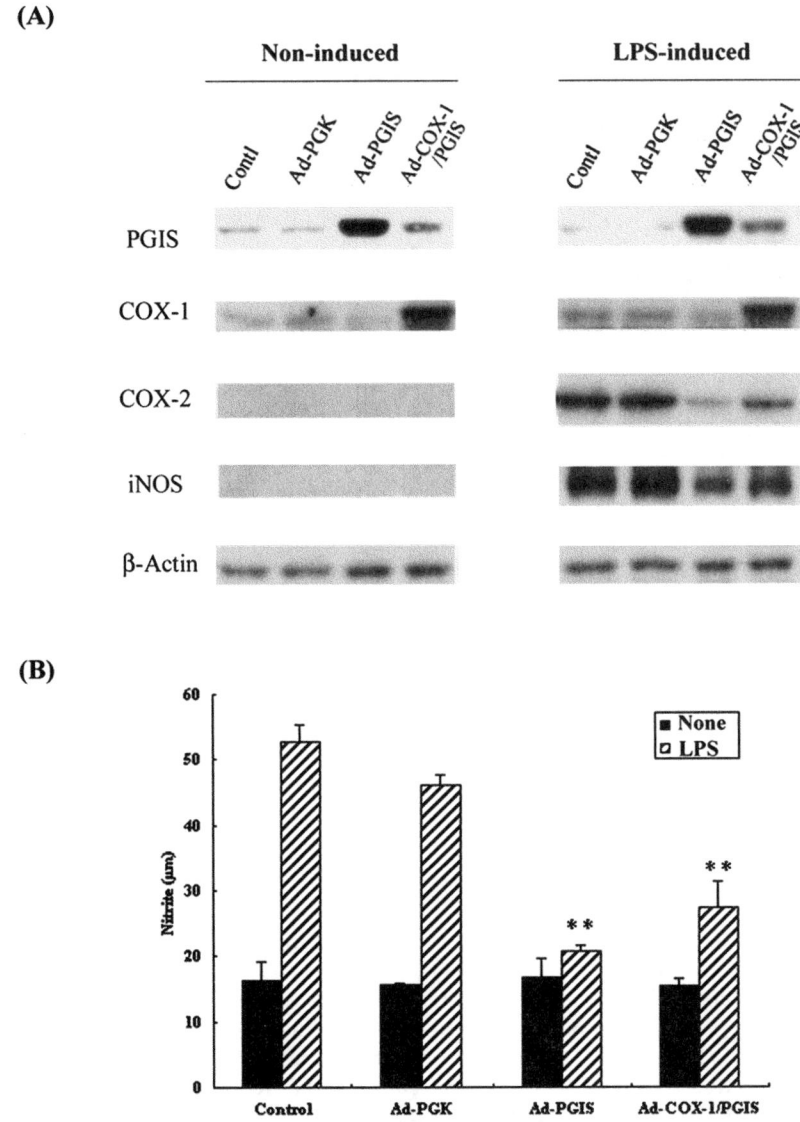

FIGURE 3. NO synthesis and expressions of iNOS, COX-2, and PGIS in Ad-PGK–, Ad-PGIS–, and Ad-COX-1/PGIS–infected neuron-glia cultures before and after LPS challenge. (**A**) Western blot analysis of protein levels of PGIS, COX-1, COX-2, and iNOS in neuron-glia cultures receiving Ad-PGK, Ad-PGIS, or Ad-COX-1/PGIS infection. After 2 days of Ad infection, cultures were treated with LPS (100 ng/mL) for 2 days and harvested. Equal amounts of protein (5 µg/lane) were analyzed by Western blot using anti-PGIS, anti-COX-1, anti-COX-2, and anti-iNOS antibodies. Protein bands were visualized using horseradish peroxidase-conjugated secondary antibodies and electrochemiluminescence (ECL). One experiment, which represents two experiments, is shown with similar results. (**B**) Inhibition of LPS-induced NO

affected microglia. Prostacyclin subsequently bound to IP receptors in cells exerting downstream signal transduction.

Overexpression of PGIS or COX-1/PGIS–Reduced LPS Stimulation in Mixed Neuron-Glia Cultures

LPS treatment can induce microglial activation and release a variety of cytokines, NO, and eicosanoids. Whether Ad infection can affect LPS stimulation is not clear. To address this question, mixed neuron-glia cultures were infected with each of Ad-PGK, Ad-PGIS, and Ad-COX-1/PGIS. Two days later, cells were further treated with LPS at a dose of 100 ng/mL for 2 days. As shown in FIGURE 3A, Ad-PGIS or Ad-COX-1/PGIS infection enhanced the expression of PGIS or COX-1/PGIS, whereas iNOS and COX-2 levels were not detectable. By contrast, LPS treatment induced both iNOS and COX-2 expression in control and Ad-PGK–infected cells. Ad-PGIS or Ad-COX-1/PGIS infection effectively reduced LPS-induced COX-2 and iNOS levels. Concurrently, LPS-stimulated nitrite releases were significantly inhibited in Ad-PGIS– and Ad-COX-1/PGIS–infected cells (FIG. 3B) compared to control and Ad-PGK–infected cells. This indicates that enhanced prostacyclin synthesis in PGIS– or COX-1/PGIS–overexpressed cells significantly inhibits iNOS expression and NO production. This inhibitory effect on NO production seems to be due to a reduction in GSA(+) microglial numbers (FIG. 4A) and/or a reduction in iNOS protein expression (FIG. 3A). Cell viabilities, measured as LDH release in medium, were not affected by Ad infection or LPS treatment. Furthermore, GFAP(+) astrocytes were not affected by either Ad infection or LPS treatment (data not shown).

We also measured TNF-α levels and lactate release in the medium. Increased levels of lactate, a product of glycolysis, is a sensitive index for impairment in energy metabolism. As shown in FIGURE 4B and C, significantly lower TNF-α levels were found in Ad-PGIS– and Ad-COX-1–infected cells compared to those of the control. LPS, at a dose of 100 ng/mL, significantly enhanced TNF-α and lactate levels in the control and Ad-PGK-infected cells (FIG. 4B and C). This LPS-stimulating effect was not seen in Ad-PGIS– or Ad-COX-1–infected cells. This indicates that overexpression of PGIS or COX-1/PGIS with enhanced prostacyclin synthesis in culture significantly inhibited LPS-induced lactate and TNF-α release. The lactate levels induced by LPS treatment could be due to the enhanced production of NO and TNF-α, which impairs oxidative phosphorylation.

Results from this study demonstrate that PGI_2 synthesis can be augmented by cultured neurons infected with either Ad-PGIS only or with bicistronic Ad-COX-1/PGIS. The mechanism by which Ad-PGIS– or Ad-COX-1/PGIS–infected cells reduces microglial numbers has not yet been clarified and is subject to hypotheses at present. In the brain, at least two distinct prostacyclin receptors, designated as IP_1 and IP_2, were reported.[25] The IP_2 receptor was found only in the central nervous sys-

---

production, as nitrite release, in the medium by Ad-PGK, Ad-PGIS, or Ad-COX-1/PGIS infection in neuron-glia cultures. Control *on x-axis,* no viral infection of naive cultures; *filled bar* in chart, unstimulated cultures. Data were expressed as means ± SEM from four independent experiments done in triplicate. *Asterisks* (**) indicate significant differences between no stimulation and LPS stimulation ($P < 0.01$ by ANOVA and *t*-test) within each Ad-infected cell.

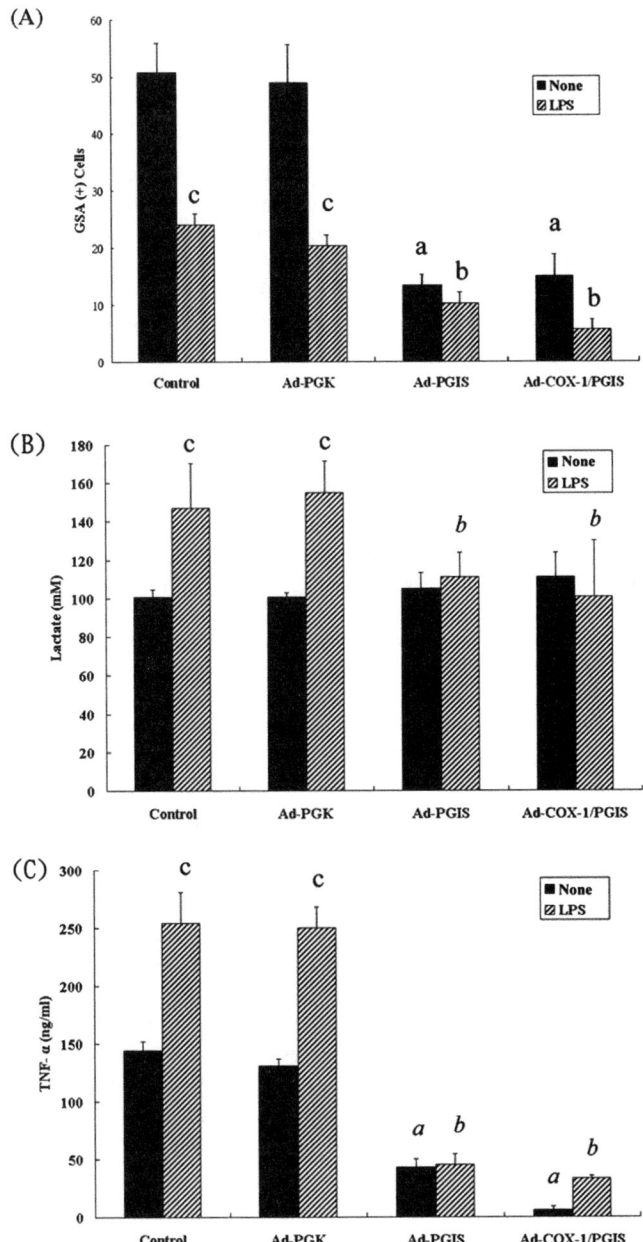

FIGURE 4. Effect of Ad-PGIS or Ad-COX-1/PGIS infection on microglial numbers, lactate, and TNF-α levels in neuron-glia cultures before and after LPS stimulation. (**A**) Lectin-reactive microglia cells expressed as GSA(+)/field (magnification ×100). (**B**) Lactate levels in the medium. (**C**) TNF-α level in the medium. Control on *x-axis,* no viral infection of naive cul-

tem and therefore termed as central-type prostacyclin receptor.[25] The IP_1 receptor is mainly coupled to G protein-coupled receptor (Gs); hence stimulation of the receptor results in cAMP production.[26] IP_2, however, had no effect on cAMP production. A selective ligand for IP_2 attenuated neuronal damage in cerebral ischemia, whereas treatment with iloprost, a selective IP_1 ligand, did not show a protective effect.[27] In the present study, microglial numbers were significantly reduced in Ad-PGIS– and Ad-COX-1/PGIS–infected cells. It remains to be determined which receptor subtype (IP_1 or IP_2) is involved in this effect. Further studies are needed to elucidate the role played by downstream molecules in the inhibition of microglial activation.

Understanding the role of prostaglandin *in vivo* is difficult because the activated COX cascade is often accompanied by the generation of a broad range of other active molecules, including cytokines, NO and the prostanoid precursor arachidonic acid itself and its metabolites. Furthermore, individual prostaglandins may also exert opposite or synergetic effects on common targets. Thus, the final contribution of prostaglandins to tissue damage or repair seems to depend on the balance between the different prostaglandins released by non-neuronal or neuronal cells in different circumstances. Here we show that enhanced prostacyclin synthesis effectively reduced microglial numbers and LPS stimulation. Inhibition of the microglial response by enhanced prostacyclin synthesis could be an effective therapeutic approach to alleviating the progression of diseases. We suggest that prostacyclin may have a neuroprotective role in modulating the immune response at the injured site.

ACKNOWLEDGMENTS

This work was supported by National Science Council Grants NSC 91-2320-B-075-017, NSC 92-3112-B-075-001, and NSC 92-3112-B-075-002 and by Taipei Veterans General Hospital Grants VGH92-387 and VGH93-371.

REFERENCES

1. LEVI, G., L. MINGHETTI & F. ALOISI. 1998. Regulation of prostanoid synthesis in microglial cells and effects of prostaglandin E2 on microglial functions. Biochimie **80:** 899–904.
2. LIU, B., L. DU & J.S. HONG. 2000. Naloxone protects rat dopaminergic neurons against inflammatory damage through inhibition of microglia activation and superoxide generation. J. Pharmacol. Exp. Ther. **293:** 607–617.
3. LIU, B. & J.S. HONG. 2003. Role of microglia in inflammation-mediated neurodegenerative diseases: mechanisms and strategies for therapeutic intervention. J. Pharmacol. Exp. Ther. **304:** 1–7.
4. McGUIRE, S.O. *et al.* 2001. Tumor necrosis factor alpha is toxic to embryonic mesencephalic dopamine neurons. Exp. Neurol. **169:** 219–230.

---

tures. [a]Significant difference between control and Ad-infected cells ($P < 0.05$ or 0.01 by ANOVA and *t*-test). [b]Significant difference between LPS-stimulated and LPS plus Ad-infected cells ($P < 0.05$ by ANOVA and *t*-test). [c]Significant difference between control and LPS within each Ad-infected cell. Data were expressed as means ± SEM from three or four independent experiments done in triplicate.

5. LIU, B. *et al.* 2002. Role of nitric oxide in inflammation-mediated neurodegeneration. Ann. N.Y. Acad. Sci. **962:** 318–331.
6. DAVIES, P. *et al.* 1984. The role of arachidonic acid oxygenation products in pain and inflammation. Annu. Rev. Immunol. **2:** 335–357.
7. MONCADA, S. & J.R. VANE. 1979. The role of prostacyclin in vascular tissue. Fed. Proc. **38:** 66–71.
8. SATOH, T. *et al.* 1999. CNS-specific prostacyclin ligands as neuronal survival-promoting factors in the brain. Eur. J. Neurosci. **11:** 3115–3124.
9. LIN, H. *et al.* 2002. Cyclooxygenase-1 and bicistronic cyclooxygenase-1/prostacyclin synthase gene transfer protect against ischemic cerebral infarction. Circulation **105:** 1962–1969.
10. CUI, Y. *et al.* 1999. Protective effect of prostaglandin I(2) analogs on ischemic delayed neuronal damage in gerbils. Biochem. Biophys. Res. Commun. **265:** 301–304.
11. NAREDI, S. *et al.* 2001. An outcome study of severe traumatic head injury using the "Lund therapy" with low-dose prostacyclin. Acta Anaesthesiol. Scand. **45:** 402–406.
12. BLOM, M.A. *et al.* 1997. NSAIDs inhibit the IL-1 beta-induced IL-6 release from human post-mortem astrocytes: the involvement of prostaglandin E2. Brain Res. **777:** 210–218.
13. BRAMBILLA, R. *et al.* 1999. Cyclo-oxygenase-2 mediates P2Y receptor-induced reactive astrogliosis. Br. J. Pharmacol. **126:** 563–567.
14. MINGHETTI, L. *et al.* 1997. Inducible nitric oxide synthase expression in activated rat microglial cultures is downregulated by exogenous prostaglandin E2 and by cyclooxygenase inhibitors. Glia **19:** 152–160.
15. ZHANG, J. & S. RIVEST. 2001. Anti-inflammatory effects of prostaglandin E2 in the central nervous system in response to brain injury and circulating lipopolysaccharide. J. Neurochem. **76:** 855–864.
16. LIOU, J.Y. *et al.* 2000. Colocalization of prostacyclin synthase with prostaglandin H synthase-1 (PGHS-1) but not phorbol ester-induced PGHS-2 in cultured endothelial cells. J. Biol. Chem. **275:** 15314–15320.
17. SHYUE, S.K. *et al.* 2001. Selective augmentation of prostacyclin production by combined prostacyclin synthase and cyclooxygenase-1 gene transfer. Circulation **103:** 2090–2095.
18. HUNG, H. *et al.* 2000. Age-dependent increase in C7-1 gene expression in rat frontal cortex. Brain Res. Mol. Brain Res. **75:** 330–336.
19. KOH, J.Y. & D.W. CHOI. 1987. Quantitative determination of glutamate mediated cortical neuronal injury in cell culture by lactate dehydrogenase efflux assay. J. Neurosci. Methods **20:** 83–90.
20. ADAMS, J.D. JR., P.W. KALIVAS & C.A. MILLER. 1989. The acute histopathology of MPTP in the mouse CNS. Brain Res. Bull. **23:** 1–17.
21. GREEN, L.C. *et al.* 1982. Analysis of nitrate, nitrite, and [15N]nitrate in biological fluids. Anal. Biochem. **126:** 131–138.
22. SANDUJA, S.K. *et al.* 1991. Differentiation-associated expression of prostaglandin H and thromboxane A synthases in monocytoid leukemia cell lines. Blood **78:** 3178–3185.
23. HORWITZ, M.S., B.R. FRIEFELD & H.D. KEISER. 1982. Inhibition of adenovirus DNA synthesis in vitro by sera from patients with systemic lupus erythematosus. Mol. Cell. Biol. **2:** 1492–1500.
24. NADEAU, I. *et al.* 2002. Low-protein medium affects the 293SF central metabolism during growth and infection with adenovirus. Biotechnol. Bioeng. **77:** 91–104.
25. TAKECHI, H. *et al.* 1996. A novel subtype of the prostacyclin receptor expressed in the central nervous system. J. Biol. Chem. **271:** 5901–5906.
26. NAMBA, T. *et al.* 1994. cDNA cloning of a mouse prostacyclin receptor. Multiple signaling pathways and expression in thymic medulla. J. Biol. Chem. **269:** 9986–9992.
27. TAKAMATSU, H. *et al.* 2002. Specific ligand for a central type prostacyclin receptor attenuates neuronal damage in a rat model of focal cerebral ischemia. Brain Res. **925:** 176–182.

Potential Mechanism of Blood Vessel Protection by Resveratrol, a Component of Red Wine

HUEI-MEI HUANG,[a] YU-CHIH LIANG,[b] TZU-HURNG CHENG,[c] CHENG-HEIEN CHEN,[c] AND SHU-HUI JUAN[d]

[a]*Graduate Institute of Cell and Molecular Biology, Center for Stem Cells Research at Wan-Fang Hospital, Taipei Medical University, Taipei, Taiwan*

[b]*Biomedical Technology, Taipei Medical University, Taipei, Taiwan*

[c]*Department of Medicine, Taipei Medical University–Wan Fang Hospital, Taipei, Taiwan*

[d]*Graduate Institute of Medical Sciences and Department of Physiology, School of Medicine, Taipei Medical University, Taipei, Taiwan*

ABSTRACT: Resveratrol-mediated heme oxgenase-1 (HO-1) induction has been shown to occur in primary neuronal cultures and has been implicated as having potential neuroprotective action. Further, antioxidant properties of resveratrol have been reported to protect against coronary heart disease. We attempted to examine the HO-1 inducing potency of resveratrol and the regulatory mechanism of its induction in rat aortic smooth muscle cells (RASMC). We showed that resveratrol-induced HO-1 expression was concentration- and time-dependent. The level of HO-1 expression and its promoter activity mediated by resveratrol was attenuated by nuclear factor-kappa B (NF-κB) inhibitors, but not by mitogen-activated protein kinase (MAPK) inhibitors. Deletion of NF-κB binding sites in the promoter region strongly reduced luciferase activity. Collectively, we suggest that resveratrol-mediated HO-1 expression occurs, at least in part, through the NF-κB pathway.

KEYWORDS: rat aortic smooth muscle cell; heme oxygenase-1; resveratrol; nuclear factor-kappa B

INTRODUCTION

Resveratrol, a polyphenolic compound, is abundant in red wine, grape skin, and seeds. It is present in the *cis* and *trans* isoforms, of which the latter is the biologically active one. Epidemiologic studies have reported that a low incidence of cardiovascular disease is associated with moderate red wine consumption in southern France—the so-called French paradox.[1] Additionally, *in vivo* and *in vitro* studies have revealed that resveratrol has significant antioxidant properties and is able to protect against coronary heart disease.[2]

Address for correspondence: Shu-Hui Juan, Ph.D., Graduate Institute of Medical Sciences and Department of Physiology, Taipei Medical University, 250 Wu-Hsing Street, Taipei 110, Taiwan. Voice: +886-2-27361661 ext. 3717; fax: +886-2-27390214.
juansh@tmu.edu.tw

Heme oxygenase (HO) is a rate-limiting catalyst in the heme degradation that produces biliverdin, free iron, and carbon monoxide. Three HO isozymes have been identified to have distinct genes.[3] Among them, HO-1, a stress-response protein, can be induced by various oxidative-inducing agents, including heme, heavy metals, UV radiation, cytokines, and endotoxin.[4] The potential use of HO-1 on therapeutic targets in various diseases has been explored. For instance, adenovirus-mediated gene transfer of HO-1 in animal models can protect against hyperoxia-induced lung injury[5] and reperfusion-induced injury of a transplanted liver.[6] We recently reported that adenovirus-mediated HO-1 gene transfer inhibits the development of atherosclerosis in apolipoprotein E–deficient mice.[7] Its notable effect prompts us to search for an alternative medicine to serve as a potent HO-1 inducer.

Activation of the transcription of HO-1 or other genes is mediated by a cascade of signal transduction pathways, mainly by modulating the activities of transcription factors. MAPKs contribute to the major pathway for regulating numerous cellular processes such as cell proliferation, differentiation, stress response, and cell death. Nevertheless, conflicting results exist in the literature: stress-related stimuli (sodium arsenite, cobalt chloride, hemin, and cadmium) might be regulated through extracellular signal-regulated kinase 1/2 (ERK1/2) or c-Jun N-terminal kinase (JNK or p38MAPK) in the transcription of HO-1 gene expression in different tissues or species. Furthermore, the redox-sensitive transcription factors of AP-1, Nrf2, and NF-κB have been shown to be involved in heavy metals' or curcumin's (a major component of the food spice turmeric) induction of HO-1 in tumor and human renal proximal tubule cells.[8,9]

Our study attempted to reveal resveratrol's HO-1 inducing effect in preventing cardiovascular diseases, and to explore its inducing potency and molecular mechanism in rat aortic smooth muscle cells (RASMC).

MATERIALS AND METHODS

Resveratrol and curcumin were purchased from Sigma Chemical Co. U0126 and SB202190 were obtained from Tocris Cookson (Bristol, UK). [γ-^{32}P]dATP (6000 Ci/mmol) was purchased from Perkin-Elmer Life Science (North Point, Hong Kong). Dulbecco's modified Eagle's medium (DMEM) medium, fetal bovine serum (FBS), and tissue culture reagents were obtained from Invitrogen (Carlsbad, CA). Protein assay agents were purchased from Bio-Rad (Hercules, CA).

Culture of Rat Aortic Smooth Muscle Cells

RASMC were isolated from thoracic aorta of male Sprague-Dawley rats (200–250 g) by explant outgrowth[10] and subculture in DMEM supplemented with 10% FBS. The purity and identity of cells were examined by immunostaining using specific antibody against smooth muscle cell α-actin. Cells from passage 5 to 12 were used for experiments.

Constructs of Plasmid Variants and Luciferase Activity Assay

The pGL3/hHO-1 reporter plasmid, which contains a 3293-bp fragment, −3106 to +186 relative to the transcription start of the human HO-1 gene, was amplified from the human BAC clone CTA-286B10[11] using the primers 5'-AGAGAACAGT-TAGAAAAGAAAG-3' (sense) and 5'-TACGGGCACAGGCAGGATCAGAA-3' (antisense). The PCR products were inserted to the pCR2.1-TOPO cloning vector (Invitrogen) and cut with KpnI/XbaI, such that the resulting PCR products contained the KpnI/XbaI site, and this was ligated into the unique KpnI/NheI site present within the pGL3 plasmid (Promega). Therefore, we obtained a pGL3/hHO-1 reporter construct containing about a 3.3-kbp region of human HO-1 promoter driving luciferase gene expression. The luciferase reporter gene constructs pHO-2920 (BsiHKAI/BglII), pHO-2739 (SspI/BglII), pHO-2520 (HindIII/BglII), pHO-1890 (BstEII/BglII), and pHO-1041 (BciVI/BglII) were generated by deletion of the −3293/−2920 fragment of HO-2920, −3293/−2520 fragment of HO-2520, etc., by the restriction enzymes indicated in parentheses. The Klenow enzyme was used to create a blunt end at the 5'-sequence, and BglII was used to generate a 3'-cohesive end. These fragments were ligated into the SmaI and BglII sites of pGL3. The identities of the sequences were confirmed using an ABI PRISM 377 DNA Analysis System (Perkin-Elmer).

For the reporter activity assay, cells were seeded in six-well plates at a density of 1×10^5 cells/well. In brief, RASMC were transiently transfected with 3 µg of plasmid DNA containing 1 µg of the Renilla luciferase construct, phRL-TK (Promega) to control transfection efficiency, and 2 µg of the appropriate HO-1 promoter firefly luciferase (FL) construct. The next day, cells were transfected with pGL3/hHO-1 and phRL-TK (Promega) as internal control plasmid using the Lipofet-AMINE 2000^{TM} (Invitrogen). After transfection (12 h), the medium was replaced with complete medium and continually incubated for another 36 h. Transfected cells were then treated with drugs for 12 h, and cell lysates were collected. Luciferase activities were recorded in a TD-20/20 luminometer (Turner Designs) using the dual luciferase assay kit (Promega) according to the manufacturer's instructions. Luciferase activities of reported plasmids were normalized to luciferase activities of the internal control plasmid.

Western Blots

Western blot analysis was carried out using the following antibodies: HO-1 and α-tubulin. To prepare whole-cell lysates, cells were washed twice with ice-cold PBS, resuspended in ice-cold RIPA buffer containing 1 mM Na_3VO_4, 1 mM PMSF, 1 µg/mL leupeptin, and 1 µg/mL aprotinin, incubated on ice for 30 min, and vortexed every 10 min followed by centrifugation at 12,000 rpm for 30 min at 4°C. Whole-cell lysates (80 µg) were electrophoresed on 10% SDS-polyacrylamide gels and then transblotted onto Hybond-P membranes Pharmacia). Membranes were blocked in PBS containing 0.1% Tween-20 and 5% skim milk at room temperature for 30 min. Blots were incubated, in blocking buffer, with the indicated antibodies (with dilutions used according to the manufacturer's instructions) for 1 h at room temperature. After three washes with PBS containing 0.1% Tween-20, blots were incubated with peroxidase-conjugated goat anti-rabbit immunoglobulin G (IgG)/anti-mouse IgG (1:5000) for 1 h at room temperature, followed by another washing. Expression of protein was detected using an enhanced chemiluminescence system.

Electrophoretic Mobility Shift Assay

To prepare nuclear protein extracts, RASMC, in 10-cm^2 dishes, were washed twice with ice-cold PBS and scraped off into 1 mL PBS after having been treated with resveratrol for 8 h in various concentrations as were indicated. After centrifugation of the cell suspension at 500 × g for 3 min, the supernatant was removed and cell pellets were subjected to NE-PER^{TM} nuclear extraction reagents (Pierce) with the addition of 0.5 mg/mL benzamidine, 2 µg/mL aprotinin, 2 µg/mL leupeptin, and 2 mM phenylmethylsulfonyl fluoride. The subsequent procedures for the nuclear protein extraction followed the manufacturer's instructions. The fraction containing the nuclear protein was used for the assay or was stored at −70°C until use. The sequence of the oligonucleotides used was AGTTGAGGGGACTTTCCAGGC for NF-κB binding (Promega). The oligonucleotides were end-labeled with [γ-^{32}P] dATP. Extracted nuclear proteins (10 µg) were incubated with 0.1 ng of ^{32}P-labeled DNA for 15 min at room temperature in 25 µL of binding buffer containing 1 µg of poly (dI-dC). For competition with unlabeled oligonucleotides, a 100-fold molar excess of unlabeled oligonucleotides relative to the radiolabeled probe was added to the binding assay. Mixtures were electrophoresed on 5% nondenaturing polyacrylamide gels. Gels were dried and imaged by means of autoradiography.

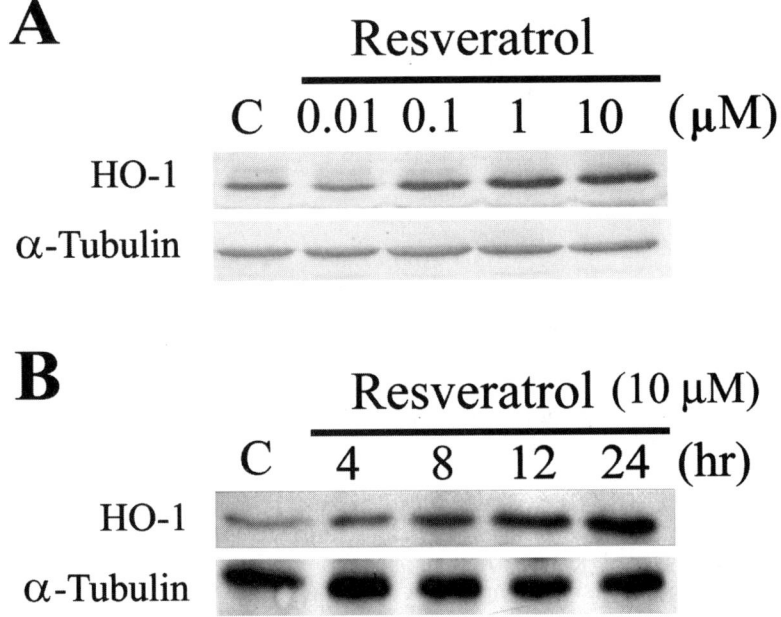

FIGURE 1. Western blot analysis of resveratrol-mediated HO-1 expression in various concentrations (**A**) and time points (**B**). Equal loading was confirmed by incubating with an anti-tubulin antibody. Results of three experiments are shown.

Statistical Analysis

Values are expressed as the mean ± SD. The significance of the difference from the control group was analyzed by Student's t-test. A value of $P < 0.05$ was considered statistically significant.

RESULTS AND DISCUSSION

Cells were treated with various concentrations of resveratrol. After 48 h in culture, cell lysates were prepared, and the protein levels of HO-1 in treated cells were assessed by Western blot analysis. A 32-kDa protein band corresponding to HO-1 was highly expressed in resveratrol-treated cells in a concentration-dependent manner (FIG. 1A). Time-course experiments demonstrated that the expression of HO-1 was observed at 4 h after resveratrol treatment and sustained ≥ 24 h after treatment (FIG. 1B). Given that resveratrol induces HO-1, it is tempting to postulate that the antioxidant and anti-inflammatory properties of resveratrol may be, at least in part, related to HO-1 induction. Resveratrol-mediated HO-1 induction is regulated at both the transcriptional and translational levels, because actinomycin D (a transcriptional inhibitor) and cycloheximide (a translational inhibitor) eliminated HO-1 mRNA and protein expression (data not shown). The mechanism of translation regulation of HO-1 induction by resveratrol remains to be established.

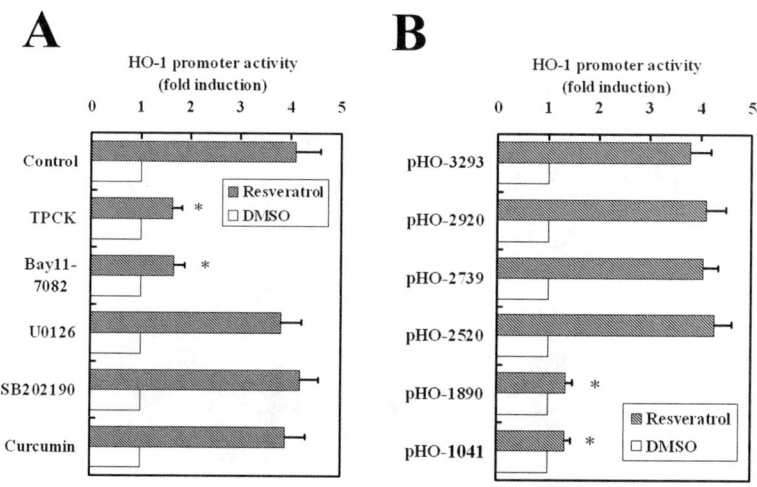

FIGURE 2. Resveratrol-mediated HO-1 promoter activity in relation to NF-κB and MAPK inhibitors (**A**) and deletion mutations in the HO-1 promoter region (**B**). RASMC were transiently transfected with pGL3/hHO-1 variants and phRL-TK as an internal control plasmid for 48 h, followed by treatment of inhibitors (10 μM) for 1 h prior to incubation with resveratrol (5 μM) for another 12 h. Luciferase activities of the reported plasmid were normalized to those of the internal control plasmid and are presented as the mean ± SD of three independent experiments. *$P < 0.01$ indicates a significant difference from the control.

To reveal the molecular mechanism of resveratrol-mediated HO-1 induction, MAPK and NF-κB inhibitors were employed. The results showed that the level of resveratrol-induced HO-1 expression was attenuated by TPCK (a protease inhibitor that blocks activation of NF-κB) and BAY 11-7082 (an inhibitor of IκB phosphorylation), but not by inhibitors of Erk1/2, JNK, and p38MAPK (data not shown).

Another approach consisting of deletion mutations was used to confirm that the importance of the NF-κB binding site is attributable to resveratrol-mediated HO-1 induction. A luciferase reporter vector containing −3293 bp of the human HO-1 promoter region was obtained. This construct was transiently transfected into RASMC for 48 h followed by treatment with MAPK and NF-κB inhibitors (10 μM) for 1 h prior to the addition of resveratrol (5 μM). After incubation for another 12 h, cells were harvested and analyzed for luciferase activity. Consistently, NF-κB inhibitors abolished the HO-1 promoter activity induced by resveratrol, but not by MAPK inhibitors (FIG. 2A). NF-κB consensus binding elements are located at −2641 to −2632 and −2451 to −2442 bp in the HO-1 promoter region. A series of deletion mutations on HO-1 promoter were carried out, and their activities were eliminated when the consensus binding sites for NF-κB were removed (FIG. 2B).

The most significant finding in this study is the demonstration of the involvement of the NF-κB pathway in resveratrol-mediated HO-1 gene induction. The electro-

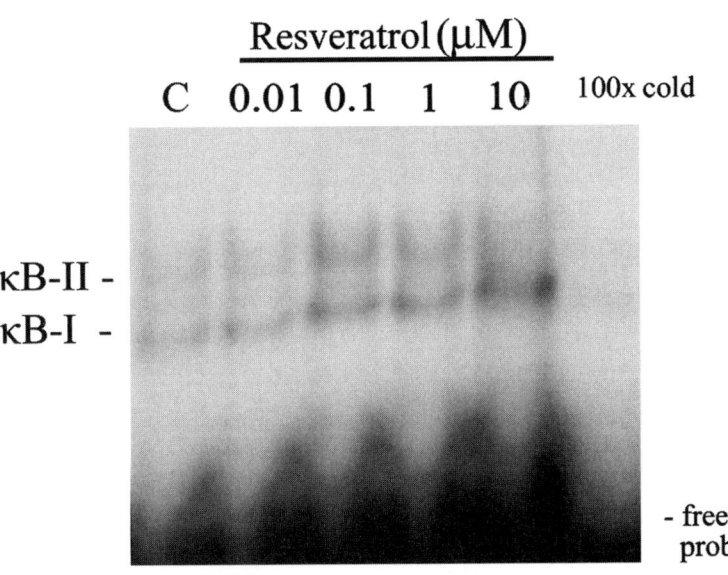

FIGURE 3. The electrophoretic mobility shift assay (EMSA) of the NF-κB binding activity by resveratrol. Binding activity of NF-κB in RASMC exposed to resveratrol. Cells were cultured and treated with increasing concentrations of resveratrol as indicated. Nuclear proteins were assayed for NF-κB binding activity by EMSA. "100× cold" denotes a 100-fold molar excess of unlabeled oligonucleotides relative to the ^{32}P-labeled probe; this was added to the binding assay for competition with the unlabeled oligonucleotides. The mobility of specific NF-κB complexes is indicated. Representative results of three separate experiments are shown.

phoretic mobility shift assay revealed that resveratrol—within a range of 0.1 to 10 µM—indeed increased NF-κB binding activity (FIG. 3). Evidence in support of this pathway was provided by two approaches. The NF-κB inhibitors, TPCK and BAY 11-7082, completely prevented resveratrol-induced HO-1 expression and the activity of the HO-1 promoter, which indicated involvement of the NF-κB pathway. In addition, serial mutants confirmed that NF-κB binding elements are responsible for resveratrol-mediated HO-1 induction. Furthermore, our study revealed that the transactivation of NF-κB by resveratrol within a range of 0.1 to 10 µM occurs by increasing the extent of IκBα phosphorylation. As a result, resveratrol facilitated the translocation of NF-κB into nucleus and then modulated HO-1 gene expression (data not shown).

NF-κB is generally considered a redox-sensitive and proinflammatory transcription factor. Interestingly, the results of our study are in accord with a recent study showing that curcumin-induced HO-1 upregulation occurs through the NF-κB pathway.[9] Both resveratrol and curcumin are natural polyphenolic compounds possessing antitumor and anti-inflammatory properties. Thus, the structural and functional similarities of resveratrol and curcumin also exert similar molecular actions on HO-1 induction. It is intriguing that resveratrol, on the one hand, induces HO-1 via a pro-oxidative and proinflammatory transcription factor (NF-κB) but, on the other hand, possesses antioxidant and anti-inflammatory properties, further emphasizing the functional importance of HO-1 as an adaptive response to oxidative stress and inflammation.

In conclusion, results from our study indicate that the activation of NF-κB is attributable to resveratrol-mediated HO-1 induction. Recently, numerous *in vitro* and *in vivo* studies have shown that the induction of HO-1 is an important cellular protective mechanism against oxidative injury.[4] Therefore, resveratrol might be a potential dietary component for protecting cells and tissues against oxidative injuries. Further studies using resveratrol will clarify the possibility of developing this new "drug" for the prevention or treatment of cardiovascular disease.

ACKNOWLEDGMENTS

This study was supported by grants from the National Science Council, Taiwan (NSC92-2320-B-038-015; NSC93-2320-B-038-022).

REFERENCES

1. RENAUD, S. & M. DE LORGERIL. 1992. Wine, alcohol, platelets, and the French paradox for coronary heart disease. Lancet **339:** 1523–1526.
2. BORS, W. & M. SARAN. 1987. Radical scavenging by flavonoid antioxidants. Free Radic. Res. Commun. **2:** 289–294.
3. MAINES, M.D. 1997. The heme oxygenase system: a regulator of second messenger gases. Annu. Rev. Pharmacol. Toxicol. **37:** 517–554.
4. PLATT, J.L. & K.A. NATH. 1998. Heme oxygenase: protective gene or Trojan horse. Nat. Med. **4:** 1364–1365.
5. OTTERBEIN, L.E. *et al.* 1999. Exogenous administration of heme oxygenase-1 by gene transfer provides protection against hyperoxia-induced lung injury. J. Clin. Invest. **103:** 1047–1054.
6. AMERSI, F. *et al.* 1999. Upregulation of heme oxygenase-1 protects genetically fat Zucker rat livers from ischemia/reperfusion injury. J. Clin. Invest. **104:** 1631–1639.

7. JUAN, S.H. *et al.* 2001. Adenovirus-mediated heme oxygenase-1 gene transfer inhibits the development of atherosclerosis in apolipoprotein E-deficient mice. Circulation **104:** 1519–1525.
8. ELBIRT, K.K. *et al.* 1998. Mechanism of sodium arsenite-mediated induction of heme oxygenase-1 in hepatoma cells. Role of mitogen-activated protein kinases. J. Biol. Chem. **273:** 8922–8931.
9. HILL-KAPTURCZAK, N. *et al.* 2001. Mechanism of heme oxygenase-1 gene induction by curcumin in human renal proximal tubule cells. Am. J. Physiol. Renal Physiol. **281:** F851–859.
10. FISHER-DZOGA, K. *et al.* 1973. Ultrastructural and immunohistochemical studies of primary cultures of aortic medial cells. Exp. Mol. Pathol. **18:** 162–176.
11. KIM, U.J. *et al.* 1996. Construction and characterization of a human bacterial artificial chromosome library. Genomics **34:** 213–218.

Pravastatin Attenuates Ceramide-Induced Cytotoxicity in Mouse Cerebral Endothelial Cells with HIF-1 Activation and VEGF Upregulation

SHANG-DER CHEN,[a] CHAUR-JONG HU,[b] DING-I YANG,[c] ABDULLAH NASSIEF,[d] HONG CHEN,[d] KEJIE YIN,[d] JAN XU,[d] AND CHUNG Y. HSU[d,e]

[a]*Department of Neurology, Chang Gung Memorial Hospital, Kaohsiung, Taiwan*

[b]*Department of Neurology, Molecular Medicine Laboratory, Taipei Municipal Jen-Ai Hospital, Taipei, Taiwan*

[c]*Institute of Neuroscience, Tzu Chi University, Hualien, Taiwan*

[d]*Department of Neurology, Washington University School of Medicine, St. Louis, Missouri, USA*

[e]*Stroke Center, Taipei Medical University, Taipei, Taiwan*

ABSTRACT: Ceramide is a pro-apoptotic lipid messenger that induces oxidative stress and may mediate apoptosis in cerebral endothelial cells (CECs) induced by TNF-α/cycloheximide, lipopolysaccharide, oxidized LDL, IL-1, and amyloid peptide. Exposure of CECs to C_2 ceramide for 12 h caused cell death in a concentration-dependent manner, with a LC_{50} of 30 μM. Statins are inhibitors of 3-hydroxyl-3-methyl coenzyme A reductase which is the rate-limiting enzyme for cholesterol biosynthesis. Pretreatment with pravastatin at 20 μM for 16 h substantially attenuated ceramide cytotoxicity in mouse CECs. Increases in vascular endothelial growth factor (VEGF) expression were detected within 1–3 h after pravastatin treatment. This pravastatin action was accompanied by the activation of hypoxia-inducible factor–1 (HIF-1), a transcription factor known to activate VEGF expression. These results raise the possibility that pravastatin may protect CECs against ceramide-induced death via the HIF-VEGF cascade.

KEYWORDS: ceramide; reactive oxidative species; apoptosis; endothelium; HMG-CoA reductase inhibitors; pravastatin; vascular endothelial growth factor; hypoxia-inducible factor

INTRODUCTION

Ceramide is generated in endothelial cells in response to stress stimuli, such as TNF-α/cycloheximide,[1,2] lipopolysaccharide,[3] oxidized LDL,[4] and IL-1.[5] A variety

Address for correspondence: Chung Y. Hsu, M.D., Ph.D., Taipei Medical University, No. 250, Wu-Hsing Street, Taipei 110, Taiwan. Voice: +886-2-27361661-2016; fax: +886-2-23787795.
hsuc@tmu.edu.tw

of physiological signals, including those initiated by cytokines and growth factors, induce changes of ceramide levels.[6] Once generated, the ceramide signal affects multiple aspects of cellular function, including apoptosis. We have previously shown that amyloid peptide increased the synthesis of ceramide via the neutral sphingomeylinase (nSMase) pathway, and we have established the causal role of the nSMase–ceramide cascade in amyloid-induced apoptosis of cerebral endothelial cells (CECs).[7–10]

The downstream mechanism by which ceramide causes endothelial dysfunction has yet to be determined. Several mechanisms, including generation of excessive reactive oxygen species (ROS) and inhibition of mitochondria complex III enzyme activity, have been implicated in the apoptotic processes.[11,12]

The atheroma-retarding properties of 3-hydroxy-3-methylglutaryl coenzyme A reductase inhibitors, or statins, in both the coronary and carotid arterial beds are well established.[13,14] Recently, a growing body of evidence suggests that statins possess important adjunctive properties that may confer additional benefit beyond the retardation of atherosclerosis.[15] These include antioxidant activity, anti-inflammatory action, improved endothelial cell function, anti-platelet effect, plaque stabilization, and anti-thrombotic effect.[16] Statins appear to be at least equipotent to vascular endothelial growth factor (VEGF) in promoting differentiation of endothelial progenitor cells.[17] VEGF is an important trophic factor supporting the survival and proliferation of endothelial cells. VEGF has also been shown to protect endothelial cells against the apoptosis induced by ceramide.[18] Upregulation of VEGF through hypoxia-inducible factor–1 (HIF-1) activation during tissue hypoxia has been well established.

In this study, we report pravastatin protection of CECs against ceramide-induced death. This pravastatin action was accompanied by HIF-1 activation and VEGF upregulation, suggesting that the HIF-1–VEGF cascade may be involved in the protective action of pravastatin in CECs against ceramide cytotoxicity.

MATERIALS AND METHODS

Reagents

All chemicals and reagents were purchased from Sigma-Aldrich (St. Louis, MO) unless specified otherwise. C_2 ceramide was from Calbiochem (La Jolla, CA). Cell culture supplies were from Invitrogen Corporation (Carlsbad, CA). Pravastatin was a generous gift from Bristol-Myers Squibb Co. (New York, NY).

Cell Culture

Mouse CECs were prepared as previously described.[2] In brief, mouse cerebral cortex was homogenized, filtered, and sequentially digested with collagenase B and collagenase/dispase (Roche Molecular Biochemicals, Indianapolis, IN). This was followed by centrifugation in a 40% Percoll solution. The second band containing microvessels was collected and plated onto collagen-coated dishes. Mouse CECs (4 to 15 passages, >95% purity based on expression of factor VIII and bradykinin receptor) were grown to 85–95% confluence before use.[19]

Assessment of Mouse CEC Death and Apoptosis

Cell survival was measured by 3-(4,5-dimethylthiazol-2-yl)-2,5-diphenyl-tetrazolium bromide (MTT) assay as described previously.[2,20] CECs grown on coverslips were washed with PBS and fixed in methanol/acetic acid (3:1) at 4°C for 5 min. After fixation, the cells were stained for 10 min with the fluorescent DNA-binding dye Hoechst 33258 at 8 µg/mL. Nuclear morphology was examined by fluorescence microscopy. Individual nuclei were visualized at 400× to distinguish the normal uniform nuclear pattern from the characteristic condensed, coalesced chromatin pattern of apoptotic cells.

Western Blot Analysis

Cytoplasmic and nuclear proteins were isolated from CECs as described previously.[20,21] The cytoplasmic proteins were subjected to 15% sodium dodecyl sulfate-polyacrylamide gel electrophoresis (SDS-PAGE) followed by electroblotting onto a polyvinylidene difluoride membrane according to standard protocols. The blot was incubated with the following primary antibodies for 1 to 2 h: mouse anti-VEGF antibody at 1:100 or mouse anti-actin antiserum at 1:500 (both from Santa Cruz Biotechnology, Inc., Santa Cruz, CA). The membrane was then incubated with the goat anti-mouse IgG secondary antibody conjugated with alkaline phosphatase at 1:5000 (Promega, Madison, WI) at room temperature for 1 h.

Reverse Transcription-Coupled Polymerase Chain Reaction (RT-PCR)

Total RNA was isolated with RNeasy Mini Kit (QIAGEN Inc., Valencia, CA). Equal amounts of total RNA (600 ng) were reverse-transcribed and the cDNAs amplified in 0.2 mM dNTP, 1 µM of each primer (forward: 5'-TCTTCAAGCCGTCCT-GTGTG-3'; reverse: 5'-AGGAACATTTACACGTCTGC-3'), 1.5 mM $MgCl_2$, and 2.5 U of Taq polymerase (Roche Diagnostics Corporation). PCR was performed for 26 cycles with denaturation at 95°C for 20 s, primer annealing at 53°C for 30 s, and extension at 72°C for 1 min. Primers were designed based on the mouse VEGF sequences. The relative mRNA level of VEGF was normalized to the internal reference cyclophilin mRNA that was co-amplified in the same reaction for each sample. The RT-PCR was conducted within the linear ranges of PCR cycles and RNA input.[20]

Electrophoretic Mobility Shift Assay (EMSA)

Gel shift assay for HIF-1 binding activity has been described in details elsewhere.[21,22] The oligonucleotide with hypoxia-response element in the promoter of EPO gene (5'-agcttGCCCTACGTGCTGTCTCAg-3' and 5'-aattcTGAGACAG-CACGTAGGGCa-3') was labeled with $\gamma[^{32}P]$-ATP. Binding reaction was performed in a total volume of 20 µL containing 1× binding buffer, 0.0175 pmol of labeled probe (>10,000 cpm), 20 µg of nuclear proteins, and 1 µg of poly(dI-dC). After incubation for 20 min at room temperature, the mixtures were subjected to electrophoresis on a nondenaturing 6% polyacrylamide gel at 180 V for 2 h under low ionic strength conditions. The gel was dried and subjected to autoradiography.

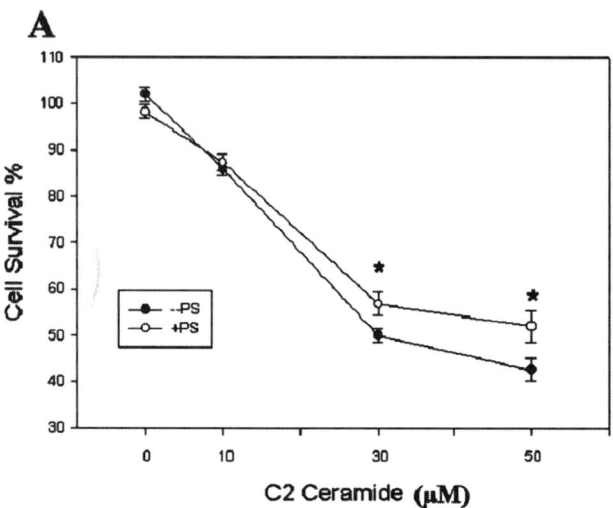

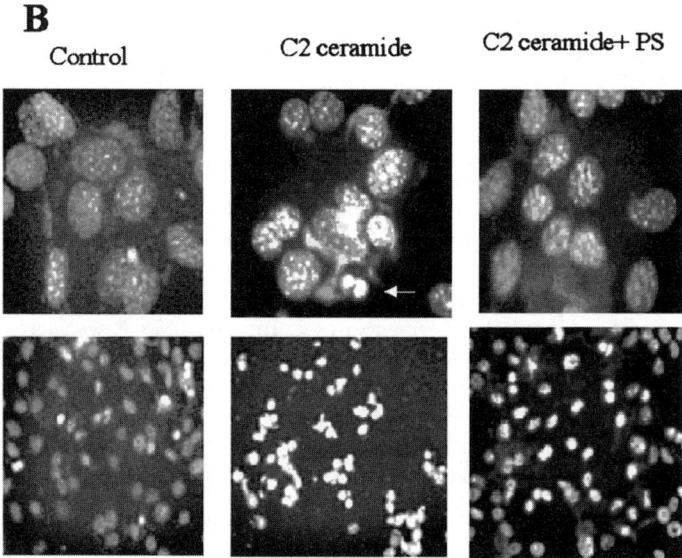

FIGURE 1. Pravastatin attenuated C_2 ceramide-induced cytotoxicity in mouse CEC. (**A**) Mouse CECs pretreated with 20 µM pravastatin for 16 h were challenged with C_2 ceramide in serum-free medium for additional 12 h before MTT assay. (**B**) CECs treated with ceramide with or without pravastatin were stained with the fluorescent DNA-binding dye Hoechst 33258. *Left panels*: control cells without any treatment; m*iddle panels*: CECs treated with C_2 ceramide for 12 h; *right panels*: CECs pretreated with 20 µM pravastatin for 16 h followed by exposure to C_2 ceramide at 30 µM for additional 12 h. *Upper panels*: 400× magnification; *lower panels*, 100× magnification. The *arrow* indicates ceramide-induced condensed, coalesced, and fragmented nuclei. Data shown are representatives of three separate experiments with similar results. PS, pravastatin. $^*P < 0.05$ as compared to the CECs without PS pretreatment.

Statistical Analyses

Data are expressed as mean ± SEM derived from triplicates of at least three separate experiments. Comparison between two experimental groups was based on the two-tailed Student's t-test. A value of $P < 0.05$ was considered significant.

RESULTS

Exposure of mouse CECs to C_2 ceramide for 12 h caused cell death in a concentration-dependent manner, with a half-lethal concentration of approximately 30 μM as determined by MTT assay. Pretreatment with pravastatin at a concentration of 20 μM for 16 h significantly attenuated the cytotoxicity of C_2 ceramide at 30 to 50 μM in mouse CECs (FIG. 1A). Results based on Hoechst staining revealed that pravastatin pre-exposure at 20 μM for 16 h also markedly reduced the numbers of apoptotic cells (FIG. 1B).

VEGF, an endothelial cell–specific mitogen, promotes endothelial cell survival and angiogenesis. The anti-apoptotic effect of VEGF against ceramide-induced apoptosis has been shown in microvascular endothelial cells.[18] We demonstrated that pravastatin increased VEGF expression at the mRNA level after pravastatin treat-

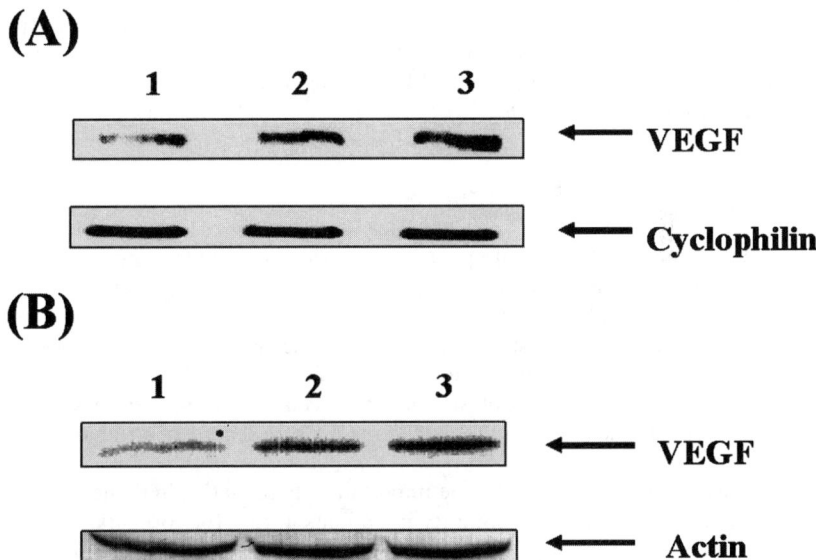

FIGURE 2. Pravastatin enhanced VEGF expression at both mRNA and protein levels. (**A**) CECs had been treated with 20 μM of pravastatin for various periods of time before VEGF mRNA was quantified via RT-PCR. *Lanes 1, 2,* and *3,* respectively, indicate control, 1 h, and 3 h after PS treatment. Cyclophilin cDNA was co-amplified as an internal standard to correct for the inherent variations of PCR. (**B**) The experimental condition was the same as described in (**A**) except that total proteins were prepared for Western analysis. Western blot for actin served as an internal control for equal loading of proteins in each lane. Results shown are representatives from three separate experiments with similar results.

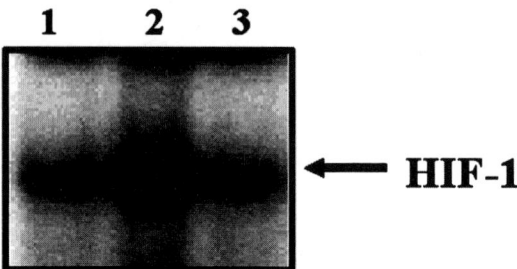

FIGURE 3. Pravastatin increased DNA binding activity of transcription factor HIF-1 in mouse CECs. Gel shift assay for HIF-1 binding activity was performed with nuclear proteins isolated from CECs with or without treatment with 20 μM pravastatin. *Lanes 1, 2,* and *3,* respectively, indicate control, 1 h, and 3 h after PS treatment. The HIF-1 binding activity was increased at 1 h after pravastatin treatment and declined to basal level in 3 h. Data shown are representative of three separate experiments with similar results.

ment for 1–3 h in mouse CECs (FIG. 2A). An in VEGF protein expression was also observed following pravastain pretreatment based on the Western analysis (FIG. 2B). VEGF is known to be regulated by HIF-1 activation under an hypoxic environment. We therefore tested whether pravastatin-induced VEGF expression involves HIF-1 activation under a non-hypoxic condition. Results based on EMSA revealed that the HIF-1 binding activity was increased at 1 h, and then declined to basal level within 3 h, after pravastatin treatment (FIG. 3).

DISCUSSION

Ceramide is a key mediator of apoptosis in various death paradigms. Signaling of the stress response through ceramide appears to play a role in the development of human diseases, including amyloid-induced vascular changes,[7,8,10] ischemia/reperfusion injury, insulin resistance in diabetes, atherogenesis, septic shock, and ovarian failure.[23] Exogenous application of short-chain ceramide analogues can induce apoptosis in various types of vascular wall cells.[24] In the present study, we showed that C_2 ceramide induced cytotoxicity in CECs, in accordance with our earlier studies.[2]

Clinical trials have demonstrated the important role of statins in reducing the incidence of coronary events and mortality in patients at risk for coronary disease.[25] Meta-analysis of statin trials has shown a lower risk of ischemic stroke in patients with a history of coronary artery disease with average and/or elevated serum cholesterol levels.[26] Subgroup analysis of the major statin trials suggests that some of the beneficial effects of statins may not be associated with cholesterol reduction. These potentially salutary actions of statins include antioxidant actions, anti-inflammatory action, improvement in endothelial cell function, anti-platelet aggregation, anti-thrombotic effect, and plaque stabilization.[15,16] In the present study, we showed that pravastatin attenuated ceramide-induced cytotoxicity in mouse CECs. This favorable action of pravastain was accompanied by an increase in VEGF expression and

HIF-1 activation. VEGF is a well-known endothelial cell mitogen that promotes endothelial cell survival and angiogenesis. The anti-apoptotic effect of VEGF on starvation- and ceramide-induced apoptosis has been shown in human dermal microvascular endothelial cells.[18] Statins stimulate VEGF expression in osteoblasts via reduction of protein prenylation and phosphatidylinositide-3 kinase pathway, which promotes osteoblastic differentiation.[27] Circulating endothelial progenitor cells (EPCs) are derived from bone marrow and are mobilized in response to endogenous tissue ischemia or exogenous cytokine stimulation. Statins can augment the circulating population of EPCs through Akt signaling pathway.[17,28] The statin effect appears to be at least equipotent to VEGF in increasing EPC differentiation.[28] The augmentation of circulating EPC might contribute to the well-established beneficial effects of statins in patients with coronary artery disease.[17] Pravastatin may attenuate the ceramide-induced CEC death via a similar pathway.

Transcriptional activation of VEGF by an increase in HIF-1 binding activity under the hypoxic condition is a well-known phenomenon.[29,30] Evidence also supports that HIF-1α expression in many types of cancer cells may be hypoxia independent. Several mechanisms have been proposed for hypoxia-independent activation of HIF-1 in cancer cells, including loss of function in VHL, p53, PTEN, p14^{ARF}, or gain of function in SRC.[31] In addition, HIF-1 can also be induced under normaxic conditions by pharmacologic reagents such as cobalt ion and iron chelators.[22] The molecular mechanisms underlying pravastatin-induced HIF-1 activation, however, remain to be fully delineated.

In conclusion, in the present study we present evidence to show that pravastatin attenuated ceramide-induced cytotoxicity in mouse CECs that may involve the HIF–VEGF cascade. Further investigations are needed to establish the causal role of HIF-1 and VEGF in statin-mediated cytoprotection in CECs.

ACKNOWLEDGMENTS

This work was supported by grants from the National Science Council in Taiwan (NSC91-2314-B-182A-057 to S.-D.C. and NSC92-2321-B-038-003 to C.Y. H.).

REFERENCES

1. LOPEZ-MARURE, R. et al. 2000. Ceramide mimics tumour necrosis factor-alpha in the induction of cell cycle arrest in endothelial cells. Induction of the tumour suppressor p53 with decrease in retinoblastoma/protein levels. Eur. J. Biochem. **267:** 4325–4333.
2. XU, J. et al. 1998. Involvement of de novo ceramide biosynthesis in tumor necrosis factor-alpha/cycloheximide-induced cerebral endothelial cell death. J. Biol. Chem. **273:** 16521–16526.
3. HAIMOVITZ-FRIEDMAN, A. et al. 1997. Lipopolysaccharide induces disseminated endothelial apoptosis requiring ceramide generation. J. Exp. Med. **186:** 1831–1841.
4. DEIGNER, H.P. et al. 2001. Ceramide induces aSMase expression: implications for oxLDL-induced apoptosis. FASEB J. **15:** 807–814.
5. MASAMUNE, A. et al. 1996. Regulatory role of ceramide in interleukin (IL)-1 beta-induced E-selectin expression in human umbilical vein endothelial cells. Ceramide enhances IL-1 beta action, but is not sufficient for E-selectin expression. J. Biol. Chem. **271:** 9368–9375.

6. KOLESNICK, R. & Z. FUKS. 1995. Ceramide: a signal for apoptosis or mitogenesis? J. Exp. Med. **181:** 1949–1952.
7. XU, J. *et al.* 2001. Amyloid beta peptide-induced cerebral endothelial cell death involves mitochondrial dysfunction and caspase activation. J. Cereb. Blood Flow Metab. **21:** 702–710.
8. YIN, K.J. *et al.* 2002. Amyloid-beta induces Smac release via AP-1/Bim activation in cerebral endothelial cells. J. Neurosci. **22:** 9764–9770.
9. LEE, J.T. *et al.* 2004. Amyloid-beta peptide induces oligodendrocyte death by activating the neutral sphingomyelinase-ceramide pathway. J. Cell Biol. **164:** 123–131.
10. YANG, D.I. *et al.* 2004. Neutral sphingomyelinase activation in endothelial and glial cell death induced by amyloid beta-peptide. Neurobiol. Dis. **17:** 99–107.
11. ANDRIEU-ABADIE, N. *et al.* 2001. Ceramide in apoptosis signaling: relationship with oxidative stress. Free Radic. Biol. Med. **31:** 717–728.
12. FOSSLIEN, E. 2001. Mitochondrial medicine—molecular pathology of defective oxidative phosphorylation. Ann. Clin. Lab. Sci. **31:** 25–67.
13. CROUSE, J.R., III *et al.* 1995. Pravastatin, lipids, and atherosclerosis in the carotid arteries (PLAC- II). Am. J. Cardiol. **75:** 455–459.
14. BROWN, G. *et al.* 1990. Regression of coronary artery disease as a result of intensive lipid-lowering therapy in men with high levels of apolipoprotein B. N. Engl. J. Med. **323:** 1289–1298.
15. VAUGHAN, C.J. & N. DELANTY. 1999. Neuroprotective properties of statins in cerebral ischemia and stroke. Stroke **30:** 1969–1973.
16. BLAUW, G.J. *et al.* 1997. Stroke, statins, and cholesterol. A meta-analysis of randomized, placebo-controlled, double-blind trials with HMG-CoA reductase inhibitors. Stroke **28:** 946–950.
17. DIMMELER, S. *et al.* 2001. HMG-CoA reductase inhibitors (statins) increase endothelial progenitor cells via the PI 3-kinase/Akt pathway. J. Clin. Invest. **108:** 391–397.
18. GUPTA, K. *et al.* 1999. VEGF prevents apoptosis of human microvascular endothelial cells via opposing effects on MAPK/ERK and SAPK/JNK signaling. Exp. Cell Res. **247:** 495–504.
19. XU, J. *et al.* 1992. Receptor-linked hydrolysis of phosphoinositides and production of prostacyclin in cerebral endothelial cells. J. Neurochem. **58:** 1930–1935.
20. XU, J. *et al.* 2000. Oxygen-glucose deprivation induces inducible nitric oxide synthase and nitrotyrosine expression in cerebral endothelial cells. Stroke **31:** 1744–1751.
21. XU, J. *et al.* 2001. Amyloid-beta peptides are cytotoxic to oligodendrocytes. J. Neurosci. **21:** RC118.
22. YANG, D.I. *et al.* 2004. Carbamoylating chemoresistance induced by cobalt pretreatment in C6 glioma cells: putative roles of hypoxia-inducible factor-1. Br. J. Pharmacol. **141:** 988–996.
23. MATHIAS, S. *et al.* 1998. Signal transduction of stress via ceramide. Biochem. J. **335:** 465–480.
24. LEVADE, T. *et al.* 2001. Sphingolipid mediators in cardiovascular cell biology and pathology. Circ. Res. **89:** 957–968.
25. SACKS, F.M. *et al.* 1996. The effect of pravastatin on coronary events after myocardial infarction in patients with average cholesterol levels. Cholesterol and recurrent events trial investigators. N. Engl. J. Med. **335:** 1001–1009.
26. CROUSE, J.R., III *et al.* 1998. HMG-CoA reductase inhibitor therapy and stroke risk reduction: an analysis of clinical trials data. Atherosclerosis **138:** 11–24.
27. MAEDA, T. *et al.* 2003. Statins augment vascular endothelial growth factor expression in osteoblastic cells via inhibition of protein prenylation. Endocrinology **144:** 681–692.
28. LLEVADOT, J. *et al.* 2001. HMG-CoA reductase inhibitor mobilizes bone marrow–derived endothelial progenitor cells. J. Clin. Invest. **108:** 399–405.
29. LIU, Y. *et al.* 1995. Hypoxia regulates vascular endothelial growth factor gene expression in endothelial cells. Identification of a 5′ enhancer. Circ. Res. **77:** 638–643.
30. RAVI, R. *et al.* 2000. Regulation of tumor angiogenesis by p53-induced degradation of hypoxia-inducible factor 1alpha. Genes Dev. **14:** 34–44.
31. SEMENZA, G.L. 2002. HIF-1 and tumor progression: pathophysiology and therapeutics. Trends Mol. Med. **8:** S62–67.

Alleviation of Oxidative Damage in Multiple Tissues in Rats with Streptozotocin-Induced Diabetes by Rice Bran Oil Supplementation

RONG-HONG HSIEH,[a,c] LI-MING LIEN,[b,c] SHYH-HSIANG LIN,[a] CHIA-WEN CHEN,[a] HUEI-JU CHENG,[a] AND HSING-HSIEN CHENG[a]

[a]*School of Nutrition and Health Sciences, Taipei Medical University, Taipei, Taiwan*

[b]*Department of Neurology, Shin Kong Wu Ho-Su Memorial Hospital, Taipei, Taiwan*

ABSTRACT: The possibility of 8-hydroxy-2′-deoxyguanosine (8-OHdG) serving as a sensitive biomarker of oxidative DNA damage and oxidative stress was investigated. Reactive oxygen species (ROS) have been reported to be a cause of diabetes induced by chemicals such as streptozotocin (STZ) in experimental animals. In this study, we examined oxidative DNA damage in multiple tissues in rats with STZ-induced diabetes by measuring the levels of 8-OHdG in the liver, kidney, pancreas, brain, and heart. Levels of 8-OHdG in mitochondrial DNA (mtDNA) and nuclear DNA (nDNA) were also determined in multiple tissues of rats treated with rice bran oil. Levels were 0.19 ± 0.07, 0.88 ± 0.30, 1.97 ± 0.05, and 9.79 ± 3.09 ($1/10^5$ dG) in the liver of nDNA of normal rats, nDNA of STZ-induced diabetic rats, mtDNA of normal rats, and mtDNA of STZ-induced diabetic rats, respectively. Levels of mtDNA of 8-OHdG were 10 times higher than those of nDNA in multiple tissues. Significant reductions in mtDNA 8-OHdG levels were seen in the liver, kidney, and pancreas of diabetic rats treated with rice bran oil compared with diabetic rats without intervention. Our study demonstrated that oxidative mtDNA damage may occur in multiple tissues of STZ-induced diabetics rats. Intervention with rice bran oil treatment may reverse the increase in the frequency of 8-OHdG.

KEYWORDS: 8-OHdG; diabetes; oxidative stress; streptozotocin

INTRODUCTION

DNA is susceptible to damage by reactive oxygen species (ROS). ROS, such as superoxide, hydrogen peroxide, the hydroxyl radical, and singlet oxygen, are produced during normal and pathophysiological processes, and production of such radicals is further enhanced by ionizing radiation, environmental mutagens, and carcinogens.[1] In terms of oxidative DNA damage, nucleic acids exposed to oxygen radicals generate various modified bases, and more than 20 oxidatively altered pu-

---

[c]R-H.H. and L-M.L. contributed equally to this work.

Address for correspondence: Hsing-Hsien Cheng, School of Nutrition and Health Sciences, Taipei Medical University, Taipei, Taiwan 110, Republic of China, Voice: +886-2-27361661 ext. 6551-112; fax: +886-2-27373112.

chenghh@tmu.edu.tw

rines and pyrimidines have been detected.[2,3] 8-Hydroxyguanine (7,8-dihydro-8-oxoguanine) is believed to be an important compound because of its role as the marker of mutagenicity and carcinogenicity in bacterial and mammalian cells.[4,5]

The diabetogenic agent streptozotocin (STZ) is a D-glucopyranose derivative of N-methyl-N-nitrosourea (MNU).[6,7] STZ has broad-spectrum antibiotic activity and is often used to induce diabetes mellitus in experimental animals through its toxic effects on pancreatic β cells. Diabetic complications in target organs arise from the chronic elevation of glucose levels. The pathogenic effect is mediated to a significant extent via increased production of ROS and/or reactive nitrogen species (RNS) and subsequent oxidative stress.[8,9] This study was undertaken to examine oxidative DNA damage in various tissues of STZ-induced diabetic rats by measuring the levels of 8-OHdG.

MATERIALS AND METHODS

STZ-Induced Diabetic Rats

At 8 weeks of age, Sprague-Dawley rats were injected intraperitoneally with STZ at 45 mg/kg body weight twice in a 3-day period. Nicotinamide (200 mg/kg) was injected before the STZ injection as a protection reagent. The development of diabetes was verified by the presence of hyperglycemia. On day 3 post–STZ treatment, rats were randomly assigned to two groups ($n = 6$ per group): diabetes with soybean oil (control) and diabetes with rice bran oil (intervention group). Rice bran oil was extracted by the supercritical CO_2 extraction method.[10] There are 33.90 mg of γ-oryzanol, 0.5 mg of γ-tocotrienol, and 3.02 mg of α-tocopherol per gram of rice bran oil, all of which were analyzed by high-performance liquid chromatography.[11] The saturated fatty acid (SFA), monounsaturated fatty acid (MUFA), and polyunsaturated fatty acid (PUFA) levels of rice bran oil were 13.61%, 50.67%, and 35.5%, respectively, as analyzed by gas chromatography.[12] On day 28, rats were anesthetized with ether, and various tissues were quickly removed and stored at –70°C until DNA was extracted.

Enzyme Digestion of DNA and ELISA for 8-OHdG

The presence of 8-OHdG was detected as reported previously.[13] Aliquots of DNA equivalent to 50 μg were freeze-dried and reconstituted in 20 mM sodium acetate, pH 4.8, containing 45 mM zinc chloride. Samples were heated in a boiling-water bath for 3 min and cooled quickly on ice prior to the addition of nuclease P1 to yield 0.1 U/μg of DNA, and then samples were incubated at 37°C for 1 h. Samples were made alkaline by the addition of 1.5 M Tris–HCl, pH 8.0, and alkaline phosphatase was added to give ~0.05–0.075 U/μg of DNA; then samples were incubated at 37°C for 30 min.

Enzymatic digests of DNA were analyzed by competitive enzyme-linked immunosorbent assay (ELISA) using a monoclonal antibody to 8-OHdG (8-OHdG Check; Genox, Baltimore, MD). Fifty microliters of DNA digest or standard 8-OHdG solution was added to the wells of a 96-well plate precoated with 8-OHdG, and then 50 μL of primary antibody (mouse anti–8-OHdG) was added; the plate was then incubated at 37°C for 1 h. After incubation, wells were washed with 200 μL/well of

TABLE 1. Contents of 8-OHdG in mtDNA and nuclear DNA from multiple tissues of normal rats and STZ-induced diabetic rats

	Liver	Kidney	Pancreas	Brain	Heart
N-nDNA	0.19 ± 0.07*	0.17 ± 0.08*	0.16 ± 0.09	0.05 ± 0.03	0.07 ± 0.03
D-nDNA	0.88 ± 0.30*	1.06 ± 0.39*	0.75 ± 0.36	0.34 ± 0.15	0.34 ± 0.19
N-mtDNA	1.97 ± 0.65†	1.77 ± 0.68†	1.73 ± 0.71†	0.64 ± 0.39†	0.63 ± 0.33†
D-mtDNA	9.79 ± 3.09†	9.20 ± 3.04†	8.03 ± 4.37†	3.28 ± 1.38†	3.76 ± 1.51†

Results are presented as the mean ± SD. *,†Statistically significant difference ($P < 0.05$). N, normal rats; D, diabetic rats; nDNA, nuclear DNA; mtDNA, mitochondrial DNA.

0.05% (vol/vol) Tween 20 in 0.01 M phosphate-buffered saline (PBS), pH 7.4, and then 100 µL/well secondary antibody (peroxidase conjugated anti–mouse immunoglobulin G) was added and the plate was incubated for 1 h at 37°C. Following incubation, peroxidase substrate (*o*-phenylenediamine/hydrogen peroxide/PBS) was added, followed 30 min later by 2 N sulfuric acid; the absorbance was then read at 450 nm.

RESULTS

Five tissues, including the liver, kidney, pancreas, brain, and heart, were collected from normal rats and from STZ-induced diabetic rats. DNA extracted from various tissues was classified into mitochondrial DNA (mtDNA) and nuclear DNA (nDNA) groups. The 8-OHdG contents of mtDNA and nDNA were determined and are listed in TABLE 1. Hepatic levels of 0.19 ± 0.07, 0.88 ± 0.30, 1.97 ± 0.05, and 9.79 ± 3.09 ($1/10^5$ dG) of nDNA of normal rats, nDNA of STZ-induced diabetic rats, mtDNA of normal rats, and mtDNA of STZ-induced diabetic rats were found, respectively. mtDNA had levels of 8-OHdG that were 10 times higher than those of nDNA in multiple tissues. There were significantly increased 8-OHdG levels of nDNA in liver and kidney of STZ-induced diabetic rats compared with the control group. Significant

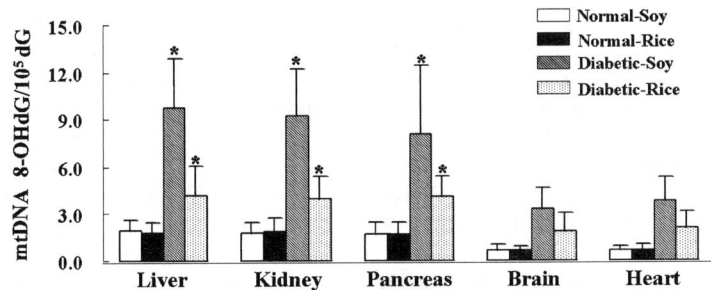

FIGURE 1. Levels of 8-OHdG in mtDNA of multiple tissues compared between normal rats and diabetic rats supplemented with rice bran oil.

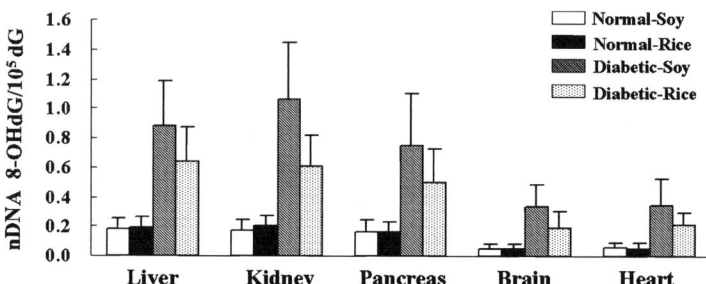

FIGURE 2. Contents of 8-OHdG in nDNA determined from diabetic rats supplemented with rice bran oil.

differences between mtDNA of diabetic rats and normal rats were detected in all examined tissues.

Protective effects of oxidative damage to DNA by supplementation with rice bran oil were determined. Levels of 8-OHdG in mtDNA of various tissues were also compared between normal and diabetic rats supplemented with rice bran oil (FIG. 1). There were significant decreases in mtDNA 8-OHdG contents in the liver, kidney, and pancreas from diabetic rats treated with rice bran oil. Intervention using rice bran oil treatment may reverse the increase in the frequency of 8-OHdG from mtDNA. Contents of 8-OHdG in nDNA from diabetic rats supplemented with rice bran oil were also determined (FIG. 2). Lower 8-OHdG proportions in the liver, kidney, and pancreas were found. However, there were no significant differences in 8-OHdG levels in nDNA between control subjects and rice bran oil–treated rats.

DISCUSSION

STZ is a well-known genotoxic agent, and it has been shown to induce DNA damage in mouse liver and kidney, which are the tumor target organs.[14,15] This study found higher levels of 8-OHdG in mtDNA from the liver, kidney, and pancreas of STZ-induced diabetic rats than in controls. The results indicate that diabetes-induced oxidative stress may cause major damage in the above-mentioned tissues. Diabetic nephropathy is the major cause of morbidity and mortality in diabetic patients. Several mechanisms have been proposed for the pathogenesis of diabetic nephropathy, such as hyperfiltration,[16] increased production of advanced glycation end products (AGEs),[17] activation of protein kinase C,[18,19] and enhanced oxidative stress.[20,21] Our findings are consistent with the notion that oxidative stress is an etiological factor in the occurrence of nephropathy.

Increased oxidative stress has been hypothesized to contribute to the pathological processes of diabetic complications. However, the detailed biomolecular mechanism has not been ascertained. In general, oxidative stress can modify nucleic acids such as 8-OHdG, affect its exact linkage, and play a crucial role in mutagenesis.[22] Increased levels of 8-OHdG have been reported in urine,[23,24] mononuclear cells,[23,25] skeletal muscles,[26] pancreas,[27] liver,[28,29] and kidney[28,29] of diabetic patients.

It is widely accepted that mtDNA is ~10–20 times more vulnerable to oxidative damage and subsequent mutations than is nDNA.[30] Higher levels of oxidative DNA damage of mtDNA than nDNA occur for the following three reasons. First, the lack of protective histones renders mtDNA more vulnerable to oxidative attack. Second, it has been argued that the proximity of mtDNA to the mitochondrial electron transport chain, a site of superoxide anion generation, predisposes it to oxidative damage. Third, it has been pointed out that oxidative DNA damage may be less efficiently repaired in mtDNA. This argument has been reiterated in many research reports.[31,32] However, there are only a few studies reporting on oxidative DNA damage in mtDNA in various tissues of diabetic rats.

In this study, we further examined the 8-OHdG levels in mtDNA and nDNA of various tissues. We found increased levels of 8-OHdG in mtDNA of the liver, kidney, and pancreas of diabetic rats at 4 weeks after the onset of diabetes. In contrast, levels of 8-OHdG in nuclear DNA were not significantly increased in diabetic rats compared with controls. These results are consistent with previous findings that mtDNA is more sensitive to oxidative stress than nDNA. Interestingly, previous studies indicated that modified mtDNA may contribute to increased levels of 8-OHdG. The possible consequence of increased formation of 8-OHdG in mtDNA is the increase in mtDNA mutation, including mispairing and deletions. Kakimoto *et al.*[33] also showed that increases in mtDNA deletion were in concert with higher levels of 8-OHdG in the kidney of diabetic rats. Deleted mtDNA may accumulate with the duration of diabetes in multiple tissues of diabetic rats.

ACKNOWLEDGMENTS

This work was supported by research grants NSC92-2320-B-038-030 from the National Science Council of the Republic of China and SKH-TMU-93-19 from the Shin Kong Wu Ho-Su Memorial Hospital.

REFERENCES

1. HENLE, E.S. & S. LINN. 1997. Formation, prevention, and repair of DNA damage by iron/hydrogen peroxide. J. Biol. Chem. **272:** 19095–19098.
2. GAJEWSKI, E., G. RAO, Z. NACKERDIEN & M. DIZDAROGLU. 1990. Modification of DNA bases in mammalian chromatin by radiation-generated free radicals. Biochemistry **29:** 7876–7882.
3. KAMIYA, H. 2003. Mutagenic potentials of damaged nucleic acids produced by reactive oxygen/nitrogen species: approaches using synthetic oligonucleotides and nucleotides: survey and summary. Nucleic Acids Res. **31:** 517–531.
4. KASAI, H. & S. NISHIMURA. 1984. Hydroxylation of deoxyguanosine at the C-8 position by ascorbic acid and other reducing agents. Nucleic Acids Res. **12:** 2137–2145.
5. KASAI, H. 1997. Analysis of a form of oxidative DNA damage, 8-hydroxy-2′-deoxyguanosine, as a marker of cellular oxidative stress during carcinogenesis. Mutat. Res. **387:** 147–163.
6. WEISS, R.B. 1982. Streptozocin: a review of its pharmacology, efficacy, and toxicity. Cancer Treat. Rep. **66:** 427–438.

7. WANG, Z. & H. GLEICHMANN. 1998. GLUT2 in pancreatic islets: crucial target molecule in diabetes induced with multiple low doses of streptozotocin in mice. Diabetes **47:** 50–56.
8. ROSEN, P., P.P. NAWROTH, G. KING, et al. 2001. The role of oxidative stress in the onset and progression of diabetes and its complications: a summary of a Congress Series sponsored by UNESCO-MCBN, the American Diabetes Association and the German Diabetes Society. Diabetes Metab. Res. Rev. **17:** 189–212.
9. BROWNLEE, M. 2001. Biochemistry and molecular cell biology of diabetic complications. Nature **414:** 813–820.
10. DOWD, M.K. & M.S. KUK. 1998. Supercritical CO_2 extraction of rice bran. J. Am. Oil Chem. Soc. **75:** 623–628.
11. QURESHI, A.A., H. MO, L. PACKER & D.M. PETERSON. 2000. Isolation and identification of novel tocotrienols from rice bran with hypocholesterolemic, antioxidant, and antitumor properties. J. Agric. Food Chem. **48:** 3130–3140.
12. MORRISON, W.R. & L.M. SMITH. 1964. Preparation of fatty acid methyl esters and dimethylacetals from lipids with boron fluoride-methanol. J. Lipid Res. **53:** 600–608.
13. COOKE, M.S., M.D. EVANS, I.D. PODMORE, et al. 1998. Novel repair action of vitamin C upon *in vivo* oxidative DNA damage. FEBS Lett. **439:** 363–367.
14. SCHMEZER. P., C. ECKERT & U.M. LIEGIBEL. 1994. Tissue-specific induction of mutations by streptozotocin *in vivo*. Mutat. Res. **307:** 495–499.
15. KRAYNAK, A.R., R.D. STORER, R.D. JENSEN, et al. 1995. Extent and persistence of streptozotocin-induced DNA damage and cell proliferation in rat kidney as determined by *in vivo* alkaline elution and BrdUrd labeling assays. Toxicol. Appl. Pharmacol. **135:** 279–286.
16. HOSTETTER, T.H., H.G. RENNKE & B.M. BRENNER. 1982. The case for intrarenal hypertension in the initiation and progression of diabetic and other glomerulopathies. Am. J. Med. **72:** 375–380.
17. BROWNLEE, M., A. CERAMI & H. VLASSARA. 1988. Advanced glycosylation end products in tissue and the biochemical basis of diabetic complications. N. Engl. J. Med. **318:** 1315–1321.
18. INOGUCHI, T., R. BATTAN, E. HANDLER, et al. 1992. Preferential elevation of protein kinase C isoform beta II and diacylglycerol levels in the aorta and heart of diabetic rats: differential reversibility to glycemic control by islet cell transplantation. Proc. Natl. Acad. Sci. USA **89:** 11059–11063.
19. KOYA, D. & G.L. KING. 1998. Protein kinase C activation and the development of diabetic complications. Diabetes **47:** 859–866.
20. WOLFF, S.P., Z.Y. JIANG & J.V. HUNT. 1991. Protein glycation and oxidative stress in diabetes mellitus and ageing. Free Radic. Biol. Med. **10:** 339–352.
21. BAYNES, J.W. 1991. Role of oxidative stress in development of complications in diabetes. Diabetes **40:** 405–412.
22. KUCHINO, Y., F. MORI, H. KASAI, et al. 1987. Misreading of DNA templates containing 8-hydroxydeoxyguanosine at the modified base and at adjacent residues. Nature **327:** 77–79.
23. HINOKIO, Y., S. SUZUKI, M. HIARI, et al. 1999. Oxidative DNA damage in diabetes mellitus: its association with diabetic complications. Diabetologia **42:** 995–998.
24. LEINONEN, J., T. LEHTIMAKI, S. TOYOKUNI, et al. 1997. New biomarker evidence of oxidative DNA damage in patients with non-insulin-dependent diabetes mellitus. FEBS Lett. **417:** 150–152.
25. DANDONA, P., K. THUSU, S. COOK, et al. 1996. Oxidative damage to DNA in diabetes mellitus. Lancet **347:** 444–445.
26. SUZUKI, S., Y. HINOKIO, K. KOMATU, et al. 1999. Oxidative damage to mitochondrial DNA and its relationship to diabetic complications. Diabetes Res. Clin. Pract. **45:** 161–168.
27. SAKURABA, H., H. MIZUKAMI, N. YAGIHASHI, et al. 2002. Reduced beta-cell mass and expression of oxidative stress-related DNA damage in the islet of Japanese type II diabetic patients. Diabetologia **45:** 85–96.
28. IMAEDA, A., T. AINIKO, T. AOKI, et al. 2002. Antioxidative effects of fluvastatin and its metabolites against DNA damage in streptozotocin-treated mice. Food Chem. Toxicol. **40:** 1415–1422.

29. PARK, K.S., J.H. KIM, M.S. KIM, et al. 2001. Effects of insulin and antioxidant on plasma 8-hydroxyguanine and tissue 8-hydroxydeoxyguanosine in streptozotocin-induced diabetic rats. Diabetes **50:** 2837–2841.
30. RICHTER, C., J.W. PARK & B.N. AMES. 1988. Normal oxidative damage to mitochondrial and nuclear DNA is extensive. Proc. Natl. Acad. Sci. USA **85:** 6465–6467.
31. KANG, C.M., B.S. KRISTAL & B.P. YU. 1998. Age-related mitochondrial DNA deletions: effect of dietary restriction. Free Radic. Biol. Med. **24:** 148–154.
32. SALAZAR, J.J. & B. VAN HOUTEN. 1997. Preferential mitochondrial DNA injury caused by glucose oxidase as a steady generator of hydrogen peroxide in human fibroblasts. Mutat. Res. **385:** 139–149.
33. KAKIMOTO, M., T. INOGUCHI, S. SONTA, et al. 2002. Accumulation of 8-hydroxy-2′-deoxyguanosine and mitochondrial DNA deletion in kidney of diabetic rats. Diabetes **51:** 1588–1595.

Curcumin Inhibits ROS Formation and Apoptosis in Methylglyoxal-Treated Human Hepatoma G2 Cells

WEN-HSIUNG CHAN,[a] HSIN-JUNG WU,[a] AND YAN-DER HSUUW[b]

[a]*Department of Bioscience Technology and Center for Nanotechnology, Chung Yuan Christian University, Chung Li, Taiwan*

[b]*Department of Nutrition and Health Science, Fooyin University, Kaohsiung, Taiwan*

ABSTRACT: Methylglyoxal (MG) is a reactive dicarbonyl compound endogenously produced mainly from glycolytic intermediates. Elevated MG levels in diabetes patients are believed to contribute to diabetic complications. MG is cytotoxic through induction of apoptosis. Curcumin, the yellow pigment of *Curcuma longa*, is known to have antioxidant and anti-inflammatory properties. In the present study, we investigated the effect of curcumin on MG-induced apoptotic events in human hepatoma G2 cells. We report that curcumin prevented MG-induced cell death and apoptotic biochemical changes such as mitochondrial release of cytochrome *c*, caspase-3 activation, and cleavage of PARP (poly [ADP-ribose] polymerase). Using the cell permeable dye 2′,7′-dichlorofluorescein diacetate (DCF-DA) as an indicator of reactive oxygen species (ROS) generation, we found that curcumin abolished MG-stimulated intracellular oxidative stress. The results demonstrate that curcumin significantly attenuates MG-induced ROS formation, and suggest that ROS triggers cytochrome *c* release, caspase activation, and subsequent apoptotic biochemical changes.

KEYWORDS: curcumin; methylglyoxal; oxidative stress; apoptosis; cytochrome *c*

INTRODUCTION

Methylglyoxal (MG) is a reactive dicarbonyl compound that is found at high levels in diabetic patients and is believed to contribute to diabetic complications.[1] MG is cytotoxic because of its triggering of apoptosis.[2,3] JNK activation, mitochondrial membrane potential change, and caspase-3 activation have been detected during MG-induced apoptosis in Jurkat cells.[4] Despite these observations, the precise details regarding the molecular mechanisms underlying MG-induced cell death are yet to be clarified.

Address for correspondence: Wen-Hsiung Chan, Department of Bioscience Technology, Chung Yuan Christian University, Chung Li, Taiwan 32023. Voice: +886-3-2653515; fax: +886-3-2653599.
whchan@cycu.edu.tw

Curcumin is a common dietary pigment and spice used in traditional Indian medicine.[5] As a potent antioxidant, curcumin has been shown to display anti-proliferative and anti-carcinogenic properties in a wide variety of cell lines and animals.[6,7] Although multiple biological functions of curcumin have been identified, the precise molecular mechanisms underlying its actions appear largely unknown.

In this study, we describe the effects of curcumin on MG-induced apoptosis. We report that curcumin inhibited apoptotic biochemical changes such as cytochrome c release and caspase-3 activation in hepatoma G2 (Hep G2) cells. In what appears to be a mechanism underlying this inhibition of apoptosis, we also found curcumin was able to suppress MG-stimulated ROS formation.

MATERIALS AND METHODS

Methylgyloxal Treatment

Hep G2 cells were incubated in medium containing various concentrations of methylglyoxal for the indicated times. Cell extracts were prepared following procedures described in a previous report.[8]

Apoptosis Assay

Oligonucleosomal DNA fragmentation in apoptotic cells was measured using the Cell Death Detection ELISA^{plus} kit according to the manufacturer's protocol (Roche Molecular Biochemicals, Mannheim, Germany).

Measurement of Caspase-3 Activity

The caspase-3 activity assay was carried out essentially as described in a previous report.[9] Caspase-3 activity was measured using the fluorogenic substrate Z-DEVD-AFC (Calbiochem, La Jolla, CA).

Measurement of Oxidative Stress

Reactive oxygen species (ROS) arbitrary units were measured using 2′7′-dichlorofluorescein diacetate (DCF-DA) dye as described in a previous report.[10]

Cytochrome c Release Assay

Mitochondrial cytochrome c release assays were performed essentially according to the method described by Yang *et al.*[11]

Statistics

Experimental data were analyzed using one-way ANOVA. $P < 0.05$ was considered significant.

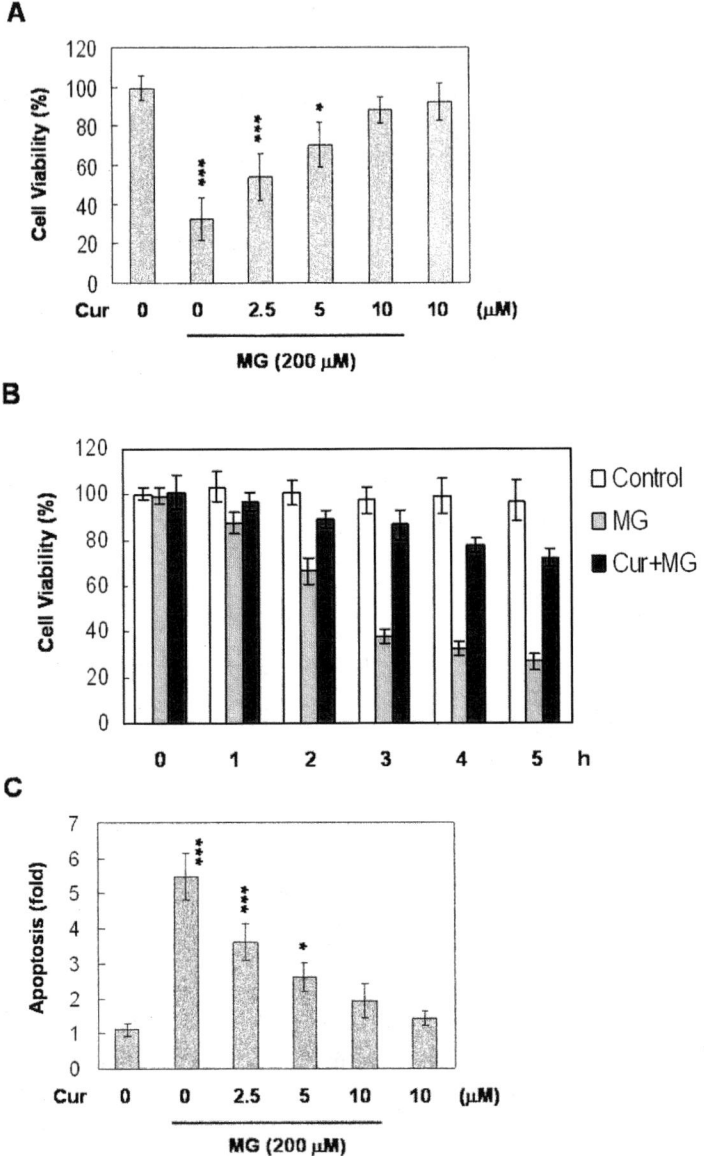

FIGURE 1. Curcumin prevents methylglyoxal (MG)-induced cell death. Hep G2 cells were incubated with or without various concentrations of curcumin (Cur) for 1 h. Cells were then treated with or without MG (200 µM) for a further 3 h. (**A**) Cell viability was determined by MTT assay. (**B**) Hep G2 cells preincubated with or without 10 µM curcumin for 1 h were either untreated or MG-treated and viable cells counted as in **A** at various time points after MG treatment, as indicated. (**C**) Apoptosis was evaluated using a commercially available cell death ELISA kit. For all graphs, $*P < 0.05$, $**P < 0.01$, and $***P < 0.0001$, respectively, from the control value.

RESULTS

Inhibition of MG-Induced Cell Death and Apoptotic Biochemical Changes in Human Hep G2 Cells by Curcumin

We examined the effect of curcumin on MG-induced cell death. Curcumin alone up to 10 µM had no effect on cell viability. Approximately 65% of cells died after MG (200 µM) treatment for 3 h, and curcumin inhibited this cell death in a dose-dependent manner (FIG. 1A). Further experiments showed a MG treatment time-dependent decrease in viable cells but a significant increase in the number of viable cells by curcumin pretreatment (FIG. 1B). We investigated whether prevention of MG-induced cell death by curcumin was due to inhibition of apoptosis by analyzing the effect of curcumin on biochemical events such as DNA fragmentation, caspase-3 activation, and cleavage of PARP (poly [ADP-ribose] polymerase). An ELISA was used to quantitatively determine the amount of histone-associated oligonucleosome DNA fragments. We found that in comparison with untreated cells, MG treatment induced a 5.5-fold increase in this apoptosis-associated parameter, but in the presence of curcumin (10 µM) this increase was only 1.9-fold (FIG. 1C). We further examined MG-stimulated activation or cleavage of casapse-3 and PARP and found that >10 µM curcumin completely blocked the effects of MG (FIG. 2A and B). Taken to-

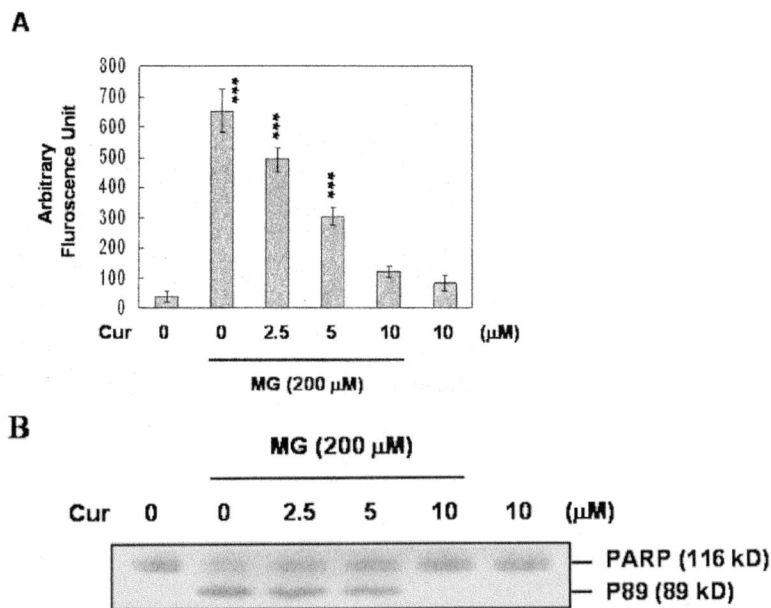

FIGURE 2. Effect of curcumin on MG-induced caspase-3 activation and cleavage of PARP. Hep G2 cells were incubated with or without various concentrations of curcumin (Cur) for 1 h. Cells were then treated with or without MG (200 µM) for a further 3 h. The cell extracts were prepared for assays. (**A**) Caspase-3 activity assay. (**B**) Cell extracts (40 µg) were immunoblotted using anti-PARP antibody. P89 (89 kDa) represents the cleavage product of PARP. $*P < 0.05$, $**P < 0.01$, and $***P < 0.0001$, respectively, from the control value.

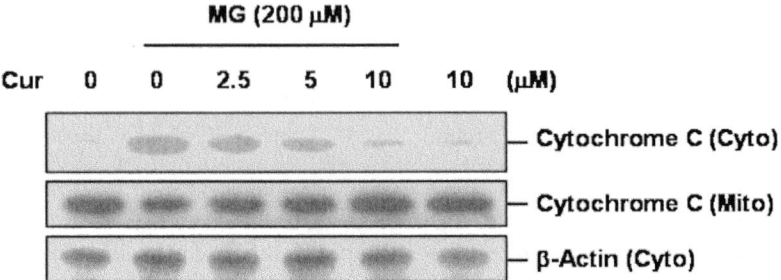

FIGURE 3. Effect of curcumin on MG-induced cytochrome c release from mitochondria. Hep G2 cells were pre-incubated with various concentrations of curcumin (Cur) for 1 h and then treated with MG (200 μM) for a further 3 h. Cytosolic (Cyto) and mitochondrial (Mito) fractions were separated. Cytosol aliquots (60 μg) were immunoblotted using anti-cytochrome c (Cyt. C) and anti–actin antibody. Mitochondrial aliquots (20 μg) were immunoblotted using anti–cytochrome c antibody.

gether, the results demonstrate that curcumin is a potent inhibitor of MG-induced apoptosis.

Curcumin Inhibits MG-Induced Mitochondrial Cytochrome c Release in Hep G2 Cells

Cytochrome c release is directly associated with apoptosis.[12] We investigated the effect of curcumin on mitochondrial cytochrome c release and found that MG caused significant cytochrome c release into the cytosol compared with control cells (without MG treatment) (FIG. 3). Curcumin pretreatment dose-dependently inhibited this cytochrome c release (FIG. 3).

Curcumin Prevents MG-Induced ROS Formation in Hep G2 Cells

We examined the effect of curcumin on ROS formation in MG-treated cells using DCF-DA as the detection reagent. Pretreatment of cells with curcumin significantly attenuated the increase in intracellular ROS content triggered by MG (FIG. 4), with 10 μM curcumin reducing intracellular ROS levels to ~30% of those in MG-treated cells.

DISCUSSION

MG, an endogenous carbonyl compound, triggers multiple intracellular responses causing a diverse array of functional changes.[3,13] Recent evidence indicates that MG-induced ROS generation plays an important role in the effect of MG on cellular responses.[14] We have previously shown that many apoptotic triggers (i.e., UV irradiation and photodynamic treatment) directly elicit ROS formation, and that antioxidants prevent these apoptotic processes.[8,9] The present study shows that curcumin at doses higher than 10 μM inhibits MG-induced apoptotic biochemical

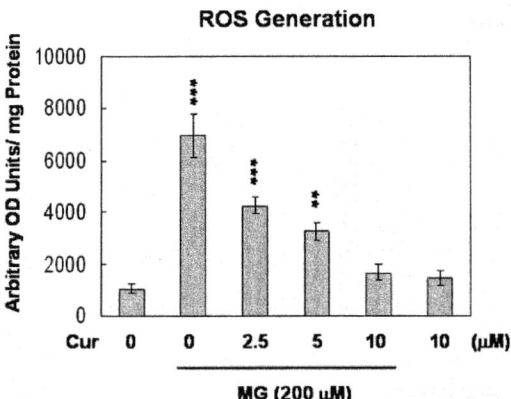

FIGURE 4. Curcumin attenuates MG-induced oxidative stress. Hep G2 cells were incubated with or without curcumin for 1 h and then treated with MG (200 µM) for a further 3 h. The generation of ROS was assayed using 2′,7′-dichlorofluorescein diacetate (DCF-DA) and is expressed as absorbance/mg of protein. $*P < 0.05$, $**P < 0.01$, and $***P < 0.0001$, respectively, from the control value.

changes in Hep G2 cells (FIGS. 1, 2, and 3). Others have shown that curcumin can both stimulate and inhibit apoptotic signaling,[15,16] suggesting that treatment protocols and cell type may determine the effect of curcumin.

The inhibitory effect of curcumin on apoptotic biochemical changes triggered by several stimuli has been attributed to its antioxidant properties.[17] Oxidative stress is now recognized as a stimulator of cell responses such as apoptosis.[18] Mechanisms underlying the antioxidant and anti-inflammatory properties of curcumin may involve glutathione (GSH)-linked detoxification. Our finding that curcumin attenuated MG-induced intracellular ROS formation supports the notion that curcumin suppresses apoptosis by quenching ROS that form after MG treatment (FIG. 4).

The present study supports the hypothesis that curcumin inhibits MG-induced apoptotic biochemical changes by blocking ROS formation and subsequent apoptotic biochemical changes. Recently, we have studied the time course of ROS formation, cytochrome c release, caspase-3 activation, and PARP cleavage during MG-treated Hep G2 cells. Our observations, taken together with the finding that ROS generation triggered by MG treatment can be blocked by curcumin, support the hypothesis that curcumin inhibits MG treatment-induced apoptotic biochemical changes by blocking ROS formation, which is an important trigger for cytochrome c release from mitochondria into the cytoplasm and subsequent activation of caspase. A model for MG-triggered apoptotic signal pathway is proposed as following: MG treatment→ ROS generation→ cytochrome c release→ caspases activation→ apoptosis. However, the precise regulatory mechanisms and signal transduction pathway of MG-induced apoptosis need further investigation.

ACKNOWLEDGMENTS

This work was supported by grants (NSC92-2320-B-033-003 and NSC 92-2120-M-033-001) from the National Science Council of Taiwan, ROC.

REFERENCES

1. THORNALLEY, P.J. 1998. Glutathione-dependent detoxification of alpha-oxoaldehydes by the glyoxalase system: involvement in disease mechanisms and antiproliferative activity of glyoxalase I inhibitors. Chem. Biol. Interact. **111–112:** 137–151.
2. OKADO, A. *et al.* 1996. Induction of apoptotic cell death by methylglyoxal and 3-deoxyglucosone in macrophage-derived cell lines. Biochem. Biophys. Res. Commun. **225:** 219–224.
3. KANG, Y.V., L.G. EDWARDS & P.J. THORNALLEY. 1996. Effect of methylglyoxal on human leukaemia 60 cell growth: modification of DNA G1 growth arrest and induction of apoptosis. Leuk. Res. **20:** 397–405.
4. DU, J. *et al.* 2000. Methylglyoxal induces apoptosis in Jurkat leukemia T cells by activating c-Jun N-terminal kinase. J. Cell. Biochem. **77:** 333–344.
5. NADKARNI, K.M. 1976. Indian Materia Medica. Popular Prakashan. Bombay.
6. HUANG, M.T. *et al.* 1994. Inhibitory effects of dietary curcumin on forestomach, duodenal, and colon carcinogenesis in mice. Cancer Res. **54:** 5841–5847.
7. JEE, S.H. *et al.* 1998. Curcumin induces a p53-dependent apoptosis in human basal cell carcinoma cells. J. Invest. Dermatol. **111:** 656–661.
8. CHAN, W.-H., C.-C. WU & J.-S. YU. 2003. Curcumin inhibits UV irradiation-induced oxidative stress and apoptotic biochemical changes in human epidermal carcinoma A431 cells. J. Cell. Biochem. **90:** 327–338.
9. CHAN, W.-H., J.-S. YU & S.-D. YANG. 2000. Apoptotic signalling cascade in photosensitized human epidermal carcinoma cells: Involvement of singlet oxygen, c-Jun N-terminal kinase, caspase-3, and p21-Activated kinase 2. Biochem. J. **351:** 221–232.
10. CHAN, W.-H. & H.-J. WU. 2004. Antiapoptotic effects of curcumin on photosensitized human epidermal carcinoma A431 cells. J. Cell. Biochem. **92:** 200–212.
11. YANG, J. *et al.* 1997. Prevention of apoptosis by bcl-2: Release of cytochrome c from mitochondria blocked. Science **275:** 1129–1132.
12. ZOU, H. *et al.* 1997. Apaf-1, a human protein homologous to *C. elegans* CED-4, participates in cytochrome c-dependent activation of caspase-3. Cell **90:** 405–413.
13. MILANESA, D.M. *et al.* 2000. Methylglyoxal-induced apoptosis in human prostate carcinoma: potential modality for prostate cancer treatment. Eur. Urol. **37:** 728–734.
14. DU, J. *et al.* 2001. Superoxide-mediated early oxidation and activation of ASK1 are important for initiating methylglyoxal-induced apoptosis process. Free Radic. Biol. Med. **31:** 469–478.
15. BUSH, J.A., K.J. CHEUNG, JR. & G. LI. 2001. Curcumin induces apoptosis in human melanoma cells through a Fas receptor/caspase-8 pathway independent of p53. Exp. Cell. Res. **271:** 305–314.
16. JARUGA, E. *et al.* 1998. Glutathione-independent mechanism of apoptosis inhibition by curcumin in rat thymocytes. Biochem. Pharmacol. **56:** 961–965.
17. RUBY, A.J. *et al.* 1995. Antitumour and antioxidant activity of natural curcuminoids. Cancer Lett. **94:** 79–83.
18. BUTTKE, T.M. & P.A. SANDSTROM. 1994. Oxidative stress as a mediator of apoptosis. Immunol. Today **15:** 7–10.

Prevention of Cellular Oxidative Damage by an Aqueous Extract of *Anoectochilus formosanus*

LENG-FANG WANG,[a] CHUN-MAO LIN,[a] CHWEN-MING SHIH,[a] HUI-JU CHEN,[a] BORCHERNG SU,[a] CHENG-CHUANG TSENG,[b] BAO-BIH GAU,[a] AND KUR-TA CHENG[a]

[a]*Department of Biochemistry, Taipei Medical University, Taipei, Taiwan, ROC*

[b]*Department of Microbiology and Immunology, Taipei Medical University, Taipei, Taiwan, ROC*

ABSTRACT: *Anoectochilus formosanus* (AF) is a popular folk medicine in Taiwan whose pharmacological effects have been characterized. In this work we investigated the antioxidant properties of an aqueous extract prepared from AF. The AF extract was capable of scavenging H_2O_2 in a dose-dependent manner. We induced oxidative stress in HL-60 cells, either by the addition of hydrogen peroxide (H_2O_2) or by the xanthine/xanthine oxidase reaction. Apoptosis caused by oxidative damage was displayed by DNA fragmentation on gel electrophoresis, and the apoptotic fraction was quantified with flow cytometry. The cell damage induced by oxidative stress was prevented by the plant extract in a concentration-dependent manner. Furthermore, the proteolytic cleavage of poly(ADP-ribose) polymerase during the apoptotic process was also inhibited by AF extract. Our results provide the basis for determining an AF extract to be an antioxidant.

KEYWORDS: *Anoectochilus formosanus*; reactive oxygen species; apoptosis; PARP

INTRODUCTION

Reactive oxygen species (ROS) play an important role in the development of tissue damage and pathological events in living organisms.[1] Imbalance of oxidation–reduction in a healthy living system may lead to cellular dysfunction and various diseases. Selected natural products from plant sources share antioxidant actions. The genus *Anoectochilus (Orchidaceae)*, comprising more than 35 species, is a perennial herb. Its distribution spread from India, the Himalayas, Southeast Asia, and Indonesia to New Caledonia and Hawaii. Of the 35 species, four were found in Taiwan, including *Anoectochilus formosanus* (AF), *A. inabai, A. koshunensis,* and *A. lanceolatus.* AF and *A. koshunensis* were first used by local people to treat snakebite. AF is still a popular folk medicine used to treat hepatitis, hypertension, and can-

---

Address for correspondence: Kur-Ta Cheng, Department of Biochemistry, Taipei Medical University, 250 Wu Hsing St., Taipei, Taiwan, ROC. Voice: +886-2-27361661 ext. 3169; fax: +886 -2-27399650.
ktbot@tmu.edu.tw

cer in Taiwan. A crude AF extract, with high antioxidant content, was found to have hepatoprotective action in a rat model of CCl_4-induced liver damage.[2] This finding prompted us to examine the protective effect of an aqueous extract of AF in H_2O_2- or superoxide-induced apoptosis in HL-60 cells.

MATERIALS AND METHODS

Preparation of Plant Extracts

AF whole plant was collected from Puli in central Taiwan. The species of the plant was authenticated, and a voucher specimen (accession no. SP 9703010) was deposited in the herbarium of the Taipei Medical University.[3] One hundred grams of the whole plant was homogenized with 900 mL of cold water and stirred at 4°C for 30 min and then centrifuged at 20,000 × g for 20 min at 4°C. The supernatant was freeze-dried to powder form. The yield of crude extracts was 1.92 g. The dry extract was maintained at −20°C until used. The extract was dissolved in the appropriate medium immediately prior to the experiments.

Cells and Cell Culture

The promyelocytic cell line HL-60 was obtained from the American Type Culture Collection (Rockville, MD) and maintained as a continuous culture in RPMI 1640 medium supplemented with 2 mM L-glutamine, 10% heat-inactivated fetal bovine serum (Life Technologies, Merelbeke, Belgium), 100 U/mL penicillin G, and 100 µg/mL streptomycin in a humidified atmosphere of 5% CO_2 in air at 37°C. Medium was refreshed every 3–4 days. Cell cultures were free of mycoplasma.

Hydrogen Peroxide Scavenging Assay

The ability of the extract to scavenge hydrogen peroxide was determined according to the method of Ruch et al.[4] Hydrogen peroxide concentration was determined spectrophotometrically with absorption at 230 nm. Because the AF extract has an absorption spectrum covering 230 nm (A_{230}), the scavenging of hydrogen peroxide activity was determined by the decrease of the A_{230} after 10 min of incubation at room temperature in the presence of various concentrations of AF and 10 mM hydrogen peroxide in 1 mL phosphate-buffered saline (PBS, pH 7.4) against a blank solution containing AF extracts in PBS without hydrogen peroxide.

DNA Fragmentation Assay

Cells under various treatments were harvested and lysed with DNA extraction buffer (50 mM Tris–HCl, 10 mM EDTA, 0.5% N-lauroylsarcosine, 0.5 mg/mL proteinase K, and 0.5 µg/mL RNase A) overnight at 50°C. The DNA samples were isolated by phenol–chloroform extraction and then analyzed by a 1.8% agarose gel electrophoresis.[5] The DNA fragmentation was visualized by UV illumination after ethidium bromide staining.

Protection of HL-60 Cells from ROS Injury

A total of 2×10^5 HL-60 cells were incubated with various concentrations of AF extracts for 10 min and then treated with 50 µM xanthine and 3.2 U/L xanthine oxidase for 4 h. The resulting cell apoptosis was characterized with a DNA fragmentation assay.

Quantification of Apoptosis by Flow Cytometry

The percentage of cells displaying DNA fragmentation was quantified using propidium iodide. Cells were washed with PBS, fixed in 70% ethanol, and then incubated in extraction buffer (0.1 M Na_2HPO_4/0.05 M citric acid, pH 7.8) at room temperature for 10 min in the dark. The cells were stained with 10 µg/mL propidium iodide containing 0.5% Triton X-100 and 0.1 mg/mL RNase A for 30 min at room temperature. The apoptotic fraction, taken as the sub-G_1 peak, was analyzed on a FACScan (Becton Dickenson, San Jose, CA).[6]

Immunoblot Analysis

Cells under various treatments were lysed in 2× sodium dodecyl sulfate (SDS)/ sample buffer (125 mM Tris–HCl, pH 6.8, 4% SDS, 10% glycerol, 0.006% bromophenol blue, 1.8% β-mercaptoethanol) and boiled for 5 min at 100°C. Samples containing 10 µg of protein were subjected to electrophoresis in an 10% SDS-PAGE

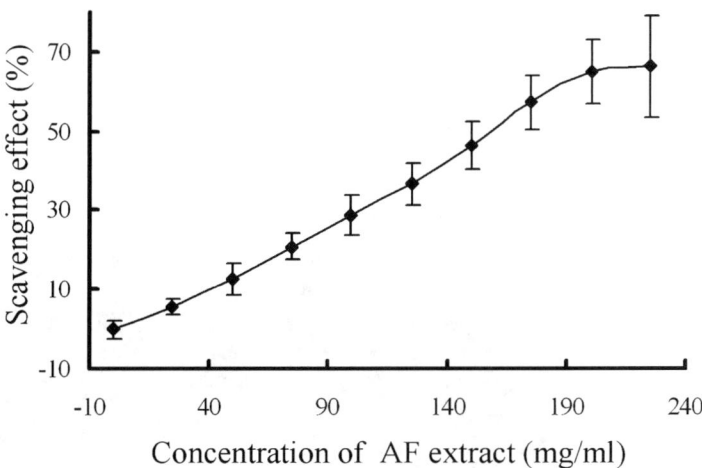

FIGURE 1. The scavenging effect of AF extract on hydrogen peroxide. Each data point represents a triplicate measurement of three preparations and was expressed as mean ± SD. Calculation of scavenging effect as following. (**A**) A_{230} (mixture of various concentration of AF and 10 mM hydrogen peroxide at room temp, 10 min) minus A_{230} (mixture of various concentration of AF at room temp, 10 min). (**B**) A_{230} (mixture of various concentration of AF and 10 mM hydrogen peroxide at room temp, 0 min) minus A_{230} (mixture of various concentration of AF at room temp, 0 min). Scavenging effect (%) = [(B − A)/B] × 100.

and transferred to a polyvinylidene difluoride membrane (Millipore, Bedford, MA). The membrane was blocked in 10 mM Tris–HCl buffer (pH 7.5) containing 100 mM NaCl, 0.05% Tween 20 and 4% skimmed milk for 1 h and then probed with the antibodies raised against PARP (Clontech, Palo Alto, CA) and β-actin (Sigma, St. Louis, MO) for 1 h. The membrane was then incubated with a horseradish peroxidase–conjugated secondary antibody, and the immunoreactive bands were visualized with enhanced chemiluminescent reagents.[7]

RESULTS AND DISCUSSION

Hydrogen Peroxide Scavenging Effect by AF Extract

The aqueous extract of AF was capable of scavenging H_2O_2 in a dose-dependent manner during 10 min of incubation (FIG. 1), suggesting its antioxidant activity.

Reduction of Hydrogen Peroxide–Induced Apoptosis by AF Extract

Oxidative stress has been implicated in several pathological processes. The antioxidant effect provides a mode of action of several selected dietary items in disease prevention. The protection of HL-60 cells from H_2O_2-induced apoptosis by the aqueous extract of AF was compared with a number of known antioxidants. HL-60 cells were exposed to 50 µM H_2O_2 for 4 h. A characteristic DNA laddering pattern suggestive of apoptosis was noted following H_2O_2 exposure. H_2O_2-scavenging en-

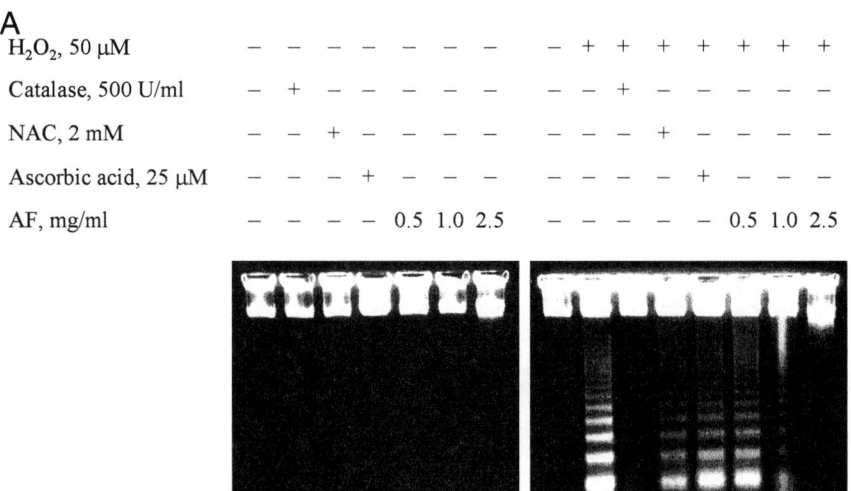

FIGURE 2. Effects of antioxidants on hydrogen peroxide induced cell apoptosis.
(A) Effects of antioxidants on DNA fragmentation of apoptotic HL-60 cells. Cells were pretreated with catalase, NAC, ascorbic acid, or AF extract for 10 min prior to 50 µM H_2O_2 treatment for 4 h. The DNA ladder was analyzed using 1.8% agarose gel electrophoresis. The results in the left panel were control without hydrogen peroxide treatment, all showing genome integrity.

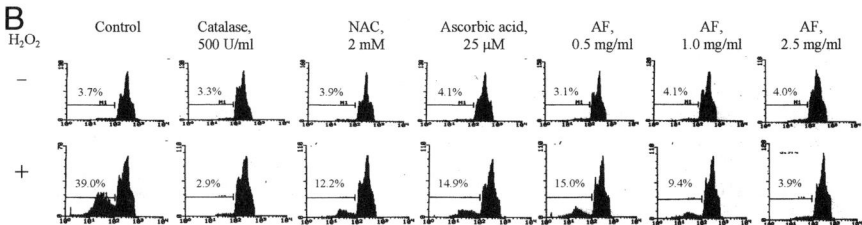

FIGURE 2 (continued). (**B**) Quantification of the apoptotic fraction upon various treatments using flow cytometry. Cells were treated as described in panel A. The cells were fixed in 70% ethanol, stained with 10 μg of propidium iodide/mL, and then the sub-G_1 peaks were analyzed on FACS. The percentage of sub-G1 peak fraction is indicated.

zyme catalase (500 U/L), or antioxidants including ascorbic acid (25 μM) and N-acetylcysteine (NAC, 2 mM), or the aqueous extract of AF (0–2.5 mg/mL) reduced the extent of H_2O_2-induced DNA laddering (FIG. 2A, right panel). The aqueous extract of AF reduced the extent of DNA laddering in a dose-dependent manner with the potency of 2.5 mg/mL comparable to that of 500 U/L catalase. To further assess the extent of apoptosis under various treatment conditions, the sub-G1 peak area was quantified using flow cytometry (FIG. 2B). Exposure to 50 μM H_2O_2 resulted in a 39% cell fraction in the sub-G1 peak area. The aqueous extract of AF dose-dependently reduced the extent of apoptosis based on flow cytometry, with the cell fraction in the sub-G1 peak area reduced to 3.9% at a concentration of 2.5 mg/mL, consistent with results shown in FIGURE 2A.

Inhibition of Xanthine/Xanthine Oxidase–Induced Cell Apoptosis by AF Extract

Xanthine/xanthine oxidase (X/XO) reaction is an important biological source of ROS contributing to the oxidative stress on the organism and is involved in many pathological processes. AF with ROS scavenging activity might be useful candidates for the prevention or suppression of X/XO reaction–induced damage. The extent of DNA laddering was suppressed when cells were treated with 0–5 mg/mL of AF extract (FIG. 3). AF effectively protected HL-60 cells from ROS-induced apoptosis in the X/XO reaction in a dose-dependent manner.

AF Extract Rescued PARP from Apoptotic Cleavage in Apoptotic Cells

When the apoptotic signaling was triggered by H_2O_2 in the HL-60 cells, the PARP was effectively cleaved from its native 113-kDa form to an 89-kDa fragment. Therefore, the cleavage pattern of PARP is used as an indication of apoptosis progression. HL-60 cells were treated with H_2O_2, and the PARP level was assessed by western blot analysis (FIG. 4). The processed 89-kDa cleavage forms of PARP appeared in samples derived from H_2O_2-treated HL-60 cells.[8] Upon the addition of various antioxidants, the levels of the 89-kDa fragment were decreased with catalase (500 U/mL), NAC (2 mM), ascorbic acid (25 μM), or AF (0.5, 1.0 and 2.5 mg/mL, respectively).

Xanthine, mM	–	50	–	–	–
Xanthine oxidase, U/L	–	3.2	–	–	–
AF, mg/ml	–	–	1.0	2.5	5.0

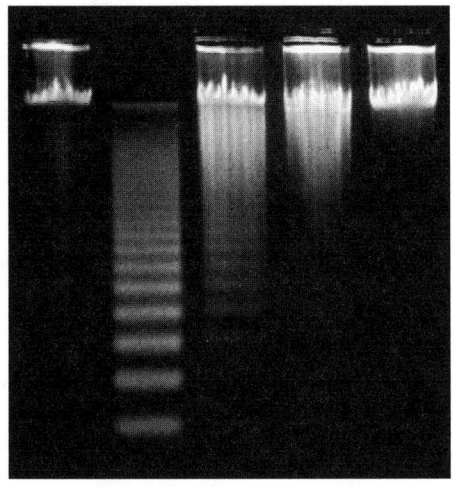

FIGURE 3. AF extract protected HL-60 cells from ROS-induced damage in the xanthine/xanthine oxidase reaction. DNA fragmentation of HL-60 cells treated with X/XO for 4 h in the presence of AF extract (0–5 mg/mL) is shown.

FIGURE 4. Effect of antioxidants on H_2O_2-induced PARP cleavage in HL-60 cells. Catalase, NAC, ascorbic acid, and an aqueous extract of AF were contained in various cell treatments, respectively, as described in FIGURE 2. Protein samples of whole-cell lysates were prepared, and 10 µg of protein for each lane was separated by 10% SDS-PAGE. Western immunoblots were performed using anti-PARP and –β-actin monoclonal antibodies to probe protein levels.

The pharmacological effects of aqueous extract of AF include hepatoprotective activity,[9] hypoglycemic and antioxidant effects in diabetic rats,[10] inhibitory action on the oxidation of low density lipoprotein,[11] and anti-inflammatory actions.[12] Excessive and persistent formation of ROS by inflammatory cells is thought to be a key factor in the ROS genotoxic effect. ROS was also proposed to be the common second messenger for activation of NF-κB, and some have suggested that the modulation of these oxidation–reduction reactions provides therapeutic benefit for controlling diseases involving inflammatory processes.[13] The aqueous AF extract exhibited the effects on protecting cells from hydrogen peroxide– or X/XO reaction–triggered apoptosis. Apoptotic cleavage of PARP in HL-60 was also reduced by the aqueous extract of AF. Results from this study provide pharmacological findings supporting the antioxidant action of this aqueous extract of AF.

ACKNOWLEDGMENTS

This work was supported by Grant NSC90-2317-B-038-001 from the National Science Council and by Grant TMU91-Y05-A124 from Taipei Medical University, Taiwan, Republic of China.

We are grateful to Dr. Sheng-Tung Huang, Department of Biochemistry, Taipei Medical University, for critically reading the manuscript.

REFERENCES

1. Cuzzocrea, S., D.P. Riley, A.P. Caputi, et al. 2001. Antioxidant therapy: a new pharmacological approach in shock, inflammation, and ischemia/reperfusion injury. Pharmacol. Rev. **53:** 135–159.
2. Lin, C.C., P.C. Huang & J.M. Lin. 2000. Antioxidant and hepatoprotective effects of Anoectochilus formosanus and Gynostemma pentaphyllum. Am. J. Chin. Med. **28:** 87–96.
3. Cheng, K.T., L.C. Fu, C.S. Wang, et al. 1998. Determination of the components in a Chinese prescription, yu-ping-feng san, by RAPD analysis. Planta Med. **64:** 46–49.
4. Ruch, R.J., S.J. Cheng & J.E. Klaunig. 1989. Prevention of cytotoxicity and inhibition of intercellular communication by antioxidant catechins isolated from Chinese green tea. Carcinogenesis **10:** 1003–1008.
5. Lin, C.M., C.S. Chen, C.T. Chen, et al. 2002. Molecular modeling of flavonoids that inhibits xanthine oxidase. Biochem. Biophys. Res. Commun. **294:** 167–172.
6. Lin, C.M., C.T. Chen, H.H. Lee, et al. 2002. Prevention of cellular ROS damage by isovitexin and related flavonoids. Planta Med. **68:** 363–365.
7. Wang, I.K., S.Y. Lin-Shiau & J.K. Lin. 1999. Induction of apoptosis by apigenin and related flavonoids through cytochrome c release and activation of caspase-9 and caspase-3 in leukaemia HL-60 cells. Eur. J. Cancer **35:** 1517–1525.
8. DiPietrantonio, A.M., T. Hsieh, & J.M. Wu. 1999. Activation of caspase 3 in HL-60 cells exposed to hydrogen peroxide. Biochem. Biophys. Res. Commun. **255:** 477–482.
9. Du, X.M., N.Y. Sun, J. Hayashi, et al. 2003. Hepatoprotective and antihyperliposis activities of in vitro cultured Anoectochilus formosanus. Phytother. Res. **17:** 30–33.
10. Shih, C.C., Y.W. Wu & W.C. Lin. 2002. Antihyperglycaemic and anti-oxidant properties of Anoectochilus formosanus in diabetic rats. Clin. Exp. Pharmacol. Physiol. **29:** 684–688.
11. Shih, C.C., Y.W. Wu & W.C. Lin. 2003. Scavenging of reactive oxygen species and inhibition of the oxidation of low density lipoprotein by the aqueous extraction of Anoectochilus formosanus. Am. J. Chin. Med. **31:** 25–36.

12. LIN, J.M., C.C. LIN, H.F. CHIU, et al. 1993. Evaluation of the anti-inflammatory and liver-protective effects of *Anoectochilus formosanus, Ganoderma lucidum* and *Gynostemma pentaphyllum* in rats. Am. J. Chin. Med. **21:** 59–69.
13. TSE, H.M., M.J. MILTON & J.D. PIGANELLI. 2004. Mechanistic analysis of the immunomodulatory effects of a catalytic antioxidant on antigen-presenting cells: implication for their use in targeting oxidation-reduction reactions in innate immunity. Free Radic. Biol. Med. **36:** 233–247.

Inhibitory Effects of a Rice Hull Constituent on Tumor Necrosis Factor α, Prostaglandin E2, and Cyclooxygenase-2 Production in Lipopolysaccharide-Activated Mouse Macrophages

SHENG-TUNG HUANG,[a] CHIEN-TSU CHEN,[a] KUR-TA CHIENG,[a] SHIH-HAO HUANG,[b] BEEN-HUANG CHIANG,[b] LENG-FANG WANG,[a] HSIEN-SAW KUO,[a] AND CHUN-MAO LIN[a]

[a]*College of Medicine, Taipei Medical University, Taipei, Taiwan*

[b]*Graduate Institute of Food Science and Technology, National Taiwan University, Taipei, Taiwan*

ABSTRACT: Isovitexin, isolated from rice hull of *Oryza sativa,* has been characterized as a potent antioxidant. Its antioxidant activity, determined on the basis of inhibition of lipid peroxidation by the Fenton reaction, was comparable with that of α-tocopherol, a well-established antioxidant. Isovitexin was able to reduce the amount of hydrogen peroxide production induced by lipopolysaccharide (LPS) in mouse macrophage RAW264.7 cells. In this study, we assessed its effects on the production of tumor necrosis factor α (TNF-α), prostaglandin E2 (PGE_2), and the expression of cyclooxygenase-2 (COX-2) in LPS-activated RAW 264.7 macrophages. Isovitexin inhibited the release of TNF-α, a proinflammatory cytokine, upon LPS activation with a 50% inhibitory concentration (IC_{50}) of 78.6 μM. Isovitexin markedly reduced LPS-stimulated PGE_2 production in a concentration-dependent manner, with an IC_{50} of 80.0 μM. The expression of COX-2 was also inhibited by isovitexin treatment. Our results suggest that suppression of ROS-mediated COX-2 expression by isovitexin is beneficial in reducing inflammation and carcinogenesis.

KEYWORDS: antioxidant; inflammation; isovitexin; COX-2; PGE_2; tumor necrosis factor

INTRODUCTION

Oxidative stress has been implicated in a variety of pathological processes, including aging, cancer, diabetes mellitus, atherosclerosis, neurological degeneration, and arthritis. Increased uptake of antioxidants may prevent organ injury associated

Address for correspondence: Chun-Mao Lin, Ph.D., College of Medicine, Taipei Medical University, No. 250, Wu-Hsing Street, Taipei 110, Taiwan. Voice: +886-2-27361661 ext. 3152; fax: +886-2-27361661 ext. 3163.
cmlin@tmu.edu.tw

with excessive generation of reactive oxygen species (ROS).[1,2] ROS have been shown to initiate a wide range of toxic oxidative reactions.[3] The effects of oxidants on signaling pathways are often characterized as resulting from heightened oxidative stress. Many antioxidants also exhibit anti-inflammatory effects. Theses observations provide a significant molecular basis for understanding the mode of actions of selected dietary ingredients in preventing diseases associated with heightened oxidative stress.

Tumor necrosis factor α (TNF-α), a proinflammatory cytokine, and prostaglandin E2 (PGE_2) are key mediators in inflammatory reaction. TNF-α generated in inflammation may induce tissue damage.[4] Prostaglandins, especially PGE_2, are involved in stimulating cell proliferation, tumor growth, and suppressing the immune response to malignant cells. Overproduction of prostaglandins from upregulation of cyclooxygenase-2 (COX-2) in cells had been associated with malignant growth.[5] COX-2 is rapidly induced by tumor promoters, growth factors, cytokines, and mitogens in various cell types.[6] Many cell types associated with inflammation, such as macrophages and endothelial cells and fibroblasts, may be induced to overexpress COX-2.[7] Treatment with TPA (12-O-tetradecanoyl-phobol-13-acetate) in mice led to edema and papilloma formation by enhancing COX-2 expression. Specific COX-2 inhibitors could counteract these biological events. Suppression of COX-2 induction and or its activity may be an effective approach for the prevention of carcinogenesis in several organs. Selected inhibitors of COX-2 may also have a therapeutic role in certain cancers.

Rice is an important dietary staple in Asia, where the incidence of breast and colon cancer is markedly lower than that in the western world.[8] It has been reported that rice constituents counteract chemical-induced mutagenicity, tumor promotion, and carcinogenicity.[9] Constituents from rice bran have been found to be beneficial for cancer prevention by epidemiological survey. Rice bran contains several classes of chemopreventive agents (e.g., flavonoids and their glycosides, tocotrienols, and γ-oryzanol). Isovitexin and related flavonoids are constituents of the rice hull of *Oryza sativa* and have been shown to exhibit potent antioxidant activity,[10] including inhibition of xanthine oxidase, protection of DNA from oxidative damage, and prevention of heavy-metal–induced cell injury.[11] Here, we further studied the antioxidative properties of isovitexin by examining its effects on LPS-induced PGE_2 and TNF-α production and COX-2 expression in the murine macrophage-like cell line RAW264.7.

MATERIALS AND METHODS

Materials

Isovitexin, a glycosylflavonoid, was isolated from rice hull as described previously.[12] The final product showed a major peak on capillary chromatography (P/ACE 5000; 75 μm × 37 cm [Beckman, Fullerton, CA]; borax buffer, 20 mM, pH = 10.0; UV absorbance at 214 nm; 25°C; UV absorbance at 214 nm, 25°C, 10K voltage; injection pressure, 80 psi; migration time: 6.64 min.). LPS (*Escherichia coli* O127:B8) and chemicals were purchased from Sigma (St. Louis, MO) unless specified.

Cell Culture

The mouse monocyte–macrophage cell line RAW 264.7 (ATCC TIB-71) was cultured in Dulbecco's modified Eagle's medium (Gibco, Grand Island, NY) supplemented with 10% heat-inactivated fetal bovine serum (Gibco). Cells were plated in 24-well plates or petri dishes before activation by LPS (100 ng/mL). Isovitexin dissolved in dimethyl sulfoxide (DMSO) with the final DMSO concentration of less than 0.2% (vol/vol) was administered with LPS. Control samples contained the same concentration of DMSO.

Lipid Peroxidation

Ethyllinoleate was oxidized by the Fenton reaction (Fe^{2+}/H_2O_2). In selected experiments, equal concentration (30 µM) of apigenin, isovitexin, and α-tocopherol were added to 0.5 mL of an aqueous solution containing 1.5 mg/L ethyl linoleate, 0.25 mM Trizma–HCl/0.75 mM KCl buffer (pH 7.4), 0.2% N-lauroyl sarcosine, 1 µM ferrous chloride, and 0.5 µM hydrogen peroxide in a 2-mL microtube. The mixture was incubated for 16 h at 37°C. The quantity of oxidation was measured by the thiobarbituric acid (TBA) assay.[11] The antioxidative activity of the samples was calculated according to the following formula: antioxidative activity (%) = 1 − (absorbance of sample/absorbance of control) × 100. The level of TBA-reactive substance (TBARS) by autooxidation of ethyllinoleate was calculated as the amount of malondialdehyde.

Determination of TNF-α Release

Murine macrophages were seeded in 24-well plates at a density of 5.0×10^5 cells/well the day before the experiment. Cells were treated with or without isovitexin (10–100 µM) and/or LPS (100 ng/mL) in 500 µL of medium containing 10% fetal bovine serum for 1 h at 37°C. TNF-α levels in the media were then determined by a quantitative sandwich enzyme-linked immunosorbent assay using the commercially available mouse TNF-α immunoassay kit (Amersham, Buckinghamshire, UK) according to the manufacturer's instructions. All experiments were done in triplicate.[13]

PGE_2 Assay

Cells were plated at a density of 1.0×10^6 cells/mL in 24-well culture plate and stimulated with LPS (50 ng/mL) in the presence or absence of various concentrations of isovitexin for 18 h. The culture medium of control and treated cells was collected and centrifuged. The level of PGE_2 released into culture medium was determined using an enzyme immunoassay according to the manufacturer's instructions (Amersham, Buckinghamshire, UK).[4]

Western Blotting

The cellular protein fraction was prepared using lysis buffer containing 10% glycerol, 1% Triton X-100, 1 mM sodium orthovanadate, 1 mM EGTA, 5 mM EDTA, 10 mM NaF, 1 mM sodium pyrophosphate, 20 mM Tris–HCl (pH 7.9), 100 µM β-glycerophosphate, 137 mM NaCl, 1 mM phenylmethylsulfonyl fluoride (PMSF), 10 µg/mL aprotinin, and 10 µg/mL leupeptin. Proteins (50 µg) were separated on sodium dodecyl sulfate–polyacrylamide gel electrophoresis (SDS-PAGE) and electrotransferred to a polyvinylidene difluoride (PVDF) membrane (Immobilon-P; Millipore, Bedford, MA). The membrane was incubated with an anti–COX-2 monoclonal antibody (Oncogene Science, Cambridge, UK) containing 1% bovine serum albumin and 0.2% NaN_3 overnight at 4°C. After incubation with horseradish peroxidase–conjugated anti–mouse immunoglobulin G antibody (Oncogene Science, Cambridge, UK), the immunoreactive bands were visualized with enhanced chemiluminescent reagents (ECL; Amersham, Buckinghamshire, UK). Relative protein content in each band was assessed using a densitometer (Alpha Innotech IS-1000; Digital Imaging System, San Leandro, CA).

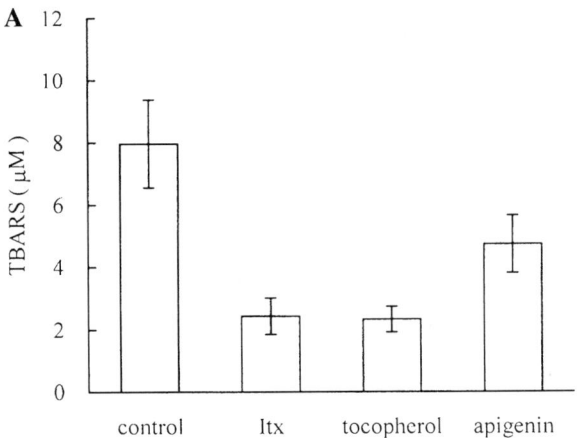

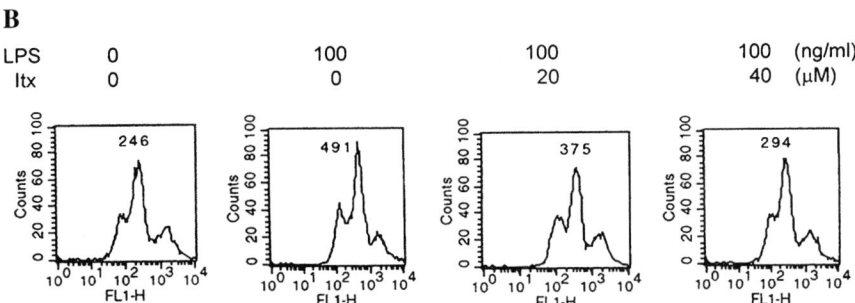

FIGURE 1. Antioxidative activity of isovitexin. (**A**) Isovitexin (Itx) inhibited the Fenton reaction–induced lipid peroxidation. Isovitexin (Itx), α–tocopherol, and apigenin (30 μM) were compared on the basis of their antioxidative activities. Quantity of oxidation was measured using the TBA assay. Antioxidative activity of the samples was measured according to levels of TBA-reactive substance (TBARS). (**B**) Isovitexin suppressed LPS-induced production of hydrogen peroxide. RAW 264.7 cells were treated with LPS (100 ng/mL) alone or LPS and isovitexin (20 and 40 μM) for 30 min. Cells were stained with DCFH-DA and subjected to flow cytometry. Peaks of FL1-H fluorescence intensity are indicated for each treatment.

Flow Cytometric Detection of Hydrogen Peroxide

RAW 264.7 cells were suspended in phenol red–free medium at a density of 10^5 cells/mL. Cells were stained with 100 μM 2′,7′-dichlorofluorescin diacetate (DCFH-DA) in the dark for 30 min and then analyzed using a FACScan (Becton Dickinson, San Jose, CA). Oxidation of green DCH fluorescence by hydrogen peroxide in living cells was detected using the FL1-H wavelength band. The fluorescence signals of 10,000 cells were processed using a logarithmic amplifier as described previously.[14]

RESULTS AND DISCUSSION

Antioxidative Activity of Isovitexin

Many antioxidants exert anti-inflammatory effects, and many polyphenolic compounds have been reported to be potent ROS scavengers. The antioxidative activity of isovitexin was evaluated by the TBA method. The inhibitory effect on ethyllinoleate oxidation by isovitexin (69.5% inhibition, TBARS value = 2.43 μM) was comparable to that of α-tocopherol (70.8% inhibition, TBARS value = 2.33 μM) and higher than that of apigenin, an aglycon flavonoid, (40.7% inhibition, TBARS value = 4.73 μM) (FIG. 1A). The results suggest that isovitexin is a potent antioxidant. Isovitexin also lowered the LPS-induced increase in the cellular content of hydrogen peroxide in RAW 264.7 macrophage cells in a dose-dependent manner (FIG. 1B).

Isovitexin Inhibition of TNF-α Production

To determine the effect of isovitexin on LPS-activated TNF-α production, we stimulated the cells with LPS in the presence or absence of isovitexin at concentra-

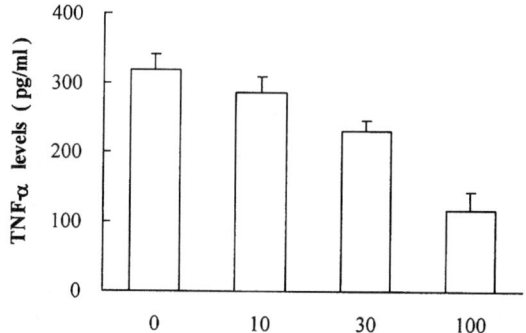

FIGURE 2. Inhibition of TNF-α production by isovitexin. RAW 264.7 cells were treated with LPS (100 ng/mL) in the presence or absence of isovitexin (Itx) (0, 10, 30, and 100 μM, respectively) for 1 h. Amount of TNF-α released into culture medium was determined. Data are presented as means ± standard error of the mean ($n = 3$).

tions ranging from 10 to 100 μM for 60 min. The production of TNF-α by the unstimulated RAW 264.7 cells was 20 pg/mL ($n = 3$). Incubation of these cells with LPS (100 ng/mL) for 1 h caused a substantial increase in TNF-α production (318.9 pg/mL; $n = 3$). When RAW 264.7 cells were stimulated with LPS in the presence of isovitexin (10–100 μM), a concentration-dependent inhibition of TNF-α production with a 50% inhibitory concentration (IC_{50}) value of 78.6 μM was observed (FIG. 2).

Isovitexin Inhibition of PGE₂ Production and COX-2 Expression

The effect of isovitexin on PGE_2 production and COX-2 protein expression following LPS stimulation was examined 18 h after treatment with LPS (50 ng/mL). PGE_2 production was reduced in a concentration-dependent manner with an IC_{50} of 80.0 μM (FIG. 3A). The inhibition was not due to general cellular toxicity (data not shown). LPS-induced COX-2 protein expression was also suppressed by isovitexin in a concentration-dependent manner after treatment for 18 h. The relative protein levels of COX-2 in the presence of 20, 40, 60, 80, and 100 μM isovitexin were 1.02,

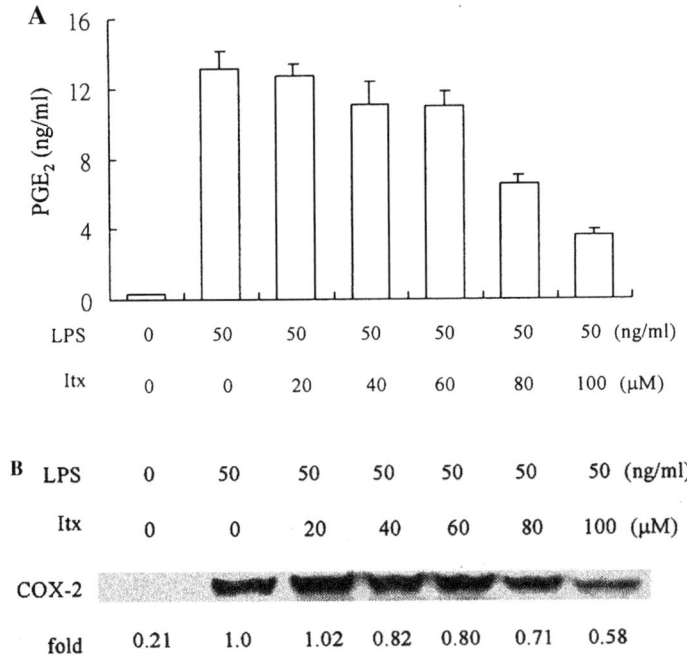

FIGURE 3. Effects of isovitexin on LPS-induced PGE_2 generation and COX-2 expression. (**A**) Cells were treated with LPS (50 ng/mL) in the presence or absence of isovitexin (Itx) for 18 h. Amount of PGE_2 released into culture medium was determined. Data are presented as means ± standard error of the mean ($n = 3$). (**B**) Values for expression of COX-2 protein upon various treatments for 18 h were quantities after resolving in 8.0% SDS-PAGE and Western blot analysis. Relative level of COX-2 expression observed with LPS alone was set at 1.0.

0.82, 0.80, 0.71, and 0.58, respectively, in reference to a value of 1.00 in samples treated with LPS alone (FIG. 3B).

A Possible Mechanism for Isovitexin Inhibition of LPS Activity

Macrophages release various mediators in cellular inflammatory response. Among those released are ROS including H_2O_2, nitric oxide, superoxide, and hydroxyl radical. The intracellular ROS production is associated with several cellular events including the activation of NAD(P)H oxidase, xanthine oxidase, and the cellular mitochondrial respiratory chain.[15,16] ROS thus formed are potent activators of NF-κB, an essential transcription factor in the inflammatory signal transduction pathway.[17] ROS plays an essential role in the initiation of the NIK/MAPK cascade to trigger a series of responses in which IκB is phosphorylated to free NF-κB from IκB inhibition.[18] The active form of NF-κB is translocated from cytoplasm to the nucleus to bind the cognate NF-κB binding site in the promoter regions of genes including COX-2.[19,20] COX-2 overexpression mediated by activation of NF-κB is a possible molecular mechanism in cellular transformation to neoplasia. Another consequence of excessive ROS generation is oxidative DNA damage leading to gene

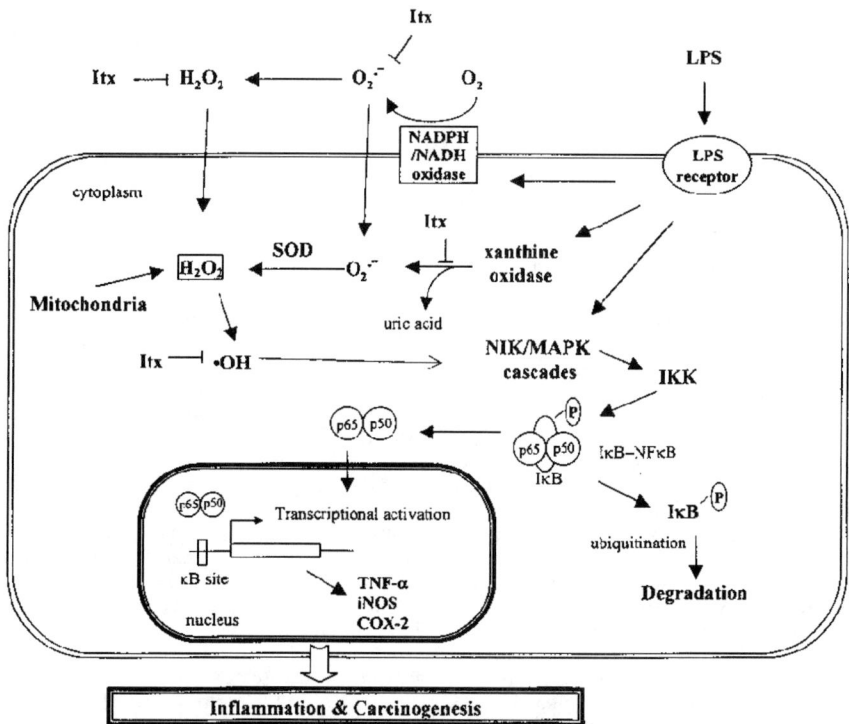

FIGURE 4. Possible mechanisms in which isovitexin (Itx) exerts its action against inflammation and carcinogenesis.

mutations and cancer formation. The inflammation-signaling transduction pathway mediated by LPS on macrophage is depicted in FIGURE 4.

In FIGURE 4, the roles of isovitexin in inhibiting inflammation and carcinogenesis are proposed. Isovitexin, with its potent antioxidant activity, is likely to suppress NF-κB activation. This contention is supported by the finding that the expression of TNF-α and COX-2, both transactivated by NF-κB, was suppressed by isovitexin. Reduced PGE_2 formation by isovitexin is likely due to its inhibition of COX-2 expression.

In summary, isovitexin is a potent antioxidant that inhibits TNF-α and COX-2 expression caused by LPS in mouse macrophage RAW264.7 cells. This effect of isovitexin is likely due to its inhibition of NF-κB activation via its antioxidant activity. Isovitexin also inhibits PGE_2 formation, likely secondary to its suppression of COX-2 expression. These antioxidant and anti-inflammatory effects of isovitexin may have implications in carcinogenesis. Isovitexin is a food phytochemical enriched in rice. Whether lower incidence of selected neoplasia such as breast and colon cancer in countries with rice as a major staple can be attributed to the antioxidant gradients in rice including isovitexin remains to be determined.

ACKNOWLEDGMENTS

This study was supported by grants from the National Science Council (NSC92-2113-M-038-005 and NSC92-2113-M-038-003).

REFERENCES

1. FITZPATRICK, F.A. 2001. Inflammation, carcinogenesis and cancer. Int. Immunopharmacol. **1:** 1651–1667.
2. PAVLICK, K.P., F.S. LAROUX, J. FUSELER, et al. 2002. Role of reactive metabolites of oxygen and nitrogen in inflammatory bowel disease. Free Radic. Biol. Med. **33:** 311–322.
3. CUZZOCREA, S., D.P. RILEY, A.P. CAPUTI, et al. 2001. Antioxidant therapy: a new pharmacological approach in shock, inflammation, and ischemia/reperfusion injury. Pharmacol. Rev. **53:** 135–159.
4. JU, H.K., S.H. BAEK, R.B. AN, et al. 2003. Inhibitory effects of nardostachin on nitric oxide, prostaglandin E2, and tumor necrosis factor-alpha production in lipopolysaccharide activated macrophages. Biol. Pharm. Bull. **26:** 1375–1378.
5. CROFFORD, L.J., R.L. WILDER, A.P. RISTIMAKI, et al. 1994. Cyclooxygenase-1 and -2 expression in rheumatoid synovial tissue. Effects of interleukin-1β, phorbol ester, and corticosteroids. J. Clin. Invest. **93:** 1095–1101.
6. INOUE, H., C. YOKOYAMA, S. HARA, et al. 1995. Transcriptional regulation of human prostaglandin-endoperoxide synthase-2 gene by lipopolysaccharide and phorbol ester in vascular endothelial cells. J. Biol. Chem. **270:** 24965–24971.
7. HLA, T., A. RISTIMAKI, S. APPLEBY, et al. 1993. Cyclooxygenase gene expression in inflammation and angiogenesis. Ann. N.Y. Acad. Sci. **696:** 197–204.
8. KOO, L.S., O.W. MANG, & J.H. HO. 1997. An ecological study of trends in cancer incidence and dietary changes in Hong Kong. Nutr. Cancer **28:** 289–301.
9. TAKESHITA, M., S. NAKAMURA, F. MAKITA, et al. 1992. Antitumor effect of RBS (rice bran saccharide) on ENNG-induced carcinogenesis. Biotherapy **4:** 139–145.
10. RAMARATHNAM, N., T. OSAWA, M. NAMIKI, et al. 1989. Chemical studies on novel rice hull antioxidants. 2. Identification of isovitexin, a C-glycosyl flavonoid. J. Agric. Food Chem. **37:** 316–319.

11. LIN, C.M., C.T. CHEN, H.H. LEE, et al. 2002. Prevention of cellular ROS damage by isovitexin and related flavonoids. Planta Medica **68:** 363–365.
12. OSAWA, T., H. KATSUZAKI, Y. HAGIWARA, et al. 1992. A novel antioxidant isolated from young green barley leaves. J. Agric. Food Chem. **40:** 1135–1138.
13. GAO, B. & M.F. TSAN. 2003. Endotoxin contamination in recombinant human heat shock protein 70 (Hsp70) preparation is responsible for the induction of tumor necrosis factor alpha release by murine macrophages. J. Biol. Chem. **278:** 174–179.
14. UBEZIO P. & F. CIVOLI. 1994. Flow cytometric detection of hydrogen peroxide production induced by doxorubicin in cancer cells. Free Radic. Biol. Med. **16:** 509–516.
15. IRANI, K., Y. XIA, J.L. ZWEIER, et al. 1997. Mitogenic signaling mediated by oxidants in RAS-transformed fibroblasts. Science **275:** 1649–1652.
16. LEE, S.F., Y.T. HUANG, W.S. WU, et al. 1996. Induction of c-jun protooncogene expression by hydrogen peroxide through hydroxyl radical generation and p60^{src} tyrosine kinase activation. Free Radic. Biol. Med. **21:** 437–448.
17. SCHRECK, R., P. RIEBER, & P.A. BAEUERLE. 1991. Reactive oxygen intermediates as apparently widely used messengers in the activation of the NF-kappa B transcription factor and HIV-1. EMBO J. **10:** 2247–2258.
18. CHEN, C.C., & J.K. WANG. 1999. p38 but not p44/42 mitogen-activated protein kinase is required for nitric oxide synthase induction mediated by lipopolysaccharide in RAW 264.7 macrophages. Mol. Pharmacol. **55:** 481–488.
19. SCHMIDT, K.N., P. AMSTAD, P. CERUTTI, et al. 1995. The roles of hydrogen peroxide and superoxide as messengers in the activation of transcription factor NF-kappa B. Chem. Biol. **2:** 13–22.
20. CHUN, K.S., H.H. CHA, J.W. SHIN, et al. 2004. Nitric oxide induces expression of cyclooxygenase-2 in mouse skin through activation of NF-kappaB. Carcinogenesis **25:** 445–454.

Raffinee in the Treatment of Spinal Cord Injury: An Open-Labeled Clinical Trial

HSIN-YING CHEN,[a] JER-MIN LIN,[b] HUNG-YI CHUANG,[c] AND WEN-TA CHIU[a,d]

[a]*Graduate Institute of Injury Prevention, Taipei Medical University, Taipei, Taiwan*

[b]*Department of Cosmetic Application and Management, Tung-Fang Institute of Technology, Ping-Tong, Taiwan*

[c]*Department of Occupational Medicine, Kaohsiung Medical University Hospital, Kaohsiung, Taiwan*

[d]*Department of Neurosurgery, Taipei Medical University–Wan Fang Hospital, Taipei, Taiwan*

> ABSTRACT: The purpose of this preliminary clinical trial was to assess the safety and therapeutic trend of Raffinee, a mixture of free radical scavengers derived from natural products, on subacute spinal cord injury (SCI). SCI patients (three females and seven males) were enrolled from three medical centers in Taiwan. Their ages ranged from 18 to 76 years, with an average of 47.6 ± 19.0 years. There were four incomplete and six complete cases in this study. Standard regimens for SCI at acute stage were given at each hospital. Approximately 7 weeks (35.4 ± 13.9 days) after SCI, 2 U of Raffinee was given orally before meals three times a day for 6 weeks. Patients received motor, sensory, and activities of daily living (ADL) assessments before and after the treatment with Raffinee every week for 6 weeks. The Wilcoxon signed-ranks test was used for statistical analysis. Significant motor and sensory recovery began in the second week of Raffinee treatment, and significant ADL recovery was also noted in the third week. Functional recovery was more prominent in incomplete cases. Raffinee appeared to be safe in the subacute stage of SCI and may be an effective adjuvant therapy for enhancing functional recovery. Further clinical studies including double-blinded randomized placebo-controlled trials with follow-up for more than 1 year are necessary to validate the effectiveness of Raffinee in SCI.
>
> KEYWORDS: spinal cord injury; free radical scavenger; lipid peroxidation; Raffinee

INTRODUCTION

The primary insult of spinal cord injury (SCI) is usually caused by compression or laceration of the spinal cord due to displacement of vertebra or disc into the spinal

---

Address for correspondence: Wen-Ta Chiu, MD, PhD, Department of Neurosurgery, Taipei Medical University–Wan Fang Hospital, 111 Hsing-Long Rd., Sec. 3, Taipei, Taiwan. Voice: +886-2-2933-5222; fax:+ 886-2-2933-5221.
wtchiu@tmu.edu.tw

canal during fracture–dislocation or bursting fracture of the spine.[1] However, mechanical damage rarely transects the core completely, even when the injury is neurologically complete.[2] In addition, biochemical and pathological changes in the spinal cord may worsen after injury. The concept of secondary injury has been proposed based on several pathophysiological mechanisms.[3–6] The major changes include: (1) reduced microcirculation due to vasospasm, platelet aggregation, and loss of autoregulation[7]; (2) increased intracellular calcium and sodium ions following activation of NMDA (N-methyl-D-aspartate) receptor by glutamate, which is released from the activated or damaged neurons[8]; (3) free radical formation and lipid peroxidation of cell membrane leading to progressive cell death[9]; and (4) apoptosis of neural cells.[10]

According to the secondary-injury theory, several neuroprotective agents with antioxidant and free radical scavenging effects, such as methylprednisolone[11] and tirilazad mesylate,[12] have been proven effective in randomized controlled clinical studies. Raffinee, derived from natural products with SOD-like antioxidant formulation, has powerful free radical scavenging actions. Raffinee was developed jointly by Ziel Enterprise in Taiwan and Toyo Hakko Company in Japan. The main components of Raffinee are GMT (Guardox), which is fermented, and infrared extracts from a mixture of rice germ, soybean, green tea, and coffee bean. It contains polyphenols (flavonols, genistein, and chlorogenic acid), proteins, minerals, and carbohydrates.[13] Each unit of Raffinee prepared for this preliminary clinical study contains 10 mL of transparent brown-colored liquid filled by aseptic vacuum technique in a plastic cup covered by a special sheet to preserve the product for 2 years without the use of preservatives. Each unit of Raffinee contains 7.5 g of GMT (Guardox), including 337.5 mg of polyphenols. The antioxidant capacity of Raffinee has been assessed by electron spin resonance (ESR) and spin trapping technique as reported previously.[14] The *in vitro* data suggested that 6 U of Raffinee, which was the total daily dose in this study, possesses 2.28×10^6 U of superoxide radical scavenging activity and 4.74×10^4 U of hydroxyl radical scavenging activity. Raffinee was tested in an open-label preliminary clinical trial to assess its safety and therapeutic trend in a small population of patients with SCI.

SUBJECTS AND METHODS

Ten SCI patients (three females and seven males) from three medical centers (Kaohsiung Medical University Hospital, Kaohsiung Veterans' General Hospital, and Chi-Mei Hospital) in Taiwan were enrolled after written informed consent had been obtained from each patient. The study protocol was approved by the institutional review board at each medical center. The age of patients ranged from 18 to 76 years (47.6 ± 19.0, mean ± SD). Four of the 10 patients had incomplete injury based on the American Spinal Injury Association (ASIA) impairment scale[15] for completeness, which closely parallels that of the Frankel scale. The remaining six patients had complete SCI. Causes of SCI included traffic accident and fall (TABLE 1). At the onset of the insult, each of the 10 patients was treated at the emergency department of the hospital, following the NASCIS 3 guidelines.[12] To not interfere with functional recovery on methylprednisolone therapy, Raffinee was given as an adjunct intervention for these cases 35.4 ± 13.9 days (range, 19–60 days) after SCI. In this study,

TABLE 1. List of patients enrolled in the Raffinee trial

No.	Sex	Age	Dx	Cause	Length of Tx (days)	Length of hospital stay (days)
1	F	25	C6 quadriparesis	fall	45	109
2	M	56	T5 paraplegia	TA	60	127
3	F	38	T7 paraplegia	TA	36	112
4	M	62	T10 paraplegia	fall	23	73
5	F	63	C7 quadriparesis	fall	30	22
6	M	37	C6 quadriplegia	TA	27	79
7	M	18	T12 paraplegia	TA	50	39
8	M	38	C5 quadriparesis	TA	44	64
9	M	63	T3 paraplegia	TA	20	55
10	M	76	C6 quadriparesis	fall	19	35

Dx, diagnosis; TA, traffic accident; Tx date, interval between the date of injury and the date of initiating Raffinee treatment.

2 U of Raffinee was given orally three times a day before a meal for 6 weeks. None of the enrolled patients had a history of chronic infectious processes such as granulomatosis or leprosy. Administration of free radical scavengers to patients with chronic infection may reduce their immunity due to interference with phagocytic functions.[16,17]

Clinical motor, sensory, and activities of daily living (ADL) assessments were done before and every week after initiation of Raffinee treatment. The motor score was recorded according to the standards of the ASIA.[15] The motor function below the neck (ranging from 0 to 100) was divided into 10 myotomes bilaterally, and the major muscle of each myotome was examined. Score 0 was for complete paralysis; 1 for visible muscle contraction without joint movement; 2 for joint movement without antigravity; 3 for antigravity without resistance; 4 for partial resistance; and 5 for normal response. The sensory score was recorded by testing light touch and pain sensation by pin prick over 29 dermatomes from C2 to S5. Score 1 was for complete sensory loss; 2 for partial loss; and 3 for normal response. The sum of sensory score ranged from 29 to 87. The ADL score was recorded according to the Barthel Index.[18] The subject was required to perform feeding, transfer, grooming, toileting, bathing, walking, climbing stairs, dressing, and voiding for assessment of the extent of assistance needed.

The Wilcoxon signed-ranks test was used to analyze the extent of functional recovery in motor, sensory, and ADL scores.

RESULTS

Every patient enrolled in this clinical trial completed the Raffinee regimen. A few subjects experienced adverse side effects, including occasional mild diarrhea (case 8) and insomnia (case 4), during the course. All these possible side effects were self-limited and resolved at the end of the course without need for medical interventions.

TABLE 2. Changes in ASIA motor scores during Raffinee treatment

Pt.No.	Pre	Week 1	Week 2*	Week 3*	Week 4*	Week 5*	Week 6*
1	77	78	86	88	94	95	95
2	17	17	17	19	19	19	20
3	50	50	54	54	54	56	56
4	50	50	50	50	50	50	50
5	71	71	76	82	–	–	90
6	50	50	50	53	56	59	60
7	61	61	62	64	65	65	65
8	85	85	87	90	92	92	92
9	45	45	45	45	45	45	45
10	67	67	80	90	96	98	98

Pre: Scores prior to Raffinee treatment. Wilcoxon signed ranks test, $*P < 0.05$.

TABLE 3. Changes in sensory scores during Raffinee treatment

Pt.No.	Pre	Week 1	Week 2*	Week 3*	Week 4*	Week 5*	Week 6*
1	62	62	70	72	80	83	83
2	33	33	33	35	35	35	35
3	69	69	73	73	73	73	73
4	63	63	63	63	63	63	63
5	64	64	67	69	–	–	75
6	67	67	67	70	72	75	75
7	69	69	69	69	69	69	69
8	73	73	77	77	79	79	79
9	49	49	49	50	50	52	52
10	51	51	68	75	80	85	85

Pre: Scores prior to Raffinee treatment. Wilcoxon signed ranks test, $*P < 0.05$.

Changes in motor scores are shown in TABLE 2. Significant motor recovery was noted at the second week of intervention and persisted until the end of the course. Changes in sensory scores are presented in TABLE 3. Similar to motor function, significant recovery was also noted at week 2 and continued throughout the course. Significant ADL recovery was also noted, starting in the third week of the course (TABLE 4). Functional recovery was more prominent in incomplete cases. In cases 1, 5, and 8, the treatment was started 4–6 weeks after SCI. Before Raffinee treatment, motor function in these three patients had been stationary, and physical activities were limited to those conducted in bed. In the second week after Raffinee treatment, the motor function started to improve throughout the course with eventual recovery of the ability to walk. Case 6 with C6 quadriplegia was rated Frankel A. Two weeks after the intervention, his left leg began to move and the muscle strength improved to grade 3, but motor function in his upper limbs improved only slightly. The results of

TABLE 4. Changes in ADL scores during Raffinee treatment

Pt. No.	Pre	Week 1	Week 2	Week 3*	Week 4*	Week 5*	Week 6*
1	15	15	15	25	40	55	75
2	0	0	5	10	15	15	15
3	15	15	15	15	20	20	20
4	15	15	15	20	20	20	25
5	20	20	20	30	–	–	50
6	15	15	15	15	15	20	20
7	70	70	70	70	70	80	80
8	25	25	25	30	30	50	65
9	0	0	0	5	10	10	20
10	45	45	65	70	85	90	90

Pre: Scores prior to Raffinee treatment. Wilcoxon signed ranks test, *$P < 0.05$.

the MRI study of cervical spine before and after the intervention revealed complete remission of spinal cord edema, but a large central syringomyelia was noted at the C7 level.

DISCUSSION

In 1990, the NASCIS 2 led to the broad use of methylprednisolone as an acceptable treatment for patients with acute SCI, assuming the therapy can be given within 8 h of injury.[11] Since then, high-dose methylprednisolone, administered within 8 h of onset, has been a worldwide "gold standard" for treatment of acute SCI. The possible mechanism of action of methylprednisolone via its antioxidant action was supported by the result of NASCIS 3. In NASCIS 3, the efficacy of methylprednisolone was compared with that of tirilazad, an aminosteroid with potent antioxidant action but without glucocorticoid effects. Tirilazad given for 48 h showed comparable therapeutic efficacy to that of methylprednisolone given for 24 h.[12] NASCIS 2[11] and NASCIS 3[12] together support the contention that it is the antioxidant action of methylprednisolone or tirilazad, but not the glucocorticoid effects, that improves neurological recovery after SCI. These two clinical trials also substantiate the concept of a secondary-injury mechanism involving excessive free radical formation after SCI that is amenable to therapies based on antioxidant strategies.

Antioxidant enzyme activities are lower, whereas the content of oxidizing substances are higher in the central nervous system than other organs.[19,20] For instance, there was a 10-fold difference in vitamin C concentration between the cerebrospinal fluid (CSF) and serum.[21] After SCI, an excessive amount of iron is released into the CSF. Iron is oxidized to ferrous ion by vitamin C, leading to the generation of hydroxyl radicals in the presence of hydrogen peroxide and superoxide (iron-dependent Fenton reaction and Harber–Weiss reaction). Another pathway for hydroxyl radical production after SCI may be mediated by glutamate (an iron-independent pathway). An excessive amount of glutamate is released after SCI to activate the NMDA receptor, resulting in increased calcium influx.[22] Calcium influx activates neuronal nitric oxide synthase (nNOS), which catalyzes the formation of nitric oxide radical from oxygen and L-arginine. The nitric oxide radical and superoxide radical react to form peroxynitrite in acid circumstance.[23] The highly reactive oxygen species such as peroxynitrite and hydroxyl radicals may bind unsaturated fatty acids in the cell membrane to cause lipid peroxidation, triggering a cascade of cell destruction.[24] The mechanism of hydroxyl radical–induced secondary injury begins within hours of SCI and may last for days.[3,25]

The results of this preliminary study suggest that natural antioxidants in Raffinee are not deleterious to the injured cord during a subacute stage. None of the patients suffered from serious adverse side effects. The observation that the patients show significant improvement during the course of Raffinee treatment is encouraging and raises the possibility that extended antioxidant therapies during the subacute stage of SCI may still be effective in improving the neurological function. Double-blinded placebo-controlled randomized trials in larger populations of patients with SCI are needed to confirm the therapeutic efficacy of Raffinee.

In conclusion, results of an open-label trial of Raffinee, a preparation rich in natural antioxidants, in 10 patients during a subacute phase of SCI showed that Raffinee was not deleterious and might improve neurological recovery. Further studies applying double-blinded randomized placebo-controlled protocols with long-term follow-up are needed to validate the effectiveness of Raffinee in treating SCI patients.

REFERENCES

1. TATOR, C.H. 1983. Spine-spinal cord relationships in spinal cord trauma. Clin. Neurosurg. **30:** 479–494.
2. KAKULAS, B.A. 1984. Pathology of spinal injuries. Cent. Nerv. Syst. Trauma **1:** 117–129.
3. TATOR, C.H. & M.G. FEHLINGS. 1991. Review of the secondary injury theory of acute spinal cord trauma with emphasis on vascular mechanism. J. Neurosurg. **75:** 15–26.
4. SIESJO, B.K. 1992. Pathophysiology and treatment of focal cerebral ischemia. Part I: pathophysiology. J. Neurosurg. **77:** 169–184.
5. SIESJO, B.K. 1992. Pathophysiology and treatment of focal cerebral ischemia. Part II: mechanisms of damage and treatment. J. Neurosurg. **77:** 337–354.
6. PETTY, M.A., P. POULET, A. HAAS, et al. 1996. Reduction of traumatic brain edema by a free radical scavenger. Eur. J. Pharmacol. **307:** 149–155.
7. GUHA, A., C.H. TATOR & J. ROCHON. 1989. Spinal cord blood flow and systemic blood pressure after experimental spinal cord injury in rats. Stroke **20:** 372–377.
8. FADEN, A.L. & R.P. SIMON. 1988. A potential role for excitotoxins in the pathophysiology of spinal cord injury. Ann. Neurol. **23:** 623–626.
9. GUTTERIDGE, J.M. 1995. Hydroxyl radicals, iron, oxidative stress, and neurodegeneration. Ann. N.Y. Acad. Sci. **765:** 201–213.
10. PARK, E., Y. LIU & M.G. FEHLINGS. 2003. Changes in glial cell white matter AMPA receptor expression after spinal cord injury and relationship to apoptotic cell death. Exp. Neurol. **182:** 35–48.
11. BRACKEN, M.B., M.J. SHEPARD, W.F. COLLINS, et al. 1990. A randomized, controlled trial of methylprednisolone or naloxone in the treatment of acute spinal cord injury. N. Engl. J. Med. **322:** 1405–1411.
12. BRACKEN, M.B., M.J. SHEPARD, T.R. HOLFORD, et al. 1997. Administration of methylprednisolone for 24 or 48 hours or tirilazad mesylate for 48 hours in the treatment of acute spinal cord injury: results of the third national acute spinal cord injury randomized controlled trial. JAMA **277:** 1597–1604.

13. NAITO, M. *et al.* 2003. Anti-atherogenic effects of fermented fresh coffee bean, soybean and rice bran extracts. Food Sci. Technol. Res. **9:** 170–175.
14. CHEN, H.Y., J.M. LIN & C.C. LIN. 1998. "Raffinee," a free radical scavenger, in the treatment of subacute stage brain and spinal cord lesions: a case report. Am. J. Chin. Med. **XXVI:** 98–103.
15. AMERICAN SPINAL INJURY ASSOCIATION. 1996. International Standards for Neurological and Functional Classification of Spinal Cord Injury. ASIA. Chicago.
16. SEGAL, A.W. & O.T. JONES. 1980. Absence of cytochrome b reduction in stimulated neutrophils from both female and male patients with chronic granulomatous disease. FEBS Lett. **110:** 111–114.
17. YUKIE, N., S. TSUYOSHI, M. YOSHIKI & O. MOTOAKI. 1984. Oxygen metabolism in phagocytes of leprotic patients: enhanced endogenous superoxide dismutase activity and hydroxyl radical generation by clofazimine. J. Clin. Microbiol. **20:** 837–842.
18. MAHONEY, F.I. & D.W. BARTHEL. 1965. Functional evaluation: the Barthel Index. Md. State Med. J. **14:** 61–65.
19. HARRISON, W.W., M.G. NETSKY & M.D. BROWN. 1968. Trace elements in human brain: copper, zinc, iron and magnesium. Clin. Chim. Acta **21:** 55–60.
20. STOCKS, J., J.M. GUTTERIDGE, R.J. SHARP & T.L. DORMANDY. 1974. Assay using brain homogenate for measuring the antioxidant activity of biological fluids. Clin. Sci. Mol. Med. **47:** 215–222.
21. GUTTERIDGE, J.M. 1992. Ferrous ions detected in cerebrospinal fluid by using bleomycin and DNA damage. Clin. Sci. **82:** 315–320.
22. CHOI, D.W. 1990. Cerebral hypoxia: some new approaches and unanswered questions. J. Neurosci. **10:** 2493–2501.
23. LIPTON, S.A., Y.B. CHOI, C.H. PAN, *et al.* 1993. A redox-based mechanism for the neuroprotective and neurodestructive effects of nitric oxide and related nitroso-compounds. Nature **364:** 626–632.
24. AIKENS, J. & T.A. DIX. 1991. Perhydroxyl radical (HOO•) initiated lipid peroxidation. J. Biol. Chem. **266:** 15091–15098.
25. YOUNG, W. 1993. Secondary injury mechanism in acute spinal cord injury. J. Emerg. Med. **11:** 13–22.

Induction of Thioredoxin and Mitochondrial Survival Proteins Mediates Preconditioning-Induced Cardioprotection and Neuroprotection

CHUANG C. CHIUEH,[a] TSUGUNOBU ANDOH,[b] AND P. BOON CHOCK[c]

[a]*School of Pharmacy, Taipei Medical University, Taipei 100, Taiwan, and Laboratory of Clinical Science, National Institute of Mental Health, National Institutes of Health, Bethesda, Maryland 20892-1264, USA*

[b]*Department of Applied Pharmacology, Toyama Medical and Pharmaceutical University, Toyama 930-0194, Japan*

[c]*Laboratory of Biochemistry, Department of Biochemistry and Biophysics, National Institute of Heart, Lung and Blood, National Institutes of Health, Bethesda, Maryland 20892-8012, USA*

ABSTRACT: Delayed cardio- and neuroprotection are observed following a preconditioning procedure evoked by a brief and nontoxic oxidative stress due to deprivation of oxygen, glucose, serum, trophic factors, and/or antioxidative enzymes. Preconditioning protection can be observed *in vivo* and is under clinical trials for preservation of cell viability following organ transplants of liver. Previous studies indicated that ischemic preconditioning increases the expression of heat-shock proteins (HSPs) and nitric oxide synthase (NOS). Our pilot studies indicate that the treatment of neuronal NOS inhibitor (7-nitroindazole) and 6Br-cGMP blocks and mimics, respectively, preconditioning protection in human neuroblastoma SH-SY5Y cells. This minireview focuses on nitric oxide–mediated cellular adaptation and the related cGMP/PKG signaling pathway in a compensatory mechanism underlying preconditioning-induced hormesis. Both preconditioning and 6Br-cGMP increase the induction of human thioredoxin (Trx) mRNA and protein for cytoprotection, which is largely prevented by transfection of cells with Trx antisense but not sense oligonucleotides. Cytosolic Trx1 and mitochondrial Trx2 suppress free radical formation, lipid peroxidation, oxidative stress, and mitochondria-dependent apoptosis; knock out/down of either Trx1 or Trx2 is detrimental to cell survival. Other recent findings indicate that a transgenic increase of Trx in mice increases tolerance against oxidative nigral injury caused by 1-methyl-4-phenyl-1,2,3,6-tetrahydropyridine (MPTP). Trx1 can be translocated into nucleus and phosphoactivated CREB for a delayed induction of mitochondrial anti-apoptotic Bcl-2 and antioxidative MnSOD that is known to increase vitality and survival of cells in the brain and the heart. In conclusion, preconditioning adaptation or a brief oxidative stress induces a delayed nitric oxide–mediated compensatory mech-

Current address for correspondence: Professor C.C. Chiueh, Dean, School of Pharmacy, Taipei Medical University, 250 Wu-Hsing Street, Taipei 100, Taiwan. Voice/fax: +886-22-739-0671.
chiueh@tmu.edu.tw

anism for cell survival and vitality in the central nervous system and the cardiovascular system. Preconditioning-induced adaptive tolerance may be signaling through a cGMP-dependent induction of cytosolic redox protein Trx1 and subsequently mitochondrial proteins such as Bcl-2, MnSOD, and perhaps Trx2 or HSP70.

KEYWORDS: adaptation; ATP-sensitive potassium channel/K(ATP); Bcl-2; cyclic GMP; MnSOD; nitric oxide (NO); nitric oxide synthase (NOS); thioredoxin

INTRODUCTION

Hypoxic Preconditioning

The first discovery of preconditioning protection in the heart suggests that brief periods of ischemia could prevent or reduce cytotoxic damage caused by a subsequent prolonged period of ischemic/reperfusion stress.[1–4] A prolonged ischemic stress is known to produce free radicals, oxidative stress, and delayed apoptotic and necrotic cell death in oxygen-sensitive cells and organs. Reactive oxygen, nitrogen, and lipid peroxyl species are generated at the time of significant cell death in the heart or the brain. A brief episode of preconditioning stress consisting of a mild and nonlethal hypoxic period can increase cell survival for days when this preconditioning stress is applied at least 12 to 24 h prior to a prolonged and severe stress. Free radicals may play a pivotal role in the initiation of preconditioning hormesis mechanisms because the inhibition of oxidative and nitrosyl stress by respective antioxidants and NOS inhibitors block the compensatory cytoprotection.[5–7]

This phenomenon of preconditioning has received most attention in the cardiovascular field; a similar phenomenon is now induced in the central nervous system. Ischemic preconditioning-induced cytoprotection can be recognized in cultures, in brain slices, and *in vivo*. A brief deprivation of oxygen, glucose, serum, and/or trophic factors from culture media would lead to the initiation of preconditioning gene induction, delayed protein synthesis, and associated cellular adaptation for tolerance and survival. For *in vivo* preparations, a single or repeated brief hypoxia challenges would protect brain neurons, hepatic cells, and other myocardial cells from subsequent severe hypoxia/reperfusion damage.[8–13] There are reports of developing a cross-tolerance. For example, human SH-SY5Y cells preconditioned with a brief serum withdrawal could develop tolerance against oxidative stress and apoptosis caused by the toxic metabolite of MPTP, 1-methyl-4-phenylpyridinium (MPP^+).[14,15]

In addition to free radical scavengers, inhibition of new protein synthesis also prevents the cytoprotective phenomenon following preconditioning stress in both *in vitro* and *in vivo* preparations.[4,5,7,16–26] Apparently, preconditioning-induced adaptive cytoprotection is largely mediated by a brief gene induction and followed by a sustained protein synthesis with a lag time of several hours that lasts for days. Several stress proteins [HSP, NOS, erythropoietin (EPO), and redox factor 1 (Ref-1)], trophic factors [vascular endothelial growth factor (VEGF) and brain-derived neurotrophic factor (BDNF)], redox protein (thioredoxin/Trx), mitochondrial proteins (HSP70, Bcl-2, and MnSOD), adenosine, ceramide, and mitochondrial ATP-sensitive potassium [K(ATP)] channels are expressed or activated following preconditioning induction by brief episodes of either oxygen glucose deprivation or serum deprivation. A DNA microarray study revealed that more than 49 genes with altered expression are found in the preconditioned heart. However, most of these genes have not been shown to be involved in the mechanism of preconditioning.[27] The reprogramming of survival proteins peaks about 24 h post preconditioning and lasts for at least 3 but not more than 7 days later.[4]

Preconditioning activates protein kinases and signaling pathways including the nitric oxide–mediated cGMP/PKG pathway.[7,15,28–33] It is interesting to note that NOS inhibitors prevent both cardio- and neuroprotection induced by most if not all preconditioning procedures. The activation of this nitric oxide–mediated preconditioning signaling pathway by 6Br-cGMP also results in neuroprotection via prolonged expression of survival proteins.[14,15] Preconditioning-induced compensatory mechanisms that help cells in different organs avoid oxidative damage caused by cytotoxic reactive oxygen species and/or toxic agents are largely unknown. It is now recognized that the preconditioning phenomenon could be induced in many organs containing different cell types. Although the nature of preconditioning-induced adaptive mechanisms is not understood, several recent reports indicate that preconditioning of donor organs may improve clinical outcomes of organ transplantation, especially liver and heart transplants.[11,34–36] A hypothermic preconditioning procedure has also been investigated concerning its efficacy in enhancing recovery of cardiac functioning and minimizing neurodegenerative aftereffects following a prolonged cardio- and/or neurosurgery.[34,37,38] With therapeutic uses of preconditioning-induced cytoprotection in mind, we decided to address this critical issue of the signaling pathways of compensatory gene induction for enhancing cellular tolerance against subsequent severe oxidative stress and damage. We investigated which stress genes and mitochondrial survival proteins are reprogrammed by a brief period of preconditioning stress for protecting against oxidative stress caused by serum deprivation and dopaminergic neurotoxin, MPTP/MPP$^+$, in the human neurotrophic SH-SY5Y cell line.[7,14,15,24]

Chemical Preconditioning

In addition to the above-described preconditioning phenomenon induced by a brief episode of hypoxic ischemia, serum withdrawal, oxygen-glucose deprivation, and/or hypothermia stress, the cytoprotective phenomenon of preconditioning could also be triggered by other distinct chemical stimuli.[19,39–48] Endotoxin (lipopolysaccharide/LPS) and its nontoxic fragment diphosphoryl lipid A, at small doses/concentrations, can activate either NOS or TNFα that could initiate the preconditioning process and associated cytoprotection in cellular and animal models. Nontoxic levels of metabolic inhibitors such as 3-nitropropionic acid (3-NP), 2-deoxyglucose (2-DG), cyanide, gas anesthetics (i.e., desflurane, isofurane, sevoflurane, halothane), and excitatory toxins are also useful in developing preconditioning tolerance. Potassium-induced cortical spraying depression may also lead to preconditioning neuroprotection. Moreover, hormesis induced by small nontoxic concentrations/doses of selective neurotoxins could also result in cellular adaptation and tolerance in the brain. Furthermore, brief episodes of mild irradiation could also prime the dermal cells leading to tolerance against subsequent exposures to harmful UV and cosmic rays. Free radical scavengers could prevent chemical preconditioning and thus block reprogramming of genes and associated adaptive cytoprotection.[7,49,50] Apparently, the preconditioning-induced cytoprotective mechanism is a primitive cellular adaptation in response to harmful external stimuli; an adaptive cytoprotective mechanism could be activated by brief episodes of free radical attack inflicted by nonlethal external stimuli.

NITRIC OXIDE MEDIATES THE PRECONDITIONING-INDUCED ADAPTIVE CYTOPROTECTIVE MECHANISM

Preconditioning-Induced Mitochondrial Survival/Stress Proteins

In vivo studies demonstrated that a family of stress proteins, heat-shock proteins (HSPs), is induced in the preconditioned brain.[4,16,20,26,51] Some of HSP factors such as HSP70 are located in the mitochondria. This is the first sign to suggest that the preconditioning procedure may induce mitochondrial proteins for enhancing cellular defense or tolerance against subsequent lethal oxidative damage. Recent studies infer that the anti-apoptotic mitochondrial protein Bcl-2 is often expressed in brain regions following preconditioning procedures.[5,7,14,24] Moreover, we recently demonstrated that preconditioning, S-nitrosoglutathione (GSNO), and 6Br-cGMP promote the expression of not only Bcl-2, but also mitochondrial antioxidative enzyme MnSOD and the redox protein Trx.[14,52] Based on this new finding, we propose that preconditioning may activate AP-1, NOS, and associated cGMP/PKG signaling pathway in human neurotrophic cells for inducing mitochondrial survival proteins and inducing tolerance against subsequent severe oxidative damage and/or toxic insults. In addition to early genes (c-fos and c-Jun), Ref-1[7] and hypoxia-inducible factor 1α (HIF-1α) are also upregulated following preconditioning in neuronal and glial preparation, respectively.[21,53]

Preconditioning Hormesis: Role of Nitric Oxide and cGMP

The potential cytoprotective role of endogenously synthesized nitric oxide and 6Br-cGMP has been repeatedly demonstrated in both cardiovascular and neuronal preparations.[2,7,14,15,32,33,54–60] However, nonphysiological concentrations of nitric oxide release from man-made nitric oxide donors have often obscured these beneficial antioxidative and cytoprotective actions of nitric oxide at physiological levels.[61] For example, sodium nitroprusside is an iron complex, which releases nitric oxide and concomitantly generates reactive hydroxyl radicals.[52,62] Sodium nitroprusside-evoked reactive oxygen species are responsible for its cytotoxicity; light-irradiated and nitric oxide–exhausted solution is far more toxic than the freshly prepared. Low concentrations of nitric oxide and S-nitrosothiols (GSNO and SNAP) prevent iron-catalyzed generation of reactive oxygen and lipid peroxyl radicals, leading to interruption of lipid peroxidation and oxidative stress.[63,64] Moreover, nitric oxide binds to heme moiety of guanylyl cyclase and increases cGMP levels for signaling through the PKG pathway, yielding several important biological functions (TABLE 1) including vasorelaxation, long-term depression, cell-to-neuron communications, anti-inflammation, protein transcription (BDNF, Trx), phosphoactivation of K(ATP) channels, and preconditioning adaptation.[14,28,31,58,65–67] In the neurotrophic SH-SY5Ycells, preconditioning by a brief serum withdrawal is mediated by

TABLE 1. Biological factors and functions induced by cGMP

EDRF	endothelial-derived relaxation factor[66,119]
K(ATP)	mitochondrial ATP-sensitive potassium channel[77,120]
LTP	long-term depression[121,122]
Preconditioning adaptation	cardioprotection neuroprotection
Trx	expression of survival/redox protein[24,67]

nitric oxide–cGMP-PKG-dependent expression of the redox protein Trx because the transfection of cells with Trx antisense, but not sense, oligonucleotides blocks Trx expression and associated neuroprotection.[14,15,24] Furthermore, in this human cell model, 6Br-cGMP mimics preconditioning-evoked PKG signaling, expression of Trx, Bcl-2, and MnSOD, and associated neuroprotection.

It is now recognized that the nitric oxide–cGMP-PKG pathway may mediate gene reprogramming acutely and protein synthesis chronically for preconditioning adaptation. Pharmacological inhibition of NOS, guanylyl cyclase, and/or PKG prevents preconditioning protection caused by a brief episode of serum withdrawal, oxygen glucose deprivation, ischemia, anoxia, isoflurane, halothane, and endotoxin lipopolysaccharide (LPS).[7,14,28,32,33,44,47,56,60,68–70] Nitric oxide mediates preconditioning-induced cellular adaptation produced both by LPS and ischemic preconditioning, whereas mitochondrial HSP70 is involved in cytoprotection evoked by ischemic preconditioning but not by LPS preconditioning, which may be mediated by ceramide synthesis triggered by TNFα.[19,71] Different signaling pathways could be reprogrammed by preconditioning activation by subtypes of NOS in different cell preparations.

There is conflicted information about which inducible NOS is involved in the mediation of preconditioning-induced cellular adaptation. In fact, all three human NOS isoforms, including iNOS, eNOD, and nNOS, are inducible under oxidative stress and they are also under a feedback control. Therefore, human iNOS behaves differently from that of murine iNOS; because of feedback regulation, human iNOS may not produce persistently a large amount of nitric oxide in pathophysiological conditions. Upregulation of murine iNOS may activate Cox-2 and Ras via a cGMP-independent pathway.[68] Recently, it has been inferred in publications that all three NOS isoforms may participate in preconditioning adaptation in different cell and/or animal models.[7,14,32,68,70,72–75] In the brain, both eNOS and nNOS are involved in preconditioning adaptation. Ischemic preconditioning activates eNOS that leads to the expression of HIF-1α, which in turn upregulates the expression of EPO in astrocytes.[53] It has not been determined yet whether the expression of HIF-1α in glials is mediated by the nitric oxide–cGMP-PKG pathway.[21] It is highly likely that the proposed nitric oxide–cGMP-PKG pathway may also mediate the expression of astroglial EPO in preconditioned brain since preconditioning also induces NOS in astroglias. The activation of neuronal EPO receptors leads to deactivation of BAD via AKT-mediated phosphorylation and thus inhibits hypoxia-induced apoptosis in neurons. Furthermore, acting through the cGMP-dependent pathway, the activation of eNOS in endothelial cells could increase VEGF levels,[22,76] while the activation of nNOS in brain neurons may increase cGMP-dependent Trx expression for enhancing cell survival and thus neuroprotection.[7,14,60]

cGMP/PKG-DEPENDENT PRECONDITIONING PATHWAYS

Phosphoactivation of Mitochondrial K(ATP) Channels

The notion of involvement of NOS, nitric oxide, and cGMP in mitochondrial ATP-sensitive potassium channels [K(ATP)] for preconditioning cardioprotection has received more attention because of its apparent therapeutic importance.[16,17,28,35,58,77,78] Ischemic preconditioning in the cardiovascular system increases eNOS and/or iNOS expression, nitric oxide release, cGMP production, and the associated PKG-signaling pathway for opening mitochondrial K(ATP), for maintaining mitochondrial proton gradient, electron flow, and energy supply. Excessive build-up of mitochondrial protonic potential is harmful to cells by disrupting the structure-function of the inner membrane and the transfer of electron/energy. Exogenous nitric oxide mimics while guanylyl cyclase inhibition blocks concomitantly the activation of K(ATP) and preconditioning protection in both *in vitro* and *in vivo* myocardial preparations. Interestingly, antioxidants and NOS inhibitors also block ischemic preconditioning activation of mitochondrial K(ATP). Moreover, the nitric oxide–cGMP-PKG pathway contributes to phosphoactivation of K(ATP) channels that may contribute to the adaptive mechanism of preconditioning-induced delayed cardioprotection.[28,77] Some gas anesthetic agents, such as sevoflurane-induced cardioprotection, may be mediated via the activation of these K(ATP) channels located in the inner membrane of the mitochondria.[48] In addition to preconditioning cardioprotection induced by mild free radical stress during a brief episode of either ischemia or gas anesthesia, improved cardiac functions also arise when potassium channel openers or protonophores (2,4-dinitrophenyl, diazoxide, pinacidil) are employed to activate these mitochondrial K(ATP) channels.[9,41,56,79–81]

Induction of Cytosolic and Mitochondrial Thioredoxin (Trx1, Trx2)

We recently used the human neuroblastoma SH-SY5Y cell model to investigate whether mitochondrial proteins play a pivotal role in the mediation of the cGMP/PKG-dependent preconditioning pathway (FIG. 1).[14] When these human neurotrophic cells are subjected to a brief 2-h serum deprivation, nNOS and Trx levels are elevated, and these preconditioned cells are subsequently more tolerant against apoptosis caused by prolonged oxidative stress.[7] Active PKG promotes the phosphorylation of MAPK/ERK1/2, and c-Myc for elevating the expression of the redox protein Trx.[15,24] Nitric oxide may protect cells against oxidative stress through the upregulation of the multifunctional redox protein Trx that exists in two distinct isoforms, cytosolic Trx1 and mitochondrial Trx2. Knockout/-down of either Trx1 or Trx2 is detrimental to cell vitality and survival.[82–89] Food restriction also induces this redox protein Trx for improving longevity in rodents.[90] The activation of PKG is required to elevate levels of Trx and Trx peroxidase (Tpx-1). Trx per se could convert superoxide anion to H_2O_2, suppress the generation of cytotoxic hydroxyl and

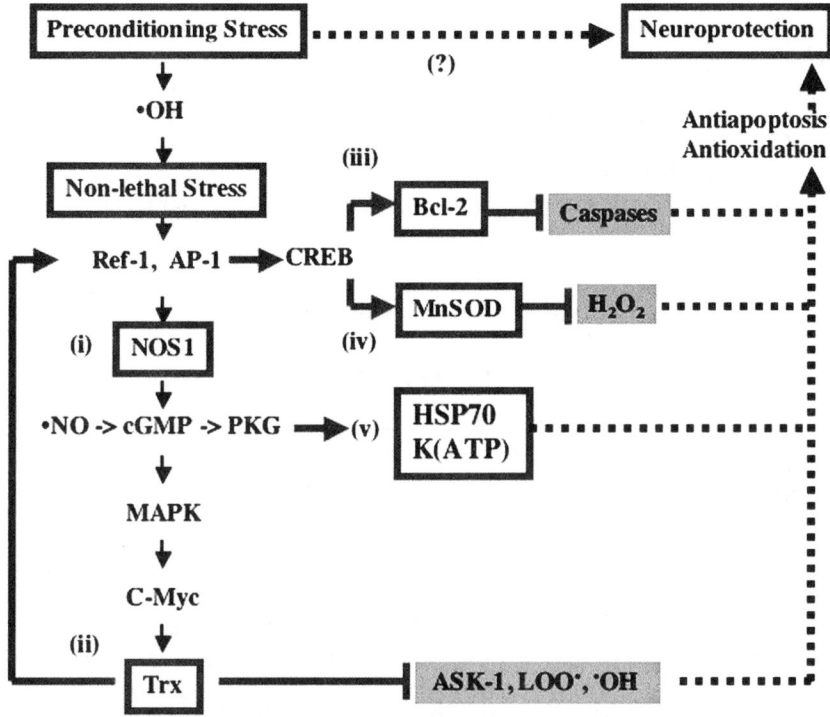

FIGURE 1. Preconditioning induction of survival proteins: Trx and mitochondrial proteins including Bcl-2, MnSOD, HSP70, and K(ATP).

lipid peroxyl radicals, inhibit the inflammatory process, and translocate into nucleus for regulation of transcription factors and synthesis of additional survival proteins for cytoprotection.[15,24,25,91–97] The antioxidative activity of Trx is at least three orders of magnitude more potent than the classic antioxidant glutathione (GSH).[24,61,64] Both the preconditioning-induced and the exogenously administered Trx increase cellular tolerance and prevent or reduce oxidative damage produced by the neurotoxin MPP+.[14,15,24,96] Preconditioning by irradiation could also induce Trx in preparations both *in vitro* and *in vivo*. Finally, transfection of human neurotrophic cells with Trx mRNA antisense, but not sense, oligonucleotides blocks concomitantly Trx biosynthesis and associated preconditioning tolerance/neuroprotection, indicating Trx may become a key molecule in mediating cGMP-dependent adaptive neuroprotection[14,15,24] and cardioprotection[25,50] induced by preconditioning.

Upregulation of Mitochondrial Survival Proteins (Bcl-2 and MnSOD) by Trx

It is known that preconditioning protection is mediated by synthesis or upregulation of new proteins. Preconditioning activation of PKG elevates not only the redox protein Trx but also mitochondrial survival proteins, MnSOD and Bcl-2, in human SH-SY5Y cells (FIG. 1). The expression of Trx is upstream of the induction of anti-apoptotic Bcl-2 and antioxidative MnSOD.[14,15,24] Cytosolic Trx1 at elevated levels enters the nucleus, activates the transcription factor[1] (phosphoactivation of CREB), and leads to the increased synthesis of Bcl-2 and MnSOD in mitochondria. Bcl-2 decreases the release of cytochrome c from mitochondria, inhibits capsase-9 and caspase-3, and thus prevents mitochondria-dependent apoptosis. Increased synthesis of MnSOD reduces superoxide anion radicals and protect mitochondria from oxidative stress. Other studies demonstrated that upregulation of either Bcl-2 or MnSOD in the mitochondria results in protection against oxidative stress caused by numerous neurotoxins.[14,62,98–102]

CONCLUDING REMARKS

Free Radicals Trigger Preconditioning Hormesis

Despite the fact that preconditioning protection is repeatedly demonstrated in different cell types both *in vitro* and *in vivo* using different external stimuli or chemicals, a considerable amount of evidence suggests that preconditioning adaptation may be triggered by free radicals, reactive oxygen and/or nitrogen species. Most preconditioning procedures are prevented by free radical scavengers including hydroxyl radical scavengers (i.e., GSNO, salicylate, *N*-(2-mercaptopropionyl)-glycine, *N*-acetylcysteine, and superoxide dismutase mimetics).[7,25,49,103,104] As discussed above, many preconditioning phenomena, if not all, are also blocked by pretreatment of inhibitors of iNOS, eNOS, and/or nNOS, indicating that nitric oxide may participate in preconditioning mechanisms.[7,11,55,73,105,106] Exogenous nitric oxide donors including GSNO trigger preconditioning-induced signaling pathways.[7,14,15,24,32,33,107] In the human SH-SY5Y cell model, a brief 2-h serum withdrawal immediate increases the release of hydroxyl radicals while there is a delayed increase in nitric oxide levels with a lag time of several hours.[7] A mild and brief preconditioning stress increases reactive oxygen species that immediately activate c-fos and c-jun in the nucleus, elevating not only transcription factor AP-1 but also Ref-1 for upregulation of nNOS mRNA. A two-fold increase in nNOS expression elevates the nitric oxide level approximately four-fold and the cGMP level 20-fold when assayed 24 h post a brief 2-h preconditioning stress in human SH-SY5Y cells. Therefore, these new findings indicate that reactive oxygen species (i.e., superoxide anion and/or hydroxyl radicals) may serve as a trigger for initiating the preconditioning process by activating early genes, AP-1 and Ref-1. The activation of AP-1 leads to the expression of the NOS gene and protein for elevating nitric oxide–cGMP-PKG signaling pathway and delayed gene and protein expression for maintaining a prolonged period of preconditioning adaptation for survival (1 to 3 days). Our new data support a notion that free radicals trigger a preconditioning adaptive mechanism; they (e.g., superoxide anions and hydroxyl radicals) activate Ref-1, Ap-1, and the nitric oxide–cGMP-PKG pathway, for reprogramming of survival genes/proteins [Trx, Bcl-2, MnSOD, and K(ATP)] following preconditioning procedures.

cGMP-Dependent Preconditioning Pathways: Trx and K(ATP)

A brief, mild oxidative stress activates nuclear transcription factor AP-1 or NFκB to increase NOS mRNA and synthesis of NOS proteins. All three isoforms of NOS are inducible following mild and severe oxidative stress. In human SH-SY5Y cells, preconditioning activates the nitric oxide–cGMP-PKG pathway, phosphoactivates MAPK/ERK1/2 and c-Myc, resulting in the expression of a redox protein Trx.[7,14,15,24,88] Both cytosolic Trx1 and mitochondrial Trx2 inhibits the generation of reactive oxygen species and lipid peroxyl radicals.[24,85,108–111] Their significant roles in cell viability have been demonstrated both in *in vitro* and *in vivo* models. In addition to Trx, Tpx-1 is also upregulated following preconditioning adaptation. In the presence of NADPH, Trx reductase, and Tpx-1, the active moiety of Trx, -Cys-Gly-Pro-Cys- is redox cycling for maintaining vital cellular proteins and their functions. Moreover, Trx1 could translocate into nucleus redox, regulate functional thiol groups of nuclear transcription factors (Ref-1, AP-1) and activate CREB, leading to elevated levels of mitochondrial survival proteins including Bcl-2 and MnSOD. Elevated mitochondrial Bcl-2 prevents the release of pro-apoptotic cytochrome *c* and the activation of caspases 9 and 3.[14,15] Upregulation of MnSOD reduces superoxide anion radicals and thus decreases the generation of cytotoxic hydroxyl radicals and peroxynitrite. Increased tolerance against MPP^+ neurotoxin by at least 30-fold has been reported in preconditioned human SH-SY5Y cells.[15,24] Therefore, preconditioning-induced neuroprotective adaptation may be due to a cGMP-mediated reprogramming of survival genes and proteins such as redox proteins (Trx1, Trx2), anti-apoptotic protein Bcl-2, and antioxidative enzyme MnSOD. This notion of mitochondria-mediated preconditioning protection through the proposed induction of mitochondrial survival proteins such as HSP70, Trx2, Bcl-2, and MnSOD is consistent with early reports that nitric oxide/cGMP mediates also the phosphoactivation of mitochondrial K(ATP) channels for supporting cellular vitality.

Drugs Mimic Preconditioning or Awaken Adaptive Expression of Survival Proteins

Preconditioning-induced cytoprotective genes and associated signaling mechanisms offer attractive targets for developing cardio- and/or neuroprotective therapies. Based on the cGMP-dependent protein expression obtained from the investigation of endogenous cytoprotective adaptation produced by preconditioning procedures, new drugs could be developed for the activation of the preconditioning signaling pathway and for achieving cardio- and neuroprotection. These potential preconditioning, mimicking agents include *S*-nitrosothiols (GSNO), cGMP agonists (6Br-cGMP), redox proteins (Trx), sphigolipids (ceramide), adenosine, EPO, and K(ATP) channel openers. For example, several K(ATP) channel openers have been developed and evaluated for their potential therapeutic approaches, mostly for cardioprotection.[35,56,80,112] An early report suggests that adenosine may activate K(ATP) for cardioprotection.[16] Recently identified protonphoric property of potassium channel openers infers a new component in their cardioprotective mechanism for suppressing excessive build-up of mitochondrial proton gradient and maintaining the normal functioning of inner mitochondrial membrane.[112] These potassium channel openers could have neuroprotective indications if they could pass through the blood–brain barrier.

Based on recent discoveries that the redox protein Trx[113] may play a pivotal role in mediating cGMP-dependent preconditioning cytoprotective mechanism[14,25]; several lead compounds have been identified for their ability in awakening Trx mRNA

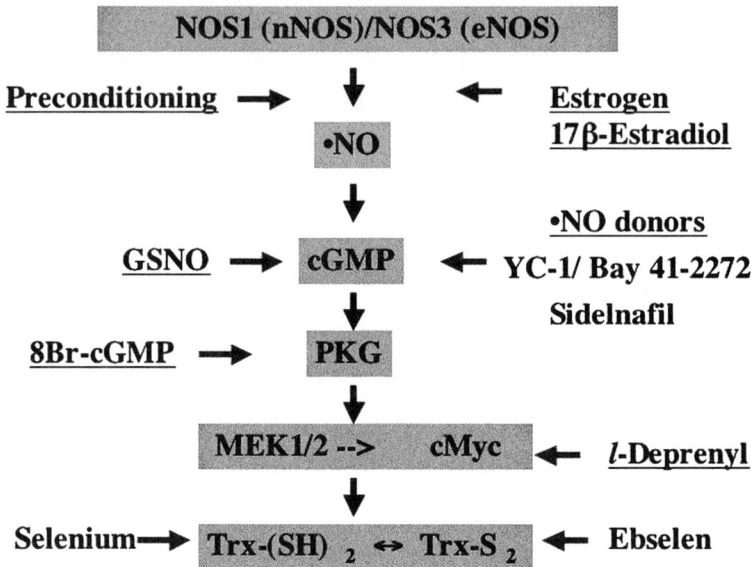

FIGURE 2. Trx-inducing pro-drugs for mimicking preconditioning hormesis.

and synthesis of Trx for enhancing cellular tolerance against oxidative damage (FIG. 2). The neuroprotective mechanism of 17β-estradiol at nanomolar concentrations may be mediated by new protein synthesis of Trx, MnSOD, and Bcl-2 in a way similar to the pathway elicited by preconditioning stress.[29,102] Another group has identified another lead compound ebselen, which increases Trx expression and protects brain cells and neurons against ischemia/reperfusion damage in stroke models and patients.[114,115] Therefore, 17β-estradiol and ebselen are the prototype of Trx-inducing agents. S-Nitrosothiols including SNAP has been employed to trigger preconditioning activation of NOS, PKG, and thus K(ATP) channels in cardiovascular system. GSNO and 6Br-cGMP protect brain cells and neurons against severe oxidative stress and neurodegeneration, possibly through the induction of mitochondrial survival proteins. New subtypes of phosphodiesterase inhibitors and 3-(5′-hydroxymethyl-2′-furyl)-1-benzylindazole (YC-1) may be useful in augmenting cytoprotective effects of cell membrane permeable 6Br-cGMP because they prevent enzymatic degradation of cGMP.[31,32,69,116–118] The investigation of preconditioning adaptation in reprogramming of survival genes/proteins and associated signaling pathways would provide insights of endogenous mechanisms of cardio- and neuroprotection. Identification of these preconditioning compensatory changes in mitochondria could lead to development of new therapeutics that enhance cellular tolerance against oxidative damage caused by free radicals and/or external stimuli such as oxidant-producing neurotoxins.

ACKNOWLEDGMENTS

This work was supported, in part, by grants from the National Science Council (to C.C.C.), the Japanese Society for Promoting Science (JSPS; to T.A.), and the National Institutes of Health Intramural Research Program (IRP; to P.B.C.).

REFERENCES

1. VEGH, A., J.G. PAPP, L. SZEKERES, et al. 1992. The local intracoronary administration of methylene blue prevents the pronounced antiarrhythmic effect of ischaemic preconditioning. Br. J. Pharmacol. **107:** 910–911.
2. KATO, H., K. KOGURE, Y. LIU, et al. 1994. Induction of NADPH-diaphorase activity in the hippocampus in a rat model of cerebral ischemia and ischemic tolerance. Brain Res. **652:** 71–75.
3. SZILVASSY, Z., P. FERDINANDY, P. BOR, et al. 1994. Loss of preconditioning in rabbits with vascular tolerance to nitroglycerin. Br. J. Pharmacol. **112:** 999–1001.
4. CHEN, J., S.H. GRAHAM, R.L. ZHU, et al. 1996. Stress proteins and tolerance to focal cerebral ischemia. J. Cereb. Blood Flow Metab. **16:** 566–577.
5. MAULIK, N., R.M. ENGELMAN, J.A. ROUSOU, et al. 1999. Ischemic preconditioning reduces apoptosis by upregulating anti-death gene Bcl-2. Circulation **100**(19 Suppl.): II369–375.
6. XI, L. 1999. Nitric oxide-dependent mechanism of anti-ischemic myocardial protection induced by monophosphoryl lipid A. Zhongguo Yao Li Xue Bao **20:** 865–871.
7. ANDOH, T., S.Y. LEE & C.C. CHIUEH. 2000. Preconditioning regulation of bcl-2 and p66shc by human NOS1 enhances tolerance to oxidative stress. FASEB J. **14:** 2144–2146.
8. ILIODROMITIS, E.K., C.C. PAPADOPOULOS, M. MARKIANOS, et al. 1996. Alterations in circulating cyclic guanosine monophosphate (c-GMP) during short and long ischemia in preconditioning. Basic Res. Cardiol. **91:** 234–239.
9. DOMOKI, F., J.V. PERCIACCANTE, R. VELTKAMP, et al. 1999. Mitochondrial potassium channel opener diazoxide preserves neuronal-vascular function after cerebral ischemia in newborn pigs. Stroke **30:** 2713–2718; discussion 2718–2719.
10. YU, Z.F. & M.P. MATTSON. 1999. Dietary restriction and 2-deoxyglucose administration reduce focal ischemic brain damage and improve behavioral outcome: evidence for a preconditioning mechanism. J. Neurosci. Res. **57:** 830–839.
11. YIN, D.P., H.N. SANKARY, A.S. CHONG, et al. 1998. Protective effect of ischemic preconditioning on liver preservation-reperfusion injury in rats. Transplantation **66:** 152–157.
12. KLEIN, M., J. GEOGHEGAN, R. WANGEMANN, et al. 1999. Preconditioning of donor livers with prostaglandin I2 before retrieval decreases hepatocellular ischemia-reperfusion injury. Transplantation **67:** 1128–1132.
13. FUNG, J.J. 2001. Ischemic preconditioning: application in clinical liver transplantation. Liver Transpl. **7:** 300–301.
14. ANDOH, T., P.B. CHOCK & C.C. CHIUEH. 2002. Preconditioning-mediated neuroprotection: role of nitric oxide, cGMP, and new protein expression. Ann. N.Y. Acad. Sci. **962:** 1–7.
15. ANDOH, T., C.C. CHIUEH & P.B. CHOCK. 2003. Cyclic GMP-dependent protein kinase regulates the expression of thioredoxin and thioredoxin peroxidase-1 during hormesis in response to oxidative stress-induced apoptosis. J. Biol. Chem. **278:** 885–890.
16. HEURTEAUX, C., I. LAURITZEN, C. WIDMANN, et al. 1995. Essential role of adenosine, adenosine A1 receptors, and ATP-sensitive K+ channels in cerebral ischemic preconditioning. Proc. Natl. Acad. Sci. USA **92:** 4666–4670.
17. ELLIOTT, G.T. 1996. Monophosphoryl lipid A: a novel agent for inducing pharmacologic myocardial preconditioning. J. Thromb. Thrombolysis **3:** 225–237.
18. RAVATI, A., B. AHLEMEYER, A. BECKER, et al. 2001. Preconditioning-induced neuroprotection is mediated by reactive oxygen species and activation of the transcription factor nuclear factor-kappaB. J. Neurochem. **78:** 909–919.

19. ZIMMERMANN, C., I. GINIS, K. FURUYA, et al. 2001. Lipopolysaccharide-induced ischemic tolerance is associated with increased levels of ceramide in brain and in plasma. Brain Res. **895:** 59–65.
20. CURRIE, R.W., J.A. ELLISON, R.F. WHITE, et al. 2000. Benign focal ischemic preconditioning induces neuronal Hsp70 and prolonged astrogliosis with expression of Hsp27. Brain Res. **863:** 169–181.
21. DIGICAYLIOGLU, M. & S.A. LIPTON. 2001. Erythropoietin-mediated neuroprotection involves cross-talk between Jak2 and NF-kappaB signalling cascades. Nature **412:** 641–647.
22. WICK, A., W. WICK, J. WALTENBERGER, et al. 2002. Neuroprotection by hypoxic preconditioning requires sequential activation of vascular endothelial growth factor receptor and Akt. J. Neurosci. **22:** 6401–6407.
23. TRUETTNER, J., R. BUSTO, W. ZHAO, et al. 2002. Effect of ischemic preconditioning on the expression of putative neuroprotective genes in the rat brain. Brain Res. Mol. Brain Res. **103:** 106–115.
24. ANDOH, T., P.B. CHOCK & C.C. CHIUEH. 2002. The roles of thioredoxin in protection against oxidative stress-induced apoptosis in SH-SY5Y cells. J. Biol. Chem. **277:** 9655–9660.
25. DAS, D.K. & N. MAULIK. 2003. Preconditioning potentiates redox signaling and converts death signal into survival signal. Arch. Biochem. Biophys. **420:** 305–311.
26. DHODDA, V.K., K.A. SAILOR, K.K. BOWEN, et al. 2004. Putative endogenous mediators of preconditioning-induced ischemic tolerance in rat brain identified by genomic and proteomic analysis. J. Neurochem. **89:** 73–89.
27. ONODY, A., A. ZVARA, L. HACKLER, JR., et al. 2003. Effect of classic preconditioning on the gene expression pattern of rat hearts: a DNA microarray study. FEBS Lett. **536:** 35–40.
28. HORIMOTO, H., G.R. GAUDETTE, A. E. SALTMAN, et al. 2000. The role of nitric oxide, K(+)(ATP) channels, and cGMP in the preconditioning response of the rabbit. J. Surg. Res. **92:** 56–63.
29. CHIUEH, C., S. LEE, T. ANDOH, et al. 2003. Induction of antioxidative and antiapoptotic thioredoxin supports neuroprotective hypothesis of estrogen. Endocrine **21:** 27–31.
30. SUZUKI, E., H. KAGEYAMA, T. NAKAKI, et al. 2003. Nitric oxide induced heat shock protein 70 mRNA in rat hypothalamus during acute restraint stress under sucrose diet. Cell. Mol. Neurobiol. **23:** 907–915.
31. CHIEN, W.L., K.C. LIANG, C.M. TENG, et al. 2003. Enhancement of long-term potentiation by a potent nitric oxide-guanylyl cyclase activator, 3-(5-hydroxymethyl-2-furyl)-1-benzyl-indazole. Mol. Pharmacol. **63:** 1322–1328.
32. KIM, J.S., S. OHSHIMA, P. PEDIADITAKIS, et al. 2004. Nitric oxide protects rat hepatocytes against reperfusion injury mediated by the mitochondrial permeability transition. Hepatology **39:** 1533–1543.
33. QIN, Q., X.M. YANG, L. CUI, et al. 2004. Exogenous nitric oxide triggers preconditioning by a cGMP- and mKATP-dependent mechanism. Am. J. Physiol. Heart Circ. Physiol. **287:** H712–718.
34. KARCK, M., P. RAHMANIAN & A. HAVERICH. 1996. Ischemic preconditioning enhances donor heart preservation. Transplantation **62:** 17–22.
35. KEVELAITIS, E., A. OUBENAISSA, J. PEYNET, et al. 1999. Preconditioning by mitochondrial ATP-sensitive potassium channel openers: an effective approach for improving the preservation of heart transplants. Circulation **100**(19 Suppl.): II345–350.
36. TORRAS, J., I. HERRERO-FRESNEDA, N. LLOBERAS, et al. 2002. Promising effects of ischemic preconditioning in renal transplantation. Kidney Int. **61:** 2218–2227.
37. HOLZER, M. & F. STERZ. 2003. Therapeutic hypothermia after cardiopulmonary resuscitation. Expert Rev. Cardiovasc. Ther. **1:** 317–325.
38. WANG, H., W. OLIVERO, G. LANZINO, et al. 2004. Rapid and selective cerebral hypothermia achieved using a cooling helmet. J. Neurosurg. **100:** 272–277.
39. MAULIK, N., R.M. ENGELMAN, Z. WEI, et al. 1995. Drug-induced heat-shock preconditioning improves postischemic ventricular recovery after cardiopulmonary bypass. Circulation **92:** II381–388.
40. RIEPE, M.W. & A.C. LUDOLPH. 1997. Chemical preconditioning: a cytoprotective strategy. Mol. Cell. Biochem. **174:** 249–254.

41. NOVALIJA, E., S. FUJITA, J.P. KAMPINE, et al. 1999. Sevoflurane mimics ischemic preconditioning effects on coronary flow and nitric oxide release in isolated hearts. Anesthesiology **91:** 701–712.
42. GUO, Z.H. & M.P. MATTSON. 2000. In vivo 2-deoxyglucose administration preserves glucose and glutamate transport and mitochondrial function in cortical synaptic terminals after exposure to amyloid beta-peptide and iron: evidence for a stress response. Exp. Neurol. **166:** 173–179.
43. HE, S.Y., H.W. DENG & Y.J. LI. 2001. Monophosphoryl lipid A-induced delayed preconditioning is mediated by calcitonin gene-related peptide. Eur. J. Pharmacol. **420:** 143–149.
44. WANG, Y.P., C. SATO, K. MIZOGUCHI, et al. 2002. Lipopolysaccharide triggers late preconditioning against myocardial infarction via inducible nitric oxide synthase. Cardiovasc. Res. **56:** 33–42.
45. GARNIER, P., N. BERTRAND, C. DEMOUGEOT, et al. 2002. Chemical preconditioning with 3-nitropropionic acid: lack of induction of neuronal tolerance in gerbil hippocampus subjected to transient forebrain ischemia. Brain Res. Bull. **58:** 33–39.
46. RAEBURN, C.D., M.A. ZIMMERMAN, J. ARYA, et al. 2002. Ischemic preconditioning: fact or fantasy? J. Card. Surg. **17:** 536–542.
47. KAPINYA, K.J., D. LOWL, C. FUTTERER, et al. 2002. Tolerance against ischemic neuronal injury can be induced by volatile anesthetics and is inducible NO synthase dependent. Stroke **33:** 1889–1898.
48. ZAUGG, M., E. LUCCHINETTI, D.R. SPAHN, et al. 2002. Volatile anesthetics mimic cardiac preconditioning by priming the activation of mitochondrial K(ATP) channels via multiple signaling pathways. Anesthesiology **97:** 4–14.
49. BOUWMAN, R.A., R.J. MUSTERS, B.J. VAN BEEK-HARMSEN, et al. 2004. Reactive oxygen species precede protein kinase C-delta activation independent of adenosine triphosphate-sensitive mitochondrial channel opening in sevoflurane-induced cardioprotection. Anesthesiology **100:** 506–514.
50. DAS, D.K. & N. MAULIK. 2004. Conversion of death signal into survival signal by redox signaling. Biochemistry (Mosc). **69:** 10–17.
51. LIU, Y., H. KATO, N. NAKATA, et al. 1993. Correlation between induction of ischemic tolerance and expression of heat shock protein-70 in the rat hippocampus. No. To. Shinkei. **45:** 157–162.
52. RAUHALA, P., T. ANDOH, K. YEH, et al. 2002. Contradictory effects of sodium nitroprusside and S-nitroso-N-acetylpenicillamine on oxidative stress in brain dopamine neurons in vivo. Ann. N.Y. Acad. Sci. **962:** 60–72.
53. RUSCHER, K., D. FREYER, M. KARSCH, et al. 2002. Erythropoietin is a paracrine mediator of ischemic tolerance in the brain: evidence from an in vitro model. J. Neurosci. **22:** 10291–10301.
54. KONOREV, E.A., J. JOSEPH, M.M. TARPEY, et al. 1996. The mechanism of cardioprotection by S-nitrosoglutathione monoethyl ester in rat isolated heart during cardioplegic ischaemic arrest. Br. J. Pharmacol. **119:** 511–518.
55. CENTENO, J.M., M. ORTI, J.B. SALOM, et al. 1999. Nitric oxide is involved in anoxic preconditioning neuroprotection in rat hippocampal slices. Brain Res. **836:** 62–69.
56. OCKAILI, R., V.R. EMANI, S. OKUBO, et al. 1999. Opening of mitochondrial KATP channel induces early and delayed cardioprotective effect: role of nitric oxide. Am. J. Physiol. **277:** H2425–2434.
57. OGAWA, T., A.K. NUSSLER, E. TUZUNER, et al. 2001. Contribution of nitric oxide to the protective effects of ischemic preconditioning in ischemia-reperfused rat kidneys. J. Lab. Clin. Med. **138:** 50–58.
58. HAN, J., N. KIM, H. JOO, et al. 2002. ATP-sensitive K(+) channel activation by nitric oxide and protein kinase G in rabbit ventricular myocytes. Am. J. Physiol. Heart Circ. Physiol. **283:** H1545–1554.
59. FISCUS, R.R. 2002. Involvement of cyclic GMP and protein kinase G in the regulation of apoptosis and survival in neural cells. Neurosignals **11:** 175–190.
60. WIGGINS, A.K., P.J. SHEN & A.L. GUNDLACH. 2003. Neuronal-NOS adaptor protein expression after spreading depression: implications for NO production and ischemic tolerance. J. Neurochem. **87:** 1368–1380.

61. CHIUEH, C.C. 1999. Neuroprotective properties of nitric oxide. Ann. N.Y. Acad. Sci. **890:** 301–311.
62. PAN, S.L., J.H. GUH, Y.L. CHANG, et al. 2004. YC-1 prevents sodium nitroprusside-mediated apoptosis in vascular smooth muscle cells. Cardiovasc. Res. **61:** 152–158.
63. RAUHALA, P., A.M. LIN & C.C. CHIUEH. 1998. Neuroprotection by S-nitrosoglutathione of brain dopamine neurons from oxidative stress. FASEB J. **12:** 165–173.
64. RAUHALA, P. & C.C. CHIUEH. 2000. Effects of atypical antioxidative agents, S-nitrosoglutathione and manganese, on brain lipid peroxidation induced by iron leaking from tissue disruption. Ann. N.Y. Acad. Sci. **899:** 238–254.
65. HANSSON, G.K., H. JORNVALL & S.G. LINDAHL. 1998. The Nobel Prize 1998 in physiology or medicine. Nitrogen oxide as a signal molecule in the cardiovascular system. Ugeskr. Laeger. **160:** 7571–7578.
66. MURAD, F. 1999. Cellular signaling with nitric oxide and cyclic GMP. Braz. J. Med. Biol. Res. **32:** 1317–1327.
67. CHIUEH, C.C. & P. RAUHALA. 1999. The redox pathway of S-nitrosoglutathione, glutathione and nitric oxide in cell to neuron communications. Free Radic. Res. **31:** 641–650.
68. XI, L., D. TEKIN, E. GURSOY, et al. 2002. Evidence that NOS2 acts as a trigger and mediator of late preconditioning induced by acute systemic hypoxia. Am. J. Physiol. Heart Circ. Physiol. **283:** H5–12.
69. SALLOUM, F., C. YIN, L. XI, et al. 2003. Sildenafil induces delayed preconditioning through inducible nitric oxide synthase-dependent pathway in mouse heart. Circ. Res. **92:** 595–597.
70. JIANG, X., E. SHI, Y. NAKAJIMA, et al. 2004. Inducible nitric oxide synthase mediates delayed cardioprotection induced by morphine in vivo: evidence from pharmacologic inhibition and gene-knockout mice. Anesthesiology **101:** 82–88.
71. CUI, J., R.M. ENGELMAN, N. MAULIK, et al. 2004. Role of ceramide in ischemic preconditioning. J. Am. Coll. Surg. **198:** 770–777.
72. ZHAO, L., P.A. WEBER, J.R. SMITH, et al. 1997. Role of inducible nitric oxide synthase in pharmacological "preconditioning" with monophosphoryl lipid A. J. Mol. Cell. Cardiol. **29:** 1567–1576.
73. LAUDE, K., J. FAVRE, C. THUILLEZ, et al. 2003. NO produced by endothelial NO synthase is a mediator of delayed preconditioning-induced endothelial protection. Am. J. Physiol. Heart Circ. Physiol. **284:** H2053–2060.
74. WANG, Y., N. AHMAD, M. KUDO, et al. 2004. Contribution of Akt and endothelial nitric oxide synthase to diazoxide-induced late preconditioning. Am. J. Physiol. Heart Circ. Physiol. **287:** H1125–1131.
75. HASHIGUCHI, A., S. YANO, M. MORIOKA, et al. 2004. Up-regulation of endothelial nitric oxide synthase via phosphatidylinositol 3-kinase pathway contributes to ischemic tolerance in the CA1 subfield of gerbil hippocampus. J. Cereb. Blood Flow Metab. **24:** 271–279.
76. BERNAUDIN, M., Y. TANG, M. REILLY, et al. 2002. Brain genomic response following hypoxia and re-oxygenation in the neonatal rat. Identification of genes that might contribute to hypoxia-induced ischemic tolerance. J. Biol. Chem. **277:** 39728–39738.
77. HAN, J., N. KIM, E. KIM, et al. 2001. Modulation of ATP-sensitive potassium channels by cGMP-dependent protein kinase in rabbit ventricular myocytes. J. Biol. Chem. **276:** 22140–22147.
78. GARLID, K.D., P. DOS SANTOS, Z.J. XIE, et al. 2003. Mitochondrial potassium transport: the role of the mitochondrial ATP-sensitive K(+) channel in cardiac function and cardioprotection. Biochim. Biophys. Acta **1606:** 1–21.
79. LIU, D., C. LU, R. WAN, et al. 2002. Activation of mitochondrial ATP-dependent potassium channels protects neurons against ischemia-induced death by a mechanism involving suppression of Bax translocation and cytochrome c release. J. Cereb. Blood Flow Metab. **22:** 431–443.
80. SORIMACHI, T. & T.S. NOWAK, JR. 2004. Pharmacological manipulations of ATP-dependent potassium channels and adenosine A1 receptors do not impact hippocampal ischemic preconditioning in vivo: evidence in a highly quantitative gerbil model. J. Cereb. Blood Flow Metab. **24:** 556–563.

81. SEBILLE, S., P. DE TULLIO, S. BOVERIE, et al. 2004. Recent developments in the chemistry of potassium channel activators: the cromakalim analogs. Curr. Med. Chem. **11:** 1213–1222.
82. TAKAGI, Y., A. MITSUI, A. NISHIYAMA, et al. 1999. Overexpression of thioredoxin in transgenic mice attenuates focal ischemic brain damage. Proc. Natl. Acad. Sci. USA **96:** 4131–4136.
83. KAHLOS, K., Y. SOINI, M. SAILY, et al. 2001. Up-regulation of thioredoxin and thioredoxin reductase in human malignant pleural mesothelioma. Int. J. Cancer **95:** 198–204.
84. YOON, B.I., Y. HIRABAYASHI, T. KANEKO, et al. 2001. Transgene expression of thioredoxin (TRX/ADF) protects against 2,3,7,8-tetrachlorodibenzo-p-dioxin (TCDD)-induced hematotoxicity. Arch. Environ. Contam. Toxicol. **41:** 232–236.
85. TANAKA, T., F. HOSOI, Y. YAMAGUCHI-IWAI, et al. 2002. Thioredoxin-2 (TRX-2) is an essential gene regulating mitochondria-dependent apoptosis. EMBO J. **21:** 1695–1703.
86. MITSUI, A., J. HAMURO, H. NAKAMURA, et al. 2002. Overexpression of human thioredoxin in transgenic mice controls oxidative stress and life span. Antioxid. Redox Signal. **4:** 693–696.
87. OKUYAMA, H., H. NAKAMURA, Y. SHIMAHARA, et al. 2003. Overexpression of thioredoxin prevents acute hepatitis caused by thioacetamide or lipopolysaccharide in mice. Hepatology **37:** 1015–1025.
88. NONN, L., R.R. WILLIAMS, R.P. ERICKSON, et al. 2003. The absence of mitochondrial thioredoxin 2 causes massive apoptosis, exencephaly, and early embryonic lethality in homozygous mice. Mol. Cell. Biol. **23:** 916–922.
89. DAMDIMOPOULOS, A.E., A. MIRANDA-VIZUETE, M. PELTO-HUIKKO, et al. 2002. Human mitochondrial thioredoxin. Involvement in mitochondrial membrane potential and cell death. J. Biol. Chem. **277:** 33249–33257.
90. CHO, C.G., H.J. KIM, S.W. CHUNG, et al. 2003. Modulation of glutathione and thioredoxin systems by calorie restriction during the aging process. Exp. Gerontol. **38:** 539–548.
91. BJORNSTEDT, M., J. XUE, W. HUANG, et al. 1994. The thioredoxin and glutaredoxin systems are efficient electron donors to human plasma glutathione peroxidase. J. Biol. Chem. **269:** 29382–29384.
92. AOTA, M., K. MATSUDA, N. ISOWA, et al. 1996. Protection against reperfusion-induced arrhythmias by human thioredoxin. J. Cardiovasc. Pharmacol. **27:** 727–732.
93. CASAGRANDE, S., V. BONETTO, M. FRATELLI, et al. 2002. Glutathionylation of human thioredoxin: a possible crosstalk between the glutathione and thioredoxin systems. Proc. Natl. Acad. Sci. USA **99:** 9745–9749.
94. UEDA, S., H. MASUTANI, H. NAKAMURA, et al. 2002. Redox control of cell death. Antioxid. Redox Signal. **4:** 405–414.
95. YODOI, J., H. NAKAMURA & H. MASUTANI. 2002. Redox regulation of stress signals: possible roles of dendritic stellate TRX producer cells (DST cell types). Biol. Chem. **383:** 585–590.
96. BAI, J., H. NAKAMURA, I. HATTORI, et al. 2002. Thioredoxin suppresses 1-methyl-4-phenylpyridinium-induced neurotoxicity in rat PC12 cells. Neurosci. Lett. **321:** 81–84.
97. KOBAYASHI-MIURA, M., H. NAKAMURA, J. YODOI, et al. 2002. Thioredoxin, an antioxidant protein, protects mouse embryos from oxidative stress-induced developmental anomalies. Free Radic. Res. **36:** 949–956.
98. OFFEN, D., P.M. BEART, N.S. CHEUNG, et al. 1998. Transgenic mice expressing human Bcl-2 in their neurons are resistant to 6-hydroxydopamine and 1-methyl-4-phenyl-1,2,3,6-tetrahydropyridine neurotoxicity. Proc. Natl. Acad. Sci. USA **95:** 5789–5794.
99. KLIVENYI, P., D. ST. CLAIR, M. WERMER, et al. 1998. Manganese superoxide dismutase overexpression attenuates MPTP toxicity. Neurobiol. Dis. **5:** 253–258.
100. YANG, L., R.T. MATTHEWS, J.B. SCHULZ, et al. 1998. 1-Methyl-4-phenyl-1,2,3,6-tetrahydropyride neurotoxicity is attenuated in mice overexpressing Bcl-2. J. Neurosci. **18:** 8145–8152.
101. FLANAGAN, S.W., R.D. ANDERSON, M.A. ROSS, et al. 2002. Overexpression of manganese superoxide dismutase attenuates neuronal death in human cells expressing mutant (G37R) Cu/Zn-superoxide dismutase. J. Neurochem. **81:** 170–177.

102. LEE, S.Y., T. ANDOH, D.L. MURPHY, *et al.* 2003. 17beta-estradiol activates ICI 182,780-sensitive estrogen receptors and cyclic GMP-dependent thioredoxin expression for neuroprotection. FASEB J. **17:** 947–948.
103. MORI, T., H. MURAMATSU, T. MATSUI, *et al.* 2000. Possible role of the superoxide anion in the development of neuronal tolerance following ischaemic preconditioning in rats. Neuropathol. Appl. Neurobiol. **26:** 31–40.
104. CARINI, R. & E. ALBANO. 2003. Recent insights on the mechanisms of liver preconditioning. Gastroenterology **125:** 1480–1491.
105. GIDDAY, J.M., A.R. SHAH, R.G. MACEREN, *et al.* 1999. Nitric oxide mediates cerebral ischemic tolerance in a neonatal rat model of hypoxic preconditioning. J. Cereb. Blood Flow Metab. **19:** 331–340.
106. BOLLI, R. 2001. Cardioprotective function of inducible nitric oxide synthase and role of nitric oxide in myocardial ischemia and preconditioning: an overview of a decade of research. J. Mol. Cell. Cardiol. **33:** 1897–1918.
107. NAKANO, A., G.S. LIU, G. HEUSCH, *et al.* 2000. Exogenous nitric oxide can trigger a preconditioned state through a free radical mechanism, but endogenous nitric oxide is not a trigger of classical ischemic preconditioning. J. Mol. Cell. Cardiol. **32:** 1159–1167.
108. SAITOH, M., H. NISHITOH, M. FUJII, *et al.* 1998. Mammalian thioredoxin is a direct inhibitor of apoptosis signal-regulating kinase (ASK) 1. EMBO J. **17:** 2596–2606.
109. KOJIMA, S., O. MATSUKI, T. NOMURA, *et al.* 1999. Elevation of antioxidant potency in the brain of mice by low-dose gamma-ray irradiation and its effect on 1-methyl-4-phenyl-1,2,3,6-tetrahydropyridine (MPTP)-induced brain damage. Free Radic. Biol. Med. **26:** 388–395.
110. RYBNIKOVA, E., A.E. DAMDIMOPOULOS, J.A. GUSTAFSSON, *et al.* 2000. Expression of novel antioxidant thioredoxin-2 in the rat brain. Eur. J. Neurosci. **12:** 1669–1678.
111. ZHANG, R., R. AL-LAMKI, L. BAI, *et al.* 2004. Thioredoxin-2 inhibits mitochondria-located ASK1-mediated apoptosis in a JNK-independent manner. Circ. Res. **94:** 1483–1491.
112. HOLMUHAMEDOV, E.L., A. JAHANGIR, A. OBERLIN, *et al.* 2004. Potassium channel openers are uncoupling protonophores: implication in cardioprotection. FEBS Lett. **568:** 167–170.
113. HIROTA, K., H. NAKAMURA, H. MASUTANI, *et al.* 2002. Thioredoxin superfamily and thioredoxin-inducing agents. Ann. N.Y. Acad. Sci. **957:** 189–199.
114. MOUSSAOUI, S., M.C. OBINU, N. DANIEL, *et al.* 2000. The antioxidant ebselen prevents neurotoxicity and clinical symptoms in a primate model of Parkinson's disease. Exp. Neurol. **166:** 235–245.
115. ZHAO, R. & A. HOLMGREN. 2004. Ebselen is a dehydroascorbate reductase mimic, facilitating the recycling of ascorbate via mammalian thioredoxin systems. Antioxid. Redox Signal. **6:** 99–104.
116. TENG, C.M., C.C. WU, F.N. KO, *et al.* 1997. YC-1, a nitric oxide-independent activator of soluble guanylate cyclase, inhibits platelet-rich thrombosis in mice. Eur. J. Pharmacol. **320:** 161–166.
117. HWANG, T.L., C.C. WU & C.M. TENG. 1999. YC-1 potentiates nitric oxide-induced relaxation in guinea-pig trachea. Br. J. Pharmacol. **128:** 577–584.
118. MARTIN, E., Y.C. LEE & F. MURAD. 2001. YC-1 activation of human soluble guanylyl cyclase has both heme-dependent and heme-independent components. Proc. Natl. Acad. Sci. USA **98:** 12938–12942.
119. MURAD, F. 2003. The excitement and rewards of research with our discovery of some of the biological effects of nitric oxide. Circ. Res. **92:** 339–341.
120. GOMMA, A.H., H.J. PURCELL & K.M. FOX. 2001. Potassium channel openers in myocardial ischaemia: therapeutic potential of nicorandil. Drugs **61:** 1705–1710.
121. DANIEL, H., N. HEMART, D. JAILLARD, *et al.* 1993. Long-term depression requires nitric oxide and guanosine 3':5' cyclic monophosphate production in rat cerebellar Purkinje cells. Eur. J. Neurosci. **5:** 1079–1082.
122. SHIBUKI, K. & D. OKADA. 1991. Endogenous nitric oxide release required for long-term synaptic depression in the cerebellum. Nature **349:** 326–328.

Mitochondrion-Targeted Photosensitizer Enhances the Photodynamic Effect–Induced Mitochondrial Dysfunction and Apoptosis

TSUNG-I PENG,[a] CHENG-JEN CHANG,[b] MEI-JIN GUO,[c] YU-HUAI WANG,[c] JAU-SONG YU,[d] HONG-YUEH WU,[c] AND MEI-JIE JOU[c]

Departments of Neurology[a] and Plastic Surgery,[b] Lin-Kou Medical Center, Chang Gung Memorial Hospital, Tao-Yuan, Taiwan

[c]*Department of Physiology and Pharmacology, Chang Gung University, Tao-Yuan, Taiwan*

[d]*Department of Cell and Molecular Biology, Chang Gung University, Tao-Yuan, Taiwan*

ABSTRACT: Recently, the mitochondrion has been considered as a novel pharmacological target for anticancer therapy due to its crucial role involved in arbitrating cell apoptosis. We have previously demonstrated that 488-nm laser irradiation induced a specific mitochondrial reactive oxygen species (mROS) formation and apoptotic death. In this study, we used a second generation of photosensitizers, the benzoporphyrin-derivative monoacid ring A (BPD-MA). We investigated specifically mechanisms at the mitochondrial level for BPD-MA coupled with 690-nm laser irradiation, the photodynamic effect (PDE) of BPD-MA, using conventional and laser scanning imaging microscopy in intact C6 glioma cells. We demonstrated BPD-MA localized mainly in the mitochondrial area. The phototoxicity induced by 1~10 J 690-nm laser irradiation was minor as compared to that induced by 488-nm laser irradiation. Unlike other mitochondrion-targeted photosensitizers, the dark toxicity induced by BPD-MA (0.05~5 mg/mL, effective doses used for the PDE) was relatively low. Nevertheless, the PDE of BPD-MA using 0.5 mg/mL coupled with 5J 690- nm irradiation induced profound and rapid (< 1 min) mitochondrial swelling, mROS formation, and severe plasma membrane blebing as compared to that induced by 488-nm laser irradiation (< 10 min). Later, the PDE of BPD-MA resulted in positive propidium iodide cell-death stain and positive TUNEL apoptotic nuclear stain and DNA laddering. Finally, the PDT of BPD-MA also instantaneously promoted the mitochondrion to diminish its covalent binding with a mitochondrial marker, MitoTracker Green. We conclude that the PDT of BPD-MA targeted primarily and compellingly the mitochondrion to induce effective mitochondria-mediated apoptosis and thus may serve as a powerful photosensitizer for clinical cancer therapy.

KEYWORDS: mitochondria; reactive oxygen species; BPD-MA; photodynamic therapy; apoptosis

Address for correspondence: Mei-Jie Jou, Department of Physiology and Pharmacology, Chang Gung University, 259 Wen-Hwa 1st Rd., Kwei-Shan, Tao-Yuan, Taiwan.
mjjou@mail.cgu.edu.tw

INTRODUCTION

The mitochondrion, due to its pivotal role in arbitrating cell apoptosis,[1–4] has recently been considered as a novel pharmacological target for various clinical applications including cancer therapy.[5–7] Mitochondrial dysfunction leads not only to the interruption of the energy supply, but also to the activation of the mitochondria-mediated apoptotic pathway.[8] Key mitochondrial mechanisms involved include the opening of the mitochondrial permeability transition pore (MPTP), which results in mitochondrial membrane potential depolarization, mitochondrial swelling, and release of apoptotic lethal proteins to the cytosol. Subsequently, downstream caspases-dependent or -independent nuclear proteases are activated to cleave nuclear DNA to small fragments.[9,10] Intriguingly, mitochondria-associated apoptotic mechanisms, particularly the opening of the MPTP can be initiated and augmented by elevated reactive oxygen species (ROS), especially the ROS that originated from the mitochondrion (mROS).[11,12] We have previously demonstrated that mROS formation can be stimulated by the 488-nm laser irradiation, which leads cells to an apoptotic cell death.[1,13,14] The photo-killing phenomenon consequently provides a unique clinical application for photodynamic therapy in the treatment of a variety of tumors.[15,16] Photodynamic therapy employs photosensitizers coupled with specific wavelengths for irradiation, that is, the photodynamic effect (PDE) of photosensitizers.[17,18] In the presence of oxygen, the PDE of photosensitizers causes significant formation of singlet oxygen as well as other ROS, which produces peroxidative reactions as a major contribution to photo-induced cell damage and death.[19–22] Among these photosensitizers, porphrins, and porphrin-related macrocycles, Photofrin (manufactured by Lederle Parenterals, Carolina, Puerto Rico, under license from Quadra Logic Technologies, Inc., Vancouver, BC, Canada) is the one that has been studied most extensively. In this study, we conducted a second generation of photosensitizers, benzoporphyrin-derivative monoacid ring A (BPD-MA, verteporfin),[23,24] for investigation. BPD-MA is a semisynthetic porphrin prepared by the reaction of protoporphrin with dimethyl acetylene dicarboxylate and has been shown to be more efficient than hematoporphyrin compounds. With the application of conventional and laser-scanning imaging microscopy, coupled with mitochondria-targeted fluorescent probes, we investigated the cellular distribution of BPD-MA and mechanisms at the mitochondrial level for the PDE of BPD-MA in intact single C6 glioma cells. Toxicities induced by either BPD-MA alone (the dark toxicity) or by 690-nm laser irradiation alone were measured and compared to that induced by the PDT of BPD-MA. In addition, mitochondrial dysfunction induced by the PDE of BPD-MA including mitochondrial morphology, mROS formation, and mitochondrial-mediated apoptosis were explored and compared to our previous study of 488-nm laser irradiation–induced mitochondrial damage and apoptosis.

MATERIALS AND METHODS

Cell Preparation

A tumor line, C6 glioma cells, was used for this study. All cells were grown in medium consisting of Dulbecco's modified Eagle's medium (Life Technologies,

Grand Island, NY) supplemented with 10% (v/v) fetal bovine serum. The cells were plated onto glass coverslips coated with poly-L-lysin for fluorescent measurement (Model No. 1, VWR Scientific, San Francisco, CA).

Chemicals and Fluorescent Dyes

The BPD-MA was provided by Quadra Logic Technologies Inc. (Vancouver, BC, Canada). The drug was protected from light and maintained at 4°C until use. All fluorescent dyes were purchased from Molecular Probes Inc. (Eugene, OR) and chemicals were obtained from Sigma (St. Louis, MO). Loading concentrations of fluorescent probes were as follows: mitochondrial fluorescent dyes: MitoTracker green (Mito G) 100 nM; MitoTracker Red (Mito R) 100 nM; tetramethyl rhodamine ethyl ester (TMRM) 100 nM; ROS fluorescent dye: 6-carboxy-2',7'- dichlorodihydrofluorescein diacetate (C-2938) 1 µM; nucleus of dead cell: propidium iodide (PI) 1 µM. Fluorescent probes were all loaded at room temperature for 20–30 min. After loading, cells were rinsed three times with HEPES-buffered saline (140 mM NaCl, 5 mM KCl, 1 mM $MgCl_2$, 2 mM $CaCl_2$, 10 mM glucose, 5 mM HEPES, pH 7.4). Cells loaded with C-2938 required additional 30–40 min of incubation after the dye loading to allow intracellular deacetylation of the ester form of dye. All experiments were performed in HEPES-buffered saline.

Conventional and Laser Scanning Imaging Microscopy

All phase-contrast images and conventional fluorescence images were obtained using a Zeiss inverted microscope (AxioVert 200M; Carl Zeiss, Jena, Germany) equipped with a mercury lamp (HBO 103), a cool CCD camera (Coolsnap fx; Roper Scientific, Tucson, AZ), and Zeiss objectives (Plan-NeoFluar 100×, N.A. 1.3 oil). Filters used for detecting Mito G and C-2938 were No. 10 (Exi: BP 450–490 nm; Emi: BP 515–565 nm) and for Mito R and TMRM, No. 15 (Exi: BP 546/12 nm; Emi: LP 590 nm). Multiphoton fluorescence images were obtained using a Leica SP2 MP (Leica-Microsystems, Leica, Mannheim, Germany) fiber-coupling system equipped with a Ti:Sa-Laser system (model : Millenia/Tsunami; Spectra-Physics, Mountain View, CA), providing pulse repetition rate at 82 MHz, laser pulse width of 1.2 ps, spectral bandwidth of 1 nm and object pulse width of 1.3 ps). Wavelength at 800 nm with average laser power of 600 mW was selected for illumination. During fluorescence imaging, the illumination light was reduced to the minimal level by selecting suitable neutral density filters to prevent the photosensitizing effect from the interaction of light with fluorescent probes. In some experiments, confocal images of cells and mitochondria were collected on a Bio-Rad Radiance 2100 (Bio-Rad, London, UK) using 488-nm argon-laser illumination. All images were processed and analyzed using MetaMorph software (Universal Imaging Corp., West Chester, PA). Intensity levels were analyzed from the original images and graphed by using Microsoft EXCEL software and Photoshop.

The PDE and Cellular Distribution of BPD-MA

To assess PDE of BPD-MA, BPD-MA was prepared from a stock solution of 1 mg/mL BPD-MA in HEPES-buffered solution stored at 4°C. Cells were incubated with different concentrations (0.05 mg/mL, 0.5 mg/mL, and 5 mg/mL in HEPES-buffered solution) separately for 30 min, and then were exposed to a Coherent 690-nm diode laser with laser power of 15 mW. Laser irradiation strengths of 10 J (11 min and 6 s), 5 J (5 min and 30 s) and 1 J (1 min and 6 s) were compared. To investigate the precise cellular distribution of BPD-MA, cells were treated with 0.5 mg/mL BPD-MA in HEPES at 37°C. To identify the cellular distribution of BPD-MA, cells were co-loaded with Mito G and NBD C6-ceramide for identifying the mitochondrion and the Golgi apparatus, respectively. Confocal images of BPD-MA, Mito-G and NBD C6-ceramide were collected simultaneously.

Fluorescence Measurement of Intracellular ROS with Fluorescent-Imaging Microscopy

Intracellular ROS was detected using an improved version of dichlorohydrofluorescein diacetate, C-2938, which has two negative charges at physiological pH that allow it to pass through membranes more easily during cell loading. In addition, C-2938 was used to test whether it is less light sensitive than a classic ROS probe, dichlorodihydrofluorescein, during fluorescence measurement as seen in our previous study.[13,25] The nonfluorescent C-2938 is oxidized by intracellular ROS to form the highly fluorescent C-2938. Fluorescent images of C-2938 were obtained before and after the treatment with 690-nm laser irradiation alone, BPD-MA alone, or the PDT of BPD-MA. Results were collected and displayed with a pseudocolor or black and white scale from 0 to 255 units. In C-2938 images, mROS was calculated from areas based on a mitochondrial marker Mito R distribution from the same cell. Intensity values were analyzed per pixel from the cytosolic, the mitochondrial, and the nuclear area of the cells. Fluorescence-intensity changes in the experiments with C-2938 were normalized with the control. Differences were evaluated using Student's t-test.

Apoptotic TUNAL Assay and DNA Ladders

Apoptotic DNA ladders were analyzed using the ApoAlert LM-PCR™ Ladder Assay Kit from Clontech (Palo Alto, CA), which specially amplified nucleosomal ladders generated during apoptosis. In brief, cellular DNA was collected from control cells and cells treated with the PDT of BPD-MA. The extracts were treated with RNase A and proteinase K. The LM-PCR assay was performed according to the user manual (15 PCR cycles). Each PCR product (15L) was electrophoresed on a 1.2% agarose/EtBr gel at 6V/cm for approximately 2.5 h, and DNA was detected by ethidium bromide under UV light.

Apoptotic cell death was also confirmed by using an Apo-BrdU *In Situ* DNA Fragmentation Assay Kit (BioVision, Mountain View, CA). The terminal deoxynucleotidyl transferase (TdT) dUTP nick-end labeling (TUNEL) identifies apoptotic cells *in situ* using TdT to transfer BrdU-dUTP to the free 3'-OH of cleaved DNA. The BrdU-labeled cleavage sites were then visualized by reaction with fluorescein-conjugated anti-BrdU monoclonal antibody. After the incubation period, cells were washed twice with PBS, fixed for 1 min with ice-cold ethanol/acetic acid (1:1) solution, and then washed three times with PBS. The fixed cells were permeabilized in ice-cold 0.2% Triton X-100 detergent for 5 min and then washed three times with PBS. Finally, the photomicrographs were obtained with a Zeiss inverted-fluorescence microscope.

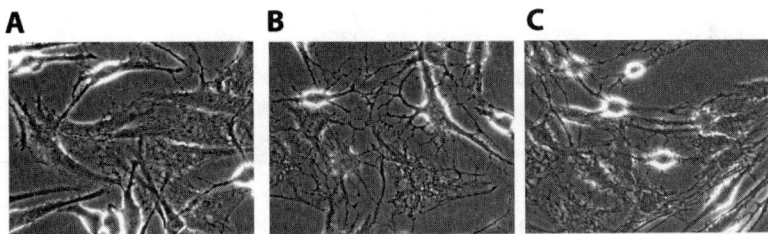

FIGURE 1. Low phototoxicity and low dark toxicity induced by laser irradiation alone and BPD-MA alone. Phase-contrast images of C6 glioma cells taken by a 40× objective. (**A**) Controls, (**B**) cells received 10 J 690-nm laser irradiation, and (**C**) cells loaded with 5 μg/mL BPD-MA. Note the low phototoxicity induced by 10 J 690-nm laser irradiation and the low dark toxicity induced by BPD-MA alone. Nuclear size was about 10 μm.

RESULTS AND DISCUSSION

Low Phototoxicity and Low Dark Toxicity Induced by Laser Irradiation Alone and BPD-MA Alone, Respectively

A successful PDE relies considerably on the low dark toxicity of photosensitizers as well as the low phototoxicity of the coupled laser to ensure a minor nonspecific damage prior to the two being coupled. Therefore, the sole toxic effect of 1~10 J 690 nm laser irradiation (the phototoxicity) and that of 0.05~5 mg/mL BPD-MA (the dark toxicity) were cautiously examined in C6 glioma cells. As shown in FIGURE 1, unlike the phototoxicity induced by 488-nm irradiation seen in our previous studies,[1,13,14] neither 690-nm laser irradiation (FIG. 1B) nor the treatment with BPD-MA (FIG. 1C) affected cell morphology and viability long after the treatment (>60 min). The low phototoxicity induced by 690-nm laser and by the low dark effect of BPD-MA make the PDE of BPD-MA superior to other photosensitizers.

Potent PDE of BPD-MA Using a 690-nm Laser Resulted in Rapid Apoptotic Cell Death in C6 Glioma Cells

The potent PDE of BPD-MA using 690-nm laser irradiation is demonstrated in FIGURE 2. In contrast to the low toxicity carried by either laser alone or BPD-MA alone, the PDE of BPD-MA (0.5 mg/mL) coupled with 5 J 690-nm laser irradiation resulted in profound and rapid (< 1 min) mitochondrial swelling and severe plasma membrane blebing. These damages soon resulted in positive nuclear PI stain in all of the PDE of BPD-MA treated cells. Nuclear condensation and positive PI staining were detected 30 min after the PDE of BPD-MA (FIG. 2A and B). The PDE of BPD-MA–induced apoptotic cell death was confirmed by both positive TUNEL apoptotic nuclear assay (FIG. 2C) and by DNA gel laddering (FIG. 2D). The potent apoptosis induced by the PDT of BPD-MA confirm that mitochondria are potential pharmacological targets for clinical application.

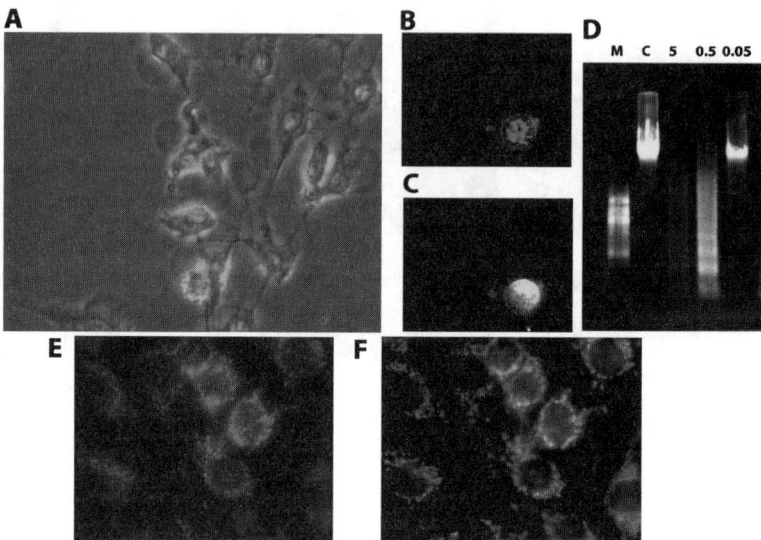

FIGURE 2. The potent PDE of BPD-MA using a 690-nm laser resulted in rapid plasma-membrane blebing and apoptotic cell death in C6 glioma cells. (**A**) A phase-contrast image superimposed with a red fluorescent PI image. Note all the cells received the PDT effect of 0.5 μg/mL BPD-MA coupled with 5 J 690-nm laser irradiation PDE formed plasma-membrane blebbing and positive PI stain. (**B**) and (**C**) Images taken from the same cells that double stained with PI and TUNEL-FITC, respectively. (**D**) DNA gel ladders assay from cells received the PDT of different concentrations of BPD-MA (0.05, 0.5, 5 μg/mL) and 5 J 690-nm laser irradiation. (**E**) Cellular distribution of BPD-MA. (**F**) Mitochondrial distribution detected using Mito G. Note the co-localization of BPD-MA and the Mito G. Images in (**E**) and (**F**) were taken using a 100× fluorescent objective. The size of the nucleus was about 10 μm.

BPD-MA Is Localized Mainly in the Mitochondrial Area

The potency of the photodynamic cell-killing effect often relies on the cellular distribution of the photosensitizers used. For instance, photosensitizers that localized to plasma membrane, lysosome, and Golgi apparatus resulted in a different photo-killing effect.[26] Intracellular studies carried by Rousset *et al.* has revealed stronger cytoplasmic than nuclear fluorescence for both BPD and Photofrin.[27,28] Because mitochondria play a pivotal role in arbitrating apoptosis, we focused on whether the BPD-MA can localize to the mitochondrial area by time-lapse imaging the emitted red fluorescence of BPD-MA after cells were loaded with BPD-MA. Intriguingly, BPD-MA localized mainly in the mitochondria (FIG. 2E) as the red fluorescent image of BDP-MA resembled closely the green fluorescent image of a mitochondrial marker, Mito G (FIG. 2F). BMP-MA, although diminished slightly in the mitochondrial area, did not, however, localize into the nucleus even when plasma membrane blebing and apoptotic cell death occurred (data not shown). These results imply that the potent photo-killing effect induced by the PDT of BPT-MA may contribute directly from the original target site for BPD-MA, the mitochondrial stress. Thus, we

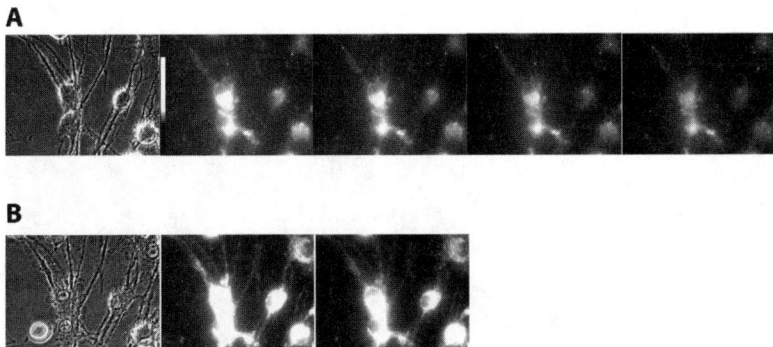

FIGURE 3. The PDT of BPD-MA coupled with 690-nm laser irradiation enhanced mROS formation, plasma membrane blebing, and apoptotic cell death. (**A**) Cells received irradiation without BPD-MA. A phase-contrast image followed with four continuously recorded fluorescent images of C-2938. (**B**) Cells received irradiation with BPD-MA. Note the potent mROS formation and plasma membrane blebing induced by the PDT of BPD-MA. Autooxidation of the C-2938 was minor during time-lapse imaging. 40× objective was used.

hypothesized that the mitochondrial-associated apoptotic pathway may be involved in the mechanism of action for the PDT of BPD-MA.

PDE of Mitochondrial-Targeted BPD-MA Enhanced Mitochondrial Dysfunction

To investigate mitochondrial mechanisms involved in the PDE of BPD-MA, we measured the dynamic changes in mitochondrial-associated events, including mitochondrial morphology and mitochondrial ROS, as well as the capability for mitochondria to retain a covalent binding of a mitochondrial fluorescent marker, Mito G, using conventional and laser-scanning imaging microscopy as previous described.[1,13,14] As shown in FIGURE 3A, the effect of visible light irradiation alone did not induce any mROS formation (FIG. 3A) when the ROS level was detected using an improved ROS probe, C-2938 (see MATERIALS AND METHODS). The autooxidation of C-2938 during the ROS measurement was minor as shown by the four continuous images recorded after the irradiation, which showed a slight decrease instead of an increase in the fluorescent intensity (FIG. 3A). These results indicate that C-2938 may be a better candidate with less light-induced autooxidation for ROS measurement as compared to the ROS probe, dichloroflurescien, used in our previous study.[13]

FIGURE 3B, in contrast to FIGURE 2A, shows that the PDE of BPD-MA induced significant and almost instantaneous increases in mROS formation (FIG. 3B) right after the BPD-MA–loaded cell received irradiation. ROS formation was particularly profound in the identified mitochondrial area using a mitochondrial marker, Mito R (not shown). The elevated mROS was about 3~4-fold higher than that in the control cells. Soon, the mROS formation led to severe mitochondiral swelling and plasma membrane blebing and apoptotic cells death. Intriguingly, the PDT of BPD-MA also induced a sweep and significant loss of Mito G stain. As shown in FIGURE 4, the first

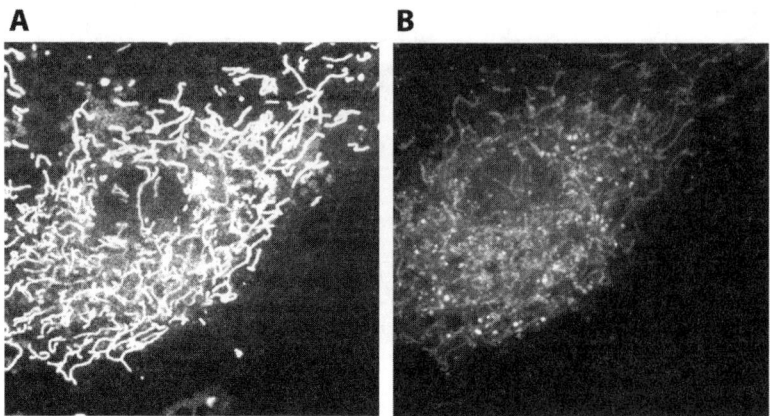

FIGURE 4. The PDT of BPD-MA induced rapid loss of the fluorescent intensity of Mito G. One second before (**A**) and 1 s after (**B**) the PDT of BPD-MA.

image taken after BPD-MA–loaded cells received visible laser irradiation, the fluorescent intensity of Mito G diminished dramatically. Note that mitochondria at this stage were still thread-like in shape. Since Mito G has been reported to form a covalent bond with the mitochondrial membrane, these results imply that the photo damage induced by the PDE of BPD-MA must be rather strongly and specifically targeted on the mitochondrion. In contrast, mROS stress, induced by 10 mM H_2O_2 as in our previous study, did not, however, reduce the fluorescent signal from a Mito G–loaded cells until the mitochondrial were fully swollen.[30]

All together, this study demonstrates that the mitochondrial-targeted photosensitizer, BPD-MA, provides a compelling and specific photodynamic reaction via the induction of mitochondrial stress. The severe damage induced by the BPD-MA on the mitochondria is mainly contributed from the oxidation carried by the freshly formed mROS upon the photodynamic reaction between the photosensitizer and the light irradiation. A viciously initiated and amplified mROS cycle can further induce more rigorous nonspecific dysregulation or dysfunction of mitochondria so that pathological apoptosis or even necrosis of cells becomes obvious and inevitable. The potent mitochondria-mediated apoptosis induced by the PDT of BPD-MA, together with its high selective localization in tumors,[29] consequently provide a novel and potential pharmacological target for the photodynamic therapy in the clinical treatment of cancer.

ACKNOWLEDGMENTS

This research work was supported by grants from the Chang Gung Research Foundation (CMRP 1009, CMRPD 32054 to M-J.J. and CMRPG 33011 to T.-IP.) and from the National Science Council (NSC 91-2320-B-182-056 to M -J.J. and 91-2314-B-182A-021 to T.-IP.)

REFERENCES

1. Jou, M.J. et al. 2002. Critical role of mitochondrial reactive oxygen species formation in visible laser irradiation-induced apoptosis in rat brain astrocytes (RBA-1) J. Biomed. Sci. **9:** 507–516.
2. Liu, X. et al. 1996. Induction of apoptotic program in cell-free extracts: requirement for dATP and cytochrome c. Cell **86:** 147–157.
3. Yang, J. et al. 1997. Prevention of apoptosis by Bcl-2: release of cytochrome c from mitochondria blocked. Science **275:** 1129–1132.
4. Reed, J.C. 1997. Cytochrome c: can't live with it—can't live without it. Cell **91:** 559–562.
5. Szewczyk, A. & L. Wojtczak. 2002. Mitochondria as a pharmacological target. Pharmacol. Rev. **54:** 101–127.
6. Modica, N. & K. Singh. 2002. Mitochondria as targets for detection and treatment of cancer. Expert Rev. Mol. Med. **2002:** 1–19.
7. Decaudin, D. et al. 1998. Mitochondria in chemotherapy-induced apoptosis: a prospective novel target of cancer therapy (review). Int. J. Oncol. **12:** 141–152.
8. Wang, X. 2001. The expanding role of mitochondria in apoptosis. Genes Dev. **15:** 2922–2933.
9. Zou, H. et al. 1997. Apaf-1, a human protein homologous to C. elegans CED-4, participates in cytochrome c-dependent activation of caspase-3. Cell **90:** 405–413.
10. Chauhan, D. et al. 1997. Cytochrome c-dependent and -independent induction of apoptosis in multiple myeloma cells. J. Biol. Chem. **272:** 29995–29997.
11. Atlante, A. et al. 2000. Cytochrome c is released from mitochondria in a reactive oxygen species (ROS)-dependent fashion and can operate as a ROS scavenger and as a respiratory substrate in cerebellar neurons undergoing excitotoxic death. J. Biol. Chem. **275:** 37159–37166.
12. Backway, K.L. et al. 1997. Relationships between the mitochondrial permeability transition and oxidative stress during ara-C toxicity. Cancer Res. **57:** 2446–2451.
13. Peng, T.I. & M.J. Jou. 2004. Mitochondrial swelling and generation of reactive oxygen species induced by photoirradiation are heterogeneously distributed Ann. N.Y. Acad. Sci. **1011:** 112–122.
14. Jou, M.J. et al. 2004. Mitochondrial reactive oxygen species generation and calcium increase induced by visible light in astrocytes Ann. N.Y. Acad. Sci. **1011:** 45–56.
15. Castro, D.J., R.E. Saxton & J. Soudant. 1996. The concept of laser phototherapy. Otolaryngol. Clin. North Am. **29:** 1011–1029.
16. Schuitmaker, J.J. et al. 1996. Photodynamic therapy: a promising new modality for the treatment of cancer. J. Photochem. Photobiol. B **34:** 3–12.
17. Dougherty, T.J. et al. 1998. Photodynamic therapy. J. Natl. Cancer Inst. **90:** 889–905.
18. Oleinick, N.L., R.L. Morris & I. Belichenko. 2002. The role of apoptosis in response to photodynamic therapy: what, where, why, and how. Photochem. Photobiol. Sci. **1:** 1–21.
19. Friedmann, H. et al. 1991. A possible explanation of laser-induced stimulation and damage of cell cultures. J. Photochem. Photobiol. B **11:** 87–91.
20. Luo, Y., C.K. Chang & D. Kessel. 1996. Rapid initiation of apoptosis by photodynamic therapy. Photochem. Photobiol. **63:** 528–534.
21. Oleinick, N.L. & H.H. Evans. 1998. The photobiology of photodynamic therapy: cellular targets and mechanisms. Radiat. Res. **150:** S146–S156.
22. Lam, M., N.L. Oleinick & A.L. Nieminen. 2001. Photodynamic therapy-induced apoptosis in epidermoid carcinoma cells. Reactive oxygen species and mitochondrial inner membrane permeabilization. J. Biol. Chem. **276:** 47379–47386.
23. Richter, A.M. et al. 1987. Preliminary studies on a more effective phototoxic agent than hematoporphyrin. J. Natl. Cancer Inst. **79:** 1327–1332.
24. Lui, H. & R.R. Anderson. 1993. Photodynamic therapy in dermatology: recent developments. Dermatol. Clin. **11:** 1–13.
25. Quillet, M. et al. 1997. Implication of mitochondrial hydrogen peroxide generation in ceramide-induced apoptosis. J. Biol. Chem. **272:** 21388–21395.

26. WATERFIELD, E.M. *et al.* 1994. Wavelength-dependent effects of benzoporphyrin derivative monoacid ring A in vivo and in vitro. Photochem. Photobiol. **60:** 383–387.
27. ROUSSET, N. *et al.* 2000. Cellular distribution and phototoxicity of benzoporphyrin derivative and Photofrin. Res. Exp. Med. (Berl.) **199:** 341–357.
28. TAKEUCHI, Y. *et al.* 2003. Induction of intensive tumor suppression by antiangiogenic photodynamic therapy using polycation-modified liposomal photosensitizer. Cancer **97:** 2027–2034.
29. RICHTER, A.M. *et al.* 1990. Biodistribution of tritiated benzoporphyrin derivative (3H-BPD-MA), a new potent photosensitizer, in normal and tumor-bearing mice. J. Photochem. Photobiol. B **5:** 231–244.
30. JOU, M.J. *et al.* 2004. Visualization of the antioxidative effects of melatonin at the mitochondrial level during oxidative stress-induced apoptosis of rat brain astrocytes. J. Pineal Res. **37:** 55–70.

Attenuation of UV-Induced Apoptosis by Coenzyme Q_{10} in Human Cells Harboring Large-Scale Deletion of Mitochondrial DNA

CHENG-FENG LEE,[a,c] CHUN-YI LIU,[a,c] SHU-MEI CHEN,[a]
MARIANNA SIKORSKA,[b] CHEN-YU LIN,[a] TZU-LING CHEN,[a]
AND YAU-HUEI WEI[a]

[a]*Department of Biochemistry and Molecular Biology, and
Center for Cellular and Molecular Biology, National Yang-Ming University,
Taipei 112, Taiwan*

[b]*Institute for Biological Sciences, National Research Council of Canada,
Ottawa, Ontario, Canada*

ABSTRACT: Chronic progressive external ophthalmoplegia (CPEO) syndrome is one of the mitochondrial diseases caused by large-scale deletions in mitochondrial DNA (mtDNA) that impair the respiratory function of mitochondria and result in decreased production of ATP in affected tissues. In order to investigate whether CPEO-associated mtDNA mutations (i.e., 4,366-bp and 4,977-bp large-scale deletions) render human cells more vulnerable to apoptosis, we constructed cybrids carrying the deleted mtDNA. Assays for cell viability, DNA fragmentation, cytochrome c release, and caspase 3 activation revealed that UV irradiation at 20 J/m^2 triggered apoptosis in all the cybrids. This treatment also produced elevated intracellular levels of reactive oxygen species (ROS). The rate of UV-induced cell death was more pronounced in the cybrids harboring mtDNA deletions than in the control cybrid with wild-type mtDNA. Subsequently, we evaluated the effect of coenzyme Q_{10} on the UV-triggered apoptosis. The results showed that after pretreatment of the cybrids with 100 μM coenzyme Q_{10} the UV-induced cell damage (i.e., ROS production and activation of caspase 3) was significantly reduced. Taken together, these findings suggest that large-scale deletions of mtDNA increased the susceptibility of human cells to the UV-triggered apoptosis and that coenzyme Q_{10} mitigated the damage; hence, it might potentially serve as a therapeutic agent to treat mitochondrial diseases resulting from mtDNA deletions.

KEYWORDS: mitochondrial DNA; mutation; apoptosis; UV irradiation; coenzyme Q_{10}

---

[c] C-F.L. and C-Y.L. contributed equally to this work.

Address for correspondence: Professor Yau-Huei Wei, Department of Biochemistry and Molecular Biology, National Yang-Ming University, Taipei 112, Taiwan. Voice: +886-2-2826-7118; fax: +886-2-2826-4843.

joeman@ym.edu.tw

Ann. N.Y. Acad. Sci. 1042: 429–438 (2005). © 2005 New York Academy of Sciences.
doi: 10.1196/annals.1338.036

INTRODUCTION

In the past four decades more than 100 mitochondrial diseases have been described; most of them affect selective populations of cells in the neuromuscular and central nervous systems. Mitochondrial diseases are often caused by point mutation or deletion of mtDNA and are associated with a broad spectrum of clinical manifestations.[1] There is also strong evidence suggesting that mitochondrial function defects, resulting from the qualitative and quantitative alterations of mtDNA, contribute to the development of neurodegenerative diseases.[2,3] The abnormalities in mtDNA affect directly the oxidative phosphorylation process and result in the reduction of ATP level and increased production of reactive oxygen species (ROS) in the affected tissues. Indeed, mitochondria not only generate most of the intracellular ROS, but also serve as a key player in the initiation and execution of apoptotic cell death. This, in turn, can worsen the pathophysiology and clinical outcome of the mitochondrial diseases.

Although great progress has been made in the fundamental understanding of several mitochondrial disorders, at present there are essentially no effective treatments for these diseases. The general strategy for selecting pharmacological therapies is to use agents that can interfere with the pathological process, or preferably agents that can modify respiratory chain function and reduce the level of cytotoxic metabolites, including ROS. One agent that falls into this category is coenzyme Q_{10}. Coenzyme Q_{10} is a well-characterized electron carrier of the respiratory chain, which is mainly localized in the inner mitochondrial membrane where it serves as a highly mobile carrier of electrons and protons between the flavoproteins and the cytochrome system.[4] Several studies demonstrated that the fully reduced form of ubiquinone (ubiquinol) might function as an antioxidant.[5,6] These studies indicate that ubiquinol may prevent both the initiation and propagation phases of lipid peroxidation, because of its location in the hydrophobic region of the membrane lipid bilayer where lipid peroxidation often takes place.

In order to evaluate the effects of mtDNA mutations on cellular behavior under stress, we have established cybrids harboring large-scale mtDNA deletions (4,366 bp and 4,977 bp) using skin fibroblasts from patients with clinically proven CPEO syndrome as the mtDNA donors.[7] The cybrids were challenged with UV irradiation, a well-known trigger of apoptosis,[8] in the presence or absence of coenzyme Q_{10}. We hypothesized that the mutated mtDNA would render the cybrids more vulnerable to the UV-induced cell death and that coenzyme Q_{10}, by acting as an effective antioxidant, might protect them from the UV irradiation-triggered cellular damage.

MATERIALS AND METHODS

Cell Culture and Experimental Treatments

A series of cybrid clones harboring different proportions of mtDNA with large-scale deletions were made by fusing enucleated skin fibroblasts from two CPEO patients with mtDNA-null human osteosarcoma (ρ^0) cells.[9] The proportion of mtDNA with large-scale deletions in different cybrid clones was quantified by the Southern hybridization method developed in our laboratory.[7]

Cybrids were grown in DMEM supplemented with 5% FBS, 100 mg/mL pyruvate, and 50 mg/mL uridine in humidified 5% CO_2/95% air and were exposed to 20 J/m^2 UV irradiation when 70% confluent was reached. The cells were analyzed 25 h later. In some experiments, prior to the UV exposure, the cybrids were pretreated for 24 h with 100 µM coenzyme Q_{10} added directly to the culture media from a water-soluble formulation prepared as described below.

Cell viability was measured by trypan blue exclusion assay. In brief, cells were harvested by trypsinization, resuspended in phosphate-buffered saline (PBS, pH 7.3), stained with 0.4% trypan blue, and counted using a hemocytometer.

Analysis of DNA Fragmentation

DNA fragmentation was analyzed by electrophoresis on a 2% agarose gel as described by Herrmann et al.[10]

Assay of Caspase 3 Activation

Caspase 3 activity was measured using a fluorescent substrate, Ac-DEVD-AFC.[11] Fluorescence intensity of the cleaved substrate was determined using a Hitachi F-3000 Spectrofluorometer (Hitachi, Ltd., Tokyo) set at an excitation wavelength of 380 nm and an emission wavelength of 508 nm.

Western Blotting

Cell lysate was separated by electrophoresis on 12% SDS-PAGE, blotted onto a Hybond-P$^+$ membrane (Amersham Biosciences, Uppsala, Sweden) and immunoblotted with anti–caspase 3 antibody (Santa Cruz Biotechnology, Santa Cruz, CA).

Immunofluorescent Staining of Cytochrome c

Cells were first incubated with 100 nM MitoTracker Red and then immunostained for cytochrome *c* with an anti–cytochrome *c* antibody (Santa Cruz Biotechnology) and FITC-conjugated secondary antibody (Jackson ImmunoResearch, West Grove, PA). The coverslips were examined under a confocal microscope.[12]

Measurement of Intracellular ROS

After incubation with 80 µM 2′,7′-dichlorofluorescin diacetate (DCFH-DA; Molecular Probes, Eugene, OR)[13] at 37°C for 20 min, cells were harvested and resuspended in 0.5 mL HEPES buffer, and fluorescence intensity was measured with a flow cytometer (model EPICS XL-MCL, Beckman Coulter, Miami, FL).

Preparation of Water-Soluble Formulation of Coenzyme Q_{10}

Water-soluble formulation of coenzyme Q_{10} described by Sikorska et al.[14] was made using polyoxyethanyl 600 tocopheryl sebacate (PTS-600) as a carrier, prepared according to U.S. patent 6,045,826.[15]

TABLE 1. Summary of biological effects of UV irradiation with and without coenzyme Q_{10} pretreatment on cybrid clones harboring mutated mtDNA

mtDNA mutation	4,366-bp deletion		4,977-bp deletion	
Cybrid clones	Lin 0	Lin 2	1-3-16	51-10
Genotype	wild type	mutant	wild type	mutant
Mutant mtDNA(%)	undetectable	14.3	undetectable	79.8
Viability (%)	50.2 ± 27.7	42.1 ± 17.8	35.4 ± 3.4	14.8 ± 2.4
DNA ladder				
Control	–	–	–	–
UV	+	+ +	+	+ +
Caspase 3 activity				
Control	14.2 ± 5.7	17.2 ± 2.0	8.8 ± 2.7	11.3 ± 2.3
CoQ_{10}	9.9 ± 4.2	14.3 ± 2.9	9.1 ± 4.0	9.1 ± 2.7
UV	260.2 ± 23.3	280.4 ± 25.9	190.8 ± 5.5	222.9 ± 13.3
CoQ_{10} + UV	223.5 ± 44.6	206.6 ± 32.1	150.3 ± 40.1	158.4 ± 14.5
Relative intracellular level of H_2O_2 (%)				
Control	100.0 ± 0.0	102.1 ± 8.1	ND	ND
CoQ_{10}	76.2 ± 18.6	86.9 ± 3.3	ND	ND
UV	133.9 ± 19.2	152.8 ± 11.1	ND	ND
CoQ_{10} + UV	110.4 ± 21.5	120.7 ± 8.2	ND	ND

Note: Caspase 3 activity was assayed 25 h after exposure to UV irradiation at 20 J/m². The activity of caspase 3 is expressed in arbitrary units. The intracellular level of H_2O_2 is expressed as a relative value with respect to a control cybrid harboring wild-type mtDNA. ND, not determined.

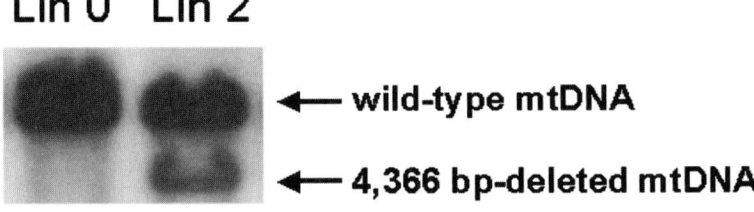

FIGURE 1. Quantification of the proportion of mtDNA with 4,366-bp deletion in the cybrids by Southern hybridization. The *upper* and *lower bands* represent the wild-type mtDNA and mtDNA with 4,366-bp deletion, respectively. The relative amount of deleted mtDNA was then determined by a laser-scanning densitometer.

RESULTS

Two sets of cybrids harboring the 4,366-bp deletion (Lin 2 with the deletion and control Lin 0 with wild-type mtDNA) and the 4,977-bp deletion (51-10 with the deletion and control 1-3-16 with wild-type mtDNA) were used in this study. The Lin 2 cybrid clone contained 14.3% of mtDNA with 4,366-bp deletion (FIG. 1) and the clone 51-10 harbored 79.8% of mtDNA with 4,977-bp deletion (TABLE 1).

The cybrids were subjected to UV irradiation at 20 J/m^2 and were analyzed 25 h later. The results are summarized in TABLE 1. After UV exposure, the viability of cybrids, especially those with mutated mtDNA, was significantly reduced. The viability of Lin 2 mutant cybrids dropped by nearly 60% after the UV exposure and the same treatment killed 85% of the 51-10 cybrids. Clearly, there was a positive corre-

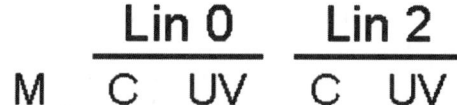
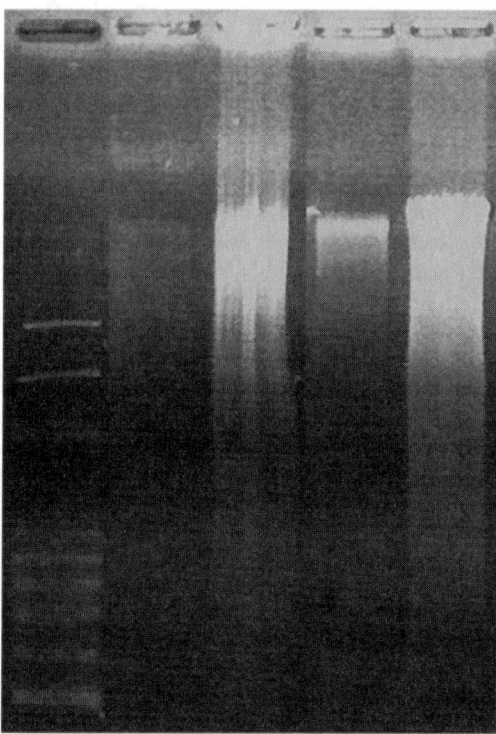

FIGURE 2. UV-triggered DNA fragmentation in cybrids harboring 4,366-bp–deleted mtDNA. DNA fragmentation was more extensive in cybrids harboring 4,366-bp mtDNA deletion than in cybrids harboring wild-type mtDNA.

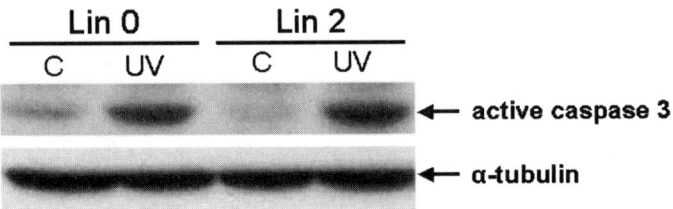

FIGURE 3. UV-induced caspase 3 activation in the cybrids harboring 4,366-bp–deleted mtDNA. The activation of caspase 3 by proteolytic cleavage was more pronounced in the cybrids harboring the deletion as compared with the cybrids harboring only the wild-type mtDNA. α-Tubulin was used as an internal standard.

lation between the degree of mitochondrial defects and cell death after exposure of cybrids to UV irradiation.

The UV-induced DNA fragmentation was very evident in the Lin cybrids (FIG. 2), and the fragmentation was more pronounced in the Lin 2 mutant cybrid. The same was true for the 51-10 and 1-3-16 cybrid clones (TABLE 1). The treatment triggered activation of caspase 3 (FIGS. 3 and 5A, TABLE 1), which was marginally higher in the mutant cybrids than in controls (Lin 2 vs. Lin 0 by 11% and 51-10 vs. 1-3-16 by 17%); it was accompanied by a release of cytochrome c from mitochondria (FIG. 4). Generation of ROS in response to the irradiation was established by measuring intracellular concentration of hydrogen peroxide (FIG. 5B, TABLE 1). The intracellular level of ROS in the mutant cybrids was approximately 14% higher than that in the relevant controls.

In a parallel set of experiments, the Lin cybrids were pretreated for 24 h with 100 μM coenzyme Q_{10} added directly to cell culture media prior to UV irradiation. This treatment lowered the ROS release by approximately 30% (FIG. 5B, $P < 0.05$) and caspase 3 activation by nearly 40% ($P < 0.05$) when analyzed 25 h after the UV exposure (FIG. 5, TABLE 1). Clearly, this pretreatment with coenzyme Q_{10} was able to offset the severity of the apoptotic response in the cybrids.

Taken together, our results showed that UV irradiation triggered, in all the cybrids, apoptotic cell death associated with ROS production and mediated by caspase 3 activation and cytochrome c release. The apoptotic markers were more pronounced in cybrids with defective mtDNA, and sensitivity of the cybrids to apoptosis and their survival after the UV exposure was related to the degree of mitochondrial defects. Coenzyme Q_{10} pretreatment quenched, to a certain degree, the ROS generation and activation of caspase 3, and thus reduced apoptotic response of the cybrids.

DISCUSSION

MtDNA mutations are responsible for a heterogeneous group of mitochondrial diseases. Their heterogeneity, however, cannot be fully explained by the proportion of mutated mtDNA. The clinical manifestations of these diseases are thought to be related not only to the severity of defects in energy metabolism, but also to apoptosis

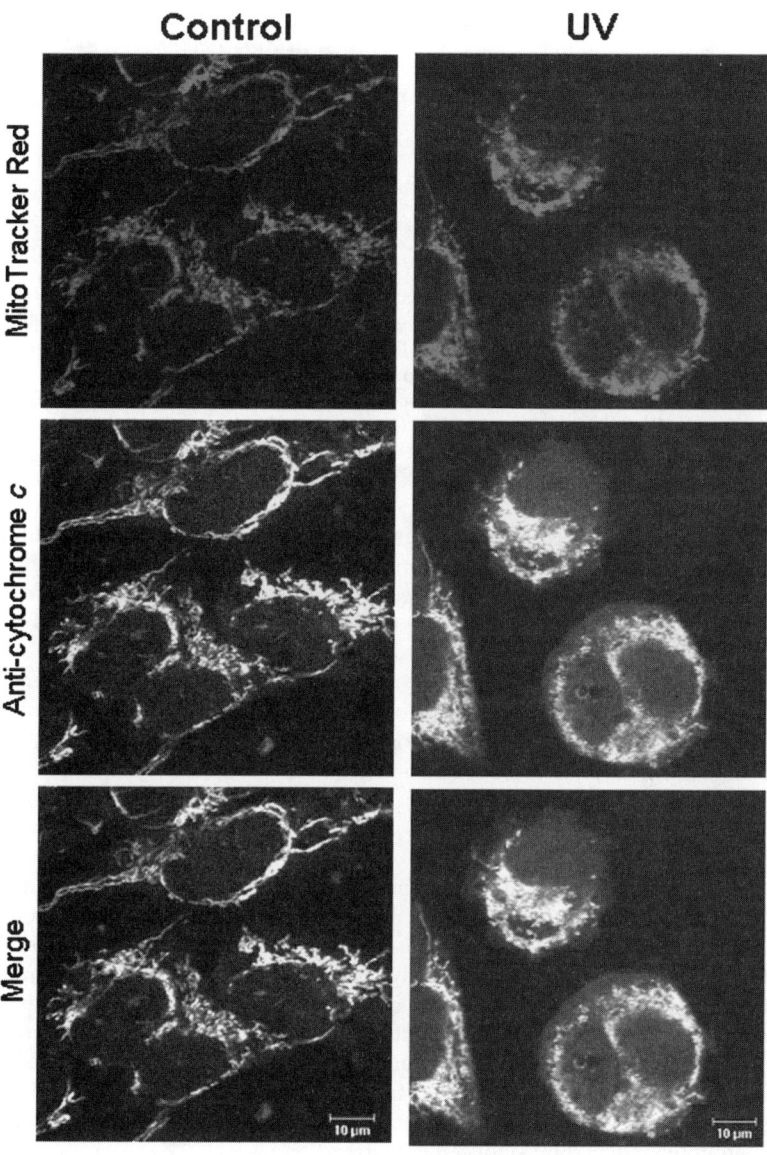

FIGURE 4. Cytochrome c release from mitochondria after UV irradiation. The cybrid (Lin 2) harboring 4,366-bp–deleted mtDNA was double stained with MitoTracker Red (*upper panel*) and anti-cytochrome c–FITC antibodies (*middle panel*). Under the control condition without exposure to UV (*left panel*), mitochondria assumed filamentous structure, and cytochrome c was localized in the mitochondria. However, after exposure to UV (*right panel*), mitochondria became swollen and moved away from the cell periphery to form aggregates around the nucleus, and cytochrome c was released from mitochondria to the cytoplasm.

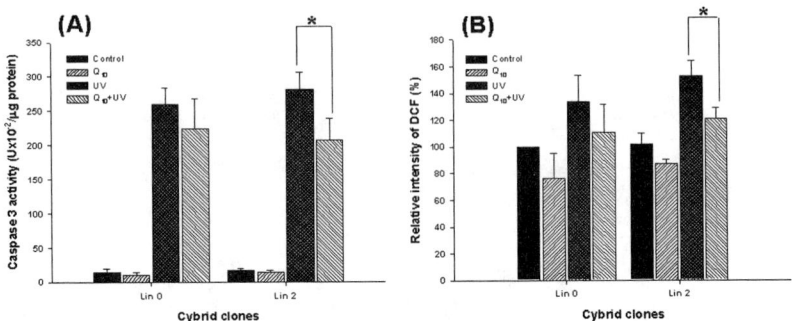

FIGURE 5. Protective effect of coenzyme Q_{10} on UV-induced apoptosis in cybrids harboring 4,366-bp–deleted mtDNA. (**A**) UV-induced caspase 3 activation was less pronounced in cybrids pretreated with 100 μM coenzyme Q_{10} than in cybrids not exposed to coenzyme Q_{10} ($^*P < 0.05$). (**B**) UV-induced increase of hydrogen peroxide was less pronounced in cybrids pretreated with 100 μM coenzyme Q_{10} compared to the cells without pretreatment with coenzyme Q_{10} ($^*P < 0.05$).

in affected tissues. Indeed, apoptotic features have been reported in muscle fibers of patients carrying a high proportion of mtDNA with large-scale deletion.[16] However, a question as to what extent the presence of mutated mtDNA renders cells more susceptible to apoptosis has not been fully answered. In this study, we used cybrids carrying two kinds of large-scale deletions in mtDNA, and we established that UV irradiation at 20 J/m^2 triggered caspase 3 and cytochrome c–mediated apoptosis. The cells' sensitivity to the UV exposure closely correlated with the degree of mitochondrial defects—that is, the higher the proportion of mutated mtDNA the more vulnerable the cybrids were to the UV irradiation (TABLE 1). These results are consistent with our previous study on UV and staurosporine-induced apoptosis in cybrids harboring mtDNA with A3243G mutation, A8344G mutation, and 4,977 bp-deletion, respectively.[17]

The molecular mechanism of UV-induced apoptosis is not yet fully understood, although previously published studies suggest that the direct UV damage to the cellular macromolecules such as nucleic acids, lipids, and proteins plays a significant role.[8] Another possible mechanism may involve the UV-elicited production of ROS such as superoxide anions and singlet oxygen that further increase oxidative stress burden and cellular damage, including the damage to mtDNA. Here we showed that the increased vulnerability to UV irradiation might, indeed, be related to the higher oxidative stress elicited by increased production of ROS in the cybrids harboring mutated mtDNA. Accordingly, modulation of the ROS production by coenzyme Q_{10} resulted in the reduced UV-induced apoptosis.

Coenzyme Q_{10} is a naturally occurring lipid-soluble antioxidant that is capable of quenching free radical production and preventing oxidative damage to cellular constituents. It is also a critical component of the mitochondrial respiratory chain and is absolutely essential for the production of ATP. And, coenzyme Q_{10} has been investigated as a potential therapeutic agent to treat cardiovascular diseases, neurodegenerative diseases, and mitochondrial disorders.[18–20] Although numerous beneficial outcomes, both experimental and clinical, have been reported in response to coenzyme Q_{10} supplementation, its full therapeutic potential is greatly limited by the lack of solubility in aqueous media. This limits its bioavailability and a scope of clinical applications. In this study, to offset the damaging effect of UV exposure, we used a unique aqueous coenzyme Q_{10} formulation developed in Dr. Sikorska's laboratory at the National Research Council, Canada.[14,15] The results showed that pretreatment of the cybrids with 100 μM coenzyme Q_{10} was protective against UV irradiation. This observation is consistent with the previously reported beneficial effects of coenzyme Q_{10}—that is, improved function of pancreatic beta-cells from patients with diabetes mellitus and MELAS syndrome,[21] offsetting a paraquat-induced neurotoxicity,[22] mitigating endotoxemic and ischemic brain damage in rodents.[23] These studies imply that water-soluble coenzyme Q_{10} could also be an effective therapeutic agent for the treatment of mitochondrial diseases. In addition to scavenging free radicals, coenzyme Q_{10} may improve mitochondrial function by stabilizing the inner membrane to facilitate mitochondrial respiration and efficiency of ATP synthesis.[24]

Taken together, we have demonstrated that large-scale deletions of mtDNA render human cells more vulnerable to apoptosis triggered by UV irradiation. The data support the notion that apoptosis might play an important role in the development of clinical symptoms such as muscle weakness and neurological disorders characteristic of the mitochondrial encephalomyopathies. We have also demonstrated that coenzyme Q_{10} can interfere with the apoptotic mechanisms, and it therefore might be an effective drug to alleviate the symptoms and slow down the progression of mitochondrial diseases.

ACKNOWLEDGMENTS

This work was supported by a grant (NSC92-2321-B-010-011-YC) from the National Science Council and a grant (EX93-9120BN) from the National Health Research Institutes, Executive Yuan, Taiwan.

REFERENCES

1. WALLACE, D.C. 1999. Mitochondrial diseases in man and mouse. Science **283:** 1482–1488.
2. LIN, F.H., R. LIN, H.M. WISNIEWSKI, et al. 1992. Detection of point mutations in codon 331 of mitochondrial NADH dehydrogenase subunit 2 in Alzheimer's brains. Biochem. Biophys. Res. Commun. **182:** 238–246.
3. IKEBE, S., M. TANAKA & T. OZAWA. 1995. Point mutations of mitochondrial genome in Parkinson's disease. Brain Res. Mol. Brain Res. **28:** 281–295.
4. MITCHELL, P. 1975. Protonmotive redox mechanism of the cytochrome b-c_1 complex in the respiratory chain: protonmotive ubiquinone cycle. FEBS Lett. **56:** 1–6.
5. FORSMARK-ANDREE, P. & L. ERNSTER. 1994. Evidence for a protective effect of endogenous ubiquinol against oxidative damage to mitochondrial protein and DNA during lipid peroxidation. Mol. Aspects Med. **15:** S73–S81.
6. SANDHU, J.K., S. PANDEY, M. RIBECCO-LUTKIEWICZ, et al. 2003. Molecular mechanisms of glutamate neurotoxicity in mixed cultures of NT2-derived neurons and astrocytes: protective effects of coenzyme Q_{10}. J. Neurosci. Res. **72:** 691–703.
7. WEI, Y.H., C.F. LEE, H.C. LEE, et al. 2001. Increases of mitochondrial mass and mitochondrial genome in association with enhanced oxidative stress in human cells harboring 4,977 BP-deleted mitochondrial DNA. Ann. N.Y. Acad. Sci. **928:** 97–112.

8. KULMS, D. & T. SCHWARZ. 2000. Molecular mechanisms of UV-induced apoptosis. Photodermatol. Photoimmunol. Photomed. **16:** 195–201.
9. KING, M.P. & G. ATTARDI. 1989. Human cells lacking mtDNA: repopulation with exogenous mitochondria by complementation. Science **246:** 500–503.
10. HERRMANN, M., H. M. LORENZ, R. VOLL, *et al.* 1994. A rapid and simple method for the isolation of apoptotic DNA fragments. Nucleic Acids Res. **22:** 5506–5507.
11. SHIAH, S.G., S.E. CHUANG, Y.P. CHAU, *et al.* 1999. Activation of c-Jun NH_2-terminal kinase and subsequent CPP32/Yama during topoisomerase inhibitor beta-lapachone-induced apoptosis through an oxidation-dependent pathway. Cancer Res. **59:** 391–398.
12. JIANG, S., J. CAI, D.C. WALLACE, *et al.* 1999. Cytochrome *c*-mediated apoptosis in cells lacking mitochondrial DNA. Signaling pathway involving release and caspase 3 activation is conserved. J. Biol. Chem. **274:** 29905–29911.
13. LEE, H.C., P.H. YIN, C.Y. LU, *et al.* 2000. Increase of mitochondria and mitochondrial DNA in response to oxidative stress in human cells. Biochem. J. **348:** 425–432.
14. SIKORSKA, M., H. BOROWY-BOROWSKI, B. ZURAKOWSKI, *et al.* 2003. Derivatised alpha-tocopherol as a CoQ_{10} carrier in a novel water-soluble formulation. Biofactors **18:** 173–183.
15. BOROWY-BOROWSKI, M., M. SIKORSKA & P.R. WALKER. 2004. Water-soluble composition of bioactive lipophilic compounds. U.S. Patents 6,045,826 and 6,191,172 B1.
16. MIRABELLA, M., S. DI GIOVANNI, G. SILVESTRI, *et al.* 2000. Apoptosis in mitochondrial encephalomyopathies with mitochondrial DNA mutations: a potential pathogenic mechanism. Brain **123:** 93–104.
17. LIU, C.Y., C.F. LEE, C.H. HONG, *et al.* 2004. Mitochondrial DNA mutation and depletion increase the susceptibility of human cells to apoptosis. Ann. N.Y. Acad. Sci. **1011:** 133–145.
18. BEAL, M.F. 2002. Coenzyme Q_{10} as a possible treatment for neurodegenerative diseases. Free Radical Res. **36:** 455–460.
19. LANGSJOEN, P.H. & A.M. LANGSJOEN. 1999. Overview of the use of CoQ_{10} in cardiovascular disease. Biofactors **9:** 273–284.
20. SOBREIRA, C., M. HIRANO, S. SHANSKE, *et al.* 1997. Mitochondrial encephalomyopathy with coenzyme Q_{10} deficiency. Neurology **48:** 1238–1243.
21. LIOU, C.W., C.C. HUANG, T.K. LIN, *et al.* 2000. Correction of pancreatic beta-cell dysfunction with coenzyme Q_{10} in a patient with mitochondrial encephalomyopathy, lactic acidosis and stroke-like episodes syndrome and diabetes mellitus. Eur. Neurol. **43:** 54–55.
22. MCCARTHY, S., M. SOMAYAJULU, M. SIKORSKA, *et al.* 2004. Paraquat induces oxidative stress and neuronal cell death: neuroprotection by water-soluble coenzyme Q_{10}. Toxicol. Appl. Pharmacol. **201:** 21–31.
23. CHUANG, Y.C., J.Y. CHAN, A.Y. CHANG, *et al.* 2003. Neuroprotective effects of coenzyme Q_{10} at rostral ventrolateral medulla against fatality during experimental endotoxemia in the rat. Shock **19:** 427–432.
24. TURUNEN, M., J. OLSSON & G. DALLNER. 2004. Metabolism and function of coenzyme Q. Biochim. Biophys. Acta **1660:** 171–199.

Antisense RNA to Inducible Nitric Oxide Synthase Reduces Cytokine-Mediated Brain Endothelial Cell Death

DING-I YANG,[a] SHAWEI CHEN,[b] UTHAYASHANKER R. EZEKIEL,[b] JAN XU,[b] YINGJI WU,[b] AND CHUNG Y. HSU[b,c]

[a]*Institute of Neuroscience, Tzu Chi University, Hualien, Taiwan*

[b]*Department of Neurology, Washington University School of Medicine, St. Louis, Missouri, USA*

[c]*Taipei Medical University, Taipei, Taiwan*

ABSTRACT: We test whether inhibition of inducible nitric oxide synthase (iNOS) can exert a cytoprotective effect on cerebral endothelial cells upon stimulation by pro-inflammatory cytokines. Mouse brain endothelial cells were stably transfected to express an antisense RNA against iNOS driven by an endothelium-specific von Willebrand factor (vWF) promoter. Upon stimulation with tumor necrosis factor-α (TNF-α) plus interferon-γ (IFN-γ), antisense transfectants showed less iNOS enzymatic activity with less nitric oxide (NO) when compared to the sense control cells. Correspondingly, the antisense cells showed a reduced LDH release and less cytosolic content of oligonucleosomes. These findings establish a cell-specific antisense strategy and confirm the cytotoxic role of iNOS expression in cultured cerebral endothelial cells.

KEYWORDS: endothelial cells; interferon-γ; tumor necrosis factor-α; von Willebrand factor

INTRODUCTION

Nitric oxide (NO), synthesized through the enzymatic conversion of L-arginine and molecular oxygen to L-citrulline by nitric oxide synthases (NOS),[1] serves a number of physiological functions. Among the three NOS isoforms that have been identified, inducible NOS (iNOS)[2] was first cloned from mouse macrophages and, unlike the other two constitutive NOS isoforms, expresses only under stimulation— mostly by inflammatory signals such as cytokines and/or endotoxin— to massively produce NO. NO produced by iNOS in a number of disease models has been demonstrated to exert either cytotoxic[3] or cytoprotective[4] effects, depending on the cell types and experimental paradigms selected.

In the central nervous system, NO synthesized by iNOS may contribute to secondary tissue damage during the inflammatory response that occurs hours to days

---

Address for correspondence: Chung Y. Hsu, M.D., Ph.D., Taipei Medical University, No. 250, Wu-Hsing Street, Taipei 110, Taiwan. Voice: +886-2-27361661-2016; fax: +886-2-23787795.
hsuc@tmu.edu.tw

following the primary insults, including brain ischemia and trauma. A major target of such an inflammatory reaction is the cerebral endothelial cell, which plays a pivotal role in sustaining blood flow and nutrient supply for brain tissues including neurons and glia. In addition to the expression of iNOS in response to inflammatory signals, cerebral endothelial cells also constitutively express eNOS,[5] which could also release NO to function as a vasodilator and which may be beneficial in maintaining optimal cerebral blood flow.[6] Therefore, selective inhibition of iNOS sparing eNOS is an attractive approach to fully delineate the cytotoxic versus cytoprotective potential of NO. Herein we used a brain endothelial (Bend) cell line as a model system in which iNOS can be induced by synergistic exposure to pro-inflammatory cytokines.[7] Our findings suggest that reduction in NO synthesis by antisense RNA to the iNOS gene protects these cells against cytokine-mediated injuries.

MATERIALS AND METHODS

Plasmid Construction

Based on the published cDNA sequence of mouse macrophage iNOS (Genebank accession number M92649), two sets of primers were designed to amplify two 145 bp fragments from mouse genomic DNA via polymerase chain reaction (PCR), one complementary (forward primer: 5'-CGGGATCCCTCCGTGGAGTGAACAAGA-3'; reverse primer: 5'-GGGGTACCTTTACAGGGAGTTGAAGAC-3') and one homologous (forward primer: 5'-CGGGATCCTTTACAGGGAGTTGAAGAC-3'; reverse primer: 5'-GGGGTACCCTCCGTGGAGTGAACAAGA-3') to the sequence between 45 bp and 189 bp of the murine iNOS transcript. These two DNA fragments derived from PCR represent the antisense (complementary) and sense (homologous) sequences covering a portion of the 5'-untranslated region 66 bp upstream of ATG initiation codon. Recognition sequences of restriction endonucleases BamH I and Kpn I were incorporated into the 5'-ends of forward and reverse primers, respectively, in both primer pairs for efficient subcloning into expression vector. The conditions used to amplify iNOS antisense/sense fragments were as follows: the preincubation was at 94°C for 2 min, reaction mixtures were cycled 30 times at 50°C for 1 min (annealing), 72°C for 1 min (extension), and 94°C for 40 s (denaturation), with a final extension of 10 min at 72°C.

To allow the expression of iNOS antisense/sense constructs exclusively in endothelial cells, an endothelial cell–specific expression vector was constructed. This was achieved by replacing the cytomegalovirus (CMV) immediate-early promoter in pcDNA3.1 (Invitrogen, Carlsbad, CA) with the promoter of von Willebrand factor (vWF). A primer pair (forward primer: 5'-GAAGATCTTTAGCCGATCCATTCAACCC-3'; reverse primer: 5'-CTAGCTAGCATACCTTCCCCTGCAAATGAG-3') was synthesized to generate a 750-bp vWF promoter out of mouse genomic DNA. Recognition sequences of restriction enzymes Bgl II and Nhe I were included in the 5'-ends of forward and reverse primers, respectively. PCR was performed using 1 µg mouse genomic DNA with reactions carried out for 30 cycles of 52°C annealing for 1 min, 72°C extension for 1 min, and 94°C denaturation for 40 s. This was followed by an extension of 10 min at 72°C. The amplified mouse vWF promoter was then subcloned into Bgl II/Nhe I sites of pcDNA3.1 vector previously digested with Bgl II/Nhe I to remove the CMV promoter.

Southern and Western Blotting

Southern blotting was conducted according to standard protocols using the DIG Labeling/Detection Kit (Boehringer Mannheim, Indianapolis, IN). To detect the iNOS protein by Western blotting, primary antibody against mouse macrophage iNOS (Transduction Lab., Lexington, KY) was applied at 1:500 dilution with a working concentration of 0.5 µg/mL. The anti-rabbit IgG secondary antibody (NA934, Amersham Life Science Inc., Arlington Heights, IL) was used at 1:7500 dilution. The immunoreactive components were then detected with ECL Western blotting detection reagent (Amersham Life Science Inc.) and exposed to Kodak XRP-1 films. To detect eNOS, primary antibody against mouse eNOS (Transduction Lab.) was applied at 1:1000 dilution at a final concentration of 0.25 µg/mL. The anti-rabbit IgG secondary antibody (NA934, Amersham Life Science Inc.) was used at 1:5000 dilution.

Statistical Analysis

Results are expressed as means ± SD. Statistical analysis was performed using Student's unpaired t-test between two experimental groups (e.g., sense vs. antisense). Multiple groups were analyzed by one-way analysis of variance followed by a post-hoc Bonferroni t-test. A P value of less than 0.05 was considered significant.

RESULTS AND DISCUSSION

Exposure of Bend cells to a combination of TNF-α and IFN-γ resulted in a marked increase in the accumulation of nitrite in the medium as compared to those cells treated with serum-free medium only (FIG. 1A). Neither cytokine alone was capable of inducing NO production in Bend cells (data not shown). Additional NOS enzyme activity assay indicates that the accumulation of nitrite in the transfectants following cytokine induction was most likely due to increased expression of iNOS proteins, but not constitutive eNOS (data not shown). Exposure of Bend cells to cytokines led to cell death that was apparent after 24 h based on the LDH assay (FIG. 1B) and persisted until at least 48 h by MTT reduction assay (FIG. 1C).

To establish the causal relationship between TNF-α/IFN-γ–induced iNOS expression and its cytotoxicity toward brain endothelial cells, an antisense approach was adopted. A recombinant plasmid was constructed to drive the expression of an iNOS antisense RNA by vWF promoter to constitutively inhibit iNOS translation. In pilot experiments, transient transfection with a reporter gene construct (LacZ) driven by this vWF promoter has confirmed that this endothelial-specific expression vector was indeed functional in Bend cells (data not shown). After stable transfection with iNOS antisense construct, one transfectant, designated as "antisense," was isolated for further characterization. In addition, another colony of Bend cells transfected with corresponding iNOS sense construct was also isolated (designated as "sense") to serve as a negative control. Genomic DNA isolated from both antisense and sense transfectants, as well as that from parental Bend cells, were subjected to Southern

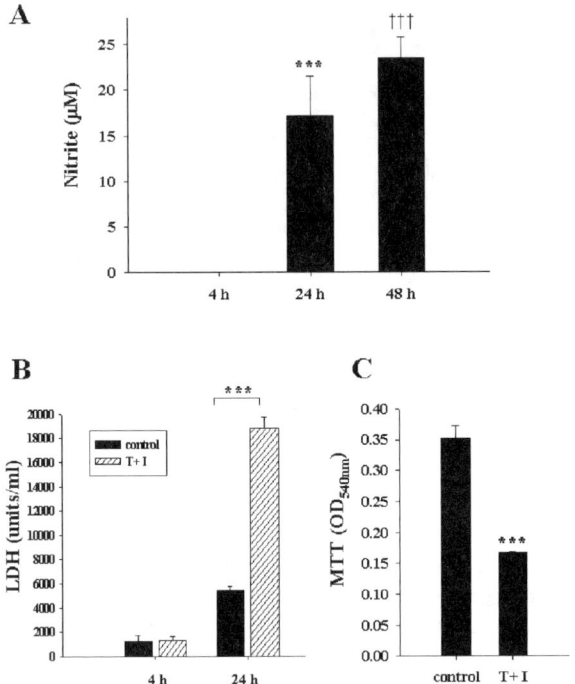

FIGURE 1. Increase in nitrite content and cell death induced by TNF-α/IFN-γ in Bend cells. (**A**) The nitrite contents in the medium from parental Bend cells following TNF-α (20 ng/mL)/IFN-γ (400 U/mL) treatment for 4, 24, and 48 h are shown. ***$P < 0.001$ compared to 4 h; †††, $P < 0.001$ compared to 4 h. (**B**) The extent of cell death based on LDH released in the parental Bend cells with (T + I) and without (control) TNF-α/IFN-γ. ***$P < 0.001$. (**C**) The extent of cell death based on the MTT assay with (T + I) or without (control) TNF-α/IFN-γ treatment for 48 h. Higher OD_{540nm} absorbency on the *ordinate* indicates higher survival rates. ***$P < 0.001$. Mean ± SD in triplicate. Data shown are representative of three separate experiments with similar results.

blotting, using a 2.2-kb fragment containing the ampicillin resistance (Amp^r) gene as a probe, to ensure the stable integration of transgene sequences into the genome of transfectants. The left panel of FIGURE 2A shows the Hinc II and Sty I digestion patterns of genomic DNA isolated from both transfectants and parental Bend cells. The same gel was then subjected to Southern hybridization (right panel, FIG. 2A). A DIG-labeled probe only recognized the bands in genomic DNA isolated from antisense and sense cells, but not that from parental Bend. Western blotting demonstrated that, under identical experimental conditions with equal amounts of proteins (20 µg) loaded in each lane, antisense transfectants expressed substantially fewer iNOS proteins compared to sense controls (FIG. 2B). Densitometry scanning of the blot revealed approximately 60 to 70% inhibition of iNOS protein expression in antisense cells compared to sense or parental Bend cells. The blockade of iNOS expression in antisense cells is significant yet incomplete because an increase in iNOS

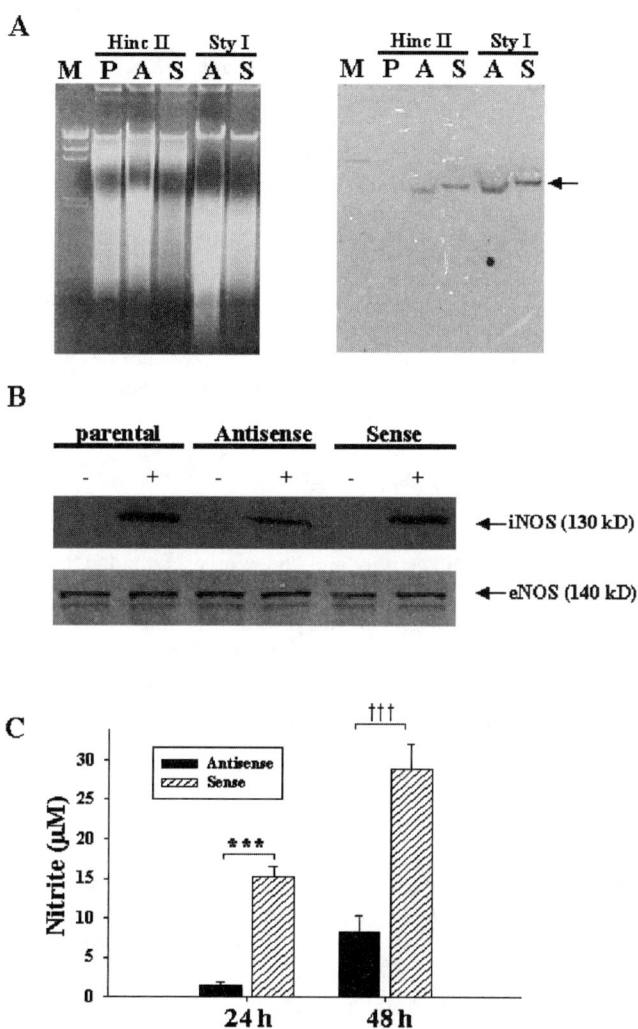

FIGURE 2. Characterizations of iNOS antisense and sense transfectants. (**A**) *(Left panel)* Agarose gel electrophoresis of genomic DNA digested with Hinc II and Sty I. *(Right panel)* The same gel blotted to a nylon membrane and probed with a DIG-labeled Amp^r sequence. M, molecular weight markers of λ DNA digested with Hind III; P, genomic DNA of parental Bend cells; A, genomic DNA of antisense transfectants; S, genomic DNA of sense transfectants. The *arrow* indicates the hybridization signal. (**B**) Western blots of iNOS and eNOS with (+) or without (−) TNF-α (20 ng/mL)/IFN-γ (400 U/mL) treatment for 24 h. (**C**) The nitrite content in the medium from transfectants treated with TNF-α/IFN-γ for 24 and 48 h. ***$P < 0.001$; †††, $P < 0.001$. Results shown in (**C**) are representative of three separate experiments. Mean ± SD in triplicate.

expression could still be detected after TNF-α/IFN-γ treatment. As shown in the lower panel of FIGURE 2B, the eNOS level, however, did not show significant changes either in antisense or sense transfectants exposed to cytokines. Moreover, the eNOS levels between uninduced antisense and sense transfectants appear similar, indicative of a specific inhibitory effect of the antisense RNA against endogenous iNOS gene without affecting eNOS expression. Results in FIGURE 2C demonstrate that nitric oxide production, as reflected by the accumulation of nitrite contents in the medium, was partially suppressed in the stable transfectants of iNOS antisense construct upon cytokine induction, as compared to those transfected with sense iNOS construct. At the end of 24-h treatment with TNF-α/IFN-γ antisense cells produced significantly less nitrite than sense controls (antisense: 1.49 ± 0.37 μM; sense: 15.24 ± 1.32 μM, $P < 0.001$). After 48 h, culture media from antisense cells contained 8.31 ± 1.99 μM nitrite whereas those from sense contained 28.95 ± 3.07 μM ($P < 0.001$). In addition to the Griess reaction, NOS activity assay was performed with cytosolic extracts from transfectants following exposure to cytokines for 24 h. We found antisense cells did contain less calcium-independent NOS activity, presumably derived from iNOS, than that from sense controls (data not shown). In contrast, neither antisense nor sense transfectants showed significant changes in calcium-dependent NOS activities (presumably from eNOS) upon stimulation with cytokines (data not shown). Altogether, results in FIGURE 2 demonstrate that stable integration and overexpression of this 145-bp antisense gene is capable of suppressing, at least in part, iNOS enzymatic activity and NO synthesis upon cytokine induction.

We examined whether antisense suppression of iNOS expression may confer cytoprotection to Bend cells challenged with cytokines. A quantitative indicator of cell death is the oligonucleosomes released into cytosol of apoptotic cells.[8] Therefore, a sandwich ELISA kit was employed to quantitatively measure the contents of cytosolic oligonucleosomes in antisense and sense transfectants following cytokine exposure. As shown in FIGURE 3A, the relative oligonucleosome content in sense transfectants is much higher than those in the antisense cell line, although both cell lines had comparable background levels of cell death without cytokine induction (1.09 ± 0.17 in antisense vs. 0.98 ± 0.20 in sense, $P = 0.4331$). This observation implies that excessive nitric oxide produced from iNOS as a result of cytokine induction may be detrimental to Bend cells. Antisense cells also demonstrated a relatively higher survival rate than did sense cells following cytokine stimulation based on LDH (FIG. 3B) and MTT (FIG. 3C) assays. We chose PI-staining of dead cells as a morphological approach to further analyze cell death (FIG. 3D). Under fluorescent microscope, very few cells were stained with PI in the antisense (upper left) and sense (lower left) transfectants without cytokines. With cytokine treatment, antisense cells (upper right) maintained a much higher cell viability as compared to sense cells (lower right) that show intense nuclear PI-staining, indicative of extensive cell loss.

Despite the constitutive expression of antisense RNA against iNOS, the antisense transfectants still expressed a small pool of iNOS proteins upon exposure to TNF-α/IFN-γ (FIG. 2B). However, TNF-α/IFN-γ did not cause an additional increase in cellular NO contents based on measurement of medium nitrite accumulation (FIG. 2C). The discrepancy between the extent of iNOS protein expression and medium nitrite accumulation remains unknown. Notably, expression of iNOS proteins may not necessarily cause massive NO production with resultant nitrite accumulation. Changes in intracellular NO contents also depend on other factors. These include substrate availability, enzymatic activities of iNOS proteins, expression levels and/or enzymatic activities of other constitutive NOS isoforms. Interestingly, iNOS activity is subject to positive feedback regulation by NO. For example, pretreatment of rat hepatocytes with S-nitroso-N-acetyl penicillamine (SNAP), an NO donor, enhanced

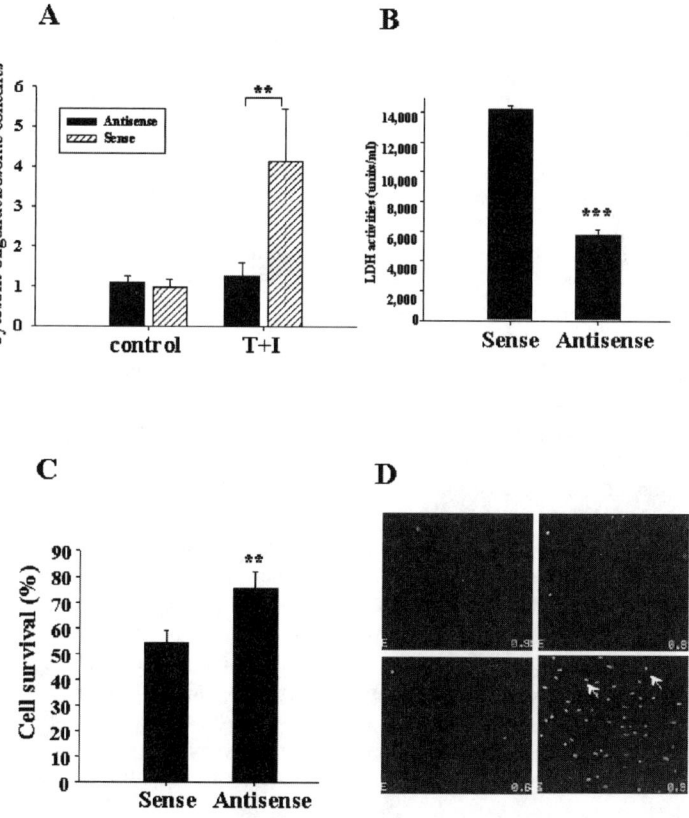

FIGURE 3. The extent of cell death in antisense and sense transfectants following cytokine exposure. (**A**) The DNA fragmentation ELISA assay with (T + I) or without (control) TNF-α (20 ng/mL)/IFN-γ (400 U/mL) treatment for 24 h. The results are expressed as relative cytosolic oligonucleosome contents in arbitrary units with serially diluted apoptotic HL60 cell lysate as reference standards. A higher value on the ordinate reflects a greater extent of cell death. **$P < 0.01$. (**B**) The LDH assay following cytokine treatment for 24 h. ***$P < 0.001$. (**C**) The MTT assay following cytokine treatment for 48 h. ** $P < 0.01$. Mean ± SD in triplicate. Data shown in **A**, **B**, and **C** are representatives of at least three separate experiments with similar results. (**D**) *(Left panels)* The fluorescent images of the antisense *(upper)* and sense *(lower)* transfectants without cytokine treatment. *(Right panels)* The images of antisense *(upper)* and sense *(lower)* transfectants with cytokine treatment for 24 h. *Arrows* indicate the dead or dying cells with propidium iodide stain.

NO production by promoting iNOS dimerization without affecting expression level of iNOS mRNA or protein.[9] Partial suppression of iNOS expression may therefore abolish this potentiation effect of NO, rendering the remaining small portion of iNOS proteins incapable of massive NO production. This may in part explain the observation that, despite the presence of remaining iNOS proteins as detected by Western blotting (FIG. 2B), antisense RNA against iNOS effectively conferred cytoprotection to Bend cells upon exposure to TNF-α plus IFN-γ, based on LDH release assay (FIG. 3B). Another possibility is that the small amount of iNOS that remained in antisense transfectants, as well as the endogenous eNOS, contributed to the basal levels of NO that may be beneficial against the insults. NO at physiological concentrations is known to protect the human monocytic U937 cells against TNF-α–induced apoptosis by reducing the generation of ceramide.[10] Antisense suppression of iNOS with resultant reduction of excessive NO formation may therefore modulate cellular NO content to a lower level that was beneficial to Bend cells, causing cytoprotective effects against inflammatory cytokines such as TNF-α and IFN-γ.

Previously, we have successfully suppressed iNOS mRNA and protein expression induced by TNF-α/IFN-γ in Bend cells using a NF-κB trap.[7] This strategy entails exogenous introduction of a hairpin oligonucleotide decoy with the consensus sequence for NF-κB binding site. However, NF-κB is a potent transactivator for the expression of a number of other genes besides iNOS; downregulation of NF-κB thus appears to be a nonspecific—hence, a less ideal—approach to elucidate the cytoprotective or cytotoxic effect of iNOS.[11] In order to circumvent these problems, we transfected the murine brain endothelial cells with recombinant constructs that constitutively express the antisense RNA to iNOS gene driven by the promoter of endothelium-specific vWF. This 750-bp vWF promoter has previously been demonstrated to target the expression of LacZ to a subpopulation of endothelial cells in yolk sac and adult brain *in vivo*.[12] Therefore, this promoter may be especially valuable in gene therapy for treating cerebral disorders that involve endothelial cells. This vWF promoter has also been engineered to confer tissue-specific expression of a suicidal gene in human umbilical vascular endothelial cells as an anti-angiogenesis approach for cancer treatment.[13] Another attractive approach recently available for gene knockdown is the small interference RNA (siRNA). The endothelium-specific promoter may also be applied to drive the expression of siRNA *in vivo* for clinical use in the near future. Overall, results reported here and by others lay the foundation for future *in vivo* studies, using alternative gene delivery techniques such as retroviral or adenoviral vector, to suppress the iNOS activity in a cell-specific manner.

ACKNOWLEDGMENTS

This work was supported by a Tzu Chi University Research Grant (TCMRC93119A-01) and a National Science Council grant (NSC93-2314-B-320-002) to D.-IY. and by Washington University School of Medicine–Pharmacia Biomedical Research Support Program, Lowell B. Miller Memorial Research Grant (National Brain Tumor Foundation), and a National Science Council grant (NSC92-2314-B-038-030) to C.Y.H.

REFERENCES

1. PALMER, R.M. et al. 1987. Nitric oxide release accounts for the biological activity of endothelium-derived relaxing factor. Nature **327:** 524–526.
2. XIE, Q.W. et al. 1992. Cloning and characterization of inducible nitric oxide synthase from mouse macrophages. Science **256:** 225–228.
3. HSU, C.Y. et al. 1997. Enhancement of apoptosis in cerebral endothelial cells by selected inflammatory signals. Ann. N.Y. Acad. Sci. **843:** 148–153.
4. RAUHALA, P. et al. 1998. Neuroprotection by S-nitrosoglutathione of brain dopamine neurons from oxidative stress. FASEB J. **12:** 165–173.
5. TOMIMOTO, H. et al. 1994. Distribution of nitric oxide synthase in the human cerebral blood vessels and brain tissues. J. Cereb. Blood Flow Metab. **14:** 930–938.
6. HUANG, P.L. et al. 1995. Hypertension in mice lacking the gene for endothelial nitric oxide synthase. Nature **377:** 239–242.
7. XU, J. et al. 1997. Regulation of cytokine-induced iNOS expression by a hairpin oligonucleotide in murine cerebral endothelial cells. Biochem. Biophys. Res. Commun. **235:** 394–397.
8. XU, J. et al. 1998. Involvement of de novo ceramide biosynthesis in tumor necrosis factor-α/cycloheximide-induced cerebral endothelial cell death. J. Biol. Chem. **273:** 16521–16526.
9. PARK, J.H. et al. 2002. Nitric oxide (NO) pretreatment increases cytokine-induced NO production in cultured rat hepatocytes by suppressing GTP cyclohydrolase I feedback inhibitory protein level and promoting inducible NO synthase dimerization. J. Biol. Chem. **277:** 47073–47079.
10. DE NADAI, C. et al. 2000. Nitric oxide inhibits tumor necrosis factor-alpha-induced apoptosis by reducing the generation of ceramide. Proc. Natl. Acad. Sci. USA **97:** 5480–5485.
11. WRIGHTON, C.J. et al. 1996. Inhibition of endothelial cell activation by adenovirus-mediated expression of I kappa B alpha, an inhibitor of the transcription factor NF-kappa B. J. Exp. Med. **183:** 1013–1322.
12. AIRD, W.C. et al. 1995. Human von Willebrand factor gene sequences target expression to a subpopulation of endothelial cells in transgenic mice. Proc. Natl. Acad. Sci. USA **92:** 4567–4571.
13. OZAKI, K. et al. 1996. Use of von Willebrand factor promoter to transduce suicidal gene to human endothelial cells, HUVEC. Hum. Gene Ther. **7:** 1483–1490.

2,6-Diisopropylphenol Protects Osteoblasts from Oxidative Stress-Induced Apoptosis through Suppression of Caspase-3 Activation

RUEI-MING CHEN,[a,b,f] GONG-JHE WU,[c,f] HWA-CHIA CHANG,[b] JUE-TAI CHEN,[b] TZENG-FU CHEN,[d] YI-LING LIN,[a] AND TA-LIANG CHEN[e]

[a]*Graduate Institute of Medical Sciences, College of Medicine, Taipei Medical University, Taipei, Taiwan*

[b]*Department of Anesthesiology, Wan-Fang Hospital, College of Medicine, Taipei Medical University, Taipei, Taiwan*

[c]*Department of Anesthesiology, Shin Kong Wu Ho-Su Memorial Hospital, Taipei, Taiwan*

[d]*Department of Pharmacology, School of Medicine, College of Medicine, Taipei Medical University, Taipei, Taiwan*

[e]*Taipei City Hospital and Taipei Medical University, Taipei, Taiwan*

ABSTRACT: 2,6-Diisopropylphenol is an intravenous anesthetic agent used for induction and maintenance of anesthesia. Since it is similar to α-tocopherol, 2,6-diisopropylphenol may have antioxidant effects. Osteoblasts play important roles in bone remodeling. In this study, we attempted to evaluate the protective effects of 2,6-diisopropylphenol on oxidative stress-induced osteoblast insults and their possible mechanisms, using neonatal rat calvarial osteoblasts as the experimental model. Clinically relevant concentrations of 2,6-diisopropylphenol (3 and 30 μM) had no effect on osteoblast viability. However, 2,6-diisopropylphenol at 300 μM time-dependently caused osteoblast death. Exposure to sodium nitroprusside (SNP), a nitric oxide donor, increased amounts of nitrite in osteoblasts. 2,6-Diisopropylphenol did not scavenge basal or SNP-releasing nitric oxide. Hydrogen peroxide (HP) enhanced levels of intracellular reactive oxygen species in osteoblasts. 2,6-Diisopropylphenol significantly reduced HP-induced oxidative stress. Exposure of osteoblasts to SNP and HP decreased cell viability time-dependently. 2,6-Diisopropylphenol protected osteoblasts from SNP- and HP-induced cell damage. Analysis by a flow cytometric method revealed that SNP and HP induced osteoblast apoptosis. 2,6-Diisopropylphenol significantly blocked SNP- and HP-induced osteoblast apoptosis. Administration of SNP and HP increased caspase-3 activities. However, 2,6-diisopropylphenol significantly decreased SNP- and HP-enhanced caspase-3 activities. This study shows that a therapeutic concentration of 2,6-diisopropylphenol can protect osteoblasts from SNP- and HP-induced cell insults, possibly via suppression of caspase-3 activities.

[f]R-M.C. and G-J.W. contributed equally to this paper.

Address for correspondence: Professor Ta-Liang Chen, Department of Anesthesiology, Wan-Fang Hospital, or Ruei-Ming Chen, Graduate Institute of Medical Sciences, College of Medicine, Taipei Medical University, No. 111, Hsing-Lung Rd., Sec. 3, Taipei 116, Taiwan. Voice: +886-2-29307930 ext. 2159; fax: +886-2-86621119.

rmchen@tmu.edu.tw or tlc@tmu.edu.tw

Ann. N.Y. Acad. Sci. 1042: 448–459 (2005). © 2005 New York Academy of Sciences.
doi: 10.1196/annals.1338.038

KEYWORDS: 2,6-diisopropylphenol; osteoblasts; nitric oxide; hydrogen peroxide; apoptosis; caspase-3

INTRODUCTION

2,6-Diisopropylphenol, an intravenous anesthetic agent, has the advantages of rapid onset, a short duration of action, and rapid elimination.[1] This intravenous anesthetic agent is widely used for induction and maintenance of anesthesia in surgical procedures. Structurally, 2,6-diisopropylphenol is similar to α-tocopherol with a phenol group so it has been implicated as having antioxidant roles.[2] Previous studies reported that 2,6-diisopropylphenol can protect thymocytes and erythrocytes from peroxinitrite- and 2,2′-azo-bis(2-amidinopropane) dihydrochloride–induced apoptosis.[3,4] Studies in human plasma and rat liver mitochondria revealed that 2,6-diisopropylphenol has antioxidant effects against lipid peroxidation.[2,5] Our previous study further demonstrated that 2,6-diisopropylphenol can protect macrophages from nitric oxide (NO)-induced damage.[6]

2,6-Diisopropylphenol is also used as an inducing or maintaining agent for patients undergoing orthopedic surgery.[7] Osteoblasts play critical roles in bone remodeling.[8] Reactive oxygen species (ROS) are one of the important factors that contribute to osteoblast metabolism.[8] NO and hydrogen peroxide (HP) are two typical ROS. In response to stimulation by inflammatory cytokines, osteoblasts synthesize massive amounts of NO and HP.[9,10] Elevation of NO and HP leads to cell injury.[11,12] Our previous study showed that NO can induce osteoblast apoptosis.[12] Caspase-3 is a proteolytic enzyme.[13] Activation of caspase-3 can drive cells to undergo apoptosis. In this study, we attempted to verify the protective effects of propofol on NO- and HP-induced osteoblast insults and their possible mechanisms.

MATERIALS AND METHODS

Cell Culture and Drug Treatment

Osteoblasts were prepared from 3-day-old Wistar rat calvaria following an enzymatic digestion method described previously.[14] Osteoblasts were seeded in Dulbecco's modified Eagle's medium (Gibco-BRL, Grand Island, NY) supplemented with 10% heat-inactivated fetal bovine serum, 100 U/mL penicillin, and 100 mg/mL streptomycin in tissue culture flasks at 37°C in a humidified atmosphere of 5% CO_2.

2,6-Diisopropylphenol was purchased from Aldrich (Milwaukee, WI). 2,6-Diisopropylphenol was stored under nitrogen, protected from light, and freshly prepared by dissolving it in dimethyl sulfoxide (DMSO) for each independent experiment. DMSO in the medium was kept to less than 0.1% to avoid the toxicity of this solvent to osteoblasts. Sodium nitroprusside (SNP) and HP were purchased from Sigma (St. Louis, MO). SNP and HP were freshly prepared by dissolving them in phosphate-buffered saline (0.14 M NaCl, 2.6 mM KCl, 8 mM Na_2HPO_4, and 1.5 mM KH_2PO_4) for each independent experiment. Osteoblasts were exposed to various concentrations of 2,6-diisopropylphenol, SNP, and HP for different time intervals.

Assay of Cell Viability

Cell viability was determined by a colorimetric 3-(4,5-dimethylthiazol-2-yl)-2,5-diphenyltetrazolium bromide (MTT) assay as described previously.[15] In brief, osteoblasts (2×10^4) were seeded in 96-well tissue culture plates overnight. After drug treatment, osteoblasts were cultured with new medium containing 0.5 mg/mL MTT for 3 more hours. The blue formazan products in osteoblasts were dissolved in DMSO and spectrophotometrically measured at a wavelength of 570 nm.

Assay of Nitrite Production

After drug treatment, amounts of nitrite, an oxidative product of NO, in the culture medium of osteoblasts were detected following a procedure described in a technical bulletin of Promega's Griess Reagent System (Promega, Madison, WI).

Quantification of Intracellular Reactive Oxygen Species

Intracellular levels of ROS were quantified following a method as described previously.[16] In brief, osteoblasts (5×10^5) were cultured in 12-well tissue culture plates overnight. 2′,7′-Dichlorofluorescin diacetate, an ROS-sensitive dye, was co-treated with 2,6-diisopropylphenol or HP for different time intervals. Osteoblasts were harvested and suspended in phosphate-buffered saline. Fluorescence intensities of osteoblasts were quantified using a flow cytometer (FACS Calibur, Becton Dickinson, San Jose, CA).

Analysis of Apoptotic Cells

Apoptotic osteoblasts were determined using propidium iodide to detect DNA fragments in nuclei according to a method as described previously.[17] After drug administration, osteoblasts were harvested and fixed in cold 80% ethanol. Following centrifugation and washing, fixed cells were stained with propidium iodide and analyzed by a FACScan flow-cytometer (Becton Dickinson) on the basis of 560-nm dichroic mirror and 600-nm band-pass filter.

Assay of Caspase-3 Activity

Activity of caspase-3 was determined by a fluorogenic substrate assay. In brief, cell extracts were prepared by lysing osteoblasts in a buffer containing 1% Nonidet P-40, 200 mM NaCl, 20 mM Tris/HCl, pH 7.4, 10 mg/mL leupeptin, 0.27 U/mL aprotinin, and 100 mM PMSF. Caspase-3 activity is determined by incubating cell lysates (25 mg total protein) with 50 mM fluorogenic substrate in a 200 mL cell-free system buffer comprising 10 mM Hepes, pH 7.4, 220 mM mannitol, 68 mM sucrose, 2 mM NaCl, 2.5 mM KH_2PO_4, 0.5 mM EGTA, 2 mM $MgCl_2$, 5 mM pyruvate, 0.1 mM PMSF, and 1 mM dithiothreitol. Intensities of fluorescent products in cells were measured by a spectrofluorometer.

Statistical Analysis

Statistical differences between the drug-treated groups were considered significant when the P value of Duncan's multiple-range test was less than 0.05. Statistical

TABLE 1. Concentration- and time-dependent effects of 2,6-diisopropylphenol (DIP) on osteoblast viability

DIP (µM)	Cell viability (O.D. values at 570 nm)		
	1 h	6 h	24 h
0	0.724 ± 0.083	0.768 ± 0.062	0.868 ± 0.072
3	0.760 ± 0.079	0.783 ± 0.081	0.842 ± 0.071
30	0.746 ± 0.086	0.760 ± 0.079	0.885 ± 0.063
300	0.717 ± 0.068	0.614 ± 0.063*	0.564 ± 0.065*

Each value represents the mean ± SEM for $n = 9$. *Values significantly differ from the respective control, $P < 0.05$.

TABLE 2. Effects of 2,6-diisopropylphenol (DIP) and sodium nitroprusside (SNP) on nitrite production

Treatment	Nitrite (µM)		
	1 h	6 h	24 h
Control	5.3 ± 1.1	4.2 ± 0.6	4.6 ± 0.6
DIP	4.7 ± 0.7	5.0 ± 0.9	3.9 ± 0.8
SNP	28.1 ± 5.6*	40.6 ± 7.1*	56.5 ± 8.2*
DIP + SNP	26.8 ± 6.4*	31.3 ± 6.9*	35.2 ± 6.6*

Each value represents the mean ± SEM for $n = 6$. *Values significantly differ from the respective control, $P < 0.05$.

analysis between the drug-treated groups over time was carried out by two-way ANOVA.

RESULTS

Treatment with 3 and 30 µM 2,6-diisopropylphenol for 1, 6, and 24 h was not cytotoxic to osteoblasts (TABLE 1). Osteoblast viability was not affected following exposure to 300 µM 2,6-diisopropylphenol for 1 h. However, when the administered time intervals reached 6 and 24 h, 2,6-diisopropylphenol significantly decreased osteoblast viability by 20% and 35%, respectively (TABLE 1).

Administration of 2 mM SNP in osteoblasts for 1, 6, and 24 h enhanced amounts of nitrite in the culture medium by 5.3-, 9.7-, and 12.3-fold, respectively (TABLE 2). Exposure to 30 µM 2,6-diisopropylphenol for 1, 6, and 24 h did not affect nitrite production. Co-treatment of osteoblasts with 2,6-diisopropylphenol and SNP did not influence the amounts of nitrite in the culture medium.

Treatment of osteoblasts with 25 µM HP for 1, 6, and 24 h significantly increased intracellular levels of ROS by 4.5-, 4.4-, and 3.4-fold, respectively (TABLE 3). Administration of 2,6-diisopropylphenol did not affect intracellular ROS levels. Co-treatment with 2,6-diisopropylphenol and HP for 1 and 6 h significantly reduced HP-

TABLE 3. Effects of 2,6-diisopropylphenol (DIP) and hydrogen peroxide (HP) on levels of intracellular reactive oxygen species (ROS)

Treatment	ROS, fluorescence intensities		
	1 h	6 h	24 h
Control	16 ± 3	19 ± 5	15 ± 2
DIP	13 ± 2	4 ± 3	2 ± 1
HP	72 ± 16*	82 ± 17*	51 ± 10*
DIP + HP	37 ± 11*†	51 ± 9*†	43 ± 10*

Each value represents the mean ± SEM for $n = 6$. *Values significantly differ from the respective control, $P < 0.05$. †Values significantly differ from the HP-treated groups, $P < 0.05$.

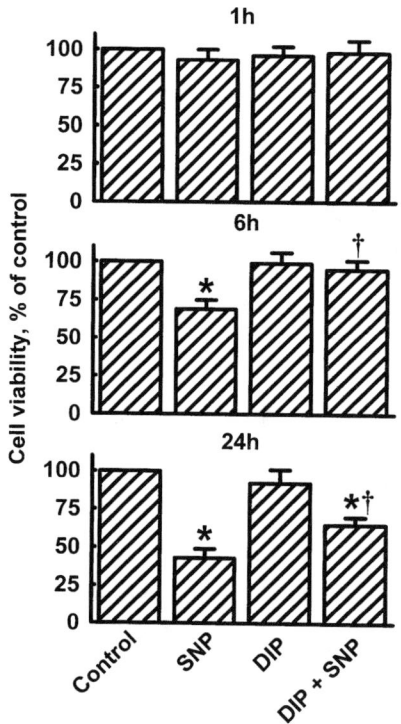

FIGURE 1. Protective effects of 2,6-diisopropylphenol (DIP) on sodium nitroprusside (SNP)-induced osteoblast death. Osteoblasts prepared from neonatal rat calvaria were exposed to 30 μM DIP, 2 mM SNP, and a combination of DIP and SNP. Cell viability was analyzed by the MTT assay. Each value represents the mean ± SEM for $n = 12$. *Values significantly differ from the respective control, $P < 0.05$. †Values significantly differ from the SNP-treated groups, $P < 0.05$.

induced intracellular ROS production by 49% and 38%, respectively. Administration of 2,6-diisopropylphenol and HP for 24 h had no effect on HP-induced intracellular ROS production (TABLE 3).

Administration of osteoblasts with SNP for 1 h was not cytotoxic to osteoblasts (FIG. 1). Viability of osteoblasts was respectively reduced by 31% and 57% following administration of SNP for 6 and 24 h. Exposure to 2,6-diisopropylphenol for 1, 6, and 24 h did not affect cell viability. Co-treatment with 2,6-diisopropylphenol and SNP for 6 h completely ameliorated SNP-induced cell death. In 24-h-treated osteo-

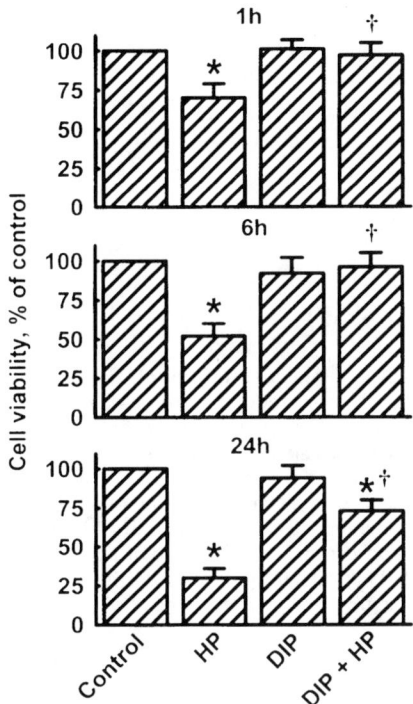

FIGURE 2. Protective effects of 2,6-diisopropylphenol (DIP) on hydrogen peroxide (HP)-induced osteoblast death. Osteoblasts prepared from neonatal rat calvaria were exposed to 30 μM DIP, 25 μM HP, and a combination of DIP and HP. Cell viability was analyzed by the MTT assay. Each value represents the mean ± SEM for $n = 12$. *Values significantly differ from the respective control, $P < 0.05$. †Values significantly differ from the SNP-treated groups, $P < 0.05$.

blasts, 2,6-diisopropylphenol caused a partial 22% reduction in SNP-induced cell death (FIG. 1).

Exposure of osteoblasts to HP for 1, 6, and 24 h significantly decreased osteoblast viability by 30%, 52%, and 70%, respectively (FIG. 2). 2,6-Diisopropylphenol did not affect cell viability. Co-treatment with 2,6-diisopropylphenol and HP for 1 and 6 h completely ameliorated HP-induced osteoblast death. Administration of 2,6-diisopropylphenol led to a 43% decrease in HP-induced cell death (FIG. 2).

Analysis by a flow cytometric method revealed that administration of SNP to osteoblasts for 1 h did not induce cell apoptosis (FIG. 3). When the administered time intervals reached 6 and 24 h, SNP significantly induced osteoblast apoptosis by 25% and 40%, respectively. Exposure to 2,6-diisopropylphenol had no effect on osteoblast apoptosis. Co-treatment with 2,6-diisopropylphenol and SNP for 6 and 24 h significantly blocked SNP-induced osteoblast apoptosis by 40% and 35%, respectively (FIG. 3).

Administration of HP to osteoblasts for 1, 6, and 24 h significantly induced osteoblast apoptosis by 15%, 35%, and 56%, respectively (FIG. 4). 2,6-Diisopropylphenol did not influence cell apoptosis. Co-treatment with 2,6-diisopropylphenol and HP, respectively, decreased HP-induced osteoblast apoptosis by 11%, 31%, and 25% (FIG. 4).

In untreated osteoblasts, low levels of caspase-3 activities were detected (FIG. 5). Administration of 2,6-diisopropylphenol had no effect on caspase-3 activity. Expo-

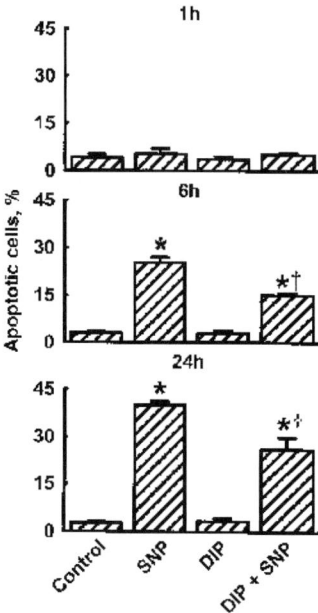

FIGURE 3. Protective effects of 2,6-diisopropylphenol (DIP) on sodium nitroprusside (SNP)-induced osteoblast apoptosis. Osteoblasts prepared from neonatal rat calvaria were exposed to 30 μM DIP, 2 mM SNP, and a combination of DIP and SNP. Apoptotic cells were determined using flow cytometry. Each value represents the mean ± SEM for $n = 6$. *Values significantly differ from the respective control, $P < 0.05$. †Values significantly differ from the SNP-treated groups, $P < 0.05$.

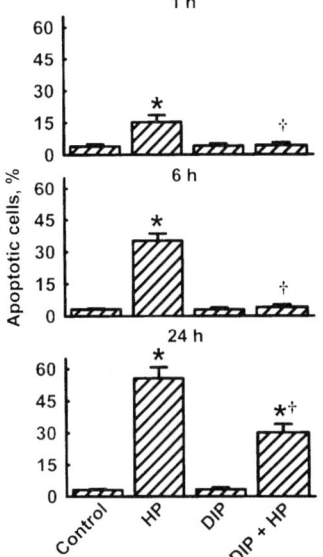

FIGURE 4. Protective effects of 2,6-diisopropylphenol (DIP) on hydrogen peroxide (H_2O_2)-induced osteoblast apoptosis. Osteoblasts prepared from neonatal rat calvaria were exposed to 30 μM DIP, 25 μM HP, and a combination of DIP and HP. Apoptotic cells were determined using flow cytometry. Each value represents the mean ± SEM for $n = 6$. *Values significantly differ from the respective control, $P < 0.05$. †Values significantly differ from the SNP-treated groups, $P < 0.05$.

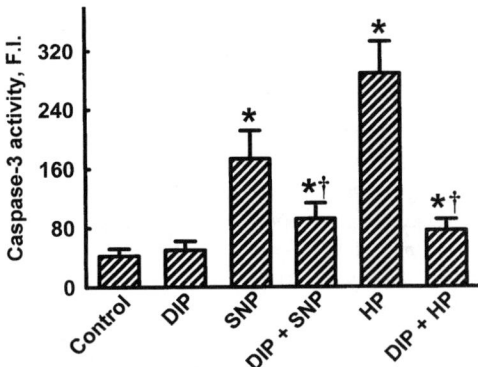

FIGURE 5. Effects of 2,6-diisopropylphenol (DIP) on sodium nitroprusside (SNP)- and hydrogen peroxide (HP)-induced caspase-3 activities. Osteoblasts prepared from neonatal rat calvaria were exposed to 30 μM DIP, 2 mM SNP, 25 μM HP, and a combination of DIP and SNP or DIP and HP. Caspase-3 activity was determined using a fluorogenic substrate assay. Each value represents the mean ± SEM for $n = 6$. *Values significantly differ from the respective control, $P < 0.05$. †Values significantly differ from the SNP- or HP-treated groups, $P < 0.05$. F.I., fluorescence intensities.

sure to SNP and HP significantly increased caspase-3 activity by 131% and 246%, respectively. 2,6-Diisopropylphenol caused a significant 47% decrease in SNP-enhanced caspase-3 activities. The HP-induced caspase-3 activity was significantly reduced by 74% following co-treatment with 2,6-diisopropylphenol and HP (FIG. 5).

DISCUSSION

2,6-Diisopropylphenol is one of the common intravenous anesthetic agents used for induction and maintenance of anesthesia during surgical procedures, including orthopedic operations.[1,7] Osteoblasts are involved in bone remodeling.[8] This study showed that administration of 3 and 30 μM 2,6-diisopropylphenol was not cytotoxic to osteoblasts. However, viability of osteoblasts was decreased time-dependently following administration of 300 μM 2,6-diisopropylphenol. Our previous study presented similar results of 2,6-diisopropylphenol at therapeutic concentrations having no effect on macrophage viability.[16] However, 2,6-diisopropylphenol at a high concentration (300 μM) increased lactate dehydrogenase release and led to an arrest of the cell cycle in the G1/S phase. 2,6-Diisopropylphenol at 30 μM is within the range of clinical plasma concentrations.[18] Therefore, clinically relevant concentrations of 2,6-diisopropylphenol are not harmful to osteoblasts and macrophages.

SNP significantly increased nitrite production in osteoblasts. SNP can be decomposed to NO under light exposure or in the presence of a biological reducing system.[12,19] The use of SNP in this study had a biochemical advantage because it provides NO for an investigation of signaling transduction pathways without interfering with NO synthase-involved second messenger systems. Nitrite is an oxidative product of NO. Increases in amounts of nitrite correspond to enhanced levels of NO in cells. Thus, administration of SNP to osteoblasts significantly increases levels of intracellular NO. Osteoblasts can constitutively produce NO.[20] Low levels of nitrite were detectable in untreated osteoblasts. 2,6-Diisopropylphenol did not decrease basal or SNP-released NO levels. Our previous study demonstrated that 2,6-diisopropylphenol downregulates NO biosynthesis via inhibition of inducible NO synthase in lipopolysaccharide-activated macrophages.[21] Therefore, 2,6-diisopropylphenol cannot directly scavenge exogenous NO, but reduces endogenous NO production through inhibition of the NO-synthesizing enzyme.

Administration of HP to osteoblasts increased intracellular levels of ROS. This study used DCFH-DA dye to quantify ROS. The major forms of ROS reacting with this dye are peroxides.[22] HP is one of the cellular peroxides. Thus, HP itself can enhance intracellular ROS levels in osteoblasts. A previous study reported that HP leads to depolarization of the mitochondrial membrane potential and promotes apoptotic factor release from the mitochondria to the cytoplasm.[23] ROS are typical mitochondrial apoptotic factors.[24] The other possible reason that explains the enhancement of intracellular ROS in osteoblasts following HP administration may be the release of mitochondrial ROS via depolarization of the mitochondrial membranes. 2,6-Diisopropylphenol significantly decreased HP-enhanced intracellular ROS levels. 2,6-Diisopropylphenol has a phenol group that can scavenge ROS.[2] Previous studies revealed that 2,6-diisopropylphenol can directly scavenge lipid peroxyryl radicals, hydroxyl radicals, and superoxide.[2,25] The present study further shows that 2,6-diisopropylphenol can also directly scavenge HP.

SNP decreased osteoblast viability time-dependently. NO can be decomposed from SNP.[19] An elevation of NO levels increases cellular oxidative stress and results in cell damage.[9,12] Apoptotic analysis revealed that SNP induced osteoblast apoptosis time-dependently. NO is a critical bioregulator which induces apoptosis in different types of cells.[26] Thus, the SNP-induced osteoblast death mainly occurs via an apoptotic mechanism. This study shows that a therapeutic concentration of 2,6-diisopropylphenol (30 μM) can protect osteoblasts from SNP-induced cell insults and apoptosis. Because 2,6-diisopropylphenol did not decrease basal or SNP-released NO, this intravenous anesthetic agent protects osteoblasts against apoptosis through means other than the direct scavenging of NO. Following SNP administration, caspase-3 activity was significantly increased. Activation of caspase-3 causes proteolytic cleavage of key proteins and induces cell apoptosis.[13] Thus, the apoptosis induced in osteoblasts by SNP involves activation of caspase-3 activity. Co-treatment with 2,6-diisopropylphenol and SNP significantly reduced SNP-enhanced caspase-3 activity. In parallel with the suppression of caspase-3 activity, 2,6-diisopropylphenol significantly reduced SNP-induced osteoblast apoptosis. Therefore, 2,6-diisopropylphenol can block SNP-induced osteoblast apoptosis via suppression of caspase-3 activity.

HP decreased osteoblast viability time-dependently. HP is an ROS. HO enhances oxidative stress and leads to cell death.[11] In parallel with cell damage, HP induced osteoblast apoptosis in a time-dependent manner. Thus, HP induces osteoblast injury through an apoptotic pathway. Co-treatment of 2,6-diisopropylphenol and HP significantly decreased HP-induced osteoblast insults and apoptosis. 2,6-Diisopropylphenol can reduce HP-enhanced intracellular ROS. ROS are apoptotic factors involved in the regulation of cell death.[24] Thus, 2,6-diisopropylphenol can directly scavenge intracellular ROS and protect osteoblasts from HP-induced insults. Administration of HP significantly increased osteoblast caspase-3 activity. Thus, HP induces osteoblast apoptosis mainly through activation of caspase-3 activity. 2,6-Diisopropylphenol significantly decreased HP-enhanced caspase-3 activity. Therefore, the other mechanism to explain the protective effects of 2,6-diisopropylphenol against HP-induced osteoblast apoptosis may occur via a reduction in caspase-3 activity.

The protective effects of 2,6-diisopropylphenol on osteoblasts from SNP- and HP-induced cell damage decreased with time. This time-dependent recovery was also observed in the 2,6-diisopropylphenol–involved reduction of intracellular ROS levels. 2,6-Diisopropylphenol can be progressively decomposed in aerobic conditions or after exposure to visible light.[1] This characteristic might explain why the 2,6-diisopropylphenol–induced protection of osteoblasts from SNP- and HP-induced injuries decreased with time. Another possible reason which explains the time-dependent decrease in the 2,6-diisopropylphenol–caused osteoblast protection is the metabolism of this anesthetic agent by cytochrome P450-dependent monooxygenase or UDP glucuronosyltransfease.[27] These two enzymes are detectable in osteoblasts.[28,29] Therefore, the lowered contents of 2,6-diisopropylphenol in osteoblasts, due to its metabolism by these related enzymes, may explain the decrease in the protective effects of this intravenous anesthetic agent on SNP- and HP-induced osteoblast insults.

CONCLUSION

This study shows that clinically relevant concentrations of 2,6-diisopropylphenol are not cytotoxic to osteoblasts. Administration of SNP and HP increases cellular oxidative stress and leads to osteoblast death via an apoptotic mechanism. 2,6-Diisopropylphenol at a therapeutic concentration can protect osteoblasts from SNP- and HP-induced cell insults and apoptosis. 2,6-Diisopropylphenol can directly scavenge HP-enhanced intracellular ROS without affecting SNP-releasing NO. Both SNP and HP activate capase-3 activities. 2,6-Diisopropylphenol can significantly reduce SNP- and HP-enhanced caspase-3 activities. Therefore, 2,6-diisopropylphenol can block HP-induced cell apoptosis through directly scavenging intracellular ROS and suppressing caspase-3 activity. However, a decrease in caspase-3 activity may be the major mechanism contributing to the protection of osteoblasts from SNP-induced cell apoptosis by 2,6-diisopropylphenol.

ACKNOWLEDGMENTS

This work was supported by the National Science Council (NSC93-2745-B-038-002-URD) and the Shin Kong Wu Ho-Su Memorial Hospital (SKH-TMU-92-32), Taipei, Taiwan.

REFERENCES

1. SEBEL, P.S. & J.D. LOWDON. 1989. Propofol: a new intravenous anesthetic. Anesthesiology **71:** 260–277.

2. ARTS, L., R. VAN DER HEE & I. DEKKER. 1995. The widely used anesthetic agent propofol can replace alpha-tocopherol as antioxidant. FEBS Lett. **357:** 83–85.
3. SALGO, M.G. & W.A. PRYOR. 1996. Trolox inhibits peroxinitrite-mediated oxidative stress and apoptosis in rat thymocytes. Arch. Biochem. Biophys. **333:** 482–488.
4. MURPHY, P.G., M.J. DAVIES, M.O. COLUMB, et al. 1996. Effect of propofol and thiopentone on free radical mediated oxidative stress of the erythrocyte. Br. J. Anaesth. **76:** 536–543.
5. ERIKSSON, O., P. POLLESELLO & E.N. SARIS. 1992. Inhibition of lipid peroxidation in isolated rat liver mitochondria by the general anesthetic propofol. Biochem. Pharmacol. **44:** 391–393.
6. CHANG, H., S.Y. TSAI, T.L. CHEN, et al. 2002. Therapeutic concentrations of propofol protects mouse macrophages from nitric oxide-induced cell death and apoptosis. Can. J. Anesth. **49:** 477–480.
7. LEBENBOM-MANSOUR, M.H., S.K. PANDIT, S.P. KOTHARY, et al. 1993. Desflurane versus propofol anesthesia: a comparative analysis in outpatients. Anesth. Analg. **76:** 936–941.
8. COLLIN-OSDOBY, P., G.A. NICKOLS & P. OSDOBY. 1995. Bone cell function, regulation, and communication: a role for nitric oxide. J. Cell. Biochem. **57:** 399–408.
9. MODY, N., F. PARHAMI, T.A. SARAFIAN, et al. 2001. Oxidative stress modulates osteoblastic differentiation of vascular and bone cells. Free Radic. Biol. Med. **31:** 509–519.
10. MOGI, M., K. KINPARA, A. KONDO, et al. 1999. Involvement of nitric oxide and biopterin in proinflammatory cytokine-induced apoptotic cell death in mouse osteoblastic cell line MC3T3-E1. Biochem. Pharmacol. **58:** 649–654.
11. LIU, H.C., R.M. CHEN, W.C. JEAN, et al. 2001. Cytotoxic and antioxidant effects of the water extract of traditional Chinese herb gusuibu (*Drynaria fortunei*) on rat osteoblasts. J. Formos. Med. Assoc. **100:** 383–388.
12. CHEN, R.M., H.C. LIU, Y.L. LIN, et al. 2002. Nitric oxide induces osteoblast apoptosis through the de novo synthesis of Bax protein. J. Orthop. Res. **20:** 295–302.
13. GOYAL, L. 2001. Cell death inhibition: keeping caspases in check. Cell **104:** 805–808.
14. PARTRIDGE, N.C., D. ALCORN, V.P. MICHELANGELI, et al. 1981. Functional properties of hormonally responsive cultured normal and malignant rat osteoblastic cells. Endocrinology **108:** 213–219.
15. WU, C.H., T.L. CHEN, T.G. CHEN, et al. 2003. Nitric oxide modulates pro- and antiinflammatory cytokines in lipopolysaccharide-activated macrophages. J. Trauma **55:** 540–545.
16. CHEN, R.M., C.H. WU, H.C. CHANG, et al. 2003. Propofol suppresses macrophage functions and modulates mitochondrial membrane potential and cellular adenosine triphosphate levels. Anesthesiology **98:** 1178–1185.
17. NICOLETTI, I., G. MIGLIORATI, M.C. PAGLIACCI, et al. 1991. A rapid and simple method of measuring thymocyte apoptosis by propidium iodide staining and flow cytometry. J. Immunol. Methods **139:** 271–279.
18. GEPTS, E., F. CAMU, I.D. COCKSHOTT, et al. 1987. Disposition of propofol administered as constant rate intravenous infusions in humans. Anesth. Analg. **66:** 1256–1263.
19. BATES, J.N., M.T. BAKER, R. GUERRA, JR., et al. 1991. Nitric oxide generation from nitroprusside by vascular tissue. Evidence that reduction of the nitroprusside anion and cyanide loss are required. Biochem. Pharmacol. **42:** s157–s165.
20. HEFLICH, M.H., D.E. EVANS, P.S. GRABOWSKI, et al. 1997. Expression of nitric oxide synthase isoforms in bone and bone cell cultures. J. Bone Miner. Res. **12:** 1108–1115.
21. CHEN, R.M., G.J. WU, Y.T. TAI, et al. 2003. Propofol downregulates nitric oxide biosynthesis through inhibiting inducible nitric oxide synthase in lipopolysaccharide-activated macrophages. Arch. Toxicol. **77:** 418–423.
22. SIMIZU, S., M. IMOTO, N. MASUDA, et al. 1997. Involvement of hydrogen peroxide production in erbstatin-induced apoptosis in human small cell lung carcinoma cells. Cancer Res. **56:** 4978–4982.
23. TADA-OIKAWA, S., Y. HIRAKU, M. KAWANISHI, et al. 2003. Mechanism for generation of hydrogen peroxide and change of mitochondrial membrane potential during rotenone-induced apoptosis. Life Sci. **73:** 3277–3288.

24. LI, N., K. RAGHEB, G. LAWLER, et al. 2003. Mitochondrial complex I inhibitor rotenone induces apoptosis through enhancing mitochondrial reactive oxygen species production. J. Biol. Chem. **278:** 8516–8525.
25. DEMIRYUREK, A.T., I. GINEL, S. KAHRAMAN, et al. 1998. Propofol and intralipid interact with reactive oxygen species: a chemiluminescence study. Br. J. Anaesth. **68:** 13–18.
26. XIE, K. & S. HUANG. 2003. Contribution of nitric oxide-mediated apoptosis to cancer metastasis inefficiency. Free Radic. Biol. Med. **34:** 969–986.
27. SIMONS, P.J., I.D. COCKSHOTT, E.J. DOUGLAS, et al. 1988. Disposition in male volunteers of a subanaesthetic intravenous dose of an oil in water emulsion of ^{14}C-propofol. Xenobiotica **18:** 429–440.
28. NISHIMURA, A., T. SHINKI, C.H. JIN, et al. 1994. Regulation of messenger ribonucleic acid expression of 1α,25-dihydroxyvitamin D3-24-hydroxylase in rat osteoblasts. Endocrinology **134:** 1794–1799.
29. ZHANG, W., C.E. WATSON, C. LIU, et al. 2000. Glucocorticoids induce a near-total suppression of hyaluronan synthase mRNA in dermal fibroblasts and in osteoblasts: a molecular mechanism contributing to organ atrophy. Biochem. J. **349:** 91–97.

Nitric Oxide Induces Osteoblast Apoptosis through a Mitochondria-Dependent Pathway

WEI-PIN HO,[a] TA-LIANG CHEN,[b] WEN-TA CHIU,[c] YU-TING TAI,[d] AND RUEI-MING CHEN[d,e]

Departments of Orthopedics[a] and Anesthesiology,[d] Wan-Fang Hospital, College of Medicine, Taipei Medical University, Taipei, Taiwan

[b]Taipei City Hospital and Taipei Medical University, Taipei, Taiwan

[c]Division of Neurosurgery, Department of Surgery, Wan-Fang Hospital, College of Medicine, Taipei Medical University, Taipei, Taiwan

[e]Graduate Institute of Medical Sciences, College of Medicine, Taipei Medical University, Taipei, Taiwan

ABSTRACT: Osteoblasts contribute to bone remodeling. Nitric oxide can regulate osteoblast activities. In this study, we attempted to evaluate the pathophysiological effects of nitric oxide on osteoblasts and its possible mechanism using neonatal rat calvarial osteoblasts as the experimental model. Exposure of osteoblasts to sodium nitroprusside, a nitric oxide donor, decreased alkaline phosphatase activities and cell viability in a concentration- and time-dependent manner. Apoptotic analysis revealed that sodium nitroprusside time-dependently increased the percentages of osteoblasts undergoing apoptosis. Administration of sodium nitroprusside reduced the mitochondrial membrane potential of osteoblasts. In parallel with the mitochondrial dysfunction, levels of intracellular reactive oxygen species and cytochrome c were significantly elevated following sodium nitroprusside administration. Exposure of osteoblasts to sodium nitroprusside significantly increased caspase-3 activity. Results of this study show that nitric oxide, decomposed from sodium nitroprusside, can induce osteoblast apoptosis through a mitochondrion-dependent cascade that causes mitochondrial dysfunction, release of intracellular reactive oxygen species and cytochrome c from mitochondria to cytoplasm, and activation of caspase-3.

KEYWORDS: nitric oxide; osteoblasts; apoptosis; mitochondrial functions; caspase-3

INTRODUCTION

Nitric oxide (NO), synthesized from L-arginine by constitutive or inducible NO synthases, contributes to the regulation of vasodilation, neurotransmission, immuno-

Address for correspondence: Ruei-Ming Chen, Ph.D., Graduate Institute of Medical Sciences, College of Medicine, Taipei Medical University, No. 250, Wu-Hsing St., Taipei 110, Taiwan. Voice: +886-2-29307930 ext. 2159; fax: +886-2-86621119.
rmchen@tmu.edu.tw; rmchen@wanfang.gov.tw

Ann. N.Y. Acad. Sci. 1042: 460–470 (2005). © 2005 New York Academy of Sciences.
doi: 10.1196/annals.1338.039

responses, and death control.[1] Osteoblasts play important roles in bone remodeling.[2] NO has been implicated to be a critical effector to modulate osteoblast activities and bone remodeling.[2,3] Osteoblasts can constitutively produce NO.[4] Following stimulation of inflammatory cytokines, NO will be massively induced by osteoblasts.[5] Constitutive NO can regulate osteoblast proliferation and differentiation.[6,7] However, overproduced NO will damage osteoblasts.[8,9] Our previous study has shown that the NO-induced osteoblast death is mainly via an apoptotic pathway.[10]

A variety of apoptotic factors is involved in programmed cell death.[11,12] Mitochondria are critical energy-producing organelles. Depolarization of mitochondrial membrane potential will lead to release of apoptotic factors, including reactive oxygen species and cytochrome c, from mitochondria to cytoplasm.[13,14] These mitochondrial apoptotic factors can sequentially transduce cell death signals. Caspase-3 is one of the cysteine proteases. Activation of caspase-3 will drive cells undergoing apoptosis.[12] However, in osteoblasts, the roles of mitochondrial apoptotic factors and caspase-3 in the NO-induced osteoblast apoptosis are still unknown. This study aimed to evaluate whether NO can modulate mitochondrial function, apoptotic factors, and caspase-3 activity to induce osteoblast apoptosis.

MATERIALS AND METHODS

Cell Culture and Drug Treatment

Osteoblasts were prepared from 3-day-old Wistar rat calvariae following an enzymatic digestion method as described previously.[15] Osteoblasts were seeded in Dulbecco's modified Eagle medium (DMEM) (Gibco-BRL, Grand Island, NY) supplemented with 10% heat-inactivated fetal bovine serum and 100 U/mL penicillin and 100 mg/mL streptomycin in tissue culture flasks at 37°C in a humidified atmosphere of 5% CO_2.

Sodium nitroprussie (SNP) was purchased from Sigma Corporation (St. Louis, MO). In each independent experiment, SNP was freshly dissolved in phosphate-buffered saline (0.14 M NaCl, 2.6 mM KCl, 8 mM Na_2HPO_4, and 1.5 mM KH_2PO_4) and protected from light. Osteoblasts were treated with various concentrations of SNP for different time intervals.

Assay of Cell Viability

Cell viability was determined by a colorimetric 3-(4,5-dimethylthiazol-2-yl)-2,5-diphenyltetrazolium bromide (MTT) assay as described previously.[16] Ten thousand osteoblasts were seeded in 96-well tissue plates culture plates overnight. After incubation with SNP, osteoblasts were cultured with new medium containing 0.5 mg/mL MTT for another 3 h. Blue formazan product in osteoblasts was dissolved in dimethyl sulfoxide (DMSO) and spectrophotometrically measured at a wavelength of 570 nm.

Quantification of Mitochondrial Membrane Potential

Levels of mitochondrial membrane potential were determined according to the method as described previously.[16] In brief, osteoblasts (5×10^5) were seeded in 12-well tissue culture plates overnight and then treated with SNP for different time intervals. After administration of SNP, osteoblasts were harvested and incubated with 3,3′-dihexyloxacarbocyanine ($DiOC_6(3)$), a positively charged dye, at 37°C for 30 min in a humidified atmosphere of 5% CO_2. After washing and centrifuging, cell pellets were suspended with phosphate-buffered saline. Fluorescence intensities in osteoblasts were analyzed by a flow cytometer (FACS Calibur, Becton Dickinson, San Jose, CA).

Determination of Intracellular ROS

Amounts of intracellular ROS were quantified following the method as described previously.[17] Briefly, osteoblasts (5×10^5) were cultured in 12-well tissue culture plates overnight and then co-treated with SNP and 2′,7′-dichlorofluorescin diacetate, an ROS sensitive dye, for different time intervals. Osteoblasts were harvested and suspended in phosphate-buffered saline. Fluorescence intensities in osteoblasts were quantified by a flow cytometer (FACS Calibur).

Immunoblotting Analysis

Immunoblotting analysis was carried out to determine levels of cytochrome c in osteoblasts. After pretreatment with SNP, osteoblasts were washed with phosphate-buffered saline, and cell lysates were collected after dissolving cells in 50 mL of ice-cold radioimmunoprecipitation assay buffer (25 mM Tris-HCl pH 7.2, 0.1% SDS, 1% Triton X-100, 1% sodium deoxycholate, 0.15 M NaCl, and 1 mM EDTA). Protein concentrations were quantified by a bicinchonic acid (BCA) protein assay kit (Pierce, Rockford, IL). Cytosolic proteins (100 μg) were resolved on 12% polyacrylamide gels and electrophoretically blotted onto nitrocellulose membranes. Cytochrome c protein was immunodetected using a mouse monoclonal antibody against rat cytochrome c protein (Transduction Laboratories, Lexington, KY). β-actin was immunodetected by a mouse monoclonal antibody against mouse β-actin (Sigma) as an internal control. Intensities of the immunoreactive bands were determined using a UVIDOCMW version 99.03 digital imaging system (Uvtec, Cambridge, England).

Fluorogenic Substrate Assay for Caspase-3 Activity

For determining caspase-3 activity, cell extracts are prepared by lysing osteoblasts in buffer containing 1% Nonidet P-40, 200 mM NaCl, 20 mM Tris/HCl, pH 7.4, 10 μg/mL leupeptin, 0.27 U/mL aprotinin, and 100 μm PMSF. Caspase-3 activity is determined by incubating cell lysate (25 μg total protein) with 50 μm fluorogenic substrate in 200 μL cell-free system buffer comprising 10 mM Hepes, pH 7.4, 220 mM mannitol, 68 mM sucrose, 2 mM NaCl, 2.5 mM KH_2PO_4, 0.5 mM EGTA, 2 mM $MgCl_2$, 5 mM pyruvate, 0.1 mM PMSF, and 1 mM dithiothreitol). Intensities of fluorescent products were measured by a spectrofluorometer.

Statistical Analysis

Statistical difference between the control and SNP-treated groups was considered significant when the P value of the Duncan's multiple range test was less than 0.05.

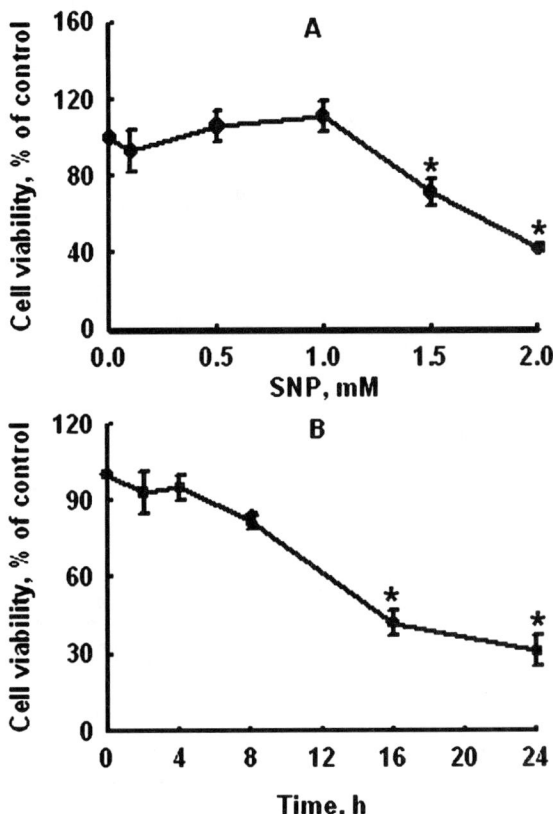

FIGURE 1. Concentration- and time-dependent effects of sodium nitroprusside (SNP) on osteoblast viability. Osteoblasts were exposed to 0.5, 1, 1.5, and 2 mM SNP (**A**) or to 2 mM SNP for 2, 4, 8, 16, and 24 h (**B**). Cell viability was determined by the MTT assay. Each value represents the mean ± SEM for $n = 12$. *Values significantly differ from the respective control, $P < 0.05$.

Statistical analysis between groups over time was carried out by the two-way ANOVA.

RESULTS

Treatment with 0.1, 0.5, and 1 mM SNP for 16 h was not cytotoxic to osteoblasts (FIG. 1A). SNP at 1.5 mM significantly reduced cell viability by 22%. After administration of 2 mM SNP, cell viability was decreased by 58% (FIG. 1A). Exposure to 2 mM SNP for 2 and 4 h did not affect cell viability (FIG. 1B). In 8-h–treated osteoblasts, cell viability was decreased by 19%. Administration of 2 mM SNP in osteoblasts for 16 h significantly decreased cell viability by 59%. Viability of osteoblasts was suppressed by 72% following SNP administration for 24 h (FIG. 1B).

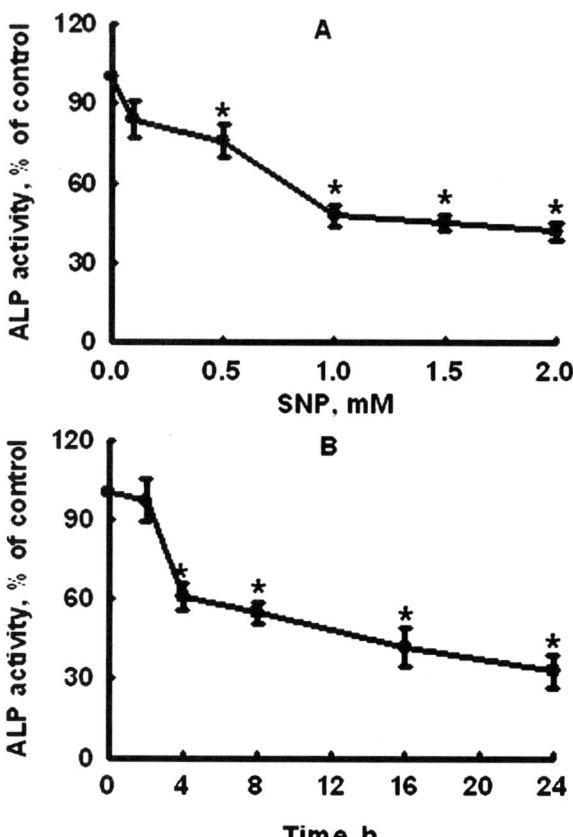

FIGURE 2. Concentration- and time-dependent effects of sodium nitroprusside (SNP) on alkaline phosphatase (ALP) activities. Osteoblasts were exposed to 0.5, 1, 1.5, and 2 mM SNP (**A**) or to 2 mM SNP for 2, 4, 8, 16, and 24 h (**B**). ALP activity was determined by a colorimetric method. Each value represents the mean ± SEM for $n = 12$. *Values significantly differ from the respective control, $P < 0.05$.

Exposure to 0.1 mM SNP for 16 h did not influence alkaline phosphatase (ALP) activity (FIG. 2A). SNP at 0.5 mM significantly decreased ALP activity by 21%. Administration of osteoblasts with 1 and 1.5 mM SNP significantly reduced ALP activity by 49% and 58%, respectively. ALP activity was suppressed by 62% after administration of 2 mM SNP (FIG. 2A). In 2-h–treated osteoblasts, 2 mM SNP did not affect ALP activity (FIG. 2B). After administration of 2 mM SNP in osteoblasts for 4 h, ALP activity was decreased by 39%. Exposure to 2 mM SNP for 8 and 16 h significantly reduced ALP activity by 45% and 59%, respectively. After exposure to 2 mM SNP for 24 h, ALP activity was suppressed by 69% (FIG. 2B).

Analysis by a flow cytometric method revealed that administration of 2 mM SNP for 2 and 4 h did not induce osteoblast apoptosis (FIG. 3). In 8-h–treated osteoblasts, SNP significantly increased apoptotic percentage by 15%. After administration of

TABLE 1. Effects of sodium nitroprusside on the mitochondrial membrane potential and intracellular reactive oxygen species

	Mitochondrial membrane potential	Intracellular ROS
Time (h)	(% of control)	(% of control)
0	100	100
1	98 ± 6	181 ± 33*
2	82 ± 5*	303 ± 85*
4	62 ± 8*	528 ± 74*

NOTE: Osteoblasts were exposed to 2 mM sodium nitroprusside for 1, 2, and 4 h. Levels of mitochondrial membrane potential and intracellular reactive oxygen species (ROS) were determined by a flow cytometer. Each value represents the mean ± SEM for $n = 6$. *Values significantly differ from the respective control, $P < 0.05$.

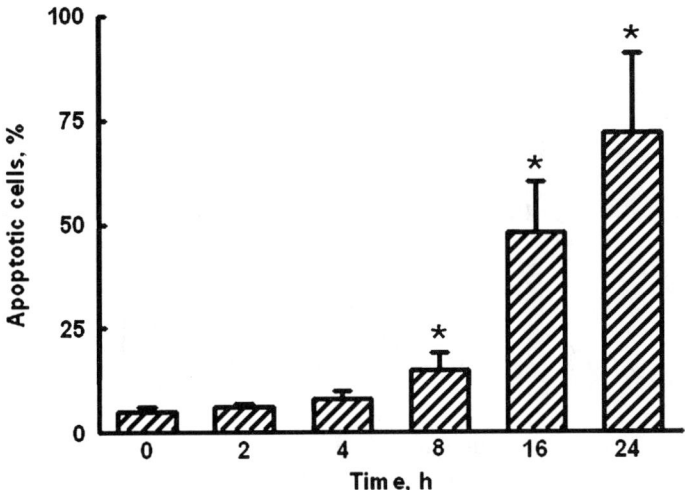

FIGURE 3. Time-dependent effects of sodium nitroprusside (SNP) on osteoblast apoptosis. Osteoblasts were exposed to 2 mM SNP for 2, 4, 8, 16, and 24 h. Apoptotic cells were determined by a flow cytometric method. Each value represents the mean ± SEM for $n = 9$. * Values significantly differ from the respective control, $P < 0.05$.

SNP for 16 h, almost 50% of osteoblasts underwent apoptosis. In 24-h–treated osteoblasts, SNP increased 68% of osteoblasts undergoing apoptosis (FIG. 3).

In 1-h–treated osteoblasts, SNP did not affect mitochondrial membrane potential (TABLE 1). Administration of osteoblasts with SNP for 2 h significantly decreased mitochondrial membrane potential by 18%. After administration of SNP for 4 h, the mitochondrial membrane potential was reduced by 38%. Exposure to SNP for 1 h significantly increased levels of intracellular ROS by 81% (TABLE 1). After administration of SNP for 2 h, levels of intracellular ROS were augmented by 203%. In 4-h–treated osteoblasts, SNP significantly enhanced levels of intracellular ROS by 428% (TABLE 1).

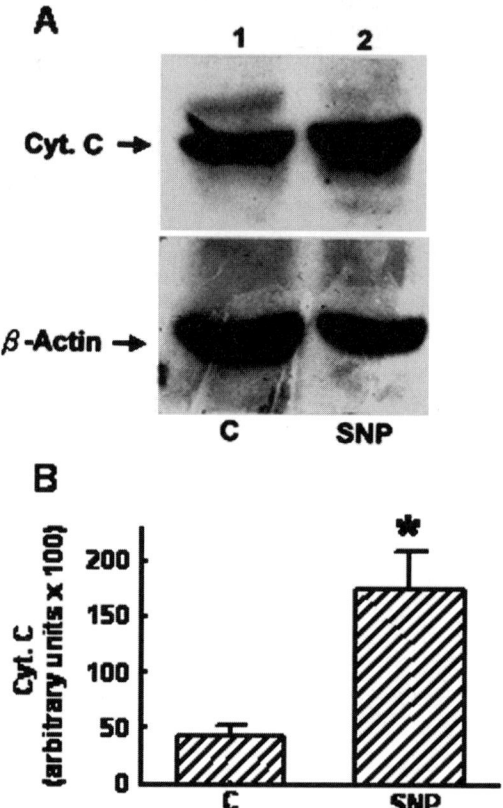

FIGURE 4. Effect of sodium nitroprusside (SNP) on cytochrome c (Cyt. C) production. Osteoblasts were exposed to 2 mM SNP for 8 h. Cytochome c in osteoblasts was immunodetected using a monoclonal antibody against rat cytochrome c protein (**A**). β-actin was immunodetected to be an internal control. Intensities of immunorelated protein bands were quantified by a digital system (**B**). Each value represents the mean ± SEM for $n = 4$. *Values significantly differ from the respective control, $P < 0.05$.

Immunoblotting analysis revealed that cytochrome c was detectable in untreated osteoblasts (FIG. 4A, lane 1). After administration of osteoblasts with SNP for 8 h, amounts of cytochrome c protein were significantly enhanced (lane 2). Intensities of immunorelated protein bands were quantified by a digital imaging system, and data are shown in FIGURE 4B. Exposure to SNP significantly increased amounts of cytochrome c protein by 3.4-fold.

Activity of caspase-3 was determined by a fluorogenic substrate assay. In untreated osteoblasts, low levels of caspase-3 activity were detected (FIG. 5). Administration of osteoblasts with SNP significantly increased caspase-3 activity by 5.5-fold.

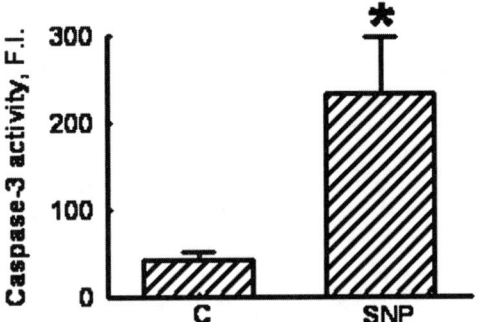

FIGURE 5. Effect of sodium nitroprusside (SNP) on caspase-3 activity. Osteoblasts were exposed to 2 mM SNP for 8 h. Caspase-3 activity was determined by a fluorogenic substrate assay. Each value represents the mean ± SEM for $n = 6$. *Values significantly differ from the respective control, $P < 0.05$.

DISCUSSION

SNP significantly decreased osteoblast viability in a concentration- and time-dependent manner. SNP can be decomposed to NO under light exposure or a biological reducing system.[18] Analysis by the Griess reaction method revealed that administration of osteoblasts with SNP concentration- and time-dependently increased nitrite production (data not shown). Nitrite is one of oxidative metabolites of NO.[1] Elevation of nitrite corresponds to the increase in NO levels. ALP is a mark protein synthesized in osteoblasts.[2] This study showed that SNP significantly decreased ALP activity. Thus, data from the present study demonstrates that NO decomposed from SNP can induce osteoblast insults, even death. Apoptotic analysis revealed that SNP time-dependently increased the percentage of osteoblasts undergoing apoptosis. Thus, the SNP-induced osteoblast death is mainly through an apoptotic pathway. Our preliminary study used a specific inhibitor of cyanide to co-treat osteoblasts with SNP; and the data revealed that this inhibitor could not block SNP-caused osteoblast insults (data not shown). In addition, administration of GS-NO, another NO donor, in osteoblasts could also induce cell apoptosis. Therefore, NO, rather than cyanide, plays a major role in SNP-induced osteoblast apoptosis.

SNP time-dependently decreased mitochondrial membrane potential. Maintenance of the mitochondrial membrane potential is critical to the synthesis of cellular adenosine triphosphate.[19] Depletion of cellular adenosine triphosphate will lead to cell dysfunction or apoptosis. Thus, one of possible mechanisms to explain the SNP-induced osteoblast apoptosis may be via the suppression of the mitochondrial membrane potential and cellular adenosine phosphate synthesis by this NO donor. In addition, depolarization of the mitochondrial membrane potential has been reported to increase the release of apoptotic factors from mitochondria to cytoplasm.[14] Enhancement of cytosolic apoptotic factors can drive cells to apoptosis. This study shows that SNP increased levels of intracellular ROS and cytochrome c in osteoblasts. ROS and cytochrome c are two typical apoptotical factors.[13,14] Therefore, the other mechanism to explain the SNP-induced osteoblast apoptosis may be through the sequentially modulating effects of this NO donor on the mitochondrial membrane potential and apoptotic factor release.

Administration of osteoblasts with SNP increased levels of intracellular ROS time-dependently. In mitochondria, ROS is a by-product synthesized during the respiratory chain reaction. Disruption of the mitochondrial membrane potential will lead to the release of ROS from mitochondria to cytoplasm.[13,14] Thus, one of the possible sources for the SNP-enhanced intracellular ROS production in osteoblasts is due to the release of ROS from mitochondria. This study used DCFH-DA dye to catch ROS. A previous study showed that NO and peroxynitrite, an oxidative product of NO and superoxide, can directly react with DCFH-DA.[20] Thus, NO decomposed from SNP may be another source of ROS in osteoblasts after exposure to this donor. Elevation of ROS will increase cellular oxidative stress. Cumulative evidence suggests that ROS can trigger cell apoptosis by upregulating pro-apoptotic Bax protein.[21] Our previous study showed that NO can increase *de novo* synthesis of Bax protein in osteoblasts.[10] Therefore, the SNP-induced ROS is involved in osteoblast apoptosis.

SNP increased cytochrome *c* release from mitochondria to cytoplasm. Depolization of the mitochondrial membrane potential can increase cytochrome *c* release.[13,14] Several lines of evidence show that Bcl-2 family members, including Bax and Bak, can bind to mitochondrial membrane and promote cytochrome *c* release.[22] Thus, NO decomposed from SNP may modulate mitochondrial function and Bcl-2 family production to enhance the release of cytochrome *c* from mitochondria. Another previous study showed that cytochrome *c* release will result in caspase activation and cell apoptosis.[23] Therefore, the SNP-induced release of cytochrome *c* is involved in osteoblast apoptosis.

Caspase-3 activity was augmented following SNP administration. A variety of apoptotic factors contribute to the activation of capase-3. This study demonstrated that SNP significantly increased intracellular ROS in osteoblasts. ROS has been reported to be capable of activating caspase-3.[19] Release of cytochrome *c* was enhanced following SNP administration. After binding to the adaptor Apaf-1, cytochrome *c* can activate procaspase-9 to mature caspase-9.[23] Sequential cleavage of procaspase-3 by caspase-9 will activate caspase-3. Thus, an increase in capase-3 activity following SNP administration may derive from enhancement of intracellular ROS and cytochrome *c* release. Active caspase-3 plays an important role in cell death control.[12] Caspase-3 can degrade key proteins in cells and drive cells undergoing apoptosis. Therefore, activation of caspase-3 plays a critical role in SNP-induced osteoblast apoptosis.

CONCLUSION

This study has shown that SNP increases levels of NO in osteoblasts, and decreases ALP activity and cell viability. The SNP-induced osteoblast death is mainly via an apoptotic pathway. SNP can reduce the mitochondrial membrane potential. In parallel to mitochondrial dysfunction, SNP enhances release of intracellular ROS and cytochrome *c* from mitochondria to cytoplasm. Following an increase in levels of mitochondrial apoptotic factors, caspase-3 activity is augmented after SNP administration. According to the present data, we suggest NO decomposed from SNP can induce osteoblast apoptosis via the sequential modulation of mitochondrial function, release of mitochondrial apoptotic factors, and caspase-3 activity.

ACKNOWLEDGMENTS

This study was supported by Grant 93TMU-WFH-07 from the Wang-Fang Hospital and Taipei Medical University, Taipei, Taiwan.

REFERENCES

1. MONCADA, S., R.M. PALMER & E.A. HIGGS. 1991. Nitric oxide: physiology, pathophysiology and pharmacology. Pharmacol. Rev. **43:** 109–142.
2. COLLIN-OSDOBY, P., G.A. NICKOLS & P. OSDOBY. 1995. Bone cell function, regulation, and communication: a role for nitric oxide. J. Cell. Biochem. **57:** 399–408.
3. EVANS, D.M. & S.H. RALSTON. 1996. Nitric oxide and bone. J. Bone Miner. Res. **11:** 300–305.
4. HEFLICH, M.H., D.E. EVANS, P.S. GRABOWSKI, et al. 1997. Expression of nitric oxide synthase isoforms in bone and bone cell cultures. J. Bone Miner. Res. **12:** 1108–1115.
5. DAMOULIS, P.D. & P.V. HAUSCHKA. 1994. Cytokines induce nitric oxide production in mouse osteoblasts. Biochem. Biophys. Res. Commun. **201:** 924–931.
6. CHAE, H.J., R.K. PARK, H.T. CHUNG, et al. 1997. Nitric oxide is a regulator of bone remodelling. J. Pharm. Pharmacol. **49:** 897–902.
7. HIKIJI, H., W.S. SHIN, S. OIDA, et al. 1997. Direct action of nitric oxide on osteoblastic differentiation. FEBS Lett. **410:** 238–242.
8. DAMOULIS, P.D. & P.V. HAUSCHKA. 1997. Nitric oxide acts in conjunction with proinflammatory cytokines to promote cell death in osteoblasts. J. Bone Miner. Res. **12:** 412–422.
9. MOGI, M., K. KINPARA, A. KONDO, et al. 1999. Involvement of nitric oxide and biopterin in proinflammatory cytokine-induced apoptotic cell death in mouse osteoblastic cell line MC3T3-E1. Biochem. Pharmacol. **58:** 649–654.
10. CHEN, R.M., H.C. LIU, Y.L. LIN, et al. 2002. Nitric oxide induces osteoblast apoptosis through the de novo synthesis of Bax protein. J. Orthop. Res. **20:** 295–302.
11. CHUNG, K.C., J.H. PARK, C.H. KIM, et al. 1999. Tumor necrosis factor-alpha and phorbol 12-myristate 13-acetate differentially modulate cytotoxic effect of nitric oxide generated by serum deprivation in neuronal PC12 cells. J. Neurochem. **72:** 1482–1488.
12. GOYAL, L. 2001. Cell death inhibition: keeping caspases in check. Cell **104:** 805–808.
13. HORTELANO, S., A.M. ALVAREZ & L. BOSCA. 1999. Nitric oxide induces tyrosine nitration and release of cytochrome c preceding an increase of mitochondrial transmembrane potential in macrophages. FASEB J. **13:** 2311–2317.
14. BLOM, W.M., H.J. DE BONT & J.F. NAGELKERKE. 2003. Regional loss of the mitochondrial membrane potential in the hepatocyte is rapidly followed by externalization of phosphatidylserines at that specific site during apoptosis. J. Biol. Chem. **278:** 12467–12474.
15. PARTRIDGE, N.C., D. ALCORN, V.P. MICHELANGELI, et al. 1981. Functional properties of hormonally responsive cultured normal and malignant rat osteoblastic cells. Endocrinology **108:** 213–219.
16. WU, C.H., T.L. CHEN, T.G. CHEN, et al. 2003. Nitric oxide modulates pro- and anti-inflammatory cytokines in lipopolysaccharide-activated macrophages. J. Trauma **55:** 540–545.
17. CHEN, R.M., C.H. WU, H.C. CHANG, et al. 2003. Propofol suppresses macrophage functions and modulates mitochondrial membrane potential and cellular adenosine triphosphate levels. Anesthesiology **98:** 1178–1185.

18. KOWALUK, E.A., P. SETH & H.L. FUNG. 1992. Metabolic activation of sodium nitroprusside to nitric oxide in vascular smooth muscle. J. Pharmacol. Exp. Ther. **262:** 916–922.
19. YU, X.H., T.D. PERDUE, Y.M. HEIMER, *et al.* 2002. Mitochondrial involvement in tracheary element programmed cell death. Cell Death Differ. **9:** 189–198.
20. RAO, K.M., J. PADMANABHAN, D.L. KILBY, *et al.* 1992. Flow cytometric analysis of nitric oxide production in human neutrophils using dichlorofluorescein diacetate in the presence of a calmodulin inhibitor. J. Leukoc. Biol. **51:** 496–500.
21. KUMAR, D. & B.I. JUGDUTT. 2003. Apoptosis and oxidants in the heart. J. Lab. Clin. Med. **142:** 288–297.
22. SCORRANO L. & S.J. KORSMEYER. 2003. Mechanisms of cytochrome c release by proapoptotic BCL-2 family members. Biochem. Biophys. Res. Commun. **304:** 437–444.
23. BORUTAITE, V. & G.C. BROWN. 2003. Mitochondria in apoptosis of ischemic heart. FEBS Lett. **541:** 1–5.

Bcl-2 Gene Family Expression in the Brain of Rat Offspring after Gestational and Lactational Dioxin Exposure

SHWU-FEN CHANG,[a,b] YU-YO SUN,[c,d] LIANG-YO YANG,[b,d] SSU-YAO HU,[b,c] SHIH-YING TSAI,[b,c,d] WEN-SEN LEE,[c,d] AND YI-HSUAN LEE[c,d]

[a]*Graduate Institute of Cell and Molecular Biology,* [b]*Brain Diseases and Aging Research Center,* [c]*Graduate Institute of Medical Sciences, and* [d]*Department of Physiology, Taipei Medical University, Taipei, Taiwan*

ABSTRACT: Recent epidemiological studies have shown that dioxin, a persistent organic pollutant, is related to cognitive and behavioral abnormalities in the offspring of exposed cohort. In order to investigate the possible impact of dioxin in survival gene expression during brain development, we established an animal model of gestational and lactational dioxin-exposed rat offspring. The expressions of dioxin-responsive gene cytochrome P450 1A1 (CYP1A1), apoptotic gene Bax, and anti-apoptotic genes Bcl-2 and Bcl-x_L were examined in rat liver and brains using Western blot analysis and RT-PCR. The results showed that treatment of pregnant rats with 2,3,7,8-tetrachlorodibenzo-*p*-dioxin (TCDD) (2 μg/kg body weight through oral delivery) at gestation day 15 resulted in an increase of Bcl-x_L in offspring male liver and cerebral cortex, but a decrease in female offspring. In contrast, the expression of Bcl-x_L in the cerebellum was decreased in male, but increased in female. Bcl-2, another anti-apoptotic gene, was also downregulated in P0 female liver, cerebral cortex, but was not observed in male. In the 4-month-old offspring, however, the Bcl-2 protein levels in the liver and cerebellum of both male and female pups were higher in the TCDD group as compared with the control group. However, the Bcl-2 level in the cerebral cortex of TCDD-treated groups was higher than the control group only in female but not male offspring at 4 months old. The expression of Bax showed no significant changes upon TCDD exposure at P0 stage, but was significantly reduced in the 4-month-old male cortex. These results indicate that early exposure of dioxin could affect the development of certain brain regions with gender difference, in terms of its differential effect on expressions of Bcl-x_L, Bcl-2, and Bax.

KEYWORDS: dioxin; brain; Bcl-2; Bcl-x_L; Bax; cerebral cortex; cerebellum; liver

Address for correspondence: Yi-Hsuan Lee, Ph.D., Department of Physiology, or Wen-Sen Lee, Graduate Institute of Medical Sciences, Taipei Medical University, 250 Wu-Hsing Street, Taipei 110, Taiwan. Voice: +886-2-2736-1661 ext. 3188 (Y-H.L.), ext. 3103 (W-S.L.); fax: +886-2-2378-1073.
hsuan@tmu.edu.tw or wslee@tmu.edu.tw

Ann. N.Y. Acad. Sci. 1042: 471–480 (2005). © 2005 New York Academy of Sciences.
doi: 10.1196/annals.1338.040

INTRODUCTION

2,3,7,8-Tetrachlorodibenzo-*p*-dioxin (TCDD) is an environmental pollutant and a toxic member of a family of compounds known as "endocrine-disrupting chemicals" (EDCs). Maternal TCDD exposure can lead to early pre- and postimplantation embryo loss.[1] TCDD-related toxicity is mainly mediated by the aryl hydrocarbon receptor (AhR) protein, which binds TCDD and heterodimerizes with its partner protein, the aryl hydrocarbon receptor nuclear translocator (ARNT). The AhR–ARNT complex binds to dioxin-responsive elements (DREs) and enhances the transcription of the phase I drug-metabolizing (*cytochrome P4501A1*) gene and phase II (*UDP-glucronosyltransferase, glutathione S-transferase*) gene.[2] As such, the function of the CYP1A1 product is metabolic activation. However, TCDD undergoes oxygenation by the CYP1A1 product, and ultimately leads to be the highly mutagenic derivatives. An alternative mechanism of CYP1A1 activation leads to oxidative stress by the production of H_2O_2.[3,4] Oxidative stress is known to cause DNA damage and P53-dependent apoptosis.[5] Oxidative stress and inappropriate apoptosis may be involved in abnormal neuron development and neurodegeneration. Recent epidemiological studies have shown that dioxin exposure is related to cognitive and behavioral abnormalities in the offspring of exposed cohort.[6,7] In this study, we established an animal model to demonstrate that early exposure of dioxin could affect the development of certain brain regions in a gender-different manner in terms of its differential effect on Bcl-2 gene family expressions. Apoptotic gene expressions, such as anti-apoptotic–promoting protein Bcl-2 and Bcl-x_L and pro-apoptotic protein Bax in the offspring of TCDD-treated rat brain tissues were examined.

MATERIALS AND METHODS

Animal Model of Gestational and Lactational Exposure of TCDD in Rats

Pregnant Sprague-Dawley (SD) rats were received from the National Institute of Experimental Animal Research, Taipei, Taiwan, on day 13 of gestation (GD13). TCDD was orally administered to the maternal rats on GD15 by gavage with a single dose of TCDD (2 µg/kg/2 mL corn oil) or vehicle (corn oil 2 mL/kg). Liver, cerebral cortex, and cerebellum were collected from offspring of TCDD-treated (T) or corn oil-treated (C) SD rats on the day of birth (P0) and at 4 months of age (115–125 days) for Western blot and RT-PCR analysis. The protocol of this animal treatment was reviewed by the Experimental Animal Committee at Taipei Medical University. All efforts were made to minimize animal suffering and to use only the number of animals necessary to produce reliable scientific data.

Western Blot Analysis

Tissues collected were homogenized with ice-cold homogenizing medium (0.32 M sucrose, and 50 mM Tris-HCl containing 1mM EDTA, 1mM sodium orthovanadate, 1 mM PMSF 10 µg/mL trypsin/chymotrypsin inhibitor, 1 mM bezamidine, and 0.1 mM leupeptin, pH 7.4). Each sample protein (30–90 µg) was separated onto 7.5% sodium dodecylsulfate–polyacrylamide gel electrophoresis (SDS-PAGE), transferred to a nitrocellulose membrane, and probed with proper diluted specific primary antibody. The immune complex was further probed with HRP-conjugated goat anti-rabbit IgG, visualized by HRP-reactive chemiluminescence reagents (Pierce, Rockford, IL), and developed on autoradiographic film (Kodak BioMax film, Eastman Kodak Co., Rochester, NY). The relative density of the protein band in the Western blot was further analyzed with an electrophoresis image analysis system (Eastman Kodak Co.). Net intensity of each protein band was normalized with the intensity of housekeeping GAPDH protein, and was then divided by the normalized intensity of the control group to obtain the ratio-to-control value. Data was further analyzed by unpaired t-test, and the graph was plotted using SigmaPlot 2001 (SPSS Inc., Chicago, IL).

Reverse Transcribed–Polymerase Chain Reaction (RT-PCR)

Total RNA was prepared from the liquid nitrogen quick-frozen tissues by directly homogenizing the tissues in extraction buffer (Trizol/phenol/chloroform), and mRNAs were reversed transcribed into cDNA using oligo-dT and reverse transcriptase (Invitrogen). Primer sets of rat β-actin or GAPDH genes and 2.5 μg of total RNA extracted from liver, cortex, or cerebellum were used. The level of housekeeping genes β-actin and GAPDH expressions were analyzed and used to demonstrate the presence of the same amount of total cDNA in each RNA sample. The expression levels of Bcl-2, Bcl-x_L, Bax, and internal control β-actin and GAPDH were detected from the same samples using designed primers. Sequences of the primer pair for amplification of each gene were 5'-CTA GAA GCA TTG CGG TGG ACG ATG GAG GG-3' and 5'-TGA CGG GGT CAC CCA CAC TGT GCC CAT CTA-3' (for β-actin gene); 5'-GAC CCC TTC ATT GAC CTC AAC-3'and 5'-GAT GAC CTT GCC CAC AGC CTT-3' (for glyceraldehydes-3-phosphate dehydrogenase; GAPDH gene); 5'-CAC CAG CTC TGA ACA GYY CAT GA-3' and 5'-TCA GCC CAT CTT CTT CCA GAT GGT-3' (for Bax gene); 5'-CAC CCC TGG CAT CTT CTC CTT-3' and 5'- AGC GTC TTC AGA GAC AGC CAG-3' (for Bcl-2 gene); 5'-AGGCTG GCGATG AGTTTG AA-3' and 5'-TGA AAC GCT CCT GGC CTT TC-3' (for Bcl-x_L gene). All amplification primers were synthesized bt GIBCO-BRL Technology. PCRs included 30 cycles of denaturation for 1 min at 94°C, annealing for 1 min at 55–60°C (various for each primer set) and extending for 1 min at 72°C. Reaction products were separated on 2% agarose gel. Band density was quantitated using Image analysis software as described in the Western blot analysis section, MATERIALS AND METHODS.

RESULTS AND DISCUSSION

Bcl-2 Gene Family Expression in the Liver of Gestational and Lactational Dioxin-Exposed Rat Offspring

Since liver is the primary target of dioxin, we first examined the expression profile of the Bcl-2 gene family in the liver of P0 pups and 4-month-old offspring. FIGURE 1 shows that the primary responsive gene CYP1A1 was strongly induced in the liver of both male and female P0 pups, assuring the positive response of the offspring rats to TCDD. The protein levels of CYP1A1 in TCDD-exposed male offspring at

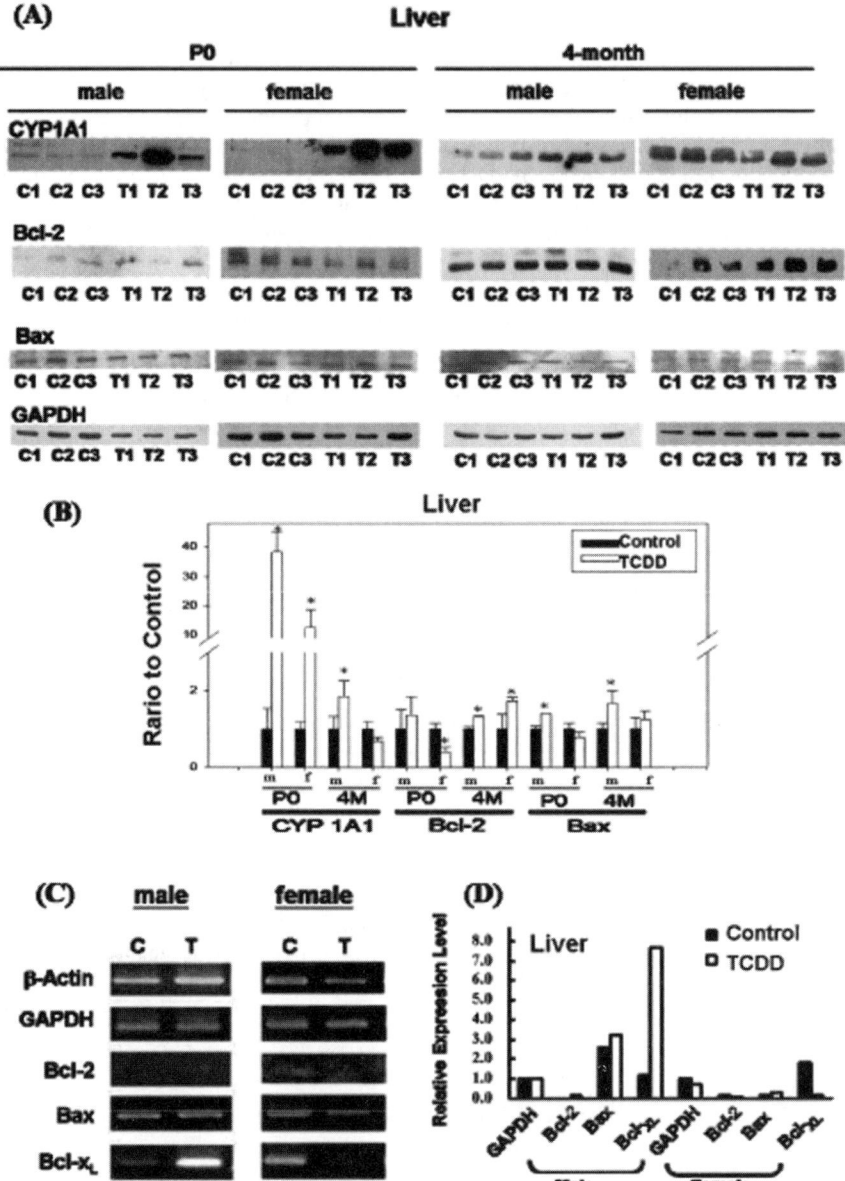

FIGURE 1. Western blot and RT-PCR analysis of expressions of CYP1A1, Bcl-2, Bcl-x_L, and Bax genes in liver of gestational and lactational dioxin-exposed rat offspring. Pregnant SD rats were orally administered with a single dose of TCDD (2 μg/kg/2 mL) or vehicle (corn oil 2 mL/2 kg) on GD15. All lobes of the liver were collected from offspring of TCDD-treated (T) or corn oil–treated (C) rats at P0 and at 4 months of age for Western blot (**A** and **B**) and RT-PCR analysis (**C** and **D**). **A** and **B**: The protein levels of CYP1A1, Bcl-2, and Bax

4 months were still higher than the control group. However, there were no significant changes of the CYP1A1 protein levels observed in 4-month-old female offspring, indicating that the primary effect of TCDD was only observed in the newborn stage, but did not sustain to the adult stage. These observations suggest that CYP1A1 may function as an acute factor for the TCDD-elicited toxicity in the newborn stage. Subsequent signals triggered by long-term exposure of TCDD, such as the desensitization of AhR, might also participate in TCDD neurotoxicity.

Examination of the Bcl-2 gene family expression showed that Bcl-x_L was much higher in the P0 TCDD-exposed male liver, but much lower in the P0 TCDD-exposed female liver as compared with their respective controls. The expression of Bcl-2 does not show not much difference between the control and the TCDD groups of P0 male, but seems to be lower in the P0 female of the TCDD group. Interestingly, TCDD-exposed offspring at 4 months had higher Bcl-2 expression in liver of both male and female offspring. The expression of pro-apoptotic gene Bax in liver was not significantly affected by the TCDD exposure in the age and gender of the offspring investigated. These results suggest that gestational exposure to dioxin, as shown in the responses of P0 pups, may induce a protective signaling in male livers, but not in female, in terms of the profound increased of Bcl-x_L expression. On the other hand, livers of female offspring might be more susceptible to dioxin cytotoxicity because both Bcl-2 and Bcl-x_L were profoundly downregulated. The higher expression of Bcl-2 in the liver of the TCDD-exposed 4-month-old female offspring suggests a possibility of carcinogenesis as a long-term effect, as reported in other literature on dioxin-induced carcinogenesis in liver.[7,8] Another possibility is that neonatal exposure to TCDD may cause a pre-stress to the liver, which in turn induces a self-protective mechanism, such as the induction of Bcl-2 expression for survival of the tissue.

Bcl-2 Gene Family Expression in the Cerebral Cortex of Gestational and Lactational Dioxin-Exposed Rat Offspring

The cerebral cortex controls the cognitive function that has been reported to be impaired in TCDD-exposed offspring in human. We, therefore, examined the expression profile of the Bcl-2 gene family in the cerebral cortex of P0 pups and 4-month-old offspring. FIGURE 2 shows that the primary responsive gene CYP1A1 was not greatly induced in the cerebral cortex of both male and female offspring at either P0 or 4 months old. However, in another study of our lab using primary cultured cortical neurons, we showed that TCDD could indeed induce CYP1A1 expression in a transient manner (unpublished data). Therefore, it is possible that the response of cerebral cortex to TCDD is rather transient as compared with the response in the liver.

---

in both P0 and 4-month-old male (m) and female (f) offspring upon TCDD exposure are shown. Data were calculated as described in MATERIALS AND METHODS, and were expressed as ratio to control (mean ± SEM, $n = 4$). $^*P < 0.05$, as compared with their respective controls. **C** and **D**: Expression levels of Bcl-2, and Bax, and Bcl-x_L in P0 male and female offspring upon TCDD exposure were analyzed by RT-PCR. For quantitation, the levels both of protein amount and gene expression were normalized with the level of housekeeping GAPDH gene expression.

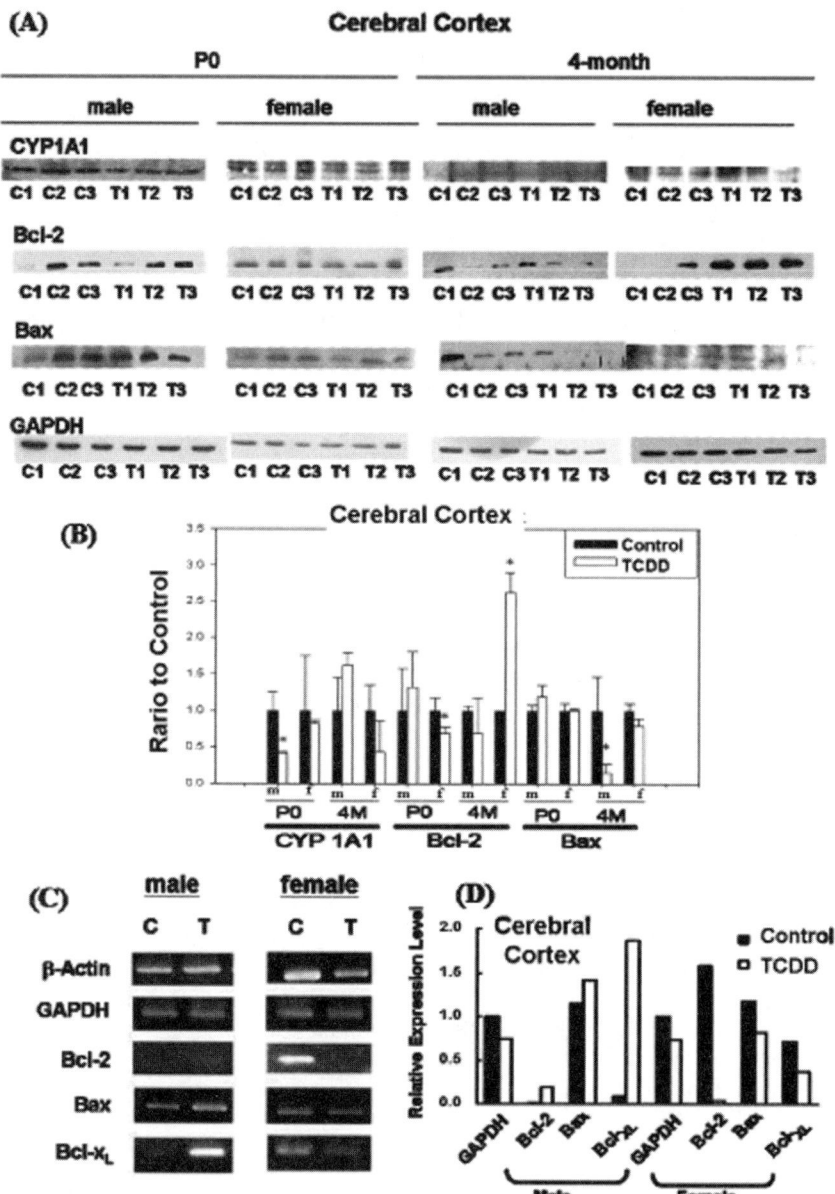

FIGURE 2. Western blot and RT-PCR analysis of expression of CYP1A1, Bcl-2, Bcl-x_L, and Bax genes in cerebral cortex of gestational and lactational dioxin-exposed rat offspring. Pregnant SD rats were orally administered with a single dose of TCDD (2 μg/kg/2 mL) or vehicle (corn oil 2 mL/2 kg) on GD15. Cerebral cortex was collected from offspring of TCDD-treated (T) or corn oil–treated (C) rats at P0 and at the 4 months of age for Western

Alternatively, it could reflect the fact that cerebral cortex is less responsive to TCDD in terms of the induction of CYP1A1 gene expression. There are several reports showing that dioxin or other PCBs have non-AhR-mediated effect in central neurons,[9] which subsequently results in a low induction level of CYP1A1.

In terms of the Bcl-2 gene family expression, expression profiles of both Bcl-2 and Bcl-x_L genes in the P0 pup cortex were similar to that in the P0 pup liver. Bcl-x_L was much higher in the P0 TCDD-exposed male cortex, but much lower in the P0 TCDD-exposed female cortex as compared with their respective controls. The expression of Bcl-2 was not much different in P0 male cortex, but was lower in the P0 female of the TCDD group. Similar to its expression profile in liver, the Bcl-2 expression of the TCDD group of 4-month-old female cerebral cortex was also much higher than the control group, but not in 4-month-old male offspring. These results suggest that cerebral cortex of female offspring at neonate stage might be more susceptible to the dioxin cytotoxicity since both Bcl-2 and Bcl-x_L were profoundly downregulated at P0. In addition, the expression of Bax in cerebral cortex was significantly reduced by the TCDD exposure in 4-month-old male but not female offspring. These results suggest that gestational exposure to dioxin may induce differential protective machinery in male and female cortex, with decreased Bax and increased Bcl-2 expression, respectively. Moreover, this observation raised the possibility that cerebral cortex of female rats might develop a long-term protective mechanism due to TCDD exposure in the fetal stage. This possibility comes from the fact that in our animal model system, TCDD-exposed offspring had a much higher mortality rate than the vehicle-exposed offspring (unpublished observation). Since the 4-month-old TCDD-exposed offspring rats were the "survivors" of the toxin insult, it is likely that they survived by means of upregulating protective machineries, such as Bcl-2 against apoptosis. These results also imply that there might be gender difference in the cognitive deficit observed in offspring of cohort upon gestation and lactation exposure to dioxin.

Bcl-2 Gene Family Expression in the Cerebellum of Gestational and Lactational Dioxin-Exposed Rat Offspring

The cerebellum serves functions related to working memory and motor planning, which also has been reported to be impaired in TCDD-exposed offspring in human. Accordingly, we examined the expression profile of the Bcl-2 gene family in the cerebellum of P0 pups and 4-month-old offspring. FIGURE 3 shows that the primary responsive gene CYP1A1 was not induced much in the cerebellum of both male and female offspring either at P0 or 4 months old. This result was similar to the one ob-

---

blot (**A** and **B**) and RT-PCR analysis (**C** and **D**). **A** and **B**: The protein levels of CYP1A1, Bcl-2, and Bax both in P0 and 4-month-old male (m) and female (f) offspring upon TCDD exposure are shown. Data were calculated as described in MATERIALS AND METHODS, and were expressed as ratio to control (mean ± SEM, $n = 4$). *$P < 0.05$, as compared with their respective controls. **C** and **D**: Expression levels of Bcl-2, and Bax, and Bcl-x_L in P0 male and female offspring upon TCDD exposure were analyzed by RT-PCR. For quantitation, the levels both of protein amount and gene expression were normalized with the level of housekeeping GAPDH gene expression.

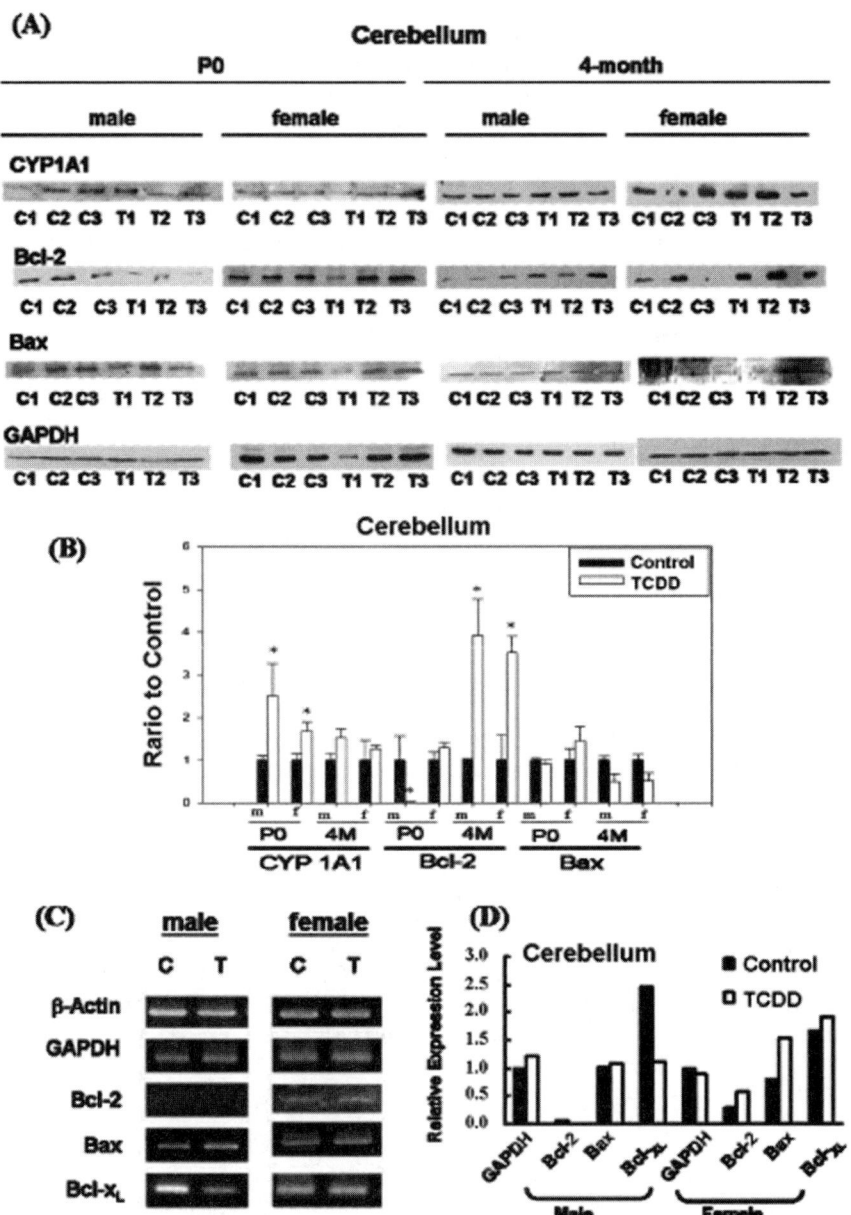

FIGURE 3. Western blot and RT-PCR analysis of expressions of CYP1A1, Bcl-2, Bcl-x_L, and Bax genes in the cerebellum of gestational and lactational dioxin-exposed rat offspring. Pregnant SD rats were orally administered with a single dose of TCDD (2 µg/kg/2 mL) or vehicle (corn oil 2 mL/2 kg) on GD15. Cerebellum was collected from offspring of TCDD-treated (T) or corn oil–treated (C) rats at P0 and at 4 months of age for Western blot

tained in the cerebral cortex, again implicating that the primary gene induction of TCDD in the brain might be transient in general as compared with the persistent elevation in liver.

The Bcl-2 gene family expression of the cerebellum of TCDD-exposed offspring has a distinct pattern different from those in liver and cerebral cortex. Bcl-x_L was much lower in the P0 TCDD-exposed male cerebellum, but no significant difference was observed in the P0 TCDD-exposed female cerebellum as compared with their respective controls. Bcl-2 was significantly lower in the P0 male cerebellum of the TCDD group, but was lower in the P0 female of the TCDD group. Similar to the expression profile in liver, the Bcl-2 expression in the cerebellum of the TCDD group was higher than the control group of both male and female offspring at 4 months. The expression of Bax in cerebellum was not significantly affected by the TCDD exposure in the age and gender of the offspring investigated. These results suggest that gestational exposure to dioxin induced a downregulation of both Bcl-2 and Bcl-x_L gene expressions in the cerebellum of male but not female offspring, rather than upregulation observed in the cerebral cortex. On the other hand, cerebellum of the female offspring might not be as susceptible as the liver and cerebral cortex to the dioxin cytotoxicity because both Bcl-2 and Bcl-x_L are not downregulated much. Higher expression of Bcl-2 was also observed in the cerebellum of both male and female TCDD-exposed offspring at 4 months. Similar explanations to the ones proposed for the Bcl-2 upregulation in the cerebral cortex are under investigation. Moreover, these results imply that the anti-apoptotic genes in different brain region are differentially regulated by dioxin.

CONCLUSION

The information obtained from this study suggests that the male and female offspring rats respond to the gestational dioxin exposure differently. The impact of the profound alteration of Bcl-x_L, Bcl-2, and Bax expressions in cerebral cortex and cerebellum of the dioxin-exposed offspring sheds a new light on the study of tissue-specific, brain region–specific, and gender-dependent neurodegeneration, possibly derived from environmental pollution. The differences may be due to the differential distribution and expression level of the Ah receptor in individual tissue, as well as the dioxin-elicited differential interruption of sex hormone functions in male and female.

(**A** and **B**) and RT-PCR analysis (**C** and **D**). **A** and **B**: The protein levels of CYP1A1, Bcl-2, and Bax both in P0 and 4-month-old male (m) and female (f) offspring upon TCDD exposure are shown. Data were calculated as described in MATERIALS AND METHODS , and were expressed as ratio to control (mean ± SEM, $n = 4$). *$P < 0.05$, as compared with their respective controls. **C** and **D**: Expression levels of Bcl-2, and Bax, and Bcl-x_L in P0 male and female offspring upon TCDD exposure were analyzed by RT-PCR. For quantitation, the levels of both protein amount and gene expression were normalized with the level of housekeeping GAPDH gene expression.

ACKNOWLEDGMENTS

This work was supported by grants from the National Science Council in Taiwan (NSC 91-2745-P038-005).

REFERENCES

1. COUTURE, L.A., B.D. ABBOTT & L.S. BIRNBAUM. 1990. A critical review of the developmental toxicity and teratogenicity of 2,3,7,8-tetrachlorodibenzo-p-dioxin: recent advances toward understanding the mechanism. Teratology **42:** 619–627.
2. WHITLOCK, J.P., JR. 1999. Induction of cytochrome P4501A1. Annu. Rev. Pharmacol. Toxicol. **39:** 103–125.
3. LIU, L., R.J. BRIDGES & C.L. EYER. 2001. Effect of cytochrome P450 1A induction on oxidative damage in rat brain. Mol. Cell. Biochem. **223:** 89–94.
4. SENFT, A.P. et al. 2002. Mitochondrial reactive oxygen production is dependent on the aromatic hydrocarbon receptor. Free Radic. Biol. Med. **33:** 1268–1278.
5. RENZING, J., S. HANSEN & D.P. LANE. 1996. Oxidative stress is involved in the UV activation of p53. J. Cell Sci. **109:** 1105–1112.
6. GOETZ, C.G., K.I. BOLLA & S.M. ROGERS. 1994. Neurologic health outcomes and Agent Orange: Institute of Medicine report. Neurology **44:** 801–809.
7. SCHANTZ, S.L. et al. 1997. Long-term effects of developmental exposure to 2,2′,3,5′,6-pentachlorobiphenyl (PCB 95) on locomotor activity, spatial learning and memory and brain ryanodine binding. Neurotoxicology **18:** 457–467.
8. STINCHCOMBE, S. et al. 1995. Inhibition of apoptosis during 2,3,7,8-tetrachlorodibenzo-p-dioxin-mediated tumour promotion in rat liver. Carcinogenesis **16:** 1271–1275.
9. Hanneman, W.H. et al. 1996. Stimulation of calcium uptake in cultured rat hippocampal neurons by 2,3,7,8-tetrachlorodibenzo-p-dioxin. Toxicology **112:** 19-28.

IL-5 Inhibits Apoptosis by Upregulation of c-myc Expression in Human Hematopoietic Cells

SHU-HUI JUAN,[a] JEFFREY JONG-YOUNG YEN,[b] HORNG-MO LEE,[c] AND HUEI-MEI HUANG[c]

[a]*Graduate Institute of Medical Sciences and Department of Physiology, Taipei Medical University, Taipei, Taiwan*

[b]*Institute of Biomedical Sciences, Academia Sinica, Taipei, Taiwan*

[c]*Graduate Institute of Cell and Molecular Biology, Center for Stem Cells Research at Wan-Fang Hospital, Taipei Medical University, Taipei, Taiwan*

ABSTRACT: Interleukin 5 (IL-5) inhibition of apoptosis is required throughout many hematopoietic lineages to regulate proliferation and differentiation. It is not clear how IL-5 mediates the intracellular molecular events regulating the anti-apoptotic effect. The cell lines TF-1 and JYTF-1 expressed different amounts of the IL-5 receptor α (IL-5Rα) subunit, which caused contrasting effects in response to IL-5. IL-5 supported the survival but not the anti-apoptotic activities of TF-1 cells, which have a low expression of IL-5Rα. In contrast, IL-5 supported both the survival and the anti-apoptotic activities of JYTF-1 cells, which overexpressed IL-5Rα compared to TF-1 cells. In this study, IL-5 stimulation increased Annexin V binding (indicative of apoptosis) in TF-1 cells and decreased apoptosis in JYTF-1 cells. The proto-oncogenes c-fos, fosB, and c-jun were not detected, whereas junB was induced by IL-5 stimulation in both types of cells. Most importantly, IL-5 significantly induced c-myc expression in JYTF-1 cells, but did not in TF-1 cells. These results are consistent with the possibility that IL-5 inhibits apoptosis in JYTF-1 cells via the upregulation of c-myc expression.

KEYWORDS: interleukin 5; IL-5 receptor α subunit; c-myc; anti-apoptosis

INTRODUCTION

In the human hematopoietic system, interleukin 5 (IL-5) plays an important role in the proliferation and the terminal differentiation of eosinophil precursors. IL-5 also regulates the growth, survival, and functions of eosinophils. In addition, IL-5 participates in regulating the differentiation of basophils and mast cells (reviewed in Ref. 1). IL-5 acts through its receptor, the IL-5 receptor (IL-5R), which is composed

---

Address for correspondence: Huei-Mei Huang, Graduate Institute of Cell and Molecular Biology, Taipei Medical University, No. 250 Wu-Shing Street, Hsinyi District, Taipei 110, Taiwan. Voice: +886-2-2930-7930 ext. 2542; fax: +886-2-8663-8283.
cmbhhm@tmu.edu.tw

of two membrane proteins, the α and β subunits.[2] The IL-5Rα subunit alone specifically binds IL-5 with low affinity. The β subunit is the common β subunit (βc) for receptors for IL-3 and granulocyte-macrophage colony-stimulating factor (GM-CSF) as well as IL-5. Although βc does not bind IL-5 by itself, it facilitates the high-affinity binding of the IL-5 ligand in combination with IL-5Rα.[2] The βc subunit is not only required for high-affinity binding, but it is also crucial for signal transduction. IL-5Rα also has an important role in transmitting a proliferative signal through IL-5R.[3] The specific binding of IL-5 to IL-5Rα recruits the βc subunit; this complex results in rapid tyrosine phosphorylation of βc and lead to the induction of many nuclear transcription factors such as c-Fos, c-Jun, and c-Myc.[3,4] The components of activator protein 1 (AP-1), c-Fos and c-Jun, as well as c-Myc are known to be able to regulate cell proliferation and anti-apoptosis.[5,6]

Previous studies have demonstrated that TF-1 cells, a human hematopoietic progenitor cell line established from the bone marrow of an erythroleukemic patient,[7] expressed high levels of GM-CSF receptor α subunit, but very low levels of IL-5Rα.[8] The growth of TF-1 cells is GM-CSF dependent. TF-1 cells can replicate DNA and undergo mitogenesis, but cannot prevent apoptosis in the presence of added IL-5. The cell line JYTF-1 is modified from TF-1 to overexpress IL-5Rα; JYTF-1 cells can replicate DNA, undergo mitogenesis, and prevent apoptosis, in IL-5 medium.[8] Furthermore, studies have demonstrated that the IL-5R signals through both MEK/MAPK and PI-3K/Akt pathways to regulate Mcl-1 expression and to prevent apoptosis.[9]

In the present study, we identified the nuclear transcription factors involved in the anti-apoptotic activity of IL-5 by comparing the differential responses of TF-1 and JYTF-1 cells to IL-5 stimulation. Our results suggest that c-myc expression is closely associated with the anti-apoptotic effect of IL-5 in JYTF-1 cells.

MATERIALS AND METHODS

Cell Lines, Culture Conditions, and Cytokines

The TF-1 cell line and its variant, JYTF-1,[8] were both maintained in RPMI1640 supplemented with 10% fetal bovine serum (FBS, GIBCO, Grand Island, NY), 50 μM β-mercaptoethanol, 2 mM L-glutamine, 100 U/mL penicillin G, 100 μg/mL streptomycin, and 2 ng/mL GM-CSF. Recombinant human GM-CSF and IL-5 were purchased from R & D Systems (Minneapolis, MN).

Detection of Apoptosis

Apoptotic cells were determined by Annexin V staining of phosphatidylserine (PS) on the outer surface of the cell membrane in apoptotic cells. Cells were cultured in GM-CSF, IL-5, or cytokine-free medium for 24 h, collected by centrifugation, and washed with PBS. The cells were then stained with Annexin V-FITC (Apoptosis Detection Kit, Oncogene Research Products, San Diego, CA) and incubated for 10–15 min at RT in the dark. After washing, the cells were analyzed using a FACScan flow cytometer (Becton Dickinson, Mountain View, CA).

Northern Blot Analysis

Total RNA was isolated from using TRIzol (GIBCO). Twenty µg of each RNA sample was separated in 1% formaldehyde agarose gel and transferred to a Hybond N$^+$ membrane. The RNA blot was probed with [α-^{32}P]-dCTP labeled cDNA for the c-fos, fosB, c-jun, jun B, c-myc, and gapdh genes. Hybridization was carried out at 42°C using standard buffer containing 50% formamide. After hybridization overnight, the blot was washed once in 2X SSC + 0.1% SDS at RT, twice in 0.2X SSC + 0.1%SDS at 55°C for 15 min, and subjected to autoradiography.

RESULTS

IL-5 Inhibited Apoptosis of JYTF-1 Cells

TF-1 and JYTF-1 cells were cultured in cytokine-free, GM-CSF, or IL-5 medium for 24 h. The cells were then stained with Annexin V-FITC and quantified by flow cytometry. As shown in FIGURE 1, about 3% and 5% of GM-CSF–treated cells were labeled with Annexin V-FITC in TF-1 and JYTF-1 cells, respectively. The maximum number of apoptotic cells was observed in cytokine-free medium, 78% in TF-1 cells and 74% in JYTF-1 cells. IL-5–treated JYTF-1 cells did not show significant changes in the percentage of dead cells compared to the GM-CSF–treated cells. However, the Annexin V-FITC–positive TF-1 cells reached 52% in the presence of IL-5.

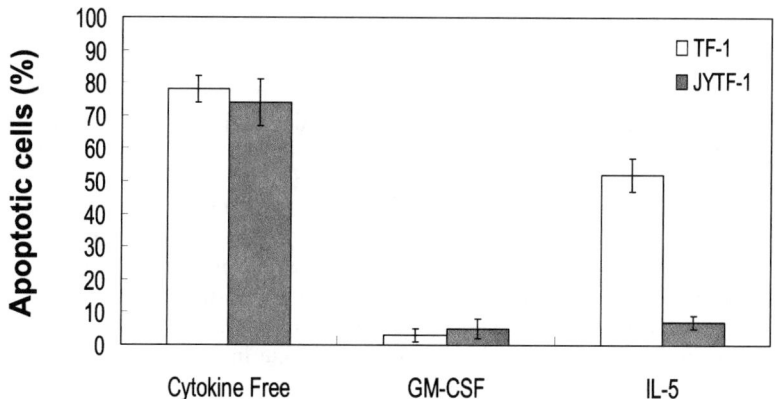

FIGURE 1. Apoptosis was inhibited by IL-5 in JYTF-1 cells, but not in TF-1 cells. Both TF-1 and JYTF-1 cells were cultured in cytokine-free, GM-CSF, or IL-5 media for 24 h. The cells were incubated with Annexin V-FITC and then analyzed using a FACScan® flow cytometer. The results are expressed as the percentage of apoptotic Annexin V binding cells and are shown as the percentage of the initial viable cell number. The data are means of three independent experiments ± standard deviations. The difference in IL-5–treated group between TF-1 cells and JYTF-1 cells was statistically significant ($P < 0.005$).

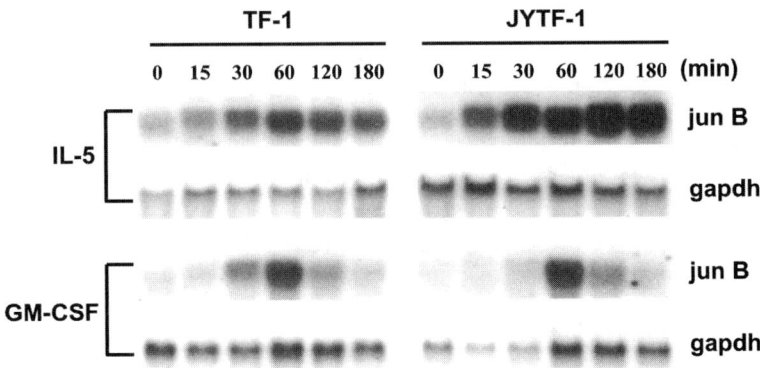

FIGURE 2. IL-5 induced expression of junB genes in TF-1 and JYTF-1 cells. Both TF-1 and JYTF-1 cells were depleted of GM-CSF for 24 h prior to stimulation either with IL-5 or GM-CSF. Cells were collected and the total RNA was prepared at each time point. Total RNA (20 μg) was fractionated in a 1% formaldehyde agarose gel and subjected to a standard Northern blot analysis. The RNA blots were hybridized with ^{32}P-labeled cDNA probes of junB and gapdh. The expression level of gapdh was used as an internal control of RNA loading.

IL-5 Induced c-myc Expression Correlated with Anti-Apoptotic Activity of JYTF-1 Cells

Nuclear oncogenes play crucial roles in the regulation of proliferation, differentiation, and anti-apoptosis. Since it has been shown that IL-5 was able to induce the expressions of c-fos, c-jun, and c-myc,[3,4] we investigated the correlation between the induction of the nuclear oncogenes and the anti-apoptotic effect of IL-5. The Fos family includes c-Fos and FosB, and the Jun family of proteins includes c-Jun and JunB. First, we measured the induction of c-fos, fosB, c-jun, and junB by stimulation with GM-CSF and IL-5 in TF-1 and JYTF-1 cells, using Northern blot analysis. The expressions of c-fos, fosB, and c-jun were undetectable in response to IL-5 stimulation (data not shown). However, junB was highly inducible by both GM-CSF and IL-5 in TF-1 and JYTF-1 cells, as shown in FIGURE 2. Second, we determined the expression of c-myc in response to GM-CSF and IL-5 (FIG. 3A). The expression of c-myc was induced very well by GM-CSF both in TF-1 and JYTF-1 cells. In contrast, c-myc was not induced by IL-5 in TF-1 cells, but induced very well by IL-5 in JYTF-1 cells. The slight increase of RNA level of c-myc gene at time points of 60 min and 180 min by IL-5 treatment in TF-1 cells were not reproducible. The c-myc induction in TF-1 and JYTF-1 cells by IL-5 was determined in triplicate, and the results were quantified by densitometry. The data are summarized in FIGURE 3B. On average, there was a threefold increase in c-myc expression induced by IL-5 in JYTF-1 cells, but no increase observed in TF-1 cells. Therefore, we conclude that IL-5 is unable to induce the expression of c-myc in TF-1 cells, and the defect is corrected in JYTF-1 cells.

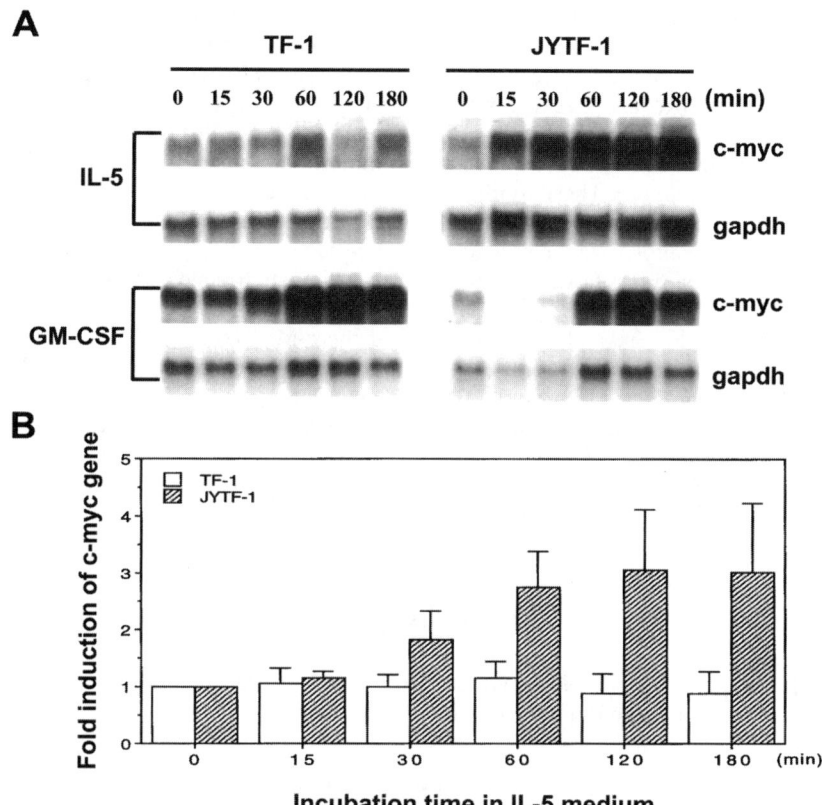

FIGURE 3. Upregulation of c-myc expression was induced by IL-5 in JYTF-1 cells, but not in TF-1 cells. (A) Defective induction of the c-myc gene by IL-5 in TF-1 cells. Both TF-1 and JYTF-1 cells were stimulated either with IL-5 or GM-CSF. The RNA blots were hybridized with ^{32}P-labeled cDNA probes of c-myc and gapdh. (B) Induction of c-myc expression in TF-1 and JYTF-1 by IL-5. Three independent Northern blot analyses of IL-5–stimulated TF-1 and JYTF-1 cells were performed as described in **panel A**. These blots were probed with c-myc and gapdh cDNA and autoradiograms were quantitated by densitometry. The arbitrary optical densities of the c-myc RNA at various time intervals were first normalized to the gapdh internal loading controls, then divided by the amount of c-myc at "0" time point to obtain "Fold induction of c-myc gene." The value at each time point represents the average of three experiments. The difference of c-myc induction in the IL-5–treated group between TF-1 cells and JYTF-1 cells was statistically significant ($P < 0.05$).

DISCUSSION

IL-5 stimulates the mitogenesis of TF-1 and JYTF-1 cells, but only prevents apoptosis of JYTF-1 cells. The present study showed that the expression of junB gene was induced by IL-5 in TF-1 and JYTF-1 cells (FIG. 2). The JunB transcription factor has been implicated in the control of mitogenesis.[10] These data suggest an important role for JunB in mitogenesis of both types of cells. By taking advantage of the differential responses of TF-1 and JYTF-1 cells to IL-5, we explored whether nuclear transcriptional factors are responsible for the anti-apoptotic effect of IL-5. We found that IL-5 was unable to induce c-myc expression in the TF-1 cell line with low expression of IL-5Rα and that IL-5 was unable to suppress apoptosis in this cell line (FIGS. 1 and 3). Increased expression of IL-5Rα in JYTF-1 cells resulted in augmentation of the induction level of c-myc and in inhibition of apoptosis on the addition of IL-5 (FIGS. 1 and 3). These results suggest that c-myc expression is closely associated with the death inhibition ability of IL-5. We will further investigate whether IL-5 inhibits apoptosis via the upregulation of c-myc expression in JYTF-1 cells using antisense inhibitors targeted against c-myc. Other studies have shown that inappropriate downregulation of c-Myc can induce apoptosis under certain circumstances. For example, the reduction of the c-Myc level by DNA damaging agents is closely correlated with cellular apoptosis. The overexpression of c-Myc can inhibit apoptosis induced by DNA damaging agents.[11] Several studies found that lowered c-Myc content in different cell systems such as Jurkat cells or B-lymphoid cells led to apoptosis.[12,13] Therefore, the level of c-Myc is carefully controlled in hematopoietic cells and is important in the control of cell activities, including proliferation, differentiation, and apoptosis.

ACKNOWLEDGMENTS

This work was supported by Grant NSC 92-2320-B-038-050 from the National Science Council of the Republic of China.

REFERENCES

1. TAKATSU, K. 1992. Interleukin-5. Curr. Opin. Immunol. **4:** 299–306.
2. TAVERNIER, J., R. DEVOS, S. CORNELIS, et al. 1991. A human high affinity interleukin-5 receptor (IL5R) is composed of an IL5-specific α chain and a β chain shared with the receptor for GM-CSF. Cell **66:** 1175–1184.
3. TAKAKI, S., H. KANAZAWA, M. SHIIBA, et al. 1994. A critical cytoplasmic domain of the interleukin-5 (IL-5) receptor alpha chain and its function in IL-5-mediated growth signal transduction. Mol. Cell. Biol. **14:** 7404–7413.
4. MIGITA, M., N. YAMAGUCHI, S. KATOH, et al. 1990. Elevated expression of proto-oncogenes during interleukin-5-induced growth and differentiation of murine B lineage cells. Microbiol. Immunol. **34:** 937–952.
5. SHAULIAN, E. & M. KARIN. 2001. AP-1 in cell proliferation and survival. Oncogene **20:** 2390–2400.
6. PELENGARIS, S. & M. KHAN. 2003. The many faces of c-MYC. Arch. Biochem. Biophys. **416:** 129–136.
7. KITAMURA, T., T. TANGE, T. TERASAWA, et al. 1989. Establishment and characterization of a unique human cell line that proliferates dependently on GM-CSF, IL-3, or erythropoietin. J. Cell. Physiol. **140:** 323–334.
8. HUANG, H.M., J.C. LI, Y.C. HSIEH, et al. 1999. Optimal proliferation of a hematopoietic progenitor cell line requires either costimulation with stem cell factor or increase of receptor expression that can be replaced by overexpression of Bcl-2. Blood **93:** 2569–2577.
9. HUANG, H.M., C.J. HUANG & J.J. YEN. 2000. Mcl-1 is a common target of stem cell factor and interleukin-5 for apoptosis prevention activity via MEK/MAPK and PI-3K/Akt pathways. Blood **96:** 1764–1771.

10. MATHAS, S., M. HINZ, I. ANAGNOSTOPOULOS, *et al.* 2002. Aberrantly expressed c-Jun and JunB are a hallmark of Hodgkin lymphoma cells, stimulate proliferation and synergize with NF-kappa B. EMBO J. **21:** 4104–4113.
11. JIANG, M.R., Y.C. LI, Y. YANG, *et al.* 2003. c-Myc degradation induced by DNA damage results in apoptosis of CHO cells. Oncogene **22:** 3252–3259.
12. WOOD, A.C., C.M. WATERS, A. GARNER & J.A. HICKMAN. 1994. Changes in c-myc expression and the kinetics of dexamethasone-induced programmed cell death (apoptosis) in human lymphoid leukaemia cells. Br. J. Cancer **69:** 663–669.
13. SONENSHEIN, G.E. 1997. Down-modulation of c-myc expression induces apoptosis of B lymphocyte models of tolerance via clonal deletion. J. Immunol. **158:** 1994–1997.

Detection of Apoptosis and Necrosis in Normal Human Lung Cells Using ^{1}H NMR Spectroscopy

CHWEN-MING SHIH,[a] WUN-CHANG KO,[b] LIANG-YO YANG,[c]
CHIEN-JU LIN,[a,d] JUI-SHENG WU,[a,d] TSUI-YUN LO,[a]
SHWU-HUEY WANG,[a] AND CHIEN-TSU CHEN[a]

[a]*Department of Biochemistry, Taipei Medical University, Taipei 110, Taiwan*

[b]*Graduate Institute of Pharmacology, Taipei Medical University, Taipei 110, Taiwan*

[c]*Department of Physiology, Taipei Medical University, Taipei 110, Taiwan*

[d]*Graduate Institute of Medical Science, Taipei Medical University, Taipei 110, Taiwan*

ABSTRACT: This study aimed to detect apoptosis and necrosis in MRC-5, a normal human lung cell line, by using noninvasive proton nuclear magnetic resonance (^{1}H NMR). Live MRC-5 cells were processed first for ^{1}H NMR spectroscopy; subsequently their types and the percentage of cell death were assessed on a flow cytometer. Cadmium (Cd) and mercury (Hg) induced apoptosis and necrosis in MRC-5 cells, respectively, as revealed by phosphatidylserine externalization on a flow cytometer. The spectral intensity ratio of methylene (CH_2) resonance (at 1.3 ppm) to methyl (CH_3) resonance (at 0.9 ppm) was directly proportional to the percentage of apoptosis and strongly and positively correlated with PI staining after Cd treatment ($r^2 = 0.9868, P < 0.01$). In contrast, this ratio only increased slightly within 2-h Hg treatment, and longer Hg exposure failed to produce further increase. Following 2-h Hg exposure, the spectral intensity of choline resonance (at 3.2 ppm) was abolished, but this phenomenon was absent in Cd-induced apoptosis. These findings together demonstrate that ^{1}H NMR is a novel tool with a quantitative potential to distinguish apoptosis from necrosis as early as the onset of cell death in normal human lung cells.

KEYWORDS: cadmium; mercury; apoptosis; necrosis; NMR

INTRODUCTION

When cells are exposed to cytotoxic agents, there are two major types of cell death: apoptosis and necrosis. Cell shrinkage, DNA damage, chromatin condensation and blebbing of the plasma, and alteration of plasma membrane phospholipids organization with phosphatidylserine externalization are major characteristics of

Address for correspondence: Dr. Chien-Tsu Chen, Department of Biochemistry, School of Medicine, Taipei Medical University, 250 Wu-Hsing Street, Taipei, Taiwan 110, ROC. Voice: +886-2-27361661 ext. 2400; fax: +886-2-27387348.
chenctsu@tmu.edu.tw

apoptosis,[1] whereas necrosis is generally characterized by swelling of cells and mitochondria, scattered chromatin condensation, and loss of plasma membrane integrity due to an overwhelmingly physical cell injury.[2] Proton nuclear magnetic resonance spectroscopy (^{1}H NMR) has been applied to study apoptotic cell death *in vitro*,[3–10] and the onset of apoptosis revealed by ^{1}H NMR is accompanied by an increase in the signal intensity of the membrane lipid methylene (CH_2) resonance (at 1.3 ppm). In this study, we investigated whether ^{1}H NMR can distinguish apoptosis from necrosis in normal human lung fibroblasts, MRC-5, triggered by cadmium (Cd) or mercury (Hg). Our results showed that ^{1}H NMR could detect and quantify different degrees of apoptosis. These findings strongly suggest that ^{1}H NMR has a great potential to become a noninvasive tool for detection of cell death in humans.

MATERIALS AND METHODS

Cell Culture

MRC-5 cells, normal human fetal lung fibroblasts, were obtained from American Tissue Culture Collection (ATCC CCL-171) and grown at 37°C in Dulbecco's Modified Eagle's Medium (DMEM) supplemented with 10% heat-inactivated fetal bovine serum (FBS), 100 U/mL penicillin, and 100 mg/mL streptomycin (pH 7.4) in a humidified atmosphere containing 5% CO_2. Because MRC-5 cells are normal human cells, all of the experiments were performed at 25–35 passages.

^{1}H NMR Spectral Analysis

^{1}H NMR spectroscopy was performed using methods published by Francis *et al.*[4,5] In brief, 5×10^7 MRC-5 cells were harvested and washed twice with D_2O-made PBS, suspended in a final volume of 500 μL, and placed immediately on ice until data acquisition. Samples were analyzed on a 500-MHz high-resolution Bruker spectrometer (Bruker; Karlsruhe, Germany) with the following settings: pulse-acquire, 90° flip angle, repetition time 10 s, 64 or 128 excitations (depending on desired signal to noise), 8 k points, and 5-kHz bandwidth. A coaxial tube filled with trimethysialoproponic acid (TSP), 0.1% solution in D_2O was used as reference (0.0 ppm) for each experiment. The relative areas underneath the CH_2 and methyl (CH_3) resonances (at 1.3 and 0.9 ppm, respectively) were calculated by integration of the proton spectrum using the trough between the CH_2 and CH_3 resonances as a baseline reference.

Measurement of Phosphatidylserine Externalization

Phosphatidylserine (PS) externalization was examined with a two-color analysis of FITC-labeled Annexin V binding and propidium iodine (PI) uptake using flow cytometry.[11] For this analysis, 1×10^6 MRC-5 cells were stained according to the manufacturer's protocol (Annexin-VFLUOS staining kit, Roche, Mannheim, Germany) and analyzed on a Becton Dickinson (San Jose, CA) FACSCalibur flow cytometer. Cell debris, characterized by a low FSC/SSC, was excluded from analysis. Cells labeled with FITC-Annexin V or PI were used to adjust the compensation. Data acquisition and analysis were performed using the CellQuest program (Becton

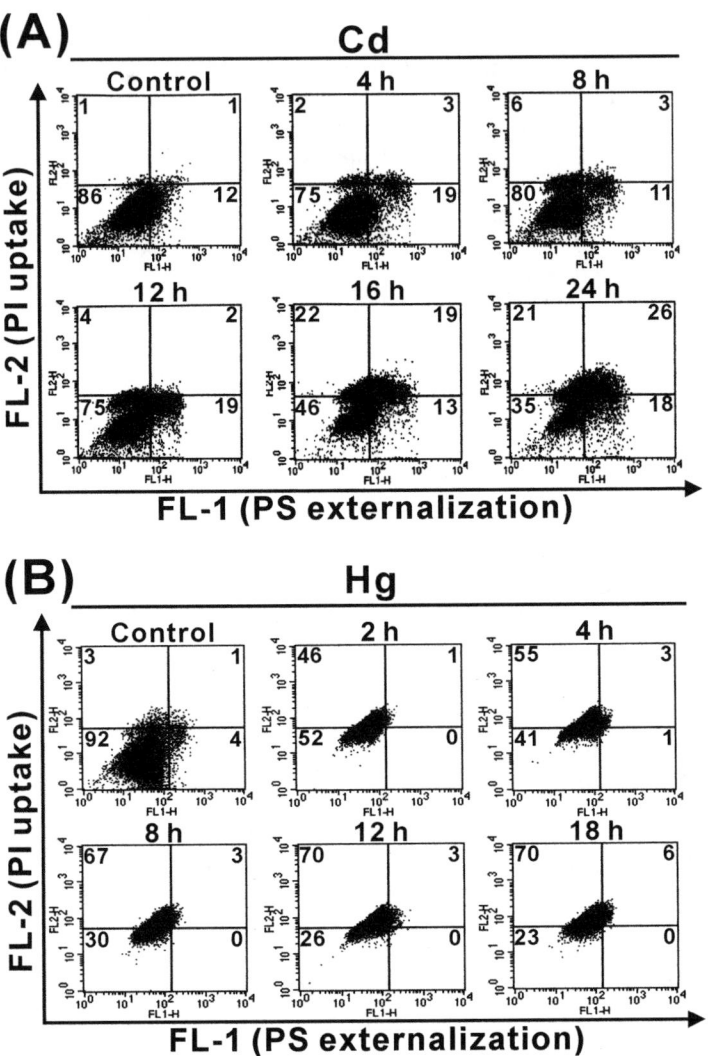

FIGURE 1. Time-course experiments of Cd-induced apoptosis and Hg-induced necrosis in MRC-5 cells. MRC-5 cells were treated with 100 μM $CdCl_2$ (**panel A**) or 100 μM $HgCl_2$ (**panel B**) for the indicated time periods, collected, and stained with Annexin-V-FLUOS staining kit (Roche), and then immediately subjected to analysis of phosphatidylserine externalization (FL-1 level of FITC-Annexin V fluorescence, X-axis) and PI uptake (FL-2 level of PI fluorescence, Y-axis) using flow cytometry. The Arabic number in each corner indicates the proportion of each quadrant. The cytogram of four quadrants was used to distinguish the normal, primary apoptotic, late apoptotic, and necrotic cells by the criteria of Annexin V^-/PI^-, Annexin V^+/PI^-, Annexin V^+/PI^+, and Annexin V^-/PI^+, respectively (see MATERIALS AND METHODS for details). The proportion of total apoptosis was summed up from that of primary (Annexin V^+/PI^-) and late apoptosis (Annexin V^+/PI^+).

Dickinson). Positioning of quadrants on Annexin V/PI dot plots was performed as reported,[12] and this method can be used to distinguish between living cells (Annexin V^-/PI^-), early apoptotic/primary apoptotic cells (Annexin V^+/PI^-), late apoptotic/secondary necrotic cells (Annexin V^+/PI^+), and necrotic cells (Annexin V^-/PI^+).

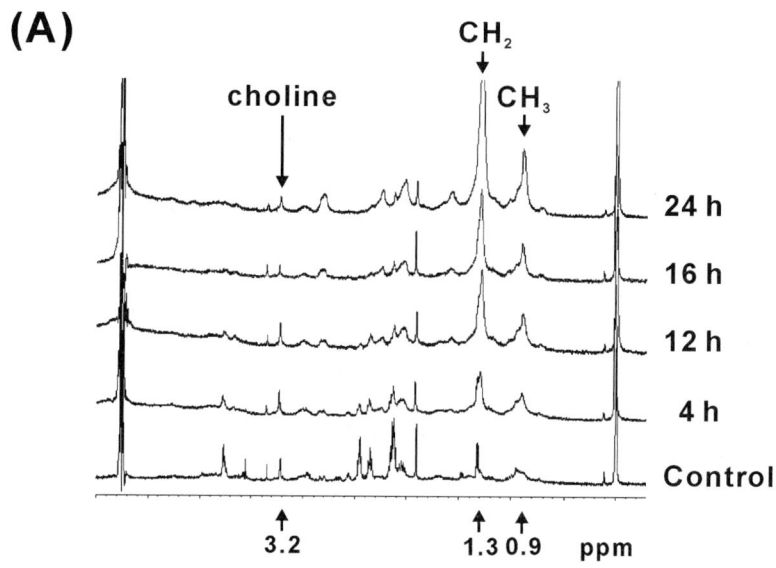

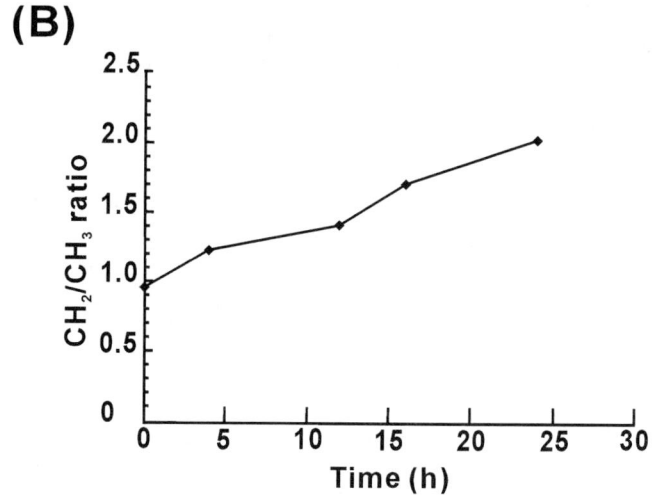

FIGURE 2. ^{1}H NMR spectra of Cd-treated MRC-5 cells. (**A**) MRC-5 cells were treated with 100 μM $CdCl_2$ for the indicated time periods and resuspended in 90% D_2O/PBS before measurement of ^{1}H NMR spectra obtained at 500 MHz. The spectral resonances of choline protons (-N(CH$_3$)) at 3.2 ppm, methylene protons (-CH$_2$-) at 1.3 ppm, and methyl protons (-CH$_3$) at 0.9 ppm are indicated. The CH$_2$/CH$_3$ signal intensity ratios were 0.96, 1.21, 1.40, 1.70, and 2.02 at 0, 4, 12, 16, and 24 h after Cd exposure, respectively (**B**).

Determination of Hypodiploid DNA Content

To measure the loss of DNA, MRC-5 cells were harvested at 1×10^6 cells/mL, washed with PBS, and fixed in ice-cold 70% ethanol for 30 min at 4°C. After centrifugation, cells were resuspended, incubated in PBS containing 0.5 mg/mL RNase A and 40 µg/mL PI at room temperature for 30 min, and analyzed using a Becton Dickinson FACSCalibur flow cytometer as described by Ormerod et al.[13] Cells with sub-G1 (hypodiploid DNA) PI incorporation were considered apoptotic.[13]

RESULT AND DISCUSSION

Time Course of Cd- or Hg-Induced Cell Death in MRC-5 Cells

MRC-5 cells were incubated with 100 µM $CdCl_2$ for 0, 2, 4, 8, 12, 16, and 24 h or cultured with 100 µM $HgCl_2$ for 0, 2, 4, 8, 12, and 18 h. To investigate the types of cell death induced by Cd or Hg, PS externalization and PI uptake in intact MRC-5 cells following Cd or Hg treatment were analyzed with a flow cytometer. FIGURE 1 is a dot plot of four quadrants scaled with logarithm as fluorescence level of FITC-labeled Annexin V (FL-1) and PI (FL-2), respectively. Cd-treated cells showed increased PS externalization with time elapsed (FIG. 1A), indicating that apoptosis was induced by Cd. In contrast, Hg-treated cells showed loss of plasma membrane integrity without PS externalization (FIG. 1B) which is a typical characteristic of necrosis.

^1H NMR Spectral Analysis of Cd- and Hg-Treated MRC-5 Cells

MRC-5 cells were incubated with 100 µM $CdCl_2$ and were harvested at 0, 4, 12, 16, and 24 h later. After resuspension in 90% D_2O-made $1 \times PBS$, cells were immediately processed for acquisition of the ^1H NMR spectra that is shown in FIG. 2A. It is worth noting that there was a progressive decrease in the choline signal (3.2 ppm) after Cd treatment. Most importantly, the CH_2/CH_3 signal intensity ratio increased from 0.96 (control) to 2.02 (at 24 h) (FIG. 2B).

Following incubation with 100 µM $HgCl_2$ for 0, 2, 4, 12, 16, or 24 h, MRC-5 cells were processed for acquisition of the ^1H NMR spectra. The CH_2/CH_3 signal intensity ratio rose from 0.92 (control) to 1.31 within 2 h after Hg exposure and reached a plateau (FIG. 3B), which was different from the pattern induced by Cd. Crucially, unlike Cd treatment, Hg was unable to evoke the choline signal in MRC-5 cells measured by ^1H NMR (FIG. 3A). These data demonstrate that the ^1H NMR can differentiate necrosis from apoptosis in MRC-5 cells treated with Hg or Cd, respectively, by differential CH_2/CH_3 signal intensity ratio and choline signal.

Correlation between the Hypodiploid DNA Content and CH_2/CH_3 Ratio

To characterize the nuclear events, Cd-treated MRC-5 cells were assessed by hypodiploid DNA assay (FIG. 4A). The percentage of cells with hypodiploid DNA (denoted by M1) (FIG. 4B) was similar to that of the cells undergoing apoptosis as revealed by the PS externalization assay showed in FIGURE 1. FIGURE 4C showed a linear regression fit of the percentage of apoptosis versus the spectral intensity ratio of the CH_2/CH_3 (1.3/0.9 ppm) resonances. The percentage of apoptosis was highly

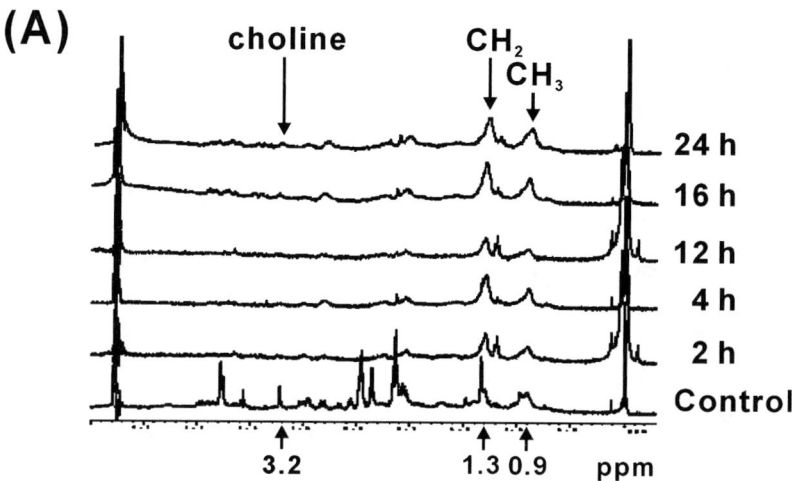

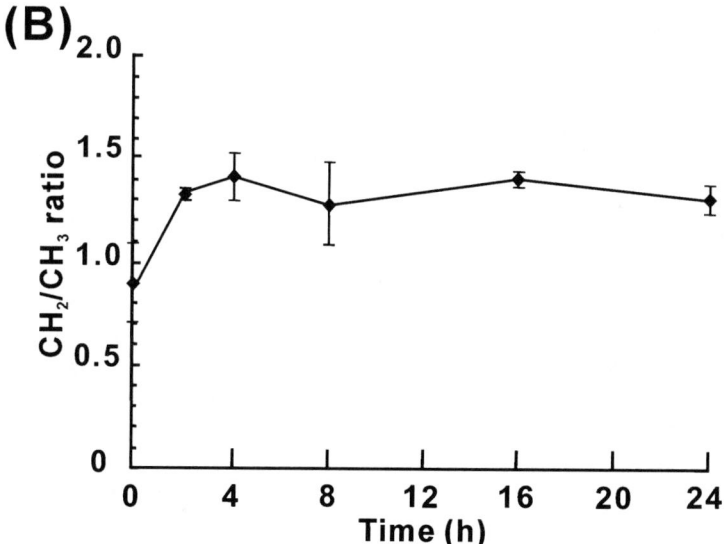

FIGURE 3. ^{1}H NMR spectra of Hg-treated MRC-5 cells. (**A**) MRC-5 cells were treated with 100 μM HgCl$_2$ for the indicated time periods and resuspended in 90% D$_2$O/PBS before measurement of ^{1}H NMR spectra obtained at 500 MHz. The spectral resonances of choline protons (-N(CH$_3$)) at 3.2 ppm, methylene protons (-CH$_2$-) at 1.3 ppm, and methyl protons (-CH$_3$) at 0.9 ppm are indicated. (**B**) The CH$_2$/CH$_3$ signal intensity ratios were 0.92, 1.31, 1.43, 1.28, 1.41, and 1.31 at 0, 2, 4, 12, 16, and 24 h following Hg exposure, respectively. It is worth noting that the CH$_2$/CH$_3$ signal intensity ratios reached the plateau within 2 h after Hg treatment.

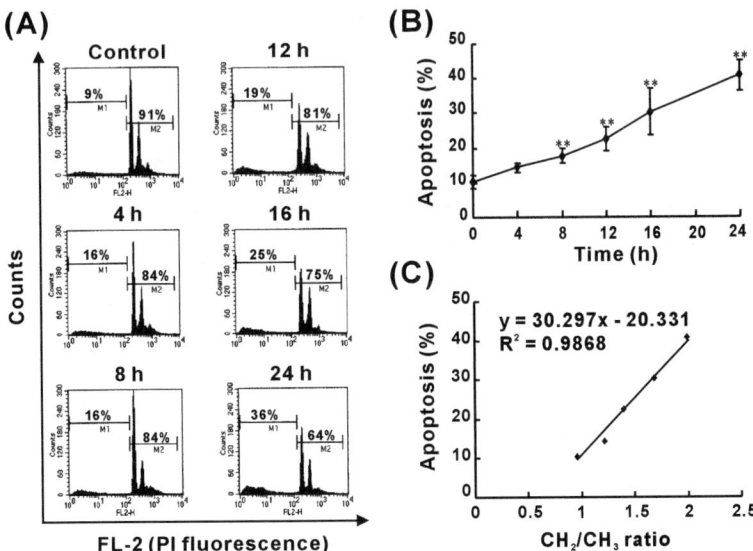

FIGURE 4. Correlation between the hypodiploid DNA content and CH_2/CH_3 ratio in Cd-induced apoptosis. (**A**) MRC-5 cells were treated with 100 μM $CdCl_2$ for the indicated time periods and then subjected to flow cytometric analysis with PI staining as described in MATERIALS AND METHODS. M1 was presented as the percentage of hypodiploid DNA in total DNA content, indicating the apoptotic percentage. Data presented in (**A**) are representative of three independent experiments. (**B**) The percentage of hypodiploid DNA content (denoted as M1) in Cd-induced MRC-5 apoptosis increased with time (**, $P < 0.01$) and data represented mean ± SD. (**C**) The linear regression analysis showed a strong and positive correlation between the percentage of apoptosis and the CH_2/CH_3 (1.3/0.9 ppm) NMR spectral ratio in Cd-induced MRC-5 apoptosis.

and positively correlated with the spectral intensity ratio of the CH_2/CH_3 with a very high correlation coefficient ($r^2 = 0.9868$), suggesting that the simple spectral intensity ratio of the CH_2/CH_3 can be used to estimate the extent of apoptosis, a very complicated process.

Comparison of ^1H NMR Spectra between Apoptosis and Necrosis

The ^1H NMR spectra between apoptosis and necrosis induced by Cd and Hg, respectively, were compared as shown in FIGURE 5. Please note that the CH_2/CH_3 intensity ratio around 1.4 generated either by Cd (12 h)- or by Hg (4 h)-treatment is listed. The intensity of the CH_3 resonance (0.9 ppm) increased in Cd-induced apoptosis, but did not change in Hg-induced necrosis. The decrease in the choline intensity (3.2 ppm) was much more obvious in necrotic cell death than in apoptosis. Compared with the control, the resonances of apoptosis or necrosis in the region between 3.4 and 3.9 ppm (consistent with myoinositol and ethanolamine) and the resonances between 2.1 and 2.9 ppm (consistent with glutamine and glutamic acid) were either reduced distinctly or completely disappeared. Recently, a glioma study

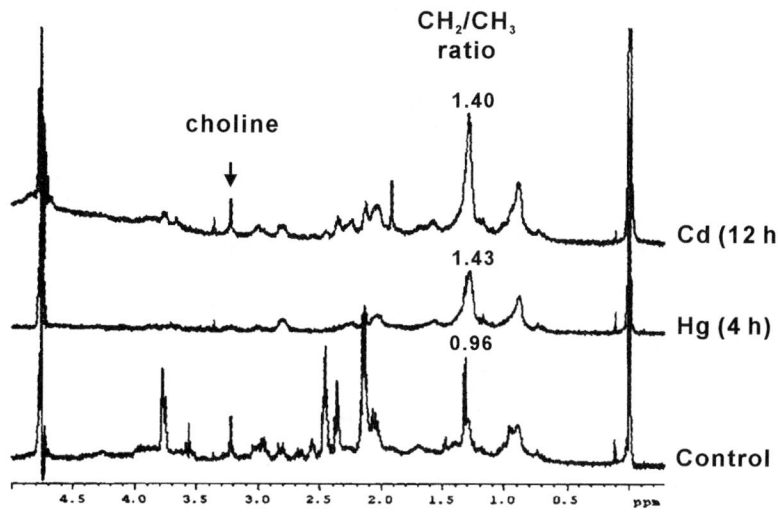

FIGURE 5. Comparison of ^{1}H NMR spectra of MRC-5 cells between Cd treatment and Hg treatment.

model implicates the possibility for detection of the apoptotic tissue *in vivo*.[14] Our current data are consistent with this earlier finding and strongly support the notion that ^{1}H NMR might provide a simple and convenient method for distinguishing apoptosis from necrosis *in vivo*, which might have a significant impact on clinical application.

CONCLUSION

One of the most significant findings in this study is that the spectral intensity ratio of CH_2/CH_3 resonances measured by ^{1}H NMR is highly and positively correlated with the percentage of apoptosis. Moreover, apoptosis and necrosis induced different ^{1}H NMR spectral patterns. These findings together strongly suggest that ^{1}H NMR is an easy and reliable tool that can distinguish apoptosis from necrosis as early as the onset of cell death and support the idea that ^{1}H NMR can be used to detect apoptosis and necrosis *in vivo* in the future.

ACKNOWLEDGMENTS

This work was supported by grants NSC 92-2320-B-038-055 and NSC 93-2320-B-038-055 (to C-M.S.) from the National Science Council, Taiwan, Republic of China.

REFERENCES

1. ROBERTSON, J. D. & S. ORRENIUS. 2000. Molecular mechanisms of apoptosis induced by cytotoxic chemicals. Crit. Rev. Toxicol. **30:** 609–627.
2. NICOTERA, P. & M. LESIA. 1997. Energy supply and the shape of death in neuron and lymphoid cells. Cell Death Differ. **4:** 435–442.
3. MOYEC, L.L. et al. 1992. Cell and membrane lipid analysis by proton magnetic resonance spectroscopy in five breast cancer cell lines. Br. J. Cancer **66:** 623–628.
4. BLANKENBERG, F.G. et al. 1996. Detection of apoptosis cell death proton nuclear magnetic resonance spectroscopy. Blood **87:** 1951–1956.
5. BLANKENBERG, F.G. et al. 1997. Quantitative analysis of apoptotic cell death using proton nuclear magnetic resonance spectroscopy. Blood **89:** 3778–3786.
6. FERRETTI, A. et al. 1999. Biophysical and structural characterization of ^{1}H-NMR-detectable mobile lipid domains in NIH-3T3 fibroblasts. Biochim. Biophys. Acta **438:** 329–348.
7. MILLIS, K. et al. 1999. Classification of human liposarcoma and lipoma using ex vivo proton NMR spectroscopy. Magn. Reson. Med. **41:** 257–267.
8. JUHANA, M. et al. 2000. ^{1}H NMR visible lipids in the life and death of cells. Trend Biochem. Sci. **25:** 357–361.
9. VITO, M.D. et al. 2001. ^{1}H NMR-visible mobile lipid domains correlate with cytoplasmic lipid bodies in apoptotic T-lymphoblastoid cells. Biochim. Biophys. Acta **1530:** 47–66.
10. BEZABEH, T. et al. 2001. Detection of drug-induced apoptosis and necrosis in human cervical carcinoma cells using ^{1}H NMR spectroscopy. Cell Death Differ. **8:** 219–224.
11. VERMES, I. et al. 1995. A novel assay for apoptosis. Flow cytometric detection of phosphatidylserine expression on early apoptotic cells using fluorescein-labeled Annexin V. J. Immunol. Methods **184:** 39–51.
12. VAN, M.E. et al. 1996. A novel assay to measure loss of plasma membrane asymmetry during apoptosis of adherent cells in culture. Cytometry **24:** 131–139.
13. ORMEROD, M.G. et al. 1992. Apoptosis in interleukin-3-dependent haemopoietic cells. Quantification by two flow cytometric methods. J. Immunol. Methods **153:** 57–65.
14. HAKUMAKI, J.M. et al. 1999. ^{1}H MRS detects polyunsaturated fatty acid accumulation during gene therapy of glioma: implication for the *in vivo* detection of apoptosis. Nat. Med. **5:** 1323–1327.

Cadmium Toxicity toward Caspase-Independent Apoptosis through the Mitochondria–Calcium Pathway in mtDNA-Depleted Cells

YUNG-LUEN SHIH,[a,e] CHIEN-JU LIN,[b,c,e] SHENG-WEI HSU,[b,c] SHENG-HAO WANG,[b,c] WEI-LI CHEN,[b,c] MEI-TSU LEE,[b] YAU-HUEI WEI,[d] AND CHWEN-MING SHIH[b]

[a]*Department of Pathology and Laboratory Medicine, Shin Kong Wu Ho-Su Memorial Hospital, Taipei 111, Taiwan, ROC*

[b]*Department of Biochemistry, Taipei Medical University, Taipei 110, Taiwan, ROC*

[c]*Graduate Institute of Medical Science, Taipei Medical University, Taipei 110, Taiwan, ROC*

[d]*Department of Biochemistry and Center for Cellular and Molecular Biology, National Yang-Ming University, Taipei 112, Taiwan, ROC*

ABSTRACT: Mitochondria are believed to be integrators and coordinators of programmed cell death in addition to their respiratory function. Using mitochondrial DNA (mtDNA)-depleted osteosarcoma cells (ρ^0 cells) as a cell model, we investigated the apoptogenic signaling pathway of cadmium (Cd) under a condition of mitochondrial dysfunction. The apoptotic percentage was determined to be around 58.0% after a 24-h exposure to 25 μM Cd using flow cytometry staining with propidium iodine (PI). Pretreatment with Z-VAD-fmk, a broad-spectrum caspase inhibitor, failed to prevent apoptosis following Cd exposure. Moreover, Cd was unable to activate caspase 3 using DEVD-AFC as a substrate, indicating that Cd induced a caspase-independent apoptotic pathway in ρ^0 cells. JC-1 staining demonstrated that mitochondrial membrane depolarization was a prelude to apoptosis. On the other hand, the intracellular calcium concentration increased 12.5-fold after a 2-h exposure to Cd. More importantly, the apoptogenic activity of Cd was almost abolished by ruthenium red, a mitochondrial calcium uniporter blocker. This led us to conclude that mtDNA-depleted cells provide an alternative pathway for Cd to conduct caspase-independent apoptosis through a mitochondria-calcium mechanism.

KEYWORDS: cadmium; caspase; apoptosis; miotochondria

---

[e]Y.-L.S. and C.-J.L. contributed equally to this work.

Address for correspondence: Dr. Chwen-Ming Shih, Department of Biochemistry, School of Medicine, Taipei Medical University, 250 Wu-Hsing Street, Taipei, Taiwan 110, ROC. Voice: +886-2-27361661 ext. 3151; fax: +886-2-86421158.
 cmshih@tmu.edu.tw

INTRODUCTION

Apoptosis, so-called programmed cell death, is important in development and tissue homeostasis of multicellular organisms. Apoptosis is associated with cell shrinkage, plasma membrane blebbing, chromatin condensation, DNA fragmentation, and formation of apoptotic bodies that can be taken up and degraded by neighboring cells without producing an inflammatory response.[1] Furthermore, the strong relationship between a dysfunction in apoptosis and diseases such as AIDS, neurodegenerative diseases, and oncogenesis is well known. In the apoptotic signaling pathway, mitochondria, in addition to their respiratory function, play a crucial role. It has been demonstrated that apoptotic proteins (cytochrome c, AIF, endonuclease G, and Smac/DIABLO) are released from the mitochondrial intermembrane space to their new destination to complete the apoptotic process.[2]

Cadmium (Cd) is a toxic metal with an extremely long biological half-life of 15~30 years in humans.[3] It has been known for decades that Cd exposure can cause a variety of adverse health effects, among which kidney dysfunction, lung diseases, disturbed calcium metabolism, and bone effects are most prominent.[4] Exposure to Cd causes loss of bone mass and increased incidence of bone fractures, leading to osteoporosis and osteomalacia as observed in itai-itai patients and laboratory animals.[5,6] The mechanisms of Cd-induced damage include the production of free radicals that alter mitochondrial activity and trigger apoptosis.[7–9] Therefore, the toxicity of cadmium is thought to occur through the induction of apoptosis. However, the apoptotic signaling induced by this toxicity is still unclear. In this report, using mitochondrial DNA (mtDNA)-depleted cells (rho zero cells, ρ^0 cells)[10–11] as a cell model, we suggest that Cd induces a caspase-independent apoptosis through the mitochondria–calcium pathway. It was noted, however, that no ROS were produced and no pro-apoptotic factors were released from mitochondria, such as cytochrome c, apoptosis-inducing factor (AIF), endonuclease G (Endo G).

MATERIALS AND METHODS

Cell Culture

The ρ^0 cells derived from a human osteosarcoma cell line, 143BTK (ATCC CRL 8303), were cultured in Dulbecco's modified Eagle's medium (DMEM) with 10% fetal bovine serum, 50 µg/mL uridine, 100 µg/mL pyruvate, 50 ng/mL ethidium bromide, 100 units/mL penicillin, and 100 units/mL streptomycin in 5% CO_2, and 95% air at 37°C in an incubator with a humidified atmosphere.[12] Serum starvation was achieved by incubation in DMEM containing 1% FBS for at least 16 h. Following this, unless otherwise stated, ρ^0 cells were treated with 25 µM Cd for the indicated time periods.

Assessment of Cell Death

As described previously, cell death was determined using a Becton Dickinson (San Jose, CA) FACSCalibur flow cytometer using propidium iodine (PI) single staining and Annexin V/PI double staining for assessment of hypodiploid DNA content and phosphatidylserine (PS) externalization, respectively.[13]

Analysis of Caspase 3 Activity

Caspase 3 activity was measured using a Caspase-3/CPP32 Fluorometric Assay Kit according to the manufacturer's instructions (BioVision, Mountain View, CA). In brief, 5×10^6 cells were incubated with 50 µL cell lysis buffer for 10 min on ice, harvested, and centrifuged at $16,000 \times g$ for 1 min at 4°C. The supernatant (200 µg proteins) was incubated with 50 µL of 2X reaction buffer and 5 µL of the 1 mM DEVD-AFC (benzyloxy-carbonyl-Asp-Glu-Val-Asp-7-amino-4-trifluoromethyl-coumarin) substrate for 1 h at 37°C in a 96-well plate. The fluorescence of the cleaved product, AFC, was measured at 405-nm excitation and 535-nm emission wavelengths on a fluorescence reader, Fluoroskan Ascent (Thermoelectron, Waltham, MA). Cleaved AFC was quantified by a calibration curve using known AFC concentrations.

Detection of the Mitochondrial Membrane Potential ($\Delta\Psi_m$)

The dual emission dye, JC-1, was used as a measure of $\Delta\Psi_m$ according to the methods described previously.[14] In brief, cells were incubated with 2.5 mg/mL JC-1 (dissolved in DMSO) for 15 min at room temperature in darkness. After centrifugation for 5 min at $200 \times g$, cells were washed twice with PBS at 4°C, resuspended in 0.5 mL PBS, and analyzed on a FACSCalibur flow cytometer. JC-1 is a lipophilic cationic fluorescence dye and is capable of selectively entering mitochondria, which will change color from red (FL-2) to greenish (FL-1) once the $\Delta\Psi_m$ declines.

Measurement of Intracellular Calcium

Cells were treated with or without Cd and harvested at indicated time periods. Before data acquisition, cells were incubated with 3 µM Fluo-3 AM dye for a total of

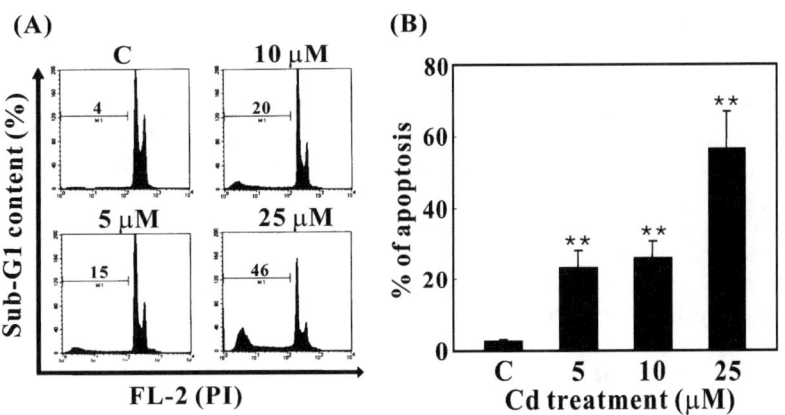

FIGURE 1. Dose-response and time course of cell death induced by Cd in ρ^0 cells. (**A**) Cells were treated with 5–25 µM CdCl$_2$ for 24 h and then analyzed by PI staining using flow cytometry to determine their hypodiploid DNA (sub-G1) proportion. The percentage of M1 indicates the cell proportion of the sub-G1 peak. Data presented in **A** are representative of three independent experiments, and their statistical results for the apoptosis are presented in **B**. **, $P < 0.01$.

30 min at 37°C and then immediately analyzed on a flow cytometer using FL1 as a detector.[15] For relative intracellular calcium concentrations, the geographic mean values of the FL-1 peak generated from Cd-treated cells over each one's own negative control group (without Cd treatment) were calculated.

Statistics

Data are expressed as the mean ± standard deviation (SD) from a minimum of three independent experiments, unless otherwise indicated. Statistical analysis was performed using Student's t-test, with $P < 0.01$ as a criterion of significance.

RESULTS AND DISCUSSION

Cd Induced Caspase-Independent Apoptosis in ρ^0 Cells

As shown in FIGURE 1, Cd induced apoptosis in a dose-responsive manner and did not seem to alter the cell cycle based on the consistent ratio of G1 (FL-1 = 200) versus G2 (FL1 = 400) using PI staining for assessing hypodiploid DNA content with a flow cytometer. To reveal the involvement of caspase, Z-VAD-fmk [Z-Val-Ala-DL-Asp(OMe)-fluoro-methylketone], a broader spectrum of caspase inhibitor, was employed to examine its ability to prevent apoptosis by Cd toxicity (FIG. 2). In the left panel of FIGURE 2A, we assumed that the Z-VAD-fmk was able to prevent HL-60 apoptosis suffered from H_2O_2 treatment, which has been demonstrated to be a caspase-dependent apoptotic system.[18] In ρ^0 cells, pretreatment of Z-VAD-fmk failed to reduce the apoptotic percentage after exposure to different concentrations of Cd, suggesting that Cd might trigger ρ^0 cells to undergo caspase-independent apoptosis (see right panel of FIG. 2A and B). Furthermore, in contrast to H_2O_2-treated HL-60 cells, Cd treatment did not activate caspase 3 activity in ρ^0 cells (TABLE 1), which obviously demonstrated that the apoptogenic activity of Cd is caspase-independent in mtDNA-depleted cells

The apoptotic pathway of ρ^0 cells is still being debated. Most of them are caspase dependent, such as anoxia- and TNFα-treated ρ^0 cells derived from human A549 lung epithelial cells,[19] saturosporine-treated ρ^0 cells derived from human D238 medulloblastoma cells,[20] saturosporine-treated ρ^0 cells derived from human WAL-3A lymphocytes,[21] and saturosporine-treated ρ^0 cells derived from human 143BTK osteosarcoma cells.[22] However, consistent with our observations, C2- and C8-ceramide–induced human D238 medulloblastoma-derived ρ^0 cells underwent caspase-independent apoptosis.[23] Therefore, further investigation is warranted to reveal the apoptotic mechanism of ρ^0 cells.

Mitochondrial Membrane Depolarization Is a Prelude to Apoptosis

Emerging evidence has suggested that Cd might exert its cell toxicity through induction of an ROS burst that accompanies collapse of the mitochondria.[7–9] Using JC-1 as an indicator of $\Delta\Psi_m$, we demonstrated that after 8 h of Cd exposure, cells with normal $\Delta\Psi_m$ dropped from 97% to 76% to 57% after 16 h of Cd treatment (see upper-left quadrant in FIG. 3A and the statistical results in FIG. 3B). Nevertheless, we failed to detect any ROS burst after Cd treatment, including hydrogen peroxide,

TABLE 1. Caspase 3 activity of Cd-treated cells[a]

Cell line	Cd treatment	Caspase 3 activity (pmole/mg/min)
HL-60	Control	1.00 ± 0.49
	100 µM, 12 h	63.66 ± 11.51
ρ^0	Control	1.00 ± 0.13
	25 µM, 8 h	0.46 ± 0.19
	25 µM, 16 h	0.99 ± 0.20
	25 µM, 24 h	0.52 ± 0.17

[a]HL-60 and ρ^0 zero cells were treated with 100 µM and 25 µM Cd, respectively, for the indicated time periods and then collected for the capase 3 activity assay as described in Materials and Methods. Data were calculated from three independent experiments and represented as mean ± SD.

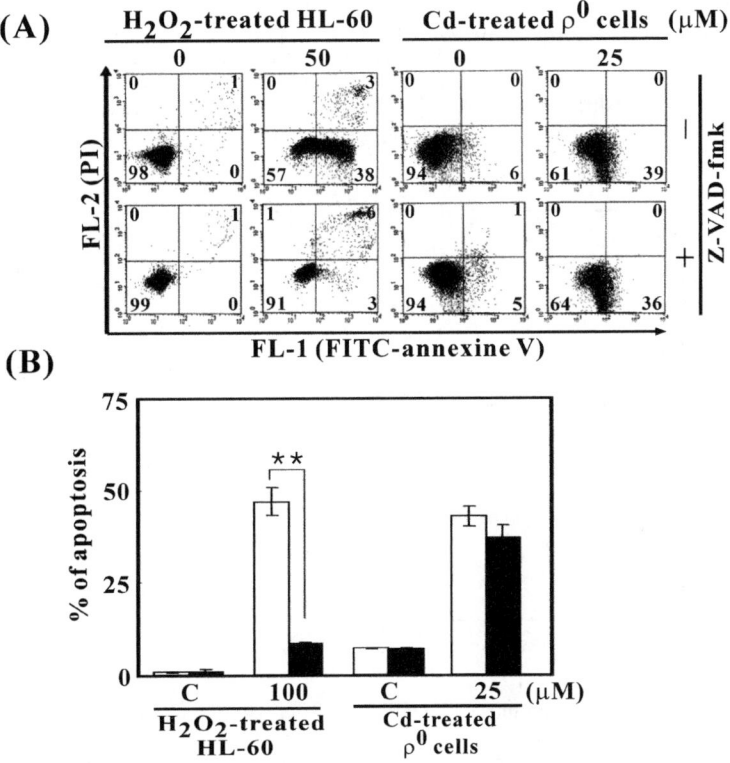

FIGURE 2. Inability of the broad-spectrum caspase inhibitor, Z-VAD-fmk, to prevent apoptosis. (**A**) Pretreatment with 40 M Z-VAD-fmk could rescue HL-60 cells from the effects of 50 µM H_2O_2 treatment. However, Z-VAD-fmk was unable to protect ρ^0 cells from Cd toxicity. Three independent experiments were performed, and their statistical results are presented in **B**. **, $P < 0.01$.

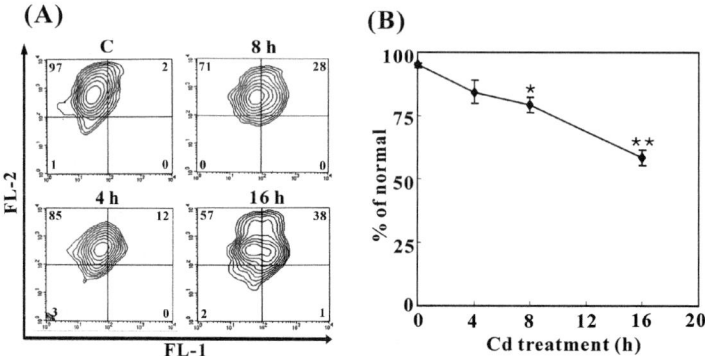

FIGURE 3. Mitochondrial membrane potential declined after Cd treatment. Cells were incubated with 2.5 μg/mL JC-1 dye for 15 min before the end of Cd treatment and were subsequently analyzed using flow cytometry. Red fluorescence (FL-2 channel) emitted from the J-aggregate form of JC-1 and green fluorescence (FL-1 channel) emitted from its monomer form increased when $\Delta\Psi_m$ was normal and depolarized, respectively. Percentages in the upper-left quadrant and right two quadrants indicate proportions of cells with normal and depolarized mitochondria, respectively. Data presented in **A** are representative of three independent experiments, and their statistical results are presented in **B**. Asterisks (*, **) indicate a significant difference from control at $P < 0.05$ and < 0.01, respectively.

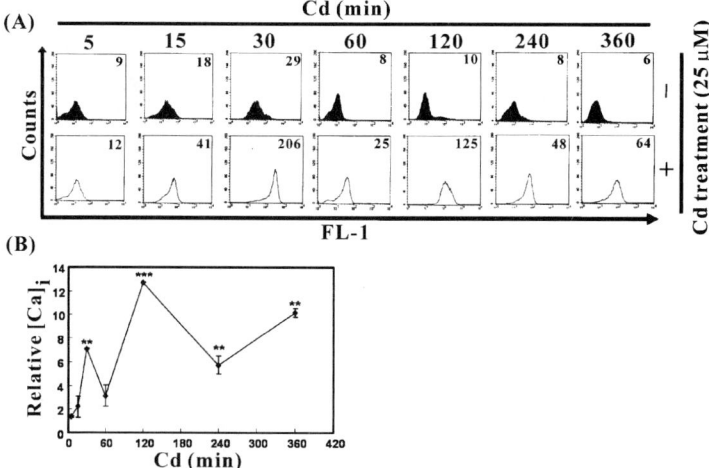

FIGURE 4. Cd induced intracellular calcium oscillation in ρ^0 cells. Cells were treated with Cd for the indicated time periods and incubated with 0.4 μM Fluo-3 AM dye for a total of 30 min before analysis by flow cytometry. The fluorescence of Fluo-3 AM (FL-1 channel) increased as the intracellular calcium increased. Data presented in **A** are representative of three separate experiments. The Arabic numeral in the upper-right conner of each cytogram indicates the geographic (GEO) mean of each peak. Panel **B** was generated from the GEO mean of the Cd-treated cytogram over its respective control. Asterisks (**, ***) indicate a significant difference from control at $P < 0.01$ and < 0.001, respectively.

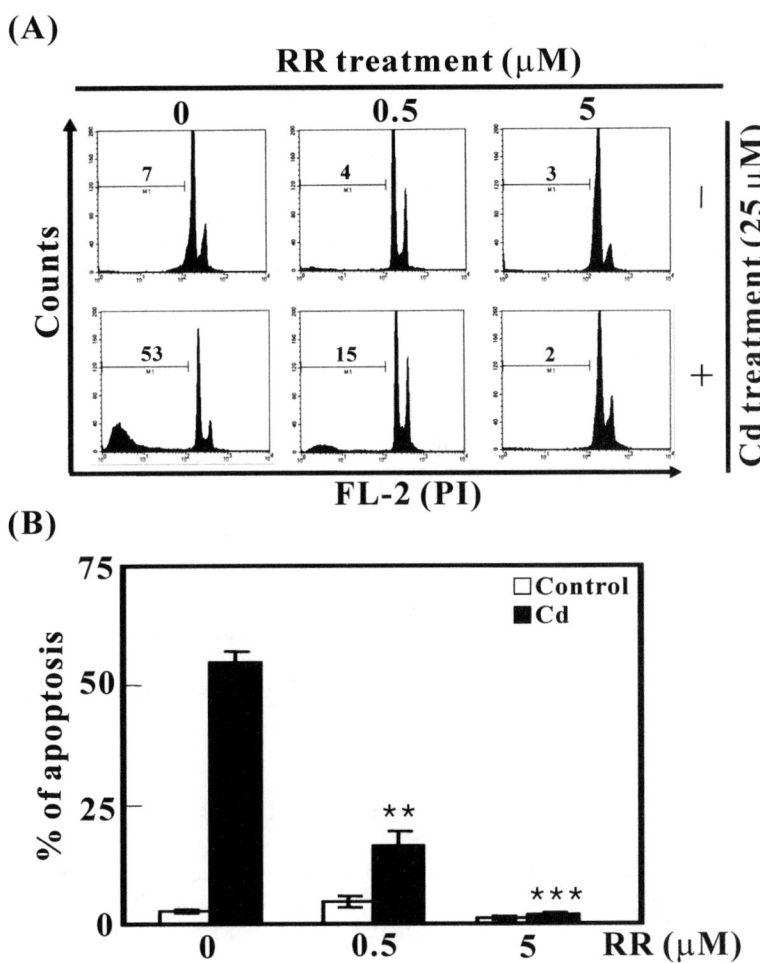

FIGURE 5. The apoptogenic activity of cadmium was suppressed by an inhibitor of the mitochondrial calcium uniporter. Cells were pretreated with various concentrations of RR, an inhibitor of the mitochondrial calcium uniporter, for 1 h, followed by treatment with Cd for another 24 h, and then were analyzed by PI staining to examine their hypodiploid DNA (sub-G1) proportion. M1 denotes the percentage of cells at the sub-G1 proportion. Three independent experiments were performed, and their statistical results are presented in **B**. Asterisks (**, ***) indicate a significant difference from control at $P < 0.01$ and < 0.001, respectively.

superoxide anions, and hydroxyl radicals (data not shown). Also, using confocal microcopy, we were unable to detect the translocation of AIF, Endo G, and cytochrome c from mitochondria to their destination (data not shown). This result suggested that the Cd-induced decline in $\Delta\Psi_m$ was irrelevant to the oxidative stress and the release of pro-apoptotic factors from mitochondria.

Elevation of Intracellular Calcium Is Involved in Apoptosis

Calcium signals have been identified as one of the major signals that converge on mitochondria to trigger mitochondria-dependent apoptosis.[24] In this report, using Fluo-3-AM as an indicator of intracellular calcium ($[Ca^{2+}]i$), Cd induced a $[Ca^{2+}]i$ oscillation which made it apparent that calcium signaling was a crucial mediator of Cd-triggered caspase-independent apoptosis in ρ^0 cells (FIG. 4). Calpains, Ca^{2+}-dependent cysteine proteases, are associated with both caspase-dependent and -independent apoptosis and are located downstream of calcium.[25] We are currently investigating the role of calpain in this system.

Ruthenium red (RR) is one of the most potent inhibitors of the mitochondrial calcium uniporter. As shown in FIGURE 5, RR can totally abolish Cd-induced apoptosis. The most likely hypothesis assumes that uptake of calcium by mitochondria might affect the mitochondrial permeability transition pore (MPTP), leading to a decline in mitochondrial $\Delta\Psi_m$, and then apoptosis. Alternatively, Cd might directly damage mitochondria through the calcium uniporter. Further investigation is required to examine mitochondrial Cd concentrations after Cd treatment.

CONCLUSIONS

In this report, our results support the notion that Cd induces caspase-independent apoptosis through induction of $[Ca^{2+}]i$ oscillation and disruption of mitochondrial ΔY_m. In this process, there is no ROS burst or release of proapoptotic factors from the mitochondria.

ACKNOWLEDGMENTS

This study was sponsored by the Shin Kong Wu Ho-Su Memorial Hospital (Grant SKH-TMU-93-35) and the National Science Council, Taiwan, ROC (Grants NSC 92-2320-B-038-055 and NSC 93-2320-B-038-047) to C.M.S.

REFERENCES

1. ROBERTSON, J.D. & S. ORRENIUS. 2000. Molecular mechanisms of apoptosis induced by cytotoxic chemicals. Crit. Rev. Toxicol. **30:** 609–627.
2. RAVAGNAN, L. *et al.* 2002. Mitochondria, the killer organelles and their weapons. J. Cell. Physiol. **192:** 131–137.
3. GOYER, R.A. & M.G. CHERIAN. 1995. Renal effects of metals. *In* Metal Toxicology. R.A. Goyer, C.D. Klaassen & M.P. Waalkes, Eds.: 389–412. Academic Press. San Diego, CA.

4. SATARUG, S. *et al.* 2003. A global perspective on cadmium pollution and toxicity in non-occupationally exposed populations. Toxicol. Lett. **137:** 65–83.
5. BHATTACHARYYA, M.H. *et al.* 1995. Metal-induced osteotoxicities. *In* Metal Toxicology. R.A. Goyer, C.D. Klaassen & M.P. Waalkes, Eds.: 465–510. Academic Press. San Diego, CA.
6. LIU, J. *et al.* 1998. Susceptibility of MT-null mice to chronic $CdCl_2$-induced nephrotoxicity indicates that renal injury is not mediated by the CdMT complex. Toxicol. Sci. **46:** 197–203.
7. ACHANZAR, W.E. *et al.* 2000. Cadmium induces c-myc, p53, and c-jun expression in normal human prostate epithelial cells as a prelude to apoptosis. Toxicol. Appl. Pharmacol. **164:** 291–300.
8. HARSTAD, E.B. & C.D. KLAASSEN. 2002. Tumor necrosis factor-α-null mice are not resistant to cadmium chloride-induced hepatotoxicity. Toxicol. Appl. Pharmacol. **179:** 155–162.
9. SHEN, H.M. *et al.* 2001. Critical role of calcium overloading in cadmium-induced apoptosis in mouse thymocytes. Toxicol. Appl. Pharmacol. **171:** 12–19.
10. MARUSICH, M.F. *et al.* 1997. Expression of mtDNA and nDNA encoded respiratory chain proteins in chemically and genetically-derived Rho0 human fibroblasts: a comparison of subunit proteins in normal fibroblasts treated with ethidium bromide and fibroblasts from a patient with mtDNA depletion syndrome. Biochim. Biophys. Acta **1362:** 145–159.
11. DEY, R. & C.T. MORAES. 2000. Lack of oxidative phosphorylation and low mitochondrial membrane potential decrease susceptibility to apoptosis and do not modulate the protective effect of Bcl-x_L in osteosarcoma cells. J. Biol. Chem. **275:** 7087–7094.
12. JIANG, S. *et al.* 1999. Cytochrome c-mediated apoptosis in cells lacking mitochondrial DNA. J. Biol. Chem. **274:** 29905–29911.
13. SHIH, C.M. *et al.* 2004. Mediating of caspase-independent apoptosis by cadmium through the mitochondria-ros pathway in MRC-5 fibroblasts. J. Cell Biochem. **91:** 384–397.
14. CASTEDO, M. *et al.* 2002. Quantitation of mitochondrial alterations associated with apoptosis. J. Immunol. Methods **265:** 39–47.
15. MONTEIRO, M.C. *et al.* 1999. A flow cytometric kinetic assay of platelet activation in whole blood using Fluo-3 and CD41. Cytometry **35:** 302–310.
16. VAN ENGELAND, M. *et al.* 1996. A novel assay to measure loss of plasma membrane asymmetry during apoptosis of adherent cells in culture. Cytometry **24:** 131–139.
17. PIETRA, G. *et al.* 2001. Phases of apoptosis of melanoma cells, but not of normal melanocytes, differently affect maturation of myeloid dendritic cells. Cancer Res. **61:** 8218–8226.
18. DIPIETRANTONIO, A.M. *et al.* 1999. Activation of caspase 3 in HL-60 cells exposed to hydrogen peroxide. Biochem. Biophys. Res. Commun. **255:** 477–482.
19. SANTORE, M.T. *et al.* 2002. Anoxia-induced apoptosis occurs through a mitochondria-dependent pathway in lung epithelial cells. Am. J. Physiol. Lung Cell. Mol. Physiol. **282:** L727–734.
20. LUETJENTS, C.M. *et al.* 2000. Delayed mitochondria dysfunction in excitotoxic neuron death: cytochrome *c* release and a secondary increase in superoxide production. J. Neurosci. **20:** 5715–5723.
21. CAI, J. *et al.* 2000. Separation of cytochrome c-dependent caspase activation from thiol-disulfide redox change in cells lacking mitochondria DNA. Free Radical Biol. Med. **29:** 334–342.
22. DEY, R. & C.T. MORAES. 2000. Lack of oxidative phosphorylation and low mitochondrial membrane potential decrease susceptibility to apoptosis and do not modulate the protective effect of Bcl-x_L in osteosarcoma cells. J. Biol. Chem. **275:** 7087–7094.
23. POPPE, M. *et al.* 2002. Ceramide-induced apoptosis of D283 medulloblastoma cells requires mitochondrial respiratory chain activity but occurs independently of caspases and is not sensitive to Bcl-xL overexpression. J. Neurochem. **82:** 482–494.
24. DUCHEN, M.R. 2000. Mitochondria and calcium: from cell signaling to cell death. J. Physiol. **529:** 57–68.
25. WANG, K.K.W. 2000. Calpain and caspase: Can you tell the difference? Trends Neurosci. **23:** 20–26.

Pectinesterase Inhibitor from Jelly Fig (*Ficus awkeotsang* Makino) Achene Induces Apoptosis of Human Leukemic U937 Cells

JIA-HUEI CHANG,[a] YUH-TAI WANG,[b] AND HUNG-MIN CHANG[a]

[a]*Graduate Institute of Food Science and Technology, National Taiwan University, Taipei 106-17, Taiwan*

[b]*Life Science Center, Hsing Wu College, No. 11-2, Fen-Liao Road, Lin-Kou, Taipei 112-44, Taiwan*

ABSTRACT: The antitumor activity of pectinesterase inhibitor (PEI), a group of cationic polypeptides, from jelly fig (*Ficus awkeotsang* Makino) achene was first examined as a treatment for leukemia in this study. PEI displayed strong growth inhibition against human leukemic U937 cells via induction of apoptosis in a dose- and time-dependent manner. At a level of 50 μg/mL, PEI inhibited 90% of cell growth, and the concentration of PEI required to induce 50% of cell viability (LC_{50}) was about 180 μg/mL. Meanwhile, cell cycle arrest at G2/M phase was observed when cells were incubated with 100 μg PEI/mL for 24 h. PEI displayed a dose-dependent influence on mitochondria transmembrane potential (MTP, $\Delta\Psi m$) of cells when detected by a flow cytometry. MTP of more than 50% cells was reduced when cells were incubated with PEI at levels higher than 50 μg PEI/mL for 24 h. In addition, PEI upregulated caspase-3 activity. Taken together, PEI potently inhibited the proliferation of human leukemic U937 cells via cell cycle arrest and apoptosis in association with MTP reduction and caspase-3 activation, respectively, and showed therapy potential for U937 cells.

KEYWORDS: PEI; U937; cell cycle arrest; caspase 3

INTRODUCTION

In current cancer chemoprevention/chemotherapy, one of the attractive strategies to devitalize malignant cells through apoptosis is the application of dietary or pharmaceutical treatment.[1] Apoptosis plays an important role in multicellular organisms' development and homeostasis and is a physiologically programmed cell death.[2] It is triggered by the activation[3,4] of caspases in the early stages of apoptosis to break down or cleave key cellular substrates, which are required for cellular functions, such as structural proteins in the cytoskeleton, DNA repair enzymes, and DNases in the nucleus, and to induce biochemical and morphological changes in cells.[5–7]

Address for correspondence: Dr. Hung-Min Chang, Graduate Institute of Food Science and Technology, National Taiwan University. Voice: +886-2-2363-0231 ext. 2776; fax: +886-2-2362-0849.
changhm@ntu.edu.tw

Recently, polypeptides with molecular weights of 3.5–4.5 kDa, consisting of more than 50% histidine, from jelly fig (*Ficus awkeotsang* Makino) achene exhibited marked competitive inhibition on the activity of pectinesterases (PEs) from fruits and vegetables.[8] Chitosan suppresses the bacteria growth by the strong adhesion to the cell membranes as a result of the positively charged molecules.[9–10] Consequently, the novel feature of these strongly positive-charged polypeptides might be associated with the growth inhibition of tumor cells.

The present results demonstrated for the first time that natural polypeptides strongly induced apoptosis in human leukemic cells. Effects of PEI on cell proliferation were achieved by examining the induction of apoptosis, mitochondrial transmembrane potential (MTP, $\Delta\Psi m$), surface exposure of phosphatidylserine (PS) by Annexin V–FITC staining, and caspase-3 activity using a flow cytometry.

MATERIALS AND METHODS

Preparation of Pectinesterase Inhibitor (PEI) from Jelly Fig Achenes

PEI solution was prepared from pectin-depleted jelly fig (*Ficus awkeotsang* Makino) achenes as reported by Jiang *et al.*[8] In brief, the achenes were washed with saline to remove PE and pectin, followed by homogenization and centrifugation to obtain supernatant and gel permeation chromatographies to purify PEI. The pooled fractions with PE inhibitory activity were lyophilized and stored at –20°C before use.

Cell Culture

Human acute monocytic leukemic suspension cell line of U937 was obtained from the American Type Culture Collection (ATCC) (Rockville, MD). Cells were cultured in RPMI 1640 medium supplemented with 1% glutamine/10% FCS and maintained in an exponential growth status. Cultures were subcultured by total medium replacement using centrifugation every 2–3 days, depending on cell number, and maintained at a cell concentration between 1×10^5 and 1×10^6 cells/mL by incubation at 37°C in a 5% humidified incubator.

PEI Treatment

U937 cells were incubated in 35-mm petri dishes at an initial concentration of 1×10^5 cells/mL in the presence of 25–200 (mg/mL) PEI at 37°C in a humidified 5% CO_2 incubator for 3 days with the untreated cells acting as a control. The numbers of adherent cells in day-3 cultures were collected by gently rubbing the dishes with a rubber policeman (Bellco Glass, Vineland, NJ).

Growth Inhibition and Cell Viability Assay

Viable cells and cell count were assessed by Trypan blue (0.04%) exclusion dye using a hemocytometer to determine the growth inhibition (%). Growth inhibition (%) = (1–cell number of PEI treatment/cell number of control group) × 100%. A cell viability assay, based on Trypan blue dye exclusion by viable cells, was used in order to establish an LC_{50} (concentration required to induce 50% of cell viability). Three separate experiments were each tested in duplicate.

Cell Cycle Arrest Analysis

After PEI treatment, cells were centrifuged for 5 min ($300 \times g$) and washed twice with RPMI 1640/10% FBS and PBS. The pellet was suspended, fixed, followed by addition of 1 mL PBS containing 50 mg/mL propidium iodide (PI), 0.1% Triton X-100, and 0.2 mg/mL RNase A, and incubated at room temperature for 30 min. Fifteen thousand events were collected with a FACScan flow cytometry (Becton Dickinson, San Jose, CA), and the data were analyzed using the Modfit LT® software package (Becton Dickinson). Each data was collected as an average of 15,000 counting cells. Three separate experiments were each tested in duplicate.

Analysis of Mitochondrial Transmembrane Potential

MTP (DYm) of U937 cells was determined according to the method described by Ludovico et al.,[11] with minor modifications. Cell suspensions (1×10^5 cells/mL) were stained by 1 mg/mL Rh123 (Sigma) at 37°C for 30 min, and 15,000 events were collected with a FACScan flow cytometry (Becton Dickinson, San Jose, CA). Flow cytometry data were analyzed using the CellQuest software package (Becton Dickinson, San Jose, CA). The loss of $\Delta\Psi m$ was expressed as the percentage of rhodamine 123–negative cells. Three separate experiments were each tested in duplicate.

Annexin V Flow Cytometric Assay

Surface exposure of phosphatidylserine (PS) by apoptotic U937 cells was measured by a flow cytometry with Annexin-V-FITC apoptosis kit obtained from Strong Biotech Corp. (Taiwan) to identify cells undergoing apoptosis and death. Cells were stained with PI. Excitation was set at 488 nm and the emission filters used were 515–545 nm (green, FITC) and 600 nm (red, PI). Ten thousand events were collected with a CyFlow® SL flow cytometry (Partec, Germany). Three separate experiments were each tested in duplicate.

Activity of Caspase 3

After a 3-day incubation with PEI, U937 cells (1×10^5 cells/mL) were harvested, washed with PBS, and analyzed by a PhiPhiLux-G_1D_1 kit (OncoImmunin, Inc., Gaithersburg, MD) for caspase-3 activity determination. Ten thousand events were collected with a CyFlow· SL flow cytometry (Partec, Germany). Three separate experiments were each tested in duplicate.

RESULTS AND DISCUSSION

Effect of PEI on Cell Survival

U937 cells treated with different concentrations (5–200 mg/mL) of PEI from jelly fig achenes were cultivated for 3 days, and the percentage of living cells was determined. Growth inhibition reached a plateau (about 90%) at a dose of 50 mg PEI/mL

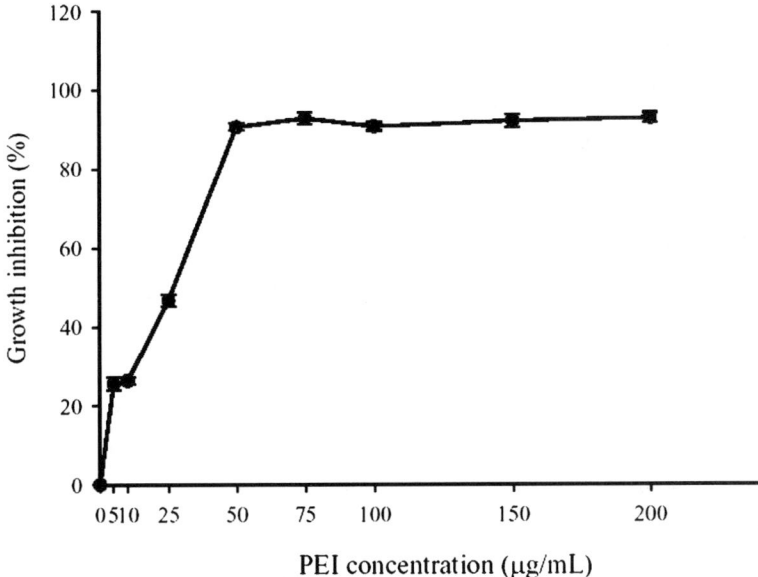

FIGURE 1. Growth inhibition of pectinesterase inhibitor on U937 cells. U937 cells (1×10^5 cells/mL) were incubated with various levels of PEI and counted after 3-day cultivation. Growth inhibition (%) = (1−cell number of PEI treatment/cell number of control group) × 100%. *Bars* refer to standard deviation. Each value represents the average of three determinations.

medium (FIG. 1). Apparently, PEI induced a dose-dependent decrease in cell viability and appeared to be a potent inhibitor of cell viability with an LC_{50} value of about 180 mg/mL (FIG. 2). PEI is composed of polypeptides that are about 53% histidine and is highly positively charged in neutral aqueous solution.[8] The adhesion to cell membranes as a result of electrostatic forces could lead to the disruption of cell membranes and the subsequent cell apoptosis.

Effect of PEI on Cell Cycle Arrest

U937 cells treated with graded concentrations of PEI for up to 72 h were analyzed for cell cycle distribution by a flow cytometry. The number of cells cultured for 24 h showed a dose-dependent reduction at Go/G1 phase and increase at G2/M phase (TABLE 1). The G2/M population increased from 14.3 ± 3.3% in the control group to 56.5 ± 5.5% and 44.3 ± 3.0% in cells treated with 100 and 200 mg PEI/mL, respectively. However, U937 cells treated with 100 mg PEI/mL and cultured for 48 h showed a reduction at G2/M phase and an increase at S phase. The percentage of S phase cells increased from 31.6 ± 4.0% at 24-h cultivation to 59.9 ± 4.9% at 48-h cultivation. It reveals that PEI induces cell cycle perturbations (G2/M and S phase arrest).

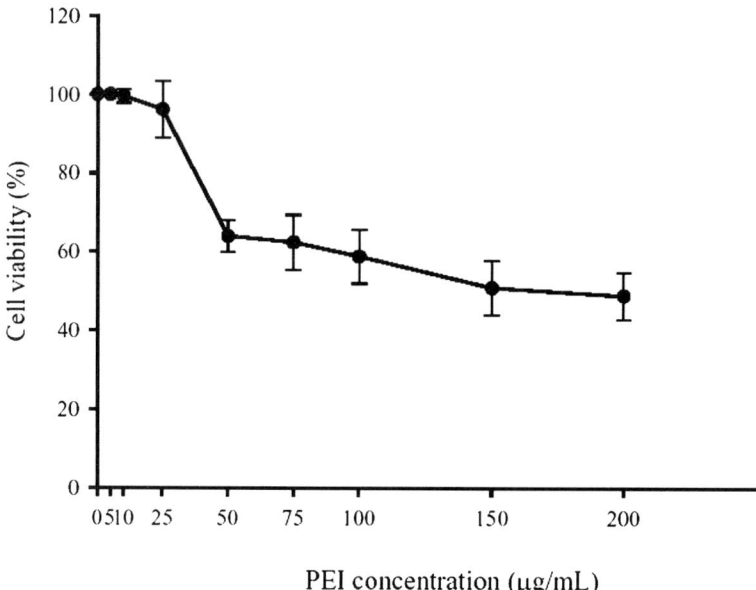

FIGURE 2. Viability of U937 cells after treatment with various concentrations of pectinesterase inhibitor. Cells incubated with various levels of PEI were counted after 3-day cultivation. *Bars* in the figures refer to standard deviation. Each value represents the average of three determinations.

In addition, cell cycle analysis showed that the increase of the apoptotic cells (sub-G1 cell fraction) by PEI treatment was time- and dose-dependent (TABLE 1). After incubation for 24 h, apoptotic cells increased from $5.4 \pm 0.9\%$ to $28.3 \pm 1.9\%$ when U937 cells were cultivated with 50 to 200 mg PEI/mL, respectively. Of note, the appearance of the sub-G1 population was accompanied by characteristic morphological changes (data not shown). Hence, there may be a direct relationship between the doses required to induce programmed cell death and to inhibit cell growth of the U937 cells. That is, the arrest stage of cell cycle is both affected by PEI concentration and incubation time. Since analysis of cell cycle distribution by a flow cytometry was inappropriate while cell membrane was damaged, cells incubated with 100 and 200 mg PEI/mL for a time longer than 48 and 24 h, respectively, were not used in this analysis.

Effect of PEI on MTP ($\Delta\Psi m$)

The altered mitochondrial function is linked to apoptosis, and a decreased MTP is associated with mitochondrial dysfunction.[12] A fluorescent probe Rh123 and flow cytometry were used to monitor the changes of fluorescence intensity during cell incubation, and the decline of Rh123 fluorescence intensity in U937 cells indicated the loss of MTP.[12] After being treated with PEI (50–200 mg/mL) for 24 h, U937 cells showed a sharp decline in Rh123 fluorescence intensity while the RH123-negative

TABLE 1. Effects of incubation time and level of pectinesterase inhibitor on cell cycle distribution of U937 cells

Time (h)	PEI concentration (μg/mL)	Cell cycle phase (%)			Apoptosis(%)
		G0/G1[2]	G2/M	S	
0	0[1]	45.1 ± 1.7	13.0 ± 4.0	41.9 ± 3.4	0.3 ± 0.2
3	0	42.6 ± 4.2^a	14.5 ± 4.9^a	42.9 ± 3.6^a	0.4 ± 0.2^c
	50	36.8 ± 4.5^{a,b}	19.7 ± 4.0^a	43.6 ± 5.0^a	0.6 ± 0.8^{b,c}
	100	36.2 ± 2.0^b	16.8 ± 1.0^a	47.0 ± 2.5^a	1.3 ± 0.4^{a,b}
	200	40.0 ± 2.3^{a,b}	14.9 ± 1.9^a	45.1 ± 1.5^a	1.8 ± 0.3^a
6	0	46.0 ± 3.6^a	15.1 ± 3.4^b	39.8 ± 2.8^a	0.3 ± 0.3^b
	50	31.9 ± 8.5^b	26.1 ± 4.3^a	42.0 ± 5.6^a	0.5 ± 0.3^b
	100	27.1 ± 6.6^b	28.0 ± 4.3^a	44.8 ± 5.2^a	2.6 ± 0.8^a
	200	32.3 ± 4.7^b	22.6 ± 4.9^a	45.1 ± 3.9^a	3.2 ± 1.4^a
9	0	44.1 ± 3.3^a	17.7 ± 5.7^c	38.2 ± 3.6^{a,b}	0.2 ± 0.2^c
	50	17.7 ± 6.9^c	31.0 ± 6.7^b	41.8 ± 4.7^{a,b}	1.2 ± 0.8^c
	100	12.8 ± 0.1^c	44.4 ± 7.5^a	35.3 ± 6.2^b	3.3 ± 0.4^b
	200	27.2 ± 5.4^b	27.5 ± 2.5^{b,c}	45.3 ± 5.2^a	4.6 ± 0.6^a
12	0	47.5 ± 3.9^a	16.7 ± 6.5^c	35.1 ± 4.2^b	0.9 ± 1.0^b
	50	41.3 ± 7.5^a	25.5 ± 6.5^{b,c}	34.9 ± 3.0^b	1.9 ± 0.9^b
	100	6.1 ± 0.3^b	45.5 ± 1.8^a	44.3 ± 7.2^a	4.3 ± 0.4^b
	200	14.2 ± 5.3^b	36.3 ± 6.7^{a,b}	44.3 ± 4.4^a	12.8 ± 4.6^a
24	0	51.6 ± 4.2^a	14.3 ± 3.3^c	34.0 ± 4.1^b	0.7 ± 0.3^a
	50	48.0 ± 4.4^a	20.3 ± 5.6^c	31.7 ± 5.5^b	5.4 ± 0.9^b
	100	12.0 ± 3.0^b	56.5 ± 5.5^a	31.6 ± 4.0^b	11.3 ± 3.2^c
	200	5.0 ± 2.7^c	44.3 ± 3.0^b	50.6 ± 5.8^a	28.3 ± 1.9^d
48	0	57.6 ± 3.8^a	9.5 ± 3.5^b	33.0 ± 2.2^c	0.6 ± 0.3^c
	50	49.6 ± 4.3^b	10.7 ± 2.7^b	41.9 ± 2.8^b	8.0 ± 2.5^b
	100	14.5 ± 3.1	24.9 ± 10.9^a	59.9 ± 4.9^a	17.5 ± 1.2^a

[1]Controls.
[2]Values are means ± SEM, $n = 6$. Those mean values in the same row with a superscript letter (a, b, c, and d) in common do not differ significantly ($P > 0.05$).

cells (%) increased markedly (FIG. 3). Cells incubated with 50 mg PEI/mL for 72 h or with 100 mg PEI for 96 h showed an apparent dose-dependent mitochondrial potential loss, similar to the cell cycle arrest at G2/M phase of cells treated with PEI for 24 h (TABLE 1).

Annexin V Flow Cytometric Assay

To further characterize the functional activity of PEI against U937 cells, PEI-stimulated cells undergoing apoptosis was determined by an Annexin-V-FITC apoptosis kit. Interestingly, when U937 cells were mixed with PEI (> 50 mg/mL) and cul-

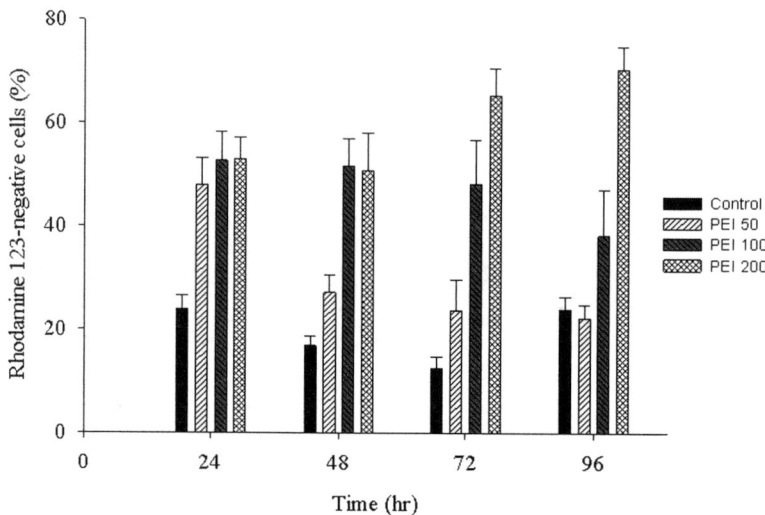

FIGURE 3. Effect of pectinesterase inhibitor level on mitochondria function of U937 cells. Flow cytometric analysis of mitochondrial transmembrane potential ($\Delta\Psi m$) of U937 cells incubated with PEI for various periods of time was estimated by the rhodamine 123 intensity. The loss of MNP was expressed as the percentage of rhodamine 123-negative cells.

tured for 24 and 48 h, the percentages of cells undergoing apoptosis increased remarkably and reached a maximum at 100 mg PEI/mL (FIG. 4A). Apparently, the change of apoptotic cells (%) was dose-dependent (25–100 mg/mL). On the other hand, the dead cell percentage was also dose-dependent (FIG. 4B) which demonstrated that PEI exerted cytotoxicity against leukemic cells. A significant increase in PEI-induced cytotoxicity was observed while U937 cells were mixed with PEI at a level of 100 mg/mL for 48 h or of 200 mg/mL for 24 h (FIG. 4B).

Compared to the 7.72% of control group (FIG. 4C), Q4 (59.72%) in FIGURE 4D was much higher, suggesting the marked increase in early apoptotic cells when U937 cells were incubated with 200 mg PEI/mL for 24 h. Similarly, Q2 (22.67%) in FIG. 4E reached a value of 22.67%, revealing a sharp increase in late apoptotic and dead cells as a result of the treatment with 200 mg PEI/mL for 48 h.

Activity of Caspase 3

Caspases are believed to play critical roles in mediating various apoptotic responses.[13–17] Augmented activity of caspase 3, an effector caspase, may be indicative of drug sensitivity due to increased cellular apoptosis.[18] In an effort to directly address the involvement of caspase 3 in PEI-induced apoptosis, caspase activity was determined by a flow cytometry using a PhiPhiLux-G_1D_1 kit. After exposure of the U937 cells to graded concentrations of PEI, the caspase 3 activity increased in a dose-dependent manner (FIG. 5) and inversely correlated with loss of cell viability, which was in agreement with the results of Majlessipour et al.[18] Kwon et al.[3] revealed that caspase 3 could be activated by hydrogen peroxide generation, which eventually

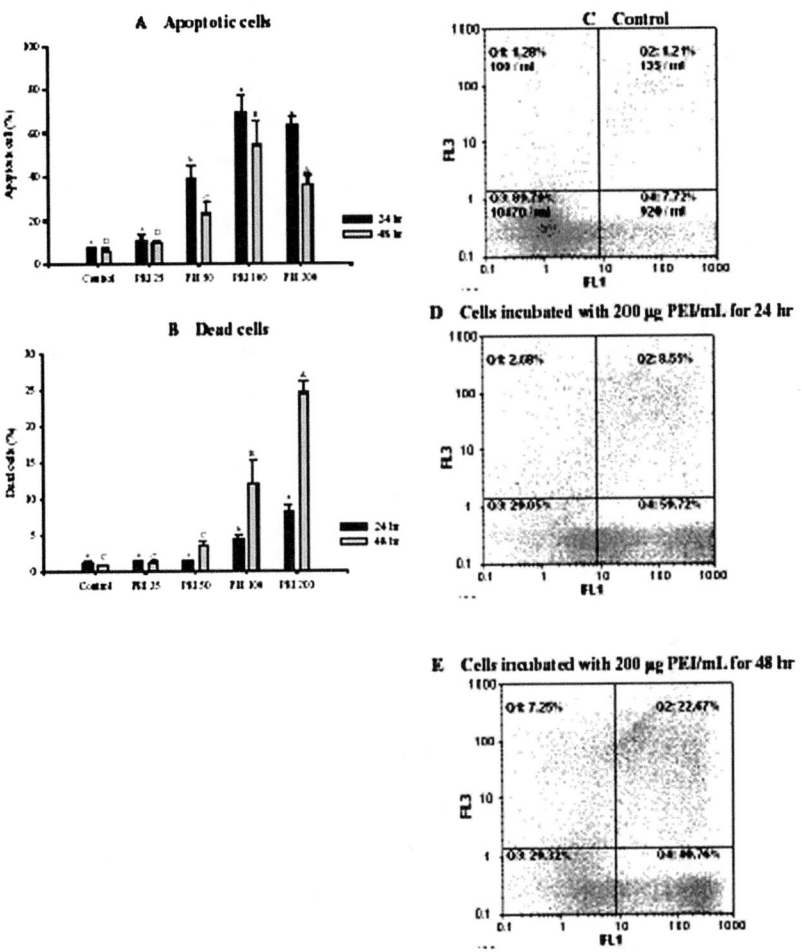

FIGURE 4. Annexin V labeling of U937 cells by flow cytometric analysis. Cells cultured with various levels of PEI for 24 and 48 h were stained with a combination of Annexin V-FITC and PI. Four subpopulations of cells were made on the basis of Annexin V-FITC binding and PI uptake: Q1, damaged cells (Annexin V-FITC negative/PI positive); Q2, late apoptotic and necrotic cells (annexin V-FITC positive/PI positive); Q3, live cells (Annexin V-FITC negative/PI negative); Q4, early apoptotic cells (annexin V-FITC positive/PI negative).

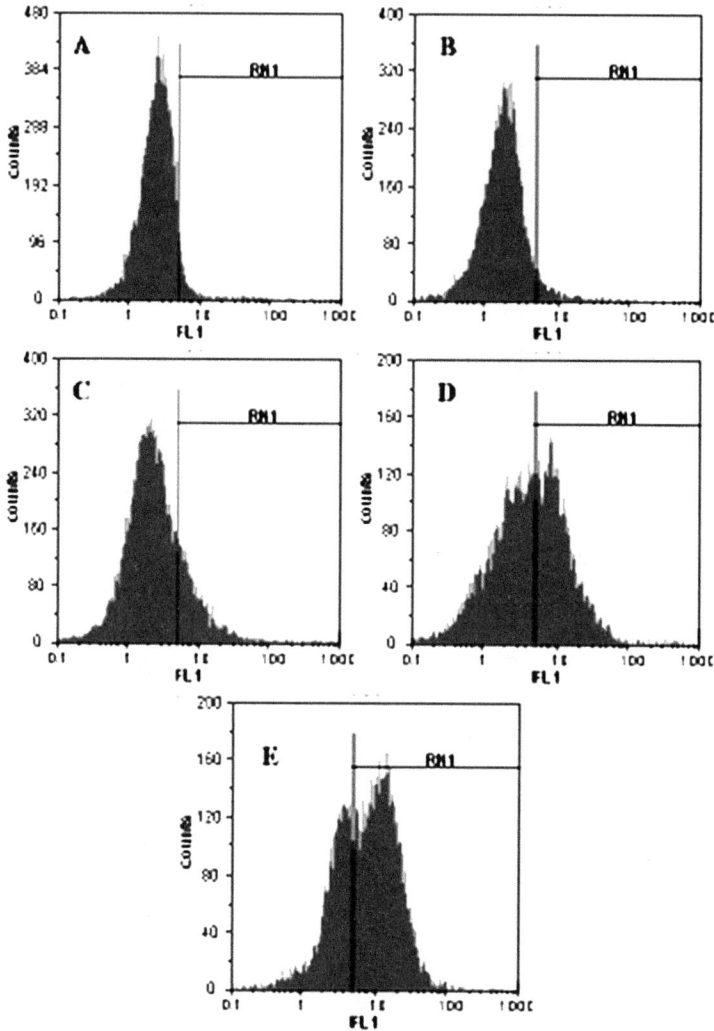

FIGURE 5. Effect of pectinesterase inhibitor on caspase-3 activity of U937 cells treated with various levels of PEI for 36 h. (**A**) Control, (**B**) 25 mg/mL, (**C**) 50 mg/mL, (**D**) 100 mg/mL, and (**E**) 200 mg/mL.

led to apoptotic cell death; this suggested that the apoptotic mechanisms were ROS generation activation of caspase 3 → cleavages of its substrate → apoptotic cell death in TRSDY-treated HL-60 cells.[3]

In conclusion, PEI inhibited the proliferation of U937 cells, affected the cell cycle distribution, increased the surface exposure of PS, decreased the mitochondrial membrane integrity (loss of mitochondrial MTP), and thus led to cell apoptosis in this study. Our findings suggest that PEI induces apoptosis through a mitochondria-dependent pathway and upregulation of caspase 3 activity. Therefore, PEI from jelly fig achenes is probably one alternative treatment for leukemia.

REFERENCES

1. KWON, K.B. et al. 2003. Induction of apoptosis by takrisodokyeum through generation of hydrogen peroxide and activation of caspase-3 in HL-60 cells. Life Sci. **73:** 1895–1906.
2. ARENDS, M.J. & A.H. WYLLIE. 1991. Apoptosis: mechanisms and roles in pathology. Int. Rev. Exp. Pathol. **32:** 223–254.
3. KWON, K.B. et al. 2002. Induction of apoptosis by diallyl disulfide through activation of caspase-3 in human leukemia HL-60 cells. Biochemical Pharmacol. **63:** 41–47.
4. POLVERINO, A.J. & S.D. PATTERSON. 1997. Selective activation of caspases during apoptotic induction in HL–60 cells. Effects of a tetrapeptide inhibitor. J. Biol. Chem. **272:** 7013–7021.
5. BELLOC, F. et al. 2000. Flow cytometry detection of caspase 3 activation in preapoptotic leukemic cells. Cytometry **40:** 151–160.
6. SCHLOFFER, D. et al. 2003. Induction of cell cycle arrest and apoptosis in human cervix carcinoma cells during therapy by cisplatin. Cancer Detect. Prevent. **27:** 481–493.
7. SU, S. et al. 2003. Effects of soy isoflavones on apoptosis induction and G2-M arrest in human hepatoma cells involvement of caspase-3 activation, Bcl-2 and Bcl-XL downregulation, and Cdc2 kinase activity. Nutrit. Cancer **45:** 113–123.
8. JIANG, C.M. et al. 2002. Characterization of pectinesterase inhibitor in jelly fig (*Ficus awkeotsang* Makino) achenes. J. Agric. Food Chem. **50:** 4890–4894.
9. TSAI, G.H. & W.H. SU. 1999. Antibacterial activity of shrimp chitosan against *Escherichia coli*. J. Food Prot. **62:** 239–243.
10. WANG, G.H. 1992. Inhibition and inactivation of five species of foodborne pathogens by chitosan. J. Food Prot. **55:** 916–919.
11. LUDOVICO, P. et al. 2001. Assessment of mitochondrial membrane potential in yeast cell populations by flow cytometry. Microbiology **147:** 3335–3343.
12. PAN, M.H. et al. 2001. Induction of apoptosis by garcinol and curcumin through cytochrome c release and activation of caspases in human leukemia HL-60 cells. J. Agric. Food Chem. **49:** 1464–1474.
13. BARKETT, M. et al. 1997. Phosphorylation of IkappaB-alpha inhibits its cleavage by caspase CPP32 in vitro. J. Biol. Chem. **272:** 29419–29422.
14. FREDERICH, M. et al. 2003. Isostrychnopentamine, an indolomonoterpenic alkaloid from Strychnos usambarensis, induces cell cycle arrest and apoptosis in human colon cancer cells. J. Pharmacol. Exp. Ther. **304:** 1103–1110.
15. LI, Y.C. et al. 2004. Baicalein induced in vitro apoptosis undergo caspases activity in human promyelocytic leukemia HL-60 cells. Food Chem. Toxicol. **42:** 37–43.
16. REUTHER, J.Y. & A.S. BALDWIN, JR. 1999. Apoptosis promotes a caspase-induced amino-terminal truncation of IkappaBalpha that functions as a stable inhibitor of NF-kappaB. J. Biol. Chem. **274:** 20664–20670.
17. ITO, K. et al. 2003. Caffeine induces G2/M arrest and apoptosis via a novel p53-dependent pathway in NB4 promyelocytic leukemia cells. J. Cell. Physiol. **196:** 276–283.
18. MAJLESSIPOUR, F. et al. 2002. The combination regimen of idarubicin and taxotere is effective against human drug-resistant leukemic cell lines. Anticancer Res. **22:** 1361–1368.

Enhancement of Cisplatin-Induced Apoptosis and Caspase 3 Activation by Depletion of Mitochondrial DNA in a Human Osteosarcoma Cell Line

HSIU-CHUAN YEN,[a] YI-CHIA TANG,[a] FAN-YI CHEN,[a] SHIH-WEI CHEN,[a] AND HIDEYUKI J. MAJIMA[b]

[a]*Graduate Institute of Medical Biotechnology and School of Medical Technology, Chang Gung University, Tao-Yuan, Taiwan*

[b]*Department of Oncology, Kagoshima University Graduate School of Medical and Dental Sciences, Kagoshima, Japan*

ABSTRACT: Cisplatin is an anticancer drug that can induce apoptosis. In this study, we investigated the effect of mitochondrial DNA (mtDNA) depletion on cisplatin-induced cell death using a human osteosarcoma cell line (143B) and mtDNA-depleted 143B cells (143B-ρ^0). Results showed that cisplatin decreased cell survival in 143B-ρ^0 cells. Moreover, cisplatin induced a greater extent of apoptosis-associated DNA fragmentation and caspase 3 activation in 143B-ρ^0 cells. The release of mitochondrial cytochrome c into cytosol by cisplatin was enhanced more obviously in 143B cells than in 143B-ρ^0 cells; however, in the control group of 143B-ρ^0 cells, it was already dramatically greater. Depletion of mtDNA may increase sensitivity of cells to cisplatin-induced apoptosis by enhancing caspase 3 activation via both cytochrome c–dependent and –independent pathways.

KEYWORDS: cisplatin; mitochondrial DNA; apoptosis; caspase 3; cytochrome c

INTRODUCTION

Cisplatin (cis-diamminedichloroplatinum [II]; CDDP) is a chemotherapeutic agent that is known to inhibit DNA replication by cross-linking DNA.[1] It has been shown that reactive oxygen species could be important in CDDP-induced apoptosis.[2] It has also been indicated that suppression of apoptosis was important in CDDP resistance in cancer cells,[3] and DNA platination or p53 status did not necessarily determine the sensitivity of cancer cells to CDDP-induced apoptosis.[4] In the mechanisms leading to apoptosis, activation of caspase 3 can result from different upstream caspases or release of the Smac/DIABLO protein from mitochondria independent of other caspases.[5] It has been indicated that CDDP induced apoptosis via release of mitochondrial cytochrome c into cytosol with subsequent activation of

---

Address for correspondence: Hsiu-Chuan Yen, Ph.D., Graduate Institute of Medical Biotechnology, Chang Gung University, 259 Wen-Hwa 1st Rd., Kwei-Shan, Tao-Yuan 333, Taiwan. Voice: +886-3-2118800 ext. 5207; fax: +886-3-2118692.
yen@mail.cgu.edu.tw

caspase 9 and then caspase 3.[3] However, activation of caspase 8 upstream of caspase 3 has also been demonstrated.[6]

Human mitochondrial DNA (mtDNA) is a circular DNA with 16,569 bp that encodes 13 subunits of complexes I, III, IV, and V of oxidative phosphorylation.[7] Although it has been shown that CDDP can cause mtDNA damage,[8] the role of mtDNA in CDDP-induced apoptosis is not clear. In this study, we compared cell survival, extent of apoptosis, activation of caspase 3, and cytochrome c release in 143B and 143B-ρ^0 cells following CDDP treatment.

MATERIALS AND METHODS

The human osteosarcoma cell line, 143B cells (thymidine kinase deficient, TK), and 143B-ρ^0 cells (clone 143B-87) lacking mtDNA were obtained from Dr. Douglas Wallace at the University California, Irvine, CA.[9,10] The absence of mitochondrial gene expression has also been previously confirmed.[11] Both cell lines were cultured in high-glucose (4.5 g/L) Dulbecco's modified Eagle's medium supplemented with 10% fetal bovine serum, 110 μg/mL pyruvate, and 50 μg/mL uridine.

Evaluation of Cell Survival following CDDP Treatment

The colony formation assay was used to test the clonogenic ability of cells. Approximately 200 cells were plated on 60-mm dishes for 24 h and exposed to various doses of CDDP for another 24 h. The number of visible colonies grown from single cells was counted after staining cells with crystal violet to evaluate cell survival following CDDP treatment.[12] The threshold to define a colony was set at 50 cells per colony, which was much less than the number of cells in regular colonies of control dishes.

Detection of Apoptosis-Associated Nucleosomal Ladder

Total genomic DNA was isolated from cells using the Puregene DNA Isolation Kits from Gentra Systems (Minneapolis, MN). The ApoAlert ligation-mediated (LM)–PCR Ladder Assay kit from BD Biosciences Clontech (Palo Alto, CA) was used to determine the presence of nucleosomal ladder. The PCR products were electrophoresed in a 1.5% agarose gel containing ethidium bromide.

Measurement of Caspase 3 Activity

The ApoAlert Caspase Fluorescent Assay kit from BD Biosciences Clontech was used. A total of 10^6 cells were harvested, and supernatant from lysed cells was incubated with the substrate, DEVD-AFC (7-amino-4-trifluoromethyl coumarin), for 1 h at 37°C. Cleavage of the substrate by active caspase 3 resulted in the increase of fluorescence (excitation = 400 nm and emission = 505 nm), which was recorded by a 96-well fluorometer. The reading from no-substrate control was subtracted.

Western Blot Analysis for Detecting Cytochrome c Release

Cells were homogenized in mitochondrial isolation solution (0.225 M mannitol, 0.075 M sucrose, 1 mM EGTA, pH 7.4) containing the protease inhibitor cocktail from Roche (Mannheim, Germany) and centrifuged at 600 × g. The supernatant was centrifuged at 10,000 × g, and the final supernatant was the cytosolic fraction without nuclear and mitochondrial fractions. The final pellet was washed and resuspended in 50 mM potassium phosphate buffer (pH 7.4) containing 0.5% Triton X-100 and saved as the mitochondrial fraction. Proteins were size-separated by SDS–polyacrylamide gel electrophoresis and analyzed for cytochrome c, actin, and Mn superoxide dismutase (MnSOD) using antibodies obtained from BD Pharmingen (San Diego, CA), Chemicon (Temecula, CA), and Upstate (Charlottesville, VA), respectively.

Statistical Analysis

Analysis of variance (ANOVA) was used to compare the difference among multiple groups using SASfor Windows version 8.1 (SAS Institute, Cary, NC). Data are presented as mean ± standard deviation.

RESULTS AND DISCUSSION

Results of cell survival assay are shown in FIGURE 1. The doses of drug that inhibited 50% and 90% of colony formation, IC_{50} and IC_{90}, were calculated from three independent experiments.[12] The IC_{50} for 143B and 143B-ρ^0 cells was 0.37 ± 0.01 and 0.24 ± 0.01 μg/mL, respectively; and the IC_{90} for 143B cells and 143B-ρ^0 cells was 0.77 ± 0.12 and 0.42 ± 0.03 μg/mL, respectively. Those results indicated that CDDP resulted in much lower cell survival in 143B-ρ^0 than in 143B cells, which could be caused by enhanced apoptosis or necrosis.

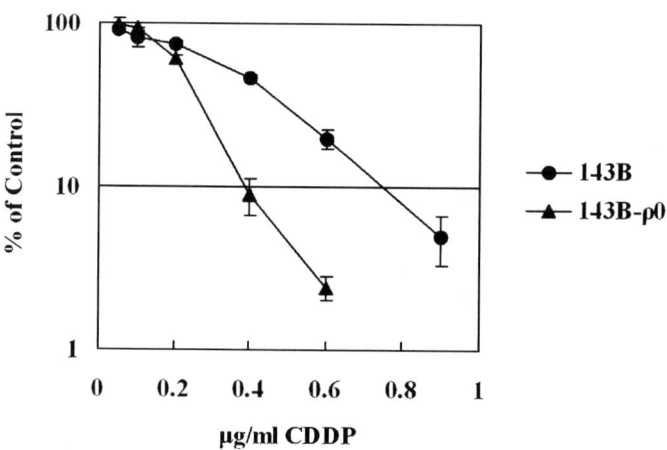

FIGURE 1. Cell survival curve following CDDP treatment obtained from colony formation assay. There are three dishes per group. The y-axis is in log scale.

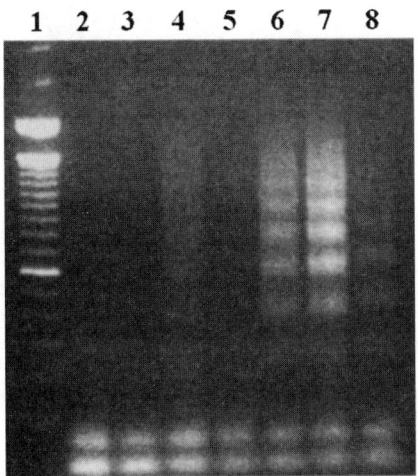

FIGURE 2. Detection of nucleosomal DNA ladder by LM-PCR. Lanes 1, 100-bp ladder; 2, 134B-control; 3, 143B-0.5 µg/mL CDDP; 4, 143B-1 µg/mL CDDP; 5, ρ^0-control; 6, ρ^0-0.5 µg/mL; 7, ρ^0-1 µg/mL; 8, positive control (calf thymus DNA).

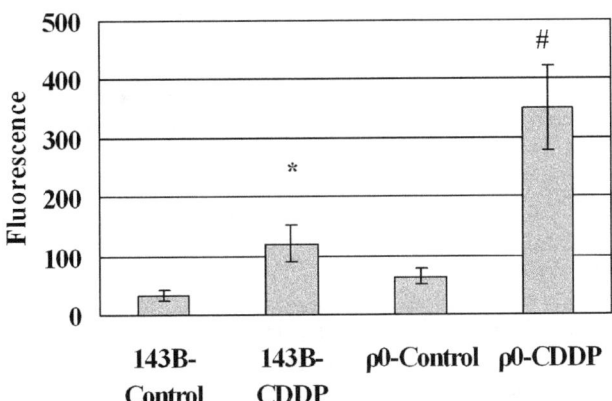

FIGURE 3. Caspase 3 activities following 24-h CDDP treatment (0.5 µg/mL). *P = .01 for 143B-control versus 143B-CDDP; #P < 0.0001 for ρ^0-control versus ρ^0-CDDP. There are four replicates for each group.

We next investigated the effect of CDDP on apoptosis in these two cell lines. Results of LM-PCR (FIG. 2) demonstrated that 24-h CDDP treatment (0.5 and 1 µg/mL) induced a greater extent of apoptosis-associated DNA fragmentation in 143B-ρ^0 cells. Moreover, FIGURE 3 shows that 24-h CDDP treatment (0.5 µg/mL) increased caspase 3 activities in both cells. Two-way ANOVA revealed that the increase in 143B-ρ^0 cells was greater (P = 0.0003). Therefore, enhancement of CDDP-induced cell death by depletion of mtDNA was correlated with the increase in DNA fragmentation and caspase 3 activation.

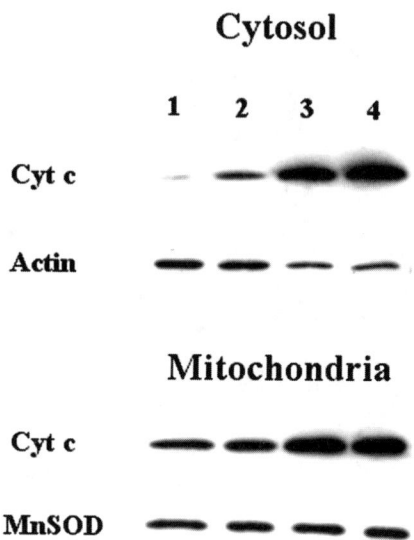

FIGURE 4. Western blot analysis for detection of cytochrome c release. Lanes 1, 2, 3, and 4 represent 143B-control, 143B-CDDP, ρ^0-control, and ρ^0-CDDP, respectively.

Furthermore, results of Western blot analysis (FIG. 4) showed that the amount of cytochrome c released into cytosol, relative to that of actin, was markedly increased by 24-h CDDP treatment (0.5 µg/mL) in 143B cells. The level of cytochrome c in cytosol was significantly higher in the control of 143B-ρ^0 cells than in 143B cells, although the increase after CDDP treatment was less obvious in 143B-ρ^0 cells. That could be due to the status of mitochondrial injury or a compensatory increase in cytochrome c expression because its level in mitochondria was also augmented in 143B-ρ^0 cells compared with that of MnSOD (FIG. 4). However, this increase did not lead to a proportional increase in the basal caspase 3 activity (FIG. 3), suggesting that 143B-ρ^0 cells could have become less sensitive to the death signal of cytochrome c, such as by suppression of apoptosome formation and caspase 9 activation,[5] during the stage without apoptotic insult. Persistent presence of higher levels of cytosolic cytochrome c in 143B-ρ^0 cells could result in greater caspase 3 activation by CDDP (FIG. 3), although cytochrome c–independent pathways might be also involved.

It has been shown that apoptosis and cytochrome c release induced by staurosporine were not altered following mtDNA depletion, demonstrating that those pathways were not impaired in ρ^0 cells.[10,13] However, apoptosis and translocation of cytochrome c were suppressed in ρ^0 cells after the treatment of tumor necrosis factor-related apoptosis–induced ligand (TRAIL),[14] whereas mtDNA depletion enhanced CDDP-induced apoptosis as shown by this study. Therefore, the use of ρ^0 cells has revealed diverse roles of mtDNA in apoptosis depending on the apoptotic stimuli used. Liang and Ullyatt also showed that CDDP-induced apoptosis was increased in ρ^0 cells derived from a lymphoma cell line (U937 cells) that was assessed by morphological changes, but they did not investigate any related apoptotic pathways.[15]

Our results first indicate that enhanced caspase 3 activation plays an important role in increased CDDP-induced cell death in 143B-ρ^0 cells. A higher degree of cytochrome c release in 143B-ρ^0 cells before treatment may be an important role, but mechanisms independent of cytochrome c release could be also enhanced, such as augmented activation of caspase 8 and release of Smac/DIABLO protein from mitochondria. Other mechanisms related to cytochrome c, such as the formation of apoptosome with apoptotic protease activating factor 1 (Apaf-1) and nitrosylation of cytochrome c, could be also important.[5,16] On the other hand, the existence of caspase-independent pathways in CDDP-induced apoptosis has been indicated.[17] Therefore, the role of caspase-independent mechanisms, such as release of apoptosis-inducing factor (AIF) from mitochondria, remains to be elucidated.[5] Furthermore, because mitochondrial dysfunction is known to induce oxidative stress and alter expression of nuclear genes,[7] chronic perturbation of redox status may result in different signaling pathways of cell death in response to CDDP in ρ^0 cells because CDDP can also increase oxidative stress.[1,2] However, as shown by this study, it is clear that the caspase-dependent pathway is, at least, a mechanism leading to increased extent of apoptosis by CDDP following mtDNA depletion in 143B cells.

ACKNOWLEDGMENTS

This work is supported by Grant CMRP1230 from Chang Gung Memorial Hospital, Taiwan.

REFERENCES

1. SIDDK, Z.H. 2003. Cisplatin: mode of cytotoxic action and molecular basis of resistance. Oncogene **22:** 7265–7279.
2. BAEK, S.M., C.H. KWON, J.H. KIM, *et al.* 2003. Differential roles of hydrogen peroxide and hydroxyl radical in cisplatin-induced cell death in renal proximal tubular epithelial cells. J. Lab. Clin. Med. **142:** 178–186.
3. MUELLER, T., W. VOIGT, H. SIMON, *et al.* 2003. Failure of activation of caspase-9 induces a higher threshold for apoptosis and cisplatin resistance in testicular cancer. Cancer Res. **63:** 513–522.
4. BURGER, H., K. NOOTER, A.W. BOERSMA, *et al.* 1997. Lack of correlation between cisplatin-induced apoptosis, p53 status and expression of Bcl-2 family proteins in testicular germ cell tumor cell lines. Int. J. Cancer **73:** 592–599.
5. HENGARTNER, M.O. 2000. The biochemistry of apoptosis. Nature **407:** 770–776.
6. HOTTA, T., H. SUZUKI, S. NAGAI, *et al.* 2003. Chemotherapeutic agents sensitize sarcoma cell lines to tumor necrosis factor-related apoptosis-inducing ligand-induced casepase-8 activation, apoptosis and loss of mitochondrial membrane potential. J. Orthop. Res. **21:** 949–957.
7. WALLACE, D.C. 1999. Mitochondrial diseases in man and mouse. Science **283:** 1482–1488.
8. OLIVERO, O.A., P.K. CHANG, D.M. LOPEZ-LARRAZA, *et al.* 1997. Preferential formation and decreased removal of cisplatin-DNA adducts in Chinese hamster ovary cell mitochondrial DNA as compared to nuclear DNA. Mutat. Res. **391:** 79–86.
9. TROUNCE, I., S. NEILL & D.C. WALLACE. 1994. Cytoplasmic transfer of the mtDNA nt 8993 T→G (ATP6) point mutation associated with Leigh syndrome into mtDNA-less cells demonstrates cosegregation with a decrease in state III respiration and ADP/O ratio. Proc. Natl. Acad. Sci. USA **91:** 8334–8338.
10. JIANG, S., J. CAI, D.C. WALLACE & D.P. JONES. 1999. Cytochrome c-mediated apoptosis in cells lacking mitochondrial DNA. J. Biol. Chem. **274:** 29905–29911.

11. YEN, H.C., C.Y. NIEN, H.J. MAJIMA, *et al.* 2003. Increase of lipid peroxidation by cisplatin in WI38 cells but not in SV40-transformed WI38 cells. J. Biochem. Mol. Toxicol. **17:** 39–46.
12. FRESHNEY, R.I. 2000. Culture of Animal Cells. A Manual of Basic Technique. 4th edit. John Wiley & Sons. New York.
13. CAI, J., D.C. WALLACE, B. ZHIVOTOVSKY & D.P. JONES. 2000. Separation of cytochrome c-dependent caspase activation from thiol-disulfide redox change in cells lacking mitochondrial DNA. Free Radic. Biol. Med. **29:** 334–342.
14. KIM, J., Y.H. KIM, Y. CHANG, *et al.* 2002. Resistance of mitochondrial DNA-deficient cells to TRAIL: role of Bax in TRAIL-induced apoptosis. Oncogene **21:** 3139–3148.
15. LIANG, B.C. & E. ULLYATT. 1998. Increased sensitivity to cis-diamminedichloroplatinum induced apoptosis with mitochondrial DNA depletion. Cell Death Differ. **5:** 694–701.
16. SCHONHOFF, C.M., B. GASTON & J.B. MANNICK. 2003. Nitrosylation of cytochrome c during apoptosis. J. Biol. Chem. **278:** 18265–18270.
17. HENKELS, K.M. & J.J. TURCHI. 1999. Cisplatin-induced apoptosis proceeds by caspase-3-dependent and -independent pathways in cisplatin-resistant and -sensitive human ovarian cancer cell lines. Cancer Res. **59:** 3077–3083.

Thallium Acetate Induces C6 Glioma Cell Apoptosis

CHEE-FAH CHIA,[a,b] SOUL-CHIN CHEN,[a] CHIN-SHYANG CHEN,[a] CHUEN-MING SHIH,[b] HORNG-MO LEE,[c] AND CHIH-HSIUNG WU[a]

[a]*Department of Surgery, School of Medicine, Taipei Medical University and Hospital, Taipei, Taiwan*

[b]*Graduate Institute of Medical Sciences, School of Medicine, Taipei Medical University and Hospital, Taipei, Taiwan*

[c]*Graduate Institute of Cell and Molecular Biology, School of Medicine, Taipei Medical University adn Hospital, Taipei, Taiwan*

ABSTRACT: Thallium acetate is a known neurotoxic agent. In this study, we investigated the mechanisms by which thallium acetate induces cell cycle arrest and cell apoptosis in cultured LC6 glioma cells. Exposure of C6 glioma cells to thallium acetate decreased cell viability as demonstrated by the MTT assay. Incubation of thallium acetate arrested cell cycle progression at the G_2/M phase and caused cellular apoptosis at 300 μM as determined by trypan blue exclusion and flow cytometric analysis. The G2/M arrest was associated with a decrease in expression of CDK2 protein and an upregulation of p53 and the CDK inhibitor p21^{Cip1}, but not p27^{Kip1}. Thallium acetate did not alter the protein levels of cyclin A and B; cyclin D1, D2, and D3; and CDK4 expression in C6 glioma cells. Incubation of C6 glioma cells with thallium acetate upregulated the expression of proapoptotic proteins Bad and Apaf and downregulated the expression of anti-apoptotic proteins Bcl-xL and Bcl-2. In conclusion, these data suggest that thallium acetate inhibits cell cycle progression at G2/M phase by suppressing CDK activity through the p53-mediated induction of the CDK inhibitor p21^{Cip1}. Impairment of cell cycle progression may trigger the activation of a mitochondrial pathway and shifts the balance in the Bcl-2 family toward the proapoptotic members, promoting the formation of the apoptosome and, consequently, apoptosis.

KEYWORDS: C6 glioma cells; G_2/M arrest; mitochondria; p21^{Cip1}; p27^{Kip1}; thallium acetate

INTRODUCTION

Thallium, a toxic heavy metal originally found in coal combustion and cement manufacturing, has become an important environmental pollutant.[1] Thallium salts were once used as medicine to treat tuberculosis, sexually transmitted diseases, and

---

Address for correspondence: Chih-Hsiung Wu, M.D., Ph.D., Department of Surgery, School of Medicine, Taipei Medical University and Hospital, 252 Wu-Hsing St., Taipei, Taiwan. Voice: +00886-2-27372181 ext. 3311; fax: +00886-2-27369438.
  chwu@tmu.edu.tw

ringworm of the scalp.[2] Thallium has also been used as a poison in criminal plots to hurt or kill people. A case of acute thallium poisoning in a suspected murder attempt has been reported in Taiwan recently.[3] The manifestations of thallium toxicity include extremely painful sensory neuropathy, motor paralysis alteration of mental status, and dysfunction and damage of such organs as the heart, liver, and kidney.[4] However, the exact mechanism of thallium toxicity has not been completely elucidated. Thallium interferes with energy production by inhibiting sodium–potassium–ATPase.[5] Thallium-induced brain damage is associated with increases in lipid peroxidation in the brain, suggesting that free radical production plays a pivotal role in thallium toxicity.[6,7] An increase in the generation of reactive oxygen species (ROS) and the disruption of cellular energy production may block cell cycle progression, which in turn may induce cell apoptosis.[8]

Cell cycle progression is controlled by a growing family of cell cycle regulatory proteins, cyclins, and their catalytic subunit cyclin-dependent kinases (CDKs).[9] Arrest of the cell cycle requires a group of CDK inhibitors (CKIs) that interact with the cyclin–CDK complexes and block cell cycle progression.[10] Recent reports have demonstrated that CKI expression is upregulated upon exit from the cell cycle or the maintenance of irreversible growth arrest.[11]

Apoptosis is an endogenous mechanism that regulates cell mass in various organ systems. Apoptosis can be regulated by several Bcl-2 family members. The Bcl-2 family comprises both anti-apoptotic proteins (e.g., Bcl-2 and Bcl-x_L) and proapoptotic proteins (e.g., Bad and Bax).[12] Bcl-2 and Bcl-x_L are mitochondrial membrane proteins that interact with Bax, another mitochondrial membrane-bound proapoptotic protein, to inhibit the Bax homo-oligomeric channel. The proapoptotic protein, Bad, binds and prevents the inhibitory constraint of Bcl-2/Bcl-x_L, resulting in ion flux across the mitochondrial outer membrane and subsequently leading to the release of cytochrome c from the mitochondrial intermembrane space into the cytosol. Cytochrome c binds to the adaptor protein Apaf-1 to promote autoactivation of procaspase-9 and the downstream caspase cascade, causing apoptosis executed by the terminal caspase-3.[13] Apoptosis can be triggered by downregulation of the anti-apoptotic homolog Bcl-2 and Bcl-x_L with or without upregulating the expression of proapoptotic proteins such as Bax and Bak or other apoptosis-related gene products such as p53 and p21^{Cip1}.[12]

In this study, we investigated whether thallium acetate regulates cell cycle progression and cell apoptosis in C6 glioma cells. We demonstrated that thallium inhibits DNA replication, leading to cell cycle arrest and cell death. The induction of cell cycle arrest is associated with upregulation of the CDK inhibitor p21^{Cip-1}. On the other hand, thallium-induced apoptosis is associated with elevation of proapoptotic proteins, Apaf and Bad, and downregulation of the anti-apoptotic proteins, Bcl-2 and Bcl-x_L.

MATERIALS AND METHODS

Materials

Affinity-purified mouse polyclonal antibodies to cyclin D1, D2, and D3, as well as CDK2, CDK4, p21^{Cip1}, and p27^{Kip1}, were obtained from Transduction Laboratory (Lexington, KY); Dulbecco's modified Eagle's medium, heat-inactivated fetal bovine serum, and other reagents required for cell cultures from Life Technologies (Gaithersburg, MD); protease inhibitor cocktail tablets from Boehringer Mannheim (Mannheim, Germany); thallium acetate and other chemicals from Sigma (St. Louis, MO).

Cell Culture, Preparation of Cell Lysates, Polyacrylamide Gel Electrophoresis, and Western Blotting

C6 glioma cells (from Food Industry Research and Development Institute, Hsin-Chu, Taiwan) were cultured as described previously.[14] Cells were lysed by adding lysis buffer containing 10 mM Tris–HCl (pH 7.5), 1 mM EGTA, 1 mM $MgCl_2$, 1 mM sodium orthovanadate, 1 mM dithiothreitol, 0.1% mercaptoethanol, 0.5% Triton X-100, and the protease inhibitor cocktails (final concentrations: 0.2 mM phenylmethylsulfonyl fluoride, 0.1% aprotinin, 50 mg/mL leupeptin). Cells adhering to the plates were scraped using a rubber policeman and stored at $-70°C$ for further measurements. Electrophoresis was carried out using 7.5% sodium dodecyl sulfate–polyacrylamide gel. Following electrophoresis, proteins on the gel were electrotransferred onto a polyvinylidene difluoride (PVDF) membrane. The PVDF membrane was incubated with a solution containing appropriate primary antibodies in the blocking buffer. Finally, the PVDF membrane was incubated with peroxidase-linked anti–mouse immunoglobulin G antibodies for 1 h and then developed using a LumiGLO chemiluminescence kit (Amersham, UK).

MTT Assay

The effect of thallium acetate on cell viability was measured by a modified MTT assay, based on the ability of live cells to utilize thiazolyl blue and convert it into dark blue formazan. In brief, 6,000 cells/well (in 96-well microtiter plates) were incubated with different concentrations of thallium acetate in 0.2 mL of culture medium overnight. The supernatant was then removed, and 0.1 mL of 40 mM HCl in isoprotanol was added to each well at 37°C for 2 h. The absorbance at a wavelength of 550 nm was measured with a micro-ELISA reader (Bio-Rad, Hercules, CA). Each assay was performed in triplicate.

Cell Cycle Synchronization and Flow Cytometry

Cells were synchronized at the G_1 phase by serum starvation for 24 h at 37°C. The cells were collected and fixed with 70% ethanol and then washed with phosphate-buffered saline. RNA was lysed with RNAse for 20 min at 37°C and stained with propidium iodide for 2 h at 4°C. The DNA content was measured using a flow cytometry analysis (Becton-Dickinson, San Jose, CA); 15,000 events were analyzed for each sample.

Statistical Analysis

Results are expressed as mean ± standard error of the mean from the number of independent experiments performed. One-way analysis of variance (ANOVA) and Student's two-tailed t test were used to determine the statistical significance of the difference between means. A P value of less than 0.05 was considered significant.

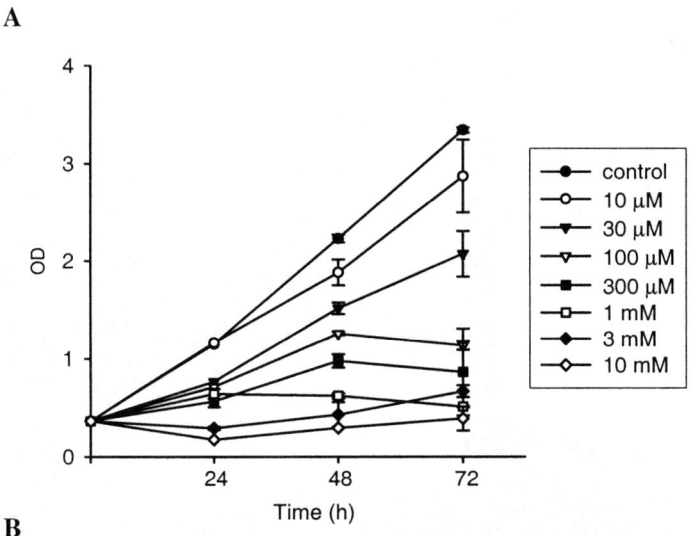
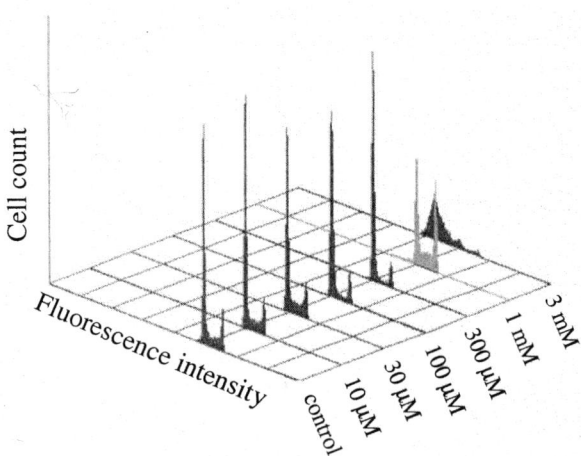

FIGURE 1. Effects of thallium acetate on cell viability and cell cycle progression. (**A**) MTT assay of cell viability revealed a dose- and time-dependent reduction in cell viability following thallium acetate exposure. (**B**) Cell cycle analysis showed an increase in cell population at G_2/M phase after thallium acetate exposure.

RESULTS

Effects of Thallium Acetate on Cell Viability and Cell Cycle Progression

Thallium has been shown to interfere with energy production and increase ROS generation.[6,7] Because oxidative stress is a deleterious event leading to cell death, we investigated whether thallium alters cell viability. C6 glioma cells were incubated with different concentrations of thallium acetate and cell viability determined by the MTT assay. FIGURE 1A shows that cell viability was reduced by thallium acetate treatment in a dose- and time-dependent manner. To further address whether thallium acetate halts cell cycle progression, we synchronized C6 glioma cells by serum starvation for 24 h; serum was then added to the quiescent cells to activate the cell cycle progression. Different concentrations of thallium acetate were added to the growth-stimulated cells for 24 h, and the percentage of the cells in the G_0/G_1 phase and G2/M phase were determined by flow cytometry. As shown in FIGURE 1B, the cell population at G_2/M phase was increased after thallium acetate exposure, suggesting cell cycle arrest at G_2/M phase. Because cell proliferation and death are closely linked to several regulatory proteins, we analyzed the protein levels of cyclins, CDK, and pro- and anti-apoptotic proteins in C6 glioma cells.

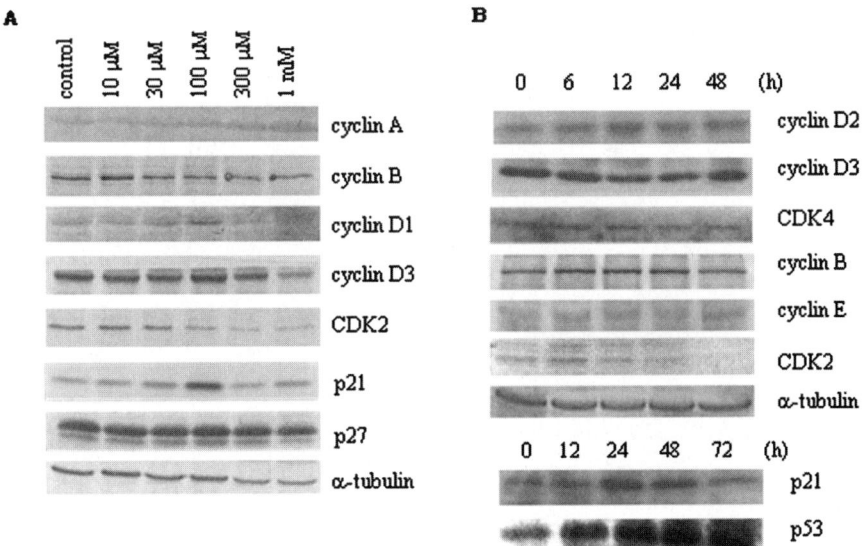

FIGURE 2. Effects of thallium acetate on the expression of cell cycle regulatory proteins. (**A**) Western blot analysis for cyclins and CDKs showed a significant reduction in the protein level of CDK-2 following thallium acetate exposure. (**B**) Western blot analysis revealed that treatment of cells with thallium acetate increased the expressions of p21^{Cip1} and p53.

Effects of Thallium Acetate on Expression of Cell Cycle Regulatory Proteins

To investigate changes in the expression of cell cycle regulatory proteins during these processes, we carried out Western blot analysis for cyclins A, B, D1, D2, D3, E, and CDK2 and CDK4, using specific monoclonal antibodies. CDK2 expression was significantly suppressed by thallium acetate (300 μM) (FIG. 2A). In contrast to CDK2, the expression of CDK4 was not affected. Cell cycle arrest was characterized by increased expression of CDK inhibitors. Treatment of cells with thallium acetate (100 μM) increased the expression of p21^{Cip1} but not p27^{kip1}, with the maximum induction of p21^{Cip1} seen after 24 h (FIG. 2B). The induction of p21^{Cip1} was in parallel with an increase in p53 expression.

Effects of Thallium Acetate on Expression of Pro- and Anti-Apoptotic Proteins

The increase of sub-G_1 population (FIG. 1B) suggests that thallium acetate induces C6 glioma cell apoptosis. In an effort to delineate the mechanism, we examined the effect of thallium on the expression of Bcl-2, Bcl-x_L, Bad, and Apaf. Western blot analysis revealed that treatment with thallium acetate for 24 h downregulated the expression of two anti-apoptotic proteins, Bcl-2 and Bcl-x_L (FIG. 3A). Thallium acetate upregulated the expression of proapoptotic proteins Bad and Apaf in C6 glioma cells (FIG. 3B).

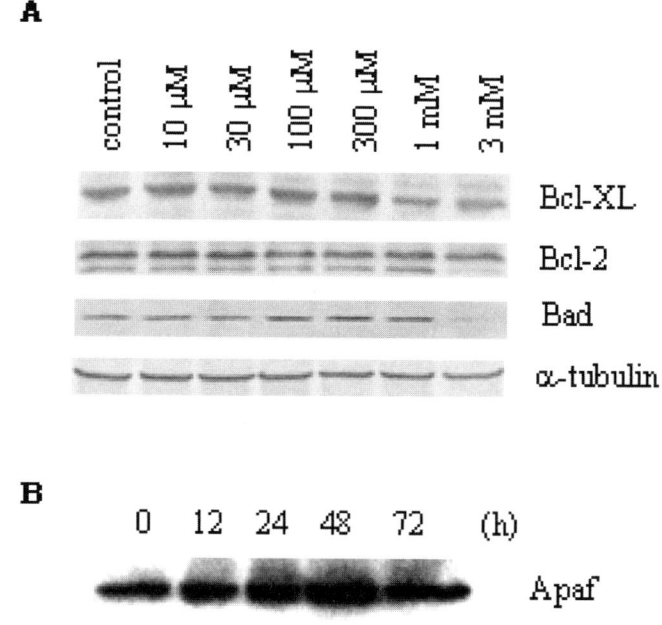

FIGURE 3. Effects of thallium acetate on the expression of apoptosis regulatory proteins. (**A**) Western blot analysis showed that treatment of C6 glioma cells with thallium acetate for 24 h downregulated the expression of anti-apoptotic protein Bcl-2 and Bcl-XL. (**B**) Thallium acetate also upregulated the Bad and Apaf protein levels in C6 glioma cells.

DISCUSSION

Thallium is known to damage the nervous system and organs such as the heart, liver, and kidney. However, little is known regarding the precise mechanisms of thallium toxicity at molecular and cellular levels. This study demonstrated that thallium acetate inhibits cell cycle progression at G_2/M phase by suppressing CDK2 expression and by inhibiting the CDK activity, likely through induction of the CDK inhibitor p21^{Cip1}. Impairment of cell cycle progression probably is the mechanism leading to C6 glioma cell apoptosis. Thallium crosses the blood–brain barrier and accumulates in the brain after exposure.[6] Results from this study suggest that thallium exerts its toxic effects in cells in the nervous system via the influence of cell cycle progression and induction of cell apoptosis.

Several mechanisms of thallium toxicities have been proposed. Thallium may substitute for potassium in the Na, K-ATPase pump, possibly due to the similar ionic charges and radii.[5,15] Thallium induces lipid peroxidation, presumably due to an increase in ROS generation.[6] In this study, we demonstrated that thallium toxicity may be associated with induction of cell cycle arrest and apoptosis in C6 glioma cells. Increased ROS generation leads to DNA damage. C6 glioma cells possess normal p53 status,[16] which plays an important role in regulating DNA repair. Induction of p53, probably secondary to ROS-induced DNA damage, may be an initiating event to cause G_1 and G_2 arrest.[17] Thallium acetate increased the expression of both p53 and p21^{Cip1} in C6 glioma cells, suggesting that cell cycle arrest and cell death occurs through a p53-dependent induction of p21^{Cip1}. Concomitant with the expression of p53 and p21^{Cip1}, downregulation of Bcl-2 and Bcl-x_L (anti-apoptotic proteins) and upregulation of Bad and Apaf, (proapoptotic proteins) were also noted. These apoptosis-regulatory proteins play prominent roles in several cell death paradigms. Bcl-2 and Bcl-x_L suppress, whereas Bad promotes, apoptosis through their actions in preserving and disrupting, respectively, the integrity of the mitochondrial membrane.[18] Downregulation of Bcl-2 and Bcl-x_L and upregulation of Bad may thereby facilitate the disintegration of the mitochondrial membrane, leading to the release of mitochondrial proteins such as cytochrome c. Once cytochrome c is released from mitochondria into the cytosol, it binds to the adaptor protein Apaf-1 and promotes activation of the caspase cascade to execute the cell death program. Thus, thallium acetate–induced cell cycle arrest at G_2/M followed by the induction of p53 and p21^{Cip1} may be the triggering events. Prolonged cell cycle arrest that resulted in C6 glioma cell apoptosis is likely due to an imbalance in the expression of anti-apoptotic (Bcl-2 and Bcl-x_L) and proapoptotic (Bad) proteins.

A relatively high concentration of thallium acetate is required to induce cell apoptosis. The 50% lethal dose of thallium has been estimated to be 32 mg/kg of body weight in rats (given intraperitoneally).[6] However, because thallium acetate tends to accumulate in rat brain regions,[6] it is difficult to estimate the dose required to exert neurological toxicity *in vivo*.

In summary, we demonstrated that treatment of C6 glioma cells with thallium acetate results in cell cycle arrest and apoptosis induction. Thallium is known to heighten oxidative stress to cause DNA damage. DNA damage may in turn upregulate p53 expression, leading to the induction of CDK inhibitors such as p21^{Cip1}. Induction of p21^{Cip1} expression may cause cell cycle arrest and lead to apoptosis.

REFERENCES

1. HOFFMAN, R.S. 2003. Thallium toxicity and the role of Prussian blue in therapy. Toxicol. Rev. **22:** 29–40.
2. HOFFMAN, R.S. 2000. Thallium poisoning during pregnancy: a case report and comprehensive literature review. J. Toxicol. Clin. Toxicol. **38:** 767–775.
3. HUANG, J. *et al.* 1998. Thallium poisoning a clinical analysis of 5 cases. Chin. Med. J. **78:** 610–611.
4. RUSYNIAK, D.E., R.B. FURBEE & M.A. KIRK. 2002. Thallium and arsenic poisoning in a small midwestern town. Ann. Emerg. Med. **39:** 307–311.
5. PEDERSEN, P.A., J.M. NIELSEN, J.H. RASMUSSEN & P.L. JORGENSEN. 1998. Contribution to Tl+, K+, and Na+ binding of Asn776, Ser775, Thr774, Thr772, and Tyr771 in cytoplasmic part of fifth transmembrane segment in alpha-subunit of renal Na, K-ATPase. Biochemistry **37:** 17818–17827.
6. GALVAN-ARZATE, S., A. MARTINEZ, E. MEDINA, *et al.* 2000. Subchronic administration of sublethal doses of thallium to rats: effects on distribution and lipid peroxidation in brain regions. Toxicol. Lett. **116:** 37–43.
7. APPENROTH, D. & K. WINNEFELD. 1999. Is thallium-induced nephrotoxicity in rats connected with riboflavin and/or GSH?—reconsideration of hypotheses on the mechanism of thallium toxicity. J. Appl. Toxicol. **19:** 61–66.
8. ANNUNZIATO, L., S. AMOROSO, A. PANNACCIONE, *et al.* 2003. Apoptosis induced in neuronal cells by oxidative stress: role played by caspases and intracellular calcium ions. Toxicol. Lett. **139:** 125–133.
9. SHERR, C.J. 1995. D-type cyclins. Trends Biochem. Sci. **20:** 187–190.
10. SHERR, C.J. & J.M. ROBERTS. 1995. Inhibitors of mammalian G1 cyclin-dependent kinases. Genes Dev. **9:** 1149–1163.
11. HIYAMA, H., A. IAVARONE & S.A. REEVES. 1998. Regulation of cdk inhibitor p21 gene during cell cycle progression is under the control of E2F. Oncogene **16:** 1513–1523.
12. GREEN, D.R. 2003. Overview: apoptotic signaling pathways in the immune system. Immunol. Rev. **193:** 5–9.
13. CREAGH, E.M., H. CONROY & S.J. MARTIN. 2003. Caspase-activation pathways in apoptosis and immunity. Immunol. Rev. **193:** 10–21.
14. LIN, C.H., Y.F. LIN, M.C. CHANG, *et al.* 2001. Advanced glycosylation end products induce nitric oxide synthase expression in C6 glioma cells: involvement of a p38 MAP kinase-dependent mechanism. Life Sci. **69:** 2503–2515.
15. AOYAMA, H., M. YOSHIDA & Y. YAMAMURA. 1986. Acute poisoning by ingestion of thallous malonate. Hum. Toxicol. **5:** 389–392.
16. MOBIO, T.A., E. TAVAN, I. BAUDRIMONT, *et al.* 2003. Comparative study of the toxic effects of fumonisin B1 in rat C6 glioma cells and p53-null mouse embryo fibroblasts. Toxicology **183:** 65–75.
17. TAYLOR, W.R. & G.R. STARK. 2001. Regulation of the G2/M transition by p53. Oncogene **20:** 1803–1815.
18. CHAO, D.T. & S.J. KORSMEYER. 1998. BCL-2 family: regulators of cell death. Annu. Rev. Immunol. **16:** 395–419.

Effects of Gonadotrophin-Releasing Hormone Agonists on Apoptosis of Granulosa Cells

NU-MAN TSAI,[a,b] RONG-HONG HSIEH,[b,c] HENG-KIEN AU,[b,d] MING-JER SHIEH,[c] SHIH-YI HUANG,[c] AND CHII-RUEY TZENG[b,d]

[a]*Institute of Medical Sciences, Buddhist Tzu-Chi University, Hualien, Taiwan*

[b]*Center for Reproductive Medicine and Sciences, Taipei Medical University, Taipei, Taiwan*

[c]*School of Nutrition and Health Sciences, Taipei Medical University, Taipei, Taiwan*

[d]*Department of Obstetrics and Gynecology, Taipei Medical University Hospital, Taipei, Taiwan*

ABSTRACT: Granulosa cells are known to contribute to maturation of oocytes, and most of the growth factors exert their action via granulosa cells. It has been established that granulosa cell death during follicular atresia and luteolysis results from apoptosis. However, the precise mechanistic pathways of granulosa cell apoptosis have not yet been defined. In this study, we determined the proportions of apoptosis in granulosa cells treated with two kinds of gonadotrophin-releasing hormone agonists (GnRHa): buserelin and leuprorelin depot. The incidences of DNA fragmentation of human granulosa cells treated with buserelin and leuprorelin were 54.33% and 39.02%, respectively. The proportions of apoptotic bodies were 6.04% and 4.29%, respectively. There was a significant difference in the proportions of DNA fragmentation between the two kinds of GnRHa-treated granulosa cells. The apoptosis pathway and associated protein expression in granulosa cells treated with GnRHa were also determined. The Bax molecule, a pro-apoptosis protein, was expressed in granulosa cells undergoing apoptosis. In contrast, Bcl-2, an anti-apoptosis protein, could not be detected in the same group of granulosa cells. The distribution of cytochrome c determined via immunostaining showed a diffuse pattern, which most likely indicated that cytochrome c was translocated from mitochondria into the cytoplasm. Western blotting showed the expressions of caspase-9 and caspase-3 in patients' granulosa cells. The GnRHa effects on granulosa cells indicated a higher incidence of DNA fragmentation and apoptotic bodies in the buserelin-treated than in the leuprorelin depot–treated group. The granulosa cells go through the mitochondria-dependent apoptosis pathway; the indicated pro-apoptosis protein Bax was expressed and induced cytochrome c release from mitochondria, which then activated caspase-9 and caspase-3 until cell death occurred.

KEYWORDS: apoptosis; gonadotrophin-releasing hormone; granulosa cell

Address for correspondence: Chii-Ruey Tzeng, Department of Obstetrics and Gynecology, Taipei Medical University Hospital, Taipei, Taiwan 110, R.O.C. Voice: +886-2-27372181 ext. 1996; fax: +886-2-23774207.
  tzengcr@tmu.edu.tw

INTRODUCTION

During follicular development, only a few follicles are selected for ovulation, whereas the remaining follicles undergo atresia. Recent studies have suggested that follicular atresia is associated with the apoptosis of granulosa cells.[1,2] Granulosa cells play a major role in regulating ovarian physiology, including ovulation and luteal regression.[3] Granulosa cells secrete a wide variety of growth factors that may attenuate gonadotrophin's action in the ovary in paracrine–autocrine processes.[4,5] Most of these factors do not directly affect oocytes but exert their actions via granulosa cells. The presence of granulosa cells appears to be beneficial for oocyte maturation and early development.[6] The downregulating effects of a gonadotrophin-releasing hormone agonist (GnRHa) are related to the frequency of administration and the prolonged occupation of GnRH receptors by the agonist.[7]

It has been established that granulosa cell death during follicular atresia and luteolysis results from apoptosis. However, the precise mechanistic pathways of granulosa cell apoptosis have not yet been defined. Mitochondria may be intimately involved in apoptosis, and some apoptosis-associated proteins have been determined to be present in granulosa cells.[8,9] Granulosa cells are known to contribute to the maturation of oocytes, and most growth factors exert their action via granulosa cells. Researchers have also reported that the incidence of apoptotic bodies in granulosa cells can predict the developmental capacity of oocytes in an *in vitro* fertilization (IVF) program.[10] In this study, we wanted to determine whether there is a different proportion of apoptosis in granulosa cells of patients treated with two kinds of GnRHa: buserelin and leuprorelin depot. The apoptotic pathway and associated protein in GnRHa-treated granulosa cells were determined.

MATERIALS AND METHODS

Granulosa Cells in Culture

The institutional review board of Taipei Medical University Hospital approved the study, which we intended to perform on granulosa cells discarded during an IVF program. Human granulosa cells were collected from patients who were recruited into an IVF–embryo transfer (ET) program. Ovarian stimulation was performed by desensitization by using a GnRHa followed by treatment with gonadotropins (follicle-stimulating hormone [FSH] and human menopausal gonadotrophin [hMG]). Ovulation was induced using human chorionic gonadotrophin (hCG). Oocytes were then retrieved by transvaginal ultrasonography–guided aspiration 34–36 h after hCG administration. Granulosa cells were harvested from follicular aspirates by centrifugation through Ficoll–Paque, washed, and suspended in Dulbecco's modified Eagle's medium containing 50 μg/mL uridine supplemented with 10% fetal bovine serum, 100 IU/mL penicillin, and 100 μg/mL streptomycin. Human granulosa cells were grown at 37°C in a humidified atmosphere with 5% CO_2 for 3 days before experimental treatment.

Determination of Apoptosis

Apoptosis was assayed using terminal deoxynucleotidyl transferase–mediated dUDP nick-end labeling (TUNEL) and propidium iodide. DNA fragmentation and apoptotic bodies were counted under a fluorescence microscope. Immunocytochemical staining of granulosa cells was carried out with antibodies specific for Bcl-2, Bax, cytochrome c, and caspase-3, as well as an LSAB-2 kit (Santa Cruz Biotechnology, Santa Cruz, CA). The first antibodies were mouse anti–human Bcl-2, goat anti–human Bax, rabbit anti–human cytochrome c, and mouse anti–human caspase-3 (1/200 dilution), respectively. The secondary antibodies and substrate were provided by the LSAB-2 kit, and then crystal mount was added before histomounting. The staining phenomenon of granulosa cells was examined under a light microscope. Double staining of the mitochondria and cytochrome c of granulosa cells with MitoTracker Red and rabbit anti–human cytochrome c antibody (1/200 dilution), and the secondary antibody, fluorescein isothiocyanate–conjugated mouse immunoglobulin G (IgG; at a 1/200 dilution), was then carried out, respectively. The results of double staining were classified into punctate and diffuse patterns by fluorescence microscopy to assess the translocation of cytochrome c from mitochondria. Western blot analysis of granulosa cells was determined specifically for caspase-3 and caspase-9. Granulosa cells were extracted with cell lysis buffer on ice for 30 min, and the total protein was detected with bicinchoninic acid assay. Aliquots containing 10 μg of cell lysates were separated by 12.5% SDS-PAGE and transferred to polyvinylidene difluoride (PVDF) membranes. Membranes were incubated with specific antibodies: mouse anti–human caspase-3 (at a 1/200 dilution, stained with the precursor form), goat anti–human caspase-3 p20 (at a 1/200 dilution, stained with the p20 subunit and precursor form), and rabbit anti–human caspase-9 p35 (at a 1/200 dilution, stained with the p35 subunit and precursor form). The secondary antibodies were horseradish peroxidase (HRP)–conjugated mouse, goat, and rabbit IgGs (at a 1/1000 dilution). Blots were then developed using an enhanced chemiluminescence (ECL) system.

RESULTS

The incidences of DNA fragmentation of human granulosa cells treated with buserelin and leuprorelin were 54.33% and 39.02%, respectively. The proportions of apoptotic bodies were 6.04% and 4.29%, respectively. There was a significant difference in the DNA fragmentation proportions between granulosa cells treated with the two kinds of GnRHa. After buserelin depot treatment, patients had shorter oocyte recovery times than those treated with leuprorelin depot (TABLE 1). The expressions of apoptosis-associated proteins in granulosa cells were investigated. The Bax molecule, a pro-apoptosis protein, was expressed in granulosa cells undergoing apoptosis. In contrast, Bcl-2, an anti-apoptosis protein, was not detected in the same group of granulosa cells. Bax, cytochrome c, and caspase-3 were all expressed in the cytosol of patients' granulosa cells (FIG. 1). We also explored the subcellular distribution of cytochrome c, a well-known apoptogenic factor that acts upstream of caspase-3 activation, to determine whether cytochrome c participates in granulosa cell apoptosis. The distribution of cytochrome c immunostaining showed a punctate pattern that coincided with localization of mitochondria in granulosa cells without apoptosis. However, many cells showed a diffuse pattern, which mostly likely indicates that cytochrome c is translocated from the mitochondria. Western blotting was used to ex-

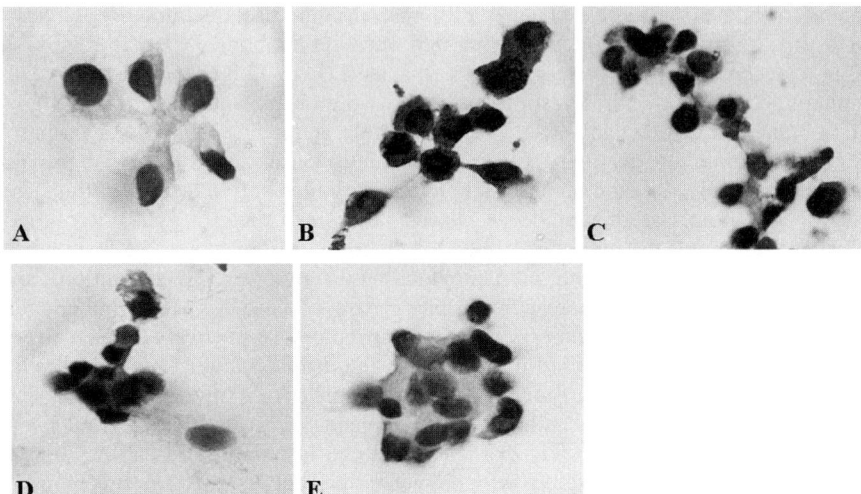

FIGURE 1. Immunocytochemical staining of granulosa cells. Granulosa cells were stained with specific antibodies and the LSAB-2 kit (red) and counterstained with hematoxylin (blue). (**A–D**) Granulosa cells were stained with specific antibodies against human Bcl-2, Bax, cytochrome c, and caspase-3, respectively. (**E**) Granulosa cells were stained without the first antibody as a negative control.

TABLE 1. Correlation of gonadotrophin-relating hormone agonists on granulosa cells and clinical factors

Pattern	Buserelin	Leuprorelin
DNA fragmentation (%)	55.55 ± 0.03*	39.02 ± 0.07
Apoptotic bodies (%)	6.04 ± 0.88	4.29 ± 0.74
Development index[a]	3.74 ± 0.15	4.06 ± 0.15
Fertilization (%)	63.85 ± 4.1	58.76 ± 6.1
Day of oocytes recovery	13.42 ± 0.3*	14.57 ± 4.2*
Age	34.04 ± 1.03	33.43 ± 1.04
Number of oocytes	7.13 ± 0.84	8.71 ± 2.16
E2 (before HCG)	1158.7 ± 204.9	1850.33 ± 63.81
P4 (before HCG)	1.14 ± 0.14	1.49 ± 0.29

*t-Test, $P < 0.05$.
[a]Development index: unfertilized oocytes were scored 0; dead embryos, 1; fully fragmented embryos, 2; arrested embryos, 3; embryos with a slow cleavage rate, 4; and embryos with a normal cleavage rate, 5. The total score of each patient was determined, and the development index (total score/total oocyte number) was calculated for each patient.

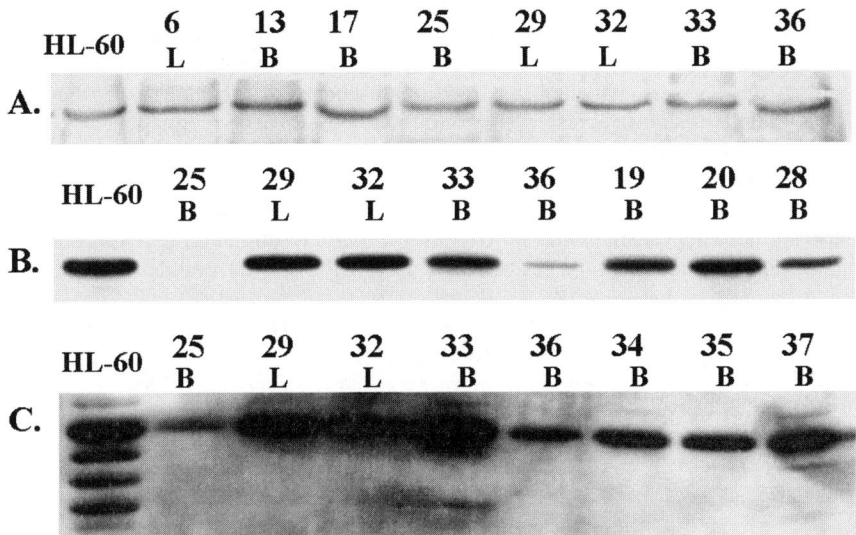

FIGURE 2. Western blot analysis of granulosa cells for caspase-3 and caspase-9. Aliquots containing 10 μg of cell lysates were separated by electrophoresis on 12% SDS-PAGE gels and transferred to PVDF membranes with HL-60 serving as a positive control. Membranes were incubated with specific antibodies and horseradish peroxidase–conjugated secondary antibodies, and blots were developed using an enhanced chemiluminescence (ECL) system. Immunoblot analysis: a 34-kDa precursor form of caspase-3 was stained (**A**), caspase-3 p20, a 34-kDa p20 subunit, and the precursor form were stained (**B**), and caspase-9 p35, a 48-kDa p35 subunit, and the precursor form were stained (**C**).

amine the expression of caspase in granulosa cells, and the results indicated that caspase-9 and caspase-3 were activated in patients' granulosa cells (FIG. 2).

DISCUSSION

In this study, granulosa cells treated with buserelin showed a higher incidence of DNA fragmentation than the group treated with leuprorelin depot. After buserelin treatment, patients also showed a shorter oocyte recovery time than those treated with leuprorelin. Apoptosis-positive nuclei were detected in some granulosa cells cultured for 48 h in the presence of FSH, and apoptosis-positive nuclei were less abundant than in granulosa cells in untreated control cultures.[11] However, in the presence of GnRHa, apoptosis-positive nuclei were obviously more abundant than in untreated granulosa cells.[11] GnRHa has attracted attention as a therapeutic agent for synchronizing follicle growth. However, the use of GnRHa agonist neither increases the number of oocytes obtained nor improves the quality of oocytes.[11,12] More atretic oocytes were acquired when GnRHa was used simultaneously with FSH than when FSH was used alone.[11,13] Those reports are consistent with this study concerning the stimulatory effect of GnRHa on granulosa cell apoptosis.

The expressions of apoptosis-associated proteins in granulosa cells were determined in this study. The data indicate that the pro-apoptosis protein Bax was expressed and that cytochrome c release was induced from mitochondria; then caspase-9 and caspase-3 were activated until cell death occurred. Incubation of granulosa cells with ATP or staurosporine was also found to possibly trigger apoptosis.[9,14] Activation of caspases works concurrently with mitochondrial dynamics during apoptosis. The mechanism of apoptosis is conserved across species and is known to occur with a cascade of sequential activation of initiator and executioner caspases.[15,16] These studies suggest that GnRHa triggers apoptosis in human granulosa cells and that the downstream apoptotic cascade may act at least in part through mitochondria.

ACKNOWLEDGMENTS

This work was supported by research grants NSC 91-2320-B-038-026 and NSC 92-2320-B-038-045 from the National Science Council of the Republic of China.

REFERENCES

1. TILLY, J.L., K.I. KOWALSKI, A.L. JOHNSON & A.J. HSUEH. 1991. Involvement of apoptosis in ovarian follicular atresia and postovulatory regression. Endocrinology **129:** 2799–2801.
2. HSUEH, A.J., H. BILLIG & A. TSAFRIRI. 1994. Ovarian follicle atresia: a hormonally controlled apoptotic process. Endocr. Rev. **15:** 707–724.
3. AMSTERDAM, A. & N. SELVARAJ. 1997. Control of differentiation, transformation, and apoptosis in granulosa cells by oncogenes, oncoviruses, and tumor suppressor genes. Endocr. Rev. **18:** 435–461.
4. NAKAMURA, T., K. TAKIO, Y. ETO, et al. 1990. Activin-binding protein from rat ovary is follistatin. Science **247:** 836–838.
5. GRAS, S., J. HANNIBAL, B. GEORG & J. FAHRENKRUG. 1996. Transient periovulatory expression of pituitary adenylate cyclase activating peptide in rat ovarian cells. Endocrinology **137:** 4779–4785.
6. CANIPARI, R. 2000. Oocyte granulosa cell interactions. Hum. Reprod. Update **6:** 279–289.
7. YEN, S.S. 1983. Clinical applications of gonadotropin-releasing hormone and gonadotropin-releasing hormone analogs. Fertil. Steril. **39:** 257–266.
8. IZAWA, M., P.H. NGUYEN, H.H. KIM & J. YEH. 1998. Expression of the apoptosis-related genes, caspase-1, caspase-3, DNA fragmentation factor, and apoptotic protease activating factor-1, in human granulosa cells. Fertil. Steril. **70:** 549–552.
9. KHAN, S.M., L.M. DAUFFENBACH & J. YEH. 2000. Mitochondria and caspases in induced apoptosis in human luteinized granulosa cells. Biochem. Biophys. Res. Commun. **269:** 542–545.
10. NAKAHARA, K., H. SAITO, T. SAITO, et al. 1997. The incidence of apoptotic bodies in membrana granulosa can predict prognosis of ova from patients participating in *in vitro* fertilization programs. Fertil. Steril. **68:** 312–317.
11. TAKEKIDA, S., H. MATSUO & T. MARUO. 2003. GnRH agonist action on granulosa cells at varying follicular stages. Mol. Cell. Endocrinol. **202:** 155–164.
12. POLSON, D.W., V. MACLACHLAN, J.A. KRAPEZ, et al. 1991. A controlled study of gonadotropin-releasing hormone agonist (buserelin acetate) for folliculogenesis in routine *in vitro* fertilization patients. Fertil. Steril. **56:** 509–514.
13. BRZYSKI, R.G., S.J. MUASHER, K. DROESCH, et al. 1988. Follicular atresia associated with concurrent initiation of gonadotropin-releasing hormone agonist and follicle-stimulating hormone for oocyte recruitment. Fertil. Steril. **50:** 917–921.

14. PARK, D.W., T. CHO, M.R. KIM, et al. 2003. ATP-induced apoptosis of human granulosa luteal cells cultured in vitro. Fertil. Steril. **80:** 993–1002.
15. CHINNAIYAN, A.M., K. ORTH, K. O'ROURKE, et al. 1996. Molecular ordering of the cell death pathway. Bcl-2 and Bcl-xL function upstream of the CED-3-like apoptotic proteases. J. Biol. Chem. **271:** 4573–4576.
16. THORNBERRY, N.A. & Y. LAZEBNIK. 1998. Caspases: enemies within. Science **281:** 1312–1316.

Index of Contributors

Andoh, T., 403–418
Au, H-K., 136–141, 177–185, 531–537

Bai, R-K., 36–47
Bai, Y., 25–35

Chan, K-H., 255–261
Chan, P., 286–293
Chan, P.H., 203–209
Chan, W-H., 372–378
Chang, A., 314–324
Chang, A.Y.W., 195–202
Chang, C-C., 76–81, 262–271
Chang, C-J., 294–302, 419–428
Chang, H-C., 262–271, 448–459
Chang, H-M., 506–515
Chang, J-H., 506–515
Chang, K-C., 64–69
Chang, P-T., 338–348
Chang, S-F., 471–480
Chang, W-Y., 255–261
Chang, Y-Y., 76–81
Chao, H-T., 148–156
Chen, C-H., 349–356
Chen, C-S., 523–530
Chen, C-T., 303–313, 387–395, 488–496
Chen, C-W., 365–371
Chen, C-Y., 109–122
Chen, F-Y., 516–522
Chen, H., 357–364
Chen, H-J., 379–386
Chen, H-L., 272–278
Chen, H-Y., 396–402
Chen, J-T., 448–459
Chen, R-M., 168–176, 262–271, 448–459, 460–470
Chen, S., 439–447
Chen, S-C., 186–194, 523–530
Chen, S-D., 64–69, 229–234, 357–364
Chen, S-M., 429–438
Chen, S-W., 516–522
Chen, T.H., 235–245
Chen, T-F., 448–459
Chen, T-G., 262–271

Chen, T-H., 286–293
Chen, T-L., 64–69, 168–176, 262–271, 429–438, 448–459, 460–470
Chen, W-L., 497–505
Chen, Y-C., 55–63
Chen, Y-P., 303–313
Chen, Y-S., 279–285
Chen, Y-Y., 82–87
Cheng, H., 314–324, 338–348
Cheng, H-H., 325–337, 365–371
Cheng, H-J., 365–371
Cheng, K-T., 379–386
Cheng, T-H., 349–356
Cheng, W-L., 70–75, 82–87
Chi, C-W., 109–122, 163–167
Chia, C-F., 523–530
Chiang, B-H., 387–395
Chiang, M-F., 48–54
Chieng, K-T., 387–395
Chin, M-Y., 294–302
Chiu, W-T., 396–402, 460–470
Chiueh, C.C., 403–418
Cho, Y.M., 1–18
Chock, P.B., 403–418
Choi, Y-S., 88–100
Chou, C-M., 303–313
Chuang, H-Y., 396–402
Chuang, Y-C., 64–69
Chung, Y., 338–348

Dai, P., 36–47
Deng, J-H., 25–35

Ezekiel, U.R., 439–447

Gau, B-B., 379–386
Guo, M-J., 419–428

Halliwell, B., 210–220
Hamasaki, N., 101–108
Ho, J-D., 157–162
Ho, W-P., 460–470
Hou, C-C., 235–245
Hou, R.C-W., 272–278, 279–285

Hsieh, R-H., 136–141, 157–162, 177–185, 186–194, 246–254, 294–302, 365–371, 531–537
Hsu, C.Y., xiii, 229–234, 357–364, 439–447
Hsu, C-H., 36–47
Hsu, M-C., 255–261
Hsu, S-W., 497–505
Hsuuw, Y-D., 372–378
Hu, C-J., 357–364
Hu, P., 25–35
Hu, S-Y., 471–480
Huang, C-C., 163–167
Huang, C-J., 303–313
Huang, C-S., 70–75
Huang, C-T., 338–348
Huang, H-C., 186–194
Huang, H-M., 272–278, 349–356, 481–487
Huang, J-R., 279–285
Huang, M-C., 338–348
Huang, S-H., 387–395
Huang, S-T., 387–395
Huang, S-Y., 136–141, 294–302, 531–537
Huang, W-C., 338–348
Hung, S-L., 235–245

Jeng, K.C.G., 272–278, 279–285
Jenner, A., 210–220
Jeyaseelan, K., 210–220
Jou, M-J., 221–228, 419–428
Juan, S-H., 349–356, 481–487

Kang, D., 101–108
Kao, L-S., 163–167
Kao, S-H., 148–156, 177–185, 186–194, 235–245
Ko, W-C., 325–337, 488–496
Kuo, C-L., 70–75
Kuo, H-S., 338–348, 387–395
Kwon, H., 36–47

Lan, M-Y., 76–81
Lee, C-F., 64–69, 70–75, 246–254, 429–438
Lee, C-H., 76–81
Lee, H-C., 109–122
Lee, H.K., 1–18

Lee, H-M., xiii , 130–135, 142–147, 148–156, 235–245, 286–293, 481–487, 523–530
Lee, M-J., 338–348
Lee, M-T., 497–505
Lee, S-W., 88–100
Lee, S-Y., 148–156
Lee, W-S., 471–480
Lee, Y.Y., 1–18
Lee, Y-H., 471–480
Li, F.C.H., 195–202
Li, Y., 25–35
Liang, Y-C., 349–356
Liao, L-P., 314–324
Liao, T-L., 148–156
Lien, L-M., 365–371
Lim, K.S., 210–220
Lin, A.M-Y., 314–324
Lin, C-C., 163–167
Lin, C-J., 325–337, 488–496, 497–505
Lin, C-L., 255–261
Lin, C-M., 325–337, 379–386, 387–395
Lin, C-Y., 429–438
Lin, H.Y., 163–167
Lin, J-C., 109–122
Lin, J-M., 396–402
Lin, J-W., 325–337
Lin, L-L., 168–176, 262–271
Lin, M-W., 130–135
Lin, S-H., 136–141, 365–371
Lin, T-K., 64–69
Lin, Y-L., 448–459
Liou, C-W., 64–69
Liou, D-Y., 338–348
Liou, K.D., 338–348
Liu, C-S., 70–75, 82–87
Liu, C-Y., 55–63, 246–254, 429–438
Liu, J-F., 255–261
Liu, J-S., 76–81
Liu, T-P., 303–313
Liu, V.W.S., 123–129
Lo, M-C., 130–135
Lo, T-Y., 488–496
Lu, C-Y., 55–63

Ma, Y-S., 55–63, 82–87
Majima, H.J., 516–522
Mau, B-L., 142–147
Min, H-K., 88–100

INDEX OF CONTRIBUTORS

Nagley, P., 123–129
Nassief, A., 357–364
Ngan, H.Y.S., 123–129

Ou, H-C., 272–278

Pak, Y.K., 1–18, 88–100
Pang, C-Y., 82–87
Park, J.S., 25–35
Park, K.S., 1–18
Peng, T-I., 221–228, 419–428
Perng, C-L., 36–47

Ryu, B-K., 88–100

Shieh, M-J., 531–537
Shih, B-F., 48–54
Shih, C-M., 303–313, 325–337, 379–386, 488–496, 497–505, 523–530
Shih, Y-L., 497–505
Shyue, S-K., 338–348
Sikorska, M., 429–438
Su, B., 379–386
Sun, Y-Y., 471–480

Tai, Y-T., 168–176, 460–470
Tam, T-N., 109–122
Tan, T-Y., 64–69
Tang, Y-C., 516–522
Thajeb, P., 48–54
Tsai, D.P., 163–167
Tsai, M-C., 186–194
Tsai, M-J., 314–324, 338–348
Tsai, N-M., 531–537
Tsai, S-Y., 471–480
Tsai, W-Y., 255–261
Tseng, C-C., 379–386
Tseng, H.P., 195–202
Tzen, C-Y., 19–24, 48–54, 130–135, 142–147
Tzeng, C-R., 136–141, 157–162, 177–185, 186–194, 531–537

Ueng, Y-F., 168–176

Wang, C-H., 314–324
Wang, H-F., 294–302
Wang, J-F., 286–293
Wang, L-F., 379–386, 387–395
Wang, P-W., 64–69
Wang, S-H., 488–496, 497–505
Wang, Y., 123–129
Wang, Y-H., 419–428
Wang, Y-T., 506–515
Wei, Y-H., xiii, 55–63, 64–69, 70–75, 82–87, 109–122, 148–156, 221–228, 246–254, 429–438, 497–505
Weng, C-F., 338–348
Whiteman, M., 210–220
Wong, L-J.C., 36–47
Wu, C-C., 109–122, 279–285
Wu, C-H., 262–271, 523–530
Wu, C-W., 109–122
Wu, G-J., 168–176, 448–459
Wu, H-J., 372–378
Wu, H-S., 76–81
Wu, H-Y., 221–228, 419–428
Wu, J-C., 325–337
Wu, J-S., 488–496
Wu, M-C., 48–54
Wu, T-Y., 19–24
Wu, Y., 439–447

Xu, J., 357–364, 439–447

Yang, D-I., 229–234, 357–364, 439–447
Yang, D-M., 163–167
Yang, L-Y., 325–337, 471–480, 488–496
Yang, V.W-C., 157–162
Yeh, T-S., 136–141, 157–162, 177–185
Yen, H-C., 516–522
Yen, J.J-Y., 481–487
Yin, J-H., 229–234
Yin, K., 357–364
Yin, P-H., 109–122
Yu, J-S., 419–428
Yuan, R-Y., 48–54

OHIO UNIVERSITY LIBRARY

Please return this book as soon as you have finished with it. In order to avoid a fine it must be returned by the latest date stamped below. All books are subject to recall after two weeks or immediately if needed.

CF